Advances in temperate grassland science and management

It is widely recognised that agriculture is a significant contributor to global warming and climate change. Agriculture needs to reduce its environmental impact and adapt to current climate change whilst still feeding a growing population, i.e. become more 'climate-smart'. Burleigh Dodds Science Publishing is playing its part in achieving this by bringing together key research on making the production of the world's most important crops and livestock products more sustainable.

Based on extensive research, our publications specifically target the challenge of climate-smart agriculture. In this way we are using 'smart publishing' to help achieve climate-smart agriculture.

Burleigh Dodds Science Publishing is an independent and innovative publisher delivering high quality customer-focused agricultural science content in both print and online formats for the academic and research communities. Our aim is to build a foundation of knowledge on which researchers can build to meet the challenge of climate-smart agriculture.

For more information about Burleigh Dodds Science Publishing simply call us on +44 (0) 1223 839365, email info@bdspublishing.com or alternatively please visit our website at www.bdspublishing.com.

Related titles:

Achieving sustainable production of milk Volume 1
Print (ISBN 978-1-78676-044-9); Online (ISBN 978-1-78676-046-3, 978-1-78676-047-0)

Advances in organic dairy cattle farming
Print (ISBN 978-1-80146-538-0); Online (ISBN 978-1-80146-539-7, 978-1-80146-540-3)

Advances in sheep production
Print (ISBN 978-1-80146-802-2); Online (ISBN 978-1-80146-803-9, 978-1-80146-804-6)

Ensuring safety and quality in the production of beef Volume 2
Print (ISBN 978-1-78676-060-9); Online (ISBN 978-1-78676-062-3, 978-1-78676-063-0)

Improving grassland and pasture management in temperate agriculture
Print (ISBN 978-1-78676-200-9); Online (ISBN 978-1-78676-202-3, 978-1-78676-203-0)

Improving organic animal farming
Print (ISBN 978-1-78676-180-4); Online (ISBN 978-1-78676-182-8, 978-1-78676-183-5)

Chapters are available individually from our online bookshop:
https://www.bdspublishing.com/webshop/

climate•SMART•publishing
IN AGRICULTURAL SCIENCE

BURLEIGH DODDS SERIES IN AGRICULTURAL SCIENCE
NUMBER 169

Advances in temperate grassland science and management

Edited by Professor Agnes van den Pol-van Dasselaar

burleigh dodds
SCIENCE PUBLISHING

Published by Burleigh Dodds Science Publishing Limited
82 High Street, Sawston, Cambridge CB22 3HJ, UK
www.bdspublishing.com

Burleigh Dodds Science Publishing, 1518 Walnut Street, Suite 900, Philadelphia, PA 19102-3406, USA

First published 2025 by Burleigh Dodds Science Publishing Limited

Library of Congress Control Number: 2025946832

British Library Cataloguing in Publication Data
A catalogue record for this book is available from the British Library

ISBN 978-1-80146-902-9 (print)
ISBN 978-1-80146-903-6 (ePub)
ISBN 978-1-80146-904-3 (PDF)
ISSN 2059-6936 (print)
ISSN 2059-6944 (online)

DOI: 10.19103/AS.2024.0154

Typeset by Deanta Global Publishing Services, Dublin, Ireland

Contents

Series list

Title	Series number
Achieving sustainable cultivation of maize - Vol 1 From improved varieties to local applications Edited by: Dr Dave Watson, CGIAR Maize Research Program Manager, CIMMYT, Mexico	001
Achieving sustainable cultivation of maize - Vol 2 Cultivation techniques, pest and disease control Edited by: Dr Dave Watson, CGIAR Maize Research Program Manager, CIMMYT, Mexico	002
Achieving sustainable cultivation of rice - Vol 1 Breeding for higher yield and quality Edited by: Prof. Takuji Sasaki, Tokyo University of Agriculture, Japan	003
Achieving sustainable cultivation of rice - Vol 2 Cultivation, pest and disease management Edited by: Prof. Takuji Sasaki, Tokyo University of Agriculture, Japan	004
Achieving sustainable cultivation of wheat - Vol 1 Breeding, quality traits, pests and diseases Edited by: Prof. Peter Langridge, The University of Adelaide, Australia	005
Achieving sustainable cultivation of wheat - Vol 2 Cultivation techniques Edited by: Prof. Peter Langridge, The University of Adelaide, Australia	006
Achieving sustainable cultivation of tomatoes Edited by: Dr Autar Mattoo, USDA-ARS, USA and Prof. Avtar Handa, Purdue University, USA	007
Achieving sustainable production of milk - Vol 1 Milk composition, genetics and breeding Edited by: Dr Nico van Belzen, International Dairy Federation (IDF), Belgium	008
Achieving sustainable production of milk - Vol 2 Safety, quality and sustainability Edited by: Dr Nico van Belzen, International Dairy Federation (IDF), Belgium	009
Achieving sustainable production of milk - Vol 3 Dairy herd management and welfare Edited by: Prof. John Webster, University of Bristol, UK	010
Ensuring safety and quality in the production of beef - Vol 1 Safety Edited by: Prof. Gary Acuff, Texas A&M University, USA and Prof. James Dickson, Iowa State University, USA	011
Ensuring safety and quality in the production of beef - Vol 2 Quality Edited by: Prof. Michael Dikeman, Kansas State University, USA	012
Achieving sustainable production of poultry meat - Vol 1 Safety, quality and sustainability Edited by: Prof. Steven C. Ricke, University of Arkansas, USA	013
Achieving sustainable production of poultry meat - Vol 2 Breeding and nutrition Edited by: Prof. Todd Applegate, University of Georgia, USA	014

Acknowledgements

We wish to acknowledge the following for their help in reviewing particular chapters:

- Chapter 16: Dr Michael O'Donovan, Teagasc, Ireland
- Chapter 26: Dr Lisa Fracchetti, Laimburg Research Centre, Italy, Dr Luc Delaby, INRAE, France, Dr Paul Newell-Price, ADAS, UK, and Dr Catherine Pfeifer, FiBL, Switzerland

Introduction

Grasslands cover a large proportion of the global land area. As a result, provision of food from grassland-based production systems has long played an important role in global food security. However, grasslands face a range of challenges, including the shift to more intensive farming methods, the impact of climate change and debates over land use change to achieve net zero emissions. Yet, research also suggests that grasslands also have a huge potential in supporting the transition to more resilient, sustainable and circular agricultural systems. This collection reviews the wealth of research addressing both these challenges and opportunities. It builds on an earlier title published by Burleigh Dodds Science Publishing in 2018: *Improving grassland and pasture management in temperate agriculture* (ed. Marshall and Collins).

Part 1 The role of grasslands in agricultural systems

Part 1 addresses the role of grasslands in agricultural systems. Chapter 1 focuses on the societal role of grasslands in agriculture. Because options for using grasslands are limited, the pressure to downsize livestock production to reduce negative environmental impacts is likely to result in a further reduction in (permanent) grasslands in regions such as North-west Europe. This risks losing important ecosystem services (ESs) provided by grasslands. This chapter compares different types of grasslands (permanent, semi-natural and temporary) with arable land in provision of key ESs: soil quality, water consumption and regulation, climate change, water quality and biodiversity. The chapter shows that grasslands can score better than arable land in ESs provision. This has important implications for future decisions about land use change.

Chapter 2 outlines the anticipated impacts of climate change on grasslands, focusing on plant productivity and composition. Grassland ecosystems are vital in global carbon cycling, biodiversity conservation and agricultural productivity. However, as ecosystems typically bordering arid or semi-arid zones, they are particularly vulnerable to the impacts of climate change which will alter key environmental drivers such as temperature and precipitation regimes. The chapter discusses the methods and techniques used to determine the impact of climate change on grassland dynamics, highlighting the potential for shifts in phenology, species composition and ecosystem functioning. Methods such as soil sampling, remote sensing, and ecosystem flux measurements are outlined. Challenges in predicting and mitigating the impacts of climate change are also discussed, emphasising the importance of researching adaptation strategies to enhance grassland resilience.

Chapter 3 discusses the role of grasslands within mixed farming/crop-livestock systems. Grasslands play a crucial role in mixed farming systems, contributing to both sustainability and productivity. As an example, they provide essential forage for livestock while offering vital ecosystem services like soil conservation, improved nutrient cycling, biodiversity support, and carbon (C) sequestration. Well-managed integrated farming systems, incorporating coupled crop-pasture rotations and in situ animal grazing, can optimize resources and enhance sustainability and resilience, particularly when managed at low to moderate grazing intensities. This integration encourages closed-loop cycles by maximizing productivity within these systems e.g. using animal manure as fertilizer while feeding crop residues to livestock. It also promotes a more resilient system less reliant on external inputs and with potentially more diverse income streams. The chapter reviews the diverse role of grasslands in mixed farming, management strategies to enhance grassland utilization within mixed systems and includes case studies on successful practices.

Part 2 Grassland dynamics and management

Part 2 discusses grassland dynamics and their implications for improved management. Chapter 4 reviews plant-soil-animal interactions and nutrient cycling in grazed grasslands. Focusing on grazed grasslands in temperate environments, this chapter describes nutrient pools and fluxes, and grassland management practices that enhance nutrient cycling efficiency. Nutrient cycling efficiency is enhanced by practices that sustain pasture cover, reduce heterogeneity of dung and urine deposition, and increase legume adoption as an alternative to nitrogen (N) fertilizer. Efficiency is also increased through presence of earthworms, dung beetles, and beneficial microbial communities, while the use of chemical and biological nitrification inhibitors shows increasing promise for reducing nitrate leaching and nitrous oxide emissions.

Chapter 5 addresses advances in grazing management. Grassland provides a cost-effective and environmentally friendly source of high-quality food for ruminants. The chapter identifies how variability in grass growth is affected by season, weather, soil type and grazing management, making grassland utilisation challenging for farmers. Improved understanding of soil and plant nutrition, plant physiology, as well as perennial ryegrass and clover variety improvement, has helped to optimize grassland production. Similarly, improved understanding of feed evaluation, ruminant nutrition, grazing management and silage technology have contributed to managing grassland utilization to match grass availability. Development and use of grassland databases, decision support systems and grass selection indices for grass/clover varieties are new innovations that promote improved grassland management.

Chapter 6 focuses on the use of precision technologies in grassland management. Maximising production efficiencies while minimising environmental and ecosystem impacts will be crucial to meet demand, while also contributing to the achievement of climate and biodiversity goals. Precision technologies offer the opportunity to map variability across grasslands and to tailor management interventions accordingly. The chapter also examines the continued role of traditional grassland measurement systems, data management and decision support systems.

Building on the previous chapter, Chapter 7 follows the long path from data collection to decisions supporting sustainable grassland management. The chapter provides an overview of techniques available to collect data about grasslands. Different spatio-temporal scales require different modes of data collection to capture grassland dynamics. The chapter discusses examples such as the use of high temporal resolution measurements of greenhouse gas exchanges to quantify soil carbon sequestration, the use of remote sensing data and models to help predict biomass production and long-term carbon sequestration rates.

Picking up from Chapter 7, Chapter 8 discusses decision support tools for the sustainable management of grasslands. As the chapter points out, decisions about grassland management involve not only farmers and their advisors but other land management groups, public policy actors and others as well. Management strategies need to take account of these different interests and their specific needs, as well as the often heterogeneous nature of data on grasslands, particularly on semi-natural grasslands. The chapter discusses how such data can be analysed in order to support sustainable multi-actor and multi-functional grassland management and decision-making. The chapter illustrates ways this can be achieved, for instance by simulating the impact of different management regimes and public policies.

Chapter 9 covers advances that have been made in managing grassland weeds. The term "weed" is both relative and subjective, defined contextually and in relation to human activities. This chapter identifies different grassland types, and how the definition of weeds can vary, ranging from cultivated feed crops in arable land to natural multi-species grasslands. As agricultural systems transition from monocultures to more diverse natural ecosystems, the perception of what constitutes a weed may shift. In this new context questions arise regarding whether weeds are always detrimental or if they can potentially contribute to ecosystem service delivery. The chapter reviews different management practices to prevent weed proliferation through nature-based solutions, particularly the role of biodiversity in natural weed control. Challenges and solutions in managing weeds as both a threat and an opportunity are explored through a case study.

Part 3 Grasslands, grazing and animal nutrition: effects on health and product quality

Part 3 covers the ways grasslands affect health and product quality. Chapter 10 highlights the contribution of forages to high nutrient use efficiency in ruminants. The chapter aims to integrate feed quality assessment and grassland science to optimise nutrient fluxes in the plant-animal system. Although animal performance is the most accurate way to define appropriate feed quality, in vitro laboratory methods are often used to evaluate feed quality by feed composition, digestibility or degradability of defined groups of substances. The estimation of the energy value of forages is very important, with the digestibility of organic matter playing an important role in determining the energy value of forages. As the chapter shows, high milk yields are possible if animals can be supplied with energy and nutrients to meet their performance requirements, a major part of which can come from forage intake.

Chapter 11 focuses on the effects of pasture grazing on milk quality. The environment and farm management practices affect herbage allowance and grazing time and - in relation to that - also milk quality. The chapter addresses milk fat composition as it relates to human health. It also covers topics such as lipids in feeds, their fate in the digestive tract, and the effects of several feedings systems–including grazing and indoor feeding. The chapter describes how ashift away from grazing can increase saturated and *trans*-10 milk FA concentrations and reduce naturally occurring fat-soluble vitamins and antioxidants in dairy products. Starch-rich fodders affect rumen retention time and alter rumen biohydrogenation pathways, and grassland species diversity can improve the transfer of polyunsaturated fatty acids (PUFA) from feed to milk.

Chapter 12 discusses grasslands and human health. The chapter reviews what we know about the main shifts in diet-health interactions during human evolution and (pre)history, with a particular focus on the role of grassland-derived foods in developing a species-specific human diet over millennia. Although the current role of grassland-derived foods (e.g. meat and dairy products) in the human diet is heavily debated, the chapter argues that there are strong evolutionary grounds for reconsidering their value in promoting good nutrition and health.

Part 4 Developments in botanical composition of grasslands

Part 4 reviews recent research results in understanding the botanical composition of grasslands. Biodiversity conservation is a rapidly advancing research area in sustainable grassland management. Chapter 13 explores evidence-based approaches to enhance biodiversity in agricultural grassland systems. The biodiversity value of these grasslands varies widely, primarily influenced by production-driven management intensity. The chapter explores different pathways for maintaining and enhancing grassland biodiversity using context-specific adaptations for different settings. The chapter includes a case study demonstrating the ecological benefits resulting from modified farm management to promote biodiversity. It also addresses ways of reconciling potential trade-offs between biodiversity and yield objectives.

Chapter 14 highlights ways of designing and developing multispecies swards (MSS) for grassland systems. The use of multispecies swards (MSS) has increasingly been recognised as a viable alternative in temperate grass-based agricultural systems to enhance forage yield, quality, and ecosystem health. Unlike traditional monocultures, MSS leverage the concept of functional diversity, typically utilising four to eight species that interact to maximise productivity. This diversity enables improved nutrient cycling, reduced weed growth, increased soil organic matter, and enhanced carbon sequestration, all of which contribute to ecosystem stability. This chapter outlines the principles, benefits and practical management strategies in developing MSS and highlights their importance in creating sustainable and resilient agricultural systems that meet both agronomic and environmental goals.

Building on the previous chapter, Chapter 15 studies how narrow-leaved plantain (*Plantago lanceolata L.*) can reduce nitrogen losses/leaching and reduce nitrous oxide emissions. As the chapter shows, including plantain in perennial ryegrass/white clover pasture can be a low-cost option to reduce leaching and emissions without negative effects on pasture production, animal production, milk quality and animal health. On the contrary, its addition can deliver a range of benefits, including optimising output from grazed dairy pastures in temperate regions.

Chapter 16 then shows how asexual *Epichloë* fungi form mutualistic endophytic associations with cool-season grasses such as perennial ryegrass (*Lolium perenne*) and tall fescue (*Festuca arundinacea*). These can produce both toxic effects, which can be detrimental to livestock health, but also produce secondary metabolites responsible for bio-protective properties that benefit livestock performance. The chapter explores the mechanisms underpinning both sets of effects and how they can be appropriately managed.

Part 5 Case studies

Part 5 contains case studies on grassland-based production systems in Europe, alpine/mountain grasslands, and New Zealand. Chapter 17 focuses on grassland-based production systems in Europe, including their future role and research needs. The chapter describes the main grassland systems used for agricultural production in Europe and explores their different functions, benefits and challenges. It addresses key issues such as adapting grasslands to climate change as well as balancing conservation with agricultural demands.

Chapter 18 addresses the particular challenges facing alpine/mountain grasslands. Climatic, topographical and soil quality constraints typically mean that grazed pasture is often the only viable option for food production. However, these alpine/mountain grasslands may also be important sources of biodiversity (including a range of predator species) and may deliver a range of ecosystem services. The chapter explores ways of reconciling these different objectives and ensuring the economic, environmental and social sustainability of mountain grasslands, particularly in the face of climate change which is expected to affect mountain areas more than other ecosystems.

Chapter 19 highlights the issues and options facing grasslands in New Zealand. New Zealand is unique among developed nations given the importance of its dairy, sheep meat and other exports from grassland systems. This chapter reviews the range of pressures facing pastoral livestock production in New Zealand, from changing consumer expectations to environmental and ethical challenges. It addresses the range of potential solutions to address these challenges, whether in terms of animal and pasture/forage management or broader changes in production management, and assesses what constitutes the best way forward for a successful and sustainable grassland livestock sector.

Part 6 Conclusion

The concluding chapter highlights the cross-cutting insights that have emerged from the preceding chapters, provides practical recommendations for practitioners and policymakers, and sets out overarching priorities for a research agenda to accelerate the transition to resilient, productive, sustainable, and biodiversity-rich temperate grasslands.

Part 1

The role of grasslands in agricultural systems

Chapter 1

The societal role of grasslands: the case of northwest Europe

N. van Eekeren, J. de Wit, M. van den Hout, M. Bruinenberg, J. Pijlman and N. Hoekstra, Louis Bolk Institute, The Netherlands

1 Introduction

Grasslands constitute 67% of global agricultural land (FAO, n.d.) and provide a wide range of ecosystem services (ES). Zhao et al. (2020), however, showed that ES research with a particular focus on grasslands is limited. Simultaneously, the importance of (semi-) permanent grasslands for livestock production is decreasing in regions such as northwest (NW) Europe and many other parts of the world. Particularly from the 1950s onwards, this has been fuelled by the intensification of beef and dairy production with a sharp increase in productivity levels. This development has been based on using increased amounts of

http://dx.doi.org/10.19103/AS.2025.0154.01

arable crops for feed, like silage maize, and (purchased) concentrates like grains and soy, and the replacement of permanent grasslands by temporary grasslands.

In most regions in NW Europe, this intensification process has been partially slowed by environmental legislation, forcing farmers to focus on efficiency. Recently, the call for downsizing livestock production to reduce negative environmental impacts, particularly greenhouse gas (GHG) and ammonia emissions, has increased. Because alternative options for utilisation of grassland products are few and with poor economic prospects, a continued reduction in the size of cattle production will result in a further reduction in (permanent) grasslands, thereby compromising ESs provided by grasslands. To avoid throwing away the proverbial 'baby with the bathwater' in societal and policy considerations concerning the cattle production sector, this chapter emphasises the often underestimated ESs of different types of grasslands compared to arable land.

2 Ecosystem services and grasslands

ESs can be defined as the goods and services provided by ecosystems to humans. The concept of ES has been productive in focusing attention on the importance of neglected outputs of systems (several 'provisioning' and nearly all 'cultural services') and of functions that are essential for maintaining agroecosystems ('regulating and supporting services'), particularly since the formalisation of the concept in the Millenium Ecosystem Assessment (Board, n.d.). The definition and valuation of ESs is complicated by a number of issues relating to final versus intermediate services or ecosystem functions, the risk of double counting versus the underestimation of ecosystem functions, indicators versus direct measures of ES, consumption/renewability, etc. (Ojea, 2010; Silvis & van der Heide, 2013).

Attempts to resolve these issues have resulted in extensive classification systems. These often have limited connection with concerns about the endangered or neglected services which are commonly discussed. The most recent Common International Classification of ESs (CICES; Haines-Young, 2023) does include simple descriptors, such as 'controlling or preventing soil or sediment loss by running water/wind' for the ES-clause 'the reduction in the loss of material by virtue of the stabilising effects of the presence of plants and animals that mitigates or prevents potential damage by water flow/wind to human use of the environment or human health and safety' (CICES clauses 2.2.1.1 and 2.2.1.2). Note that many ESs are related to the prevention of present and future damage to humans (a 'disservice') instead of being directly linked to a positive service of an ecosystem to humans, as illustrated by the ES label on erosion.

Despite the extensive theoretical work on ES classifications, the multifunctionality of grasslands is not easily captured in ecosystem service assessments. Richter et al. (2021), for example, have shown that one (direct or indirect measure of an) indicator might relate (positively and/or negatively) to many ESs, while many indicators might relate to one ES. Therefore, they and others (Ojea, 2010; Haines-Young, 2023) conclude that a classification of relevant ESs needs to be case and context specific, depending on a particular application context to avoid double-counting and other pitfalls.

Although in general, many different ESs are identified, we have selected only eight indicators listed in Table 1 as most relevant for the comparison of land use types in NW Europe with the Netherlands as a case study. These indicators cover 14 different ESs (CICES; based on Haines-Young, 2023). Indications to which ESs as an indicator relates to, some at multiple/different spatial levels, are included. Other ESs with potential differences between land use type, e.g. erosion or spiritual meaning of land use, were not selected because they are of little relevance in (most parts of) the Netherlands.

2.1 Grasslands and ecosystem services

The common characteristics of a grassland that are important for different ESs are that:

- it covers the soil all year round;
- it is a perennial crop; and
- it has a high, continuous supply of crop and root residues which serves as nutrition for soil biota.

Arable land, on the contrary, is characterised by:

- an alternation of crops with periods without ground cover;
- no continuous supply of nutrition for soil biota; and
- frequent disturbance of soil/soil biota by tillage operations.

Because agro-ecological conditions vary greatly around the world, thereby influencing the importance of ESs delivered by grasslands and other land use types, in this chapter we have focussed on ESs relevant to the situation in NW Europe. Throughout the chapter, the Netherlands will be used as a case study, illustrating the different ESs for:

- arable land;
- permanent grasslands;

Table 1 Selected relevant (indicators of) ecosystem services (Haines-Young, 2023) for comparing grasslands and arable land in NW Europe with examples from the Netherlands as a case study

Ecosystem service or indicator (simple name)	Mainly related to ... (description + reference to CICES definition number to which it relates)
Food production	Provisioning of cultivated crops that can be used (directly or indirectly) for human nutrition (CICES 1.1.1.1).
Soil quality	Regulating/maintaining soil conditions that enable (futural) human use (CICES 2.3.4.3).
Water consumption and regulation	Regulating water flows by virtue of the characteristics of ecosystems that assist people in maintaining hydrological systems (CICES 2.2.2.1 & 2.2.2.2).
Climate change	Regulating the concentrations of gases in the atmosphere that impact on global climate (CICES 2.3.6.2).
Ammonia	Regulating/maintaining ecological conditions necessary for protecting populations of species that people use or enjoy or seek to preserve (off-side effect of the land use, e.g. on terrestrial nature reserves; CICES 2.3.2.4, 2.3.2.5 & 3.4.2.1). Also related to biophysical qualities of ecosystems that are enjoyed actively or passively, or appreciated for their inherent beauty (CICES 3.1.1.1, 3.1.1.2 & 3.2.1.4). At higher concentrations, it also relates to odours that are harmful/a nuisance to people (CICES 2.1.2.1).
Water quality	Regulating/maintaining the chemical condition of fresh waters that enables human use/health (CICES 2.3.5.1) and ecological conditions necessary for protecting populations of species that people enjoy or seek to preserve (off-site effect of the land use, on aquatic systems; CICES 2.3.2.4, 2.3.2.5 & 3.4.2.1). Also related to the fixing and storage of (organic or inorganic) substances that mitigate their harmful effects (CICES 2.1.1.2).
Above-ground biodiversity	Regulating/maintaining ecological conditions necessary for protecting populations of species that people enjoy (mainly on-site effects, e.g. on meadow birds; CICES 2.3.2.3 & 2.3.2.5) or maintaining biophysical qualities of ecosystems that are appreciated for their inherent beauty (e.g. grazing animals; CICES 3.2.1.4). These on-site effects of land use are an important ingredient of the 'license to produce' that agriculture needs in the Netherlands.
Below-ground biodiversity	Regulating/maintaining ecological conditions necessary for protecting populations of species that people enjoy or seek to preserve (mainly on-site effects, e.g. on meadow birds; CICES 2.3.2.3, 2.3.2.4, 2.3.2.5 & 3.4.2.1). Also, below-ground biodiversity has a crucial role in ecosystem functioning, affecting several other ESs like soil quality, water regulation and above-ground biodiversity (on-site as well as off-side effects of the land use considered).

- temporary grasslands; and
- semi-natural grasslands.

Table 2 illustrates the development of dairy production in the Netherlands as an extreme example of the intensification process with related negative environmental impacts, like nutrient surpluses and land use change.

Permanent grasslands involve continuous grass cultivation in a field. Disturbance of the soil by tillage is minimal, only during grassland renewal which, e.g., in the Netherlands occurs on average once every 10 years (Velthof et al., 2009). Tillage in permanent grasslands is followed by the reseeding of grass ('grass-on-grass'). This is in contrast to temporary grasslands where sooner or later (but at least once every 5 years) arable land follows grassland after tillage.

Temporary grasslands are composed of annual, biennial or perennial forage species kept for a short period of time (usually only a few years) (Allen et al., 2011). Temporary grasslands in the EU are defined as grasslands included as a part of a normal crop rotation, lasting at least one crop year and less than 5 years (Lesschen et al., 2012).

Grasslands in the Netherlands do differ amongst each other in terms of intensity of use, which affects several ESs, but the difference between land use types are small:

- Both permanent and temporary grasslands are grasslands whose main purpose is the production of feed for ruminants.
- Both are dominated by perennial ryegrass (*Lolium perenne* L.) and intensively fertilised (around 350 kg of nitrogen per hectare annually of which max 170–230 kg nitrogen from organic fertilisers).
- Herbicides are used occasionally to control weeds.

Table 2 Main characteristics of Dutch dairy production

	1960	1985	2005	2020
Milk production (ton/ha/year)	5.5	12.5	12.7	16.4
N-soil surplus (kg/ha/year)	–	400	179	157
Milk production (kg/cow/year)	4200	5330	7690	8980
Concentrate purchased (kg/cow/year)	830	2280	2020	2530
	(% of total area for roughage production)			
Permanent grasslands	96	84	63	63
Temporary grasslands	4	3	17	19
Arable roughage crops	1	13	20	18

Ps. Changes in % permanent and temporary grasslands from 1985 to 2005 are exaggerated due to changed legislative definitions. Since 1975, arable roughage crops mainly consist of silage maize.
Source: Adapted from: Van Horne & Prins, 2002; BINternet, n.d.-b; Statline, n.d.

Semi-natural grasslands are a special case of permanent grasslands, defined as ecosystems with limited management dominated by indigenous or naturally occurring grasses and other herbaceous species (Allen et al., 2011). These grasslands are of low but growing importance in NW Europe, because of their use in 'transitional situations' in and around nature reserves. Their main objective is to maintain or improve biodiversity; production (both feed quantity and quality) is secondary. To foster biodiversity, these grasslands receive little or no fertilisation and have additional restrictions on livestock utilisation (e.g. delayed mowing date and/or extensive grazing).

3 Comparing arable and grassland ecosystem services: permanent, temporary and semi-natural grasslands

The ESs of three types of grasslands were compared to arable land (as the main land use type for food production). These four systems (arable land, permanent grasslands, temporary grasslands and semi-natural grasslands) were scored based on the available literature and 'expert judgement'. The land use type with the highest score is given five points, and the land use type with the lowest score one point. The other land use types are scored relative to these scores.

The main focus in this chapter is on land use type and ESs at the field level. The final impact of grasslands on ESs also depends on the entire farming system, including the number of animals and their level of production, and the use of alternative farming systems. However, such a comprehensive analysis is beyond the scope of this chapter. Here, we want to provide building blocks for an integral analysis by listing the effects of different land use types on ESs at the field level. Occasionally, we will give a slightly broader (system) focus on an ES as an illustration.

A summary of the contribution of ESs per land use type is shown in Fig. 1. These contributions will be specified in the following sections. Briefly, food production is higher from 1 hectare of arable land than from 1 hectare of grasslands while it is low from semi-natural grasslands. Thanks to very low levels of fertilisation and permanent vegetation, semi-natural grasslands score better than arable land for most ESs, except for ammonia emissions, which are comparably low in both systems.

Temporary and permanent grasslands have a position between arable land and semi-natural grasslands for many ESs, with the exception of ammonia. In addition, permanent grasslands score similar or better than semi-natural grasslands on the ESs soil quality and below-ground biodiversity, mainly associated with a high and stable organic matter supply to the soil and few soil tillage operations.

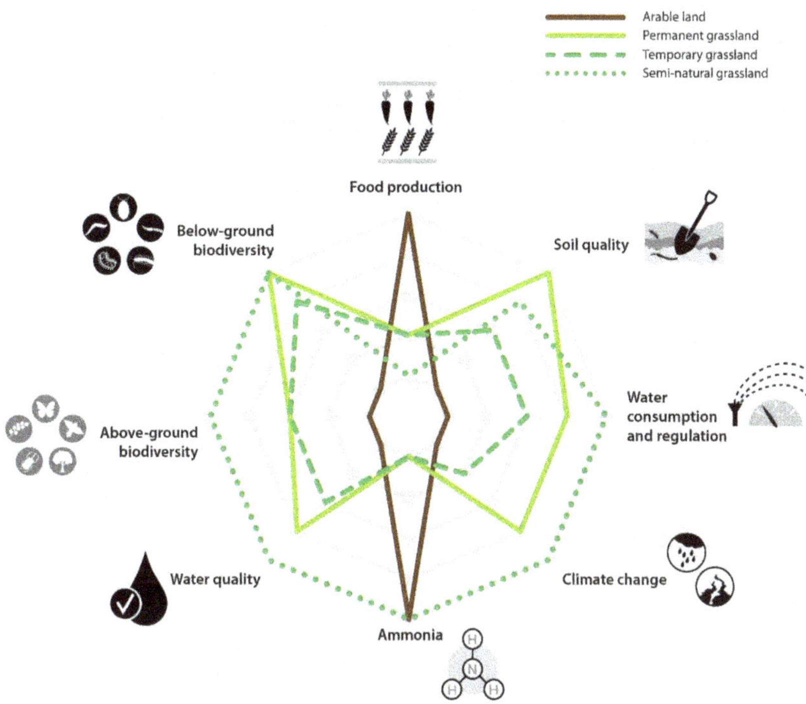

Figure 1 Ecosystem services from arable land and grassland types. The land use type with the highest ('best') score is given five points and the land use type with the lowest ('worst') score one point. The other land use types are scored relative to these scores. In Sections 4 to 10 the different scores in this figure are discussed.

4 Comparing ecosystem services: food production

The production of food or fodder is the main purpose of most agricultural land in NW Europe; this is taken as the starting point in this comparison. Grasslands produce fodder that is transformed into milk and meat by livestock. Arable land produces either feed for livestock (usually in the form of silage maize, cereals or protein-rich crops) or food that is used directly for human consumption.

On average, more people can be fed from 1 hectare of arable land than from 1 hectare of grassland (Spedding, 1988).

To give an example of this, a simplified calculation has been made from the yields of 1 hectare of wheat, potatoes and milk (from grass) in the Netherlands (see Table 3). In this table no direct comparison is made with the yield of grass, because this has to be converted into milk before it can be consumed by humans (see Box 1).

Table 3 Production of protein and energy from milk, wheat and potatoes (see Box 1 for calculation and references).

Food	Yield/ha (tonnes of product)	Protein/ha (tonnes)	Energy (million kcal/ha)
Milk (from grass)	9.3	0.33	6
Wheat	8.8	1.10	29
Potatoes	48.5	1.00	41

Box 1. Calculation of food production

Average milk production in the Netherlands is about 16 000 kg of milk per hectare (BINternet, n.d.-b). However, this production includes purchased concentrates, and therefore does not originate entirely from grasslands. Of the total feed ration, an average of 58% of the protein comes from own land, mainly grasslands and some silage maize (van der Meulen, 2022). This results in an estimated production of 9.3 tonnes of milk per hectare of own land. Multiplied by an average content of 3.58% protein (BINternet, n.d.-b) and the energy value of milk (voedingswaardetabel.nl), a hectare of grassland yields 0.33 tonnes of protein and (rounded) 6 million kcal of energy.

On average, over the years 2019–2021, the Dutch production of potatoes was 48.5 tonnes per hectare and of wheat 8.8 tonnes per hectare (BINternet, n.d.-a). These yields have been multiplied by energy and protein values of these foods (voedingswaardetabel.nl) resulting in a yield for protein and energy per hectare.

This example illustrates that arable crops produce about three times more (consumable) protein than grasslands (through milk). The energy production of arable land is a multitude of grasslands, particularly for energy because the crops in this example are very rich in energy.

There are some nuances to this quantitative comparison:

- Arable farming is generally located on the most fertile soils, resulting in higher yields. The average grass yield (and thus milk yield) is therefore naturally already somewhat lower than the yields of arable land. In addition, some soils are inherently suitable only for grasslands (heavy clay soils or peat soils; to counteract soil subsidence caused by peat oxidation when tilled).
- An arable crop rotation contains more than just potatoes and cereals. If vegetables, for example, are included, a slightly different picture might emerge, as many vegetables have lower yields in terms of energy and protein than cereals.

- Protein from milk has a higher quality (amino acid pattern) than protein from arable crops (Schaafsma, 2000) and therefore may have added value. On the other hand, a diverse vegetarian diet as a whole can also contain sufficient protein quality (Springmann et al., 2018b).
- Simultaneously with milk, some meat and manure are produced. Due to manure surpluses in the Netherlands (related to high livestock densities), manure has lost its (financial) value, but inherently organic manure is a valuable material and important to keep soil quality high, especially for arable land.

At semi-natural grasslands food production is not the main goal and only a small part of the grass can be converted into food, due to the low feeding value of the grass.

In summary, semi-natural-grasslands have the lowest score (see Fig. 1). Permanent and temporary grasslands score better than semi-natural grasslands, but arable land has by far the highest score for food production.

5 Comparing ecosystem services: soil quality

Good soil quality underpins food production but is also important for other ESs. Different intensities of soil use have different effects on factors such as soil structure, nutrient availability and soil biota. When soil quality is poor, more external inputs such as fertilisers, irrigation and pesticides are needed to achieve good yields.

Soil quality improves with grasslands and usually decreases with arable land (van Eekeren et al., 2008). Grasslands have a high supply of organic matter from crop residues and roots and a relatively low decomposition rate compared to arable land. Therefore, organic matter content increases with permanent grasslands and decreases with arable land (Vleeshouwers and Verhagen, 2002). This is illustrated in Table 4, showing the effects of land use (grasslands and arable land) on six elements of soil quality. Under grasslands, soil structure is greatly improved by an interplay of increased organic matter, rooting and soil biota. On arable land, natural recovery and soil structure formation are lower due to a lower organic matter content, no permanent rooting and less soil biota. To counter this, various agricultural management practices (like increased tillage) are often necessary on arable land.

Compared to temporary grasslands, permanent grasslands contain more organic matter (van Eekeren et al., 2008; Kayser et al., 2018) (see Table 4), allowing the soil to retain more moisture. A comparison between young (<10 years) and old (20 years) grasslands showed that old grasslands contain (in addition to more organic matter) a higher nitrogen stock, resulting in higher nutrient availability for plant growth (Iepema et al., 2020; 2022). Biological

Table 4 Six elements of soil quality from a long-term trial in Belgium

Six elements of soil quality	36 years of permanent grassland	3 years of temporary grassland after 3 years of arable land	3 years of arable land after 3 years of temporary grassland	36 years of continuous arable land
1. Organic matter (%) (0–10 cm)	5.7	3.3	3.8	2.3
2. Soil chemistry (SNS kg/ha) (0–10 cm)*	159	93	102	55
3. Soil structure (crumb + rounded%) (0–10 cm)	76	65	19	21
4. Rooting (number/m^2 at 20 cm)	1081	1813	ND	ND
5. Soil biota (earthworms kg/ha) (0–20 cm)	1660	520	140	50
6. Water management (infiltration mm/s)	2.7	1.1	ND	ND

Average of measurements over 3 years.
* SNS, soil nitrogen supply.
Source: Adapted from: van Eekeren et al. (2008).
ND = Not Determined.

soil quality (see Section 10.2 'Below-ground biodiversity') is also better on permanent grasslands. More active soil biota can give functional benefits in terms of nutrient delivery, soil structure and water balance, thus improving soil quality.

Soil quality of grasslands can decrease when grasslands are used more extensively. Organic matter content on semi-natural grasslands is often lower than or equal to permanent grasslands (Struyk et al., 2020; Lindborg et al., 2022). Extensification often reduces nutrient availability for plant growth. On semi-natural grasslands on peat soils, Deru et al. (2018) found that soil structure and rooting were also worse than on permanent grasslands. This was reflected in poorer water infiltration.

In summary, permanent grasslands have the highest score for soil quality, followed by semi-natural grasslands and then temporary grasslands (see Fig. 1). Arable land has the lowest score for soil quality.

6 Comparing ecosystem services: water consumption and regulation

6.1 Water consumption

The water consumption required by a crop to reach a harvestable product varies by crop type. In (seasonally) dry areas, irrigation supplies the major part of the water consumption. Preferably irrigation is limited as much as possible due to water scarcity.

The water consumption of a crop is closely related to production. Grass consumes 360 litres of water to produce 1 kg of dry matter of feed (Schröder and van Middelkoop, 2016). Triticale and maize are examples of forage crops that use less, i.e. 270 and 180 litres of water per kilogram of dry matter aboveground biomass production, respectively (Schröder and van Middelkoop, 2016). Most arable crops have a high water consumption per hectare of land because of high production. Take for example potatoes: with a production of 48.5 tonnes and a water consumption of 100 litres per kilo of product (van Houwelingen, 2014), the total water consumption is 4850 m^3 per hectare. For grasslands, with a consumption of 360 litres per kilo of dry matter and a yield of 10 tonnes of dry matter, the consumption results in 3600 m^3 per hectare.

Delaying or omitting irrigation in dry years or seasons contributes to lower water consumption but also to lower production because of a reduced or complete stop of crop growth. For arable land, this is not a realistic option because most crops do not recover from drought damage. However, with grasslands, the choice can be made not to irrigate and accept some drought damage. Young grasslands often have deeper roots than older grasslands, making temporary grasslands slightly more drought-tolerant than permanent grasslands (de Haan et al., 2019). In contrast, an old sward of permanent grasslands often recovers relatively easily from drought damage and will start growing again when rain falls. It can therefore be a deliberate choice (also financially) not to irrigate grasslands and accept some damage. Therefore, per land area, arable farming irrigates more than livestock farming (see Box 2).

Box 2. Per land area, arable farming irrigates more than livestock farming.

For example, in a year with average rainfall (2016), 56% of Dutch arable land was irrigated at least once, compared to 18% of feed crops (grasslands and maize; Compendium for the Environment, 2019). Looking at the water consumption of the entire agricultural sector, in the years 2018, 2019 and 2020 (all dry years), an average of 136 billion m^3 of water was used for irrigation in livestock farming and 108 billion m^3 in arable farming (CBS, 2022b). This extra 28 billion m^3 of water for irrigation in livestock farming is only a small additional consumption as the area of feed crops is about double that of arable farming (CBS, 2022a). Per land area, arable farming therefore irrigates more than livestock farming.

In semi-natural grasslands, water consumption will be significantly lower than for the other land use types due to lower production (Lindborg et al., 2022), and, more importantly, it is never irrigated.

6.2 Water regulation

Moisture retention capacity (in times of drought) and water infiltration capacity (during heavy rainfall) are important characteristics in the water-regulating capacity of soils, which can reduce the burden to the main water-regulating system in times of drought and heavy rainfall. Both are strongly related to soil organic matter content. A higher organic matter content prevents clogging of the soil and acts like a sponge; it can fill up with water and slowly release it. Especially for water infiltration, soil structure is also important: an airy, open soil structure can drain (infiltrate) water faster than a soil that is compacted (Deru et al., 2018).

Soil tillage practices such as ploughing result in loss of organic matter, less soil biota and poorer soil structure and as a consequence arable land is able to retain less water than permanent grasslands (Kodešová et al., 2011; Horel et al., 2015). Even a single ploughing in permanent grasslands can negatively affect soil structure and organic matter content, thereby reducing water infiltration and soil water retention capacity (Li et al., 2007).

The increase in organic matter, improvement in soil structure and an increase in earthworm tunnels also allow a grassland soil to drain water faster during and after intense rainfall than arable land (Bouché and Al-Addan, 1997; McLenaghen et al., 2017). These soil improvements increase over the lifetime of grasslands resulting in higher rates of water infiltration in permanent compared to temporary grasslands (van Eekeren et al., 2008). Specifically, anecic earthworms disappear on arable land, while they are maintained by permanent grasslands (van Eekeren et al., 2008; Kautz et al., 2014). These earthworms dig vertical, relatively large tunnels which can greatly improve soil water infiltration (Bouché and Al-Addan, 1997). Moreover, the roots of grasses and herbs can grow deeper through these tunnels relatively easily (little resistance). These roots thus have better access to (ground) water during times of drought (Logsdon and Linden, 1992; van Eekeren et al., 2008). On semi-natural grasslands on peat soils, the combination of lower pH, fewer earthworms, poorer rooting and poorer soil structure gave rise to significantly lower water infiltration than on permanent grasslands (Deru et al., 2018).

In summary, semi-natural grasslands score best, mainly due to the lower production and thus water consumption per hectare, though water regulation of permanent grasslands is slightly better (see Fig. 1). The water consumption of permanent and temporary grasslands is lower than of arable land per unit land, while water regulation of, particularly permanent, grasslands is better.

7 Comparing ecosystem services: carbon dioxide (CO_2) and nitrous oxide (N_2O) emissions

Among GHG emissions, carbon dioxide (CO_2) and nitrous oxide (N_2O) are particularly relevant for grasslands. Because the main focus in this chapter is at field level (see beginning of this chapter) we do not include the (large) methane (CH_4) emissions from ruminants using grasslands. Methane emissions from soil are generally limited, showing little difference between grasslands and arable land (Schils et al., 2022). Therefore in this section, the focus is on CO_2 and N_2O.

7.1 CO_2

The stock of CO_2 in soil is a balance between the formation and decomposition of organic matter. Soil organic matter consists, on average, of 50% carbon. Building up organic matter in a soil therefore contributes to CO_2 sequestration and climate change mitigation. Soil tillage can accelerate decomposition, reducing its content and thus releasing carbon as CO_2.

Permanent grasslands, in which no tillage takes place and in which the soil is covered by plants throughout the year, contain more organic matter than arable land (van Eekeren et al., 2008; Kayser et al., 2018). Kuikman et al. (2003) calculated the amount of carbon in the total plant–soil system of grasslands and arable land with different crops. On average, the soil under grasslands contains four to five times more carbon than arable land. Additionally, the average carbon content in grasslands in Europe increases annually, while it decreases on arable land (Vleeshouwers and Verhagen, 2002).

The higher organic matter content of grasslands compared to arable land is related to a higher supply of organic matter from crop and root residues on grasslands, combined with a lower decomposition due to less tillage. Because the organic matter content increases with the age of grasslands, carbon contents in old grasslands are also higher than in young or renewed (temporary) grasslands (see Fig. 2). Increasing the age of grasslands can store up to 1.8 tonnes C (= 6.6 tonnes CO_2) per hectare per year (Garcia et al., 2023). After about 40–100 years of permanent grassland, saturation occurs, after which there is not much additional sequestration (Six et al., 2002; de Wit et al., 2018; Iepema et al., 2022). The soil then contains a large carbon stock, so higher carbon losses can be expected when ploughing old permanent grasslands compared to ploughing young grasslands (Iepema et al., 2022) (Box 3).

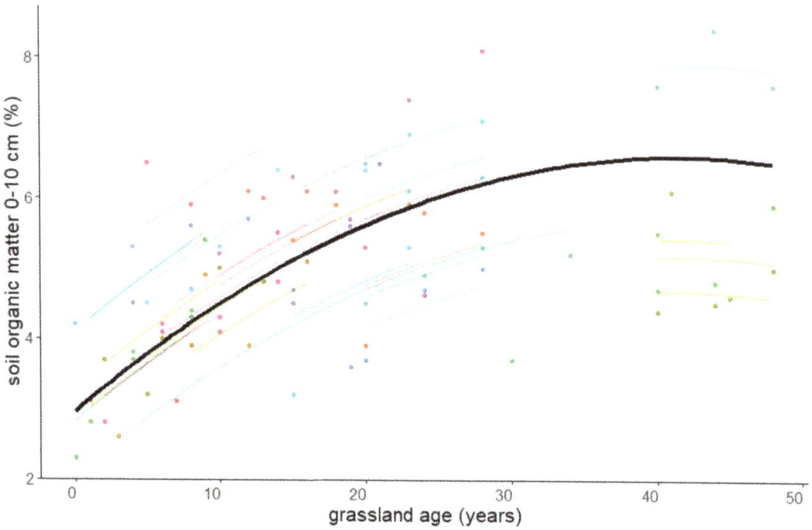

Figure 2 Increase in organic matter content with age of permanent grasslands on sandy soils. The thick black line represents the average, the coloured lines represent increases in individual grasslands. Source: Adapted from: De Wit et al. (2018).

Box 3. Grasslands organic matter and GHG emissions on a dairy farm in perspective

Including animal emissions, the average dairy farm in the Netherlands emits 992 g CO_2 equivalents per kilogram of delivered milk, or 1125 g if changes in land use and organic matter are also included (Hospers et al., 2022). At a production of 17 000 kg of delivered milk per hectare, GHG emissions are calculated to be 17–19 tonnes CO_2-equivalent per hectare per year. The organic matter build-up of young grasslands (up to 6.6 ton CO_2-equivalent per hectare per year; Garcia et al., 2023) may offset maximum 38% of this.

Carbon sequestration is strongly related to the supply of organic matter from crop and root residues, which in turn is related to plant growth and fertilisation. Higher fertilisation levels, and thus higher plant production, contribute positively to carbon sequestration (Schils et al., 2022). Semi-natural grasslands will therefore sequester less carbon than permanent grasslands (Struyk et al., 2020; Lindborg et al., 2022). Semi-natural grasslands do generally have similar or lower soil carbon stocks than productive grasslands, despite the lower carbon sequestration. This similarity in carbon stocks is mainly related to the fact that these semi-natural grasslands tend to be old permanent grasslands and thus have accumulated carbon for a longer time, in addition to lower organic matter decomposition rates due to soil conditions (such as lower pH and higher groundwater levels) (van Eekeren et al., 2022).

7.2 N_2O

N_2O is a GHG with ±273 times the global warming potential of CO_2 on a time scale of 100 years, though this factor has been slightly changing over time (Van Dijk et al., 2023). N_2O is mainly formed when there is a local lack of oxygen in the soil, due to the soil being very wet or compacted, and when sufficient easily degradable organic matter, ammonium or nitrate is available.

Apart from soil conditions, N_2O emissions are mainly linked to fertiliser levels; land use (grasslands versus arable land) only has a limited influence (Schils et al., 2022). When less fertiliser is applied, there is generally also less N_2O emission. Since the fertilisation level of grasslands in the Netherlands is higher than of arable land (350 vs 220 kg N per ha; Hoogeveen, 2022), N_2O emissions from grasslands in the Netherlands are assumed to be higher than from arable land; the difference between grasslands and arable land is approximately 1 kg N_2O-N or 0.5 ton CO_2-equivalent per hectare (based on Van Dijk et al., 2023). There are also considerable N_2O emissions associated with renewing permanent grasslands or converting temporary grasslands to arable land when a lot of nitrogen is released through mineralisation (Vellinga et al., 2004). Compared to a reference grassland, this extra N_2O emission amounts to 5.5 kg N or approx. 2.2 ton CO_2-equivalent per hectare in total after grassland renewal, measured during one whole year after renewal but mainly emitted during a short period with favourable conditions (Velthof et al., 2009).

In semi-natural grasslands, N_2O emissions are low as fertilisation is limited, except if there are changes in groundwater levels: N_2O and/or CH_4 emissions can rise sharply in case the land is converted into wetlands.

In summary, semi-natural grasslands with a low N_2O emission and reasonable potential for carbon sequestration and maintaining carbon stock has the highest score (see Fig. 1). Permanent grasslands with its high potential for carbon sequestration and maintaining carbon stock is second despite a relatively high N_2O emission. Temporary grasslands are on the third place, while arable land has the lowest score because of the low potential of carbon sequestration.

8 Comparing ecosystem services: ammonia

Losses of nitrogen to the environment in the form of ammonia can cause unwanted enrichment (eutrophication) and acidification of natural areas. This has negative effects on biodiversity, particularly in N-sensitive terrestrial areas.

Ammonia emissions are related to the level of fertilisation, fertilisation method and weather conditions, and type of (organic or inorganic) fertiliser. These factors determine the percentage of ammonia from fertiliser that is lost to air, i.e. the emission factor. On arable land, ammonia emissions from organic fertiliser application are usually lower than on permanent and temporary grasslands, because the organic fertiliser can be worked directly into the soil (van Bruggen et al., 2022).

For Dutch conditions, it is estimated that in dairy slurry manure, 48% of the nitrogen is ammonia bound (this is total ammoniacal N, or TAN) and is prone to losses through volatilisation (Van Dijk et al., 2023). The average emission factor during and after slurry application is estimated to be 17% of the TAN for grasslands (mainly superficial sod injection) and 2% for arable land (van Bruggen et al., 2022). For both grasslands and arable land an emission factor of 2.5% is estimated for inorganic fertiliser (CAN-N) (van Bruggen et al., 2022). Total ammonia emissions from grasslands (permanent and temporary) and arable land are calculated to be on average 22 kg and 4 kg per hectare, respectively (Table 5). In semi-natural grasslands, no or very little manure is applied and ammonia emissions from manure application will therefore be minimal.

In summary, semi-natural grasslands and arable land score best for ammonia emission (see Fig. 1). Permanent and temporary grasslands have a relatively high ammonia emission and have therefore the lowest score.

9 Comparing ecosystem services: water quality

9.1 Nitrate

Excess nitrate can leach into groundwater and reduces water quality. The risk of nitrate leaching is higher at higher levels of fertilisation (Schils et al.,

Table 5 Average ammonia emissions following fertilisation of grasslands and arable land in the Netherlands

	Slurry			Inorganic fertiliser		
	Application rate (kg N/ha)	TAN (kg/ha)	Emissions kg N/ ha (EF)	Application rate (kg N/ha)	Emissions kg N/ ha (EF)	Total emissions Kg N/ha
Grasslands	230[a]	110[c]	19 (17%)[d]	120[a]	3 (2.5%)[d]	22
Arable land	110[b]	52[c]	1 (2%)[d]	110[b]	3 (2.5%)[d]	4

[a]Hoogeveen (2022), [b]van Leeuwen (2022), [c]Van Dijk et al. (2023), [d]Van Bruggen et al., 2022.

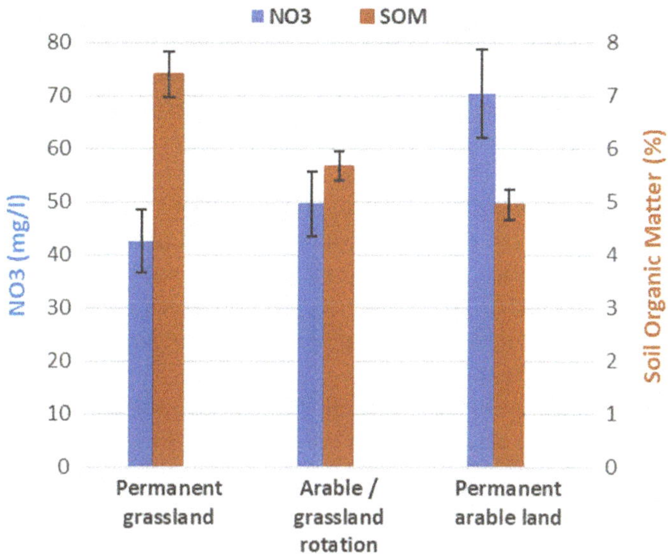

Figure 3 Average nitrate concentrations in upper groundwater (0.5–1.5 m) and percentage of organic matter by land use type. Source: Adapted from: van der Wal et al. (2019).

2022). In general, grasslands are less susceptible to nitrate leaching than arable land (Di and Cameron, 2002). Apart from differences in soil type (with effect on denitrification), nitrate leaching is lower in grasslands because grass has a long growing season and a high nitrogen uptake. Because grasslands have a high organic matter content and microbial biomass in which much nitrate can be stably sequestered (see Fig. 3), tillage for re-seeding increases nitrate leaching from temporary grasslands (Schils et al., 2022). On arable land, annual tillage causes mineralisation of organic matter, releasing a lot of nitrate. When mineralisation releases more nitrate than the crop can absorb, it can easily leach out during periods with an excess of precipitation. In the Netherlands, monitoring shows that nitrate concentrations in groundwater on sandy soils under arable land with maize are more than 2.5 times higher than under grasslands (Hooijboer et al., 2017). Especially after arable crop harvest (during autumn and winter), there is an increased risk of leaching because the soil is no longer covered (Di and Cameron, 2002). Nitrate may also leach out in spring before the roots of the field crop are sufficiently developed to absorb nitrogen.

9.2 Phosphate

Losses of phosphate through leaching are higher for arable land than for grasslands, but tend to be limited for both types of land use (Marrs et al., 1991), depending on the degree of phosphate saturation. In general, phosphate is strongly bound to the soil and does not leach out easily. However, due to long-term and high phosphate fertilisation in the Netherlands in the past, this binding capacity has been partly saturated on a part of the soils, resulting in higher leaching and leakage of phosphate for long periods (Schoumans, 2004; van Grinsven and Bleeker, 2017). The degree of phosphate saturation is almost independent of present land use and thus not discriminating between the land use types considered in this chapter.

Soil erosion can also cause phosphate bound to soil particles to run off into surface water. On slopes, there is a higher risk of erosion. Permanent and temporary grasslands prevent the soil from erosion and hence phosphate runoff (Butler & Haygarth, 2007). When renewing grasslands, losses of phosphate can be minimised by carrying out soil tillage during dry periods, thus preventing erosion and runoff (Butler & Haygarth, 2007). Erosion and runoff can also occur on flat land during heavy rainfall and poor drainage. Moreover, lack of infiltration can cause ponding, which can lead to phosphate release and runoff. On grasslands, water can infiltrate better than on arable land (see also Section 6, 'Water consumption and regulation'), limiting ponding and runoff.

As for nitrate, higher levels of fertilisation also put phosphate at risk of leaching shortly after application. When fertiliser is applied and it rains shortly afterwards, the phosphate in the organic and inorganic fertiliser can leach out even before it can attach to the soil (Magid et al., 1999; Bajouco et al., 2020).

9.3 Pesticides

Pesticides can be environmentally damaging, particularly when they leach into surface and groundwater, reducing water quality. The use of pesticides is low on grasslands compared to arable land. A study by Lageschaar et al. (2020) shows that in the Dutch province of North Brabant, the environmental impact of pesticides in water from grasslands is very small compared to other crops. Crops such as potatoes (56–169 times more) and cereals (9–13 times more) are treated with a multitude amount of pesticides compared to grasslands, where herbicides are the only pesticides used (see Table 6) (BINternet, n.d.-c).

Temporary grasslands generally require less weed control than permanent grasslands. For example, in a rotation of 3 years maize followed by 3 years grass, virtually no weed control was needed during the grassland phase (Kool

Table 6 Pesticides used per hectare in 2021 in various crops and grasslands in the Netherlands

Land use	Cultivation	Total active substance (kg/ha)	Insecticides (kg/ha)	Fungicides (kg/ha)	Herbicides (kg/ha)	Nematicides (kg/ha)	Other (kg/ha)
Arable land	Wheat	2.7		0.6	1.3		0.8
	Barley	1.9	0.1	0.2	1.6		0.1
	Potato						
	-Seed	33.8	0.2	5.7	2.4	0.5	25.0
	-Consumption	11.3	0.5	5.6	3.1	0.2	2.0
	-Starch	13.1	0.0	8.8	3.0	0.7	0.5
	Silage maize	0.9			0.9		
Grasslands	Grass[a]	0.2			0.2		

Use (kilogram active ingredient, rounded to one decimal place) is shown as total and by category (insecticides, fungicides, herbicides, nematicides and other) (BINternet, n.d.-c).
[a]Both permanent and temporary grasslands.

et al., 2004). However, this lower herbicide use in temporary grasslands is linked to more weed control during the arable phase (and therefore inherent to the system). In addition, glyphosate (RoundUp) is commonly used during the conversion of (temporary) grassland to arable land to destroy the sod. On average, the difference between permanent and temporary grasslands in the use of herbicides is small and depending on the frequency of conversion of temporary grasslands in the crop rotation.

In summary, semi-natural grasslands score best for water quality, followed by permanent grasslands and then temporary grasslands (see Fig. 1). Arable land has the lowest score.

10 Comparing ecosystem services: biodiversity

10.1 Above-ground biodiversity

Important above-ground species that use agricultural land as habitat are plants, insects, large and small mammals and birds. Permanent grasslands contribute more to the conservation of endangered species than arable land (Schils et al., 2022). A greater diversity of plant species is also found in grasslands than in arable land. On arable land, often only one type of crop is grown and weeds are removed, while grasslands usually consists of multiple species of herbs and grasses (wild or seeded). Arable land often provides little above-ground biomass that can serve as food for wild species, as a major part of the plant is usually harvested as food for humans. To ensure this harvest, in some cases the arable crop is also protected from predation by pesticides (see also Section 9.3, 'Pesticides') which can be detrimental to biodiversity (Hallmann et al., 2017).

Grasslands also offer a more stable habitat than arable land, which is tilled at least once every year. Many organisms depend on a stable habitat for, e.g., reproduction, nesting or the presence of specific host plants. On average, the biodiversity of arable land in Europe still has 10% of its original value, while the biodiversity of grasslands for agricultural use is on average 26% of its original value (Reidsma et al., 2006).

Permanent grasslands can be expected to ensure greater biodiversity than periodically renewed young grasslands (Tscharntke and Greiler, 1995). Permanent grasslands make a greater contribution to plant diversity, pollinators and the conservation of endangered species and thus deliver a greater biodiversity than temporary grasslands (Schils et al., 2022). In part, this is because permanent grasslands provide a long-term stable habitat, allowing populations to establish.

Semi-natural grasslands also form a long-term stable habitat and form a better basis for increased biodiversity through its lower land use intensity, offering plants, insects and birds sufficient time to complete their interconnected life cycle. Flowering herbs provide nectar and pollen that benefit flower-visiting insects (Ozinga et al., 2018; Stip and van Grunsven, 2018). The presence of many insects in turn contributes to the food supply of insectivorous farmland birds. Such low-productivity grasslands can also provide a favourable vegetation structure for meadow bird chicks (sufficient shelter, low crop density and high structural variation), which enables chicks to wade through the vegetation and catch insects while walking (Kentie et al., 2013; Schekkerman & Beintema, 2007).

Above-ground biodiversity is a main ingredient for landscape appreciation, with the presence of birds, insects and other animals being the most important (Buijs et al., 2019). The presence of (flowering) herbs also increases the landscape appreciation of grasslands (Schils et al., 2022). Grasslands therefore generally have a clear added value in the landscape compared to arable land (Schils et al., 2022).

In summary, semi-natural grasslands score best, followed by permanent and temporary grasslands (see Fig. 1). Arable land has the lowest score for above-ground biodiversity.

10.2 Below-ground biodiversity

The influence of land use, especially grassland versus arable land, on the numbers and diversity of soil biota is very high. Grasslands are characterised by much less soil disturbance than arable land and can therefore provide a stable habitat for soil organisms. In addition, grasslands have a year-round insulating litter and vegetation layer, and a constant supply of crop and root residues that serves as food for soil organisms (Edwards and Bohlen, 1996). Permanent grasslands can therefore contain as many as 8.5 times more earthworms with a total biomass up to 33 times higher than in arable land (van Eekeren et al., 2008; Rutgers et al., 2016). The number of earthworms in temporary grasslands depends on the percentage of grassland in the cropping pattern and grassland age at the moment of measurement. In a crop rotation of 3 years of temporary grassland and 3 years of arable land, the earthworm count starts almost from zero in the first year after conversion from arable land to grassland and after 3 years, the temporary grassland will not yet reach a similar number of earthworms that is usually found on permanent grasslands (see Table 4). Extrapolations show that it will take 5 years in temporary grasslands before it returns to the level of permanent grasslands (Van Eekeren et al., 2008). This recovery can be accelerated by leguminous plants in the grasslands that stimulate the number of earthworms in the soil (Sears, 1950; van Eekeren et al., 2009; Crotty et al., 2015).

While above-ground biodiversity increases in semi-natural grasslands, this is not always the case below ground. The numbers and biomass of earthworms in semi-natural grasslands depend heavily on pH and nutritional status. With impoverishment, nutrition and pH decreases, and hence, earthworm reproduction decreases. In a comparison between semi-natural and permanent grasslands on dairy farms on peatland, permanent grasslands contained twice as many earthworms as semi-natural grasslands because of these two processes (Deru et al., 2018). Earthworm diversity was also less per field (so-called alpha diversity) in semi-natural than in permanent grasslands (Deru et al., 2018). However, due to the greater differences in geo-botanical conditions (subsoil, etc.) between the individual semi-natural grasslands, the total earthworm diversity for all the fields combined (so-called gamma diversity) was greater for semi-natural grasslands compared to permanent grasslands (Deru et al., 2018).

The abundance and diversity of mites (*Acarina*) and springtails (*Collembola*) are highest in permanent grasslands and decrease sharply when grasslands are converted to arable land (Postma-Blaauw et al., 2010, 2012). More mites and springtails were measured among semi-natural grasslands on sandy soils compared to permanent grasslands (Rutgers et al., 2009). Van Eekeren et al. (2022) found no significant difference in the abundance of mites and springtails in a comparative study of permanent grasslands and semi-natural grasslands, but did find a higher alpha diversity in semi-natural grasslands.

Microbial biomass and activity is higher in grasslands than in arable land and increases with increasing age of grasslands (Plassart et al., 2008; van Eekeren et al., 2008, 2010; Grayston et al., 2004). Microbial activity in permanent grasslands is on average 3–5 times higher than in permanent arable land (van Eekeren et al., 2008). Even one-time ploughing to renew permanent grasslands reduces microbial biomass and even though the microbial biomass does recover, effects can be observed for up to 5 or more years after disturbance (Murugan et al., 2013).

In summary, permanent and semi-natural grasslands score best, mainly due to the higher number of years without tillage (see Fig. 1). Temporary grasslands take an intermediate position. Arable land has the lowest score because of the annual tillage and no permanent vegetation cover.

11 Discussion: land use changes

11.1 Temporal effects on ecosystem services if land use changes

In the comparison in previous paragraphs, land use types have been considered as 'permanent/stable' situations. However, if assessing changes in land use types, temporal effects also have to be included. Such temporal effects are particularly relevant for GHGs and water quality: conversion of grasslands to arable land induces break-down of soil organic matter which gives, temporarily, a huge increase in GHG emissions (13–15 times the total annual emissions of an average hectare of a dairy farm; see Box 4) and nitrate leaching (well above 100 mg nitrate/L; van den Berg and Poelen, 2003). Thus, conversion from grasslands to arable land will not only result in a permanent decline in ESs such as soil quality, water regulation, biodiversity and landscape (see Fig. 1), but it will also result in huge temporal effects on CO_2 storage and water quality.

Box 4. Greenhouse gas emissions and land use change

If permanent grasslands are converted to permanent arable land, a major part of the soil organic matter will be decomposed due to increased tillage, reduced organic matter input and, often, increased liming (required for most arable crops). Vellinga et al. (2004) calculated that 250 tonnes of CO_2-equivalent per hectare can be released over the years before the soil finds a new equilibrium. This amounts to 13–15 times the total annual GHG emissions of an average hectare, including related animal emissions, of a Dutch dairy farm (17–19 tonnes of CO_2equivalent per year). For (peat) soils with a very high organic matter content, this temporal emission peak is many times higher.

On the other hand, conversion of a field of arable land to grassland can actually turn this source of CO_2 emissions into a sink, making a significant contribution to soil carbon sequestration. A literature review concludes that after converting arable land to grassland it can store up to 1.8 tonnes C per hectare per year (Garcia et al., 2023), thus sequestrating – up to 6.6 tonnes CO_2 per hectare per year.

11.2 Policy decisions and land use change

The impact assessment of policy interventions on ESs is complicated due to the uncertainty of subsequent changes in agricultural land use. Land use changes are, commonly, not dictated as such but result from policy decisions affecting (future) legal, economic and cultural circumstances on which farmers base their decisions related to land use (see Fig. 4). This can be illustrated by the likely effects of policy-regulated reduction of dairy numbers/production. Although reduced livestock production might lead to an extensification of grassland use, this is not a plausible scenario without major additional measures. More likely, the area of grasslands will decrease and will be converted into arable land (silage maize or more intensive arable crops) as these crops are more economically attractive

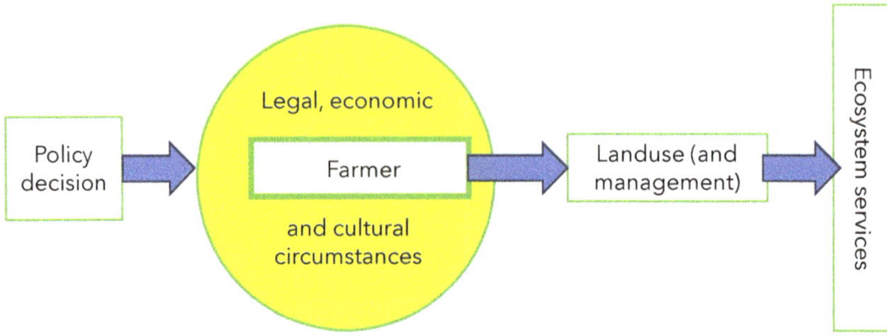

Figure 4 Relation between policy decisions and ecosystem services (simplified). Land use changes are, commonly, not dictated as such but result from policy decisions affecting (future) legal, economic and cultural circumstances on which farmers base their decisions related to land use.

(van den Hout et al., 2023). Thus a policy-regulated reduction of dairy production without additional measures will most likely result in a decline in most ESs.

The impact assessment of policy decisions becomes even more complicated if one considers that both GHG emissions and food production are a global phenomenon. This means that a regional reduction in GHG emissions is meaningless if counteracted by an increase in another region. Likewise, a reduction of dairy production in the Netherlands (of which a major part is exported) is unlikely to lead to a meaningful reduction in global dairy consumption. Without additional measures to limit consumption, there will be mainly a shift of dairy production to other regions with a comparable or higher CO_2 footprint. The combination of both characteristics means that a reduction of the Dutch dairy production is unlikely to result in a net reduction of the global warming potential, if it is not accompanied by a reduction in dairy consumption.

Such assessments, including interconnected agronomic and socio-economic effects of interventions, central to so-called consequential LCAs, are highly complex, revealing sometimes unintended or unexpected consequences of interventions compared to the more common attributional LCAs (Schaubroeck et al., 2021; Costa et al., 2020). For a German context, e.g., it is shown that substitution of cow milk with soy-based milk does not lead to significant reduction of GHG emissions, due to a connected shift from dairy-beef production to less efficient suckler-beef systems if the substitution of milk is not accompanied by drastic reductions in beef consumption (Costa et al., 2023).

12 Conclusion and future trends

A continued search for innovations which reduce the environmental impact of livestock systems (e.g. GHG and N emissions) is essential to keep the impact of an increased human population with an increased purchasing power within all planetary boundaries (Springmann et al., 2018a). This search should be accompanied by assessments of the impact of these innovations on other ESs under various agro-ecological conditions. System comparisons should include agricultural systems with less livestock production, because most studies indicate that diets with less animal products are essential to reduce the negative environmental impact of food systems (Tilman and Clark, 2014; Röös et al., 2017). For example, food system assessments show that red meat consumption, particularly beef, is minimised in sustainable healthy diets for high- and middle-income countries; dairy production is less affected (Springmann et al., 2018b; Van Zanten et al., 2023). This opens questions on suitable techniques for viable dairy systems on mainly grass from grasslands that widens crop rotations and from land less suitable for crop production. This type of questions has been hardly dealt with, but indications are that they do comprise lower productive animals than currently common in NW Europe (van Hal et al., 2019). Related to this is the question on how to attain such systems, as in most countries dairy production is still intensifying with a reduction of grass in the feed rations.

It is shown (see Fig. 1) that, particularly permanent grasslands score better than arable land on the provision of important ecosystem services such as: soil quality, water consumption and regulation, climate change, water quality, above-ground and below-ground biodiversity.

At the moment, negative environmental impact by livestock farming has received major attention, mainly related to GHG and nitrogen emissions. An emphasis on simply reducing livestock numbers risks losing sight of the positive role that grasslands play in the provisioning of ESs. Without additional measures, large-scale conversion of grasslands to arable land is likely if livestock numbers are to be reduced, which means a sharp decline in the ESs provided by agricultural land. But more importantly, conversion of grasslands into arable land also gives temporarily a huge increase in GHG emissions and nitrate leaching. Therefore, it is essential that temporal and interconnected agro- and socio-economic effects of interventions are taken into account if complex changes in food systems are envisaged.

13 Where to look for further information

- Richter, F., et al. (2021). A guide to assess and value ecosystem services of grasslands. *Ecosystem Services*, *52*, 101376.
- Schils, L. M. R., et al. (2022). Permanent grasslands in Europe: land use change and intensification decrease their multifunctionality. *Agriculture, Ecosystems & Environment, 330.*

- Van Hal, O., et al. (2019). Upcycling food leftovers and grass resources through livestock: impact of livestock system and productivity. *Journal of Cleaner Production*, *219*, 485–496.
- Wang, Y., et al. (2023). Risk to rely on soil carbon sequestration to offset global ruminant emissions. *Nature Communications*, *14*(1), 7625.

14 References

Allen, V. G., et al. (2011). An international terminology for grazing lands and grazing animals. *Grass and Forage Science*, *66*, 2–28.

Bajouco, R., et al. (2020). Risk of phosphorus losses from Andosols under fertilized pasture. *Environmental Science and Pollution Research*, *27*, 19592–19602.

BINternet (n.d.-a). *BINternet, land- en tuinbouw.* Agrimatie. https://www.agrimatie.nl/Binternet.aspx?ID=14&Lang=0 (accessed 2 March 2023).

BINternet (n.d.-b) *BINternet, land- en tuinbouw.* Agrimatie. https://www.agrimatie.nl/Binternet.aspx?ID=15&Bedrijfstype=1&SelectedJaren=2022@2021@2020@2019&GroteKlassen=Alle%20bedrijven (accessed 2 March 2023).

BINternet. (n.d.-c). *BINternet, land- en tuinbouw "gewasbescherming per gewas".* Agrimatie. https://www.agrimatie.nl/binternet.aspx?ID=21&bedrijfstype=2 (accessed 2 March 2023).

Board (n.d.). Millennium ecosystem assessment. "http://www.millenniumassessment.org" www.millenniumassessment.org (accessed 2 March 2023)

Bouché, M. B. & Al-Addan, F. (1997). Earthworms, water infiltration and soil stability: some new assessments. *Soil Biology and Biochemistry*, *29*(3–4), 441–452.

Buijs, A., et al. (2019). Resultaten Nationale Landschapsenquête; Onderzoek naar visies en waardering van de Nederlandse bevolking over het landelijk gebied in Nederland. *Wageningen Environmental Research, Rapport* 2937.

Butler, P. J. & Haygarth, P.M. (2007). Effects of tillage and reseeding on phosphorus transfers from grassland. *Soil Use and Management*, 23(1), 71–81.

CBS. (2022a, 30 november). *Landbouw; gewassen, dieren en grondgebruik naar regio.* CBS. https://opendata.cbs.nl/#/CBS/nl/dataset/80780ned/table (accessed 2 March 2023).

CBS. (2022b, 12 oktober). *De verduurzaming van de landbouw – deel I: productie en verbruik.* https://www.cbs.nl/nl-nl/longread/statistische-trends/2022/de-verduurzaming-van-de-landbouw-deel-i-productie-en-verbruik/4-verbruik-door-de-landbouw (accessed 2 March 2023).

Compendium voor de leefomgeving. (2019, 5 februari). *Waterverbruik in de land- en tuinbouw, 2001–2016.* https://www.clo.nl/indicatoren/nl001413-watergebruik-landbouw (accessed 2 March 2023).

Costa, M. P., et al. (2020). Representing crop rotations in life cycle assessment: a review of legume LCA studies. *The International Journal of Life Cycle Assessment*, *25*, 1942–1956.

Costa, M. P., et al. (2023). Environmental and land use consequences of replacing milk and beef with plant-based alternatives. *Journal of Cleaner Production*, *424*, 138826.

Crotty, F. V., et al. (2015). Assessing the impact of agricultural forage crops on soil biodiversity and abundance. *Soil Biology & Biochemistry*, *91*, 119–126.

De Haan, M., et al. (2019). Droogte op Koeien & Kansen-bedrijven in 2018. *Wageningen Livestock Research, Rapportnummer 84.* https://doi.org/10.18174/470046

Deru, J. G., et al. (2018). Soil ecology and ecosystem services of dairy and semi-natural grassland on peat. *Applied Soil Ecology*, *125*, 26–34.

De Wit, J., et al. (2018). Opbouw organische stof met blijvend grasland. *V-Focus 21 Maart, 2018*, 32–34.

Di, H. & Cameron, K. (2002). Nitrate leaching in temperate agroecosystems: sources, factors and mitigating strategies. *Nutrient Cycling in Agroecosystems*, *64*, 237–256.

Edwards, C. A. & Bohlen, P. J. (1996). *Biology and ecology of earthworms* (Vol. 3). Springer Science & Business Media.

FAO (n.d.). Land statistics 2001–2023. Global, regional and country trends.

Garcia, L. B. M., et al. (2023). *Carbon sequestration through agricultural practices – a review of international literature*. Slim Landgebruik.

Grayston, S. J., et al. (2004). Assessing shifts in microbial community structure across a range of grassland of differing management intensity using CLPP, PLFA and community DNA techniques. *Applied Soil Ecology*, *25*(1), 63–84.

Haines-Young, R. (2023). Common international classification of ecosystem services (CICES) V5.2 and guidance on the application of the revised structure. cices.eu

Hallmann, C. A., et al. (2017). More than 75 percent decline over 27 years in total flying insect biomass in protected areas. *PLoS One*, *12*, e0185809. https://doi.org/10.1371/journal.pone.0185809.

Hoogeveen, M. (2022). *Stijging van stikstofbemesting in 2020 na daling in 2019*. Agrimatie. https://www.agrimatie.nl/ThemaResultaat.aspx?subpubID=2232&themaID=2282&indicatorID=2772 (accessed 2 March 2023).

Horel, Á., et al. (2015). Effects of land use and management on soil hydraulic properties. *Open Geosciences*, 7(1), 742–754.

Hospers, J., et al. (2022). The evolution of the carbon footprint of Dutch raw milk production between 1990 and 2019. *Journal of Cleaner Production*, *380*, 134863.

Hooijboer, A. E. J., et al. (2017). Effects of crop rotation on water quality in the Netherlands: Combining the Minerals Policy Monitoring Programme and Nation-wide survey of crop data of the sandy regions of the Netherlands. *LuWQ2017, Land Use and Water Quality: Effect of Agriculture on the Environment The Hague, the Netherlands, 29 May – 1 June 2017*.

Iepema, G., et al. (2020). Productivity and topsoil quality of young and old permanent grassland: an on-farm comparison. *Sustainability*, *12*(7), 2600.

Iepema, G., et al. (2022). Extending grassland age for climate change mitigation and adaptation on clay soils. *European Journal of Soil Science*, *73*, 1–14.

Jurasinski, G., et al. (2016). *Ecosystem services provided by paludiculture – greenhouse gas emissions*. In: Wichtmann, W., et al. (eds.). *Paludiculture - productive use of wet peatlands*. Schweizerbart Science Publishers. www.schweizerbart.de/9783510652839 (accessed 25 April 2019).

Kautz, T., et al. (2014). Contribution of anecic earthworms to biopore formation during cultivation of perennial ley crops. *Pedobiologia*, *57*(1), 47–52.

Kayser, M., et al. (2018). Grassland renovation has important consequences for C and N cycling and losses. *Food and Energy Security*, *7*(4), e00146.

Kentie, R., et al. (2013). Intensified agricultural use of grasslands reduces growth and survival of precocial shorebird chicks. *Journal of Applied Ecology*, *50*(1), 243–251.

Kodešová, R., et al. (2011). Soil structure and soil hydraulic properties of Haplic Luvisol used as arable land and grassland. *Soil and Tillage Research*, *111*(2), 154–161.

Kool A., et al. (2004). Gewasbescherming binnen de grenzen. Resultaten gewasbescherming De Marke 1993–2002. *Rapport 44. CLM-Rapport 588-2004.*

Kuikman, P. J., et al. (2003). *Stocks of C in soils and emissions of CO2 from agricultural soils in the Netherlands* (No. 561). Alterra.

Lageschaar, L., et al. (2020). *Gevolgen voor milieubelasting van het water door verschuiving naar intensievere teelten in Noord-Brabant – Achtergrondrapport*. CLM.

Lesschen, J. P., et al. (2012). *Mogelijkheden voor koolstofvastlegging in de Nederlandse landbouw en natuur*. Alterra, Wageningen.

Li, X. G., Li, F. M., Zed, R., Zhan, Z. Y., & Singh, B. (2007). Soil physical properties and their relations to organic carbon pools as affected by land use in an alpine pastureland. *Geoderma*, *139*(1–2), 98–105.

Lindborg, R., et al. (2022). Ecosystem services provided by semi-natural and improved grasslands – synergies, trade-offs and bundles. *European Grassland Federation, Proceedings of the 29th General Meeting*, 55–66.

Logsdon, S. D. & Linden, D. R. (1992). Interactions of earthworms with soil physical conditions influencing plant growth. *Soil Science*, *154*(4), 330–337.

Magid, J., et al. (1999). Phosphate leaching responses from unperturbed, anaerobic, or cattle manured mesotrophic sandy loam soils. *Journal of Environmental Quality*, *28*, 1796–1803.

Marrs, R. H., et al. (1991). Soil chemistry and leaching losses of nutrients from semi-natural grassland and arable soils on three contrasting parent materials. *Biological Conservation*, *57*(3), 257–271.

McLenaghen, R. D., et al. (2017). Improvement of degraded soil physical conditions following the establishment of permanent pasture. *New Zealand Journal of Agricultural Research*, *60*(3), 287–297.

Murugan, R., et al. (2013). Specific response of fungal and bacterial residues to one-season tillage and repeated slurry application in a permanent grasslands soil. *Applied Soil Ecology*, *72*, 31–40.

Ojea, E., et al. (2010). Classifying ecosystem services for economic valuation: the case of forest water services. https://addi.ehu.es/bitstream/handle/10810/14215/BC3WP201012.pdf

Ozinga, W. A., et al. (2018). *Wilde bijen en zweefvliegen per landschapstype* (No. 2920). Wageningen Environmental Research.

Plassart, P., et al. (2008). Molecular and functional responses of soil microbial communities under grassland restoration. *Agriculture, Ecosystems & Environment*, *127*(3–4), 286–293.

Postma-Blaauw, M. B., et al. (2010). Soil biota community structure and abundance under agricultural intensification and extensification. *Ecology*, *91*(2), 460–473.

Postma-Blaauw, M. B., et al. (2012). Agricultural intensification and de-intensification differentially affect taxonomic diversity of predatory mites, earthworms, enchytraeids, nematodes and bacteria. *Applied Soil Ecology*, *57*, 39–49.

Reidsma, P., et al. (2006) Impacts of land-use change on biodiversity: an assessment of agricultural biodiversity in the European union. *Agriculture, Ecosystems and Environment*, *114*, 86–102.

Richter, F., et al. (2021). A guide to assess and value ecosystem services of grasslands. *Ecosystem Services*, *52*, 101376.

Röös, E., Bajželj, B., Smith, P., Patel, M., Little, D., & Garnett, T. (2017). Greedy or needy? Land use and climate impacts of food in 2050 under different livestock futures. *Global Environmental Change*, *47*, 1–12.

Rutgers, M., et al. (2009). Biological measurements in a nationwide soil monitoring network. *European Journal of Soil Science*, *60*, 820–832.

Rutgers, M., et al. (2016). Mapping earthworm communities in Europe. *Applied Soil Ecology*, *97*, 98–111.

Schaafsma, G. (2000). The protein digestibility–corrected amino acid score. *The Journal of Nutrition*, *130*(7), 1865S–1867S.

Schaubroeck, T., et al. (2021). Attributional & consequential life cycle assessment: definitions, conceptual characteristics and modelling restrictions. *Sustainability*, *13*(13), 7386.

Schekkerman, H. & Beintema, A. J. (2007). Abundance of invertebrates and foraging succes of Black-tailed godwit *Limosa limosa* chicks in relation to agricultural grassland management. *Ardea*, *95*(1), 39–54.

Schils, L. M. R., et al. (2022). Permanent grasslands in Europe: land use change and intensification decrease their multifunctionality. *Agriculture, Ecosystems & Environment*, *330*, 1–10.

Schoumans, O. F. (2004). *Inventarisatie van de fosfaatverzadiging van landbouwgronden in Nederland*. Wageningen, Alterra rapport 730.4.

Schröder, J. J. & van Middelkoop, J. C. (2016). *Milieukundige voetafdruk van de teelt van ruwvoergewassen*. Wageningen Plant Research.

Sears, P. D. (1950). Soil fertility and pasture growth. *Journal of the British Grassland Society*, *5*, 267–280.

Silvis, H. J. & Van der Heide, C. M. (2013). *Economic viewpoints on ecosystem services*. Wageningen, Statutory Research Tasks Unit for Nature and the Environment, WOT-report 123.

Six, J., et al. (2002). Stabilization mechanisms of soil organic matter: implications for C-saturation of soils. *Plant and Soil*, *241*, 155–176.

Spedding, C. R. W. (1988). *An introduction to agricultural systems*. Springer.

Springmann, M., et al. (2018a). Options for keeping the food system within environmental limits. *Nature*, *562*(7728), 519–525.

Springmann, M., et al. (2018b). Health and nutritional aspects of sustainable diet strategies and their association with environmental impacts: a global modelling analysis with country-level detail. *The Lancet Planetary Health*, *2*(10), e451–e461.

Statline. (n.d.). Landbouw sinds 1851. https://opendata.cbs.nl/#/CBS/nl/dataset/71904ned/table?ts=1706022043863 (accessed 23 January 2024)

Stip, A. & van Grunsven, R. H. A. (2018). *Beheermaatregelen voor insecten in graslanden in midden Friesland*. De Vlinderstichting, Rapport VS2018.018.

Struyk, P. A., Timmermans, B. G. H., & van Eekeren, N. (2020). Organisch stofgehalte in bodem natuurgraslanden niet gestegen. *Vakblad Natuur Bos Landschap 168*, 12–15.

Tilman, D. & Clark, M. (2014). Global diets link environmental sustainability and human health. *Nature*, *515*(7528), 518–522.

Tscharntke, T. & Greiler, H. J. (1995). Insect communities, grasses, and grasslands. *Annual Review of Entomology*, *40*(1), 535–558.

Vellinga, T. V., et al. (2004). The impact of grassland ploughing on CO_2 and N_2O emissions in the Netherlands. *Nutrient Cycling in Agroecosystems*, *70*(1), 33–45.

Velthof, G. L., et al. (2009). Method and timing of grassland renovation affects herbage yield, nitrate leaching, and nitrous oxide emission in intensively managed grassland. *Nutrient Cycling in Agroecosystems*, *86*(3), 401–412.

Vleeshouwers, L. M. & Verhagen, A. (2002). Carbon emission and sequestration by agricultural land use: a model study for Europe. *Global Change Biology*, *8*(6), 519–530.
Van Bruggen, C., et al. (2022). *Emissies naar lucht uit de landbouw berekend met NEMA voor 1990–2020*. Wettelijke Onderzoekstaken Natuur & Milieu, WOt-technical report 224.
Van den Berg, N. & Poelen, I. (2003). *De relatie tussen het nitraatgehalte in het grondwater en de bedrijfsvoering op melkveehouderijbedrijven*. Louis Bolk Instituut, Driebergen/ HAS Kennis Transfer, 's-Hertogenbosch.
Van den Hout, M., et al. (2023). Grasland maakt het verschil. LBI report 2023-013 LbD, Bunnik.
Van der Wal, A., et al. (2019). Bodem-en waterkwaliteit in de Nederlandse landbouw: Relatie tussen bodemorganische stof en nitraatuitspoeling op melkveebedrijven op zandgrond. *Bodem*, 5, 34–36.
Van Dijk, W., et al. (2023). *Rekenregels van de KringloopWijzer 2023*. Wageningen Research, Rapport WPR-1279.
Van der Meulen, H. (2022). *Permanent grasland in Noord-Holland lager dan in rest van Nederland; eiwit van eigen land hoger*. Agrimatie. https://www.agrimatie.nl/PublicatieRegio.aspx?subpubID=2518&themaID=2286&indicatorID=2911&sectorID=7229 (accessed 6 March 2023).
Van Eekeren, N., et al. (2008). Soil biological quality after 36 years of ley-arable cropping, permanent grasslands and permanent arable cropping. *Applied Soil Ecology*, *40*(3), 432–446.
Van Eekeren, N., et al. (2009). A mixture of grass and clover combines the positive effects of both plant species on selected soil biota. *Applied Soil Ecology*, *42*, 254–263.
Van Eekeren, N., et al. (2010). Ecosystem services in grasslands associated with biotic and abiotic soil parameters. *Soil Biology and Biochemistry*, *42*(9), 1491–1504.
Van Eekeren, N., et al. (2022). Microarthropod communities and their ecosystem services restore when permanent grasslands with mowing or low-intensity grazing is installed. *Agriculture, Ecosystems and Environment*, *323*, 107682.
Van Hal, O., et al. (2019). Upcycling food leftovers and grass resources through livestock: impact of livestock system and productivity. *Journal of Cleaner Production*, *219*, 485–496.
Van Horne, P. & Prins, H. (2002). *Development of dairy farming in the Netherlands in the period 1960–2000*. Agricultural Economics Research Institute (LEI) report 2.02.07.
Van Houwelingen, P. (2014, 9 september). *Aardappel heel efficiënt met water*. Akkerwijzer. https://www.akkerwijzer.nl/artikel/107531-aardappel-heel-efficient-met-water/#:~:text=De%20gemiddelde%20wereldwijde%20'waterfootprint'%20van,water%20per%20ton%20opbrengst%20nodig (accessed 2 March 2023).
Van Grinsven, H. & Bleeker, A. (2017). *Evaluatie Meststoffenwet 2016: Syntheserapport*. Planbureau voor de Leefomgeving, reportno. 2258.
Van Leeuwen, T. (2022). *Verschillen in stikstofgebruik tussen regio's nemen toe*. Agrimatie. https://www.agrimatie.nl/PublicatiePage.aspx?subpubID=2523&themaID=2282&indicatorID=2772&sectorID=2233 (accessed 2 March 2023).
Van Zanten, H. H. E., et al. (2023). Circularity in Europe strengthens the sustainability of the global food system. *Nature Food*, *4*, 320–330.
Zhao, Y., et al. (2020). Grassland ecosystem services: a systematic review of research advances and future directions. *Landscape Ecology*, *35*, 793–814.

Chapter 2

Grasslands and climate change

Martin Lukac[1] and Chris Reynolds[2], University of Reading, UK

1 Introduction

Grazing, as a fundamental ecological process, plays a pivotal role in shaping the structure and function of grassland ecosystems (Belsky, 1992). Through herbivores' selective consumption of vegetation, grazing influences the composition, distribution, and productivity of plant communities within grasslands. Grazing pressure and associated excreta deposition regulate plant growth, creating a mosaic of vegetation patches characterised by varying physical structure, species composition, and nutrient content (Milchunas et al., 1988; Tallowin et al., 2005). This mosaic structure not only supports biodiversity by creating diverse habitats for different species but also contributes to the resilience of grassland ecosystems by establishing a constant disturbance regime, which, in the long run, promotes dynamic responses to environmental fluctuations (Craine et al., 2013). By affecting plant biomass and diversity, grazing also influences water and nutrient cycling, soil fertility, and soil structure and function (Lavado et al., 1996; McSherry and Ritchie, 2013), thereby shaping the overall stability of grassland ecosystems. Notably, the historical co-evolution of herbivores and grasslands has resulted in mutual adaptations that underscore the integral relationship between grazing and the ecological dynamics of grassland ecosystems. The disturbance dynamics of natural grasslands are

1 m.lukac@reading.ac.uk
2 c.k.reynolds@reading.ac.uk

http://dx.doi.org/10.19103/AS.2025.0154.02

mimicked by mowing and plant biomass removal in managed grasslands, a significant management strategy in these systems. In this chapter, we will focus on natural processes. However, the reader should bear in mind that very similar climate change impact mechanisms are at play in managed grasslands.

Human-induced rapid climate change, characterised by observed and predicted shifts in temperature, precipitation patterns, and atmospheric composition (Masson-Delmotte et al., 2021), threatens the stability and productivity of grassland ecosystems worldwide. Increased frequency of extreme weather events is an early indicator of rapid change, and grasslands are particularly sensitive to it due to their unique geographic locations and ecological characteristics. Natural grasslands, often found in regions with semi-arid or seasonal climates, are susceptible to changes in water availability and seasonality (Knapp et al., 2015; Li et al., 2022). These changes can lead to plant phenology disruptions or species distribution shifts, as illustrated later in the chapter. Moreover, the intricate interplay between climate, vegetation, and soil dynamics in grasslands makes them vulnerable to the cascading effects of climate change, including increased vulnerability to invasive species and heightened risks of degradation and desertification (Zhang et al., 2018; Zhao et al., 2023; Catford et al., 2019).

However, grasslands are not passive subjects of climate change and exert strong feedback on the climate. Grasslands store substantial amounts of carbon in their soils; the deep root systems of grasses contribute to soil carbon storage by enhancing soil structure and promoting the accumulation of organic matter (Bai and Cotrufo, 2022). A warmer climate, however, may result in soil carbon loss. A recent meta-analysis of long-term warming experiments (≥5 years) has shown a 14% reduction in topsoil recalcitrant carbon (Chen et al., 2020). Grasslands stabilise regional climates by moderating surface temperature, maintaining hydrological cycles through evapotranspiration, and aiding infiltration (Xue, 1996). Understanding the two-way relationship between grasslands and climate change is imperative due to the large global distribution of grasslands. Warming and water availability strongly interact as key drivers of grassland growth, but the relationship and its effect differ between grassland types (Liu et al., 2018).

Rapid climate change can disrupt the provision of ecosystem services by grasslands, which are essential for human well-being. For example, changes in rainfall patterns may reduce water availability for agricultural production or alter the timing and magnitude of grassland vegetation growth, impacting herbage nutrient availability for livestock (Swemmer et al., 2007). In four native Great Plains grasslands, experimentally imposed drought resulted in a consistent reduction of above-ground NPP (net primary production) of 28–37% in C4 grasslands only, while below-ground NPP was reduced by 32–64% in all grasslands (Carroll et al., 2021). Conversely, excessive precipitation events are already becoming more severe and frequent (Milazzo et al., 2023), resulting in flooding, waterlogging, and increased soil erosion, all with negative consequences for grassland productivity (Leauthaud et al., 2018). This is particularly relevant to floodplains and wet grasslands, where weather extremes may severely impact typical disturbance-recovery regimes before we fully understand their dynamics (Joyce et al., 2016; Thapa et al., 2016). Thus, understanding and predicting the potential impacts of climate change on grasslands is crucial for devising effective strategies to mitigate its adverse effects and sustain ecosystem service provision.

2 Impacts of climate change on grasslands

The role of climate in grassland function and productivity is fundamental, as it directly influences various aspects of ecosystem dynamics. This chapter briefly introduces the main impacts of climate change on grasslands, while later chapters of this book will discuss them in greater detail. Temperature is crucial in regulating physiological processes in grassland plants and soils. Temperature regimes closely resembling long-term climate envelopes that have driven the development and adaptation of specific grasslands support robust growth and productivity (Zhang et al., 2023). Changes in long-term means and temperature extremes can severely inhibit metabolic functions, limiting biomass accumulation and grazer productivity. Analysis of large herbivore productivity in an 80-year study shows that variation of precipitation outside of the grazing season is the strongest predictor of grazer productivity. The stability of herbivore productivity across the years is further affected by grazing intensity, with heavy grazing resulting in a 'boom-bust' cycle threatening operations (Derner et al., 2020). Precipitation and soil water holding capacity are equally important drivers of productivity, as water availability dictates plant growth and soil moisture levels are critical for nutrient recycling (Li et al., 2022). In regions with seasonal or semi-arid climates, adequate rainfall sustains vegetation cover and productivity, whereas drought conditions can limit growth and increase the risk of degradation (Zhang et al., 2023). Climate determines the length and timing of the growing season, impacting the duration and intensity of plant growth and biomass accumulation. Past literature suggested a positive effect of cool-season warming on annual productivity. A longer growing season with favourable conditions may increase productivity, while shorter or erratic growing seasons may limit plant growth and impact plant composition (Fridley et al., 2016). However, recent observations from Southeast Australia show that anticipated benefits of cool-season warming are unlikely to be realised as optimum cool-season temperature is exceeded and indirect effects on soil water limit growth (Churchill et al., 2022).

The global level of CO_2 in the atmosphere is rising rapidly due to human activity, notably burning fossil fuels and land use change. Since CO_2 is a resource limiting the rate of photosynthesis in many plants (Lawlor and Mitchell, 1991), the continuing increase of its concentration driven by human emissions directly impacts primary productivity in grasslands. An increase in plant productivity, commonly referred to as CO_2 fertilisation, has been documented in many studies. For example, biomass was increased by 23% in a mixed species temperate meadow (Haworth et al., 2015), by 30–60% in a Swedish semi-natural grassland (Marissink et al., 2002), or by 40% (shoots) and 9% (roots) in a typical temperate grassland in northern China (Zhang et al., 2010). However, we now realise that observed increases may be transient as other factors start to limit productivity once CO_2 limitation has been alleviated (Liang et al., 2016).

Aforementioned changes in temperature have already been observed to significantly impact grassland ecosystems, affecting both the physiological processes of plants and the overall functioning of these ecosystems. Although climate change will not affect all global regions similarly (Masson-Delmotte et al., 2021), the general prediction for main grassland areas is warming. Consistently warmer temperatures can alter plant phenology, with earlier bud break, flowering, and leaf senescence (Cleland et al., 2006). Native-dominated grasslands are threatened by exotics better adapted to the new conditions; exotic plant species greened up 11 days sooner and senesced 36 days later in a latitudinal study across grasslands in the central U.S. (Wilsey et al., 2018). Shifts like this can disrupt the timing of key ecological events in a grassland, potentially affecting plant reproduction, community composition, and long-term productivity (Ren et al., 2020). Changing temperatures may thus severely influence the distribution and abundance of plant species within grassland communities. Species with narrow temperature tolerances may experience declines in abundance or shifts in distribution ranges as temperatures exceed their physiological limits (Soussana and Lüscher, 2007). Conversely, invasive species adapted to warmer conditions may proliferate (Catford et al., 2019), altering the composition and structure of grassland ecosystems.

Changes in precipitation totals and distribution will also profoundly and directly impact grasslands, altering their structure, composition, and productivity. Shifts in precipitation patterns can lead to changes in the distribution of plant species within grassland ecosystems, with drought-tolerant species potentially becoming more dominant in areas experiencing decreased rainfall (Soussana and Lüscher, 2007). Conversely, species requiring higher water availability may decline in abundance. Reduced precipitation or changes in the timing of rainfall events can limit plant growth and overall grassland productivity, impacting seed germination, plant growth, biomass accumulation, and soil functioning (Unger and Jongen, 2014). Soil function, broadly defined as the variety of roles soil plays in supporting plant growth, can also be negatively affected by excess water. Flooding is mediated by topography and does not tend to have as large-scale effects as drought does. There are several processes affected by it. Prolonged waterlogging can lead to anaerobic soil conditions, which may harm plant roots and soil microorganisms (Sánchez-Rodríguez et al., 2019). Floodwaters can physically damage plants (Sánchez-Rodríguez et al., 2019), and prolonged inundation can lead to plant stress or death (Oram et al., 2020) or can promote the growth of pests and diseases affecting plant productivity (Seidel, 2017).

This reduction in productivity can have cascading effects on ecosystem functioning, including declines in forage availability for grazing animals and alterations in nutrient cycling processes (Frank, 2007). There is a close relationship between the impact of changing temperature and precipitation, as these are the two key drivers of biome distribution on Earth. Examining these factors separately often does not yield useful information and understanding, as both strongly influence productivity and drive seasonality in grasslands.

The anticipated impacts of changing seasonality and phenology on grasslands encompass a range of ecological processes and dynamics. Alterations in seasonality, predominantely driven by shifts in temperature and precipitation patterns, can disrupt the synchronised timing of biological events such as flowering, leaf emergence, and seed maturation in grassland plants (Cleland et al., 2006). These changes can lead to mismatches in species interactions and competitive dynamics, potentially altering the composition and diversity of grassland communities. For example, non-native species may be better able to exploit new conditions and outcompete native vegetation (Wilsey et al., 2018). Furthermore, shifts in plant phenology can affect the availability and nutritional quality of forage resources for grazing animals, influencing their productivity and grazing patterns. A modelling study of future European grassland productivity shows higher than current potential annual grass forage yield, grazing capacity and livestock density, and a shift in seasonal grazing capacity in some regions but not across the continent (Chang et al., 2017). Seasonality changes will impact global animal agriculture production patterns, necessitating further investment in alternative feed resources or animal welfare (Zanon et al., 2022).

Increased CO_2 concentration stimulates photosynthesis in many grassland species. It enhances the growth and productivity of grassland vegetation (Shaw et al., 2002), leading to changes in biomass allocation, nutrient cycling, and ecosystem carbon sequestration. Elevated CO_2 levels may also influence plant–water relations, potentially affecting water use efficiency and drought tolerance in grassland species (Wullschleger et al., 2002). Carbon uptake by plants is an exercise in water loss management, elevated CO_2 lowers stomatal conductance at leaf level and has been shown to double water use efficiency in a grassland (Jackson et al., 1994). Changes in plant physiology and growth responses to elevated CO_2 can also alter species interactions, community dynamics, and the distribution of plant functional types within grassland ecosystems. Just looking at perennial grasses, species differ in apparent optimum CO_2 concentration (Zheng et al., 2018). As we increase atmospheric CO_2 content, some grassland species will experience a competitive advantage at the expense of others. Furthermore, elevated CO_2 indirectly influences other environmental factors such as soil moisture, temperature, and nutrient availability (Shaw et al., 2002), shaping grassland ecosystems' overall functioning. For example, the close coupling of carbon and nitrogen cycles in grasslands (Soussana and Lemaire, 2014) also means a strong interaction between the effects of elevated CO_2 and nitrogen deposition in these systems. Increased nitrogen availability can promote productivity but also changes in plant community composition, favouring fast-growing, nitrophilic species over native grassland flora. Plant species that can better use elevated CO_2 have been shown to outcompete others in nitrogen acquisition (Feng et al., 2015). They use additional carbon capture to speed up N uptake, putting species less responsive to elevated CO_2 at a further disadvantage.

Most, if not all, of the impacts of climate change on grasslands illustrated above are also true for other land use types, such as for arable crops or forests. One crucial difference distinguishing grasslands from other land use types is their regime of frequent disturbance, which has led to their unique regenerative capacity. The grassland plant community is dominated by species that tolerate repeated removal of above-ground structures, which hypothetically suggests a better capacity to adapt to climate change compared to other land use types. However, climate change involves a non-reversible shift of the climatic envelope and soil function driven by it rather than short-lived disturbance events. The regenerative capacity of grasslands may thus not be sufficient to maintain system stability, especially if climate change is compounded by overgrazing (Dong et al., 2020).

3 Measuring the impact of climate change on grasslands

As outlined above, climate change can profoundly affect grasslands in several ways, with many interactions between the factors and the impacts. The assessment of the impact of climate change on grasslands is not straightforward; however, several methods and metrics are used to estimate it. Agreed methodologies exist to measure several aspects of grasslands to describe their reaction to climate change. These range from measuring fundamental variables such as carbon, nutrient, or water content in the soil to describing complex characteristics such as various types of plant or diversity of soil communities and ecosystem function or ecosystem service delivery.

Measuring soil parameters in grasslands typically involves repeated soil sampling techniques to collect representative soil samples from various locations within the grassland area. Soil cores or augers are commonly used to collect samples at specified depths. The sampling depth is determined based on the parameters of interest and study objectives, typically from about 10 cm to 1 m depth (Carter and Gregorich, 2007). Once collected, soil samples undergo laboratory analysis to assess various soil properties, such as soil texture, pH, water retention capacity, infiltration rate, carbon content, and soil nutrient levels, including nitrogen, phosphorus, and potassium, which can be assessed through chemical extraction methods. Many methods exist to assess soil exploitation by vegetation by describing root mass and depth, either destructively or repeatedly using rhizotrons (Gill et al., 2002; Huot et al., 2020).

Overall, these soil measurements provide insights into the functioning and constraints of the grassland plant–soil subsystem. Repeating these measurements over time, especially in long-term studies of at least 5 or 10 years, allows for assessing changes and trends with potential association with climate change (Carter and Gregorich, 2007).

Measuring soil water fluctuations and evapotranspiration in a grassland ecosystem involves various non-destructive techniques and instruments. The key difference between the soil parameters illustrated above and those related to the water cycle is the measurement frequency. Soil moisture sensors, such as TDR (Time Domain Reflectometry (Walker et al., 2004)) or capacitance sensors, are frequently used to assess immediate soil water content affected by weather and biotic activity. These sensors are inserted into the soil at different depths to continuously monitor moisture levels, offering insights into water availability to forage plants and soil moisture dynamics. Evapotranspiration, the combined loss of water through evaporation from the soil surface and transpiration by plants, can be measured using techniques such as lysimeters, Bowen ratio systems, or eddy covariance systems (more on which later). Lysimeters collect water draining from soil columns to estimate evapotranspiration rates (Yang et al., 2000), while Bowen ratio (Perez et al., 1999) and eddy covariance systems (Aubinet et al., 2012) directly measure fluxes of water vapour and heat between the land surface and the atmosphere. Soil water content, especially during the growing season, is a key parameter constraining plant productivity and, therefore, a useful indicator of changes to the system driven by climate change. Similar to basic soil characteristics, the longer the data is recorded, the more information it may hold on climate-related impacts.

Eddy covariance systems, which measure carbon, water, and energy flux between grasslands and the atmosphere, are pivotal for understanding the dynamics of these exchanges (Baldocchi, 2003). Carbon flux measurements quantify the net ecosystem exchange (NEE) of CO_2, uncovering the overall carbon balance of the grassland. NEE represents the difference between the carbon absorbed by plants during photosynthesis and the carbon released through plant and soil respiration and is the best estimate of carbon sequestration by a grassland. Positive NEE values indicate that the grassland acts as a source of CO_2 to the atmosphere, while negative values signify a net carbon sink (Baldocchi, 2003). Continuous carbon flux measurements investigate the impacts of changes in environmental factors such as temperature, precipitation, and land management practices on carbon dynamics in grassland ecosystems. Eddy covariance sensors and data processing techniques can measure and report on energy fluxes and compounds other than CO_2, but this gas is the main variable followed and reported on due to its pivotal role in the feedback mechanism between climate change and global grasslands.

The latest developments in remote sensing techniques offer efficient, comprehensive, and increasingly accessible means of monitoring changes in grassland vegetation cover and health over large spatial scales (Wachendorf et al., 2018). Satellite or drone-mounted sensors equipped with multispectral or hyperspectral capabilities can capture data across different wavelengths of electromagnetic radiation, several of which inform on specific processes or characteristics of grassland vegetation. Vegetation indices, such as the Normalised Difference Vegetation Index (NDVI), Enhanced Vegetation Index (EVI), and others derived from these data, provide quantitative measures of vegetation activity, vigour, or composition (Goetz, 1997; Guerini Filho et al., 2020; Punalekar et al., 2021). Remote sensing enables relatively low-cost identification of vegetation stressors such as drought, disease, or nutrient deficiencies by detecting changes in spectral signatures associated with specific stress responses. Furthermore, the spatial and temporal resolution of remote sensing data allows for detecting fine-scale variations in grassland vegetation, facilitating targeted management interventions to strengthen the productivity or resilience of managed systems (Ali et al., 2016; Reinermann et al., 2020). Recent technological and methodological developments facilitate the development of time series of several grassland biophysical parameters with sufficient resolution to guide management options targeting soil degradation, grazing intensity or seasonal plant productivity (Ali et al., 2016). Projected future climate scenarios projected at various spatial scales can be combined with process-based grassland models to simulate how changes in temperature, precipitation, and other climatic factors may impact grassland ecosystems. Changes in productivity, light interception, water and nutrient use efficiency or plant composition as a result of climate change have been predicted. For example, ensemble modelling shows a significant drift of all sites towards more arid climates in France by the end of the century (Graux et al., 2013), with new opportunities for annual and seasonal herbage production, especially in spring and winter. Grassland models aim to predict the effects of climate change at the system level, an approach with significant challenges. At present, models must improve their predictions of the direct and indirect effects of climate change on the sward, climate change effects on grassland systems outputs, and – perhaps the most challenging – mediation of climate change impacts by site, system, and management (Kipling et al., 2016).

In summary, there is a range of techniques and methodologies in use that enable consistent and reliable monitoring of the status of grasslands. The methodologies range from minute observations of soil chemistry at specific soil depths to remote-sensed descriptions of the vegetation at the field or on a larger scale. The key point relating to climate change is the long-term nature of these observations, as the timescale of measurements describing grassland function or composition must be commensurate with the timescale of impacts driven by climate change. These measurements form the basis of predictive models of the impact of climate change on grassland ecosystems and their functions, which are of increasing importance in light of increased variability in global weather events associated with global warming (Wheeler and Reynolds, 2013).

4 Adaptation: challenges and opportunities

The development of successful adaptation strategies enhancing the stability and resilience of grasslands is contingent on access to realistic predictions. However, predicting the impact of climate change on grasslands presents an array of challenges, largely connected to the multifaceted nature of climate and grassland ecosystems. One significant obstacle lies in understanding and capturing the intricate web of interactions and feedback loops between various environmental factors. Grasslands are influenced by climatic variables such as temperature and precipitation, soil characteristics, and land management practices, which drive vegetation productivity, plant composition, and biotic interactions. Understanding how these factors interact and shift due to changing climate requires sophisticated modelling approaches and extensive empirical data captured across large spatial scales (e.g. see Li et al., 2020; Sándor et al., 2020; Van Oijen et al., 2018). Grasslands exhibit remarkable diversity in soil types, topography, vegetation communities, and climatic regimes, resulting in varied responses to similar climate change drivers across different regions. Predicting these diverse responses necessitates accounting for local-scale processes and patterns while considering broader regional and global climate trends. Climate projections typically involve uncertainties (Masson-Delmotte et al., 2021), making it challenging to precisely predict the timing, magnitude, and spatial distribution of future climate changes and their impacts on grasslands (Wheeler and Reynolds, 2013).

Grasslands, like many other ecosystems, exhibit the characteristics of complex systems (Schiere and Gregorini, 2023). Non-linear responses and threshold responses or abrupt shifts in ecosystem structure and function in response to relatively small or cumulative changes in climate or disturbance regimes are very difficult to predict. Needless to say, understanding grasslands as complex systems is a data-heavy exercise (Schrenk et al., 2022). New evidence shows that these systems need to be studied as socio-ecological systems where the direct (management, grazing) or indirect (nitrogen deposition, fire regime suppression) human input must be factored in with natural processes and climate (McGinlay et al., 2017). This effort must include the co-creation and shaping of policies and management strategies for grassland conservation and climate change adaptation, requiring interdisciplinary collaboration and integrated research that combines insights from science, stakeholders, and wider society (Bardgett et al., 2021).

In centuries past, many grasslands were transformed into agricultural fields or urban landscapes. Today, laws in numerous countries mandate efforts to conserve and adapt these grasslands in the face of climate change (McDonald et al., 2016). However, the starting point for these adaptation efforts varies widely. Regions with more abundant natural vegetation tend to have richer species diversity than heavily managed agricultural areas. Changes in land use, such as the suppression of natural fires and traditional grazing practices, have led to a decline in species diversity in grasslands (Stevens et al., 2017). Fire, in particular, has historically played a significant role in shaping grassland ecosystems. Therefore, restoring grasslands to their historic fire and grazing regimes is crucial for their adaptation (Tölgyesi et al., 2022). Grassland plant species generally respond positively to fire, which can help alter plant communities and reduce the dominance of unwanted species (Limb et al., 2016). Similarly, grazing can serve as an adaptation tool in some grasslands. However, the intensity of grazing is a critical factor in systems subjected to overgrazing, soil erosion, or invasive species.

Underground, grasslands rely on mutualistic relationships with soil biota to thrive, particularly arbuscular mycorrhizal fungi (AMF) (Afkhami and Strauss, 2016). Adding AMF to grasslands can enhance plant productivity, especially in degraded soils, as soil biota diversity also contributes to the stability of grassland ecosystems. Soil carbon is crucial in grassland adaptation, as it fuels soil life and improves soil functions. Grasslands rich in species tend to maintain and enhance below-ground carbon better than less diverse grasslands (Dass et al., 2018). Numerous case studies and initiatives document the challenges, solutions, and opportunities in grassland adaptation to climate change. Table 1 offers a small glimpse at some of these, even though there is no definitive 'how to' as the interventions and approaches tend to fit specific conditions in each example.

Table 1 A selection of additional online resources containing background information, research synthesis, user stories, how-to guide, and description of successful initiatives dealing with impacts and adaptation of grasslands to climate change

Natural Solutions to Climate Change: Grassland (The Wildlife Trusts). A project focusing on employing nature-based solutions to increase carbon storage while restoring habitats on land. The project emphasises the importance of grasslands in locking up carbon and boosting biodiversity.
Grasslands Report Synthesis (University of Colorado). This ongoing research synthesises the impacts of grassland restoration measures on the structure and function of grassland ecosystems in the context of global climate change.
Climate Change Risk to Global Grazing Lands (International Livestock Research Institute). A report considering the role of livestock grazing in maintaining food security, the regions likely to be most vulnerable to changing rainfall patterns, and strategies for addressing climate change-related risks.
Saving Our Grasslands (World Wildlife Fund). This initiative shows the importance of protecting and restoring grasslands, which are critical for biodiversity, nature, and climate. They also provide strategies for addressing climate change-related risks.
Managing Grazing to Improve Climate Resilience (USDA). This initiative focuses on rotational stocking systems to control the timing and intensity of grazing, giving the pastures time for rest and regrowth.
The Grassland Climate Adaptation Network (GCAN) is a collaborative initiative aimed at understanding and addressing the impacts of climate change on grassland ecosystems. It brings together scientists, land managers, policymakers, and other stakeholders to develop strategies for adapting to changing environmental conditions.
Regenerative Grazing: **A Compelling Climate Strategy** (Bio4Climate). This paper provides an overview of how regenerative grazing combats climate change and provides multiple environmental benefits.
Enhancing Biodiversity for the Future (Savoy Global). This project uses holistic planned grazing and permaculture to restore grasslands and enhance biodiversity.
Profitable and Resilient Farming Through Grassland Restoration: Winona Farm (Ellen MacArthur Foundation). This large-scale example of regenerative agriculture was adopted by over two thousand farmers in Australia and other parts of the world.
Regenerative Agriculture and Dairy Farming (Dairy Global). This initiative discusses the implementation of regenerative agriculture in large conventional dairy production systems that do not focus primarily on grazing animals on pasture.
UnderstandingAG (Gabe Brown). A learning resource that showcases the benefits of regenerative agriculture and integration of grazing animals for soil restoration.
Grasses and Climate Change (Earth at Home). A learning resource covering the impact of climate change on grasslands, considering how growers adapt to this change and how grasslands can be part of the solution.

Future research should prioritise studies to understand the specific impacts of climate change on grassland ecosystems, including assessing the vulnerability of grasslands to extreme weather events and long-term shifts in climatic patterns. Research should develop and evaluate adaptation strategies, such as altered grazing regimes, fire regimes, revegetation efforts, and water conservation practices, to maintain productivity and enhance resilience. Additionally, research should focus on identifying and conserving grassland species and assemblages to preserve genetic diversity (Hamilton et al., 2020), with the long-term view of supporting ecosystem resilience and adaptation. Moreover, policy interventions informed by research outcomes should support sustainable land management practices, incentivise grassland conservation efforts, and integrate climate change considerations into land-use planning and decision-making processes. However, grassland adaptation costs vary widely, with some strategies being more affordable than others. Current funding schemes are often insufficient to address the scale of adaptation needed to combat biodiversity loss and climate change (Richardson, 2016). Thus, exploring funding systems considering resilience and carbon credits could offer promising avenues for future grassland adaptation efforts (Lyons et al., 2023).

5 Conclusion

The strong feedback loop between climate change and grasslands, in combination with the world's huge grassland area, highlights the urgency of understanding and managing their reaction to a changing climate. As rapid climate change continues to alter the climatic envelopes of these ecosystems, the resilience and functioning of grasslands are increasingly at risk. Measuring and monitoring these impacts require methodologies from hyper-local soil analysis to remote sensing techniques covering tens of hectares or larger areas. Long-term observational data and interdisciplinary collaboration are essential for accurately assessing the vulnerability of grasslands to climate change and developing effective response prediction models and adaptation strategies. Research that elucidates the specific vulnerabilities of grassland ecosystems should be prioritised and, in collaboration with stakeholders, used to evaluate and implement adaptation strategies. Scientific insights must be integrated into policy-making processes to incentivise sustainable grassland management practices, as losing grassland ecosystems erodes our non-arable food production capacity.

6 Where to look for further information

For more high-level information on grasslands and their role in climate change, readers can explore resources provided by organisations such as the Intergovernmental Panel on Climate Change (AR6, Chapter 5: Food, Fibre, and Other Ecosystem Products), the Food and Agriculture Organisation (Climate-Smart Agriculture Sourcebook, 2013, Module 2: Climate-Smart Crop Production Systems), and the United Nations Environment Programme (UNEP).

For more in-depth information, scientific journals like *Global Change Biology*, Journal of Ecology, and Agriculture, Ecosystems & Environment regularly publish research on this topic.

7 References

Afkhami ME and Strauss SY (2016) Native fungal endophytes suppress an exotic dominant and increase plant diversity over small and large spatial scales. *Ecology* 97(5): 1159–1169.

Ali I, et al. (2016) Satellite remote sensing of grasslands: from observation to management. *Journal of Plant Ecology* 9(6): 649–671.

Aubinet M, et al. (2012) *Eddy covariance: a practical guide to measurement and data analysis.* Springer Science & Business Media.

Bai Y and Cotrufo MF (2022) Grassland soil carbon sequestration: current understanding, challenges, and solutions. *Science* 377(6606): 603–608.

Baldocchi DD (2003) Assessing the eddy covariance technique for evaluating carbon dioxide exchange rates of ecosystems: past, present and future. *Global change biology* 9(4): 479–492.

Bardgett RD, et al. (2021) Combatting global grassland degradation. *Nature Reviews Earth & Environment* 2(10): 720–735.

Belsky AJ (1992) Effects of grazing, competition, disturbance and fire on species composition and diversity in grassland communities. *Journal of Vegetation Science* 3(2): 187–200.

Carroll CJ, et al. (2021) Is a drought a drought in grasslands? Productivity responses to different types of drought. *Oecologia* 197: 1017–1026.

Carter MR and Gregorich EG (2007) *Soil sampling and methods of analysis.* CRC Press.

Catford JA, et al. (2019) Grassland invasion in a changing climate. In *Grasslands and climate change,* edited by DJ Gibson and J Newman. Cambridge University Press, pp. 149–171.

Chang J, et al. (2017) Future productivity and phenology changes in European grasslands for different warming levels: implications for grassland management and carbon balance. *Carbon Balance and Management* 12: 1–21.

Chen J, et al. (2020) Soil carbon loss with warming: new evidence from carbon-degrading enzymes. *Global Change Biology* 26(4): 1944–1952.

Churchill AC, et al. (2022) Pastures and climate extremes: impacts of cool season warming and drought on the productivity of key pasture species in a field experiment. *Frontiers in Plant Science* 13: 836968.

Cleland EE, et al. (2006) Diverse responses of phenology to global changes in a grassland ecosystem. *Proceedings of the National Academy of Sciences* 103(37): 13740–13744.

Craine JM, et al. (2013) Global diversity of drought tolerance and grassland climate-change resilience. *Nature Climate Change* 3(1): 63–67.

Dass P, et al. (2018) Grasslands may be more reliable carbon sinks than forests in California. *Environmental Research Letters* 13(7): 074027.

Derner JD, et al. (2020) Climatic and management determinants of large herbivore production in semi-arid grassland. *Agriculture, Ecosystems & Environment* 290: 106761.

Dong S, et al. (2020) Enhancing sustainability of grassland ecosystems through ecological restoration and grazing management in an era of climate change on Qinghai-Tibetan Plateau. *Agriculture, Ecosystems & Environment* 287: 106684.

Feng Z, et al. (2015) Constraints to nitrogen acquisition of terrestrial plants under elevated CO_2. *Global Change Biology* 21(8): 3152–3168.

Frank DA (2007) Drought effects on above-and below-ground production of a grazed temperate grassland ecosystem. *Oecologia* 152(1): 131–139.

Fridley JD, et al. (2016) Longer growing seasons shift grassland vegetation towards more-productive species. *Nature Climate Change* 6(9): 865–868.

Gill RA, et al. (2002) Longevity and turnover of roots in the shortgrass steppe: influence of diameter and depth. *Plant Ecology* 159: 241–251.

Goetz S (1997) Multi-sensor analysis of NDVI, surface temperature and biophysical variables at a mixed grassland site. *International Journal of Remote Sensing* 18(1): 71–94.

Graux A-I, et al. (2013) Ensemble modelling of climate change risks and opportunities for managed grasslands in France. *Agricultural and Forest Meteorology* 170: 114–131.

Guerini Filho M, et al. (2020) Estimating natural grassland biomass by vegetation indices using Sentinel 2 remote sensing data. *International Journal of Remote Sensing* 41(8): 2861–2876.

Hamilton J, et al. (2020) Evolutionary approaches to seed sourcing for grassland restorations. *The New Phytologist* 225(6): 2246–2248.

Haworth M, et al. (2015) Carbon dioxide fertilisation and supressed respiration induce enhanced spring biomass production in a mixed species temperate meadow exposed to moderate carbon dioxide enrichment. *Functional Plant Biology* 43(1): 26–39.

Huot C, et al. (2020) Root depth development in tropical perennial forage grasses is related to root angle, root diameter and leaf area. *Plant and Soil* 456: 145–158.

Jackson R, et al. (1994) CO_2 alters water use, carbon gain, and yield for the dominant species in a natural grassland. *Oecologia* 98: 257–262.

Joyce CB, et al. (2016) Future wet grasslands: ecological implications of climate change. *Ecosystem Health and Sustainability* 2(9): e01240.

Kipling RP, et al. (2016) Key challenges and priorities for modelling European grasslands under climate change. *Science of the Total Environment* 566: 851–864.

Knapp AK, et al. (2015) Differential sensitivity to regional-scale drought in six central US grasslands. *Oecologia* 177: 949–957.

Lavado RS, et al. (1996) Impact of grazing on soil nutrients in a Pampean grassland. *Rangeland Ecology & Management/Journal of Range Management Archives* 49(5): 452–457.

Lawlor D and Mitchell R (1991) The effects of increasing CO2 on crop photosynthesis and productivity: a review of field studies. *Plant, Cell & Environment* 14(8): 807–818.

Leauthaud C, et al. (2018) Modelling the growth of floodplain grasslands to explore the impact of changing hydrological conditions on vegetation productivity. *Ecological Modelling* 387: 220–237.

Li L, et al. (2022) Drought timing influences the sensitivity of a semi-arid grassland to drought. *Geoderma* 412: 115714.

Li L, et al. (2020) Modelling the effects of climate change on transpiration and evaporation in natural and constructed grasslands in the semi-arid Loess Plateau, China. *Agriculture, Ecosystems & Environment* 302: 107077.

Liang J, et al. (2016) Processes regulating progressive nitrogen limitation under elevated carbon dioxide: a meta-analysis. *Biogeosciences* 13(9): 2689–2699.

Limb RF, et al. (2016) Synthesis paper: assessment of research on rangeland fire as a management practice. *Rangeland Ecology & Management* 69(6): 415–422.

Liu D, et al. (2018) Contrasting responses of grassland water and carbon exchanges to climate change between Tibetan Plateau and Inner Mongolia. *Agricultural and Forest Meteorology* 249: 163–175.

Lyons KG, et al. (2023) Challenges and opportunities for grassland restoration: a global perspective of best practices in the era of climate change. *Global Ecology and Conservation* 46: e02612.

Marissink M, et al. (2002) Above-ground plant production under elevated carbon dioxide in a Swedish semi-natural grassland. *Agriculture, Ecosystems & Environment* 93(1–3): 107–120.

Masson-Delmotte V, et al. (2021) IPCC, 2021: summary for policymakers. In *Climate change 2021: The physical science basis. contribution of working group i to the sixth assessment report of the intergovernmental panel on climate change*. Contribution of Working Group I to the Sixth Assessment Report of the Intergovernmental Panel on Climate Change [Masson-Delmotte, V., et al., (eds.)]. Cambridge University Press, Cambridge, United Kingdom and New York, NY, USA, pp. 3–32, doi:10.1017/9781009157896.001.

McDonald T, et al. (2016) International standards for the practice of ecological restoration–including principles and key concepts. https://doi.org/10.1111/rec.13035

McGinlay J, et al. (2017) The threat of abandonment in socio-ecological landscapes: Farmers' motivations and perspectives on high nature value grassland conservation. *Environmental Science & Policy* 69: 39–49.

McSherry ME and Ritchie ME (2013) Effects of grazing on grassland soil carbon: a global review. *Global Change Biology* 19(5): 1347–1357.

Milazzo F, et al. (2023) The role of grassland for erosion and flood mitigation in Europe: a meta-analysis. *Agriculture, Ecosystems & Environment* 348: 108443.

Milchunas DG, et al. (1988) A generalized model of the effects of grazing by large herbivores on grassland community structure. *The American Naturalist* 132(1): 87–106.

Oram NJ, et al. (2020) Plant community flood resilience in intensively managed grasslands and the role of the plant economic spectrum. *Journal of Applied Ecology* 57(8): 1524–1534.

Perez P, et al. (1999) Assessment of reliability of Bowen ratio method for partitioning fluxes. *Agricultural and Forest Meteorology* 97(3): 141–150.

Punalekar SM, et al. (2021) Assessing suitability of Sentinel-2 bands for monitoring of nutrient concentration of pastures with a range of species compositions. *Agronomy* 11(8): 1661.

Reinermann S, et al. (2020) Remote sensing of grassland production and management: a review. *Remote Sensing* 12(12): 1949.

Ren S, et al. (2020) Diverse effects of climate at different times on grassland phenology in mid-latitude of the Northern Hemisphere. *Ecological Indicators* 113: 106260.

Richardson BJ (2016) Resourcing ecological restoration: the legal context for commercial initiatives. *Restoration Ecology* 24(5): 686–691.

Sánchez-Rodríguez AR, et al. (2019) Extreme flood events at higher temperatures exacerbate the loss of soil functionality and trace gas emissions in grassland. *Soil Biology and Biochemistry* 130: 227–236.

Sándor R, et al. (2020) Ensemble modelling of carbon fluxes in grasslands and croplands. *Field Crops Research* 252: 107791.

Schiere JB and Gregorini P (2023) Complexity, crash and collapse of chaos: clues for designing sustainable systems, with focus on grassland-based systems. *Sustainability* 15(5): 4356.

Schrenk H, et al. (2022) QtAC: an R-package for analyzing complex systems development in the framework of the adaptive cycle metaphor. *Ecological Modelling* 466: 109860.

Seidel P (2017) Extreme weather events and their effects on plant pests infecting potato, sugar beet, rape and grassland. *Journal für Kulturpflanzen* 69(4): 125–136.

Shaw MR, et al. (2002) Grassland responses to global environmental changes suppressed by elevated CO2. *Science* 298(5600): 1987–1990.

Soussana J-F and Lemaire G (2014) Coupling carbon and nitrogen cycles for environmentally sustainable intensification of grasslands and crop-livestock systems. *Agriculture, Ecosystems & Environment* 190: 9–17.

Soussana JF and Lüscher A (2007) Temperate grasslands and global atmospheric change: a review. *Grass and forage science* 62(2): 127–134.

Stevens N, et al. (2017) Savanna woody encroachment is widespread across three continents. *Global change biology* 23(1): 235–244.

Swemmer AM, et al. (2007) Intra-seasonal precipitation patterns and above-ground productivity in three perennial grasslands. *Journal of Ecology* 95(4): 780–788.

Tallowin J, et al. (2005) Impact of grazing management on biodiversity of grasslands. *Animal Science* 81(2): 193–198.

Thapa R, et al. (2016) The response of dryland floodplain vegetation productivity to flooding and drying. *Journal of Arid Environments* 129: 42–55.

Tölgyesi C, et al. (2022) Post-restoration grassland management overrides the effects of restoration methods in propagule-rich landscapes. *Ecological Applications* 32(1): e02463.

Unger S and Jongen M (2014) Consequences of changing precipitation patterns for ecosystem functioning in grasslands: a review. *Progress in Botany* 76: 347–393.

Van Oijen M, et al. (2018) Effects of climate change on grassland biodiversity and productivity: the need for a diversity of models. *Agronomy* 8(2): 14.

Wachendorf M, et al. (2018) Remote sensing as a tool to assess botanical composition, structure, quantity and quality of temperate grasslands. *Grass and Forage Science* 73(1): 1–14.

Walker JP, et al. (2004) In situ measurement of soil moisture: a comparison of techniques. *Journal of Hydrology* 293(1–4): 85–99.

Wheeler T and Reynolds C (2013) Predicting the risks from climate change to forage and crop production for animal feed. *Animal Frontiers* 3(1): 36–41.

Wilsey BJ, et al. (2018) Phenology differences between native and novel exotic-dominated grasslands rival the effects of climate change. *Journal of Applied Ecology* 55(2): 863–873.

Wullschleger S, et al. (2002) Plant water relations at elevated CO2–implications for water-limited environments. *Plant, Cell & Environment* 25(2): 319–331.

Xue Y (1996) The impact of desertification in the Mongolian and the Inner Mongolian grassland on the regional climate. *Journal of Climate* 9(9): 2173–2189.

Yang J, et al. (2000) A large weighing lysimeter for evapotranspiration and soil-water–groundwater exchange studies. *Hydrological Processes* 14(10): 1887–1897.

Zanon T, et al. (2022) Diverse feed, diverse benefits: the multiple roles of feed diversity at pasture on ruminant livestock production: a review. *Journal of Veterinary Science and Animal Husbandry* 10(1): 1–20.

Zhang F, et al. (2023) The legacy effects of grazing and precipitation variability on grassland productivity in a temperate steppe. *Agricultural and Forest Meteorology* 343: 109788.

Zhang G, et al. (2018) Exacerbated grassland degradation and desertification in Central Asia during 2000–2014. *Ecological Applications* 28(2): 442–456.

Zhang L, et al. (2010) Responses of a dominant temperate grassland plant (*Leymus chinensis*) to elevated carbon dioxide and nitrogen addition in China. *Journal of Environmental Quality* 39(1): 251–259.

Zhao Y, et al. (2023) Divergent trends in grassland degradation and desertification under land use and climate change in Central Asia from 2000 to 2020. *Ecological Indicators* 154: 110737.

Zheng Y, et al. (2018) The optimal CO_2 concentrations for the growth of three perennial grass species. *BMC Plant Biology* 18: 1–12.

Chapter 3

Harnessing the power of grasslands in mixed farming/crop-livestock systems

F. Taube, Christian-Albrechts-University Kiel, Germany and Wageningen University, The Netherlands; J. K. Nyameasem, Christian-Albrechts-University Kiel, Germany and University of Bonn, Germany; and S. Verma, Christian-Albrechts-University Kiel, Germany and Aarhus University, Denmark

1 Introduction

The demand for food is projected to double in the coming decades, driven by rapid population growth, urbanization, and rising incomes, factors that place enormous pressure on global agricultural systems (Elferink & Schierhorn, 2016). In response, a range of agricultural policies have been implemented to boost food production and enhance food security (Godfray et al., 2010; Poppy et al., 2014). However, the competition for land and resources for agriculture is expected to intensify significantly by 2050 (Alexandratos & Bruinsma, 2012), a challenge further compounded by the impact of climate change (Lynch et al., 2021). Unfortunately, many policy-driven strategies have promoted the specialization and intensification of farming, resulting in more simplified and high-input systems. These often involve the decoupling of crop and livestock production and the emergence of homogenized agricultural landscapes. Moreover, such systems typically rely heavily on external inputs,

http://dx.doi.org/10.19103/AS.2025.0154.04

such as synthetic fertilizers and pesticides, which contributes to environmental pollution, degrade water and air quality, disrupt ecosystems, and pose risks to human health through potential contamination of food products.

Given the drawbacks of previous approaches to ensuring food security, sustainable intensification, which focuses on increasing food production on existing cropland, has gained considerable traction (Godfray et al., 2010; Garnett et al., 2013). At the same time, the ongoing specialization and intensification of agriculture, especially in the Global North, and increasingly in the Global South, have resulted in a plethora of ecological challenges, including greenhouse gas emissions, nutrient losses, land degradation, and biodiversity decline (Giller et al., 2021; Kozicka et al., 2023). Globally, the agriculture sector is estimated to contribute 19–29% of total anthropogenic greenhouse gas emissions. If current emissions trends persist, there will be at least 0.7 degrees more warming by 2100, even if fossil fuel emissions are drastically reduced (Lee et al., 2023). Agricultural production now stands at a critical crossroads, tasked with the dual challenge of ensuring food security for a burgeoning population while also addressing the urgent threat of climate change (Koura et al., 2015). The increasing demand for animal products further intensifies pressure on land use, necessitating expansion of grazing areas. This demand is often met by producing fodder on croplands and enhancing the productivity of grasslands (Erb et al., 2016).

Despite their significant potential, the role of grasslands in ensuring food security is often overlooked (O'Mara, 2012). Beyond serving as feed sources for grazing and browsing animals, grasslands support a rich diversity of wildlife (Stypinski, 2011; Adu et al., 2017). Both natural or semi-natural grasslands play invaluable roles within agricultural systems by delivering ecosystem services that contribute to ecological stability, agricultural sustainability, and human well-being. When managed effectively, grasslands provide key regulating services, including water regulation, erosion control, climate mitigation, pollination support, and cultural value, showcasing their multifaceted contributions to both the environment and society. For instance, Doblas-Rodrigo et al. (2022) showed an annual increase of 1.26 t C ha^{-1} in soil organic carbon (SOC) within the top 30 cm of grassland soil over a 36-year simulation. Notably, grassland soils with initial SOC below 80 t C ha^{-1} have the potential to meet the "4 per 1000" C sequestration target (Minasny et al., 2017; Soussana et al., 2010).

However, the interconnected service bundles provided by grasslands, including water-related services linked to fodder production, cultural services associated with livestock traditions, and population-regulating services tied to biodiversity, are increasingly under threat in many ecosystems. For instance, while natural grasslands in South Africa maintain a delicate balance between the supply and demand of these services, semi-natural grasslands in Europe struggle to meet diverse service demands due to their smaller scale and heterogeneous landscape management (Bengtsson et al., 2019). The ongoing intensification of agriculture and land-use change (LUC) are major drivers of significant SOC losses, leading to negative impacts on biodiversity, ecosystem productivity and climate regulation (Catovsky et al., 2002). In regions characterized by intensive dairy farming, such as Northern Germany, large areas of grassland have been converted into energy-dense arable land for grain and silage crop production, particularly silage maize (Taube et al., 2014), further exacerbating these issues. In light of the growing demands and environmental challenges in modern agriculture, it is imperative to reconsider and reintegrate the vital role of grasslands within these complex agroecosystems (Godfray et al., 2010).

With future consumption of animal-source foods (ASFs) expected to rise, the demand for feed production is set to increase accordingly. At the same time, global debates around improving nutrition in low-income countries and addressing overconsumption and its related health risks in high-income countries have gained momentum. In response, several researchers advocate for a global shift toward reduced reliance on ASFs or increased adoption of plant-based diets to mitigate climate change and keep natural resource use within planetary boundaries (Kozicka et al., 2023). However, implementing such dietary transitions can be challenging, both culturally and politically (Aschemann-Witzel et al., 2021). As an alternative, a circular food systems approach has been proposed, emphasizing the development of plant-based ASF alternatives and improved resource efficiency (van Zanten et al., 2018). This strategy calls for a reduced animal husbandry in high-income regions, aligning with global food security goals. In this model, ruminants would primarily be fed on non-arable land, utilizing ecological "leftover" areas and agricultural by-products, thereby minimizing competition with land suited for direct human food production (Röös et al., 2017; van Zanten et al., 2018). Feed production on arable land should only be pursued when it supports food crop production, such as through reductions in pesticide use or the enhancement of ecosystem services, thus ensuring a more sustainable and integrated use of land resources.

A shift towards mixed farming, where crop cultivation and livestock rearing are integrated on the same agricultural land, has been proposed as a key strategy for achieving agricultural sustainability (Koura et al., 2015; Alderkamp et al., 2025). Within these systems, grasslands play a pivotal role by providing forage for livestock and delivering essential ecosystem services, such as soil conservation, biodiversity enhancement, and C sequestration (Haughey et al., 2023). When well-managed, integrated farming schemes can optimize resources and strengthen ecological resilience by leveraging the natural synergies between crops and livestock. This approach offers a practical scalable solution to the dual challeneges of meeting global food demand and addressing environmental concerns.

Ley systems, when integrated into cropping practices, play a vital role in bolstering food security in developing nations and promoting the sustainable production of both plant and animal products. Moreover, they offer a viable route towards ecological restoration in the Northern Hemisphere, helping to counterbalance the negative consequences of highly intensive and specialized agricultural systems. In both contexts, leys can contribute to economic and social resilience in the Global South while simultaneously advancing ecological sustainability in the North. This becomes especially important when considering the often-overlooked hidden costs associated with specialized systems, including greenhouse gas emissions, nitrogen (N) losses, and shifts in land usage. Although these environmental risks may not be immediately conspicuous, their long-term consequences are significant and frequently negleted in policy and decision-making processes (see Taube et al., 2023; Lord, 2023). However, the widespread adoption of mixed farming systems, including ley-based approaches, faces several hurdles, ranging from gaps in education and sociocultural barriers to insufficient policy support (Topp et al., 2023). This chapter explores the central role of grasslands within mixed farming/crop–livestock systems, highliting their diverse functions and ecological value. It also examines the benefits, and some of the trade-offs associated with with ecological intensification and de-intensification of grassland management.

2 Mixed farming/crop-livestock integration

Mixed farming practices have been implemented in diverse forms around the world, each tailored to specific ecological and economic contexts (Fig. 1; Sekaran et al., 2021). One example is agro-pastoral systems, which integrate crop cultivation with livestock grazing eto maintain a balanced nutrient cycle and support grassland ecosystem conservation through rotational use of land for grazing and crop cultivation. In general, mixed farming designates specific areas for both crop and livestock production, enabling synergies that enahnce overal system productivity. Silvopastoral systems take this integration further by incorporating trees or agroforestry components with grazing, thereby promoting biodiversity, enhancing C sequestration, and increasing climate resilience. Similarly, integrated aquaculture-agriculture systems merge fish farming with crop production, improving land productivity while offering opportunities for income diversification. Lastly, integrated crop–livestock–forest systems blend crop cultivation, animal husbandary and forest components in a cohesive framework, optimizing resource use and delivering both economic and environmental benefits.

To better understand the potential benefits, trade-offs, and barriers to adoption across different crop and livestock systems along integration gradients, several classification frameworks have been proposed. These classifications

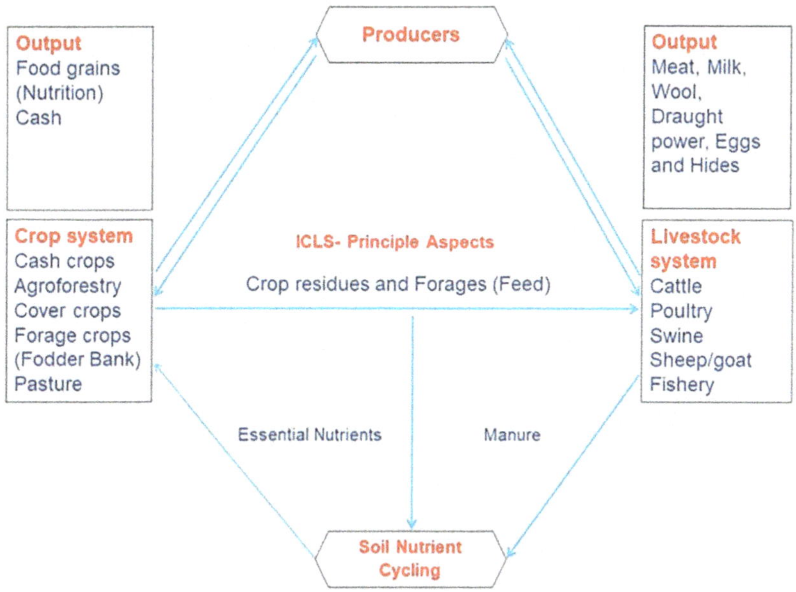

Figure 1 Principle aspects of the integrated crop–livestock system (ICLS). Source: Adopted from Sekaran et al. (2021).

consider factors such as inputs, capital, labour, space, time, ownership, and management practices, as well as the degree of interaction among these factors (Bell and Moore, 2012; Moraine et al., 2017; Garrett et al., 2020). Garrett et al. (2020), for example, outlined a spectrum ranging from segregated high-input systems to highly integrated crop-livestock systems (ICLS), including traditional, new, and territorial forms of integration. At the segregated end of the spectrum, crop and livestock operations function largely independently, interacting mainly through market exchanges. These systems typically achieve high yields but rely heavily on synthetic fertilizers, pesticides and off-farm feed sources. Such dependence often leads to concentrated livestock densities and excessive manure accumulation, contributing to environmental degradation. Conversely, highly integrated systems feature tightly coupled crop–pasture rotations and in situ animal grazing. These practice improving nutrient cycling, enhance soil structure, and support sustainability, especially when grazing is managed at low-to-moderate intensities.

Moraine et al. (2017) conceptualized the dynamic interactions within integrated crop–livestock systems through a visual framework (Fig. 2) that captures the relationships among biophysical components, material flows, and ecosystem processes. The model features three overlapping spheres, representing crops, livestock, and the environment, illustrating their interconnectedness within these systems. The relative sizes and degrees of

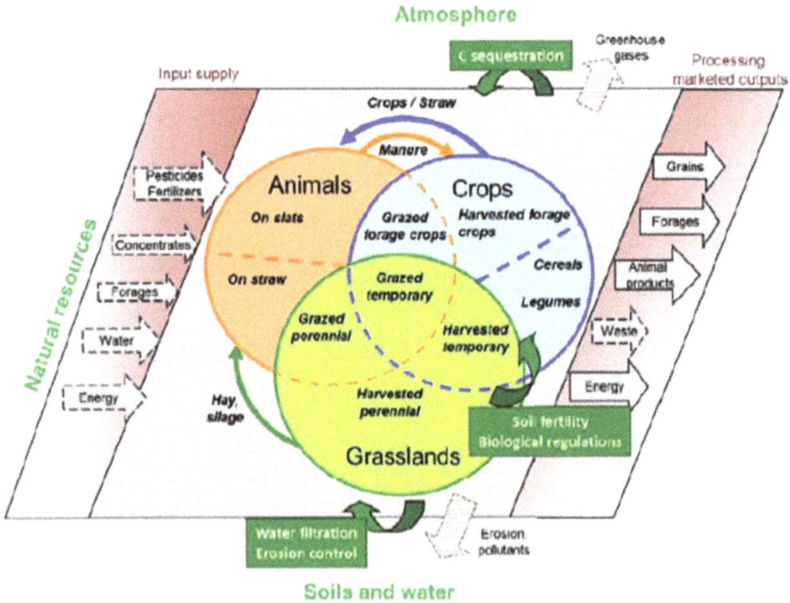

Figure 2 Interactions among biophysical components of a typical integrated crop–livestock system. Source: Adapted from Moraine et al. (2017).

overlap among the spheres can be adjusted to reflect the structural diversity of crop–livestock systems across different contexts. Crops and livestock are shown as distinct yet interconnected entities, with material flows – such as nutrients, water, and biomass – depicted as linking these components. These flows highlight the critical role of resource management in maintaining system sustainability. Surrounding the spheres, large semi-circular arrows symbolize ecosystem processes, including biological, physical, and chemical functions, which collectively drive ecosystem services. A core challenge in achieving effective crop–livestock integration lies in minimizing external inputs and reducing emissions to the environment. Sustainable integration, therefore, requires optimizing internal resource use and enhancing ecological efficiency. Ultimately, the configuration and interaction among these spheres determine the system's capacity to deliver resilient ecosystem services essential for both agricultural productivity and environmental health.

Traditional Integrated Crop–Livestock Systems (ICLS), which fall under the category of highly integrated systems, rely primarily on on-farm resources rather than external inputs to manage nutrient supply and weed control (Fig. 2). These systems offer several advantages by addressing the limitations of conventional farming through more efficient and sustainable resource use. Integration fosters closed-loop cycles, where animal manure is used as fertilizer and crop residues are fed to livestock, thereby maximizing productivity within the system (Fig. 2). This closed-loop approach supports sustainable nutrient cycling, reduces dependence on synthetic inputs, and contributes to improved soil health and overall system resilience (Dumont et al., 2020).

One of the key advantages of integrated systems is their strong contribution to food security. By incorporating a diverse mix of crops and livestock, these systems support greater dietary diversity, which plays a critical role in improving nutrition and reducing malnutrition and associated health issues. Additionally, Integrated Crop–Livestock Systems (ICLS) enhance economic stability by generating multiple income streams, thereby reducing household vulnerability to market fluctuations. These systems offer a variety of livelihood opportunities–including crop production, livestock rearing, aquaculture, and agroforestry–which are especially valuable in rural areas. This income diversification strengthens community resilience, enabling households to better adapt to unforeseen challenges while promoting long-term sustainability and the conservation of natural resources. By minimizing the risk of total income loss from crop failure or price volatility, ICLS can also contribute to poverty alleviation and improved rural well-being.

Another major advantage of integrated systems is their resilience to the impacts of climate change. Their inherent adaptability and flexibility allow farmers to respond effectively to shifting climatic conditions. By optimizing land use and resource allocation, these systems enhance overall productivity while reducing the risk of food shortages, particularly in rural and climate-vulnerable regions. Integrated systems help buffer the effects of extreme weather events on both crop and livestock production, thereby supporting a more stable and reliable food supply. Farmers can tailor their practices to evolving conditions, such as prioritizing drought-tolerant crops during periods of water scarcity or diversifying crop and livestock species to spread risk and build resilience against climate-related disruptions.

ICLSs also offer important environmental benefits. When managed effectively, they can improve soil health, enhance grassland biodiversity, and contribute to climate change mitigation through practices such as agroforestry, cover cropping, and rotational grazing. These systems provide a holistic approach to sustainable agriculture, supporting rural livelihoods, strengthening food security, and increasing resilience to climate-related stresses. Despite these advantages, ICLSs still represent a relatively small proportion of farming systems globally. Higher adoption rates–exceeding 20% of farms or land area–are primarily observed in regions such as the northern United States, Canada, Argentina, Australia, and New Zealand (Fig. 3).

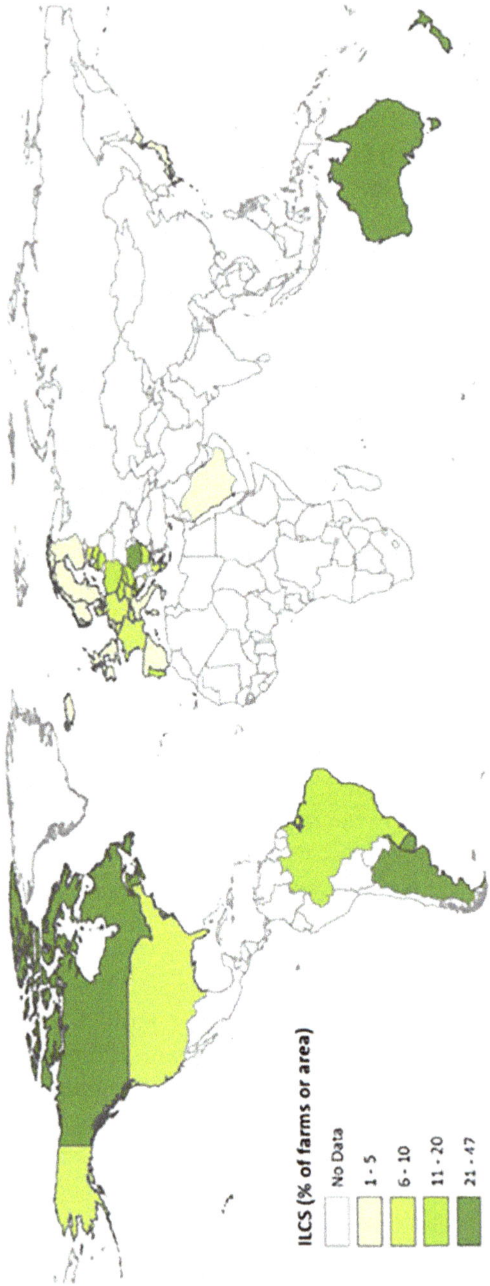

Figure 3 Integrated crop and livestock systems as a percentage of agricultural area or farms by country. Source: Adapted from Garret et al. (2020)

3 The role of grasslands in the context of mixed farming

3.1 Economic benefits

Grasslands play a vital role in supporting livestock health and productivity within mixed farming systems. The integration of grasslands, crops, and livestock creates a dynamic network of ecological interactions that enhances system sustainability. Well-managed grasslands provide high-quality forage, which supports improved weight gain, increased milk yield in dairy animals, and overall better animal health. Effective grazing management ensures a nutritionally balanced diet rich in proteins, minerals, and essential nutrients, which are critical for optimal growth and reproductive performance. Additionally, improved feed efficiency contributes to lower enteric methane (CH_4) emissions by reducing the time required for animals to reach maturity, thereby supporting both productivity and environmental goals (Franzluebbers, 2020).

The nutritional quality of grasslands is influenced by various factors, including forage species composition, phenological stage, and prevailing environmental conditions–all of which directly affect the diet and performance of grazing livestock. Under well-managed systems, grass-fed diets can significantly enhance the nutritional profile of livestock products. For example, milk from grass-fed cows is widely recognized for its superior quality. Research indicates that dairy products from grass-fed cattle contain elevated levels of antioxidants, such as vitamin E and beta-carotene, along with a more favorable omega-6 to omega-3 fatty acid ratio (Gómez-Cortés et al., 2018). Similarly, grass-fed meat, praised for its distinctive flavor (Priolo et al., 2001), is enriched with beneficial nutrients, including beta-carotene, vitamin E, and omega-3 fatty acids, making it nutritionally superior to meat from conventionally raised livestock (van Vliet et al., 2021).

Grass-fed products are particularly rich in omega-3 polyunsaturated fatty acids, such as eicosapentaenoic acid and docosahexaenoic acid, which are known to promote cardiovascular health, reduce inflammation, and support cognitive function (Swanson et al., 2012; Khalid et al., 2022). Additionally, these products often contain higher levels of conjugated linoleic acid, a compound linked to potential anticarcinogenic effects and enhanced immune function (Dachev et al., 2021). Growing consumer interest in sustainable food production and environmental stewardship, combined with the nutritional advantages of grass-fed systems, including increased concentrations of health-promoting fatty acids, vitamins, and antioxidants (Prache et al., 2020), presents a valuable opportunity for farmers to develop high-value, premium meat and dairy products.

Despite the well-documented benefits of livestock grazing for crop production, many producers across diverse environments and management contexts remain hesitant to adopt ICLSs. Their concerns often center on the potential negative impacts of grazing animals, such as soil compaction from trampling and excessive N deposition, which may affect the agronomic quality of their land (Peterson et al., 2020; Bilotta et al., 2007). However, a recent meta-analysis (Peterson et al., 2020), which compared crop yields in ICLS with those in unintegrated controls across three continents, involving 12 crops and four livestock species, revealed encouraging findings. On average, crop yields in ICLS were comparable to those in conventional systems, ranging from −7% to +2% (Fig. 4a). Interestingly, dual-purpose crops, such as rye, wheat, oats, and barley, managed simultaneously for both grazing and grain production, showed an average yield reduction of 20% compared to single-purpose crops. However, when dual-purpose systems were excluded, crops within ICLS outperformed those in unintegrated systems, particularly on loamy soils, and demonstrated equivalent yields across most other soil types. These findings suggest that regions with intermediate soil textures may be particularly suitable for implementing ICLS. Among the key variables influencing yield outcomes were crop type, soil characteristics, and weather anomalies, while climate, livestock species, and the specific type of ICLS played secondary roles (Fig. 4b).

3.2 Environmental benefits

Integrating crop cultivation with livestock rearing offers a strategic approach to grassland management, promoting adaptability and resource efficiency in the face of mounting environmental challenges. This synergy between plant and animal components optimizes the multifunctionality of grasslands, unlocking their potential for sustainable agriculture (Godfray et al., 2010; Koura et al., 2015). Grasslands serve as critical providers of ecosystem services that underpin agricultural sustainability, with soil conservation being one of their most significant contributions. The extensive root systems of grasses play a pivotal role in preventing soil erosion and nutrient loss, thereby preserving the long-term fertility of agricultural lands (Loges et al., 2018). These root structures also improve soil porosity, enhancing water infiltration and reducing surface runoff, which supports sustainable water use and land management (Yeates et al., 1998).

In addition to their soil-stabilizing properties, grasslands contribute substantially to climate change mitigation through carbon sequestration. The deep-rooted vegetation and high biomass inputs characteristic of grassland systems enhance the capture and long-term storage of atmospheric carbon dioxide (Mbow et al., 2017). Microbial activity within these systems promotes soil aggregate formation, which stabilizes SOC over extended periods (Tisdall and Oades, 1982; Bai and Cotrufo, 2022). Moreover, the continuous vegetative

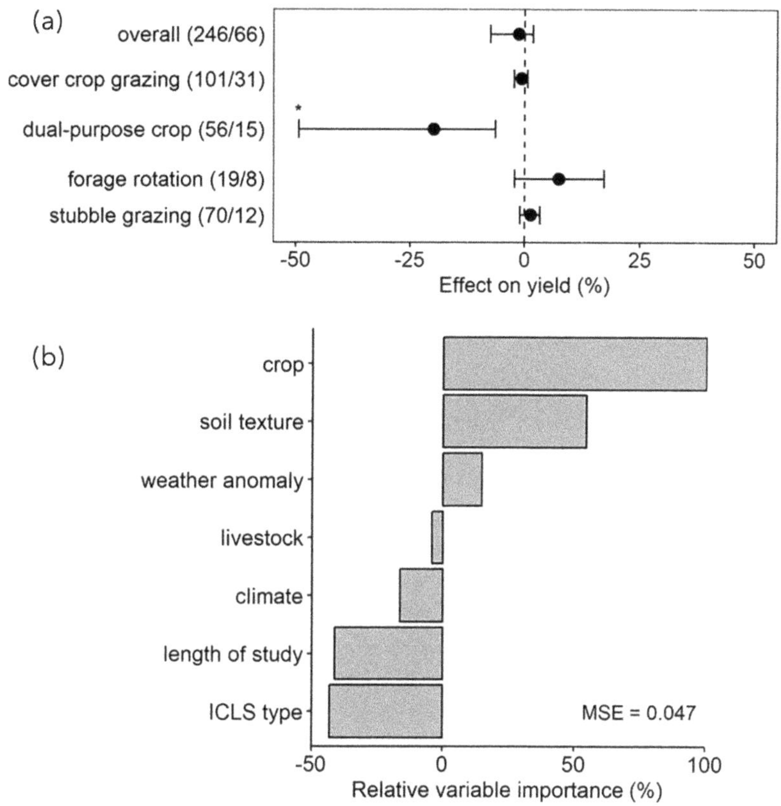

Figure 4 (a) Overall effect of ICLS on crop yield and effect within different ICLS system types; number of observations/number of studies for each category appears in parentheses. Points represent grazed system yield effect, while the dotted vertical line represents the paired ungrazed system yield. Error bars represent 95% bias-corrected-accelerated bootstrap confidence intervals. Asterisks (*) represent a significant yield response in grazed systems relative to ungrazed systems at the 95% confidence level. (b) Relative variable importance ranking. Rankings were derived from a random forest permutation procedure, with the effect size of ICLS relative to unintegrated systems as the response variable. Model variables were ranked according to the percent increase in mean squared error (MSE, or out-of-bag error) when they were removed from the model (Type 1 classification) and normalized relative to the most important variable. Negative variable importance indicates variables that were no more informative to the model than random chance. Source: Adapted from Peterson et al. (2020).

cover facilitates the cycling of key nutrients such as carbon (C), nitrogen (N), and phosphorus (P), supporting ongoing organic matter formation. Combined with minimal soil disturbance, these processes create favourable conditions for SOC accumulation, underscoring the ecological importance of grasslands in sustaining soil health and maintaining the global carbon balance.

The accumulation of SOC in both natural and managed agroecosystems depends on the balance between carbon inputs derived from photosynthesis and outputs driven by microbial decomposition, a dynamic influenced by various biotic and abiotic factors (Rodrigues et al., 2023). In grassland systems, decaying plant roots and aboveground biomass are the primary sources of organic matter. Among these, root-derived carbon tends to be retained longer in the soil due to its higher lignin content and enhanced protection within the soil matrix (Rasse et al., 2005; Schmidt et al., 2011). Grazing animals play a key role in nutrient cycling within these systems, contributing to improved soil fertility through manure deposition and the stimulation of plant regrowth (Yeates et al., 1998; Bai and Cotrufo, 2022). Beyond serving as vital sources of forage and supporting livestock productivity, grasslands deliver a suite of ecosystem services that are integral to the functionality of mixed farming systems. Grazing management, in particular, has a significant influence on SOC sequestration, as it affects root biomass, litter inputs, and the accumulation of soil organic matter–all of which are critical for sustaining grassland productivity. In addition to their soil conservation functions, grasslands are important biodiversity reservoirs, offering habitats for a wide array of plant and animal species. This biodiversity supports ecological resilience and balance, reinforcing the sustainability of integrated crop–livestock systems (Badenhausser et al., 2009; Lemaire et al., 2015).

Improper grazing practices can lead to soil compaction and erosion, undermining the long-term productivity of grasslands and adjacent croplands. Inadequate management strategies–such as continuous or excessive grazing without sufficient rest periods–can accelerate grassland degradation, resulting in diminished forage availability, compromised livestock nutrition, and the overall decline in ecosystem health (Fleischner, 1994; Bai and Cotrufo, 2022). To counter these effects, rotational grazing has emerged as an effective strategy for sustaining grassland health. By periodically relocating livestock across different pasture sections, this method promotes vegetation recovery, ensures uniform forage utilization, and contributes to habitat diversity, thereby enhancing biodiversity (Taube et al., 2014). However, while rotational grazing is beneficial in many respects, continuous grazing has been associated with greater root biomass accumulation compared to rotational systems (Hoogsteen et al., 2020). This has led to the development of innovative systems such as the 'Pure-Graze' approach, which aims to merge the benefits of both systems. By supporting uninterrupted root development and minimizing the stress periods associated with rotational rest phases, Pure-Graze offers a potential pathway to enhancing soil organic matter (SOM) levels and overall grassland resilience (Hoogsteen et al., 2020).

4 Management to enhance grassland utilization within mixed systems

4.1 Legume effects

Forage legumes, members of the Fabaceae (Leguminosae) family, are widely used as feed for ruminant livestock. Unlike cereal grains, which primarily provide seeds, forage legumes offer multiple edible plant parts, including leaves and stems. These legumes are commonly grazed, ensiled, or harvested as hay and are frequently cultivated either as monocultures or in mixtures with grasses. The Fabaceae family is notably diverse, comprising approximately 16 000 to 17 000 species across 750 genera (Allen and Allen, 1981). As protein-rich sources of forage and seeds, legumes fulfill multiple agronomic roles, serving as cover crops, green manure, and soil structure enhancers. A key attribute of legumes is their ability to fix atmospheric nitrogen (N_2) through symbiotic N-fixation, reducing dependence on synthetic N fertilizers and promoting cost-effective nutrient availability for subsequent crops. When incorporated into grass-legume mixtures, forage legumes significantly enhance overall pasture productivity and quality. For example, Loges (1998) demonstrated that the inclusion of red clover (*Trifolium pratense*) with perennial ryegrass (Lolium perenne) increased total N yield, crude protein (CP) content, N-fixation rates, and N content in the harvested biomass. This legume–grass mixture also improved N uptake and grain CP content in a following wheat crop, leading to increased grain yield across all tested clover–grass seeding rates. Furthermore, the red clover–perennial ryegrass combination outperformed mixtures with Italian ryegrass (Lolium multiflorum), highlighting the importance of species complementarity in optimizing sward performance. Outcomes from such mixtures, however, are influenced by additional factors, including sward age, ploughing timing, and fertilizer application rates. Therefore, site-specific management strategies must be developed and tailored to local conditions to maximize the agronomic and environmental benefits of forage legumes within mixed swards.

Integrating legumes with grasses can significantly enhance the nutritional profile and palatability of forage, offering livestock a more balanced and diverse diet. While red clover (*Trifolium pratense*) is recognized for its high nutritional value, it presents specific challenges during the ensiling process. These difficulties stem from its low concentration of water-soluble carbohydrates, high buffering capacity, and relatively low dry matter content at the time of harvest (McDonald et al., 1991). However, these limitations can be effectively mitigated by combining red clover with grasses, particularly when adequate wilting is performed prior to ensiling (Krawutschke, 2011). This approach not only facilitates a more successful fermentation process but also yields a higher-quality silage. Moreover, increasing the proportion of ryegrass (Lolium spp.) in the mixture has been shown to further improve silage fermentation characteristics, enhancing both the nutritional value and storability of the final product (Krawutschke, 2011).

4.2 Effects of diverse swards

Multi-species swards, also referred to as herb- and legume-rich leys, are grassland systems composed of six or more co-existing plant species. Increasing botanical diversity in grasslands provides a wide range of benefits for both livestock and the environment. Diverse, nutrient-rich vegetation better meets the dietary needs of grazing animals, often delivering forage of higher nutritional quality compared to grass-dominated swards (Katsvairo et al., 2006; Jerrentrup et al., 2020). Depending on regional conditions and target ecosystem services, these leys may consist of monocultures or mixtures of grasses, legumes, and forbs.

Grasses such as perennial ryegrass and Italian ryegrass are widely favored for their adaptability, versatility (suitable for silage, hay, and grazing), and ability to suppress weeds and improve soil structure. A Dutch study examining monocultures and grass–clover mixtures within an organic arable crop rotation revealed that grass monocultures, particularly perennial ryegrass, were associated with high root density, effective weed control, and improved soil structure (De Haas et al., 2019).

Conversely, clover monocultures contributed to increased soil mineral nitrogen, higher earthworm abundance, elevated herbage dry matter, nitrogen yield, and enhanced digestibility. When combined, grass-clover mixtures synergized these benefits, often outperforming even the best-performing monocultures.

Temporary grasslands composed of species with complementary functional traits offer greater adaptability to variable environmental conditions and demonstrate exceptional resilience under climate stress. Deep-rooted species enhance soil structure, promote water infiltration, and reduce erosion, mitigating the effects of extreme weather events such as droughts and heavy rainfall. This resilience ensures a consistent forage supply, bolstering system stability and farmer livelihoods (Lemaire et al., 2014). Although perennial ryegrass monocultures remain dominant across Western European grasslands, integrating species such as tall fescue and white clover alongside perennial ryegrass has proven highly productive (Hoogsteen et al., 2020). Tall fescue contributes substantial belowground biomass, while white clover offers nitrogen-fixing capacity, together forming N-efficient, high-yielding swards.

Beyond forage productivity, multispecies swards offer additional environmental services. They support pollinators throughout the growing season and help mitigate carbon losses (Steinbeiss et al., 2008). Enhanced plant diversity is linked to increased soil carbon (C) sequestration in both topsoil and subsoil layers (Lange et al., 2023). However, realizing this potential requires a deeper understanding of the regulatory mechanisms involved. Small-scale experimental studies have shown that greater plant diversity boosts SOC storage via mechanisms such as increased belowground C inputs, enhanced microbial diversity and activity, and reduced decomposition losses (Fig. 5) (Fornara and Tilman, 2008; Lange et al., 2015; Chen et al., 2018). For instance Fornara and Tilman (2008) demonstrated that high-diversity mixtures of perennial grassland species stored up to 500% more soil C and 600% more N compared to monocultures. Notably, plots containing C4 grasses and legumes

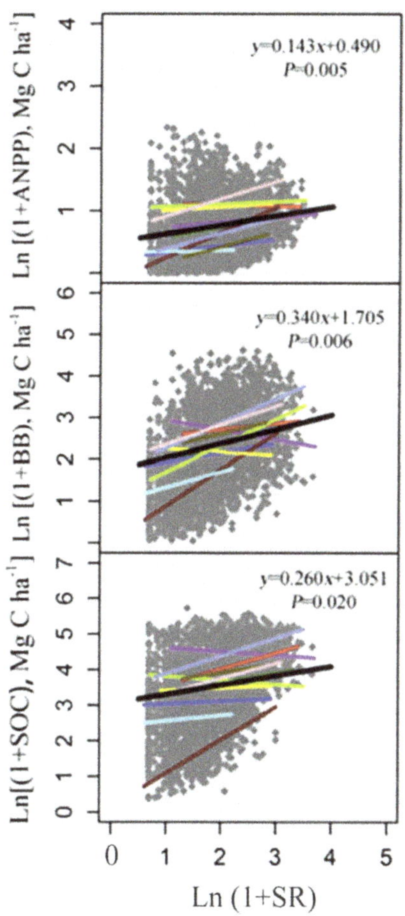

Figure 5 Relationships of species richness (SR) with aboveground net primary productivity (ANPP), belowground biomass (BB), and SOC storage at 0–30 cm soil depth in grasslands. Coloured lines present regression plots of different vegetation types, and bold black lines indicate SR–ANPP, SR–BB, and SR–SOC relationships at the biome-type level. Source: Adapted from Chen et al. (2018).

exhibited significantly greater soil C accumulation (193% and 522% increases, respectively), attributed to deeper root biomass and synergistic interactions between complementary functional groups.

Similarly, a study by Lange et al. (2015), derived from the long-term grassland biodiversity experiment (The Jena Experiment) and by radiocarbon (^{14}C) modelling, demonstrated that higher plant diversity contributes to increased rhizosphere C inputs. This increase in root-derived carbon stimulated microbial activity and soil carbon storage, resulting in the greater accumulation of recently fixed carbon in high-diversity plots. Furthermore, De Deyn et al. (2011) showed that long-term biodiversity restoration practices significantly enhance soil C and N storage, particularly when combined with an augmented abundance of specific legume species (e.g. *T. pratense*). These restoration efforts were also linked to reduced ecosystem respiration, elevated SOM content, and improved soil structure.

4.3 Bioactive herbs/forages

Bioactive forages refer to plants that are suitable as feed for livestock and contain active compounds with potential health benefits for animals and the environment. These bioactive compounds – such as phytochemicals, essential oils, tannins, flavonoids, and terpenoids – play functional roles in pasture-based livestock systems. These metabolites are found to be present in various prominent forage species, including alfalfa (*Medicago sativa*), chicory (*Cichorium intybus*), ribwort plantain (*Plantago* spp.), clovers (*Trifolium* spp.), and several other tropical legume species (Nyameasem et al., 2020; Verma et al., 2022). These compounds have been shown to exhibit anthelmintic properties, enhance nutrient utilization in ruminants, and promote soil carbon stabilization in grasslands (Adamczyk et al., 2017; Mueller-Harvey et al., 2019; Clemensen et al., 2022). Incorporating such species into multispecies leys can deliver additional ecosystem benefits, including the reduction of nitrous oxide (N_2O) emissions through biological nitrification inhibition and the mitigation of enteric methane emissions from livestock (Subbarao et al., 2012; Beauchemin et al., 2020). Furthermore, these diverse swards can enhance biodiversity by supporting a wide range of insect species critical to the food web, including those important to birds and other wildlife (Badenhausser et al., 2009; Lemaire et al., 2015; Beye et al., 2022). Despite these advantages, several challenges remain. These include the low agronomic potential of some species, variable bioactivity under differing environmental conditions, high variability across and within species and cultivars, and limited breeding programs aimed at enhancing these traits (Verma et al., 2021). Additionally, designing multispecies grasslands that fit within crop rotations while supporting diverse pollinator communities, particularly flower-visiting insects, remains a complex task, especially when striving to maintain forage productivity and feed quality (Malisch et al., 2023).

4.4 Management options for successful implementation of integrated crop-livestock system

System management should avoid heavy reliance on external inputs to preserve the balance of on-farm resources critical for sustaining soil fertility and crop yields. Excessive input use can disrupt nutrient cycles, potentially leading to environmental pollution and aggravating the effects of climate change. To achieve both high productivity and the delivery of key ecosystem services, effective management strategies are essential for optimizing grassland performance within crop rotations. Additionally, increased research investment is necessary to guide the development and implementation of these practices (Malisch et al., 2023). One promising approach involves terminating the grassland phase using no-till practices before establishing the succeeding crop. This method can help maintain high SOC stocks (Struck et al., 2019) while promoting biodiversity and ecological balance across the farmland. In contrast, frequent ploughing accelerates SOC decomposition, leading to significant carbon losses associated with conventional tillage. From an agronomic perspective, ploughing grassland may still offer advantages in certain rotational contexts, particularly when leys extend over several years, though this must be balanced against potential environmental trade-offs (Reinsch et al., 2018). For example, the incorporation of a 3-year-old grassland has been shown to provide greater agronomic benefits than annual grass–clover leys, arable legumes like field beans, or even manure applications from previous crops (Kayser et al., 2010). Over the long term, including a 2-year ley phase within a 6-year crop rotation can enhance SOC stocks by 5–6 tons per hectare over a 20-year period (Guillaume et al., 2022; Antony et al., 2022).

Undersowing arable crops with catch crops during the rotation is an effective strategy to reduce nitrate leaching through enhanced N uptake in autumn (Eriksen et al., 2015; Biernat et al., 2020). For instance, Eriksen et al. (2015) found that barley (*Hordeum vulgare*) undersown with Italian ryegrass significantly reduced nitrate leaching to less than 10 kg N ha^{-1} $year^{-1}$. In contrast, maize used as a subsequent crop led to elevated nitrate leaching, exceeding 50 kg N ha^{-1} $year^{-1}$ for five consecutive years. Similarly, Böldt et al. (2024) found that the choice of grass species as a cover crop plays a crucial role in closing the nitrogen cycle and enhancing on-farm N budgets by combining high N uptake with low N_2O emissions. These practices offer valuable insights into sustainable grassland management within crop rotations.

Grasslands possess significant potential for C sequestration; however, this capacity is not limitless, and effective long-term C storage requires a combination of complementary management practices (Smith, 2014). The mineralization of soil SOC tends to accelerate following grassland termination, especially during transitions to arable land within rotational systems. Therefore, implementing strategies that minimize SOC losses during these conversions is essential. Recommended approaches include the adoption of no-tillage practices during grassland conversion (Struck et al., 2019), the use of diverse plant species, particularly tannin-rich swards, for silage production, and the extension of ley phases with the inclusion of cover crops to increase SOM inputs. While fertilization can enhance plant biomass and promote soil C sequestration in non-tilled, permanent grasslands (Cenini et al., 2015), the application of N and P must be carefully managed. These inputs may alter soil microbial communities, potentially disrupting belowground ecological functions (Leff et al., 2015). Thus, it is crucial to consider both fertilizer type and application rates to support SOM accumulation while maintaining soil microbial integrity and ecosystem health.

5 Case studies

In this section, we briefly summarize three noteworthy case studies from northern Germany (Taube et al., 2023), Canterbury region of New Zealand (Peel et al., 2014), and Woburn ley–arable experiment in England (Johnston et al., 2017) that epitomize the transformative potential of ICLSs reliant on grasslands to promote sustainability, enhance ecosystem services, and mitigate environmental impacts. In other jurisdictions, such as Australia, mixed farmers have successfully implemented ley-arable systems integrating grasslands into their crop rotations (Bell and Moore, 2012; Bell et al., 2014). These systems capitalize on the benefits of grass–clover leys, demonstrating substantial N carry-over to subsequent cash crops, thereby reducing the risk of groundwater contamination. Additionally, incorporating legumes and diverse species into grasslands enhanced agro-biodiversity. These varied success stories highlight the adaptability and effectiveness of such systems across diverse geographical regions, further advocating for their adoption and widespread implementation within mixed farming landscapes.

5.1 The Lindhof approach in northern Germany

The Lindhof research farm in Schleswig-Holstein operates as an ICLS, managing a spring-calving Jersey herd within a low-input framework. The farm is dedicated to demonstrating the high eco-efficiency of pasture-based milk production by utilizing N self-sufficient, high-yielding grass-clover leys that are characterized by low N_2O emissions and considerable C sequestration potential. The overarching goal of the project is to maximize milk yield per hectare while simultaneously delivering a range of ecosystem services, including food production, water quality protection, climate change mitigation, biodiversity enhancement, improved animal welfare, and the maintenance of aesthetically valuable landscapes. Operating under organic farming standards, Lindhof places strong emphasis on the strategic use of grass-clover leys within ICLS to optimize N fixation and promote sustainable agricultural practices.

Grazing management at the farm incorporates rotational grazing, strip grazing, and silage supplementation, enabling over 90% of the cows' energy requirements to be met from pasture during the grazing season. To maintain consistent herbage availability, the farm employs cover crops and grass–clover swards. Each paddock's cycle concludes with oat cultivation, followed by two successive cash crops – typically wheat and mustard – before new leys are established.

A comparative analysis between the Lindhof system and 356 conventional indoor dairy farms in the region highlights the Lindhof model's superior productivity and profitability (Table 1). The spring-calving Jersey herd at Lindhof, managed intensively on pasture, demonstrates strong performance indicators, including superior milk composition, greater milk yield from forage, reduced dependence on purchased concentrates, and a lower adjusted reproduction rate. The system showcases efficient resource use, with approximately 48% of the annual energy demand met by grazed pasture, followed by grass silage at 35%, and concentrates at just 16%. In terms of crude protein (CP) contributing to milk production, grazed pasture accounts for around 47%, grass silage 41%, and grain-based concentrates only 11%. These figures highlight the central role of pasture and silage in supporting milk output under the Lindhof system.

Although forage production at Lindhof was only about 9% lower than that of conventional farms during the 2019/2020 accounting year – a trend consistent with long-term averages – the farm achieved a notable economic advantage. Total feed costs were approximately 32% lower compared to similar farms as illustrated in Table 1.

Several additional factors contribute to Lindhof's lower operating costs. Annual veterinary and hoof care expenses per cow are approximately 11% lower than those of comparable farms. Effective fertility management, such as achieving an age at first calving just over two years and maintaining a calving interval close to one year, further supports economic efficiency. Detailed documentation allows for precise labour allocation across business operations, resulting in a low annual working hour requirement per cow. Despite its scale, Lindhof maintains operational efficiency through well-planned infrastructure that ensures easy access to pastures from the milking parlour. Moreover, the farm benefits from significant cost savings in manure management compared to conventional indoor systems.

Table 1 Production parameters, economic results, and nitrogen balance (2019/2020) of the experimental farm Lindhof from Kiel University compared to the average of 356 dairy farms fully evaluated by *Vereinigung für Rinderspezialberatung* (VRS), a chamber of agriculture (extension service) in Schleswig-Holstein, Germany.

	Unit	Lindhof	Average of 356 by BZA
Dairy herd	number of cows	94	166
Body weight	kg cow^{-1}	470	670[a]
Milk yield	kg ECM cow^{-1}	7007	9433
Milk yield natural	kg cow^{-1}	5728	9257
Milk yield per kg body weight	kg ECM kg^{-1} BW	14.90	14.08
Milk solids production (fat plus protein)	kg cow^{-1}	592	702
Fat	%	5.59	4.2
Protein	%	3.99	3.45
Concentrate feeding	dt cow^{-1} $year^{-1}$	8.00	28.10
Concentrate feeding efficiency	g kg^{-1} ECM	120	295
Milk production per ha MFA on farm[b]	kg ECM ha^{-1} MFA	10,946	14,866
Milk produced from forage[c]	kg ECM cow^{-1}	5,284	3,767
Proportion of milk produced from forage[c]	%	75.41	39.93
Adjusted reproduction rate	%	18.20	33.40
First calving age (LKV-SH, 2021)	months	24.6	28.4
Calving interval (LKV-SH, 2021)	days	362	400
Costs for veterinary, medicines + hoof care	ct kg^{-1} ECM	1.48	1.64
Total feed costs[d]	ct kg^{-1} ECM	16.81	22.12
Costs of producing forage	ct kg^{-1} ECM	12.17	13.35
Concentrate feed costs	ct kg^{-1} ECM	3.83[e]	8.77
Mineral N fertilizer input	kg N ha^{-1} MFA	0	99
N balance[f]	kg N ha^{-1} MFA	88	149

Abbreviations: SH = Schleswig-Holstein, ECM = energy-corrected milk, MFA = main forage area, LK-SH = Landwirtschaftskammer (Chamber of Agriculture) Schleswig-Holstein.

[a]Estimated value based on the average of the breeds, [b]without area requirements for imported feed; [c]milk from concentrates excluded according to LK-SH (2021) calculation, [d]rearing replacement heifers included, [e]Farm-gate N balance of only the dairy operation, [f]from organic production at a 63% higher price. Source: Reported by Taube et al. (2023).

Field measurements at Lindhof (Smit et al., 2021) revealed low nitrate leaching in fertilized permanent grasslands and moderate levels in grazed grass-clover swards. Grazing was found to increase NO_3–N losses compared to cutting, though Lindhof's values remained below the typical range reported in the literature (Smit et al., 2021), possibly due to the effectiveness of its root systems (Chen et al., 2018). A notable N carry-over effect from grass-clover swards to subsequent cash crops was observed, potentially reducing the risk of groundwater contamination and lowering the need for additional N fertilizer in subsequent crops. Another study at Lindhof (Beye et al., 2022) highlighted the role of legume mixtures, particularly those with *Trifolium* species, in supporting bumblebee populations. Grazed binary and tertiary mixtures of grass-clover leys (including perennial ryegrass, white clover, and red clover) had significant positive effects on pollinator abundance. This research underscores the benefits of enriching species diversity and incorporating legumes into grass–clover leys, which enhanced agro-biodiversity compared to conventional grasslands lacking flowering legumes. Introducing clover (20–40%) into grasslands was found to notably increase wild bee abundance.

Jersey cows that grazed on the multi-species swards (*Lolium perenne*, *Trifolium repens*, *T. pratense*, *Lotus corniculatus*, *Cichorium intybus*, *Plantago lanceolata*, *Carum carvi*, and *Sanguisorba minor*) exhibited remarkably low enteric methane emissions per unit of milk, measuring less than 10 g CH^4 kg ECM^{-1}. This represents a significant reduction compared to the typical range of 10–20 g CH_4 kg ECM^{-1} for Jersey cows. This improvement is likely due to the high-quality feed consumed by the cows, as supported by studies by Loza et al. (2021), Nyameasem et al. (2022), and Lorenz et al. (2020). Additionally, Nyameasem et al. (2022) observed that the N_2O N emission factor from cow urine and dung at Lindhof was less than half of the IPCC default value (0.30% compared to 0.77%). This reduced emission factor can be attributed to the pastures' enhanced ability to capture and utilize N inputs from excreta patches. The greenhouse gas emissions associated with milk from the Lindhof grazing system, reflected in the carbon footprint (PCF), are approximately 0.6 kg CO_2eq kg ECM^{-1}. In contrast, conventional indoor systems producing milk year-round have a significantly higher PCF of more than 1.0 kg CO_2eq kg ECM^{-1} (Taube et al., 2023).

Using grass-clover mixtures in crop rotations at Lindhof results in significant soil carbon accumulation, which substantially reduces the carbon footprint of milk production compared to conventional year-round indoor farming. The ley system at Lindhof also provides valuable environmental services, helping to mitigate social costs associated with N, P, and CO_2 pollution, which are estimated at 25 cents per kg ECM, with phosphorus pollution accounting for the largest share at 13 cents per kg ECM (Taube et al., 2023). The 2.5-year grass–clover ley phase, essential for Lindhof's milk production, also benefits its cash crop production. Nitrogen-rich grass-clover roots and stubble stored in the soil prevent leaching and provide about 150 kg N ha^{-1} to the succeeding crop, such as edible oats. Additionally, Lindhof's manure resources enhance winter cereal yields. This integrated system helps balance the excess nitrogen from milk production (approximately 88 kg N ha^{-1} from grass-clover forage), reduces N losses, and supports a positive humus balance (Taube et al., 2023).

5.2 Canterbury region of New Zealand

A study by Peel et al. (2014) on seven mid-Canterbury farm businesses showcases the region's growing trend of integrating crop and dairy systems. The study aimed to explore the reasons behind this transition and the associated benefits, summarized below. The ICLS farms included a variety of cash crops such as milling wheat, vegetables, ryegrass seed, peas, potatoes, Asian brassicas, malting barley, hemp, and oats. All the dairy farms within the ICLS system used supplemented feed (600–1200 kg per cow per lactation, or 2–5 kg daily), mainly consisting of grains, placing them above average in terms of stocking rate and production per cow compared to district averages in Ashburton. Farmers in mid-Canterbury transitioned from intensive cropping systems to integrated crop and dairy operations using three primary methods: first, by converting a single cropping farm into separate, adjacent crop and dairy farms; second, by transforming a single cropping farm into combined crop and dairy units with flexible boundaries; and third, in some cases, by acquiring additional land and establishing independent, non-adjacent plots to develop supplementary dairy units.

According to the authors, some farmers shifted their land use to achieve greater profitability, while others sought improvements in their lifestyle. They found dairy farming to be simpler and less stressful compared to crop cultivation, prompting some to partially transition to dairy farming to reignite their passion for agriculture and gain more leisure time. Introducing lower-order share milkers or managers allowed farmers to step back from direct labor management, aligning with their preferences and reducing their workload. In all cases, the cropping area was reduced. Farmers employed various wintering practices, using crop farms for full or partial dairy herd wintering. Some incorporated grazing and transferred cut-and-carry feed to the milking platform, while others practiced individual drying-off of cows to extend the milking season. Self-wintering was seen as beneficial for ensuring financial security, guaranteeing profit, and supporting intensive crop rotations. Additionally, keeping cows on crop farms offered advantages, such as enhancing manure utilization in maize cultivation and reducing fertilizer requirements.

All farmers recognized the synergies associated with supplementary feed. Some sourced all their cow supplements from their cropping farm, while others evaluated the comparative costs before making decisions. Internal feed trading helped reduce transaction costs, with one farmer purchasing grain from neighboring dairy platform farms at a lower price. Farmers also sold silage, baleage, pea, and barley straw and utilized grazing on seed crops as supplements for their herds. Although crop farms did not directly subsidize feed costs for the dairy platform, information sharing facilitated the supply of appropriate feed quantities at reduced transaction costs. In these case

Table 2 Case study farms compared to Ashburton district farms

	Average (and range) for the case study farms	Averages for the Ashburton district 2011/2012
Stocking rate, cows ha^{-1}	3.9 (3.5–4.5)	3.5
Cow herd size	881 (500–1900)	859
Milking platform, ha	247 (138–420)	243
Milk solid production, kg cow^{-1}	485 (457–550)	406
Milk solid production, kg ha^{-1}	1892 (1600–2182)	1421

Source: Adapted from Peel et al. (2014).

studies, farmers shared specialized machinery between their crop and dairy enterprises, promoting the exchange of information and knowledge. They believed they had a better understanding of crop markets and feed value than single-enterprise dairy farmers, enabling more informed decisions about feed costs. Ultimately, the case-study farms recorded superior production metrics compared to the average farms (Table 2). While potential environmental benefits were acknowledged, they were not quantified in the study.

5.3 Woburn ley-arable experiment in England

Incorporating cropping land into dairy farming presents potential benefits for reducing environmental issues associated with the dairy industry, offering a promising avenue for industry growth. One key strategy is the increased sequestration of carbon in SOM, which can help mitigate the rising concentrations of atmospheric CO_2. This aligns with global initiatives such as the 4‰ initiative (Soussana et al., 2010). Specifically, it involves altering cropping patterns to include more herbaceous plants on agricultural soils (Paustian et al., 2016). However, the success of these strategies, especially in temperate climates, depends on several complex factors, including the relatively slow rate of SOC change in most cropping systems.

A long-term ley-arable experiment initiated in 1938 on sandy loam soil offers valuable insights (Johnston et al., 2017). Initially, the soil had 0.98% SOC after more than 60 years of continuous arable cropping. Over time, a decline in SOC was observed, particularly in areas with continuous arable cropping, especially when root crops or fallows were included. However, the introduction of 3-year grass or grass + clover leys within a 5-year ley-arable rotation (3 years of grass-clover ley followed by 2 years of arable crops, such as winter wheat and winter barley) significantly increased SOC in the topsoil over a period of 30–40 years, eventually stabilizing around 1.3% SOC (Fig. 6). Extending the ley phase to 8 years in a 10-year rotation (8 years of grass-clover leys followed by spring barley and winter wheat) revealed that SOC had not yet reached

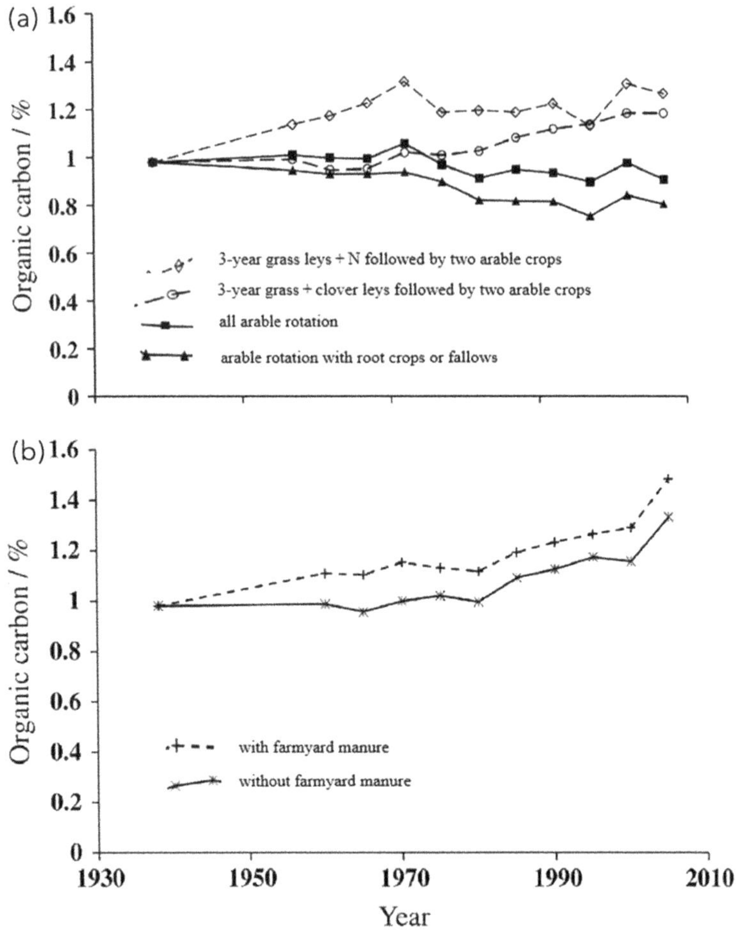

Figure 6 Percentage organic C in soil to 25 cm for selected treatments of (a) 5-year rotation experiment (data are the mean of soil samples with or without farmyard manure) and (b) 8-year ley rotations (data are the mean of the first and second cycles). Each point is the mean of data from each of five blocks. Source: Adapted from Johnston et al. (2017).

equilibrium after more than 30 years, highlighting the dynamic nature of SOC responses. Additionally, periodic farmyard manure application every fifth year had a substantial positive impact on SOC levels, underscoring the role of organic inputs.

Modeling has shed light on the fate of added SOC, showing that only 5% is retained in the soil, while a significant 95% is lost. This highlights the challenges of maintaining sequestered carbon over the long term. Both observed and modeled changes suggest that annual increases of 4% or more in SOC are achievable through the cultivation of grass or grass + clover leys. However, this improvement is sustainable only for a limited time until a new equilibrium for SOC is reached, which depends on the continuous maintenance of SOC inputs from the leys. These findings underscore the potential of grass + clover leys in helping mitigate atmospheric CO_2 levels. However, under similar conditions, arable grasslands within mixed farms may show lower C sequestration benefits compared to permanent grasslands. Nyameasem et al. (2024) found that 12 years land-use change, SOC stocks were were significantly higher in permanent grasslands compared to arable systems with ley, a finding consistent with earlier studies (Guillaume et al., 2022; Elias et al., 2023).

Factors such as vegetation type, organic input quality, and management practices play an critical role in C cycling, influencing biological communities and, consequently, C sequestration (Garnett et al., 2017). Permanent grasslands benefit from minimal soil disturbance, which promotes slower turnover rates for both native SOC and newly added organic matter. This process supports the formation of stable aggregates and prolongs the residence time of organic matter. In contrast, arable grassland soils have lower potential for SOC storage due to several factors, including reduced biomass C input (De Los Rios et al., 2022), increased leaching of inorganic and dissolved organic N and C, and the cyclical depletion of SOM due to crop growth and harvesting (Antony et al., 2022). Additionally, lower belowground biomass allocation and crop harvest further contribute to reduced litter input and organic matter accumulation in arable grasslands (Tang et al., 2019). Understanding these dynamics is essential for developing effective strategies to enhance C sequestration in agricultural systems.

6 Future trends in research

The evolution of precision agriculture has significantly had a profound impact on grassland management within mixed farming systems (Taube, 2022). Political initiatives, such as the European Commission's goal to fully integrate farmers into the digital economy for sustainable food production, highlight the growing importance of agricultural digital transformation (Dayioğlu and Turker, 2021). Agri-tech solutions are playing a key role in enabling data-driven decision-making, promoting more efficient and sustainable grassland management practices that optimize productivity while minimizing environmental impacts. Achieving this will require leveraging tools such as sensors, robotic automation, and digital data, all of which depend on robust connectivity (Berger, 2019; Fielke et al., 2020). Cutting-edge technologies, including remote sensing, eddy covariance systems, satellite imagery, and geographic information systems (GIS), are now instrumental in monitoring and managing grasslands. These technologies provide precise insights into pasture health, biomass production, environmental footprint, and grazing patterns, which can empower farmers to make informed decisions about grazing intensity, nutrient management, and rotational grazing practices. For instance, real-time monitoring using global positioning system-enabled collars on livestock helps optimize grazing management, preventing overgrazing and enhancing forage utilization.

Like all farming systems, agri-tech solutions offer considerable potential for optimizing grassland management and productivity within mixed farming systems. Integrated sensor technologies and Internet of Things (IoT) devices enable continuous monitoring of key parameters such as soil moisture, nutrient levels, and plant health, facilitating more precise and efficient resource management. AI-powered predictive models further support decision-making by providing insights into optimal grazing patterns, fertilization timing, and pest control strategies. For instance, automated drones equipped with multispectral cameras can map vegetation health, helping identify zones that require intervention or adjustments in grazing intensity. Additionally, the development of innovative irrigation systems tailored specifically for grasslands promotes efficient water use, an increasingly vital factor in regions experiencing water scarcity or irregular rainfall patterns (Stevens et al., 2021). Future research should prioritize strategies to enhance the accessibility and affordability of these technologies for farmers to ensure widespread adoption and long-term sustainability.

Simultaneously, integrating diverse forages can play a pivotal role in restoring degraded soils and enhancing C sequestration. Continued innovation in forage crop genetics is essential, particularly in the context of climate change, to develop varieties with improved nutritional value, higher yields, and enhanced traits such as drought tolerance, resource-use efficiency, and disease resistance (Sustek-Sánchez et al., 2023). These advancements will support the development of more resilient and sustainable grassland systems. Selecting forage species that can withstand extreme weather events and adopting climate-smart practices to reduce greenhouse gas emissions are key steps forward. One such practice is agroforestry, integrating trees or shrubs into grazing systems, which not only provides supplementary forage and shade for livestock but also improves soil structure and ecosystem health. Agri-tech solutions will be instrumental in designing and managing these systems by assessing soil and climate data to identify tree species best suited to local conditions and complementary to existing grassland vegetation. Increasingly, emphasis is being placed on resilience to extreme weather, resource-use efficiency, and multifunctionality in grasslands, with growing recognition of the role that plant diversity in leys can play in delivering these outcomes.

Organic N application can enhance SOC storage in grasslands; however, it is often associated with substantial ammonia and N_2O losses (Nyameasem et al., 2022, 2023, 2024), which are further intensified by climate change-driven increases in N mineralization. To better understand these complex interactions between soil processes, environmental conditions, and C and N cycling, process-based models are indispensable. These models provide a cost-effective and scalable alternative to large-scale field experiments by enabling the simulation of gaseous nitrogen emissions across different spatial and temporal scales. However, the accuracy of such models hinges on their rigorous validation against empirical data, as emphasized by the Intergovernmental Panel on Climate Change. Enhancing the development of simulation platforms that are more generic, adaptable, and robust will improve their applicability and reliability (Enders et al., 2023). This, in turn, will deepen our understanding of the key system dynamics influencing agricultural sustainability and support more informed, evidence-based policymaking.

7 Conclusion

The strategic management of grasslands emerges as a crucial element in fostering sustainable agricultural landscapes globally. Integrating grasslands into mixed farming and crop–livestock systems fosters synergies, mitigates trade-offs between ecosystem services, and extends benefits well beyond food production. Recognizing the complex and multifunctional nature of grasslands underscores the need for targeted research, supportive policies, and adaptive management strategies. As agricultural practices evolve and technology continues to advance, grasslands hold significant potential to enhance ecosystem services, lower environmental footprints, and strengthen the resilience of agricultural systems. Examples such as the Lindhof Farm's low-input model highlight the transformative potential of well-managed grasslands. By promoting multifunctionality and leveraging innovations in grassland management, integrated systems can achieve more efficient resource use, minimize negative environmental impacts, and bolster long-term sustainability. Ultimately, embedding grasslands into mixed farming systems is central to a resilient, productive, and ecologically responsible agricultural future.

8 Acknowledgement

The Federal Ministry of Food and Agriculture (BMEL), based on a decision of the Parliament of the Federal Republic of Germany via the Federal Office for Agriculture and Food (BLE) (grant number 2822ABS010), partially funded this work.

9 Where to look for further information

- Kadžiulienė, Ž., et al. (n.d.). The future role of ley-farming in cropping systems. https://www.europeangrassland.org/fileadmin/documents/Infos/Printed_Matter/Proceedings/EGF2023.pdf.

- Livestock in Mixed Farming Systems of the Hindu Kush-Himalayas (n.d.). Available at: https://www.fao.org/3/x5862e/x5862e04.htm (Accessed February 5, 2024).
- Nikzaad, R. M. and Nusrathali, N. (2023). Integrating livestock and crop systems for enhanced productivity and grassland conservation in developing countries. *IntechOpen*. https://doi.org/10.5772/intechopen.113109.
- Reinsch, T., et al. (2021). Toward specialized or integrated systems in Northwest Europe: On-farm eco-efficiency of dairy farming in Germany. *Frontiers in Sustainable Food Systems* 5. Available at: https://www.frontiersin.org/articles/10.3389/fsufs.2021.614348 (Accessed June 17, 2023).
- Taube, F., et al. (2024). Closed-loop combined crop–livestock farming systems. In B. Amon (Ed.), *Developing circular agricultural production systems*. Cambridge: Burleigh Dodds Science Publishing, pp. 35–64. https://doi.org/10.19103/AS.2023.0120.08.

10 References

Adamczyk, B., et al. (2017). Tannins can slow down and speed up soil enzymatic activity in boreal forests. *Soil Biology and Biochemistry* 107, 60–67. https://doi.org/10.1016/j.soilbio.2016.12.027

Adu, E. K., et al. (2017). The grasscutter: an untapped resource of Africa's grasslands. *Livestock Research for Rural Development* 29. Available at: http://www.lrrd.org/lrrd29/3/jnya29047.html (Accessed January 29, 2024).

Alderkamp, L. M., et al. (2025). Integrating crop and dairy production systems: Exploring different strategies to achieve environmental targets. Science of The Total Environment, 958, 177990. https://doi.org/https://doi.org/10.1016/j.scitotenv.2024.177990

Alexandratos, N. and Bruinsma, J. (2012). *World Agriculture towards 2030/2050: the 2012 revision*. Rome: FAO.

Allen, O. N. and Allen, C. K. (1981). *The leguminosae: a source book of characteristics, uses and modulation*. London, UK: Macmillan Publishers Ltd.

Antony, D., et al. (2022). Soil organic matter storage in temperate lowland arable, grassland and woodland topsoil and subsoil. *Soil Use and Management* 38, 1532–1546. https://doi.org/10.1111/sum.12801

Aschemann-Witzel, J., et al. (2021). Plant-based food and protein trend from a business perspective: markets, consumers, and the challenges and opportunities in the future. *Critical Reviews in Food Science and Nutrition* 61, 3119–3128. https://doi.org/10.1080/10408398.2020.1793730

Badenhausser, I., et al. (2009). Reconciling agricultural and environmental landscapes in cereal production areas. *Biofutur*, 27–30.

Bai, Y. and Cotrufo, M. F. (2022). Grassland soil carbon sequestration: current understanding, challenges, and solutions. *Science* 377(6606): 603–608.

Beauchemin, K. A., et al. (2020). Review: fifty years of research on rumen methanogenesis: lessons learned and future challenges for mitigation. *Animal* 14, s2–s16. https://doi.org/10.1017/S1751731119003100

Bell, L. W. and Moore, A. D. (2012). Integrated crop–livestock systems in Australian agriculture: trends, drivers and implications. *Agricultural Systems* 111, 1–12. https://doi.org/10.1016/j.agsy.2012.04.003

Bell, L. W., et al. (2014). Evolution in crop–livestock integration systems that improve farm productivity and environmental performance in Australia. *European Journal of Agronomy* 57, 10–20. https://doi.org/10.1016/j.eja.2013.04.007

Bengtsson, J., et al. (2019). Grasslands—more important for ecosystem services than you might think. *Ecosphere* 10, e02582. https://doi.org/10.1002/ecs2.2582

Berger, R. (2019). Global focus report. In *Farming 4.0: How precision agriculture might save the world.* Germany: Publisher Roland Berger GMBH.

Beye, H., et al. (2022). Species-enriched grass-clover mixtures can promote bumblebee abundance compared with intensively managed conventional pastures. *Agronomy* 12, 1080. https://doi.org/10.3390/agronomy12051080

Biernat, L., et al. (2020). Is organic agriculture in line with the EU-Nitrate directive? On-farm nitrate leaching from organic and conventional arable crop rotations. *Agriculture, Ecosystems & Environment* 298, 106964. https://doi.org/10.1016/j.agee.2020.106964

Bilotta, G., et al. (2007). The impacts of grazing animals on the quality of soils, vegetation, and surface waters in intensively managed grasslands. Advances in agronomy, 94, 237–280. https://doi.org/10.1016/S0065-2113(06)94006-1

Böldt, M. J., et al. (2024). Evaluating nitrous oxide emissions in low input systems using different cover crop strategies over the winter period. *Agriculture, Ecosystems & Environment* 364, 108895. https://doi.org/10.1016/j.agee.2024.108895

Catovsky, S., et al. (2002). Biodiversity and ecosystem productivity: implications for carbon storage. *Oikos* 97, 443–448. https://doi.org/10.1034/j.1600-0706.2002.970315.x

Cenini, V. L., et al. (2015). Chronic nitrogen fertilization and carbon sequestration in grassland soils: evidence of a microbial enzyme link. *Biogeochemistry* 126, 301–313. https://doi.org/10.1007/s10533-015-0157-5

Chen, S., et al. (2018). Plant diversity enhances productivity and soil carbon storage. *Proceedings of the National Academy of Sciences of the United States of America* 115, 4027–4032. https://doi.org/10.1073/pnas.1700298114

Clemensen, A., et al. (2022). Potential benefits of tanniferous forages in integrative crop-livestock agroecosystems. *Frontiers in Agronomy* 4. Available at: https://www.frontiersin.org/articles/10.3389/fagro.2022.911014 (Accessed January 30, 2024).

Dachev, M., et al. (2021). The effects of conjugated linoleic acids on cancer. *Processes* 9, 454. https://doi.org/10.3390/pr9030454

Dayioğlu, M. A. and Turker, U. (2021). Digital transformation for sustainable future - Agriculture 4.0 : a review. *Tarım Bilimleri Dergisi.* https://doi.org/10.15832/ankutbd.986431

De Deyn, G. B., et al. (2011). Additional carbon sequestration benefits of grassland diversity restoration. *Journal of Applied Ecology* 48, 600–608. https://doi.org/10.1111/j.1365-2664.2010.01925.x

De Haas, B. R., et al. (2019). Combining agro-ecological functions in grass-clover mixtures. *AIMS Agriculture and Food* 4, 547–567. https://doi.org/10.3934/agrfood.2019.3.547

De Los Rios, J., et al. (2022). Integrating crop-livestock system practices in forage and grain-based rotations in Northern Germany: potentials for soil carbon sequestration. *Agronomy* 12, 338. https://doi.org/10.3390/agronomy12020338

Doblas-Rodrigo, Á., et al. (2022). Grassland contribution to soil organic carbon stock under climate change scenarios in Basque Country (Spain). *Regional Environmental Change* 22, 34. https://doi.org/10.1007/s10113-022-01877-4

Dumont, B., et al. (2020). Incorporating diversity into animal production systems can increase their performance and strengthen their resilience. *Frontiers in Sustainable Food Systems* 4. Available at: https://www.frontiersin.org/articles/10.3389/fsufs.2020.00109 (Accessed January 19, 2024).

Elferink, M. and Schierhorn, F. (2016). Global demand for food is rising. Can we meet it? *Harvard Business Review*. Available at: https://hbr.org/2016/04/global-demand-for-food-is-rising-can-we-meet-it (Accessed January 29, 2024).

Elias, D. M. O., et al. (2023). The potential to increase grassland soil C stocks by extending reseeding intervals is dependent on soil texture and depth. *Journal of Environmental Management* 334, 117465. https://doi.org/10.1016/j.jenvman.2023.117465

Enders, A., et al. (2023). SIMPLACE–a versatile modelling and simulation framework for sustainable crops and agroecosystems. *In Silico Plants* 5, diad006. https://doi.org/10.1093/insilicoplants/diad006

Erb, K.-H., et al. (2016). Livestock grazing, the neglected land use. In H. Haberl, M. Fischer-Kowalski, F. Krausmann, and V. Winiwarter (Eds.), *Social ecology: Society-nature relations across time and space human-environment interactions*. Cham: Springer International Publishing, pp. 295–313. https://doi.org/10.1007/978-3-319-33326-7_13

Eriksen, J. (2015). Nitrate leaching and residual effect in dairy crop rotations with grass–clover leys as influenced by sward age, grazing, cutting and fertilizer regimes. *Agriculture, Ecosystems & Environment* 212, 75–84. https://doi.org/10.1016/j.agee.2015.07.001

Fielke, S., et al. (2020). Digitalisation of agricultural knowledge and advice networks: a state-of-the-art review. *Agricultural Systems* 180, 102763. https://doi.org/10.1016/j.agsy.2019.102763

Fleischner, T. (1994). Ecological costs of livestock grazing in Western North America. *Conservation Biology* 8, 629–644. https://doi.org/10.1046/j.1523-1739.1994.08030629.x

Fornara, D. A. and Tilman, D. (2008). Plant functional composition influences rates of soil carbon and nitrogen accumulation. *Journal of Ecology* 96, 314–322. https://doi.org/10.1111/j.1365-2745.2007.01345.x

Franzluebbers, A. J. (2020). Chapter 2 - Cattle grazing effects on the environment: greenhouse gas emissions and carbon footprint. In M. Rouquette and G. E. Aiken (Eds.), *Management strategies for sustainable cattle production in Southern Pastures*. New York, NY: Academic Press, pp. 11–34. https://doi.org/10.1016/B978-0-12-814474-9.00002-5

Garnett, T., et al. (2013). Sustainable intensification in agriculture: premises and policies. *Science* 341, 33–34. https://doi.org/10.1126/science.1234485

Garnett, T., et al. (2017). *Ruminating on cattle, grazing systems, methane, nitrous oxide, the soil carbon sequestration question – and what it all means for greenhouse gas emissions*. Oxford: University of Oxford.

Garrett, R. D., et al. (2020). Drivers of decoupling and recoupling of crop and livestock systems at farm and territorial scales. *E&S* 25, art24. https://doi.org/10.5751/ES-11412-250124

Giller, K. E., et al. (2021). The future of farming: Who will produce our food? *Food Security* 13, 1073–1099. https://doi.org/10.1007/s12571-021-01184-6

Godfray, H. C. J., et al. (2010). The future of the global food system. *Philosophical Transactions of the Royal Society B: Biological Sciences* 365, 2769–2777. https://doi.org/10.1098/rstb.2010.0180

Gómez-Cortés, P., et al. (2018). Milk fatty acids and potential health benefits: an updated vision. *Trends in Food Science & Technology* 81, 1–9. https://doi.org/10.1016/j.tifs.2018.08.014

Guillaume, T., et al. (2022). Carbon storage in agricultural topsoils and subsoils is promoted by including temporary grasslands into the crop rotation. *Geoderma* 422, 115937. https://doi.org/10.1016/j.geoderma.2022.115937

Haughey, E., et al. (2023). Sustainable intensification and carbon sequestration research in agricultural systems: a systematic review. *Environmental Science & Policy* 143, 14–23. https://doi.org/10.1016/j.envsci.2023.02.018

Hoogsteen, M. J. J., et al. (2020). Do grazing systems and species composition affect root biomass and soil organic matter dynamics in temperate grassland swards? *Sustainability* 12, 1260. https://doi.org/10.3390/su12031260

Jerrentrup, J. S., et al. (2020). Diverse swards and mixed-grazing of cattle and sheep for improved productivity. *Frontiers in Sustainable Food Systems* 3. Available at: https://www.frontiersin.org/articles/10.3389/fsufs.2019.00125 (Accessed January 30, 2024).

Johnston, A. E., et al. (2017). Changes in soil organic matter over 70 years in continuous arable and ley–arable rotations on a sandy loam soil in England. *European Journal of Soil Science* 68, 305–316. https://doi.org/10.1111/ejss.12415

Katsvairo, T. W., et al. (2006). Sod–livestock integration into the peanut–cotton rotation: a systems farming approach. *Agronomy Journal* 98, 1156–1171. https://doi.org/10.2134/agronj2005.0088

Kayser, M., et al. (2010). Nitrogen management in organic farming: comparison of crop rotation residual effects on yields, N leaching and soil conditions. *Nutrient Cycling in Agroecosystems* 87, 21–31. https://doi.org/10.1007/s10705-009-9309-0

Khalid, W., et al. (2022). Functional behavior of DHA and EPA in the formation of babies brain at different stages of age, and protect from different brain-related diseases. *International Journal of Food Properties* 25, 1021–1044. https://doi.org/10.1080/10942912.2022.2070642

Koura, B. I., et al. (2015). Determinants of crop-livestock integration by small farmers in Benin. *International Journal of Biological and Chemical Sciences* 9, 2272–2283. https://doi.org/10.4314/ijbcs.v9i5.2

Kozicka, M., et al. (2023). Feeding climate and biodiversity goals with novel plant-based meat and milk alternatives. *Nature Communications* 14, 5316. https://doi.org/10.1038/s41467-023-40899-2

Krawutschke, M. (2011). Qualitätsveränderungen im Zuwachsverlauf und bei der Gärfutterbereitung von Rotklee (Trifolium pratense L.) unter besonderer Berücksichtigung der Rohproteinfraktionen.

Lange, M., et al. (2023). Increased soil carbon storage through plant diversity strengthens with time and extends into the subsoil. *Global Change Biology* 29, 2627–2639. https://doi.org/10.1111/gcb.16641

Lange, M., et al. (2015). Plant diversity increases soil microbial activity and soil carbon storage. *Nature Communications* 6, 6707. https://doi.org/10.1038/ncomms7707

Lee, H., et al. (2023). IPCC, 2023: Climate Change 2023: Synthesis Report, Summary for Policymakers. Contribution of Working Groups I, II and III to the Sixth Assessment Report of the Intergovernmental Panel on Climate Change. H. Lee and J. Romero (Eds.), IPCC, Geneva, Switzerland.

Leff, J. W., et al. (2015). Consistent responses of soil microbial communities to elevated nutrient inputs in grasslands across the globe. *Proceedings of the National Academy of Sciences* 112, 10967–10972. https://doi.org/10.1073/pnas.1508382112

Lemaire, G., et al. (2014). Integrated crop–livestock systems: strategies to achieve synergy between agricultural production and environmental quality. *Agriculture, Ecosystems & Environment* 190, 4–8. https://doi.org/10.1016/j.agee.2013.08.009

Lemaire, G., et al. (2015). Grassland–cropping rotations: an avenue for agricultural diversification to reconcile high production with environmental quality. *Environmental Management* 56, 1065–1077. https://doi.org/10.1007/s00267-015-0561-6

Loges, R. (1998). Ertrag, Futterqualität, N2-Fixierungsleitung und Vorfruchtwert von Rotklee- und Rotkleegrasbeständen.

Loges, R., et al. (2018). Forage production in rotational systems generates similar yields compared to maize monocultures but improves soil carbon stocks. *European Journal of Agronomy* 97, 11–19. https://doi.org/10.1016/j.eja.2018.04.010

Lord, S. (2023). Hidden costs of agrifood systems and recent trends from 2016 to 2023. FAO.

Lorenz, H., et al. (2020). Does the Admixture of Forage Herbs Affect the Yield Performance, Yield Stability and Forage Quality of a Grass Clover Ley? Sustainability, 12(14), 5842. https://www.mdpi.com/2071-1050/12/14/5842

Loza, C., et al. (2021). Methane Emission and Milk Production from Jersey Cows Grazing Perennial Ryegrass–White Clover and Multispecies Forage Mixtures. Agriculture, 11(2), 175. https://www.mdpi.com/2077-0472/11/2/175

Lynch, J., et al. (2021). Agriculture's contribution to climate change and role in mitigation is distinct from predominantly fossil CO2-emitting sectors. *Frontiers in Sustainable Food Systems* 4. Available at: https://www.frontiersin.org/articles/10.3389/fsufs.2020.518039 (Accessed January 29, 2024).

Malisch, C. S., et al. (2023). The importance of multi-species grassland leys to enhance ecosystem services in crop rotations. In *The future role of ley-farming in cropping systems.* Vilnius: Grassland Science in Europe, pp. 192–206. Available at: https://ira.agroscope.ch/en-US/Page/Publikation/Index/53389 (Accessed January 30, 2024).

Mbow, H.-O. P., et al. (2017). Special report on climate change, desertification, land degradation, sustainable land management, food security, and greenhouse gas fluxes in terrestrial ecosystems (SR2). *Ginevra, IPCC* 650. Available at: https://www.ipcc.ch/site/assets/uploads/2018/07/sr2_background_report_final.pdf (Accessed January 30, 2024).

McDonald, P., et al. (1991). *The biochemistry of silage*. Marlow, UK: Chalcombe Publications.

Minasny, B., et al. (2017). Soil carbon 4 per mille. Geoderma, 292, 59–86. https://doi.org/https://doi.org/10.1016/j.geoderma.2017.01.002

Moraine, M., et al. (2017). A social-ecological framework for analyzing and designing integrated crop–livestock systems from farm to territory levels. *Renewable Agriculture and Food Systems* 32, 43–56. https://doi.org/10.1017/S1742170515000526

Mueller-Harvey, I., et al. (2019). Benefits of condensed tannins in forage legumes fed to ruminants: importance of structure, concentration, and diet composition. *Crop Science* 59, 861–885. https://doi.org/10.2135/cropsci2017.06.0369

Nyameasem, J. K., et al. (2024). Incorporating leys in arable systems as a mitigation strategy to reduce soil organic carbon losses during land-use change. *Frontiers in Environmental Science* (in press). https://doi.org/10.3389/fenvs.2024.1399197

Nyameasem, J. K., et al. (2020). Nitrogen availability determines the long-term impact of land use change on soil carbon stocks in grasslands of southern Ghana. *SOIL* 6, 523–539. https://doi.org/10.5194/soil-6-523-2020

Nyameasem, J. K., et al. (2023). Effect of slurry application techniques on nitrous oxide emission from temperate grassland under varying soil and climatic conditions. *Grass and Forage Science* 78, 338–358. https://doi.org/10.1111/gfs.12612

Nyameasem, J. K., et al. (2022). Impact of cattle slurry application methods on ammonia losses and grassland nitrogen use efficiency. *Environmental Pollution* 315, 120302. https://doi.org/10.1016/j.envpol.2022.120302

Nyameasem, J. K., et al. (2024). Incorporating leys in arable systems as a mitigation strategy to reduce soil organic carbon losses during land-use change. *Frontiers in Environmental Science*, 12. https://doi.org/10.3389/fenvs.2024.1399197

O'Mara, F. P. (2012). The role of grasslands in food security and climate change. *Annals of Botany* 110, 1263–1270. https://doi.org/10.1093/aob/mcs209

Paustian, K., et al. (2016). Climate-smart soils. Nature, 532(7597), 49–57. https://doi.org/10.1038/nature17174

Peel, S., et al. (2014). Integration of crop and dairy farms. *Primary Industry Management* 18, 36–37.

Peterson, C. A., et al. (2020). Commercial integrated crop-livestock systems achieve comparable crop yields to specialized production systems: a meta-analysis. *PLoS One* 15, e0231840. https://doi.org/10.1371/journal.pone.0231840

Poppy, G. M., et al. (2014). Food security in a perfect storm: using the ecosystem services framework to increase understanding. *Philosophical Transactions of the Royal Society B: Biological Sciences* 369, 20120288. https://doi.org/10.1098/rstb.2012.0288

Prache, S., Martin, B., & Coppa, M. (2020). Authentication of grass-fed meat and dairy products from cattle and sheep. Animal, 14(4), 854–863. https://doi.org/10.1017/S1751731119002568

Priolo, A., D. Micol and J. Agabriel (2001). "Effects of grass feeding systems on ruminant meat colour and flavour. A review." Animal research 50(3): 185–200. https://doi.org/10.1051/animres:2001125

Rasse, D. P., et al. (2005). Is soil carbon mostly root carbon? Mechanisms for a specific stabilisation. *Plant and Soil* 269, 341–356. https://doi.org/10.1007/s11104-004-0907-y

Rodrigues, C. I. D., et al. (2023). Soil carbon sequestration in the context of climate change mitigation: a review. *Soil Systems* 7, 64. https://doi.org/10.3390/soilsystems7030064

Röös, E., et al. (2017). Greedy or needy? Land use and climate impacts of food in 2050 under different livestock futures. *Global Environmental Change* 47, 1–12. https://doi.org/10.1016/j.gloenvcha.2017.09.001

Schmidt, M. W. I., et al. (2011). Persistence of soil organic matter as an ecosystem property. *Nature* 478, 49–56. https://doi.org/10.1038/nature10386

Sekaran, U., et al. (2021). Role of integrated crop-livestock systems in improving agriculture production and addressing food security – a review. *Journal of Agriculture and Food Research* 5, 100190. https://doi.org/10.1016/j.jafr.2021.100190

Smit, H. P., et al. (2021). "Very low nitrogen leaching in grazed ley-arable-systems in northwest europe." Agronomy 11(11): 2155. https://doi.org/10.3390/agronomy11112155

Smith, P. (2014). Do grasslands act as a perpetual sink for carbon? *Global Change Biology* 20, 2708–2711. https://doi.org/10.1111/gcb.12561

Soussana, J.-F., et al. (2010). Mitigating the greenhouse gas balance of ruminant production systems through carbon sequestration in grasslands. Animal, 4(3), 334–350.

Steinbeiss, S., et al. (2008). Plant diversity positively affects short-term soil carbon storage in experimental grasslands. *Global Change Biology* 14, 2937–2949. https://doi.org/10.1111/j.1365-2486.2008.01697.x

Stevens, D. R., et al. (2021). Integrating digital technologies to aid grassland productivity and sustainability. *Frontiers in Sustainable Food Systems* 5, 602350.

Struck, I. J., et al. (2019). Yield potential and nitrogen dynamics of no-till silage maize (Zea mays L.) under maritime climate conditions. *European Journal of Agronomy* 107, 30–42. https://doi.org/10.1016/j.eja.2019.04.009

Stypinski, P. (2011). The effect of grassland-based forages on milk quality and quantity. *Agronomy Research* 9, 479–488.

Subbarao, G. V., et al. (2012). Biological nitrification inhibition—a novel strategy to regulate nitrification in agricultural systems. In *Advances in agronomy*. Elsevier, pp. 249–302. https://doi.org/10.1016/B978-0-12-394275-3.00001-8

Sustek-Sánchez, F., et al. (2023). Improving abiotic stress tolerance of forage grasses – prospects of using genome editing. *Frontiers in Plant Science* 14, 1127532. https://doi.org/10.3389/fpls.2023.1127532

Swanson, D., et al. (2012). Omega-3 fatty acids EPA and DHA: health benefits throughout life1. *Advances in Nutrition* 3, 1–7. https://doi.org/10.3945/an.111.000893

Tang, S., et al. (2019). Synthesis of soil carbon losses in response to conversion of grassland to agriculture land. *Soil and Tillage Research* 185, 29–35. https://doi.org/10.1016/j.still.2018.08.011

Taube, F. (2022). From ecological intensification to hybrid agriculture—The future domain of digital farming. In J. Dörr & M. Nachtmann (Eds.), *Handbook digital farming–Digital transformation for sustainable agricul- ture*. Cham: Springer-Verlag GmbH, pp. 357–364.

Taube, F., et al. (2014). Grassland and globalization – challenges for north-west European grass and forage research. *Grass and Forage Science* 69, 2–16. https://doi.org/10.1111/gfs.12043

Taube, F., et al. (2023). Eco-efficiency of leys—the trigger for sustainable integrated crop-dairy farming systems. *Grass and Forage Science* n/a, 1–12. https://doi.org/10.1111/gfs.12639

Tisdall, J. M. and Oades, J. M. (1982). Organic matter and water-stable aggregates in soils. *Journal of Soil Science* 33, 141–163. https://doi.org/10.1111/j.1365-2389.1982.tb01755.x

Topp, E., et al. (2023). Perceptions and sociocultural factors underlying adoption of conservation agriculture in the Mediterranean. *Agriculture and Human Values*. https://doi.org/10.1007/s10460-023-10495-7.

van Vliet, S., et al. (2021). Health-promoting phytonutrients are higher in grass-fed meat and milk. *Frontiers in Sustainable Food Systems* 4. Available at: https://www.frontiersin.org/articles/10.3389/fsufs.2020.555426 (Accessed January 29, 2024).

Van Zanten, H. H. E., et al. (2018). Defining a land boundary for sustainable livestock consumption. *Global Change Biology* 24, 4185–4194. https://doi.org/10.1111/gcb.14321

Verma, S., et al. (2021). Examining the variables leading to apparent incongruity between antimethanogenic potential of tannins and their observed effects in ruminants–a review. *Sustainability* 13, 2743. https://doi.org/10.3390/su13052743

Verma, S., et al. (2022). Linking metabolites in eight bioactive forage species to their in vitro methane reduction potential across several cultivars and harvests. *Scientific Reports* 12, 10454. https://doi.org/10.1038/s41598-022-14424-2

Yeates, G. W., et al. (1998). Contrasting response to cropping of populations of earthworms and predacious nematodes in four soils. *Soil and Tillage Research* 48, 255–264. https://doi.org/10.1016/S0167-1987(98)00134-2

Part 2

Grassland dynamics and management

Chapter 4

Plant-soil-animal interactions and nutrient cycling in grazed grasslands

Lynn E. Sollenberger, University of Florida, USA; Marta M. Kohmann, University of Wisconsin-Madison, USA; K. Roger Trumpp and Jose C.B. Dubeux, Jr, University of Florida, USA

1 Introduction

Grasslands are an abundant and important global ecosystem, occupying more than 40% of Earth's ice-free terrestrial surface (Hewins et al., 2018). Grasslands grazed by livestock occur on ~40 million km^2 (Lambin and Meyfroidt, 2011). Interactions among plants, soils, and animals are inherent to grazed grasslands (Sollenberger and Wallau, 2020) and affect a wide range of ecosystem functions. Nutrient cycling is one of the most consequential (Dubeux and Sollenberger, 2020).

In extensively managed grasslands where fertilizer inputs are low, senescent plant litter is an important nutrient pool and plays a significant role in

http://dx.doi.org/10.19103/AS.2025.0154.10

nutrient cycling (Dubeux et al., 2007). Forage intake by herbivores, especially at greater stocking rates, reduces the proportion of nutrients cycling in plant litter while increasing that in excreta (Thomas, 1992). Eighty percent or more of most nutrients consumed by herbivores are excreted (Haynes and Williams, 1993), and nutrients in excreta are more labile and cycle at a greater rate than those in plant litter (Bardgett and Wardle, 2003).

Plant recovery of excreted nutrients can be negatively affected by heterogeneous distribution of excreta in the grassland (Dubeux et al., 2009) and by leaching, volatilization, and immobilization (Haynes and Williams, 1993). Although elemental cycles in perennial grasslands are generally more strongly coupled than in annual cropping systems (Soussana and Lemaire, 2014), animal consumption, digestion, and metabolism can result in decoupling of these cycles, leading to potentially negative impacts on air and water quality (Dungait et al., 2012). Decoupling occurs because dung is enriched in C and P and urine in N and K. These nutrients can be lost, recycled, or stored in plants or soils depending on how grazing animals change nutrient distribution in the landscape (Vertès et al., 2019). Decoupling can result in accelerated rates of pasture degradation or limited pasture growth due to nutrient scarcity (Boddey et al., 2004). Thus, understanding nutrient pools and cycles and the effect of management practices on nutrient cycling efficiency is a requisite for improving pasture productivity, reducing off-farm inputs, and minimizing negative consequences to the environment (Dubeux and Sollenberger, 2020).

The focus of this chapter is nutrient cycling in grazed pasture in temperate climates. A limited number of references from subtropical environments are included to provide additional depth and breadth. Within the grazed grassland context, our objectives are to:

- describe the most important nutrient pools and the nutrient fluxes among pools;
- discuss the roles of management practices, soil fauna, and the soil microbial community in nutrient cycling;
- provide example nutrient budgets for grazed swards;
- present a case study which quantifies nutrient pools and fluxes in an N-fertilized grass monoculture and a grass-legume mixed pasture; and
- identify current knowledge gaps and future research needs.

These are discussed in more detail in this chapter.

2 Nutrient pools and fluxes in grazed pastures

Others have described in detail the nutrient pools and fluxes in grazed pasture (Dubeux et al., 2007). Here, we briefly overview N, P, and K pools, providing greater emphasis on nutrient fluxes among pools.

The largest N pool is the atmosphere, but because of the high stability of N_2, it is only available to plants through biological or industrial N fixation. The atmospheric N pool is ~16 000 times greater than the sum of the soil and biotic N pools (Russelle, 1996). Thus, N is the most limiting nutrient in many grasslands (Russelle, 1992). Second in size to the atmospheric N pool is soil, with the size of this pool being site-specific as it is determined by the quantity of organic matter, microbial biomass, adsorbed NH_4^+, and plant-available inorganic N (Stevenson and Cole, 1999). Soil mesofauna (invertebrates, 0.1–2.0 mm in size, e.g. nematodes and mites) and macrofauna (e.g. earthworms, snails, beetle larvae) are important components of the soil N pool and represent far greater amounts of N than the sum of aboveground and belowground plant N (Dubeux et al., 2007). Plant litter is a relatively small N pool, but it is important because litter and soil microbiota constitute the link between N in living plant tissue and that available for subsequent plant uptake (Thomas and Asakawa, 1993).

The soil is the greatest reservoir of K. Because K is not found primarily as a component of organic compounds, its chemistry in soils is based on cation exchange reactions. In sandy soil environments, the absence of organic matter or clay decreases effective cation exchange capacity and makes K vulnerable to leaching (Dubeux et al., 2007). Animals represent a relatively small K pool, but they play an oversized role in K cycling because of the relatively high K concentration in ingested herbage and the large proportion of ingested K that is excreted.

Compared with N and K, both plant and animal tissue P concentrations are relatively small, resulting in soil being the largest and most important P pool in grasslands (Haynes and Williams, 1993). The availability of soil P depends not only on turnover of organic P but also on the chemistry of inorganic P (Rao et al., 1999). The potential retention of inorganic P in unavailable oxide forms, a function of soil pH and chemical composition, is a constraint to efficient P cycling.

The proportion of K, P, and N in live herbage, aboveground plant litter (belowground plant litter was not measured), live root-rhizome biomass, and soil (to a 20-cm depth) nutrient pools is shown for rotationally stocked bermudagrass (*Cynodon* spp.) pastures (Table 1; Liu et al., 2011c; 2017). Expressed as a proportion of the sum of nutrients in the four pools, the soil N and P pools contained >80% of these nutrients, regardless of grazing or N fertilizer management, while for K, the proportions in soil and live herbage were nearly the same and always in a range of 31–49% (Table 1). Across all nutrients, the plant litter pool was the smallest. Increasing N fertilization increased the proportion of K and N in plant pools and decreased the proportion in soil (Liu et al., 2017). As grazing stubble height increased, the size of the plant nutrient pools increased, mainly as a function of greater herbage mass (Liu et al., 2011c).

Table 1 Ranges in potassium, phosphorus, and nitrogen in each of four nutrient pools as a proportion of the total amount of that nutrient present across all measured pools of rotationally stocked Tifton 85 bermudagrass pastures

	N fertilization	Polynomial contrast: N[a]	Stubble height	Polynomial contrast: height[b]
Nutrient/Pool	% in pool % in pool			
Potassium				
Live herbage	31–42	$L^{*\uparrow c}$	35–42	$L^{**\uparrow c}$
Plant litter	2–4	$L^{*\uparrow}$	3–4	$L^{*\uparrow}$
Root-rhizome	18–22	Q^{*}	16–20	NS
Soil to 20 cm	36–49	$L^{**\downarrow}$	36–46	$L^{**\downarrow}$
Phosphorus				
Live herbage	7–10	NS	6–8	$L^{\uparrow}$
Plant litter	2	NS	2	NS
Root-rhizome	6–9	NS	6–7	NS
Soil to 20 cm	79–85	NS	83–86	$L^{\downarrow}$
Nitrogen				
Live herbage	2–4	$L^{**\uparrow}$	3–4	Q^{**}
Plant litter	1–2	$L^{**\uparrow}$	1–2	$L^{**\downarrow}$,Q^{**}
Root-rhizome	3–4	$L^{*\uparrow}$,Q^{*}	4	NS
Soil to 20 cm	90–93	$L^{**\downarrow}$,Q	90–92	Q^{**}

* and ** indicate the significance at the $P \leq 0.05$ and 0.01 levels, respectively. L or Q not followed by an asterisk, $P \leq 0.10$; NS = not significant, $P > 0.10$. ↑ indicates the linear response was increasing as the N level or stubble height level increased. ↓ indicates the linear response was decreasing as the N level or stubble height level decreased.

[a]Polynomial contrast for the effect of N level (50, 150, and 250 kg ha^{-1} $year^{-1}$) within nutrient and pool.

[b]Polynomial contrast for the effect of post-grazing stubble height (8, 16, and 24 cm) within nutrient and pool.

[c]L = linear and Q = quadratic effects of N level or stubble height.

Ranges for each nutrient in each pool bound the responses observed across different levels of pasture N fertilizer rate and post-grazing stubble height. Source: Data adapted from Liu et al. (2011c, 2017).

Nutrient pool size is a snapshot in time as nutrients cycle among pools and undergo biochemical processes that alter their chemical structure and biological availability (Dubeux and Sollenberger, 2020). Grazing animals affect grassland nutrient cycling by modifying spatial distribution of nutrients (Mathews et al., 1994a) and altering nutrient lability (Vertès et al., 2019). Within the soil–plant–animal continuum, nutrient cycling is a critical supporting ecosystem service because grassland productivity is often nutrient limited (Sollenberger et al., 2019).

Although soil is generally the largest nutrient pool in grazed pastures, mineralization rates of soil nutrients (20–50 g kg^{-1} of that nutrient in soil $year^{-1}$) are much lower than for nutrients in livestock excreta and plant residues

Published by Burleigh Dodds Science Publishing Limited, 2025.

(650–850 g kg^{-1} $year^{-1}$; Brady and Weil, 2002), limiting the quantity made available from the soil for plant growth (Dubeux et al., 2007). As a result, plant litter and animal excreta play an important role in nutrient cycling in grazed grasslands (Boddey et al., 2004).

Aboveground deposition and decomposition of plant litter occur continuously in grasslands and influence plant nutrient supply. Litter chemical composition affects the balance between nutrient mineralization and immobilization processes that are especially important for N, P, and S (Myers et al., 1994). Increasing N fertilization and stocking rate increase litter quality (i.e. greater N concentration, lesser C:N ratio) and rate of litter turnover and nutrient release (Dubeux et al., 2006a). Species composition of grasslands also affects litter dynamics, with the presence of legumes increasing litter quality, decomposition rate, and net nutrient mineralization compared with grasslands devoid of legumes (Kohmann et al., 2018).

Most mineral nutrients consumed by herbivores are excreted in dung and urine (Rotz et al., 2005). Nutrients in urine are either in inorganic forms and immediately plant-available or are mineralized in a few days and rapidly available for plant uptake (Mathews et al., 1996). Nutrients in dung vary in form and rate of availability, and the release of nutrients from dung is generally slower than from urine but more rapid than from plant litter (Bardgett and Wardle, 2003). Because the quantity of nutrients contained in excreta patches greatly exceeds the immediate plant requirement, particularly in areas of the animal congregation, significant nutrient losses to the environment can occur (Cai et al., 2017; Mitchel et al., 2021).

3 Excreta deposition and nutrient return in grazed pastures

This theme explores deposition patterns, excreta effects on plant responses and soil characteristics, nutrient losses from excreta, and the use of nitrification inhibitors to minimize loss of excreta N. In terms of excreta nutrient composition, we can generalize that P, Ca, Mg, and several micronutrients (e.g. Fe, Cu, Mn, and Zn) are excreted primarily in the dung, while K and to a lesser extent Na are excreted primarily in the urine. The proportion of N and S excreted in dung versus urine is a function of diet composition (Mathews et al., 1996), with their proportion in urine increasing as dietary N and S concentrations increase (Rotz et al., 2005). Braz et al. (2002) reported that 93% of the N ingested by animals grazing a low-protein grass pasture was excreted in dung, while 80% of N was excreted in urine when forage N concentration was high (Whitehead, 2000).

3.1 Patterns of excreta deposition

Forage consumption by grazing herbivores and subsequent excreta deposition redistributes and concentrates nutrients in the landscape, with effects on sward composition and productivity and on the environment (Auerswald et al., 2010). Excreta patch distribution is generally not homogeneous, and there are positive correlations between the time spent in a location by herbivores and excreta patch density (Draganova et al., 2016). Understanding patterns of excreta deposition and what affects them is important for enhancing our understanding of nutrient cycling within a pasture.

Grassland topography affects excreta deposition patterns. On hilly topography of southern Bavaria, Germany, aggregation of excreta occurred on the lowlands and crests due to increased time spent grazing and resting, respectively (Auerswald et al., 2010). In New Zealand hill country, annual excreta N load on low slopes was two- and four-fold that of medium and high slopes due to livestock behavior, specifically the choice of low slopes as primary areas for congregating (Hoogendorn et al., 2016). In a sloping pasture in southern Germany, nutrients accumulated in a flat crest zone and were depleted in the steeper areas (Schnyder et al., 2010). These studies indicate that slope and/or landscape position affect where cattle congregate and the quantity of excreta loading, with implications for soil nutrient concentrations, nutrient cycling, and potential nutrient loss.

Structural features of pastures affect patterns of excreta deposition. Herd concentration areas such as laneways, water troughs, and shade have greater nutrient loading from excreta than areas where such structural features are not present (De Rosa et al., 2020; Mitchell et al., 2021). Excreta deposition in locations where animal treading occurs accentuates nutrient losses due to compaction, and these areas increase spatial concentration of soil nitrate (NO_3^-) and nitrous oxide (N_2O) fluxes (Mitchell et al., 2021). In Japan, the spatial distribution of cow dung was well-predicted considering green herbage mass and distance from a water trough (Yoshitoshi et al., 2020). In warm climates or seasons characterized by warm temperatures, well-distributed shade caused dung distribution to be more homogeneous; the absence of trees resulted in the concentration of excreta around gates, water troughs, and mineral feeders (Carpinelli et al., 2020).

Shade affected excreta deposition patterns under high-temperature conditions in Hawaii, USA (Mathews et al., 1999). Authors reported that N, P, and K accumulation was greater around shade than waterers, but in paddocks without shade, substantial amounts of P and K accumulated near the water points. In a high-temperature environment in Florida, USA, soil N, P, and K accumulated in zones closest to shade and water, suggesting that positioning of structural features plays a major role in excreta deposition and soil nutrient redistribution in warm environments (Mathews et al., 1994a; 1994b). Likewise, soil P, K, and Mg concentrations in tall fescue [*Lolium arundinaceum* (Schreb.) S.J. Darbyshire] pastures were ~2–8, 3–15, and 1–1.5 times greater, respectively, near shade and water than at greater distances (Schomberg et al., 2000).

Fence line position may affect the pattern of excreta deposition. In an isolated pasture, excreta density decreased with proximity to a fence line (Auerswald et al., 2010), but deposition along fence lines increased if animals occupied adjoining pastures (Dubeux et al., 2014). In strip-grazed dairy paddocks in Australia, soil K was greater near the gate end than at the back of the paddock (Aarons et al., 2015). These studies support that structural features and arrangement of pastures are important determinants of excreta deposition patterns with consequent effects on soil nutrient concentrations and that shade has an especially important role in deposition patterns in warm environments or seasons.

3.2 Nutrient return from excreta and plant response

Nutrient input to pastures from excreta affects forage production and composition. Nutrients in urine are immediately available for plant uptake, but rate of nutrient availability from dung varies (Mathews et al., 1996). Most K in dung is water soluble and readily available, while N, Ca, Mg, and S release is slower and negatively affected by lignin and polyphenols that reduce mineralization rates (Mathews et al., 1996). Nutrient returns (kg ha^{-1} $year^{-1}$) in dung for year-round grazing systems ranged from 5 kg to 9 kg P, 10 kg to 26 kg K, 10 kg to 16 kg Ca, 9 kg to 13 kg Mg, and 23 kg to 40 kg N (Garcia et al., 2021). Total annual N excretion in dung plus urine ranged from 69 kg ha^{-1} to 89 kg ha^{-1} (Garcia et al., 2021), demonstrating the importance of nutrient return in excreta to subsequent plant response.

Urine deposition increased herbage accumulation by 15% and herbage N and K concentrations by 19% and 17%, respectively, over a no-excreta control in perennial grasslands in Germany (Scheile et al., 2018). The effect on herbage accumulation was measurable 7 months after urine deposition. Forage accumulation in perennial ryegrass (*Lolium perenne* L.)-white clover (*Trifolium repens* L.) swards in New Zealand increased linearly with increasing urine-N application rates (Di and Cameron, 2007). Forage accumulation in urine patches increased from 3040 kg ha^{-1} to 4820 kg ha^{-1} as deposition frequency increased in bahiagrass (*Paspalum notatum* Flugge) pastures in Florida, USA (White-Leech et al., 2013b). Forage accumulation and nutritive value increased to distances of 15 cm and 30 cm beyond urine deposits, respectively, with increases in forage accumulation observed for ≥84 days following single or multiple urine applications (White-Leech et al., 2013a).

Published by Burleigh Dodds Science Publishing Limited, 2025.

Forage responses to dung deposition have been less consistent. In the northern Czech Republic, herbage N, P, and K concentrations were greater at dung patches in intensively grazed swards (Pavlu et al., 2019), but herbage accumulation at dung patches was not different than the control in Solling Uplands, Germany (Scheile et al., 2018). Bahiagrass forage accumulation decreased linearly from 3040 kg ha^{-1} to 2800 kg ha^{-1} with increasing dung application frequency, apparently due to grass smothering associated with crusting of the exposed surface of dung pads and minimal dung beetle activity (White-Leech et al., 2013b). In contrast, perennial ryegrass herbage accumulation increased for 112 days after the application of dung from grazing dairy cattle in temperate Australia (Aarons et al., 2009). Overall, the literature suggests that urine deposition will almost always be accompanied by increasing forage accumulation and nutritive value, but plant responses to dung are less predictable and depend significantly on the presence of soil fauna like dung beetles and on weather conditions, especially rainfall, following deposition.

3.3 Excreta effects on soil characteristics

Excreta deposition generally changes soil nutrient status sufficiently to affect fertilizer recommendations (Carran and Theobold, 2000). Grazed New Zealand pastures with natural excreta deposition for 23 years were compared with those protected from excreta deposition and grazing (Carran and Theobold, 2000). Where excreta was deposited, total soil N was 20% greater in the top 15-cm soil layer and more N was mineralized in the 7.5–15 cm layer than in excluded areas. Exchangeable K was much greater in areas accessed by cattle than where they were excluded.

In the northern Czech Republic, the presence of dung in grazed, mixed-species perennial pastures did not affect soil nutrient concentrations, likely because nutrients not utilized by plants were leached, volatilized, or transformed into unavailable forms (Pavlu et al., 2019). Conversely, in temperate dairy pastures in Australia, extractable P and K increased in the 0–5 cm soil layer under dung, total organic P was unaffected, and microbial soil P increased (Aarons et al., 2004a; 2004b; Aarons et al., 2009). In Switzerland, cattle urine application stimulated nitrification and denitrification enzyme activities in soil and increased NH_4^+ and NO_3^- concentrations in species-rich temperate pastures (Hartmann et al., 2013).

Pasture species composition can affect nutrient cycling via animal excreta. Dung from cattle grazing a smooth bromegrass (*Bromus inermis* Leyss.)-legume mixture had 5–12% greater N concentration than that of cattle grazing a bromegrass monoculture receiving 90 kg N ha^{-1} $year^{-1}$ (Schick et al., 2019). Seven days after dung deposition, the soil below dung patches in the mixed-species pasture had 20–42% greater inorganic N than the soil under dung

Published by Burleigh Dodds Science Publishing Limited, 2025.

patches in the N-fertilized grass monoculture. Greater diet nutritive value in the grass-legume mixture increased dung N concentration, resulting in faster dung decomposition and more rapid nutrient movement into the soil.

3.4 Nutrient losses from excreta

Nutrients exit grazed grasslands in animal products or through losses including volatilization, leaching, and runoff (Dubeux and Sollenberger, 2020). Increasing intensity of grassland management leads to greater amounts of reactive N and P in the soil, increasing the likelihood of losses to the environment (Soussana and Lemaire, 2014). Because of its importance to productivity and the ease at which it can be lost from the system, N is a primary nutrient of concern. Pathways of N loss vary with climate, with gaseous losses predominating in arid environments and NO_3^- leaching predominating in high-rainfall environments (Russelle, 1992).

Deposition of either dung or urine increases N_2O emissions by enhancing N mineralization, nitrification, denitrification, or nitrifier denitrification. Nitrification is the microbe-mediated aerobic process resulting in oxidation of NH_4^+ to NO_2^- and then to NO_3^-, with the release of N_2O as a byproduct (Cai et al., 2017). Increased NH_4^+ availability after excreta deposition enhances soil nitrification by stimulating its first step (Hartmann et al., 2013). Denitrification is the reduction of NO_3^- to NO_2^-, then to NO, N_2O and N_2, in association with the increased presence of ammonia-oxidizing bacteria and denitrifying bacteria (Cai et al., 2017). Nitrifier denitrification is a process mediated by ammonia-oxidizing bacteria, where NO_2^- is reduced to N_2, but N_2O is produced as an intermediate or final product (Wrage et al., 2004; Wrage-Mönnig et al., 2018). Denitrification activity is expected to increase after dung deposition due to increasing C and decreasing O_2, and after urine deposition because of increasing soil moisture, NO_3^-, and labile C (Wrage-Mönnig et al., 2018).

Temporal and spatial fluctuations in N_2O emissions occur. Increased fluxes immediately after grazing may be due to increased N deposition from animal excreta, soil compaction, and reduced plant N uptake (De Rosa et al., 2020). Spatial differences in N_2O fluxes and soil NO_3^- concentrations occur because of greater animal traffic in some areas of the pasture and resultant greater excreta N deposition. As an example of the latter, N_2O fluxes in the proximity of pasture gates were 11 times greater than the field average (De Rosa et al., 2020). Similarly, herd congregation areas had 3–14 times greater N_2O emissions compared with remaining pasture areas (Mitchell et al., 2021). Authors concluded that 3% of farm area was responsible for 28% of soil N_2O emissions (Mitchell et al., 2021). Thus, spatial management of N loading may reduce N_2O emissions and N losses from grazing systems.

Urine patches are hotspots of N_2O emissions from grazed pasture. There is variation in emissions among patches, and in Sweden most was accounted for by cumulative precipitation 20 days after application ($R^2 = 0.60$) and average soil water-filled pore space 30 days after application ($R^2 = 0.45$) (Barczyk et al., 2023). In smooth bromegrass pastures in Nebraska, USA, patches that received a single urination event lost on average six times more N as N_2O than areas that received no urine (Snell et al., 2014). The N_2O emissions were positively associated with measurements of soil water-filled pore space and were much greater in a year with average precipitation versus drought.

Nitrate–N leaching losses from urine are generally a function of the amount of N deposited and live plant cover. On perennial ryegrass-white clover pastures in New Zealand, NO_3^-–N leaching losses increased with increasing amounts of urine N (Di and Cameron, 2007). Similar results were found on orchardgrass (*Dactylis glomerata* L.) swards in Pennsylvania, USA (Stout, 2003). Rates of 0 kg, 300 kg, and 700 kg of urine N ha^{-1} were applied to lysimeters with perennial ryegrass present or absent (Rayner et al., 2020). Presence of perennial ryegrass reduced leachate from 28 kg N ha^{-1} (plants absent) to 2 kg N ha^{-1} for the no-urine control and from 375 kg N ha^{-1} to 93 kg N ha^{-1} for the 700 kg N ha^{-1} treatment.

Nitrogen in dung is primarily in an organic form requiring mineralization for release, with less leaching occurring than from urine (Wachendorf et al., 2008). Conversely, K in both dung and urine is primarily in a soluble form. While leaching of K from grasslands is usually low (Kayser and Isselstein, 2005), it can occur in sandy soil environments if high levels of available soil K are present, for example, at urine patches (Dubeux et al., 2007). In New Zealand, grazing dairy cows were estimated to be directly or indirectly responsible for 74–92% of all K losses in pastures over a 30-year period (Williams et al., 1990).

4 Mitigating nitrous oxide emissions: nitrification inhibitors

Because livestock urine patches are an important source of N_2O emissions in grazed grasslands, nitrification inhibitors have been investigated as a N_2O mitigation strategy (Adhikari et al., 2021). Mitigation of N_2O emissions from urine patches must target the production of NO and NO_2^- during nitrification by inhibiting NH_3 oxidation (Clough et al., 2020). Research into nitrification inhibitors has evaluated chemical additives and plant species with observed biological nitrification inhibition. The effects of nitrification inhibitors and selected management interventions on N_2O emissions and other categories of N losses are described in the following text and summarized in Table 2.

In a review of 196 datasets, the chemical nitrification inhibitors dicyandiamide, 3,4-dimethylpyrazole phosphate, and nitrapyrin reduced N_2O emissions from urine patches by 44%, 28%, and 28%, respectively (Adhikari et al. 2021). A meta-analysis found the three inhibitors reduced N_2O emission

Table 2 Selected management interventions and their effects on N losses in grazed pasture

Management intervention	Treatments	Responses	References
Biological nitrification inhibitors (BNIs)	Urine applied to pasture grass with either high or low BNI potential	High BNI pasture grass ↓ N_2O emissions from urine spots 60% compared with low BNI grass	Byrnes et al., 2017
BNI	Urine applied from cattle with diet containing fodder beet or with no fodder beet	Beet diet urine ↓ ammonia-oxidizing bacteria population 36%, soil NO_3^-–N concentration 31%, and NO_3^-–N leaching losses from urine 64% versus pasture diet	Talbot et al., 2020
BNI	Standard urine applied to pastures varying in % plantain in herbage mass	Total urine N_2O emissions ↓ linearly as proportion of plantain in the sward increased ($r^2 > 0.96$)	Simon et al., 2019
Chemical nitrification inhibitor (CNI)	0 kg, 300 kg, 700 kg, and 1000 kg urine N ha^{-1} with or without dicyandiamide	CNI ↓ NO_3^-–N leaching from urine by 63%; ↑ N offtake in forage by 32%	Di & Cameron, 2007
CNI	Various meta-analysis	CNI ↓ N_2O emissions from excreta 28–44%;↑ forage accumulation (15%) and N uptake (13%)	Adhikari et al., 2021
CNI	Incubated urea + manure with or without nitrapyrin	CNI ↓ N_2O emissions from excreta by 64–71%	Tao et al., 2021
CNI	Meta-analysis (N sources: dung, urine, slurry, fertilizer)	CNI ↓ N_2O emissions from various excreta sources by 52–66%	Soares et al., 2023
Inoculation with arbuscular mycorrhizal fungi (AMF)	Grass microcosms fertilized with NO_3^- or NH_4^+ and inoculated with AMF or a non-mycorrhizal control	Presence of AMF in pasture soils ↓ N_2O fluxes by 62% and ↓ NH_4^+ leaching by 75% (pasture wide, not only excreta); plant N content ↑ 13% due to AMF	Bender et al., 2015
Legumes versus N fertilizer	Legume–grass mixture with 44% legume and no N fertilizer versus mixture with 4% legume + fertilizer	Substituting legume for N fertilizer ↓ N_2O emissions 39–54% (pasture wide, not only from excreta)	Fuchs et al., 2018
Legumes versus N fertilizer	Legume–grass mixture (~30% legume) versus grass with N (224 kg N ha^{-1} $year^{-1}$)	Legume integration ↓ NO_3^- leaching (pasture wide, not only from excreta) by 51% versus N-fertilized grass monoculture	Garcia et al., 2021; Trumpp, 2024
Minimize excreta deposition in livestock traffic areas	Excreta deposition in herd concentration areas (HCA), pasture, and riparian zones	Excreta deposits in HCA ↑ N_2O emissions 9× compared with those on pasture; 3% of farm area accounts for 28% of greenhouse gas budget	Mitchell et al., 2021
Plant cover	Presence/absence of perennial ryegrass and application of 0 kg, 300 kg, or 700 kg N ha^{-1} as bovine urine	Presence of perennial ryegrass ↓ NO_3^-N leaching losses from urine by 75%	Rayner et al., 2020

factors for urine, dung, and slurry by 57% (Soares et al., 2023). Mechanistically, dicyandiamide reduced the ammonia-oxidizing bacteria population in soil at cattle urine patches, but there was no effect on the denitrifying bacteria population (Cai et al., 2017). In a calcareous soil in China, nitrapyrin reduced total N_2O emissions from a urea plus cattle manure mixture by 65–71% compared with the no-inhibitor control, and the reduction was associated with lesser ammonia-oxidizing bacteria gene abundance (Tao et al., 2021). Application of dicyandiamide to a perennial ryegrass-white clover mixture growing in lysimeters in New Zealand reduced NO_3^-–N leaching losses by 63% and increased N harvested in forage by 32% (Di and Cameron, 2007). Across multiple studies, the application of dicyandiamide increased herbage accumulation and N uptake by 13% and 15%, respectively (Adhikari et al., 2021).

Small amounts of dicyandiamide residues in milk products have been reported from its use on grazed pasture, and 3,4-dimethylpyrazole phosphate and nitrapyrin can enter the food chain via grazing livestock (Adhikari et al., 2021). Despite these observations, no negative impacts on non-targeted organisms have been documented from the use of chemical nitrification inhibitors at recommended rates (Adhikari et al., 2021). Continued research is needed to confirm these findings.

Biological nitrification inhibitors (BNIs) may be released as root exudates of plants and can decrease nitrifier activity in agricultural soils, improving N recovery from fertilizer and excreta and enhancing N-use efficiency (Zhang et al., 2022). Two *Urochloa* spp. grasses were compared, one which had reported BNI activity (cv. Tully) and one that did not (cv. Mulato) (Byrnes et al., 2017). Following urine application, soil under Tully showed suppression of nitrification and denitrification and reduced presence of ammonia-oxidizing archaea relative to soils under Mulato. Cumulative N_2O fluxes from urine application were greater on pastures under Mulato than for Tully. Inhibitor effects also occurred following the application of urine from animals fed fodder beet (*Beta vulgaris* L.) (Talbot et al., 2020) and when urine from cattle consuming non-fodder beet diets was applied to fields where fodder beet was growing (Yao et al., 2018). Secondary plant metabolites in urine following beet consumption or from root exudates of live beets may inhibit bacteria growth and nitrification, leading to lower N_2O emissions. Biological nitrification inhibition remains under-researched as a mitigation option in pastures (Clough et al., 2020). Readers are referred to Zhang et al. (2022) for a consideration of mechanisms of BNI.

Diet species composition can affect urine volume and urine N concentration. Chicory (*Cichorus intybus* L.) and plantain (*Plantago lanceolata* L.) have high water and mineral concentrations and may modulate urination behavior and urinary N excretion (Mangwe et al., 2019). Even with similar N intake, increased urination events for animals fed chicory diluted urine N concentration compared with animals consuming white clover, decreasing N loading at individual urine patches. Similarly, urine N-loading rates of animals fed plantain decreased with increasing proportions of plantain in the diet, lowering N_2O emissions (Simon et al., 2019).

5 Litter deposition and decomposition in grazed pastures

In grazed perennial grasslands, deposition and decomposition of aboveground plant litter continuously influence plant nutrient supply compared with periodic pulses of litter deposition in annual cropping systems (Eilittä et al., 2003). Litter composition affects the net balance between nutrient mineralization and immobilization, influencing nutrient availability of N, P, and S (Myers et al., 1994).

5.1 Litter deposition

Pasture herbage mass, grazing management, N fertilization, and species composition affect litter mass and deposition rate. Smooth bromegrass litter deposition was 48% greater in N-fertilized versus unfertilized rotationally stocked pastures, and the greatest litter mass and litter deposition rate coincided with periods of peak herbage mass (Guretzky et al., 2014). Bermudagrass litter mass and litter deposition rate increased as target grazing height and residual herbage mass increased (Liu et al., 2011a).

Greater N fertilizer rates combined with greater stocking rates of bahiagrass pastures caused a larger nutrient contribution from litter than in extensively managed swards (Dubeux et al., 2006b). Comparing grass–legume mixtures with N-fertilized grass monocultures, litter mass was greater for heavily fertilized grasses than for the mixture (Silva et al., 2021). In contrast, grass and grass–legume litter deposition was similar when the grass received a relatively low N rate (Kohmann et al., 2018). Overall, management practices that increase pasture herbage mass are likely to increase litter deposition rates.

5.2 Litter decomposition

Decomposition of plant litter depends in part on initial N, soluble sugar, and recalcitrant compound concentrations and whether it has already partially senesced (Rumpel et al., 2015). When added to soil, grass litter of different stages of maturity decomposes at different rates (Sanaullah et al., 2010). Litter decomposition can also vary significantly for grasses receiving different N rates, managed under different grazing treatments, or associated with various proportions of legumes.

Greater N fertilization and stocking rate of bahiagrass pastures increased litter turnover and nutrient release compared with lesser management intensity (Dubeux et al., 2006a). Litter composition for the high and low intensities, respectively, were 22.9 versus 14.1 g N kg^{-1}, 1.3 versus 0.8 g P kg^{-1}, a C:N ratio of 22 versus 40, and a lignin:N ratio of 4.4 versus 5.8 (Dubeux et al., 2006b). In bermudagrass pastures, grazing intensity did not affect litter decomposition, but increasing N fertilization increased the decomposition rate (Liu et al., 2011b). Smooth bromegrass litter N concentration was greater when N fertilizer was applied compared with a no-N control (18.9 g N kg^{-1} and 16.5 g N kg^{-1}, respectively), and N release through decomposition was 46 g N kg^{-1} and 27 kg N ha^{-1} for the same treatments, respectively (Guretsky et al., 2014).

A comparison of grass–legume mixtures versus a grass monoculture receiving 112 kg N ha^{-1} showed similar amounts of N released through plant litter decomposition (Jaramillo et al., 2021a). When litter N returns were compared for legume–grass mixtures and grass monocultures that were unfertilized or fertilized at 50–60 kg N ha^{-1}, N return from mixture litter was greater than that from litter of unfertilized grass pastures and as great or greater than litter from the N-fertilized pastures (Kohmann et al., 2018; 2019). These data suggest the amount of N fertilizer applied to a grass monoculture affects comparisons of litter dynamics with grass–legume mixtures. This was supported further by a comparison of year-round forage systems, where those including legumes contributed important quantities of N via aboveground litter decomposition, but amounts were less than for grass systems receiving large N fertilizer inputs (290 kg N ha^{-1}; Silva et al., 2021).

6 Efficiency of nutrient cycling in grazed pastures

Grassland species composition and grazing management are important factors driving nutrient cycling efficiency. Species diversity or introduction of legumes influence nutrient budgets and soil nutrient availability through plant physiological attributes as well as intra- or interspecific interactions (Rumpel et al., 2015). Different rooting depths and root architecture among species in a diverse sward may reduce N losses. The presence of legumes, due to biological N fixation (BNF), can influence nutrient cycling. This N can be made available to associated non-fixing species through excreta of grazing animals, decomposition of legume litter, sloughing of legume roots and nodules in response to defoliation, and rhizodeposition pathways including decomposition and decay of nodules and root cells and exudation of soluble N compounds by legume roots (Fustec et al., 2010). Associating legumes with non-fixing species in pastures leads to greater efficiencies in N use than legumes grown as a component of sequential crop rotations. This is due to the niche separation effect which stimulates the BNF of legumes and increases the herbage accumulation rate of the non-fixing species (Corre-Hellou et al., 2006; Fustec et al., 2010).

Replacing fertilizer N with BNF from legumes was found to reduce N_2O emissions by 39–54% (Fuchs et al., 2018; Table 2), with N_2O emissions from clover residue decomposition and rhizodeposition less than N_2O emissions from N fertilization. A summary of whole-system and life-cycle assessment analyses for dairy farms indicated that at similar total N inputs, clover-grass pasture systems can be more efficient than N-fertilized grass systems per kilogram of milk produced from both energy-use and greenhouse gas emissions perspectives (Ledgard et al., 2009).

Livestock grazing modifies ecosystem properties and can elicit cascading effects, with ultimate impacts on soil nutrient cycling and primary production (Sun et al., 2017). The selection of an optimum stocking rate is important because it affects soil function significantly, even more so than stocking method (Read et al., 2016). Pasture decline is often caused in part by a lack of maintenance fertilization, but it can be hastened by greater than optimal stocking rates. Overgrazing results in a large proportion of nutrients cycling in dung and urine which are disproportionately deposited in high-traffic areas where plant cover and ability to take up nutrients are minimal (Boddey et al., 2004). In the Hongyuan Alpine Meadow of China, heavy grazing also decreased root exudation rate and soil N mineralization rate, while moderate grazing effects on these parameters were not different from ungrazed areas (Sun et al., 2017). During the recovery period after grazing, moderately grazed swards had greater root exudation rate, soil inorganic N concentration, and net soil N mineralization rate than heavily grazed pastures. On the Tibetan Plateau in China, heavy grazing reduced gross N mineralization rates by 19%, suppressing microbial N mining and overall soil N availability (Sun et al., 2018). Similar to the recommendations of Boddey et al. (2004), they suggested that overgrazing is likely to amplify plant N limitation, hampering productivity and pasture recovery, but a reduction in grazing pressure can be a sustainable way to maintain soil fertility, C sequestration, efficient nutrient recycling, and overall ecosystem stability.

Enhancing uniformity of excreta deposition is an important goal of grazing management with potential benefits to nutrient-use efficiency, but achieving this goal is challenging (Dubeux et al., 2009). After 3 years of cattle grazing bahiagrass at three intensities, soil nutrient concentrations were greatest close to shade and water regardless of grazing intensity (Dubeux et al., 2009). In a study comparing rotational and continuous stocking of bahiagrass during summer, soil nutrient concentration was generally not affected by stocking method, but soil nutrients accumulated in pasture zones closest to shade and water (Dubeux et al., 2014). Similarly, in a 2-year study with bermudagrass in a warm climate, stocking method had less effect on soil nutrient distribution than position of shade and water points (Mathews et al., 1994a). Of these, shade was a greater attractant than water points (Mathews et al., 1999). Thus, the position of shade and water may have a greater effect on the efficiency of nutrient cycling than stocking method in warm environments.

Grazing exclusion or duration of grazing during a season affect nutrient pools and cycling. In a subalpine pasture, longer periods of grazing during the warm season increased soil total and extractable N, almost doubled microbial biomass C, increased extracellular enzyme activity, and enhanced nutrient cycling mobilization (Gavrichkova et al., 2022). Grazing exclusion in the Tibetan Plateau enhanced net CO_2 uptake by 17% during the growing season, but it aggravated mineral nutrient limitations to plant growth by decreasing soil NO_3^-- N concentration by 50% and foliar concentrations of N by 10%, Ca by 9%, S by 10%, Fe by 39%, and Mn by 20% (Liu et al., 2020). Likewise in karst alpine grasslands in China, rotational or continuous stocking reduced soil organic C after 17 years compared with grazing exclusion, but grazing promoted soil total N (Yang et al., 2023). While exclusion provided improvements in plant recovery and ecosystem C stock, grazing promoted greater ecosystem nutrient stocks.

7 Role of soil fauna and microbial communities in pasture-nutrient cycling

Invertebrates exert a major influence on soil N availability, particularly in soils of low N fertility (Schon et al., 2011). They increase plant uptake of N from surface-applied plant litter, and in compacted soils, tunneling organisms such as earthworms improve soil structure and N availability (Schon et al., 2011).

Much of the available research assessing the role of invertebrates in pasture-nutrient cycling focuses on dung beetles. They exhibit tunneling behavior and process herbivore dung, burying it deeply in the soil profile (Ma et al., 2023). Dung beetles have been referred to as ecosystem engineers, altering physical and chemical properties of the soil in ways that affect availability of resources to other organisms (Doube, 2018). Dung beetles also aerate soil, reduce greenhouse gas emissions, aid in control of flies and nematodes that affect livestock, and support secondary seed dispersal (Noriega et al., 2023). Conservation of dung beetle populations is also valued because their activity can reduce the survival of gastrointestinal parasites on pastures (Sands and Walls, 2017).

Dung decomposition and nutrient cycling is advantaged by a functionally diverse dung beetle community (Noriega et al., 2023). In South Dakota, USA, a diverse dung beetle community was credited with more effective dung degradation in grasslands, and early beetle colonization following dung deposition was key to maximizing success (Pecenka and Lundgren, 2018). The introduction of exotic dung beetles across southeastern Australia, where native dung beetles are not efficient in processing livestock dung, reduced the quantity of dung on the soil surface (Ma et al., 2023). Arthropods that feed on dung accelerate the cycling of nutrients in grazed pasture (Tixier et al., 2016) and can increase herbage accumulation. For example, dung beetle presence increased soil N content and herbage accumulation of pearl millet [*Pennisetum glaucum* (L.) R. Br.] (Garcia et al., 2023). Likewise, herbage accumulation of oat (*Avena sativa* L.)-vetch (*Vicia benghalensis* L.) mixture in temperate Australia improved with dung beetle activity (Ma et al., 2023). To date, over 60 species of exotic dung beetles from both temperate and tropical environments have been introduced to Australian grasslands, with important positive impacts (Ma et al., 2023).

Increased use of N and P fertilizers on pastures may decrease belowground soil faunal diversity (Donnison et al., 2000). In New Zealand, greater quantity and quality of inputs, including P fertilization, disadvantaged fungal-based communities but increased the number of bacterial-feeding nematodes (Parfitt et al., 2010). Similarly, soils with greater N status may cause a shift from fungal-to-bacterial populations, increasing the rate of nutrient cycling and the risk of N leakage compared with soils containing fungal-based food webs, which dominate low-fertility pastures (van der Heijden et al., 2008).

Long-term differences in grazing management can influence microbial communities (Pan et al., 2018). Restriction from grazing in the semiarid grasslands of Inner Mongolia increased microbial biomass but did not affect microbial diversity (Liu et al., 2016). In the same study, ammonia-oxidizing bacteria and ammonia-oxidizing archaea had greater growth rates in soils of grazed versus grazing-restricted swards, but defoliation management effects were limited to the top 20 cm of soil. In Inner Mongolia, China, grazing intensity affected the distribution of nitrifying communities by impacting soil characteristics including bulk density and NH_4^+-N (Pan et al., 2018). For grasslands in Poland, the microbial community structure was sensitive to different systems of defoliation management (Musial et al., 2020), with actinobacteria growth increasing in mowed versus grazed swards, creating favorable conditions for greater release of inorganic N from its organic soil pool. Thus, microbial communities are sensitive to a range of pasture management practices, with implications for nutrient cycling in grassland ecosystems (Liu et al., 2016).

There are limited data on the contributions of arbuscular mycorrhizal fungi to nutrient cycling, but it is generally considered they enhance plant nutrition and growth. In microcosms containing Italian ryegrass [*Lolium* multiflorum (Lam.)], mycorrhizae reduced reactive and unreactive P leaching by 31%, enhanced plant P content by 15%, and soil P mobilization by 18%, reduced N_2O fluxes by 62% and NH_4^+ leaching by 75%, and increased plant N concentration by 13% (Bender et al., 2015; Table 2). Soil nutrient status affects root cell membrane integrity, quantity of root exudates, and mycorrhizal activity. When no N and

Published by Burleigh Dodds Science Publishing Limited, 2025.

K were applied to three perennial pasture grasses, plant root exudations of citrate and malate increased leading to increased mycorrhizal colonization relative to grasses receiving N and K fertilizer (Tshewang et al., 2020). The literature suggests that increasing arbuscular mycorrhizae colonization may promote sustainable nutrient cycling, but there is a need to conduct process-based studies to better understand the mechanisms involved and how they are affected by management (Bender et al., 2015).

8 The role of nutrient cycling in the sustainability of grazed grasslands

The sustainability of grazed grasslands is threatened by a range of factors including overgrazing, soil erosion, depletion of soil organic matter, inadequate nutrient supply, soil compaction, and nutrient losses to the environment (Boddey et al., 2004; Franzluebbers et al., 2020). Efficient nutrient cycling greatly enhances sustainability of grazed grasslands (Sollenberger et al., 2019), but achieving this goal is particularly challenging because the presence of animals affects nutrient availability and spatial distribution (Dubeux et al., 2006b; Vertès et al., 2019). Concurrently, the size of nutrient pools and magnitude of nutrient fluxes vary widely in grazed grasslands and can be associated with divergent outcomes ranging from grassland degradation, when nutrient inputs and availability are inadequate, to significant nutrient losses to the environment when inputs are greater than uptake and storage capacity.

Several important strategies have been suggested for enhancing the efficiency of nutrient cycling and advancing the cause of grazed grassland sustainability. Avoiding overstocking is critical because greater stocking rates increase the proportion of nutrients cycling in excreta (Thomas, 1992), often increasing gaseous or leaching losses from the system (Dungait et al., 2012). Overstocking also reduces plant cover and above- and belowground biomass, triggering cascading effects including soil erosion, nutrient runoff and leaching, and decreasing soil organic matter which reduces nutrient storage (Sollenberger et al., 2012). The use of perennials is an important sustainability practice because they maintain year-round soil cover and root biomass, minimize runoff and leaching losses, and reduce a need for soil tillage (Picasso et al., 2022). Increased species richness has been associated with greater N retention and soil nitrate concentrations (Cong et al., 2014), increased resource-use efficiency, and greater resilience when grasslands encounter disturbance (Tracy et al., 2018). Increased use of legumes allows a reduction in synthetic fertilizer use in some cases and increases the amount of N cycling (Kohmann et al., 2018; 2019) and N cycling efficiency (Garcia et al., 2021). As a result, year-round grass–legume systems were able to sustain similar animal productivity as N-fertilized grasses despite receiving only 15% as much N via fertilizer (Jaramillo et al., 2021b).

Published by Burleigh Dodds Science Publishing Limited, 2025.

Considering grassland sustainability broadly, optimization of agricultural practices is essential to close biogeochemical cycles and minimize the negative impacts of nutrient cycle decoupling on the environment (Vertès et al., 2019). Ecologically oriented practices (e.g. integrated crop–livestock systems, grassland species diversification, and greater landscape complexity) can increase productivity and ecosystem services through diversification and greater nutrient-use efficiency (Franzluebbers et al., 2020). Conservation nutrient management (e.g. minimize soil disturbance, maximize soil cover and use of perennial plant species, and integrate legumes) can improve soil health, increase soil biological activity, and provide essential nutrients to forage crops such that nutrient input via fertilizers is considered a secondary choice (Franzluebbers et al., 2020).

9 Use of nutrient budgets to guide management of grazing systems: the case of nitrogen and phosphorus

A nutrient budget, also called nutrient mass balance, is the difference between inputs and outputs of nutrients of interest within specific spatial and time scales (Soberon et al., 2015). Results expressed on a per area basis indicate environmental impact, while on a per unit of product (meat, milk) basis indicate production efficiency (Soberon et al., 2015). Farm-to-gate nutrient budgets have been successfully applied in dairy operations to evaluate environmental risks and improve economic feasibility and nutrient-use efficiency (Soberon et al., 2013; Soberon et al., 2015; Spears et al., 2003a). Evaluated over time, they quantify the effect of changes in management practices and regulations in reducing environmental risks (Aarts et al., 2000; Kohmann et al., 2021; Oenema and Roest, 1998). Variability in nutrient budgets among farms with different management practices can also be used as an indicative for improvement in nutrient-use efficiency (Akert et al., 2020).

Most nutrient budgets are estimated at farm-to-gate scales. Positive nutrient budgets imply nutrient-use inefficiency and soil nutrient buildup with potential losses to water bodies or the atmosphere, increasing environmental risks (Carpenter et al., 1998; Soberon et al., 2015). However, losses of nutrients and associated environmental risks are subject to site-specific management factors such as rate, timing, type, and placement of nutrients, as well as weather conditions (Gourley et al., 2012a; Soberon et al., 2015). Van Leeuwen et al. (2019) demonstrated that field-level nutrient balances are more useful than farm-level when there is considerable within-farm variation in soil type or management practices, particularly because of variations in leaching associated with soil type. Moreover, high losses of nutrients at a field level can be underestimated when only farm-scale nutrient budgets are used.

Van Leeuwen et al. (2019) indicate that hot spots and hot moments should be considered when there are large environmental risks associated with nutrient losses. Supporting this conclusion, Gourley et al. (2015) noted that areas from which dairy cows were restricted had lesser extractable nutrient concentration, but areas, where significant urine and dung were deposited, had greater environmental risk. These differences are not captured in farm-to-gate nutrient budgets (Gourley et al., 2012a). Negative budgets can indicate limitations for forage production and potential for pasture degradation (Kohmann et al., 2021; Soberon et al., 2015), but the risk of pasture degradation must be considered in the context of soil nutrient analysis. For example, negative nutrient budgets in areas with a history of soil nutrient loading might not result in degradation (Akert et al., 2020; Mu et al., 2016). Thus, the interpretation of farm-to-gate nutrient budgets is subject to factors such as soil analysis, soil nutrient saturation levels, climate, and management practices.

9.1 Nitrogen budgets

Various comparisons were made of farm-to-gate N budgets of two dairy farm systems in Ireland and four systems in Switzerland with a significant grazing component (Table 3). For systems that did not account for atmospheric deposition and had low pasture legume proportion with negligible BNF, chemical fertilizer and purchased feed made up a large proportion of N inputs (~97%; Mihailescu et al., 2014; Mu et al., 2016). When BNF and atmospheric deposition were considered, they made up 38–64% of total N inputs, suggesting the adoption of legumes can alleviate reliance on N for chemical fertilizer. Milk exports were the main N output (45–82%). Nitrogen balance was positive and highly variable across the systems (75–217 kg N ha^{-1} $year^{-1}$), and N-use efficiency was relatively low (22–57%). This suggests there is considerable unaccounted N and potential to improve both N balance estimates and N-use efficiency. It is important to note that these budgets did not consider N losses through leaching, runoff, or volatilization. Although this is a small selection of cases, it indicates improving estimates of N losses increases the total amount of N accounted for in dairy grazing systems and informs management decisions.

Management of animal excreta plays an important role in farm-to-gate N budgets. In grazing dairies, 75–90% of total excreted N is deposited directly on pasture while this proportion is far lower in confined systems (0–35%; Gourley et al., 2012a). Adjusting N and P fertilizer application to account for nutrients recycled through animal urine and dung in grazing systems is essential to avoid over-application that negates the benefits of nutrient cycling (van Leeuwen et al., 2019). Requiring manure application during spring instead of fall and adaptation of chemical N fertilizer recommendations to account for manure

Table 3 Average farm-level N inputs, outputs, balance, and efficiency (outputs divided by inputs, expressed in percentage) of various dairy systems with significant grazing in Ireland and Switzerland

Country	Ireland[a]	Ireland[b]	Switzerland[c]	Switzerland[d]	Switzerland[e]	Switzerland[f]
Number of farms	21	19	8	8	1	1
Years	3	1	3	3	3	3
			N inputs (kg ha^{-1})			
Concentrate fed	27	40.1	2.9	23.3	6.6	16.3
Roughage (hay, silage, wheat straw, alfalfa)	NA[g]	12.4	16.8	19.6	41.4	11.6
Forage	12.4			Accounted for under roughage		
Bedding material	4			Accounted for under roughage		
Chemical fertilizer	186	199	24.5	27.3	52	15.5
Organic fertilizer	0	0	17.6	33.2	21.4	17.9
Livestock	3.8	NA	6	5.5	2.4	2.3
N fixation	NA	NA	41.5	42.8	86.8	88.9
Atmospheric deposition	NA	8.1	22.5	23.8	25	25
Total inputs	**233**	**260**	**132**	**176**	**236**	**178**
			N outputs (kg ha^{-1})			
Milk	40	46.7	30.3	45.5	60	57.4
Livestock	12.8	11.4	10.7	9.8	10.5	11.4
Plant material	0	0	12.4	37.5	0	20.4
Manure	NA	NA	0	7.4	2.3	1.1
Total outputs	**53**	**58**	**53**	**100**	**73**	**90**
N balance (kg ha^{-1})	***180***	***217***	***78***	***75***	***163***	***87***
N-use efficiency (%)	***23***	***22***	***41***	***57***	***31***	***51***
Source	Mihailescu et al. (2014)	Mu et al. (2016)	Akert et al. (2020)	Akert et al. (2020)	Akert et al. (2020)	Akert et al. (2020)

Forage and grazing management for the farms in each column are described as: [a]pasture grazed and ensiled; [b]grass-based, wheat straw, grass silage, feed produced on-farm; [c]grazing for 230 days year^{-1}, minimum concentrate, commercial farms; [d]grazing for 202 days year^{-1}, hay and fresh herbage (192 days year^{-1}), minimum concentrate, commercial farms; [e]full-time grazing (24 h day^{-1}), minimum concentrate, experimental farm; [f]part-time grazing (6-8 h day^{-1}), hay and fresh herbage, minimum concentrate, experimental farm. [g]NA: not applicable, not considered, or negligible.

inputs were successful in reducing farm-level N balance in Ireland (Mihailescu et al., 2014). One of the limitations of farm-to-gate N budgets is they are often not sensitive enough to elucidate the risk of N losses from areas with concentrated N deposition.

Feeding strategies can also alter farm-to-gate N budgets, but results vary. Mu et al. (2016) reported the Irish dairy farms that rely on grazing have ~20% lower N inputs ha^{-1} and 10% lower N balance ha^{-1} than systems in the Netherlands that have a greater reliance on concentrate. In Switzerland, however, greater N inputs via concentrate increased N outputs due to greater milk yields per cow, resulting in lesser N balance and greater N-use efficiency in farms using part-time compared with full-time grazing (Akert et al., 2020).

9.2 Phosphorus budgets

Farm-to-gate P budgets for dairy systems are highly variable as shown in the examples from Germany, Northern Ireland, Ireland, and Switzerland (Table 4). They ranged from −10.6 to 17.3 kg P ha^{-1}, and P-use efficiency ranged from 268% to 37% (Table 4). Phosphorus inputs were mostly from animal feed (33–94%) or fertilizer (23–55%), with their proportions inversely related. Atmospheric P deposition (input) and runoff and leaching (output) were considered only in one study (Ohm et al., 2015). Most exports were from animal products (milk and livestock; ~85%) except for the case study in Germany where a large proportion of manure was exported outside the farm instead of re-distributed on grasslands (Ohm et al., 2015). Different from the N budgets, P budgets accounted for the majority of P and presented near zero budgets and ~100% P-use efficiencies.

Negative P budgets indicate the P exported is not being replenished at the farm level. However, the impact on forage production will depend on soil P status (Ohm et al., 2015, 2017). For example, in a dairy where soils were highly impacted and P was elevated, P mining by crops and grasslands can reduce risk of P losses to the environment in the short term, but over a longer time span, P inputs would likely be necessary (Ohm et al., 2015; 2017). Livestock operations that produce crops on-farm often have greater P-use efficiency, which has been suggested as the best management strategy to reduce P balance in dairy operations (Spears et al., 2003a). However, this is possible only if P fertilizer and manure applications are managed to match and not exceed crop requirements. For example, P-rate recommendations for forages in Florida (USA) are species-specific and consider both agronomic performance and environmental safety with a distinction between grazed and harvested systems (Silveira et al., 2011). This approach reflects the abundance of grazing systems in the region and the elevated risk of P pollution due to coarse-textured soils with low P-holding capacity (Obour et al., 2011; Silveira et al., 2011). Tailoring

Published by Burleigh Dodds Science Publishing Limited, 2025.

Table 4 Average farm-level P inputs, outputs, balance, and efficiency (outputs divided by inputs, expressed in percentage) of various dairy systems with significant grazing in several countries

Country	Germany[a]	Northern Ireland[b]	Ireland[c]	Switzerland[d]	Switzerland[e]	Switzerland[f]	Switzerland[g]
Number of farms	1	83	19	8	8	1	1
Years	3	6	1	3	3	3	3
				P inputs (kg ha^{-1})			
Minerals	0.6	NA[h]	NA	In concentrate			
Concentrate fed	0.5	In concentrate	7.4	1.1	5.7	0.9	7.5
Roughage (hay, silage, wheat straw, alfalfa)	6.0	19.7	2.4	2.7	3.1	6.3	2.6
Beeding material	0.2	NA	In roughage				
Chemical fertilizer	NA	7.5	3.2	1.4	2.3	0	0
Organic fertilizer	NA	0	0	5	7.2	3.6	3.2
Seed	0.2	NA	NA	NA	NA	NA	NA
Livestock	NA	NA	NA	1.4	1.3	0.6	0.5
Atmospheric deposition	0.2	NA	NA	NA	NA	NA	NA
Total inputs	**6.6**	**27.2**	**13**	**11.6**	**19.6**	**11.4**	**6.3**
				P outputs (kg ha^{-1})			
Milk	3.9	7.6	8.3	5.9	8.1	10	10.6
Livestock	1.4	2.0	3.3	2.6	2.4	2.5	2.7
Plant material	NA	0.3	0	2.2	6.4	0	3.5
Manure	11.6	NA	NA	0	1.3	0.1	0.1

(Continued)

Table 4 (Continued)

Country	Germany[a]	Northern Ireland[b]	Ireland[c]	Switzerland[d]	Switzerland[e]	Switzerland[f]	Switzerland[g]
Runoff, leaching	0.3	NA	NA	NA	NA	NA	NA
Total outputs	**17.2**	**9.9**	**11.6**	**10.7**	**18.2**	**12.6**	**16.9**
P balance (kg ha^{-1})	***−10.6***	***17.3***	***1.4***	***0.9***	***1.4***	***−1.2***	***−10.6***
P-use efficiency (%)	***260***	***37***	***89***	***92***	***93***	***111***	***268***
Source	Ohm et al. (2015)	Adenuga et al. (2018)	Mu et al. (2016)	Akert et al. (2020)	Akert et al. (2020)	Akert et al. (2020)	Akert et al. (2020)

Forage and grazing management for the farms in each column are described as [a]grazing management: young stock grazing full time; lactating animals grazed 7 h day^{-1}; [b]>3/4 of total grazing livestock is dairy; [c]wheat straw, grass silage, feed produced on-farm; [d]grazing for 230 days year^{-1}, minimum concentrate, commercial farms; [e]grazing for 202 days year^{-1}, hay and fresh herbage (192 days year^{-1}), minimum concentrate, commercial farms; [f]full-time grazing (24 h day^{-1}), minimum concentrate, experimental farm; [g]part-time time grazing (6–8 h day^{-1}), hay and fresh herbage, minimum concentrate, experimental farm. [h]NA, not applicable, not considered, or negligible.

P-rate recommendations in this manner has been successful in reducing farm-to-gate P budgets to near neutrality, with the greatest benefit in operations that are self-sufficient in feed production (Kohmann et al., 2021). When on-farm feed production is not possible, exporting manure from operations that import feed rather than producing feed on the farm reduces the P balance, but this may not be advantageous from a watershed perspective (Spears et al., 2003a).

Farm-to-gate P budget results must be interpreted while considering within-farm animal management practices. In Australia, forage-based dairy farms typically use rotational stocking, but areas where animals are held and fed supplements have high stocking density, leading to nutrient buildup from animal excreta (Gourley et al., 2015). Calving, feeding, and holding areas in Australian dairy farms had 10 times greater Olsen P and 15 times greater Colwell K compared with grazed pastures. Similarly, in Wisconsin (USA), soil Bray-1 P and K were much greater in areas used for exercise and feeding compared with crop fields (Powell et al., 2005). Management practices focused on reducing erosion and runoff in areas with high soil P concentration can be an efficient way to reduce environmental risks (Carpenter et al., 1998).

10 Case study: nitrogen cycling in nitrogen-fertilized grass and grass-legume pastures

Few studies of N cycling in pastures have measured pools and fluxes across multiple seasons and years. Such measurements are challenging because N is highly transient and affected by biotic and abiotic factors (Galloway et al., 2004). Nitrogen is a critical element for plant and animal growth, but it is also implicated in NO_3^- leaching to ground water (Markov, 2012; Santos et al., 2023) and N_2O emissions (Crews and Peoples, 2004; Jensen et al., 2012). Therefore, quantifying N flow among different pools of the air–soil–plant–animal interface is important.

This case study derives from a series of experiments conducted in the mild and humid temperate region of northwestern Florida, USA (30.77°N latitude, 85.23°W longitude). Details of the case study were derived from Jaramillo et al. (2021a, 2021b), Garcia et al. (2021), and Garcia et al. (2023), and the N cycle for the two grazing systems is shown in Figure 1.

The study compares two well-established, long-term grazing systems, each having cool- and warm-season forage components that allow for nearly year-round grazing. One system included N-fertilized bahiagrass monoculture pastures in the summer, overseeded in the fall with an N-fertilized mixture of oat and Italian ryegrass for winter grazing (Grass + N). Total annual N fertilization in this system was 224 kg ha^{-1} (urea and polymer-coated urea), split equally in the cool and warm seasons. The second system (Grass + Legume) included a bahiagrass-rhizoma peanut (*Arachis glabrata* Benth) mixed pasture during the

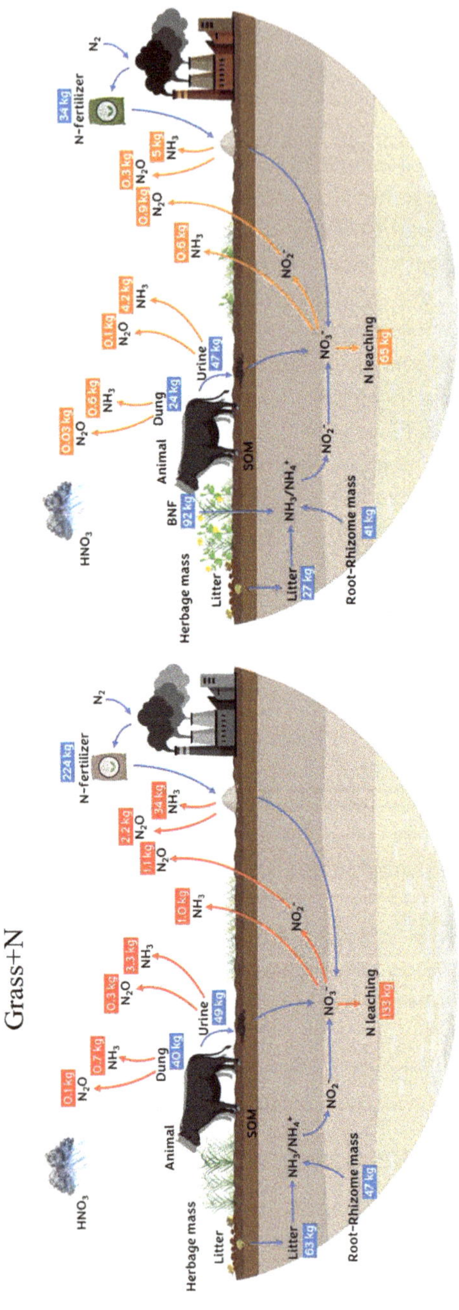

Figure 1 Nitrogen cycle in contrasting grazing systems. Multiple transformations take place in the N cycle. External N inputs (blue arrows) are industrial fertilizers (urea is usually used because it is rapidly converted to inorganic forms in the soil) and BNF (symbiosis between soil microorganisms and plant roots). Nitrogen outputs (red or orange arrows) are losses due to product harvesting (forage or animal), N in leachate and runoff, NH_3 volatilization, and N_2O emissions. Values are kg ha^{-1} $year^{-1}$.

warm season, overseeded in the fall with a mixture of oat and ryegrass plus a blend of crimson (*Trifolium incarnatum* L.), red (*T. pratense* L.), and ball clovers (*T. nigrescens* Viv.). During the cool season, Grass + Legume received 34 kg N ha^{-1} $year^{-1}$. Grass + Legume had 92 kg BNF ha^{-1} $year^{-1}$, so total N inputs were 224 and 126 kg ha^{-1} for Grass + N and Grass + Legume, respectively (Pereira Neto et al., 2024).

Pastures in both systems were continuously stocked at variable stocking rates to maintain constant forage allowance (Sollenberger et al., 2005). Nitrogen returning to pastures in dung was greater for Grass + N than Grass + Legume because of a greater dung N concentration and stocking rate in Grass + N (Garcia et al., 2021; Jaramillo et al., 2021b), while N excreted from urine was similar between systems (Table 5) (Garcia et al., 2021). The Grass + N system returned a greater amount of N in aboveground litter compared with Grass + Legume (Table 5) (Jaramillo et al., 2021a). This can be partially explained by greater herbage accumulation in Grass + N, which is often associated with greater litter deposition rates (Guretzky et al., 2014; Jaramillo et al., 2021a, 2021b).

The belowground mass in Grass + N and Grass + Legume systems returned a similar quantity of N (Table 5), and 43% of the warm-season N in Grass + Legume was from rhizoma peanut. This belowground pool acts as a nutrient source when BNF is limited, supporting aboveground growth (Dubeux et al., 2017). Belowground plant mass and microbial biomass contribute to the N cycle and are important in forming soil organic matter, the second largest N reservoir after the atmosphere (Bai and Cotrufo, 2022).

Nitrogen losses occur from fertilizers, the soil, and animal excreta via different pathways (Williams & Haynes, 1990), including ammonia (NH_3) volatilization, N_2O emission, and NO_3^- leaching. Greenhouse gas emissions from fertilizer were estimated using the IPCC emission factor (IPCC, 2019).

Ammonia loss, which is a main cause of low N-use efficiency from fertilizers, was greater for Grass + N than Grass + Legume (Table 6; Bouwman et al., 2002). After a fertilization event, there is an increase in soil N abundance

Table 5 Nitrogen fluxes from animal sources (dung and urine) and plant sources (aboveground and belowground litter) for Grass + N and Grass + Legume systems

		System	
Nitrogen fluxes, kg N ha^{-1} $year^{-1}$		Grass + N	Grass + Legume
Animal	Dung	40	24
	Urine	49	47
Plant	Aboveground litter	63	27
	Belowground litter	47	41

Source: Garcia et al. (2021); Jaramillo et al. (2021a; 2021b).

Published by Burleigh Dodds Science Publishing Limited, 2025.

Table 6 Total N losses as NH_3 and N_2O volatilization from fertilizers, soil, dung, and urine, and losses as NO_3^- from leaching in Grass + N and Grass + Legume systems

			System	
Nitrogen losses, kg N ha^{-1} year^{-1}			Grass + N	Grass + Legume
N-fertilizer		NH_3	34	5
		N_2O	2.2	0.3
Soil		NH_3	1.0	0.6
		N_2O	1.1	0.9
Animal	**Dung**	NH_3	0.7	0.6
		N_2O	0.1	0.03
	Urine	NH_3	3.3	4.2
		N_2O	0.3	0.1
Nitrate leaching			133	65

Source: Adapted from Garcia et al. (2021); Trumpp (2024).

as well as an increase in pH, resulting in NH_3 volatilization (Robertson and Vitousek, 2009). The type of fertilizers applied and weather conditions play an important role in these emissions (Robertson and Vitousek, 2009). Similarly, N_2O emissions are predicted to be greater for Grass + N than Grass + Legume (Table 6).

Excreta also contributed to N losses. Urine N losses were greater in the Grass + Legume system, particularly in the form of N_2O, but NH_3 volatilization and N_2O emissions from dung were similar between systems (Table 6). Nitrogen emissions from excreta are dependent on quantity and composition. Animals grazing in Grass + Legume have a greater percentage of N excreted in the urine (Garcia et al., 2021), and the high concentration of soluble N in urine is more susceptible to losses, whereas N in dung is in an organic form and losses are typically smaller (Williams and Haynes, 1990). Lastly, N lost as NO_3^- by leaching was greater for Grass + N (133 kg N ha^{-1} year^{-1}) in comparison with Grass + Legume (65 kg N ha^{-1} year^{-1}) (Table 6), which is associated with the high soluble-N inputs in the form of fertilizer.

Finally, animal performance demonstrates that similar average daily gain and gain per area are possible between the two systems (Jaramillo et al., 2021b) (Table 7). Based on the results of this case study, the inclusion of legumes in grazing systems allowed an 85% reduction in N-fertilizer inputs without compromising animal productivity when compared with the N-fertilized grass monoculture. As observed in Table 7, the Grass + Legume gain per N input was 1.9-fold greater than in the Grass + N system. Thus, the integration of legumes is an alternative for sustainable intensification of livestock systems.

Table 7 Animal performance during the cool and warm seasons, average daily gain, gain per area, stocking rate, and gain per kilogram of N input for Grass + N and Grass + Legume systems

	System	
Item	Grass + N	Grass + Legume
Average daily gain, kg ha^{-1} day^{-1}	0.58	0.69
Gain per area, kg ha^{-1}	620	669
[a]Stocking rate, AU ha^{-1}	4.9[a]	3.7[b]
[b]Gain per kg of N input	2.8	5.3

[a]Means within a row followed by different letters are different ($P < 0.05$).
[b]Gain per kg of N input refers to the kg of animal produced per kilogram of N.
Source: Adapted from Jaramillo et al. (2021b).

11 Conclusion and future trends

Nutrient cycling is a critical supporting ecosystem service provided by grazed grasslands. The presence of animals and their consumption, digestion, and metabolism of forage affect nutrient pools and the rate of nutrient cycling, redistribution of nutrients from excreta deposition, and the risk of nutrient losses to the environment via leaching, runoff, and gaseous losses. Choice of grazing intensity (stocking rate or sward height), stocking method, species composition (including presence of legumes), arrangement of structural features of grazed pastures (shade, watering stations, etc.), and the use of nitrification inhibitors are potentially important tools to address these challenges and enhance the efficiency of nutrient cycling.

Nutrient cycling research will continue to be an important area of scientific inquiry going forward because important knowledge gaps remain for increasing the efficiency of nutrient cycling in grazed pastures. These gaps include (i) the impact of management practices on changes in diversity and abundance of soil fauna and microbial communities, (ii) the effect of soil fauna and microbial community changes on nutrient cycling, and (iii) the efficacy and practical value of chemical and biological nitrification inhibitors in reducing N losses and increasing N-use efficiency. Future studies might be categorized within broad themes of grassland sustainability, nutrient-use efficiency, and the role of soil fauna and microbial communities.

Within grassland sustainability, research is needed to address the role of legumes as an alternative to fossil-fuel-intensive inorganic N. Legumes can be an important component of a strategy to avoid inadequate N inputs leading to degradation of grazed grasslands (Boddey et al., 2004). Ideally, this can be achieved while reducing costs by reducing N-fertilizer inputs, sustaining animal performance, and increasing delivery of critical ecosystem services (Jaramillo et al., 2021b; Sollenberger and Dubeux, 2022). Determining the optimal proportion of legume, or indeed the optimal species composition of a species-rich grassland, is critical information to achieve particular ecosystem, economic, or animal productivity goals.

Within the theme of nutrient-use efficiency, research is needed to address the efficacy, economic viability, and potential unintended consequences associated with the use of chemical and biological nitrification inhibitors. Data cited in this chapter show increases in N-use efficiency and reduction in leaching and gaseous emission losses associated with the use of nitrification inhibitors. Additional research is needed to establish optimal rates for chemical inhibitors and acceptable maximum residue levels in soil, plant, and animal products (Adhikari et al., 2021). Biological nitrification inhibition remains under-researched as a mitigation option in grazed pastures (Clough et al., 2020).

Understanding the role of soil fauna and soil microbial communities in nutrient cycling is only beginning to emerge, and the effects of management practices on their diversity and abundance are not well defined. Particularly in soils of low N fertility, invertebrates increase plant uptake of nutrients from surface-applied plant litter, and in compacted soils, tunneling organisms such as earthworms and dung beetles improve soil structure and nutrient availability (Schon et al., 2011). Much available research assessing the role of invertebrates in pasture nutrient cycling focuses on the activities of dung beetles, and studies are needed across a broader range of soil fauna. Microbial communities are sensitive to a range of pasture management practices, and the resultant changes in microbial communities have implications for nutrient cycling in grassland ecosystems that are not well understood (Liu et al., 2016).

12 Where to look for further information

The following articles provide useful additional information for subjects described in this chapter:

- Adhikari, K. P., Chibuike, G., Saggar, S., et al. (2021), Management and implications of using nitrification inhibitors to reduce nitrous oxide emissions from urine patches on grazed pasture soils – A review, *Sci. Total Environ.* 791: 148099.
- Cai, Y., Chang, S. X. and Cheng, Y. (2017), Greenhouse gas emissions from excreta patches of grazing animals and their mitigation strategies, *Earth Sci. Rev.* 171: 44–57.
- Dubeux, J. C. B., Jr., Sollenberger, L. E., Mathews, B. W., Scholberg, J. M. and Santos, H. Q. (2007), Nutrient cycling in warm-climate grasslands, *Crop Sci.* 47(3): 915–928.

Published by Burleigh Dodds Science Publishing Limited, 2025.

- Noriega, J. A., Hortal, J., de Castro-Arrazola, I., et al. (2023), Dung removal increases under higher dung beetle functional diversity regardless of grazing intensification, *Nat. Commun.* 14(1): 8070.
- Soares, J. R., Souza, B. R., Mazzetto, A. M., et al. (2023), Mitigation of nitrous oxide emissions in grazing systems through nitrification inhibitors: a meta-analysis, *Nutr. Cycl. Agroecosyst.* 125(3): 359–377.
- Zhang, M., Zeng, H., Afzal, M. R., Gao, X., Li, Y., Subbarao, G. V., and Zhu, Y. (2022), BNI-release mechanisms in plant root systems: current status of understanding, *Biol. Fert. Soils* 58: 225–233.

13 References

Aarons, S. R., et al. (2004a), Dung decomposition in temperate dairy pastures I. Changes in soil chemical properties, *Aust. J. Soil Res.* 42(1): 107–114.

Aarons, S. R., et al. (2004b), Dung decomposition in temperate dairy pastures II. Contribution to plant-available soil phosphorus, *Aust. J. Soil Res.* 42: 115–123.

Aarons, S. R., et al. (2009), Dung pads increase pasture production, soil nutrients and microbial biomass carbon in grazed dairy systems, *Nutr. Cycl. Agroecosyst.* 84: 81–92.

Aarons, S. R., et al. (2015), Between and within paddock soil chemical variability and forage production gradients in grazed dairy pastures, *Nutr. Cycl. Agroecosyst.* 102: 411–430.

Aarts, H. F. M., et al. (2000), Phosphorus (P) management in the 'De Marke' dairy farming system, *Nutr. Cycl. Agroecosyst.* 56: 218–229.

Adenuga, A. H., et al. (2018), Estimation and determinants of phosphorus balance and use efficiency of dairy farms in Northern Ireland: a within and between farm random effects analysis, *Agric. Syst.* 164: 11–19.

Adhikari, K. P., et al. (2021), Management and implications of using nitrification inhibitors to reduce nitrous oxide emissions from urine patches on grazed pasture soils – A review, *Sci. Total Environ.* 791: 148099.

Akert, F. S., et al. (2020), Farm-gate nutrient balances of grassland-based milk production systems with full- or part-time grazing and fresh herbage indoor feeding at variable concentrate levels, *Nutr. Cycl. Agroecosyst.* 117: 383–400.

Auerswald, K., et al. (2010), Coupling of spatial and temporal patterns of cattle excreta patches on a low intensity pasture, *Nutr. Cycl. Agroecosyst.* 88: 275–288.

Bai, Y. and Cotrufo, M. F. (2022),. Grassland soil carbon sequestration: current understanding, challenges, and solutions, *Science* 377(6606): 603–608.

Barczyk, L., et al. (2023), N_2O emission factors for cattle urine: effect of patch characteristics and environmental drivers, *Nutr. Cycl. Agroecosyst.* 127(2): 173–189.

Bardgett, R. D. and Wardle, D. A. (2003), Herbivore-mediated linkages between aboveground and belowground communities, *Ecology* 84(9): 2258–2268.

Bender, S. F., et al. (2015), Mycorrhizal effects on nutrient cycling, nutrient leaching and N_2O production in experimental grassland, *Soil Biol. Biochem.* 80: 283–292.

Boddey, R. M., et al. (2004), Nitrogen cycling in *Brachiaria* pastures: the key to understanding the process of pasture decline, *Agric. Ecosyst. Environ.* 103(2): 389–403.

Bouwman, A. F., et al. (2002), Estimation of global NH_3 volatilization loss from synthetic fertilizers and animal manure applied to arable lands and grasslands, *Global Biogeochem. Cycles* 16(2): 8.1–8.11.

Brady, N. C. and Weil, R. R. (2002), *The nature and properties of soil*, Prentice Hall, Upper Saddle River, USA.

Braz, S. P., et al. (2002), Aspectos quantitativos do processo de reciclagem de nutrientes pelas fezes de bovinos sob pastejo em pastagem de *Brachiaria decumbens* na Zona da Mata de Mina Gerais (In Portuguese, with English abstract.), *Rev. Bras. Zootecnia*. 31: 858–865.

Byrnes, R. C., et al. (2017), Biological nitrification inhibition by *Brachiaria* grasses mitigates soil nitrous oxide emissions from bovine urine patches, *Soil Biol. Biochem.* 107: 156–163.

Cai, Y., et al. (2017), Greenhouse gas emissions from excreta patches of grazing animals and their mitigation strategies, *Earth Sci. Rev.* 171: 44–57.

Carpinelli, S., et al. (2020), Spatial and temporal distribution of cattle dung and nutrient cycling in integrated crop-livestock systems, *Agronomy* 10(5): 672.

Carran, R. A. and Theobald, P. W. (2000), Effects of excreta return on properties of a grazed pasture soil, *Nutr. Cycl. Agroecosyst.* 56: 79–85.

Carpenter, S. R., et al. (1998), Nonpoint pollution of surface waters with phosphorus and nitrogen, *Ecol. Appl.* 8(3): 559–568.

Clough, T. J., et al. (2020), Nitrous oxide emissions from ruminant urine: science and mitigation for intensive management perennial pastures, *Curr. Opin. Environ. Sustain.* 47: 21–27.

Cong, W.-F., et al. (2014), Plant species richness promotes soil carbon and nitrogen stocks in grasslands without legumes, *J. Ecol.* 102(5): 1163–1170.

Corre-Hellou, G., et al. (2006), Interspecific competition for soil N and its interaction with N_2 fixation, leaf expansion and crop growth in pea-barley intercrops, *Plant Soil* 282: 195–208.

Crews, T. E. and Peoples, M. B. (2004), Legume versus fertilizer sources of nitrogen: ecological tradeoffs and human needs, *Agric. Ecosyst. Environ.* 102(3): 279–297.

De Rosa, D., et al. (2020), Field-scale management and environmental drivers of N_2O emissions from pasture-based dairy systems, *Nutr. Cycl. Agroecosyst.* 117: 299–315.

Di, H. J. and Cameron, K. C. (2007), Nitrate leaching losses and pasture yields as affected by different rates of animal urine nitrogen returns and application of a nitrification inhibitor - a lysimeter study, *Nutr. Cycl. Agroecosyst.* 79: 281–290.

Donnison, L. M., et al. (2000), Management influences on soil microbial communities and their function in botanically diverse haymeadows of northern England and Wales, *Soil Biol. Biochem.* 32(1): 253–263.

Doube, B. M. (2018), Ecosystem services provided by dung beetles in Australia, *Basic Appl. Ecol.* 26: 35–49.

Draganova, I., et al. (2016), The effects of temporal and environmental factors on the urination behaviour of dairy cows using tracking and sensor technologies, *Precis. Agric.* 17: 407–420.

Dubeux, J. C. B., et al. (2017), Biological N_2 fixation, belowground responses, and forage potential of rhizoma peanut cultivars, *Crop Sci.* 57(2): 1027–1038.

Dubeux, J. C. B., et al. (2006a), Litter decomposition and mineralization in grazed Pensacola bahiagrass pastures managed at different intensities, *Crop Sci.* 46(3): 1305–1310.
Dubeux, J. C. B., et al. (2006b), Litter mass, deposition rate, and chemical composition in grazed Pensacola bahiagrass pastures managed at different intensities, *Crop Sci.* 46(3): 1299–1304.
Dubeux, J. C. B., et al. (2007), Nutrient cycling in warm-climate grasslands, *Crop Sci.* 47(3): 915–928.
Dubeux, J. C. B., et al. (2009), Animal behavior and soil nutrient redistribution in continuously stocked Pensacola bahiagrass pastures managed at different intensities, *Crop Sci.* 49(4): 1503–1510.
Dubeux, J. C. B., et al. (2014), Stocking method, animal behavior, and soil nutrient redistribution: how are they linked?, *Crop Sci.* 54(5): 2341–2350.
Dubeux, J. C. B. and Sollenberger, L. E. (2020), Nutrient cycling in grazed pastures. In F. M. Rouquette and G. E. Aiken (Eds), *Management strategies for sustainable cattle production on southern pastures*, Elsevier Publishers, Amsterdam, Netherlands, pp. 59–75.
Dungait, J. A. J., et al. (2012), Advances in the understanding of nutrient dynamics and management in UK agriculture, *Sci. Total Environ.* 434: 39–50.
Eilittä, M., et al. (2003), On-farm experiments with maize–mucuna systems in the Los Tuxtlas region of Veracruz, Mexico. I. Mucuna biomass and maize grain yield, *Exp. Agric.* 39(1): 5–17.
Franzluebbers, A., et al. (2020), Focusing the future of farming on agroecology, *Agric. Environ. Lett.* 5(1): e20034.
Fuchs, K., et al. (2018), Management matters: testing a mitigation strategy for nitrous oxide emissions using legumes on intensively managed grassland, *Biogeosciences* 15(18): 5519–5543.
Fustec, J., et al. (2010), Nitrogen rhizodeposition of legumes: a review., *Agron. Sustain. Dev.* 30: 57–66.
Galloway, J. N., et al. (2004), Nitrogen cycles: past, present, and future, *Biogeochemistry* 70: 153–226.
Garcia, L., et al. (2021), Nutrient excretion from cattle grazing N-fertilized grass or grass-legume pastures, *Agron. J.* 113(4): 3110–3123.
García, C. C. V., et al. (2023), The role of dung beetle species in nitrous oxide emission, ammonia volatilization, and nutrient cycling, *Sci. Rep.* 13(1): 3572.
Gavrichkova, O., et al. (2022), Consequences of grazing cessation for soil environment and vegetation in a subalpine grassland ecosystem, *Plants* 11(16): 2121.
Gourley, C. J. P., et al. (2012), Nitrogen use efficiency and manure management practices in contrasting dairy productions systems, *Agric. Ecosyst. Environ.* 147: 73–81.
Gourley, C. J. P., et al. (2015), Soil phosphorus, potassium and sulphur excesses, regularities and heterogeneity in grazing-based dairy farms, *Agric. Ecosyst. Environ.* 201: 70–82.
Guretzky, J. A., et al. (2014), Litter deposition and nitrogen return in rotationally stocked smooth bromegrass pastures, *Agron. J.* 106(1): 175–184.
Hartmann, A. A., et al. (2013), Effects of drought and N-fertilization on N cycling in two grassland soils, *Oecologia* 171: 705–717.
Haynes, R. J. and Williams, P. H. (1993), Nutrient cycling and soil fertility in the grazed pasture ecosystem, *Adv. Agron.* 49: 119–199.

Hewins, D. B., et al. (2018), Grazing and climate effects on soil organic carbon concentration and particle-size association in northern grasslands, *Sci. Rep.* 8(1): 1336.

Hoogendoorn, C. J., et al. (2016), Nitrous oxide emission factors for urine from sheep and cattle fed forage rape (*Brassica napus* L) or perennial ryegrass/white clover pasture (*Lolium perenne* L./*Trifolium repens* L.), *Agric. Ecosyst. Environ.* 227: 11–23.

Jaramillo, D. M., et al. (2021a), Litter mass, deposition rate, and decomposition in nitrogen-fertilized or grass–legume grazing systems, *Crop Sci.* 61(3): 2176–2189.

Jaramillo, D. M., et al. (2021b), Water footprint, herbage, and livestock responses for nitrogen-fertilized grass and grass–legume grazing systems, *Crop Sci.* 61(5): 3844–3858.

Jensen, E. S., et al. (2012), Legumes for mitigation of climate change and the provision of feedstock for biofuels and biorefineries: a review, *Agron. Sustain. Dev.* 32: 329–364.

Kayser, M. and Isselstein, J. (2005), Potassium cycling and losses in grassland systems: a review, *Grass Forage Sci.* 60(3): 213–224.

Kohmann, M. M., et al. (2018), Nitrogen fertilization and proportion of legume affect litter decomposition and nutrient return in grass pastures, *Crop Sci.* 58(5): 2138–2148.

Kohmann, M. M., et al. (2019), Legume proportion in grassland litter affects decomposition dynamics and nutrient mineralization, *Agron. J.* 111(3): 1079–1089.

Kohmann, M. M., et al. (2021), Farm-scale phosphorus budgets of beef cow-calf operations, *Nutr. Cycl. Agroecosyst.* 119: 389–403.

Lambin, E. F. and Meyfroidt, P. (2011), Global land use change, economic globalization, and the looming land scarcity, *PNAS* 108(9): 3465–3472.

Ledgard, S., et al. (2009), Environmental impacts of grazed clover/grass pastures, *Irish J. Agric. Food Res.* 48(2): 209–226.

Liu, J., et al. (2016), Effects of pasture management on soil fertility and microbial communities in the semi-arid grasslands of Inner Mongolia, *J. Soils Sediments* 16: 235–242.

Liu, K., et al. (2011a), Grazing intensity and nitrogen fertilization affect litter responses in 'Tifton 85' bermudagrass pastures. I. Mass, deposition rate, and chemical composition, *Agron. J.* 103(1): 156–162.

Liu, K., et al. (2011b), Grazing intensity and nitrogen fertilization affect litter responses in 'Tifton 85' bermudagrass pastures. II. Decomposition and nitrogen mineralization, *Agron. J.* 103(1): 163–168.

Liu, K., et al. (2011c), Distribution of nutrients among soil-plant pools in 'Tifton 85' bermudagrass pastures grazed at different intensities, *Crop Sci.* 51(4): 1800–1807.

Liu, K., et al. (2017), Nutrient pools in bermudagrass swards fertilized at different nitrogen levels, *Crop Sci.* 57(1): 525–533.

Liu, Y., et al. (2020), Grazing exclusion enhanced net ecosystem carbon uptake but decreased plant nutrient content in an alpine steppe, *Catena* 195: 104799.

Ma, L., et al. (2023), Dung beetle activity is soil-type-dependent and modulates pasture growth and associated soil microbiome, *Agronomy* 13(2): 325.

Mangwe, M. C., et al. (2019), Forage herbs as an alternative to ryegrass-white clover to alter urination patterns in grazing dairy systems, *Anim. Feed Sci. Technol.* 252: 11–22.

Markov, S. A. (2012), Nitrogen cycle. In J. Spradley, et al. (Eds), *Earth's Weather, Water & Atmosphere*, pp. 347–350.

Mathews, B. W., et al. (1999), Soil macronutrient distribution in rotationally stocked kikuyugrass paddocks with short and long grazing periods, *Commun. Soil Sci. Plant Anal.* 30(3–4): 557–571.

Mathews, B. W., et al. (1994a), Impact of grazing management on soil nitrogen, phosphorus, potassium, and sulfur distribution, *J. Environ. Qual.* 23(5): 1006–1013.

Mathews, B. W., et al. (1994b), Soil sampling procedures for monitoring potassium distribution in grazed pastures, *Agron. J.* 86(1): 121–126.

Mathews, B. W., et al. (1996), Grazing systems and spatial distribution of nutrients in pastures: soil considerations. In R. E. Joost and C. A. Roberts (Eds), *Nutrient cycling in forage systems*, PPI/FAR, Columbia, USA, pp. 213–229.

Mihailescu, E., et al. (2014), Nitrogen balance and use efficiency on twenty-one intensive grass-based dairy farms in the South of Ireland, *J. Agric. Sci.* 152(5): 843–859.

Mitchell, E., et al. (2021), Herd concentration areas create greenhouse gas hotspots, *Nutr. Cycl. Agroecosyst.* 121(1): 15–26.

Mu, W., et al. (2016), Nutrient balance at chain level: a valuable approach to benchmark nutrient losses of milk production systems, *J. Clean. Prod.* 112: 2419–2428.

Musial, M., et al. (2020), Effect of pasture management system change on in-season inorganic nitrogen pools and heterotrophic microbial communities, *Agronomy* 10(5): 724.

Myers, R. J. K., et al. (1994), The synchronisation of nutrient mineralization and plant nutrient demand. In P. L. Woomer and M. J. Swift (Eds), *The biological management of tropical soil fertility*, John Wiley & Sons, Chichester, UK, pp. 81–116.

Noriega, J. A., et al. (2023), Dung removal increases under higher dung beetle functional diversity regardless of grazing intensification, *Nat. Commun.* 14(1): 8070.

Obour A. K., et al. (2011), Fluctuating water table effect on phosphorus release and availability from a Florida spodosol, *Nutr. Cycl. Agroecosyst.* 91: 207–217.

Oenema, O. and Roest, C. W. J. (1998), Nitrogen and phosphorus losses from agriculture into surface waters; the effect of policies and measures in The Netherlands, *Wat. Sci. Tech.* 37(3): 19–30.

Ohm, M., M., et al. (2015), Redistribution of soil phosphorus from grassland to cropland in an organic dairy farm, *Appl. Agric. Forestry Res.* 65(3): 193–204.

Ohm, M. M., et al. (2017), Long-term negative phosphorus budgets in organic crop rotations deplete plant-available phosphorus from soil, *Agron. Sustain. Dev.* 37: 1–9.

Pan, H., et al. (2018), Understanding the relationships between grazing intensity and the distribution of nitrifying communities in grassland soils, *Sci. Total Environ.* 634: 1157–1164.

Parfitt, R. L., et al. (2010), Effect of fertilizer, herbicide and grazing management of pastures on plant and soil communities, *Appl. Soil Ecol.* 45(3): 175–186.

Pavlu, K., et al. (2019), Effects of grazing intensity and dung on herbage and soil nutrients, *Plant Soil Environ.* 65: 343–348.

Pecenka, J. R. and Lundgren, J. G. (2018), The importance of dung beetles and arthropod communities on degradation of cattle dung pats in eastern South Dakota, *Peer J.* 6: e5220.

Pereira Neto, J. D., et al. (2024), Herbage responses and animal performance of nitrogen-fertilized grass and grass-legume grazing systems, *J. Agric. Sci.* 162: 77–89.

Picasso, V. D., et al. (2022), Diverse perennial circular forage systems are needed to foster resilience, ecosystem services, and socioeconomic benefits in agricultural landscapes, *Grassland Res.* 1(2): 123–130.

Powel, M. J., et al. (2005), Manure collection and distribution on Wisconsin dairy farms, *J. Environ. Qual.* 34(6): 2036–2044.

Rayner, S., et al. (2020), Can ruminant urine-N rate and plants affect nitrate leaching and its isotopic composition?, *N. Z. J. Agric. Res.* 63(1): 87–105.

Rao, I. M., et al. (1999), Plant adaptation to phosphorus-limited tropical soils. In M. Pessarakli (Ed), *Handbook of plant and crop stress*, Marcel Dekker, New York, NY, USA, pp. 61–95.

Read, Z. J., et al. (2016), Landscape function analysis to assess soil processes on farms following ecological restoration and changes in grazing management, *Eur. J. Soil Sci.* 67(4): 409–420.

Robertson, G. P. and Vitousek, P. M. (2009), Nitrogen in agriculture: Balancing the cost of an essential resource, *Annu. Rev. Environ. Resour.* 34: 97–125.

Rotz, C. A., et al. (2005), Whole-farm perspectives of nutrient flows in grassland agriculture, *Crop Sci.* 45(6): 2139–2159.

Rumpel, C., et al. (2015), The impact of grassland management on biogeochemical cycles involving carbon, nitrogen and phosphorus, *J. Soil Sci. Plant Nutr.* 15(2): 353–371.

Russelle, M. P. (1992), Nitrogen cycling in pasture and range, *J. Prod. Agric.* 5(1): 13–28.

Russelle, M. P. (1996), Nitrogen cycle in pasture systems. In R. E. Joost and C. A. Roberts (Eds), *Nutrient cycling in forage systems*, PPI/FAR, Columbia, MO, pp. 125–166.

Sanaullah, M., et al. (2010), How does plant leaf senescence of grassland species influence decomposition kinetics and litter compounds dynamics?, *Nutr. Cycl. Agroecosyst.* 88: 159–171.

Sands, B. and Wall, R. (2017), Dung beetles reduce livestock gastrointestinal parasite availability on pasture, *J. Appl. Ecol.* 54(4), 1180–1189.

Santos, E. R. S., et al. (2023), Integrated crop-livestock systems result in less nitrate leaching than ungrazed crop systems in North Florida, *J. Environ. Qual.* 52(4): 847–858.

Scheile, T., et al. (2018), Herbage biomass and uptake under low-input grazing as affected by cattle and sheep excrement patches, *Nutr. Cycl. Agroecosyst.* 112: 277–289.

Schick, B. D., et al. (2019), Dietary nutritive value, dung quality, decomposition, and nutrient movement into soil in smooth bromegrass pastures, *Crop Sci.* 59(3): 1294–1308.

Schomberg, H. H., et al. (2000), Spatial distribution of extractable phosphorus, potassium, and magnesium as influenced by fertilizer and tall fescue endophyte status, *Agron. J.* 92(5): 981–986.

Schon, N. L., et al. (2011), Influence of soil faunal communities on nitrogen dynamics in legume-based mesocosms, *Soil Res.* 49(2): 190–121.

Schnyder, H., et al. (2010), Nutrient redistribution by grazing cattle drives patterns of topsoil N and P stocks in a low-input pasture ecosystem, *Nutr. Cycl. Agroecosyst.* 88: 183–195.

Silva, L. S., et al. (2021), Litter mass and nitrogen disappearance in year-round nitrogen-fertilized grass and legume-grass forage systems, *Agron. J.* 113(6): 5170–5182.

Silveira, M. L., et al. (2011), The cow-calf industry and water quality in South Florida, USA: a review, *Nutr. Cycl. Agroecosyst.* 89: 439–452.

Simon, P. L., et al. (2019), The efficacy of *Plantago lanceolata* for mitigating nitrous oxide emissions from cattle urine patches, *Sci. Total Environ.* 691: 430–441.

Snell, L. K., et al. (2014), Nitrous oxide emissions and herbage accumulation in smooth bromegrass pastures with nitrogen fertilizer and ruminant urine application, *Nutr. Cycl. Agroecosyst.* 98: 223–234.

Soares, J. R., et al. (2023), Mitigation of nitrous oxide emissions in grazing systems through nitrification inhibitors: a meta-analysis, *Nutr. Cycl. Agroecosyst.* 125(3): 359–377.

Soberon, M. A., et al. (2015), Changes in nutrient mass balances over time and related drivers for 54 New Your State dairy farms, *J. Dairy Sci.*, 98(8): 5313–5329.

Soberon, M. A., et al. (2013), Whole farm nutrient balance calculator for New York dairy farms, *Nat. Sci. Edu.* 42(1): 57–67.

Sollenberger, L. E. and Dubeux, J. C. B. (2022), Warm-climate, legume-grass forage mixtures versus grass-only swards: an ecosystem services comparison, *Rev. Bras. Zootec.* 51: e20210198.

Sollenberger, L. E. and Wallau, M. O. (2020), Grassland-herbivore interactions. In K. J. Moore, et al. (Eds), *Forage, Volume II – The science of grassland agriculture*, John Wiley and Sons Publishers, West Sussex, UK, pp. 201–214.

Sollenberger, L. E., et al. (2005), Reporting forage allowance in grazing experiments, *Crop Sci.* 45(3): 896–900.

Sollenberger, L. E., et al. (2012), Prescribed grazing on pasturelands. In C. J. Nelson (Ed), *Conservation outcomes from pastureland and hayland practices: assessment, recommendations, and knowledge gaps*, Allen Press, Lawrence, USA, pp. 111–204.

Sollenberger, L. E., et al. (2019), Grassland management affects delivery of regulating and supporting ecosystem services, *Crop Sci.* 59(2): 441–459.

Soussana, J.-F. and Lemaire, G. (2014), Coupling carbon and nitrogen cycles for environmentally sustainable intensification of grasslands and crop-livestock systems, *Agric. Ecosyst. Environ.* 190: 9–17.

Spears, R. A., et al. (2003), Whole-farm phosphorus balance on western dairy farms, *J. Dairy Sci.* 86(2): 688–695.

Stevenson, F. J. and Cole, M. A. (1999), *Cycles of soil-carbon, nitrogen, phosphorus, sulfur, and micronutrients*, John Wiley & Sons, New York, USA.

Stout, W. L. (2003), Effect of urine volume on nitrate leaching in the northeast USA, *Nutr. Cycl. Agroecosyst.* 67: 197–203.

Sun, G., et al. (2017), Responses of root exudation and nutrient cycling to grazing intensities and recovery practices in an alpine meadow: an implication for pasture management, *Plant Soil* 416: 515–525.

Sun, Y., et al. (2018), Nitrogen pools and cycles in Tibetan *Kobresia* pastures depending on grazing, *Bio. Fertil. Soils* 54: 569–581.

Talbot, W. D., et al. (2020), Cattle diet and winter plant growth effects on nitrogen losses from cattle urine patches. *Nutr. Cycl. Agroecosyst.* 116: 365–379.

Tao, R., et al. (2021), Mitigating N2O emission by synthetic inhibitors mixed with urea and cattle manure application via inhibiting ammonia-oxidizing bacteria, but not archaea, in a calcareous soil, *Environmental Pollution* 273: 116478.

Thomas, R. J. (1992), The role of the legume in the nitrogen cycle of productive and sustainable pastures, *Grass Forage Sci.* 47(2): 133–142.

Thomas, R. J. and Asakawa, N. M. (1993), Decomposition of leaf litter from tropical grasses and legumes, *Soil Biol. Biochem.* 25(1): 1351–1361.

Tixier, T., et al. (2016), A four-country ring test of nontarget effects of ivermectin residues on the function of coprophilous communities of arthropods in breaking down livestock dung, *Environ. Toxicol. Chem.* 35(8): 1953–1958.

Trumpp, K. R. (2024), *Forage legumes as a tool to reduce the carbon footprint of beef cattle systems*. M.S. thesis, University of Florida, Gainesville, FL, USA.

Tshewang, S., et al. (2020), Nitrogen and potassium fertilisation influences growth, rhizosphere carboxylate exudation and mycorrhizal colonisation in temperate perennial pasture grasses growth, *Agronomy* 10: 1878.

Van Der Heijden, M. G. A., et al. (2008), The unseen majority: soil microbes as drivers of plant diversity and productivity in terrestrial ecosystems, *Ecol. Lett.* 11(3): 296–310.

van Leeuwen, M. M. W. J., et al. (2019), The relevance of spatial scales in nutrient balances on dairy farms, *Agric. Ecosyst. Environ.* 269: 125–139.

Vertès, F., et al. (2019), C-N-P uncoupling in grazed grasslands and environmental implications of management intensification. In G. Lemaire, et al. (Eds), *Agroecosystem diversity: reconciling contemporary agriculture and environmental quality*, Academic Press, Cambridge, USA, pp. 15–34.

Wachendorf, C., et al. (2008), Nitrous oxide emissions and dynamics of soil nitrogen under ^{15}N-labeled cow urine and dung patches on sandy grassland soil, *J. Plant. Nutr. Soil Sci.* 171(2): 171–180.

Whitehead, D. C. (2000), *Nutrient elements in grassland: Soil-Plant-Animal Relationships*, CABI Publishing, New York, USA.

White-Leech, R., et al. (2013a), Excreta deposition on grassland. I. Forage harvested, nutritive value, and nitrogen recovery, *Crop Sci.* 53(2): 688–695.

White-Leech, R., et al. (2013b), Excreta deposition on grassland. II. Spatial pattern and duration of forage responses, *Crop Sci.* 53(2): 696–703.

Williams, P. H., et al. (1990), Mass balance modeling of potassium losses from grazed dairy pasture, *N. Z. J. Agric. Res.* 33(4): 661–668.

Williams, P. H. and Haynes, R. J. (1990), Influence of improved pastures and grazing animals on nutrient cycling within New Zealand soils, *N. Z. J. Ecol.* 14: 49–57.

Wrage, N., et al. (2004), Nitrous oxide production in grassland soils: assessing the contribution of nitrifier denitrification, *Soil Biol. Biochem.* 36(2): 229–236.

Wrage-Mönnig, N., et al. (2018), The role of nitrifier denitrification in the production of nitrous oxide revisited, *Soil Biol. Biochem.* 123: A3–A16.

Yang, S., et al. (2023), Long-term grazing exclusion increases ecosystem carbon stock but decreases nitrogen stock in the Karst alpine grassland of China, *J. Soil Sci. Plant Nutr.* 23(3): 3714–3725.

Yao B., et al. (2018), Understanding the mechanisms for the lower nitrous oxide emissions from fodder beet urine compared with kale urine from dairy cows, *J. Soils Sediments* 18: 85–93.

Yoshitoshi, R., et al. (2020), Spatial distribution of grazing sites and dung of beef cows in a sloping pasture, *Jpn. Agric. Res. Q.* 54(4): 299–306.

Zhang, M., et al. (2022), BNI-release mechanisms in plant root systems: current status of understanding, *Biol. Fert. Soils* 58: 225–233.

Chapter 5

Advances in grazing management

Michael O'Donovan, Teagasc, Ireland

1 Introduction

Grassland management systems are evolving continually, leading to many countries placing greater emphasis on grazed grass rather than grass/maize silage or concentrates in animal production. Where grazing is included in the production system, efficient grazing is pivotal. The competitive advantage of grass-based livestock production systems is based on the efficient production and utilisation of grazed grass. Irish pasture has, for example, the potential to grow up to 10–14 t DM/ha/annum (DM, dry matter; O'Donovan et al., 2020). In the context of a changing climate, warmer and wetter winters are expected in northern Europe and increasingly warmer and drier summers in southern Europe (Ergon et al., 2018). Although grazing systems have specific challenges such as unstable feed supply and reduced individual animal intake and performance, the principal benefit of improved grazing management has been to optimise the quantity and nutritive value of the forage consumed by

http://dx.doi.org/10.19103/AS.2025.0154.11

grazing animals. Key livestock production decisions like calving date (dairy, beef) and lambing date are targeted to maximise the use of grazed grass. Such parameters will have to be continually evaluated if grass growth curves are changing based on climatic conditions.

Increasing consumer demand for more natural dairy products has resulted in the formation of 'Grass-fed' standards, stipulating minimum proportions of grass in the diet and minimum time at pasture to qualify for grass-fed status (Bord Bia, 2019; Morrisey, 2018; NZMP, 2022). The benefits of grass-fed milk and meat have been well identified (e.g. O'Callaghan et al., 2016; Noci et al., 2005; Timlin et al., 2023, 2024). Grass-fed milk has a beneficial impact on the fatty acid profile of milk and dairy products with increased concentrations of beneficial nutrients such as omega 3 fatty acids, conjugated linoleic acid, vaccenic acid and reduced levels of palmitic acid, omega 6 fatty acids and thrombogenecity index score compared to total mixed ration produced milk. This allows grass-based production systems to follow a particular marketing advantage when highlighting the 'grass-fed' benefits of animal products. This market focus is particularly positive to continue to promote the image of grassland. The key challenges in the future will be to further increase animal production by improving pasture growth and grass quality, increasing N use efficiency (NUE) and reducing greenhouse gas (GHG) and ammonia emissions. The objective of this chapter is to discuss how new advances in grazing management can further generate new opportunities to progress sustainable grassland systems. This is illustrated by Irish grass-based systems as a case study.

2 Grassland dry matter production

Grassland productivity is affected by botanical composition, soil characteristics, climatic conditions, altitude, latitude and management. Very little actual national or international annual DM production is reported, the reasons why such little data is produced is unclear. This is clearly an information gap across countries. Previously, Smit et al. (2008) presented spatial data of actual grassland productivity across regions of Europe, based on an extended set of regional, national and international statistics over a 10-year period (1995–2004). The most productive regions from this analysis were obtained in the North-western parts of Europe (Ireland, North West France, United Kingdom). In the Mediterranean countries, DM yields were much less, yielding 1.5–2 t DM/ha/annum due to moisture stress challenges.

3 Grass measurement adoption

The uptake of decision support systems (DSSs) in grassland continues to be slow by all agriculture sector actors. The role of DSS in the improvement of grassland production and utilisation has huge potential. The influence of peer-to-peer learning and the creation of grassland focused discussion groups will assist this process. The movement of grassland research onto grassland farms in Ireland has been aided by the creation of PastureBase Ireland (PBI) (Hanrahan et al., 2017), this allows a greater understanding of the performance of grass and white clover varieties in an on-farm situation giving huge insights into the farm effects on grass DM production. Such an initiative needs to be replicated in other countries particularly in temperate regions. The use of grass growth modelling, to examine and understand the effects of climate limitations, soil type, climatic variables and their impacts in the longer term, needs to be more widely used and adopted. The use of predictive real-time data in N management and a change away from date-bound decisions to more informed decision-making based on real-time data is a key prerequisite of better N management for the agricuture industry. This can be a major advancement for grazing management and nutrient efficiency improvement at the farm level.

One of the major deficiencies in grassland management at the farm level has been the lack of a measurement capability and application. O'Donovan et al. (2002) introduced the concept of grassland measurement using visual assessment to dairy farms in Ireland. The rising plate meter has been used successfully in Ireland and this technology has been improved through the incorporation of GPS and smartphone technology allowing it to be integrated with web-based grassland management tools (Murphy et al., 2021a, 2021b; McSweeney et al., 2018).

In the countries where grassland DM production is important and the main livestock feed, a number of new grassland decision support tools have been developed. PBI in Ireland (Hanrahan et al., 2017), PasturePlan (Delaby et al., 2014) in France and 'Grip op Gras' (Stienezen et al., 2018) in the Netherlands are all decision supports used extensively to assist in grazing management decision-making. The key objective of most DSS's is to increase the information available to help the decision-making process at farm level. The advancement of the internet and in particular the increase in smartphone usage has created opportunities for the development and use of web-based DSS's that facilitated the collation of large data quantities in a central data storage platform from different farms. The potential use of this information from a research perspective can be significant. In Ireland, PBI has been in operation since 2013, PBI offers farmers grassland decision support and stores grassland data from dairy, beef and sheep farmers in a central national database. PBI also provides an automated mechanism to benchmark farms across periods and to benchmark across a range of farms. Figure 1 shows the level and frequency of grassland measurement taking place in PBI (Hearn, 2024). The level of measurement has steadily increased in the last number of years, Figure 1 shows the number of farmers routinely measuring grass (>10 covers per year) at greater than 6000

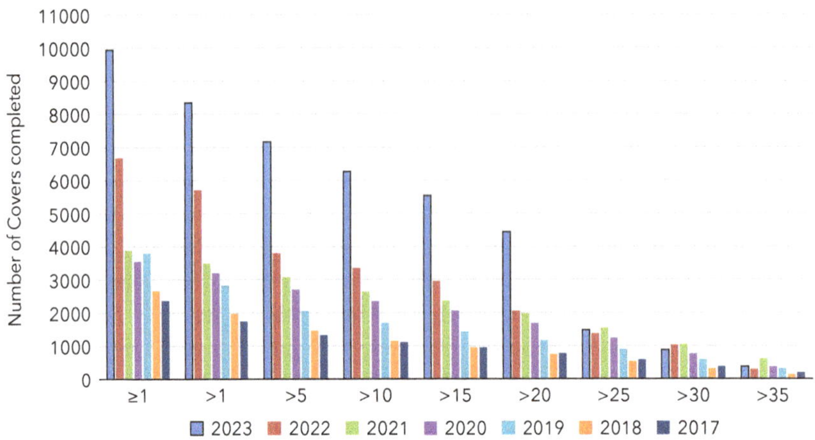

Figure 1 The level of grass covers measured from 2017 to 2023 in Pasture Base Ireland (Hearn, 2024).

farmers, this clearly shows the improved level of engagement of farmers adopting a key grassland management technology, to further improve grassland efficiency.

Without proper reporting of weekly, monthly and annual DM production, many countries are not focussing on improving grassland management. As continued levels of climate instability are being experienced, the capability to better adapt to fluctuations in grass growth is required (Ruelle et al., 2023). In 2018, Europe experienced a severe drought. When moisture deficits arrive, it is important that the livestock industries are well advised on how to manage through such events, without wasting feed and ensuring feed security for livestock. It is clear that adaptive practises need to be fully incorporated into grassland systems.

4 Grazing infrastructure

To maximise grass utilisation or indeed improve the number of days at grass, a rotational grazing system is required, whereby the grazing herd/flock is continually allocated fresh pasture as they move around the grazing area (delineated into grazing paddocks for ease of management). An integrated farm laneway network (referred to as a roadway network) is required to connect the farm yard to all areas of the grazing area, which can be defined as an area graze-able by the herd in situ (Patton et al., 2016) to use grass efficiently. This is particularly important in the spring and autumn period where suboptimal soil conditions lead to excessive treading damage to the soil structure (Kennedy et al., 2011; O'Loughlin et al., 2016). Ensuring grazing farms have adequate grazing infrastructure, farm roadways, water troughs and access to enter paddocks is important to increase the number of grazing days achieved. Hanrahan et al. (2018) estimated that every extra ton of grass DM utilised was worth around €173. The only way extra grass can be utilised is to increase the amount of grazing days in a farm system.

5 Grass growth prediction

A new innovative area in terms of grassland management decision support is grass growth modelling/prediction (Ruelle et al., 2018). Increasing the predictability of grass growth increases confidence in short-term grassland management. Figure 2 shows the average daily grass growth rates in 2021, 2022 and 2023 and the average growth rate from 2018 to 2023 (inclusive) recorded on PBI.

Numerous grass growth models have been developed, ranging from empirical models tailored to specific locations (Brereton et al., 1996) to mechanistic models aiming to understand the mechanisms influencing grass growth (e.g. the STICS model) (Brisson et al., 2009), BASGRA (Höglind et al., 2020) and T&C-BG model (Botter et al., 2020). While the majority of these models offer conceptual merits, their complexity and data requirements make them poorly suitable as practical decision support tools (Bittar et al., 2018).

In Ireland and France, the Moorepark St Gilles Grass Growth Model (MoSt GG) (Ruelle et al., 2018) has been developed as both a research tool and a

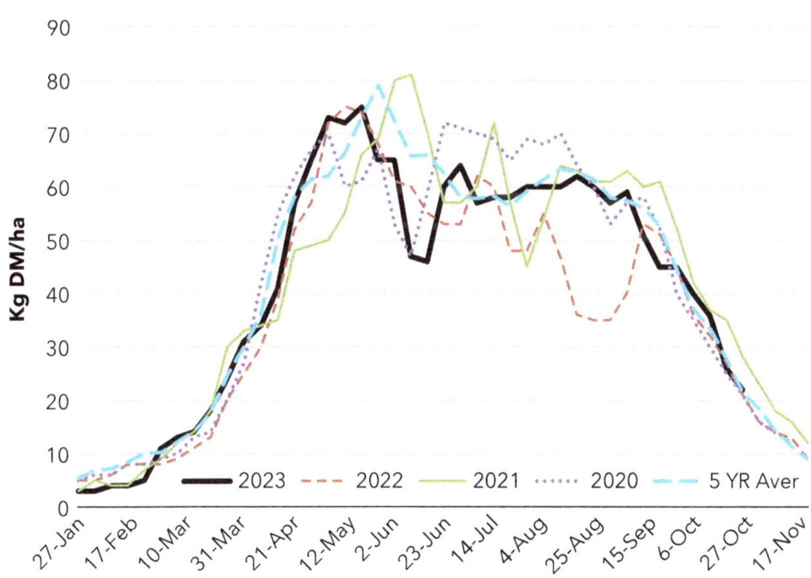

Figure 2 Daily grass growth rates (kg DM/ha) in 2020, 2021, 2022 and 2023 compared to the mean growth rate from 2018 to 2023 (inclusive) recorded on Pasture Base Ireland.

decision support tool for farmers. Since its creation, it has been used extensively in both countries at both research and farm level and is used to predict grass growth weekly on 84 Irish farms (Ruelle et al., 2021). The MoSt GG model, initially adapted from the model developed by Jouven et al. (2006) has evolved considerably over the years (Ruelle et al., 2018). The MoSt model predicts the trend in grass growth (increase, decrease, static) weekly. This information is valuable for grassland decision-making for farmers. The directional trend change in grass growth as well as the absolute value predicted are important, the directional trend takes precedence in managing the grass growth response.

The development of the MoSt GG model has been a major advancement in terms of grass growth farm modelling. For a grass growth prediction model to become a valuable grassland management decision support tool, the model must be easily adaptable to a range of soil types and management scenarios. To increase the likelihood of farm-level use, few input data should be required. The model must respond to the main factors that influence grass growth including defoliation, N fertiliser application, N returned to soil from livestock faeces, urine and soil water content. One of the main limitations of the MoSt GG model is that it does not consider pasture species other than perennial ryegrass or nutrients other than N; however, this aspect of the model will be developed in the next edition of the model (Bonnard, 2024).

6 Grass clover swards

Forage legumes, particularly white clover (*Trifolium repens* L.), are highly regarded, but over time, legume use declined mainly due to the low cost of N fertiliser (Gilliland et al., 2009), which was used in increasing amounts. White clover is assuming much more importance in the last decade due to increasing restrictions on current and future N fertiliser use. These restrictions result from policies aimed at reducing N loss, e.g. EU Water Framework Directive (2000/60/EC), the Nitrates Directive (SI 378 2006) and EU Farm to Fork Strategy 2020. Recognition of the environmental implications of grassland production systems has increased, especially over the last decade. This includes the need to not only reduce nutrient emissions from agriculture and improve water quality but also increase the role of grassland in biodiversity protection, carbon sequestration and landscape quality.

White clover (*T. repens* L.) (WC) is the most important sown legume species in grazed grassland in temperate regions (Peyraud et al., 2009). White clover needs to be present within the sward at a minimum proportion of 20% in order to realise DM production benefits. Previous research (Lüscher et al., 2014) has shown that sward WC content fluctuates over time and that a number of key soil, climatic and management factors are associated with these fluctuations. Soil factors, such as pH, are important in ensuring successful establishment, as WC has a preference for slightly acidic or neutral pH values. Increased inorganic N fertiliser has been shown to have a negative effect on the prevalence of WC in the sward through increased competition from companion grasses. This has raised doubts about how much benefit WC brings to intensively managed pastures receiving higher levels of N fertiliser (Egan et al., 2018).

The incorporation of white clover in grazing swards takes advantage of their capacity to fix atmospheric N and make it available for plant growth, allowing the reduction in chemical N fertiliser (Egan et al., 2018; Enriquez Hidalgo et al., 2018). White clover has high digestibility and has a high energy value which is attributed to its low fibre concentration which reflects the absence of structural components such as stems and sheaths (Ayres et al., 1998). A particular advantage of white clover is the reduced rate of digestibility decline in the mid-season compared to perennial ryegrass (Ulyatt, 1970). Increased animal production performance as a result of increasing the sward white clover proportion has been observed in dairy cows (Egan et al., 2018; McClearn et al., 2020), beef steers (Thomas et al., 1981) and sheep (Orr et al., 1990). Egan et al. (2017 and 2018) found in a component study that cows grazing a grass-clover sward (20%+ clover content) had greater milk yield and milk solids yield than cows grazing a grass only sward at similar N input. Spring grazing management plays an important role in maintaining WC growth and persistence. There should also be consideration taken to select companion PRG cultivars with low post-grazing capacity (approximately 40 mm) required to support the persistency of sward WC content. In autumn, grazing management needs to ensure that clover paddocks are well grazed and tillered pre-closing and to reduce winter clover death. All factors need to be balanced with the requirement to maintain sward DM productivity and ensure the clover has an additive effect (higher milk production performance and economic impact) to the farm system's performance and environmental footprint.

Research in the areas of incorporating white clover into grassland swards to optimise their contribution and persistence is a significant focus of ongoing research efforts (Egan et al., 2018). Given the increasing chemical N cost, legume usage is growing. The traits for choosing appropriate white clover cultivars need some improvement. The future development of a Clover Profit Index (CPI) in Ireland, similar to the Pasture Profit Index (PPI) (McEvoy et al., 2011; O'Donovan et al., 2016), will provide farmers with confidence in selecting the appropriate white clover varieties for their system, this development is 3–5 years from completion. Both of these indexes are designed to place economic values on the traits of importance in Irish grassland swards. The traits have to be quantified in carefully designed plot-cutting and grazing studies. In developing such agronomic indexes, it is recognised that the traits of importance are those which have the greatest impact on the grassland systems. In the Irish context, seasonal DM yield, pasture quality, silage DM production, grazing utilisation and sward persistence are the main traits of interest in the pasture profit index (Tubritt et al., 2020).

7 Alternative sward species

Improved pastures for grazing in dairy production systems have predominantly consisted of perennial ryegrass (*Lolium perenne* L., PRG) monocultures over the past several decades (Grogan and Gilliland, 2011; Lee et al., 2012). The European Union has stipulated a reduction of 20% in chemical fertiliser use over a 10-year period, 2020–2030 (EU, 2020). Alternative methods of supplying grazing swards with N are therefore required. Perennial ryegrass-dominated swards experience low growth rates in colder periods (i.e. winter) or at times of soil moisture deficit leading to a requirement for supplementary feed.

Pasture sward species complementarity has been widely promoted as a solution to the issues outlined above (Lüscher et al., 2014). Sward species complementarity occurs where multiple species, from differing plant functional groups, provide niche benefits to increase overall sward yield and resilience, such benefits include greater resource-use efficiency within and across seasons. One example of this sward complementarity is the combination of grasses with legumes where the leguminous species can supply N to all sward species via biological nitrogen fixation.

Sward botanical composition has an influence on overall sward performance where species need to be present in sufficient proportions to have a significant impact on DM production. Both defoliation strategy and N fertiliser application have been shown to affect the growth of forage herbs (Lee et al., 2015; Martin et al., 2017) and strongly influence sward clover content (Egan et al., 2017). While it is known that an annual sward WC content of ~20% is sufficient for herbage production benefits (Andrews et al., 2007; Hennessy et al., 2022), the contribution of individual species in complex MS swards under grazing is not understood as yet (Gilliland, 2022). For example, the role of red clover would appear to be more beneficial in conservation-based systems (Clavin et al., 2017).

Sward complementarity leading to increased DM production has also been observed in multispecies (MS), swards which include forage herbs along with grasses and legumes (Cranston et al., 2015; Baker et al., 2023a). There is also evidence to the contrary where little difference was observed in DM production between binary grass and legume mixtures and more complex MS swards (Bryant et al., 2017; Grange et al, 2021). Baker et al. (2023b) indicated that these differences can be subject to sward management intensity which also has a large influence on how much diverse swards can contribute to other ecosystem services and biodiversity (Hoekstra et al., 2023). The mechanisms through which forage herbs increase sward DM production are related to variations in both seasonal growth patterns (Baker et al., 2023b) and the rooting depth of species sown within the same pasture (Komainda et al., 2020). Utilising forage herbs such as ribwort plantain (*Plantago lanceolata* L., PL) and chicory in combination with grasses and legumes can increase the overall sward resource utilisation capacity where these forage herbs can continue growing in climatic conditions which may limit the growth of other forage species (Lee et al., 2015; Welten et al., 2019). Klump et al. (2018) clearly showed that grassland play a major role in the provision and regulation of important ecosystem services across many livestock production systems. Grassland acts as CO_2 sink, with evidence of positive rates of soil C accumulation across differing climates and management practises. Given that management practises can be region specific to increase C accumulation, grazing does have an advantage on grassland (especially in upland and extensively managed areas). The retention of longer term grassland swards has positive effects on C accumulation. Soil disturbance leads to associated C loss, but this can be often non linear. While the additive effects of MS swards is well argued, the evidence of their long-term additive effects is poor (Hearn et al. 2023). It is important that longer term grassland studies are undertaken in this subject area whereby the full effects of such species are properly quantified. Ecosystem services tend to be long trend additional effects in grassland swards (Klump et al., 2018).

8 Grazing management influences on grass quality

At grazing, increased frequency of defoliation results in high quality but a decrease in net herbage accumulation, whereas infrequent defoliation leads to greater herbage production, but decreased grass feed value (Tunon, 2013). High biomass yield (kg DM/ha) at grazing will limit animal performance through digestive constraints (low intake of poorly digestible matter), but high-quality swards (low pre-grazing yield) can limit animal performance through behavioural constraints if the time required to graze the required grass quantity is too great (Baumont et al., 2004). The pre-grazing yield effect at grazing may also vary across the grazing season. Tunon (2013) found no effect of pre-grazing herbage mass up to 2300 kg DM/ha (>4 cm) on milk yield from April to July but observed a reduction in milk fat plus protein yield compared to swards with pre-grazing herbage masses of 1000 and 1500 kg DM/ha from July to October. Beecher et al. (2017) observed no difference in OM digestibility between swards with a pre-grazing mass <1500 kg DM/ha and >2000 kg DM/ha in the May-to-July period, but in the July-to-October period it was observed that increasing pre-grazing yield resulted in a significant decrease in OM digestibility and in digestible DM intake (r^2 = .75) (Figure 3).

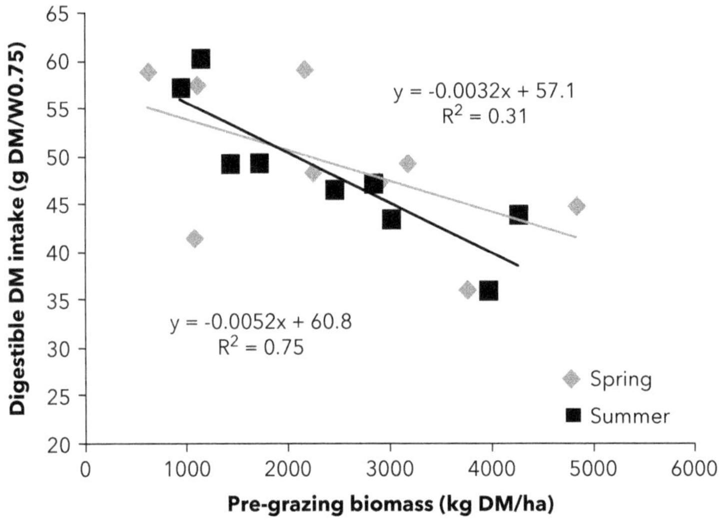

Figure 3 Effect of pre-grazing biomass on digestible dry matter intake measured in sheep. Source: Data adpated from: Beecher et al. 2017).

Animal performance at grass is highly influenced by the voluntary grass DM intake (VDMI) of the animal and the organic matter digestibility (OMD) of the grass ingested. Accurate measurements of grass VDMI and OMD are difficult at grazing where pre-grazing herbage mass is a major factor affecting both the VDMI and OMD (Wims et al., 2010). Grass OMD is higher when grass is maintained at a low herbage mass, this generally results in a greater grass VDMI been achieved. Grass quality, measured as OMD, is a key driver of animal performance in grazing systems. The chemical composition of a perennial ryegrass sward shows large variations from the top to the base of the sward, with decreasing OMD as depth increased in the sward (Delagarde et al., 2000). Perennial ryegrass varieties can vary in *in vitro* OMD due to differences in ploidy (Beecher et al., 2015) and heading date (O'Donovan and Delaby, 2005), as well as sward structure (McEvoy et al., 2010).

There are opportunities to increase the productivity and efficiency of grass-based systems by strategically modifying the ration. To select the optimal strategy, quantitative knowledge of how the diet interacts with the grazing ruminant, the nutrients it supplies and the animal's metabolic requirements are crucial. Recent research focussing on the neutral detergent fibre fraction of immature pasture comprises of a large potentially digestible pool that degrades rapidly in the rumen allowing for higher milk production performance to be achieved from grass-only diets (Dineen et al., 2021).

Ruminants consuming grass-based diets exhibit a large dependence on microbial amino acids to support metabolisable amino acid supply, how and when (season) this can be balanced is an area that requires more focussed research to increase our understanding and knowledge. The creation of high-quality, high-tannin, low methane emission swards is now a key aspect of grassland research. While we are some years from this objective been achieved, it remains a possibility. Perennial ryegrass is invariably low in tannin content, so the pursuit of high tannin species can be complimentary to the focus of reducing methane emissions of grazing ruminants.

9 Grass breeding applications

The majority of resources in forage breeding programmes in northwest Europe are committed to perennial ryegrass improvement as it is the main forage species sown in the region. White clover and red clover, the principal legumes for temperate grassland, are subject to significantly less breeding efforts. In Europe, varieties cannot be sold without first undergoing and passing a National List or Recommended List (RL) evaluation system. The RL is highly influential with almost all seed sales in Europe today consisting of RL varieties (Grogan & Gilliland, 2011). Grass and clover breeding programs have evolved to put greater emphasis on DM yield, this emphasis has changed now to put significantly greater emphasis on seasonal yield and quality. The genetic gain in annual DM yield of perennial ryegrass has been estimated at 4–6% per decade (Brummer & Caler, 2014; Gilliland et al., 2020a). This is modest in comparison with the 10–15% per decade genetic improvement in cereal seed yield. The lag in the genetic gain of forage species is due to several biological factors and limited seed industry investment. Over the 1973–2013 period, it is estimated that annual DM yield increased by 0.52% under conservation (silage harvesting) and 0.35% under simulated grazing (McDonagh et al., 2016) in the Northern Irish evaluation lists.

In many respects, it is too early to establish if there has there been an improvement in grass digestibility through grass breeding. Genetic improvement for nutritional value traits in perennial ryegrass has been modest with dry matter digestibility (DMD) increasing by 5–10 g/kg DM (0.5–1%) per decade (Gilliland et al., 2020b). Nutritional value is now a core trait for improvement in most grass breeding programmes but given it typically takes 15–20 years to develop, test and release a new variety, the effects of this increased breeding effort is only starting to be realised in new varieties.

10 Grass variety selection indices

Perennial ryegrass variety economic selection indices such as the Pasture Profit Index (PPI) in Ireland and the Forage Value Index (FVI) in New Zealand and Australia have been developed to aid variety selection decisions within the grassland industry. These indices utilise data from evaluation trials (O'Donovan et al., 2016; Chapman et al., 2017) expressing superior variety performance based on a total merit index (i.e. expressed as an increase in net profit per year). These indices increase the value of the information supplied to the end user, creating a communication tool expressed in a language (i.e. profit) that users more readily relate to. The relative difference between varieties for all traits are expressed in economic terms indicating the additional profit generated by one variety against another (O'Donovan et al., 2016). Farmers can use the PPI to select varieties based on their total economic merit (sum of all traits) or by focusing on individual traits deemed to be of greater importance (McEvoy et al., 2011).

Grazing efficiency of perennial ryegrass varieties influences the amount of herbage which can be utilised (Byrne et al., 2017). Swards that are consistently grazed to low post-grazing sward heights (approx. 4 cm) support the highest utilisation levels (McCarthy et al., 2013). Hanrahan et al. (2018) reported that utilisation was one of the single largest (controllable) variables affecting profitability of ruminant systems with each additional tonne of DM utilised increasing profitability of dairy farms by €173/ha. Utilisation is a function of herbage production and harvesting by the animal which can be influenced by perennial ryegrass varieties. Herbage DM production is already captured in the PPI, the relative differences between varieties for grazing efficiency is included as a supplementary trait (grazing utilisation). Byrne et al. (2018) reported a 0.3 cm difference for diploids relative to and tetraploids in achieved post-grazing sward height with the increased digestibility and lower tiller density of tetraploids, compared to diploids, responsible for the grazing difference. Protocols have been developed where a grazing efficiency value can be calculated for individual varieties (Tubritt et al., 2020; Byrne et al., (2018). Both authors have found important differences between varieties in their ability to be grazed by cattle. The results from these protocols have provided the data necessary to develop a grazing efficiency sub-index within the PPI. Over time, it is expected that the grazing utilisation trait will enable the elimination of poorer graze out varieties, which will benefit grazing management significantly, it will eventually remove the need for mechanical correction of pastures.

11 Sward persistence

One of the main limitations of grassland-based research is the lack of long-term DM production datasets, the key to understanding grass variety or clover persistence is having multi-year's data measured. Sward persistence is a key topic for grassland production systems. Long-term quantification of DM production of sward performance is vital.

In the past number of years, some new species (chicory and red clover) have been promoted as having significant additive benefits; however, they have persistency issues (Hearn et al., 2023). In the area of MS awards, there is little long-term research, documenting the change in sward species and DM production over time. Hearn et al. (2023) concluded there was little available data in the literature on the persistence of either plantain or chicory in MS grazing swards beyond 2 years. Baker et al. (2023a) suggested that neither forage herb persist well in the second year under intensive grazing where forage herbs accounted for less than 10% and 14% of total sward DM, respectively. Chicory is classified as biennial and persistence beyond 2 years in pasture is low (Li and Kemp, 2005; Hearn et al., 2023). Plantain has been shown to be more persistent in the second grazing season with sward DM contents of 29% and 31% reported by Vi et al. (2023) in New Zealand and Hearn et al. (2023) in Ireland, respectively.

12 Climate adaptation

Many countries are experiencing extreme changes in the climate with varying weather events (drought, high autumn rainfall, above average spring rainfall, extreme temperatures in winter and summer). In Ireland, the mean annual surface air temperature has increased by approximately 0.8°C over the last 110 years, the level of moisture deficits is likely to increase in time. Average annual national rainfall has increased by approximately 60mm or 5% in the period 1981 to 2010, compared to the 30-year period 1961 to 1990. Grassland production is likely to face more droughts in the future in many Mediterranean regions, while warming in Nordic area will increase forage DM production. There is a large variation nationally in DM production in Europe, influenced by climate conditions, soils, latitude, altitude and management. Grazing season length is predicted to increase from an observed 6.1 months in 2012 to 6.7 months in 2050 and 6.8 months in 2070 across Europe (Dillon, 2018).

The observational record suggests that the 2018–2019 European drought event is unprecedented in the last 250 years, with substantial implications for vegetation health. One of the major limitations of climate model simulation is its ability to reliably simulate the extreme events and the changes there after. Some climate models have a relative ability to simulate the past, but there remains some uncertainties in projections. Despite such limitations, climate models are the only tools available to mechanistically understand the occurrence, processes and fate of future climatic events. For instance in 2018, there were two extreme weather events: first, cold weather in spring, this prevented farmers from grazing livestock due to snow, high rainfall and low grass supply. The impact of this was that farmers needed to have an adequate conserved fodder stock and sufficient slurry storage in place to manage the situation. In the summer of 2018, an extended dry period impacted on grassland growth rates during the main grazing season, resulting in a requirement for significant supplementation to offset poor grass growth. The results of this event led to the feeding of conserved fodder and concentrate feeds during this period, which had major implications for fodder stocks subsequently. Such events have focussed the livestock sector to ensure grassland production systems are sufficiently robust to withstand the risks of climate impacts. The planning for future such events is uncertain; however, as part of climate adaptation, increased levels of feed will be required to be conserved.

This practise will have to be routinely completed to offset the climatic trends.

Europe-wide, grass and clover species are exposed to many different environmental challenges. Many recommended lists are now measuring grass diseases and are screening varieties with low disease resistance. Major diseases such as crown rust, mildew, *Drechslera*, *Rhynchosporium* and ryegrass mosaic virus have become more common and frequently found in grassland systems (British Grassland Society, 2023). Other sporadic diseases such as brown rust, stem rust and barley yellow dwarf virus can vary in their prevalence between years. The most significant effect of grass diseases are its impacts on pasture quality and its impact on the sward presented to the grazing animal, especially in autumn. In the future, national evaluation programs will have to increase their monitoring of such diseases, breeders will have to be very proactive to breed grass and clover varieties which have high disease resistance which may ultimately compromise some other agronomic traits.

13 Conclusion and future trends

The objective in successful grassland systems to achieve higher animal performance, lower nitrogen and carbon footprint without reductions in grassland vegetation or related declines in C sequestration. For grassland systems to advance, there needs to be continual improvement in performance (animal and sward). Grassland management practises have developed in the past number of years through the introduction of new decision support systems. Grasslands in all temperate zones are experiencing seasonal changes and the climate impacts are making adaptation practises take place more frequently. The increased use of legumes and particularly white clover and red clover needs to progress faster. Given current environmental constraints and challenges, there needs to be a major focus to improve legume management on farms. New decision supports and their application need to be better used across the grassland sectors, weather apps and better decision rules based on climate adjustments, sward characteristics and nutrient planning applications that will deliver tangible results in time. The main issue is ensuring wider user acceptance.

Changing sward species needs to be a fruitful exercise for the farmers, the strengths/weaknesses of such species need to be better quantified in the longer term. The role of MS in pasture-based systems needs to be better evaluated and defined. The most suitable species and combinations of species need to be evaluated. Further research work needs to focus on the added value aspect of specific species to the overall DM production, persistence, N fixation, canopy morphology and sward quality of the swards. Added benefits may bring environmental benefits (e.g. reduced N loss, increased soil C content, etc.) to grazed pasture; however, such benefits need to be better researched and quantified.

Grassland-fed systems not only enable high proportions of forage in the animal's diet but also allow the expression of the animal's natural grazing behaviours outdoors, the aspects of grassland-based systems are very positive to wider society and the likelihood is that consumer demand for more natural 'grass-fed' food products will increase. Future European and National policies in temperate countries need to fully understand, interpret and promote the economic, environmental and social structures provided by grasslands and grassland-based systems to national economies.

14 References

Andrews, M., et al. (2007). Use of white clover as an alternative to nitrogen fertiliser for dairy pastures in nitrate vulnerable zones in the UK: productivity, environmental impact and economic considerations. *Annals of Applied Biology* 151, 11–23. https://doi.org/10.1111/j.1744-7348.2007.00137.x

Ayres, J. F., et al. (1998). A study of the nutritive value of white clover in relation to different stages of phonological maturity in the primary growth in spring. *Grass and Forage Science*, 53: 250–259. https://doi.org/10.1046/j.1365-2494.1998.00137.x

Baker, S., et al. (2023a). Multispecies swards outperform perennial ryegrass under intensive beef grazing. *Agriculture, Ecosystems & Environment* 345. https://doi.org/10.1016/j.agee.2022.108335

Baker, S., et al. (2023b). Dry-matter production and botanical composition of multispecies and perennial ryegrass swards under varying defoliation management. *Grass and Forage Science*. https://doi.org/10.1111/gfs.12615

Baumont, R., et al. (2004). A mechanistic model of intake and grazing behaviour integrating sward architecture and animal decisions. *Animal Feed Science and Technology*, 112, 5–28.

Beecher, M., et al. (2015). The variation in morphology of perennial ryegrass cultivars throughout the grazing season and effects on organic matter digestibility. *Grass and Forage Science* 70, 19–29.

Beecher M., et al. (2017). Effects of harvesting perennial ryegrass at different levels of herbage mass on voluntary intake and in vivo digestibility in sheep. *Grass and Forage Science* 73, 553–561.

Bittar, A., et al. (2018). Simulation de la croissance de l'herbe et des rendements des pâturages avec ModVege. Recherche agronomique suisse 186.

Bonnard, L. (2024). *Improving. the MoSt Grass Growth model to predict environmental impact and improve on farm grass growth prediction*. M. Agr Sc Thesis. National University of Ireland.

Bord Bia. (2019). Grass Fed Dairy Standard. Accessed Aug 8, 2022. https;//www.bordbia.ie/globalassets/bordbia2020/farmers-growers/grass=fed=standard/grass-fed-daiiry-rev01-final.pdf

Brereton, A., et al. (1996). *Agrometeorology of grass and grasslands for middle latitudes*. World Meteorological Organization.

Botter, M., et al. (2020). *Impacts of fertilization on grassland productivity and water quality across the European Alps: insights from a mechanistic model*. Biogeosciences Discussions.

British Grassland Society. (2023/2024). Recommended Grass and Clover list for England and Wales (2023/24). Annual Publication (29 pages). www.bspb.co.uk/wp-content/uploads/RGCL-Merchants-Guide-2023.pdf

Brisson, N., et al. (2009). Conceptual basis, formalisations and parameterization of the STICS crop model. Conceptual basis, formalisations and parameterization of the STICS crop model, 1–298.

Brummer, E.C. and Caler, M.D. (2014). Cool-season forages. In: *Yield Gains in Major U.S. Field Crops, CSSA Special Publication 33* (eds. S. Smith, B. Diers, J. Specht and B. Carver), ASA, CSSA, and SSSA, Madison, WI, pages 33–51.

Bryant, R., et al. (2017). Milk yield and nitrogen excretion of dairy cows grazing binary and multispecies pastures. *Grass and Forage Science* 72, 806–817. https://doi.org/10.1111/gfs.12274

Byrne, N., et al. (2017). Establishing phenotypic performance of grass varieties on Irish grassland farms. *The Journal of Agricultural Science* 155, 1633–1645. https://doi.org/10.1017/S0021859617000740

Byrne, N., et al. (2018). Understanding factors associated with the grazing efficiency of perennial ryegrass varieties. *European Journal of Agronomy* 101, 101–108. https://doi.org/10.1016/j.eja.2018.09.002

Chapman, D.F., et al. (2017). An economically based evaluation index for perennial and short-term ryegrasses in New Zealand dairy farm systems. *Grass and Forage Science* 72, 1–21, https://doi.org/10.1111/gfs.12213

Clavin, D., et al. (2017). Red clover for silage: management impacts on herbage yield, nutritive value, ensilability and persistence, and relativity to perennial ryegrass. *Grass and Forage Science* 72, 414–431. https://doi.org/10.1111/gfs.12249

Cranston, L., et al. (2015). A review of the use of chicory, plantain, red clover and white clover in a sward mix for increased sheep and beef production. *Journal of New Zealand Grasslands* 89–94. https://doi.org/10.33584/jnzg.2015.77.475

Delaby, L., et al. (2014). Pâtur'Plan: Un outil dynamique pour favoriser la gestion anticipée des parcelles en système de pâturage tournant.

Delagarde, R., et al. (2000). Vertical distribution of biomass, chemical composition and pepsin-cellulase digestibility in a perennial ryegrass sward: Interaction with month of year, regrowth age and time of day. *Animal Feed Science and Technology* 84, 49–68. https://doi.org/10.1016/S0377-8401(00)00114-0

Dillon, P. (2018). The evolution of grassland in the European Union in terms of utilisation, productivity, food security and the importance of adoption of technical innovations in increasing sustainability of pasture based ruminant production systems. In Grassland Science in Europe, Vol 23 – Sustainable meat and milk production from grassland pp. 3–16. 27th General meeting of the European Grassland Federation, Cork, Ireland.

Dineen, M., et al. (2021). Characterization of the nutritive value of perennial ryegrass (*Lolium perenne* L.) dominated pastures using updated chemical methods with application for the cornell net carbohydrate and protein system. *Animal Feed Science and Technology* 272, 114752. https://doi.org/10.1016/j.anifeedsci.2020.114752

Egan, M., et al. (2018). Incorporating white clover (*Trifolium repens* L.) into perennial ryegrass (*Lolium perenne* L.) swards receiving varying levels of nitrogen fertiliser: effects on milk and herbage production. *Journal of Dairy Science* 101, 3412–3427.

Egan, M., et al. (2017). Including white clover in nitrogen fertilized perennial ryegrass swards: effects on dry matter intake and milk production of spring calving dairy cows. *Journal of Agricultural Science* 155, 657–668.

Enriquez-Hidalgo, D., et al. (2018). Production and quality benefits of white clover inclusion into ryegrass swards at different nitrogen fertilizer rates. *Journal of Agricultural Science* 156, 378–386.

Ergon, A., et al. (2018). How can forage in Nordic and Mediterranean Europe adapt to the challenges and opportunities arising from Climate change? *European Journal of Agronomy* 92, 97–106.

EU. (2020). *Farm to Fork Strategy For a fair, healthy and environmentally-friendly food system*. European Commission.

Gilliland, T.J., et al. (2009). Influence of testing procedure on evaluation of white clover (*Trifolium repens* L.). *Irish Journal of Agricultural and Food Research* 48, 227–242.

Gilliland, T.J., et al. (2020a). Opportunities and challenges for breeding perennial ryegrass cultivars with improved livestock production potential. *Irish Journal of Agricultural and Food Research*. https://doi.org/10.15212/ijafr-2020-0111

Gilliland, T.J., et al. (2020b). A proposal for enhanced EU herbage VCU and DUS testing procedures. *Grass and Forage Science* 75, 227–241.

Gilliland, T.J. (2022). *Where to next? Role of additional species in grass-white clover swards-production and environmental benefits*. Swards for the Future Conference & Workshop, 36.

Grange, G., et al. (2021). Plant diversity enhanced yield and mitigated drought impacts in intensively managed grassland communities. *Journal of Applied Ecology* 58, 1864–1875. https://doi.org/10.1111/1365-2664.13894

Grogan, D. and Gilliland, T.J. (2011). A review of perennial ryegrass variety evaluation in Ireland. *Irish Journal of Agricultural and Food Research* 50, 65–81.

Hanrahan, L., et al. (2017). PastureBase Ireland: a grassland decision support system and national database. *Computers and Electronics in Agriculture* 136, 193–201.

Hanrahan, L., et al. (2018). Factors associated with profitability in pasture-based systems of milk production. *Journal of Dairy Science* 101, 5474–5485, https://doi.org/10.3168/jds.2017-13223.

Hearn, C., et al. (2023). Seasonal variations in nutritive and botanical composition properties of multispecies grazing swards over an entire dairy grazing season. *Grassland Research*. https://doi.org/10.1002/glr2.12037.

Hearn, C. (2024). Personal communication.

Hennessy, D., et al. (2022). Better management of dairy nitrogen with grass-white clover swards. Swards for the Future Conference & Workshop, 10.

Höglind, M., et al. (2020). BASGRA_N: a model for grassland productivity, quality and greenhouse gas balance. *Ecological Modelling* 417, 108925.

Hoekstra, N.J., et al. (2023). Differences in grassland sward biodiversity and management regime lead to mixed effects on ecosystem services. *European Journal of Agronomy* 149, 126886. https://doi.org/10.1016/j.eja.2023.126886

Jouven, M., et al. (2006). Model predicting dynamics of biomass, structure and digestibility of herbage in managed permanent pastures. 1. Model description. *Grass and Forage Science* 61, 112–124.

Kennedy, E., et al. (2011). Restricting dairy cow access time to pasture in early lactation: The effects on milk production, grazing behaviour and dry matter intake. *Animal* 5, 1805–1813. https://doi.org/10.1017/S1751731111000826

Klump, K. and Fornara, D.A. (2018). The carbon sequestration of grassland soils – climate change and mitigation strategies. In *Sustainable meat and milk production from grasslands*, Vol 23, grassland Science in Europe pp. 509–535.

Komainda, M., et al. (2020). Drought tolerance is determined by species identity and functional group diversity rather than by species diversity within multi-species swards. *European Journal of Agronomy* 119, 126116. https://doi.org/10.1016/j.eja.2020.126116

Lee, J.M., et al. (2012). Perennial ryegrass breeding in New Zealand: a dairy industry perspective. *Crop and Pasture Science* 63, 107–127.

Lee, J.M., et al. (2015). Management strategies for chicory (Cichorium intybus) and plantain (Plantago lanceolata): impact on dry matter yield, nutritive characteristics and plant density. *Crop and Pasture Science* 66, 168–183. https://doi.org/10.1071/CP14181

Li, G. and Kemp, P.D. (2005). Forage chicory (Cichorium intybus L.): a review of its agronomy and animal production. *Advances in agronomy* 88, 187–222. https://doi.org/10.1016/S0065-2113(05)88005-8

Luscher, A., et al. (2014). Potential of legume-based grassland-livestock systems in Europe: a review. *Grass and Forage Science* 69, 206–228. https://doi.org/10.1111/gfs.12124

Martin, K., et al. (2017). Herbage dry-matter yield and nitrogen concentration of grass, legume and herb species grown at different nitrogen-fertiliser rates under irrigation. *Animal Production Science* 57, 1283–1288. https://www.publish.csiro.au/an/AN16455

McCarthy, B., et al. (2013). The effect of stocking rate and calving date on grass production, utilization and nutritive value of the sward during the grazing season. *Grass and Forage Science* 68, 364–377. https://doi.org/10.1111/j.1365-2494.2012.00904.x

McClearn, B., et al. (2020). The effect of perennial ryegrass ploidy and white clover inclusion on milk production of dairy cows. *Animal Production Science* 60, 143–147. https://doi.org/10.1071/AN18539

McDonagh, J., et al. (2016). Genetic gain in perennial ryegrass (*Lolium perenne*) varieties 1973 to 2013. *Euphytica* 212, 187–199. https://doi.org/10.1007/s10681-016-1754-7

McEvoy, M., et al. (2010). Effect of herbage mass and allowance on sward characteristics, milk production, intake and rumen volatile fatty acid concentration. *Grass and Forage Science* 65, 335–347. https://doi.org/10.1111/j.1365-2494.2010.00752.x

McEvoy, M., et al. (2011). Development and application of an economic ranking index for perennial ryegrass cultivars. *Journal of Dairy Science* 94, 1627–1639. https://doi.org/10.3168/jds.2010-3322

McSweeney, D., et al. (2018). Grasshopper: micro-sonic sensortechnology enables enhanced height measuremen by a rising plate meter. In: Proceedings of the 27th Genberal Meeting of the European Grassland Federation 'Sustainable meat and milk production from grasslands'. 17–21 June 2018, Cork, Ireland. *Grassland Science in Europe* 23, 879.

Morrisey, M. (2018). *Grass fed dairy* (Vol. 2023). Bord Bia. Accessed March 13, 2023 .https;///www.teagasc.ie/media/website/publications/2018/Grass-Fed-Dairy–Mary Morissery---National-Dairy-Conference-2018.pdf

Murphy, D.J., et al. (2021a). Evaluation of the precision of the rising plate meter for measuring compressed sward height on heterogeneous grassland swards. *Precision Agriculture* 22, 922–946. https://doi.org/10.1007/s11119-020-09765-9

Murphy, D.J., et al. (2021b). Utilising grassland management and climate data for more accurate prediction of herbage mass using the rising plate meter. *Precision Agriculture* (in press). https://doi.org/10.1007/s11119-020-09778-4

Noci, F., et al. (2005). The fatty acid composition of muscle fat and subcutaneous adipose tissue of pasture-fed beef heifers: Influence of the duration of grazing. *Journal of Animal Science* 83, 1167–1178.

NZMP. (2022). *Fonterra Grass and Pasture Fed Standard* (Vol. 2023). New Zealand Milk Products (NZMP).

O'Callaghan, T.F., et al. (2016). Effect of pasture versus indoor feeding systems on raw milk composition and quality over an entire lactation. *Journal of Dairy Science* 99, 9424–9440. https://doi.org/10.3168/jds.2016-10985

O'Donovan, M., et al. (2002). A comparison of four methods of herbage mass estimation. *Irish Journal of Agricultural and Food Research* 41, 17–27.

O'Donovan, M. and Delaby, L. (2005). A comparison of perennial ryegrass cultivars differing in heading date and grass ploidy with spring calving dairy cows grazed at two different stocking rates. *Animal Research* 54, 337–350.

O'Donovan, M., et al. (2016). Combining seasonal yield, silage dry matter yield, quality and persistency in an economic index to assist perennial ryegrass variety selection. *The Journal of Agricultural Science* 155, 556–568. https://doi.org/10.1017/S0021859616000587

O'Donovan, M., et al. (2020). Ruminant grassland production systems in Ireland. *Irish Journal of Agricultural and Food Research* 52. https://doi.org/10.15212/ijafr-2020-0118

O'Loughlin, J., et al. (2016). *Is Ireland ready for more milk? Page 37 in Teagasc Heavy Soils Dairy Programme Review*. Teagasc.

Orr, R.J., et al. (1990). Sward composition, animal performance and the potential production of grass/white clover swards continuously stocked with sheep. *Grass and Forage Science* 45, 325–336.

Peyraud, J.L., et al. (2009). Potential food production from forage legume-based-systems in Europe: an overview *Irish Journal of Agricultural and Food Research* 48, 115–135.

Patton, D., et al. (2016). Effect of stocking rate on milk and pasture productivity and supplementary feed use for spring calving pasture fed dairy systems. *Journal of Dairy Science* 99, 5904–5915. https://doi.org/10.3168/jds.2015-10305

Ruelle, E., et al. (2018). Development of the Moorepark St Gilles grass growth model (MoSt GG model): a predictive model for grass growth for pasture based systems. *European Journal of Agronomy* 99, 81–90.

Ruelle, E., et al. (2021). The MoSt grass growth - Predicting grass growth live on farm. In Proceedings Moorepark 21. Irish Dairying-Delivering Sustainability, pp. 70–72.

Ruelle, E., et al. (2023). Update on the MoSt Grass growth prediction model. In Proceedings Moorepark 23. Irish Dairying-Securing a Sustainable Future, pp. 84–85.

Stienszen, M.W.J., et al. (2018). Amazing grazing; Feed wedge and cutting window for getting systems with high levels of supplementation. In Grassland Science in Europe, Vol 23 - Sustainable meat and milk production from grasslands, p. 819.

Smit, H., et al. (2008). Spatial distribution of grassland productivity and land use in Europe. *Agricultural Systems* 98, 208–219.

Thomas, C., et al. (1981). Beef production for silage. 2. The performance of beef cattle given silages of either perennial ryegrass or red clover. *Animal Production* 32, 149–153.

Timlin, M., et al. (2023). Impact of varying levels of pasture allowance on the nutritional quality and functionality of milk throughout lactation. *Journal of Dairy Science* 106(10), 6597–6622. https://doi.org/10.3168/jds.2022-22921

Timlin, M., et al. (2024). Pasture feeding improves the nutritional, textural and techno-functional characteristics of butter [Manuscript submitted for publication].

Tubritt, T., et al. (2020). An investigation into the grazing efficiency of perennial ryegrass varieties. *Grass and Forage Science* 75. https://doi.org/10.1111/gfs.12481

Tunon, G.E. (2013). *Improving the use of perennial ryegrass swards for dairying in Ireland*. PhD thesis. Massey University, Palmerston North, New Zealand.

Ulyatt, M.J. (1970). Evaluation of pasture quality under New Zealand conditions. *Proceedings of the New Zealand Grassland Association* 32, 61–68.

Vi, C., et al. (2023). Effective proportion of plantain (Plantago lanceolata L.) in mixed pastures for botanical stability and mitigating nitrous oxide emissions from cow urine patches. *Agronomy* 13, 1447. https://doi.org/10.3390/agronomy13061447

Welten, B.G., et al. (2019). Efficacy of different temperate pasture species to reduce nitrogen leaching from cattle urine applied in different seasons: a soil lysimeter study. *Soil Use and Management* 35, 653–663. https://doi.org/10.1111/sum.12512

Wims, C.M., et al. (2010). Effect of pre-grazing herbage mass on methane production, dry matter intake and milk production of grazing dairy cows during the mid season period. *Journal of Dairy Science* 93, 4976–4985.

Chapter 6

The use of precision technologies in grassland management

Laurence Shalloo and Paula Molina Palma, Teagasc, Ireland; and Deirdre Hennessy, University College Cork, Ireland

1 Introduction

Grassland, or pasture, covers a large proportion of the global land area. Provision of food from grassland-based production systems plays an important role in global food security, through conversion of a low-cost human inedible food source (grassland) by ruminants into high-quality food for human consumption. Agricultural reforms and policy, increased demand for feed and

http://dx.doi.org/10.19103/AS.2025.0154.12

food, and environmental considerations and the requirement to reduce food-feed competition suggest that self-sufficient farming using locally produced or on-farm resources such as grazed grass will become more important in the future (e.g. Hofstetter et al., 2014). Dillon et al. (2005) and Hanrahan et al. (2018) showed that key drivers of profitability at farm level in temperate regions are associated with the proportion of purchased feed in the diet of the animal and with increased grass utilisation. It has been estimated that the demand for animal-derived protein may double by 2050 (Henchion et al., 2017). Maximising production efficiencies while minimising environmental and ecosystem impacts will be crucial to meet this demand, while also contributing to the achievement of climate and biodiversity goals.

In this chapter, we review the use of precision technologies in pasture-based systems as a mean to achieve these sustainability goals. Most of the examples presented in the chapter are from Ireland, as one of the few countries in Europe with predominantly grassland-based production systems.

2 Grassland-based milk production systems

Grassland-based milk production systems are common in temperate regions including, but not limited to, northwest Europe, New Zealand, parts of the United States of America, parts of southern Australia, and parts of South America where there is a long grass-growing season. In these regions, grassland-based systems have advantages over high input systems (Shalloo et al., 2004; Dillon et al., 2005), and are associated with increased product quality (O'Callaghan et al., 2016), improved animal welfare (Wagner et al., 2018), increased labour efficiency (Dillon et al., 2005; O'Brien et al., 2012) and increasing global food security (Laisse et al., 2018, Hennessy et al., 2021) relative to systems of production where the animal is eating feed that could be consumed by humans.

The potential of grass-based systems is often restrained by a lack of expertise and path dependency over the last century, which has driven most European farms towards indoor, all year-round calving systems. In temperate regions, pasture is the lowest cost feed source for milk and meat production. Finneran et al. (2010) reported the cost of grazed pasture as a feed source for livestock relative to pasture silage and concentrate to be 1:1.8:2.4, respectively. The efficiency of grassland-based systems has been found to be positively associated with grass utilisation, grazing season length and pasture management, and negatively associated with supplementation level (Shalloo et al., 2004; Macdonald et al., 2010; Läpple et al., 2012; Ramsbottom et al., 2015; Hanrahan et al., 2018).

For grassland farmers relying on pasture as their primary feed source to optimise their production systems, there is a requirement for accurate real-time measurement of pasture herbage mass and quality to optimise grazing and nutrition management. In Ireland, every 1000 kg DM increase in pasture utilisation per hectare was associated with increases in net farm profit of approximately €173 per hectare (Hanrahan et al., 2018). The availability of real-time data can increase the precision in decision making around grass feed allowance, grass quality, supplementation, fertilisation, feed budgeting, and silage harvesting.

Increasing the efficiency and sustainability of grass-based systems is thus desirable to make them even more competitive, especially in countries like Ireland and New Zealand where the grassland-based system is currently a niche production model globally (Shalloo et al., 2019). The availability of precision management systems and tools can contribute to the development of sustainable and resilient grassland-based systems and help bring those systems to the next level in overall sustainability. Maximising the utilisation of grazed grassland through improved management, which incorporates precision technologies, can contribute to increased feed self-sufficiency and farm profit, while reducing nutrient loss and GHG emissions from the overall system.

The development of precision technologies in agriculture is increasing with automation and the collection and analysis of big data. Technologies specific to grassland-based systems vary, with many technologies now available around livestock management, fertiliser management, and grassland management. Technologies designed specifically for grassland management to increase pasture utilisation are limited. Technologies for the production of other crops such as maize and soybean are more common; for example, sensors are used to measure the microclimate, spectral characteristics, and small-scale differences in soil fertility to help improve crop management and productivity (Wolfert et al., 2017). Sensors developed with a focus on improving pasture growth and management are still in development, with few currently commercially available. However, there is real potential in this area with the advent of artificial intelligence, machine learning, and its integration in grassland-based decision making. In this chapter, we review the use of precision technologies in pasture-based systems as a means to achieve these sustainability goals. Most of the examples presented in the chapter are from Ireland, as one of the few countries in Europe with predominantly grassland-based production systems.

3 Traditional grassland measurement methods

Implementing good grassland management is crucial to achieve high levels of grassland production and optimum utilisation through grazing while reducing the levels of concentrate feeds required in the system. Measuring growth rates and feed supply provides farmers with important knowledge of feed availability and allows them to make informed decisions regarding grass feed allocations and supplementation, as well as management of surplus or deficits of grass on the grazing platform. Knowledge of grass supply, as well as stocking rate, feed demand, supplementation, and annual and seasonal grass growth allows farmers to develop feed budgets, grazing plans, nutrient management plans, and reseeding/grassland renewal.

Grassland measurement has been implemented for many decades to greater or lesser extents on commercial farms. Quantifying the capacity of farmland to grow grass is an important consideration in promoting sustainable grazing systems. Measuring weekly, seasonal, and annual grass production allows farmers to determine appropriate stocking rates and grassland management strategies (O'Donovan et al., 2021). Measurement generally requires an action to be performed by the farmer and requires skill, knowledge, and confidence in the method used. Typical grassland measurement methods for grazing swards include the rising platemeter (RPM), the cut and weigh method, or visual assessment.

3.1 Platemeter

The RPM is the most established, non-destructive means of estimating grass yield (O'Donovan et al., 2002). It can be used by a wide range of users including farmers and advisors/extension officers. The RPM records sward height and combines this with sward density to estimate herbage mass. The RPM consists of a weighted disc attached to a scaled staff. This is dropped onto the sward and the plate rises to give a compressed sward height. Usually approximately 50 readings at random locations across a paddock will give reliable information. Sward density for a given grassland species/type is usually available via research but the accuracy can vary across seasons, sward species composition, weather conditions, growth stage and indeed sward height. Herbage mass is estimated using the following equation:

$$Herbage\ mass\ (kg\ DM/ha) = \begin{bmatrix} Compressed\ sward\ height\ measured\ with\ the\ RPM\ (cm) \\ -Target\ residual\ sward\ height\ (cm) \end{bmatrix} \times sward\ density\ (kg\ DM/cm)$$

where target residual sward height is the target post-grazing sward height in cm above ground level and sward density is the kg DM in each 1 cm of the sward (often defaulted to 250 kg DM/cm in temperate grassland).

The use of the RPM requires minimal training, and a large number of samples can be recorded in an evenly distributed manner throughout a paddock.

3.2 Cut and weigh

Herbage mass can be estimated by farmers and others using the cut and weigh method (O'Donovan et al., 2002). This involves selecting a random area within a paddock and cutting all of the material within a quadrat of known area, e.g. 0.5 m × 0.5 m at a desired height above ground level, e.g. 4 cm, using shears. All of the material is weighed, and the herbage mass is estimated on a per hectare basis using the following equation:

$$\begin{aligned} \textit{Herbage mass}\left(\textit{kg DM / ha}\right) = (&\textit{weight of grass harvested within the quadrate} \\ &\times \textit{dry matter content}\left(\textit{g / kg DM}\right) \\ &\times\left(10{,}000\ m^2 \div \textit{area of quadrate}\left(m^2\right)\right) \end{aligned}$$

Dry matter content is usually available from literature, advisory offices, and extension services.

3.3 Visual assessment of herbage mass

Visual estimation is a non-destructive method of grass measurement (O'Donovan et al., 2002) whereby a farmer observes the paddock/field and estimates the herbage mass (kg DM/ha). Farmers usually 'train their eye' through practising cutting or weighing or using the RPM, both described above, or through shadowing an experienced operator. This is a low-cost means of grassland measurement; however, it is subjective, requires a high level of skill, and must be practised frequently.

4 Precision technologies for grassland management

There are a number of precision technologies that can be used for grassland measurement, some commercially available, and some at lower technology readiness level (TRL)'s in development such as camera technologies combined with machine learning will take some time before they are deployed at farm level.

Digitally enabled rising plate meters streamline and automate some of the data collection process associated with pasture measurement through collecting data (sward height and/or herbage mass) required to calculate pasture availability on farm (French et al., 2015). These have an added benefit in that the skill and experience required is much less than when compared to visual assessments, while avoiding the manual tasks associated with rising plate meters. Another approach is to measure the height of the pasture canopy without a plate resting on the pasture, but this is less sensitive to pasture density and thus less sensitive to biomass. Distance measurement in both plate and canopy height is usually done via an ultrasonic sensor which measures the height of the canopy itself or of a plate resting on the canopy (Moeckel et al., 2017). The Grasshopper (www.moregrass.ie), a recent iteration of the RPM developed in Ireland, uses a Global Navigation Satellite Systems integrated ultrasonic sensor to record the height of the rising plate when it is dropped in the sward. This type of RPM has rapid data processing capabilities via automated links to online support tools such as PastureBase Ireland (www.pbi.ie) and its ability to geo-tag measurement data. Using the electronic RPM allows faster and more efficient grass measurement without the need for manually recording data and time-consuming manual calculations.

There is increasing momentum around spectral approaches to grassland measurement, be it handheld, drone or satellite-based (Sibanda et al., 2016). Spectral analysis can be completed with portable, aerial, or satellite-based systems and a common approach is to develop wavelengths to create indices such as Normalised Difference Vegetation Index which estimates if vegetation is present or not by assessing the ratio of red and near-infrared wavelengths. The challenges associated with spectral approaches include background soil effects, atmospheric effects, grazing impact and heterogeneity of species, and variation in pasture growth stages/proportion of senescent material in the canopy which disproportionately influence spectral data (Moeckel et al., 2017). Most studies have focused on measuring pastureland biomass in tropical savannahs via remote satellite sensing with less focus on temperate pasturelands (Moeckel et al., 2017).

Moeckel et al. (2017) reported combining an ultrasonic sensor with handheld hyperspectral sensor data and satellite-derived spectral data to measure canopy height. They found that the best predictions of biomass were early in the grazing season when pastures were more homogeneous. As the grazing season continued, issues such as pasture refusal post grazing resulted in variability in the sensor outputs, which could not be accounted for when predicting biomass. The R^2 values achieved with biomass, however, indicated that further development is required before practical application in pastureland management is possible (maximum $R^2 = 0.52$).

Askari et al. (2019) used an unmanned aerial vehicle to gather spectral data to assess pasture quality and achieved an R^2 of between 0.77 and 0.83 relative to chemical analysis. There has been some progress using satellite data to estimate pasture biomass using Synthetic Aperture Radar. One study of Irish pasture found that grass biomass could be predicted with an R^2 of up to 0.75 based on this approach (Ali et al., 2017). However, predictions with that level of accuracy were only achieved periodically when weather conditions were clear.

Determining herbage mass and sward botanical composition using images captured by camera and mobile phone has been investigated by Albert et al. (2021, 2022a, 2022b). Those authors reported an accuracy in terms of species composition of grass-white clover swards of 95%, and a similar accuracy in predicting herbage mass. Using a small dataset, the algorithms developed for the camera/mobile phone transferred quite well to images captured by drone (Albert et al., 2022c). This capacity creates the opportunity to both provide measurements but also to upload that information remotely to a centralised database and system that can then be integrated with other data and can be used to make informed decisions.

5 Measuring grass quality

Grass quality is traditionally quantified by collecting grass material in the field, processing it and subjecting the material to wet chemistry analysis. Processing can take a number of forms including drying at a range of temperatures (e.g. 95°C, 60°C, 40°C, or freeze drying) and then milling the sample to a desirable material size. Various laboratory methods are then applied to determine DM content, digestibility (e.g. Morgan et al. (1989)), crude protein content (e.g. AOAC method 990–03 (AOAC, 1990)), and fibre content (e.g. Van Soest et al. (1991)).

The advancement of near infrared spectroscopy (NIRS) technology allows for a more rapid screening of grassland material for various quality parameters, removing the requirement for wet chemistry. Sample processing is of course still required. Near infrared spectroscopy analysis measures the absorption rates of low energy infrared light radiation (700–2500 nm) within matter, which are then used to quantify the chemical constituents of the matter by means of empirical modelling using calibration equations, e.g. Burns et al. (2010), which must be updated on an on-going basis.

There is a requirement to develop NIRS analysis of fresh grass samples to predict herbage quality parameters, thus delivering results in real time and improving the capacity for decision making. However, NIRS analysis of fresh grass can be restricted by the presence of moisture/water on the plant, which can interfere with the prediction of quality parameters (e.g. Alomer et al., 2003; Feuerstein and Paul, 2007). NIRS has been used to predict pasture crude protein and DM content in a range of grassland swards with varying levels of accuracy (Alomar et al., 2009; Lobos et al., 2019; Dale et al., 2017; Parrini et al., 2019).

Developing a portable NIRS that would allow in-field real-time estimation of herbage quality in a non-destructive manner is highly desirable in terms of removing the requirement to collect grass material and transfer it to the laboratory as well as any requirement for pre-processing. However, the availability of such a device for grassland management at farm level is likely some time away as there are considerations around the cost, requirement for chemometric modelling techniques, and potential for errors and restrictions associated with the presence of moisture/water on the grass material, as well as restrictions associated with light noise, particle size, and wavelength range (Dos Santos et al., 2013). Constraints associated with the development of calibration data also limit the capacity for the development of portable NIRS for grassland.

Murphy et al. (2021) investigated the development of NIRS calibrations to predict grass DM content and crude protein content of fresh grass in Ireland. Grass samples were scanned using a FOSS 6500 spectrometer (FOSS-NIR System DK, Hillerød, Denmark) at 2 nm intervals in the range of 1100 ~ 2500 nm and absorption was recorded as log 1/rReflectance. Reference wet chemistry analysis was undertaken for both parameters and a partial least squares regression was applied to calibrate the spectral data with the wet chemistry data. Dry matter content was predicted with an R^2 of 0.86 with a standard error of prediction of 9.46 g/kg, while the R^2 for crude protein was lower at 0.84 with a standard error of prediction of 20.38 g/kg. While the prediction of crude protein content was not sufficient to replace wet chemistry or NIRS analysis of dried milled samples, it could potentially be used to provide approximate levels or ranges or indeed ranking of crude protein content or grass nitrogen content, providing useful information for farmers.

Development of spectral analysis using a platemeter, or similar, to assess pasture quality is a promising research area (Sibanda et al., 2016) that might provide estimates of herbage quality parameters in a quick and efficient alternative to lab-based assessments. This precision technology is at a low TRL and no commercial applications are currently available.

6 Data aggregation and processing

A database is a structured set of records acquired from a range of sources which is usually held on a server and is accessible in various ways. There are many types and structures of databases, and they can store vast quantities of data collected across a whole range of different aspects of agriculture. Most of this data resides in databases specifically designed to facilitate one form of functionality (e.g. regulation, disease epidemiology, sustainability, etc.). While that data fulfils its original purpose, linking it with other data collected for other purposes could dramatically increase the value of the data, e.g. sustainability assessment (O'Brien et al., 2014) or verification of grass-fed numbers (proportion of the diet originating from pasture) (O'Brien et al., 2018). Examples may include the use of animal identification information to help populate sustainability models or the use of activity data as an animal welfare indicator. Some databases are designed to monitor commercial farm data and, in general, are designed to fulfil a specific farm business requirement. The development of the internet allows these databases to be stored remotely and linked with other databases.

Raw data is generally of little use, so the processing and summarising of data is a key aspect of precision technologies, and these processes are generally completed by algorithms (for review, see Valletta et al., 2017). At its most basic level, algorithms can generate coefficients that can then be used to report metrics (e.g. changes in steps taken per day). Building on this and including model results, automated inferences can be made based on the data. This could, for example, include predicting grass growth. At a more developed stage, a recommendation to make decisions around fertilisation or silage harvesting could be made, taking into account factors such as weather forecast, current grass supply, and recent fertiliser application. Using live data must be the key goal of precision technology adoption in grassland-based systems.

7 Interoperability and linkages

Large quantities of data are generated on farms and stored locally or aggregated by government agencies and companies. In most cases, those databases were developed with limited applications in mind. However, there is a magnitude of potential uses for the data.

The ability of a sensor or system to work with other measurements or other *in-situ* technologies is often crucial to making them viable investments. Collaboration, protocols for interoperability, data formatting standards, open source, and open data approaches may be required to realise the full potential of precision technologies in grassland-based systems. To improve linkages between databases, there is a requirement to understand the design of existing databases including the database platform and infrastructure. The focus should centre on strategies to develop linkages between different databases. This process should identify issues around data integration and ensure that any future developments in data collection can be undertaken using a standard methodology. The integration of data across relevant databases adds value to the analysis that may be completed as not all the pertinent data resides in a unique database.

Overcoming the barriers and challenges associated with data and intellectual property and business models will be crucial. The challenges include data interoperability, sharing/access, proprietary systems, and intellectual property concerns. In response to these challenges, major vendors have responded by creating company-specific ecosystems with good intracompany interoperability. This approach includes bundling/cross-selling of products. It is also extremely important to consider data ownership, data security, and data business models, which will be required to ensure there is trust among the users.

8 Grassland management databases

An example of a grassland research database is PastureBase Ireland (PBI; Hanrahan et al., 2017). It builds on the concept of citizen science (Conrad and Hilchey, 2011). Citizen science has proved highly successful in the past in many countries, forming the bedrock of biological recording in various large research projects, particularly in ecology and environmental sciences. In the case of PBI, grassland farmers are the citizen scientists collecting information related to their grassland management and utilisation, such as grazing dates, weekly herbage mass, fertiliser application, etc., and inputting to PBI. The data stored in this database are fuelling on-farm research programmes in Ireland and are informing researchers, advisors, farmers, and other stakeholders about grassland management and grazing management across the country. One example includes on-farm grass cultivar evaluation. This function is being achieved through the creation of discussion forums within the PBI system, which allows for the roll-out of research and technology initiatives more easily and increases farmer engagement, especially through linkages to social media.

Data regarding fertiliser use on farm is generally stored with the merchant where the fertiliser is purchased. Protection of this data is important for many farmers and generally, they are unwilling to share it, harbouring concerns that any disclosure of this data might be used against them (e.g. EU Nitrates regulations, land values). The potential benefits of centrally collecting and storing that data for sustainability assessments or at individual paddock level to determine fertiliser response functions are significant. Ideally, if the farmer uses an app linked to a smartphone to record fertiliser information (spreading dates, quantities), this information can be automatically integrated with a grassland database like PBI. This simplifies recording and improves the utility of the data for the farmer.

9 Using precision technology to enhance on-farm decision support and automation

The primary objective of using precision technology is to inform humans to facilitate automated decision-making on farm. In many cases, human discernment and knowledge cannot be substituted by machines, so machine autonomous decision-making is not always possible. Precision technologies can be categorised by the number of stages of decision-making automated for the farmer. Rutten et al. (2013) described decision-making as a way to characterise how precision technologies are integrated into decision-making on farm. In many circumstances, automated decisions may be desirable where the operator's decisions may not practically be able to fully consider and interpret the data, and/or automation saves the farmer time or labour. As the quantities and types of data available in addition to the multiplicity of automated decision implementation methods develop, this will become more relevant.

In systems that rely on the operator to make and implement decisions, information is usually made available to them via an interface, report, or notification. The format and timeliness of how the information is provided affects the subsequent decision-making process. Some systems have algorithms and interfaces developed specifically to support decisions. Others only collect and report data and/or cannot link different data sources, inhibiting aggregation of relevant information in one view or report.

To date, literature regarding decision making with precision technologies in grassland systems is sparse but is growing in recent years. The robustness of data collected by precision technologies including sensors is still perceived as a priority and thus, most research focuses on sensing accuracy and systems validation, rather than the development of decision support tools or systems (DSS) that utilise the data. Decision making remains pivotal to the successful application of precision technologies (Hostiou et al., 2017). The widespread proliferation of artificial intelligence technologies will have a dramatic effect on the potential of DSS systems into the future.

10 Decision support systems (DSS) for grassland management

Some grassland management DSS are available for farmers who utilise grassland as the primary feed source, facilitating improved grassland management. These DSS can deliver tangible benefits in terms of pasture supply, quality, and utilisation (Creighton et al., 2011; Hanrahan et al., 2017) and ultimately profitability (Shalloo et al., 2011; Hanrahan et al., 2018). Grassland management tools can include the grass wedge, grass/feed budgets, and spring and autumn rotation planners which assist farmers with short to medium-term grassland management (Macdonald et al., 2010). Some DSS use historical grass growth data to estimate grass production capacity of paddocks/the farm and facilitate long-term management decisions such as paddock reseeding and determining the most appropriate farm stocking rates; both factors strongly linked to the economic performance of grass-based dairy production systems (Kennedy et al., 2007; Shalloo et al., 2011; McCarthy et al., 2013).

Many DSS for grassland management now exist in the form of web-based applications. These provide greater flexibility to the user; web-based architecture enables the collation of large quantities of data in central databases which extend the functionalities of DSS beyond farm level. Data collected on a wide number of farms can be used to benchmark performance of a single operation, but also for research purposes and policy formulation (Hanrahan et al., 2017).

11 Virtual fencing

Electric fences had a significant impact on grazing management in grassland-based systems when first introduced. Electric fencing provides a manual approach to providing an allocation of feed to the animal through allocating an area that animals can reside in for a period of time. A virtual fence (VF) is an apparatus serving an enclosure, a barrier, or a boundary without a physical barrier (Umstatter, 2011) that can be adjusted remotely. Virtual fencing offers a way to optimise grazing management of available land through linking feed available to animal performance and to the utilisation of the material offered. It has been described as having a high level of efficacy in a range of scenarios (Campbell et al., 2023). In the future, the potential to manage and herd animals creates a significant opportunity to optimise grazing management (Anderson et al., 2009, Doniec et al., 2010, Umstatter, 2011). Virtual fencing technologies use a warning cue before producing additional stimuli. Lee et al. (2009) reported in a study with five heifers that associative learning by cattle enabled effective and ethical VFs. Stimuli were delivered by remote control to GPS collars on the heifers. This study demonstrated that the appropriate use of an audio cue is an effective conditioned stimulus for the VF of cattle.

Research on virtual fencing started about 50 years ago with prototypes working 30 years ago. However, VF required two further developments: (a) a system that facilitated reasonable battery performance coupled with energy harvesting; (b) the provision of a system that animals can easily understand. During the last few years, VF technical maturity emerged from research to everyday farm conditions due to enhanced low-voltage microcomputer power, remote wireless communication, and improved battery storage, as well as machine learning technologies to optimise battery use and power. The technology is now here that would allow a collar to be put on an animal, and between battery optimisation and energy harvesting, that collar could stay on the animal for the full year. The technology is available to herd animals from the milking parlour and to the parlour, as well as to change paddocks at times other than milking time. The technology is used to individually feed animals based on their potential performance as well as their behaviour attributes. The technology is being used with grassland measurement and management software to fully optimise day-to-day grazing management. This requires measurements of pre and post herbage mass, as well as the farm wedge of the spring/autumn rotation planner. This technology can be integrated with grassland management and measurement and has the potential to revolutionise grassland management. The technology in this space is moving extremely quickly, and AI will play a huge role in this space in the future.

12 DSS for sustainability assessment

Sustainability assessment databases and DSSs will be key features of grassland management tools in the future. The key component here will be the utilisation of existing databases and data sources that have been collected and collated and that have verifiable data and linking this data with additional data that can be collected as part of any sustainability assessments. These sources of information can then be integrated into models that will utilise the information to quantify sustainability metrics (Herron et al., 2022).

A new DSS developed in Ireland called AgNav (www.agnav.ie) has been designed with the focus of delivering a centralised system that collates all sustainability information and that sits on top and utilises existing data infrastructure and databases. The system is currently generating robust farm and output-based assessments on GHG emissions, ammonia, and nutrient surpluses across dairy and beef systems. As it develops, it will include other enterprises (e.g. arable and sheep) and it will include both GHG emissions and soil carbon removals, and will ultimately include new sustainability metrics that can be captured with newly developed technologies and methodologies. The outputs of the models and applications will be used as industry sustainability metrics and benchmarks. The metrics, for example will include the carbon footprint per unit of product, ammonia emissions per unit of product, the levels of feed/food competition, biodiversity, etc.

13 Precision technologies for dairy herd management

Grassland-based dairy farmers can also benefit from the use of other precision technologies (besides precision technologies for grassland management) in the areas of reproductive, milking and calf management. These precision technologies, also called precision dairy technologies, allow continuous monitoring of grazing cows' behaviour and health and have the potential to improve animal health and welfare, increase farm efficiency, and reduce farm costs (Heins et al., 2023). This section explores the use of precision technologies for dairy herd management, including reproductive, milking, and calf management technologies.

13.1 Precision technologies for reproductive management

Digital reproductive management technologies are precision technologies that aim to improve the reproduction of dairy cows. Specifically, they are used to facilitate oestrus detection, insemination, and calving of the dairy herd. Good dairy cow fertility is especially important in seasonal calving pasture-based dairy farms where a compact period of calving is needed so dietary demand of peak lactation overlaps with optimal pasture growth (Shalloo et al., 2021). Shalloo et al. (2014) estimated that the costs of reproductive inefficiency in pasture-based dairy farms ranged from €3.86 per day (suboptimal calving dates) to €12.44 (increased replacement rates) per cow per year. By assisting and easing herd reproduction-related tasks, reproductive management technologies could have a significant role in reducing labour, increasing fertility status, and increasing farm profitability in grassland-based dairy herds.

An example of reproductive management technologies is the individual cow activity sensor. An animal health sensor is defined as 'a device that measures a physiological or behavioural parameter of an individual cow and provides information that can be utilised with other information and in combination with an algorithm to generate automated detection of changes in the condition that could be related to a health event and requires action on the part of the farmer' (Steeneveld et al., 2015).

According to Heins et al. (2023), more than 50 wearable technologies have been developed and marketed for dairy cattle, some of which are worn by the cow and some are inside the cow. These sensors can collect cow behaviour data such as changes in activity, eating time, ruminating time, laying, and standing time. Consequently, the data is processed through algorithms and can be categorised into specific behaviours or health and oestrus alerts that can be viewed on a computer system, a website or a mobile device by the farmer (Heins et al., 2023).

It has been reported that the main reason for the purchase of wearable sensors by grassland-based dairy farmers is for heat detection (Heins et al., 2023). This is because oestrus detection is key to improving submission rates (the proportion of cows which are eligible to breed that are actually bred), which is particularly important when trying to improve the reproduction status of the dairy herd. It is usually performed by visual observation with the help of heat detection aids (visual detection aids), which can be particularly time-consuming, repetitive, and require considerable skill and experience to attain a reasonable level of efficiency (Adenuga et al., 2020). Hogan et al. (2022) estimated that approximately 10% of the labour involved in a commercial pasture-based dairy farm is destined for heat detection, artificial insemination, and calving cows, as well as animal health monitoring, among other cow care tasks.

In a study by Thomas et al. (2019), sensors had an oestrus detection rate that ranged from 62.4% to 91%, compared to visual observation that has approximately 72% oestrus detection rate in pasture-based dairy systems. By improving oestrus detection, sensors are expected to shorten calving intervals (Rutten et al., 2014) and to reduce the average days to first service of the herd with a consequent increase in calving rates, milk production (Auldist et al., 2007), and the financial returns of the farm (Adenuga et al., 2020). Shalloo et al. (2014) estimated that a 1% increase in a 6-week calving rate is associated with an increase in farm profitability of €9.26/cow per year and €3.51/heifer per year. Additionally, by reducing the need for an operator to visually detect oestrus, sensors can reduce labour requirements. Increased milk production and reduced labour input would also increase the profitability and overall sustainability of dairy farms.

13.2 Precision technologies for health management

In addition to heat detection, wearable sensors can be used for dairy health management (detection of diseases such as lameness and mastitis). Accelerometers that are used to monitor locomotion activity (e.g. lying time) and detect lameness in pasture-based dairy herds are usually attached to the cow's leg. This is particularly relevant in pasture-based dairy farms where cows may have to walk long distances between the milking parlour and pasture (Shalloo et al., 2021). Early lameness detection using accelerometers could reduce the time from onset to treatment, preventing cases from becoming severe, speeding recovery, and increasing production and welfare (O'Leary et al., 2020). If we consider the significant economic losses (Dolecheck and Bewley, 2018) and environmental impacts (Chen et al., 2016) associated with lameness, the use of accelerometers could eventually help to improve the economic and environmental sustainability of dairy farms. However, studies evaluating the performance of accelerometers on lameness detection have been unable to achieve high accuracy in combination with practical implementation in an on-farm commercial setting (Borghart et al., 2021). Significant progress is still required to have a system with sufficient sensitivity (the ability of a device to identify an individual with disease as positive) and specificity (the ability of a device to identify an individual who does not have a disease as negative) for automatic lameness detection. The best performance results in pasture-based dairy settings, according to accuracy measures, are achieved using data from accelerometers in combination with production and other animal data (Borghart et al., 2021).

In addition to heat and lameness detection, different sensor systems are commercially available to detect mastitis and metabolic disorders in dairy herds (Hogeveen et al., 2021). Mastitis sensors provide data regarding the composition of milk or the physiological status of a cow, which is then processed through an algorithm to identify a deviation from normality that could be predictive of mastitis (alert). Most sensor system alerts need then to be confirmed by either clinical examination or secondary testing. Mastitis sensors have shown different detection performance levels, but so far, none of the published clinical mastitis detection models have met the target of 80% sensitivity and 99% specificity (Hogeveen et al., 2021). Khatun et al. (2018)

found that in an automatic milking rotary, detection performance for clinical mastitis achieved a sensitivity of 90% and a specificity of 90%. In automatic milking systems with online somatic cell count measurements, detection performance achieved a sensitivity of 80% and specificity of 90% (Dalen et al., 2019). Because of the graduality of mastitis, sensor performance will never be perfect. Therefore, it has been proposed that for better mastitis management, instead of using one alert algorithm, sensors should aim to monitor four different mastitis situations (cows needing immediate attention, cows not needing immediate attention, cows needing attention at drying off, and monitoring udder health at the herd level) (Hogeveen et al., 2021).

13.3 Precision technologies for calf management

Rearing pre-weaned calves is one of the most challenging tasks on the farm, especially in compact calving systems such as those in pasture-based dairy farms (Gleeson et al., 2008). Moreover, calf care duties have been found to be (after milking) one of the most time-consuming tasks in pasture-based dairy farms. Specifically, Hogan et al. (2022) estimated that milking and calf care tasks account for 48% of all labour input in the spring period. Since calf feeding is one of the most important tasks within calf care, the method of calf feeding (e.g. manual milk teat feeding, bucket milk feeding, automatic calf feeding) may have a significant impact on labour input related to calf care (Gleeson et al., 2008). Automatic calf feeders are computerised calf feeders that are capable of feeding and monitoring the performance of calves. In addition to labour saving, they can be used to detect illness in calves and track animal growth performance. A case study conducted by Hogan et al. (2023) reported that dairy farms that installed automatic calf feeders improved their calf care labour efficiency by 23%. Sinnot et al. (2021) compared the effects of manual and automatic milk feeding methods on calf health, behaviour, growth, and labour. They found no differences in days from birth to weaning, weight at weaning, average daily gain weight or in any of the behavioural patterns, but they did find that total labour input/day was consistently less for automatic feeders compared to manual feeder systems (1 minute and six seconds per calf per year). Therefore, labour saving is a distinct advantage of automatic calf feeders when compared to traditional manual calf feeders.

13.4 Other precision technologies used in herd management

Another technology used for herd management is the automatic drafting gate. Automatic drafting gates are automated systems that allow cows to be automatically drafted before or after milking. This can be done for management reasons (for example, group drafting for routine vaccinations, drying off, insemination) or to separate cows from the herd with health issues. By reducing the need for operators to manually draft the herd, automatic drafting gates are expected to improve labour efficiency (Edwards et al., 2015). They can be used in association with individual cow activity sensors to automatically identify and separate each cow for insemination or treatment. Another technology that allows for reduced labour requirement is the back latch, which essentially opens the wire to allow animals to be at the milking parlour for the start of milking.

Among dairy farm duties, milking has been found to be the most time-consuming task in pasture-based dairy settings (Hogan et al., 2022). According to Deming et al. (2018) milking and its associated tasks, such as pre- and post-milking herding, and post-milking washing of the equipment, account for 33% of total annual farm labour in Ireland. Edwards et al. (2020) estimated that at peak lactation, milking represents 43–58% of a conventional 40-hour work week in New Zealand. Considering the importance of milking on labour demand, the adoption of in-parlour automation technologies such as automatic milk meters, automatic cluster removers, automatic parlour feeders, and automatic milk washers appears to be a key strategy to reduce workload and to improve work-life balance. Additionally, they are expected to provide economic benefits in terms of labour cost savings and increased production per cow (Wade et al., 2004). Pasture-based dairy systems in New Zealand have increasingly relied on technologies to deal with the industry labour issues to improve labour productivity with a particular focus on milking tasks (Deming et al., 2018). Precision technologies that can help reduce the labour requirements around the milking and herding process will have real benefits in the areas of social and economic sustainability.

14 Case study

As alluded to, many of the examples provided relate to pasture-based dairy production systems in Ireland. Technically efficient dairy farmers use a wide range of technologies to enhance their management of their farm business. They can optimise grass production through grassland measurement and the use of Pasture Base Ireland (Hanrahan et al., 2017), nutrient management planning, weather forecasting, and grass growth predictions from the MoSt Grass Growth Model (Ruelle et al., 2018). Having grown the grass, they require animals to graze it. Using the Economic Breeding Index (Veerkamp et al., 2002) and technologies like automatic heat detection and sexed semen allows them to select the right cow for their system, while adding value to the calves from the system. Systems within the milking shed that allow selection of cows through drafting gates, collars, and indwelling sensors to provide information on cow health including rumen function, sensors to assist with animal breeding, and indeed to provide information with regard to the progress of calving ensure a herd that can efficiently and effectively utilise grazed grass. Within the milking parlour, sensors aid in ensuring efficient milking procedures can be implemented.

15 Conclusion and future trends

This review focused on precision technologies within pasture-based systems. Precision technologies offer significant potential to increase efficiency, reduce costs and labour, increase sustainability, improve sustainability verification, and raise animal welfare standards of grassland-based systems of milk production. The integration of sensor data, collated within centralised databases enriched with existing data and analysed for specific end-use requirements is recommended. Designing tools and products that have a focus on the needs of the farmer, the labour-saving potential and the return on investment will be key to ensuring that these technologies deliver in pasture-based systems. The key drivers of efficiency and profitability in grassland-based systems are pasture utilisation and dairy cow fertility. Technologies that support grass utilisation and fertility have the greatest chance of delivering positive economic returns for farmers. For the industry as a whole, the most pressing needs in addition to farm profitability are sustainability assessments and sustainability management, which technology is likely to facilitate and support. New developments (e.g. VF) have the potential to revolutionise grazing management not only for pasture utilisation but also for sustainability management and animal health management while at the same time reducing labour requirements. However, for these technologies to realise their potential, they will have to be cost-competitive and resilient, robust, and most importantly, where the farmer and technology interact will be central to the success of the technology.

The research and development of precision technologies in agriculture progresses at a rapid pace. The development of tools/sensors/systems to improve the productivity of grassland farms continues to be an ambition of many companies and researchers. Combining the expertise of technology developers with the science and data analysis skills of researchers could provide a powerful means to increasing the number and types of precision technologies available to farmers. It is clear that there is a need for an integrated approach to ensure that data capture and data integration are linked to models that can help to analyse the data and the outputs in order to provide useful outputs to end users in a time-appropriate fashion, through the most appropriate medium and in the right format.

16 Where to look for further information

- European Association for Precision Livestock Farming – https://eaplf.eu/.
- PastureBase Ireland Website – www.pbi.ie.
- Grasshopper website – www.moregrass.ie.
- EAAP – European Federation for Animal Production – https://eaap.org/.
- ADSA – American Dairy Science Association – www.adsa.org.
- Teagasc – www.teagasc.ie.

17 References

Anderson, D.M., et al. (2009). Gathering cows using virtual fencing methodologies. In Proceedings of the 62nd Society for Range Management Annual Meeting (Paper No. 2030-4).

Adenuga, A.H., et al. (2020). Economic viability of adoption of automated oestrus detection technologies on dairy farms: a review. *Animals*, 10: 1241. https://doi.org/10.3390/ani10071241

Albert, P., et al. (2021). Semi-supervised dry herbage mass estimation using automatic data and synthetic images. *Proceedings of the IEEE/CVF International Conference on Computer Vision (ICCV) Workshops*, 2021: 1284–1293. https://doi.org/10.48550/arXiv.2110.13719

Albert, P., et al. (2022a). Utilizing unsupervised learning to improve sward content prediction and herbage mass estimation. *Computer Vision and Pattern Recognition*, arXiv preprint arXiv:2204.09343. https://doi.org/10.48550/arXiv.2204.09343

Albert, P., et al. (2022b). Utilizing unsupervised learning to improve sward content prediction and herbage mass estimation. In Proceedings of the 29th General Meeting of the European Grassland Federation "Grassland at the heart of circular and sustainable food systems. *Grassland Science in Europe*, 27: 791–793. https://www.europeangrassland.org/fileadmin/documents/Infos/Printed_Matter/Proceedings/EGF2022.pdf

Albert, P., et al. (2022c). Unsupervised domain adaptation and super resolution on drone images for autonomous dry herbage biomass estimation. *Proceedings of the IEEE/CVF Conference on Computer Vision and Pattern Recognition (CVPR) Workshops*, 2022: 1636–1646. https://doi.org/10.48550/arXiv.2204.08271

Ali, I., et al. (2017). Application of repeat-pass TerraSAR-X staring spotlight interferometric coherence to monitor pasture biophysical parameters: limitations and sensitivity analysis. *IEEE Journal of Selected Topics in Applied Earth Observations and Remote Sensing*, 10: 3225–3231.

Alomar, D., et al. (2003). Effect of preparation method on composition and NIR spectra of forage samples. *Animal Feed Science and Technology*, 107: 191–200. https://doi.org/10.1016/S0377-8401(03)00124-X

Alomar, D., et al. (2009). Prediction of the composition of fresh pastures by near infrared reflectance or interactance-reflectance spectroscopy. *Chilian Journal of Agricultural Research*, 69: 198–206. https://doi.org/10.4067/S0718-58392009000200009

AOAC. (1990). *Official Methods of Analysis*. Vol. 1. 15th ed. Association of Official Analytical Chemists, Washington, DC.

Askari, M.S., et al. (2019). Evaluation of grass quality under different soil management scenarios using remote sensing techniques. *Remote Sensing*, 11: 1835.

Auldist, M.J., et al. (2007). Effects of varying lactation length on milk production capacity of cows in pasture-based dairying systems. *Journal of Dairy Science*, 90: 3234–3241.

Borghart, G.M., et al. (2021). Prediction of lameness using automatically recorded activity, behavior and production data in post-parturient Irish dairy cows. *Irish Veterinary Journal*, 74(4), 1–10. https://doi.org/10.1186/s13620-021-00182-6

Burns, G., et al. (2010). Using NIRS to predict composition characteristics of *Lolium perenne* L. cultivars. *Advances Animal Bioscience*, 1: 321–321.

Campbell, D.L., et al. (2023). Automated virtual fencing can effectively contain sheep: field trials and prospects. *Animals*, 13(4): 619.

Chen, W., et al. (2016). The effect of lameness on the environmental performance of milk production by rotational grazing. *Journal of Environmental Management*, 172: 143–150.

Conrad, C.C. and Hilchey, K.G. (2011). A review of citizen science and community-based environmental monitoring: issues and opportunities. *Environmental Monitoring and Assessment*, 176: 273–291.

Creighton, P., et al. (2011). A survey analysis of grassland dairy farming in Ireland, investigating grassland management, technology adoption and sward renewal. *Grass and Forage Science*, 66: 251–264.

Dale, A.J., et al. (2017). Impact of sampling and storage technique, and duration of storage, on the composition of fresh grass when analysed using near-infrared reflectance spectroscopy. *Grass and Forage Science*, 72: 614–625. https://doi.org/10.1111/gfs.12266

Dalen, G., et al. (2019). The detection of intramammary infections using online somatic cell counts. *Journal of Dairy Science*, 102: 5419–5429.

Deming, J., et al. (2018). Measuring labor input on pasture-based dairy farms using a smartphone. *Journal of Dairy Science*, 101(10): 9527–9543. https://doi.org/10.3168/jds.2017-14288

Dillon, P., et al. (2005). Optimising financial return from grazing in temperate pastures. Utilisation of grazed grass in temperate animal systems. In: Murphy, J.J. (Ed.), *Proceedings of a Satellite Workshop of the XXth International Grassland Congress*. Cork, Ireland, pp. 131–147.

Dolecheck, K. and Bewley, J. (2018). Animal board invited review: dairy cow lameness expenditures, losses and total cost. *Animal*, 12: 1462–1474.

Doniec, M., et al. (2010). Autonomous gathering of livestock using a multi-functional sensor network platform. *Proceedings of the 6th 2010 ACM Workshop on Hot Topics in Embedded Networked Sensors*, HotEMNETS 2010, Killarney, Ireland, pp. 1–5.

Dos Santos, T.C.A., et al. (2013). A review on the applications of portable near-infrared spectrometers in the agro-food industry. *Applied Spectroscopy*, 67: 1215–1233. https://doi.org/10.1366/13-07228

Edwards, J.P., et al. (2015). Evaluating rates of technology adoption and milking practices on New Zealand dairy farms. *Animal Production Science*, 55: 702–709. https://doi.org/10.1071/AN14065

Edwards, J.P., et al. (2020). Short communication: technologies and milking practices that reduce hours of work and increase flexibility through milking efficiency in pasture-based dairy farm systems. *Journal of Dairy Science*, 103(8): 7172–7179. https://doi.org/10.3168/jds.2019-17941

Feuerstein, U. and Paul, C. (2007). NIR-Spectroscopy of non-dried forages as a tool in breeding for higher quality – laboratory tests and online investigations on plot harvesters. In: Lübberstedt, T., et al. (Eds.). *Proc. XXVIIth EUCARPIA Symposium on Improvement of Fodder Crops and Amenity Grasses. European Association for Research on Plant Breeding*. Wageningen, Netherlands, pp. 110–114.

Finneran, E., et al. (2010). Simulation modelling of the cost of production and utilizing feeds for ruminants on Irish farms. *Journal of Farm Management*, 14: 95–116.

French, P., et al. (2015). Development and adoption of new technologies to increase the efficiency and sustainability of pasture-based systems. *Animal Production Science*, 55: 931–935.

Gleeson, D., et al. (2008). The labour input associated with calf care on Irish dairy farms. *Livestock Science*, 116: 82–89.

Hanrahan, L., et al. (2017). PastureBase Ireland: a grassland decision support system and national database. *Computers and Electronics in Agriculture*, 136: 193–201.

Hanrahan, L., et al. (2018). Factors associated with profitability in pasture-based systems of milk production. *Journal of Dairy Science*, 101: 5474–5485.

Hennessy D.P., et al. (2021). The net contribution of livestock to the supply of human edible protein: the case of Ireland. *The Journal of Agricultural Science*, 159: 463–471. https://doi.org/10.1017/S0021859621000642

Heins B.J., et al. (2023). Precision technologies to improve dairy grazing systems. *JDS Communications*, 4: 318–323. https://doi.org/10.3168/jdsc.2022-0308

Henchion, M., et al. (2017). Future protein supply and demand: strategies and factors influencing a sustainable equilibrium. *Foods*, 6: 53.

Herron, J., et al. (2022). Life cycle assessment of pasture based dairy systems: current and future performance. *Journal of Dairy Science*, 105: 5849–5869.

Hofstetter, P., et al. (2014). Dairy farming: indoor v. pasture-based feeding. *Journal of Agricultural Science*, 152: 994–1011.

Hogan, C., et al. (2022). An examination of labor time-use on spring-calving dairy farms in Ireland. *Journal of Dairy Science*, 105(7): 5836–5848. https://doi.org/10.3168/jds.2022-21935

Hogan, C., et al. (2023). Longitudinal measures of labour time-use on pasture-based dairy farms, incorporating the impact of specific facilities and technologies. *Animal*, 17: 100747.

Hogeveen, H., et al. (2021). Novel ways to use sensor data to improve mastitis management. *Journal of Dairy Science*, 104: 11317–11332.

Hostiou, N., et al. (2017). Impact of precision livestock farming on work and human-animal interactions on dairy farms. *A Review Base*, 21: 268–275.

Kennedy, E., et al. (2007). Effect of spring grazing date and stocking rate on sward characteristics and dairy cow production during mid lactation. *Journal of Dairy Science*, 90: 2035–2046.

Khatun, M., et al. (2018). Development of a new clinical mastitis detection method for automatic milking systems. *Journal of Dairy Science*, 101: 9385–9395.

Laisse, S., et al. (2018). The net feed conversion efficiency of livestock: a new approach to assess the contribution of livestock to human feeding INRA Productions *Animales*, 31: 19.

Läpple, D., et al. (2012). Extended grazing: a detailed analysis of Irish dairy farms. *Journal of Dairy Science*, 95: 188–195.

Lee, C., et al. (2009). Associative learning by cattle to enable effective and ethical virtual fences. *Applied Animal Behaviour Science*, 119(1–2): 15–22.

Lobos, I., et al. (2019). Calibration models for the nutritional quality of fresh pastures by near-infrared reflectance spectroscopy. *Ciencia e Investigacion Agraria*, 46: 234–242. https://doi.org/10.7764/rcia.v46i3.2020

Macdonald, K.A., et al. (2010). The emergence, development and effectiveness of decision rules for pasture-based dairy systems. In *Proceedings of the 4th Australasian Dairy Science Symposium* 2010, DairyNZ, Private Bag 3221, Hamilton, New Zealand, 1 Dairy Centre, Tasmanian Institute of Agricultural Research, University of Tasmania, PO Box 3523, Burnie, Tasmania, Australia.

McCarthy, B., et al. (2013). The effect of stocking rate and calving date on grass production, utilization and nutritive value of the sward during the grazing season. *Grass and Forage Science*, 68: 364–377.

Moeckel, T., et al. (2017). Fusion of ultrasonic and spectral sensor data for improving the estimation of biomass in grasslands with heterogeneous sward structure. *Remote Sensing*, 9(98): 1–14. https://doi.org/10.3390/rs9010098

Morgan, D., et al. (1989). Modified neutral detergent cellulase digestibility procedure for use with the'Fibertec'system. *Irish Journal of Agricultural Research*, 28: 91–92.

Murphy, D.J., et al. (2021). A near infrared spectroscopy calibration for the prediction of fresh grass quality on Irish pastures. *Information Processing in Agriculture*, 9(2): 243–253. https://doi.org/10.1016/j.inpa.2021.04.012

O'Brien, D., et al. (2012). A life cycle assessment of seasonal grass-based and confinement dairy farms. *Agricultural Systems*, 107: 33–46.

O'Brien, D., et al. (2014). An appraisal of carbon footprint of milk from commercial grass-based dairy farms in Ireland according to a certified life cycle assessment methodology. *International Journal of Life Cycle Assessment*, 19: 1469–1481.

O'Brien, D., et al. (2018). A national methodology to quantify the diet of a dairy cow. *Journal of Dairy Science*, 101: 8595–8604.

O'Callaghan, T.F., et al. (2016). Effect of pasture versus indoor feeding systems on raw milk composition and quality over an entire lactation. *Journal of Dairy Science*, 99: 9424–9440.

O'Donovan, M., et al. (2002). A comparison of four methods of herbage mass estimation. *Irish Journal of Agricultural and Food Research*, 41: 17–27.

O'Donovan, M., et al. (2021). Irish grassland research – main achievements and advancements in the past 60 Yrs and where to progress to next. *Irish Journal of Agricultural and Food Research*, 61: 1–11.

O'Leary, N.W., et al. (2020). Invited review: cattle lameness detection with accelerometers. *Journal of Dairy Science*, 103(5): 3895–3911.

Parrini, S., et al. (2019). Near infrared spectroscopy technology for prediction of chemical composition of natural fresh pastures. *Journal of Applied Animal Research*, 47: 514–520. https://doi.org/10.1080/09712119.2019.1675669

Ramsbottom, G., et al. (2015). Factors associated with the financial performance of spring –calving, pasture-based dairy farms. *Journal of Dairy Science*, 98: 3526–3540.

Ruelle, E., et al. (2018) Development of the Moorepark St Gilles grass growth model (MoSt GG model): a predictive model for grass growth for pasture based systems. *European Journal of Agronomy*, 99: 81–90. https://doi.org/10.1016/j.eja.2018.06.010

Rutten, C.J., et al. (2013). Invited review: sensors to support health management on dairy farms. *Journal of Dairy Science*, 96: 1928–1952.
Rutten, C.J., et al. (2014). An ex ante analysis on the use of activity meters for automated estrus detection: to invest or not to invest? *Journal of Dairy Science*, 97: 6869–6887.
Shalloo, L. and Hanrahan, L. (2019). Setting targets for the Irish dairy industry. *Animal Production Science*, 60: 159–163.
Shalloo, L., et al. (2004). Description and validation of the Moorepark dairy system model. *Journal of Dairy Science*, 87: 1945–1959.
Shalloo, L., et al. (2011). The economics of reseeding on a dairy farm. *Irish Journal of Agricultural and Food Research*, 50: 113–122.
Shalloo, L., et al. (2014). Effect of fertility on the economics of pasture-based dairy systems. *Animal*, 8: 222–231. https://doi.org/10.1017/S1751731114000615
Shalloo, L., et al. (2021). A review of precision technologies in pasture-based dairying systems. *Irish Journal of Agricultural and Food Research*, 59(2): 279–291.
Sibanda, M., et al. (2016). Comparing the spectral settings of the new generation broad and narrow band sensors in estimating biomass of native grasses grown under different management practices. *GIScience and Remote Sensing*, 53: 614–633.
Sinnot, A., et al. (2021). The effects of manual and automated milk feeding methods on group-housed calf health, behaviour, growth and labour. *Livestock Science*, 244: 104343.
Steeneveld, W., et al. (2015). Economic consequences of investing in sensor systems on dairy farms. *Computers and Electronics in Agriculture*, 119: 33–39.
Thomas, E.B., et al. (2019). A decision-support tool for investment analysis of automated oestrus detection technologies in a seasonal dairy production system. *Animal Production Science*, 59: 2280–2287.
Umstatter, C. (2011). The evolution of virtual fences: a review. *Computers and Electronics in Agriculture*, 75(1): 10–22.
Valletta, J.J., et al. (2017). Applications of machine learning in animal behaviour studies. *Animal Behaviour*, 124: 203–220.
Van Soest, P.J., et al. (1991). Methods for dietary fiber, neutral detergent fiber, and nonstarch polysaccharides in relation to animal nutrition. *Journal of Dairy Science*, 74: 3583–3597.
Veerkamp, R.F., et al. (2002). Dairy cattle breeding objectives combining yield, survival and calving interval for pasture based systems in Ireland under different milk quota scenarios. *Livestock Production Science*, 76: 137–151.
Wade, K.M., et al. (2004). Economic efficiency of automatic milking systems with specific emphasis on increases in milk production. In A. Meijering, et al. (Eds.), *Automatic Milking: A Better Understanding*. Wageningen Academic Publishers, Wageningen, the Netherlands, pp. 63–67.
Wagner, K., et al. (2018). Impact of daily grazing time on dairy cow welfare-results of the welfare quality® protocol. *Animals*, 8: 1–11.
Wolfert, S., et al. (2017). Big data in smart farming: a review. *Agricultural Systems*, 153: 69–80.

Chapter 7

The long path from data collection to sustainable grassland management

Nina Buchmann, ETH Zurich, Switzerland; Panu Korhonen, Natural Resources Institute Finland, Finland; Carsten Stefan Malisch, Aarhus University, Denmark; and Paul Newell Price, ADAS, UK

1 Introduction

Grasslands, including savannas, prairies, and steppes, are vital for ecosystems globally and particularly in Europe (White et al., 2000). These landscapes are critical for agricultural productivity as well as for biodiversity and climate change mitigation. For example, grasslands significantly contribute to Europe's agricultural landscape and provide livestock forage as well as regional revenues. But European grasslands, particularly permanent ones, also play a key role in biodiversity conservation and carbon sequestration, aiding climate change mitigation. They are essential for soil preservation and water regulation, maintaining the ecological balance in agricultural regions. Moreover, their cultural and historical values, linked with traditional agricultural practices and recreational activities, are significant. Thus, sustainable management of grasslands is crucial for preserving their ecological functions and ecosystem

http://dx.doi.org/10.19103/AS.2025.0154.34

services (Schils et al., 2022). For Europe, this means aligning food and feed production with reducing environmental footprint of permanent grassland use, while providing income to farmers and land managers. Clearly, effective management strategies rely on accurately determining grassland functions. However, which technologies and approaches to use and which strategies will be adopted by land managers is far less clear.

Many different techniques are available to observe and quantify ecosystem functions and services provided by grasslands now and in the future. Techniques vary widely, encompassing a spectrum from traditional to advanced, covering different temporal and spatial scales, and ranging from ground-based observations to airborne methods. Traditional ground surveys and manual plot sampling provide very detailed, site-specific data, while advanced methods often include not only highly scientific sophisticated but also costly approaches. For example, remote sensing technologies using new satellite missions not only offer broad-scale monitoring capabilities but also require complex image processing as well as ground-truthing. On the other hand, airborne methods like drone-based imaging can provide a balance between large-scale observation and detailed local data. These data are then often compiled in databases and integrated in models to predict future changes and potential responses of grasslands to various environmental and socio-economic stressors. This integration requires sophisticated data analysis techniques that can handle various data types and scales. Analyses and syntheses of data from different sources allow a more complete picture of grassland functionality; however, best practice examples are still scarce.

The ultimate goal of all these analyses is to translate complex agronomic and ecological data into practical management recommendations. This involves understanding the interplay between many ecosystem services and functions. To be adopted and become effective, management strategies must on the one hand balance these services and on the other hand be tailored to specific grassland conditions and thus management needs as well as to agro-political and economic constraints. This chapter aims to provide a concise overview of different techniques available to collect data to learn about grassland functions and services. We will address not only the questions 'why', 'what', 'where', 'when', and 'how to measure', but also which management options are available and what their potential to increase ecosystems functions and services might be. Finally, we briefly discuss adoption of sustainable management options and tools.

2 Assessing the pros and cons of different techniques for data collection

Based on expert knowledge as well as literature, we evaluated different tools and innovative techniques to inform sustainable management options for permanent grassland. We started with the management purpose, which ranged from feed production to reduction of the environmental footprint of grassland management but also included potential solutions to different global challenges, e.g., climate change or biodiversity loss. Different tools and techniques ('what') for those different purposes were then judged in terms of what spatial and temporal scales they are performed at ('where', 'when'), i.e. site, farm, or landscape scales. In addition, we assessed if they are collected or employed only sporadically (collected manually or in campaigns), more regularly (as with drones or in experiments), or even continuously.

Due to the heterogeneity of purposes and spatio-temporal scales, we then addressed the many different modes and techniques of data collection ('How'), which clearly depend on the grassland functions or services of interest. We further evaluated the weaknesses of these different techniques, which might limit their adoption ('Weaknesses').

To develop sustainable management options, it is crucial to select data collection techniques according to the management purpose and the respective farming and production systems, accounting for their spatial and temporal scales as well as mode of data collection and weaknesses. Some of these techniques and approaches are still being used rather in academic or experimental settings. Nevertheless, some methods and techniques have already been implemented in management options, both for assessing grassland vegetation and livestock, while others still need 'proof of concept' under real-world agricultural settings. In the following sections, we focus on new and innovative examples of data capture and analysis in grassland and livestock farming systems to illustrate the pros and cons of various techniques (Table 1).

Many of the new approaches and technologies to support sustainable management options not only provide but also require large amounts of high-quality reliable data and metadata. Within the context of academic and research institutions, and dependent on the methods or techniques used, making data known and available to others is thus key to push sustainable management of grasslands. Open science strategies and FAIR principles (Findability, Accessibility, Interoperability, and Reusability of digital assets) much depend on the willingness of the researchers to provide data and metadata to others, and on users who cite DOIs (Digital Object Identifiers) of such datasets, acknowledging the work of those who collected the data. These data then can feed into models of many kinds, which can be the basis of apps or online decision-support tools to help farmers in their day-to-day life.

Table 1 Overview of methods and techniques available to collect information about permanent grasslands

What	Where	When	How	Weaknesses	Management options
Feed production, grazing					
Feed: quantity, quality, stability over time, grazing	Site	Campaigns	Plate-meter, clipping of quadrats, plant height (ruler)	Time and space resolution, time requirement, know-how	Use of different mixtures, new species, breeds or varieties, change in management intensity (cutting, grazing, fertilization, irrigation), virtual fencing
	Animals	Campaigns	Feed intake observations, incl. tracers and holters with noseband sensor, biomass assessments	Time and space resolution, rather academic/experimental approach, expensive	
	Site, farm	Regularly	GPS collars, virtual fencing, drones, UAVs, mounted with GPS and different cameras	Know-how, privacy/permit issues, time and space resolution	
	Site, farm, landscape	Regularly	Satellites	Price, data ownership, know-how, time, and space resolution	
Resource use					
Inputs via fertilization, grazing	Site	Regularly	Farmer interviews; official statistics; legal requirements	Time demand, privacy issues	Different farming practices (e.g. stocking density, timing, animal breed), intensification/extensification, precision farming, change to different production systems (e.g. regenerative agriculture, organic agriculture, permaculture)
	Site	Regularly	Soil analyses	Time demand, price	
	Site, farm	Regularly	Drones, UAVs	Know-how, privacy/permit issues, time and space resolution	
	Site, farm, landscape	Regularly	Satellites	Price, data ownership, know-how, time and space resolution	

			Climate change: Grasslands as drivers		
Greenhouse gas emissions: magnitude of fluxes, drivers; soil carbon stocks	Site	Regular campaigns	Soil chambers, soil analyses, isotope analyses	Time and space resolution, know-how, rather academic/experimental approach	Change in fertilization (magnitude, type, timing), precision farming (e.g. variable rate fertilization)
	Site	Continuously	Eddy-covariance flux stations	Price, know-how, spatial homogeneity, ruminants, relocation difficult, rather academic approach	
	Animals	Campaigns	GPS collars	Know-how, price, time and space resolution, rather academic/experimental approach	Change in animal breed, feeding practice
			Climate change: Grasslands driven by climate change		
Resilience to climate change, e.g. extreme events	Site	Campaigns, continuously	Experiments simulating extreme weather events, long-term data sets	Academic approach, research still needed; predictability of extremes	Use of new mixtures, introduction of new grassland species, irrigation/drainage, change in farming practices (e.g. timing of management practices)
			Biodiversity		
Determining biodiversity (species and functional group diversity, plants, animals, microbiota, eDNA)	Site	Campaigns	Manual assessments, experiments, lab assessments, long-term data sets	Know-how, time, and space resolution, academic approach, available results often site-specific	Use of mixtures, new grassland species, overseeding, different animal breeds or varieties, change in farming practices (e.g. timing of management)
	Farm	Regularly	Drones, UAVs	Know-how, time and space resolution, privacy/permit issues	
	Site, farm, landscape	Regularly	Satellites	Price, data ownership, know-how, time and space resolution	

Different management purposes are considered, ranging from feed production to help solving global challenges. Different columns not only give examples of what, where, when, and how to measure but also address the weaknesses of these techniques. Relevant management options are given, although these are not exclusive.

3 Data collection on feed production, grazing behaviour, and resource use

Information about feed production in permanent grasslands and thus forage supply for animals can be collected in many different ways. While vegetation growth over time can be easily measured in meadows, such an assessment is more difficult in pastures, since not only forage supply but also animal behaviour need to be considered.

GPS collars provide an opportunity to monitor the grazing behaviour of livestock over an extended period of time. The collars typically send location data every 15–20 min via a network. Collar device data reception can be variable, depending on field topography, vegetation type, and distance to the receiving antenna (Maroto-Molina et al., 2019). However, even in uneven terrain, it is possible to generate grazing calendars and livestock use intensity maps using simple data processing, indicating grazing preferences, distances covered, habitats impacted and how this varies with pasture availability throughout the season (Chodkiewicz et al., 2023). Collars can also be used to determine where and when cows are calving and for monitoring animal health and behaviour.

However, costs of GPS collars are significant, and technical support is a very important element in their adoption (Fernández-Habas et al., 2020). Trade-offs between data resolution and data quality exist due to limited battery lifespans and the difficulty to replace batteries in grazing systems. Data quality and relevance can be enhanced in combination with additional sensors. For example, the RumiWatch holters (Itin & Hoch, Switzerland) can differentiate with high accuracy between prehension bites, rumination chews, and eating chews, by using a pressure-based sensor over the cow's nose and jaw (Li et al., 2023). While these methods enable data acquisition, further research needs to prove relevance in different grazing systems.

While GPS collars only monitor grazing behaviour, virtual fencing (VF) is a novel technology that uses a combination of audio and electrical stimuli to contain grazing livestock within a movable GPS boundary. This ability to control animal movements means that VF can provide additional productivity and conservation benefits compared with GPS collars although costs are greater. In some countries, there are animal welfare concerns to address, although Hamidi et al. (2022) found no evidence of negative impact on animal welfare of grazing heifers. VF can facilitate more flexible and tailored conservation management in high nature value grasslands. Hiron & Wahlund (2024) demonstrated that VF can be used to create areas of short, medium, and tall vegetation (creating habitat diversity), focusing grazing pressure where needed, and temporarily reducing grazing pressure for a short period (protecting wildlife). Animals quickly adapted and learned to respond to the audio signals. VF can also be combined with vegetation indices (such as RGB-VI; Red, Green, Blue) data from unmanned aerial vehicles (UAVs) to better allocate pasture resources to grazing livestock over large areas (Hamidi et al., 2024).

4 Data collection on climate change interactions

Integrating over both spatial and temporal variabilities is best done using eddy-covariance (EC) measurements of ecosystem greenhouse gas exchange, accounting for soil, vegetation and animal gas exchange with the atmosphere at EC flux stations (Eugster & Merbold, 2015). Here, micrometeorological and gas concentration measurements are combined, at 10–20 Hz (i.e. 10–20 times per second), to measure the net gas exchange of carbon dioxide (CO_2), water vapour (H_2O), methane (CH_4), and/or nitrous oxide (N_2O) over an entire grassland. All management practices can take place as normally, so EC measurements only interfere to a very limited extent with farming practices. However, grassland areas need to be at least 1.5–2 ha, following the main wind direction. So not all grasslands can be measured with this technique. Moreover, if the grassland is grazed and CH_4 fluxes are measured, the position of the animals within the footprint should be known. This requires additional observations or monitoring via GPS collars.

Prices for an EC set-up are high as is the knowledge demand for running such a station and analysing the EC data. Therefore, this technique is rather bound to academic approaches by research institutions. Nevertheless, the insights based on such EC stations cannot be gained in any other way. Responses of entire grasslands to changes in climate or management practices, as well as their sensitivities to extreme weather events like heatwaves or droughts, cannot be measured in a different way, only estimated by modelling. Moreover, combining EC measurements of greenhouse gas fluxes with management information (i.e. carbon exports with harvests; carbon inputs with organic fertilization) allows not only to quantify soil carbon sequestration (Feigenwinter et al., 2023a) but also to detect trade-offs, e.g. between maintaining carbon sequestration while reducing N_2O losses (Fuchs et al., 2018; Feigenwinter et al., 2023b).

In agronomy, process-based biogeochemistry models (BGMs) are often used to simulate growth and productivity of grasslands and, in environmental sciences, to quantify the surface exchange of greenhouse gases. These models calculate the dynamics of carbon, nutrients, and water along the soil–plant–atmosphere continuum, often on a daily or sub-daily time step, and can be helpful in predicting and understanding ecosystem functions and services, including testing the impacts of different management decisions. However, there is a great disparity among models in terms of their ability to simulate C sequestration and status of sources and sinks of greenhouse gases in different environments, which can mostly be attributed to ill-representation of effects of pedo-climatic conditions and management in the models (Brilli et al., 2017).

BGMs also usually lack detailed representation of dynamics related to plant biodiversity, reducing the capacity to utilize them effectively for decision-making (Van-Oijen et al., 2020).

Efforts to incorporate satellite and UAV-acquired data, either directly as inputs or through data assimilation, to enhance the accuracy of BGMs are highly promising. The information infrastructures for connecting models with the growing number and variety of measurements are being developed although the widespread adoption of accessible and scalable tools remains a significant challenge (Fer et al., 2020). The improvement of comprehensive data infrastructures that integrate diverse ecological measurement data, including remote sensing, with BGMs, will greatly enhance the models' functionality and effectiveness within management decision-making processes.

5 Data collection on biodiversity

Evaluating biodiversity and related ecosystem services requires precise and repeated assessment of species richness or functional group richness of the organism group of interest. With regards to acquiring georeferenced vegetation data at high temporal and spatial resolutions, a number of technologies have become available. For example, satellite data, extensively available and widely utilized in various forms of vegetation indices, are integral to numerous environmental and agricultural applications. However, many practical applications for driving management decisions in grassland farming still require laborious collection of ground truth data and often temporal and spatial resolutions are not sufficient in satellite data.

The collection of aerial spectral data by UAVs equipped with visible light, multispectral, or hyperspectral cameras presents a viable solution for applications demanding high-resolution and timely data. An example of employing UAV-derived data for making informed management decisions, such as optimizing fertilization strategies, is the precise quantification of leguminous plants within grasslands, as demonstrated in a study by Li et al. (2021). This task, along with many image segmentation applications, greatly benefits from the recent rapid advancements in deep learning methods for processing UAV-acquired data (Osco et al., 2021, Wang et al., 2022).

Models that utilize ultra-high resolution imagery from UAVs or time-lapse cameras for flower mapping have also been developed (Gallmann et al., 2022; Andreatta et al., 2023). They contribute significantly to efforts in developing new remote sensing tools for biodiversity monitoring, in addition to spectral variation assessments which have been hard to generalize over different environments (Thornley et al., 2023). More difficult remains the quantification at the vegetative stage, as required in more intensively used grasslands. Yet at close range, analyses of RGB images have been able to reliably differentiate between red and white clover at the vegetative stage (Skovsen et al., 2021), thus providing the basis for upscaling from high resolution UAV imagery. Biomass estimation using models that combine both structural (e.g. canopy height acquired using photogrammetric 3D-models) and spectral UAV-acquired data seems to provide highest accuracy in most cases (Bazzo et al., 2023).

High-resolution hyperspectral data have been utilized in assessments of forage quality with success in limited environments (e.g. Oliveira et al., 2024), but generalization across different plant species and environments is still a significant challenge. Drone-in-a-box solutions, wherein a UAV autonomously executes data collection tasks based on predefined instructions and subsequently returns to a docking station for charging and data transfer, present significant potentials, particularly for research applications. By automating the process, these systems enable more consistent and frequent data gathering, thereby yielding richer datasets that are crucial for time-sensitive research and analysis.

6 Case study: Assessment of management options

Developing sustainable management options, based on scientific evidence in the form of data, is best done using a participatory approach, collecting data not only from experiments but also from real-world farms and farm networks (Klaus et al., 2020). Within an EU project called SUPER-G (Developing SUstainable PERmanent Grassland systems and policies), we used a series of co-innovation workshops with over 20 farm networks, and more than 40 experiments on commercial and research farms across Europe to identify such promising management options (Rankin et al., 2024). We assessed the options based on the techniques discussed above for their effective potential to indeed capture grassland functions and services and to be adopted by grassland famers and thus to improve sustainability of permanent grassland farming. For permanent grasslands, options were assigned to four areas:

1 new grassland and animal species, diverse species swards: introduction of new grassland species, overseeding with diverse species or mixtures, new livestock breeds;
2 precision grassland management: use of plate meters or other yield estimation techniques, grazing and cutting management tools, rotational grazing, satellite and/or drone technologies for yield and quality estimation and to measure ecosystem services delivery, use of precision technologies for grazing (e.g. in field weighing, VF, GPS collars, apps);
3 nutrient management: quantification of nutrient balances, precision nutrient management, variable rate fertiliser application, use of slow-release fertilisers, modelling soil carbon dynamics; and

4 agri-environmental options: grazing, cutting, and other management strategies for productivity, biodiversity, and other public goods, increasing legume cover to reduce mineral N fertiliser and decrease N_2O emissions, use of locally harvested seeds for re-vegetation.

This wide range of management options, based on a broad portfolio of different methods and techniques, was subsequently ranked for their effective potential to provide increased ecosystem services, to be adopted by grassland farmers, and thus to improve sustainability of permanent grassland farming (Table 2). A set of seven criteria was used for this assessment: (1) improved provision of studied ecosystem service (yes, no); (2) potential for provisioning multiple ecosystem service (yes, no); (3) technology readiness levels (TRLs, from 7 to 9); (4) ease-of-use by farmers (low, medium, high); (5) useability for pastures and/or meadows (pasture, meadow, both types); (6) potential to improve profitability (low, medium, high); and (7) relevance and/or transfer beyond local context (yes, no).

Table 2 Identification of the most promising management options for permanent grasslands carried out in SUPER-G.

	Innovative management options and new technologies	Most promising
	i) new grassland species / diverse species swards:	
1	Introduction of new grassland species	
2	Overseeding with diverse species or mixtures	
	ii) precision grassland management:	
3	Use of other precision technologies for grazing (e.g., in field weighing / virtual fencing / GPS collars / apps)	
4	Grazing management tools, rotational grazing	
5	Platemeter, other yield estimation techniques	
6	Satellite or drone technologies for yield and quality estimation	
	iii) nutrient management:	
7	Quantification of nutrient balances	
8	Precision nutrient management, variable rate fertiliser application	
9	Use of slow-release fertilisers	
10	Modeling soil carbon (C) dynamics to increase soil C stocks	
	iv) agri-environment options:	
11	Increasing legume cover to reduce mineral N fertiliser and decrease N_2O emissions	
12	Mechanical loosening of soil for improved drainage and productivity	
13	Agroforestry, promotion of tree establishment and growth	
14	Grazing management and other management strategies for productivity, biodiversity and other public goods	
15	Cutting management and other management strategies for productivity, biodiversity and other public goods	
16	Use of locally harvested seeds for re-vegetation	
17	New livestock breeds for permanent grasslands / ecosystem services	

The following colour coding was used: dark green represents the most promising options, followed by light green and yellow. The least promising options and technologies are given in orange (based on Buchmann et al., 2024).

All options were ranked as being relevant for transfer to other regions, and most of them were seen as applicable to both pastures and meadows. Most options showed to increase the one ecosystem service under study or even to increase multiple ecosystem services. However, some of them were not seen to provide further ecosystem services as an add-on. For example, using plate meters or other yield estimation techniques such as drones and satellites were seen as rather limited to yield estimates but not to provide further advantages. Technology readiness levels for all options were typically ranked rather high (TRL 8, i.e. system complete and qualified; TRL 9, i.e. actual system proven in an operational environment), but techniques linked exclusively to drones and satellites or modelling were ranked considerably lower (TRL7, i.e. system prototype demonstration in an operational environment). This was reflected in the assessment that most of the options seemed to be easy to use by farmers, except drone and satellite technologies or modelling. In addition, precision nutrient management, variable rate fertiliser application as well as precision grazing management were rated considerably lower, indicating a large need for knowledge transfer from science to practice. The potential to improve profitability was rated medium to high for about 70% of all options. About 40% of the options were considered highly profitable (including use of plate meters, grazing management tools, rotational grazing, quantification of nutrient balances, precision farming for nutrients and grazing, and the use of legumes), while three options (modelling, agroforestry, and use of locally collected seeds) were considered as showing low potentials to improve profitability. This clearly showed a gap between scientific evidence, anticipated ease-of-use and readiness levels of innovative management options for permanent grasslands to their expected adoption by farmers, which is driven by many factors, including profitability.

Political boundary conditions, socio-economic factors as well as risk attitude of farmers can limit the adoption of scientifically proven, reliable management techniques (Hofmann et al., 2022). When agri-environmental policies support certain management options, or when front-runners successfully implement new technologies, their adoption is likely to be higher. When risks for farmers increase, e.g. to lose yields due to extreme climatic events, more diverse swards are suddenly more attractive. Knowing barriers for adoption becomes thus similarly important as collecting and providing new scientific evidence for existing knowledge gaps.

7 Conclusion

Thus, although gaps in knowledge exist, data are being collected using a wide array of methods, technologies, and approaches. Many data are openly shared, although FAIR principles are not yet everywhere implemented. Management options are being developed and tested for permanent grasslands, which might also be applicable to temporary grasslands within a crop rotation. Increased uptake of data processing technologies and inferred management options will require proof of concept, clear cost-benefit analyses, and both technical and financial support to farmers through the implementation of (inter) national policies, e.g. via common agricultural policy (CAP) Strategic Plans (CSP), and the careful design of eco-schemes and agri–environmental–climate commitments.

8 Where to look for further information

Further reading:

- A good review on hyperspectral imaging in precision agriculture is provided by Ram et al. (2024).
- A good start to read about methods to estimate aboveground biomass in grasslands using UAV is Bazzo et al. (2023).
- A short introduction into EC is provided by Eugster and Merbold (2015), a longer compendium with great overview figures and many technical details is written by Burba (2022), freely available from the LiCor webpage.

Key conferences/initiatives/projects:

- Copernicus Data Space Ecosystem Browser is a central hub to access, explore, and utilize Earth observation and environmental data. Data are open access.
- European Union Aviation Safety Agency's (EASA) website provides comprehensive and regularly updated information on regulatory aspects for drone use in Europe. .
- European Grassland Federation (EGF) holds annual meetings and workshops with international participation.
- European Open Science Cloud (EOSC) advocates FAIR and open research data (ORD) across Europe.
- FLUXNET, the global 'network of networks,' brings together global EC fluxes of carbon, water, and energy between the biosphere and atmosphere. Data are open access.
- Global Long-Term Agricultural Experiment Network (GLTEN) is a global network of over 60 long-term agricultural experiments. Meta-data are available online.
- Integrated Carbon Observation System, ICOS, provides standardized greenhouse gas flux, meteorological, and management data from close

to 180 measurement stations across 16 European countries (as of 2025). Data are open access.

9 Acknowledgements

This chapter is derived from a paper presented as an invited keynote at the 30th General Meeting of the European Grassland Federation (EGF) "Why Grasslands?" in Leeuwarden (NL) on 11 June 2024.

10 Declaration of funding/Conflict of interest statement

The authors acknowledge various funding sources, e.g. from the EU projects SUPER-G (project number 774124-2) and ForageDrone (project number 145346), as well as the SNF projects GrassGas (200021-105949) and M4P (40FA40_154245). The authors declare no conflict of interest.

11 References

Andreatta, D., et al. (2023), Extracting flowering phenology from grassland species mixtures using time-lapse cameras, *Remote Sensing Environment* 298: 113835.

Bazzo, C.O.G., et al. (2023), A review of estimation methods for aboveground biomass in grasslands using UAV, *Remote Sensing* 15: 639.

Brilli, L., et al. (2017), Review and analysis of strengths and weaknesses of agro-ecosystem models for simulating C and N fluxes, *Science of the Total Environment* 598: 445–470.

Buchmann, N., et al. (2024, October 2), SUPER-G, deliverable report 3.7 on assessment of innovative permanent grassland management options. EC Project Number 774124-2. https://www.super-g.eu/2024/02/15/deliverable-3-7-assessment-of-innovative-permanent-grassland-management-options/

Burba, G. (2022), Eddy covariance method for scientific, regulatory, and commercial applications. https://www.licor.com/products/eddy-covariance/ec-book

Chodkiewicz, A., et al. (2023), Habitat use by semi-feral Konik horses on wetlands–three-year GPS study, *Environmental Monitoring and Assessment*. 195: 1033.

Eugster, W. & Merbold, L. (2015), Eddy covariance for quantifying trace gas fluxes from soils, *Soil* 1: 187–205.

Fer, I., et al. (2020), Beyond ecosystem modeling: A roadmap to community cyberinfrastructure for ecological data-model integration, *Global Change Biology* 27:13–26.

Feigenwinter, I., et al. (2023a), Large inter-annual variation in carbon sink strength of a permanent grassland over 16 years: impacts of management practices and climate, *Agricultural and Forest Meteorology* 340: 109613.

Feigenwinter, I., et al. (2023b), N_2O and CH_4 fluxes from intensively managed grassland: the importance of biological and environmental drivers vs. management, *Science of the Total Environment* 903: 166389.

Fernández-Habas, J., et al. (2020), Preferences for grassland-management innovations in dehesa farms from Andalusia (Spain), *Grassland Science in Europe* 25: 725–727.

Fuchs, K., et al. (2018), Management matters: testing a mitigation strategy for nitrous oxide emissions using legumes on intensively managed grassland, *Biogeosciences* 15: 5519–5543.

Gallmann, J., et al. (2022), Flower mapping in grasslands with drones and deep learning, *Frontiers in Plant Science* 12: 774965.

Hamidi, D., et al. (2022), Heifers don't care: no evidence of negative impact on animal welfare of growing heifers when using virtual fences compared to physical fences for grazing, *Animal* 16: 100614.

Hamidi, D., et al. (2024, September 17), Using UAV technology combined with virtual fencing for improved information gain, In: Rankin, J., et al. (eds) *Deliverable report 3.6, synergies and tradeoffs, SUPER-G* (Sustainable Permanent Grassland Systems and Policies), EC Project Number 774124-2. https://www.super-g.eu/2024/02/15/deliverable-3-6-synergies-tradeoffs/

Hiron, M. & Wahlund, L. (2024, September 17), Virtual fencing: a flexible tool for conservation management in grazed semi-natural grasslands?, In: Rankin, J., et al. (eds) *Deliverable report 3.6, synergies and tradeoffs, SUPER-G* (Sustainable Permanent Grassland Systems and Policies), EC Project Number 774124-2. https://www.super-g.eu/2024/02/15/deliverable-3-6-synergies-tradeoffs/

Hofmann, B., et al. (2022), Barriers to evidence use for sustainability: insights from pesticide policy and practice, *Ambio* 52: 425–439.

Klaus, V.H., et al. (2020), Do biodiversity-ecosystem functioning experiments inform stakeholders how to simultaneously conserve biodiversity and increase ecosystem service provisioning?, *Biological Conservation* 245: 108552.

Li, K.Y., et al. (2021), The application of an unmanned aerial system and machine learning techniques for red clover-grass mixture yield estimation under variety performance trials, *Remote Sensing* 13: 1994.

Li, Z., et al. (2023), Validation and use of the RumiWatch noseband Sensor for monitoring grazing behaviours of lactating dairy cows, *Dairy* 2: 104–111.

Maroto-Molina, F., et al. (2019), A low-cost IOT-based system to monitor the location of a whole herd, *Sensors* 19: 2298.

Oliveira, R.A., et al. (2024), High-precision estimation of grass quality and quantity using UAS-based VNIR and SWIR hyperspectral cameras and machine learning, *Precision Agriculture* 25: 186–220.

Osco, L.P., et al. (2021), A review on deep learning in UAV remote sensing, *International Journal of Applied Earth Observation and Geoinformation* 102: 102456.

Ram, B.G., et al. (2024), A systematic review of hyperspectral imaging in precision agriculture: analysis of its current state and future prospects, *Computers and Electronics in Agriculture* 222: 109037.

Rankin, J., et al. (2024, September 17), SUPER-G, deliverable report 3.6 on synergies & trade-offs. EC Project Number 774124-2. https://www.super-g.eu/2024/02/15/deliverable-3-6-synergies-tradeoffs/

Schils, R.L., et al. (2022), Permanent grassland in Europe: land use change and intensification decrease their multifunctionality, *Agriculture, Ecosystems & Environment* 330: 107891.

Skovsen, S.K., et al. (2021), Robust species distribution mapping of crop mixtures using color images and convolutional neural networks, *Sensors* 21: 175.
Thornley, R.H., et al. (2023), Prediction of grassland biodiversity using measures of spectral variance: a meta-analytical review, *Remote Sensing* 15: 668.
Van-Oijen, M., et al. (2020), Incorporating biodiversity into biogeochemistry models to improve prediction of ecosystem services in temperate grasslands: review and roadmap, Agronomy 10: 259.
Wang, Z., et al. (2022), Review of remote sensing applications in grassland monitoring, *Remote Sensing* 14: 2903.
White, R., et al. (2000), *Pilot analysis of global ecosystems (PAGE) grassland ecosystems*, World Resource Institute.

Chapter 8

From data collection to decision support tools for the sustainable management of grasslands

Magali Jouven[1], Institut Agro Montpellier, France; Audrey Michaud[2], VetAgro Sup, France; and Pascal Carrère[3], INRAE, France

1 Introduction

From a biogeographical point of view, grasslands are ecosystems in which the vegetation is made up of herbaceous species, in particular grasses, graminoids and even forbs (Dixon et al., 2014). From an agroecological point of view, grasslands are 'land devoted to the production of forage for harvest by grazing/browsing, cutting, or both, or used for other agricultural purposes such as renewable energy production. The vegetation can include grasses, grass-like plants, legumes and other forbs. Woody species may also be present' (Peeters et al., 2014). Based on this broad definition, grasslands encompass temporary grasslands containing one or more selected herbaceous species,

1 *University of Montpellier, CIRAD, INRAE, Institut Agro Montpellier, UMR SELMET, Montpellier*
2 *Université Clermont Auvergne, INRAE, VetAgro Sup, UMR Herbivores, Saint-Genès-Champanelle*
3 *Université Clermont Auvergne, INRAE, VetAgro Sup, UREP, Clermont-Ferrand*

http://dx.doi.org/10.19103/AS.2025.0154.33

permanent grasslands containing a variety of herbaceous species and semi-natural grasslands possibly including woody plants (e.g. mountain pastures, Mediterranean rangelands, wetland pastures, shrubland and dehesas). Permanent grasslands are an administrative category defined by the European Union (EU) as 'land used permanently (for several consecutive years, normally 5 years or more) to grow herbaceous fodder, forage or energy purpose crops, through cultivation (sown) or naturally (self-seeded), and which is not included in the crop rotation on the holding' (Eurostat, 2023).

Grasslands are among the largest ecosystems in the world, covering over 26% of the global land area (40% if we exclude Greenland and Antarctica; World Resources Institute (2000) based on International Geosphere-Biosphere Programme data) and around 70% of the global agricultural area (FAO, 2019). They are an important feed source for livestock, a crucial habitat for wildlife, and they contribute to the livelihood of over 800 million people by providing food, forage and energy (FAO, 2019). However, in many favourable areas in terms of fertility and climatic conditions (especially water availability), grasslands have been converted into arable land; this occurred especially in South American Pampas, in the North American Prairie, the East European Steppe, and even in semi-arid regions of Africa. As a consequence, the surface area of semi-natural grazing land has been reduced and the traditional grassland-based breeding systems as well. In the EU-27, the total surface area covered by permanent grasslands in 2022 was estimated at 50.7 million ha (Eurostat, 2022), which corresponds to 32% of the EU's agricultural area. Based on the same source, temporary grasslands would account for 4.5% and other plants harvested green (maize, lucerne, mixtures, etc.) for 7.6%. Grasslands are thus a major component of EU agricultural landscapes. This is especially true in countries like France, Ireland, Germany, Romania, Spain and Poland, where permanent and semi-natural grasslands account for over 35% of the utilised agricultural area.

Since the end of the twentieth century (MEA, 2005), grasslands have been increasingly valued for the diversity of services provided to agriculture and, more generally, to society: forage production, protection against soil erosion, water quality, cultural value, etc. (Lindborg et al., 2022). A number of recent large-scale studies, such as the empirical mapping of ecosystem services (Brown and Fagerholm, 2015; MAES, 2020) at the EU level or Evaluation Française des Ecosystèmes et Services Ecosystémiques (EFESE) in France (Tibi and Therond, 2017), have identified and valued the variety of services provided by agricultural ecosystems (Schils et al., 2022). Semi-natural grasslands in particular have drawn much attention because their higher floristic diversity supports a variety of services beyond forage production, such as mitigation of global warming, pollination or biodiversity conservation (Manning et al., 2018; Bengtsson et al., 2019). The provision of such services is a crucial step towards the achievement of the United Nations' sustainable development goals (Mehrabi et al., 2020) or climate change mitigation through carbon storage in the soil (Beillouin et al., 2023).

To date, most of the services provided by grasslands do not imply a direct financial compensation; this is the case, for example, for flood control, nutrient cycling or landscape aesthetics. Nevertheless, in the EU, certain services such as biodiversity or habitat maintenance have been somehow recognised by the last Common Agriculture Policy (CAP) programmes: incentives are higher for farms that use large surface areas of semi-natural grasslands and/or that introduce legumes in their crop rotations (Heyl, 2022). In France, a variety of semi-natural grasslands have been eligible for substantial CAP subsidies since 2015, which has contributed to the economic sustainability of many extensive livestock farming systems (Gautier, 2017). Considering the strategic importance of the services provided by grasslands, combined with the large surface area they cover worldwide and especially in the EU, a major issue today is to find ways to maintain or enhance such services through appropriate management decisions at the field, farm, landscape and region/country scales.

Deciding on grassland management is not a simple matter. Despite the fact that the day-to-day management of grasslands is under the sole responsibility of the farmer, a number of strategic decisions may involve (Fig. 1) other farmers (e.g. in the case of crop x livestock integration at a territorial level or for collective summer pastures), agricultural advisers and even land managers (who decide on local regulations about the multi-use of grassland, especially in nature reserves and parks) and public policy officers (who deploy and implement agricultural policies). Thus, strategic decisions about grasslands and their management might be taken at a variety of levels (sub-region, country, local territory, group of farms, individual farm, plot) and by a variety of stakeholders that may interact directly or not. Within this framework, collecting, analysing and sharing data about grasslands and their contribution to agricultural landscapes is of utmost importance in order to ensure the consistency and efficiency of grassland management decisions at the various levels. According to Bahlo et al. (2019) 'there is a need for not only a new type of decision support tool, but also a consensus on data exchange standards to prove the value of shared data at farm scale (commercial benefit) and a regional scale (public good)'.

This chapter describes the diversity of data available about grasslands, and how such data may be analysed and fuel a variety of tools in order to support grassland management decisions at farm and territory scales, with a focus on EU countries. Specific attention is then drawn to the case of semi-natural grasslands that are more difficult to document but that are of high strategic importance. Based on different examples, the potential of combining different

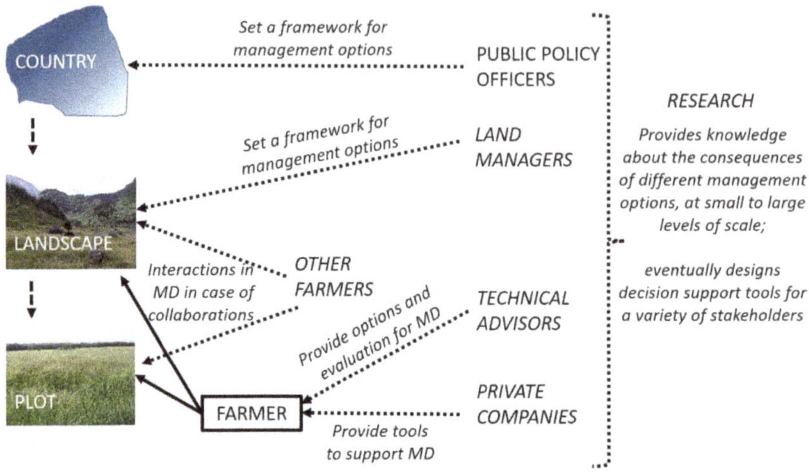

Figure 1 Contributions (dotted lines) of a variety of stakeholders to the strategic management decisions (MD; full lines) made by the farmer at plot and landscape levels. The dashed lines remind that the frameworks set at large levels of scale apply to the smaller levels.

types of data and different tools is discussed, as well as the importance of multi-stakeholder networks.

2 Heterogeneous data about grasslands: an ever-growing resource

Data collected about grasslands (i.e. both indicators of their state and the associated metadata) are heterogeneous in nature, and of variable availability in time and space. The variety of data available enables to address complex and systemic issues such as those raised by global warming or by the agro-ecological transition. For example, studies of drought resistance in grassland, which have long been carried out in semi-controlled field trials, have major limitations inherent in the experimental design and geographical location of the trials (Fraser et al., 2013; Matos et al., 2020). Data may be quantitative or qualitative and describe the amount or quality of forage production during a specific period of time, the diversity of the local flora and fauna, or assess one or more services provided by these grasslands. Data can be collected by a variety of methods: field observations or measurements carried out by humans, sensor-based collection of local conditions, smartphone or drone imagery, agricultural or ecological surveys among local stakeholders, satellite imagery, etc. Depending on the method, data collection may involve one or more stakeholders (farmers, agricultural advisers, environmental or public policy officers, scientists, technical research staff, nature management associations,

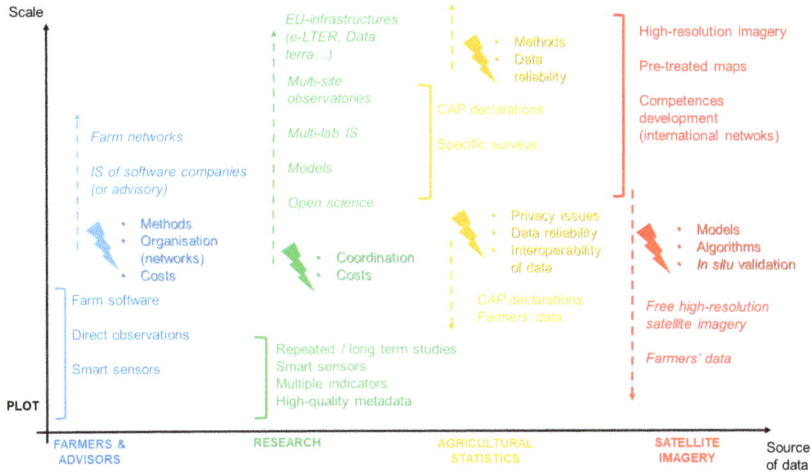

Figure 2 Data about grasslands: sources, levels of integration, characteristics and issues. The plain text refers to the characteristics of the data collected; the text in italics refers to the conditions of upscaling or downscaling the information collected, and the text in bold near the bolts refers to the issues connected to the change of scale.

private companies), as well as more or less sophisticated tools, from the eye of the local expert to the high-tech sensor. The method of data collection, management and storage might impact the availability of the data, its open or private status and the options for analysis and operational use. It also determines the precision and reliability of the data and, therefore, its possible utilisations and the decisions that depend on its analysis. Figure 2 summarises the data available from the four different sources analysed in the following paragraphs.

2.1 Data collected at farm level by farmers or technical services

Data collected at farm scale by farmers and technical services is gathered in large amounts for the purpose of supporting some of the following technical decisions about grassland management: When is the best moment to cut or graze that plot? What amount and quality of forage is available on that plot? Are the current environmental conditions (especially climatic) compatible with the predetermined management plan, or do adjustments need to be made?

In this perspective, the most widespread indicator is probably grass height, generally measured using a sward stick or rising plate metre (Murphy et al., 2021). An example of a large-scale and long-term data collection system designed by research but implemented by farmers is PastureBase Ireland[4], in operation since 2013. PastureBase Ireland (Fig. 3) is a free internet-based

4 https://www.teagasc.ie/crops/grassland/pasturebase-ireland/

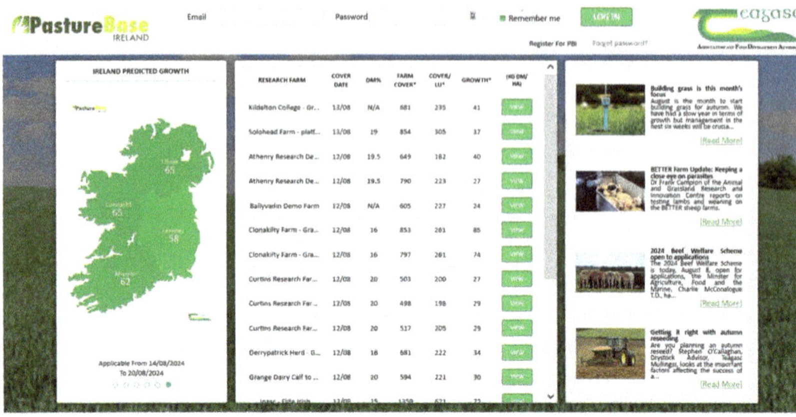

Figure 3 Screenshot of the user interface of PastureBase Ireland. On the left side, the latest predictions of grass growth are displayed for the various counties. On the right side, links to short articles with recent news about livestock farming, among which are popularised research about grasslands and herbivores. In the centre, the latest data from the network of farms. Registration to PastureBase gives access to a variety of tools at the farm scale.

grassland management programme associating a central national database of grassland data with an array of tools to support decisions about grazing rotations, feeding management, fertiliser/slurry application, etc. (Hanrahan et al., 2017). In recent years, the massive adoption of the tool (over 6000 users) has enabled the issue of reliable grass growth forecasts and has provided valuable data for grassland research (Ruelle et al., 2021a, 2021b).

Information about grassland production and utilisation may also be collected in the framework of systematic investigations of farm practices and performances coordinated by extension services. Such investigations are usually carried out in the framework of regional or national networks in order to provide technical and economic references about a variety of farming systems. For instance, in France, the INOSYS '*Réseaux d'élevage*' network, coordinated by the national technical institute for herbivore productions (IDELE), involves over 1500 livestock farms and around 250 technical advisers and engineers dispatched in regional offices. The data collected since it was set up in 1980 is used to analyse the changes occurring within the main farming systems and eventually help stakeholders from the livestock sector adapt to the evolution of climate, markets and policies. Part of such data can be freely accessed on the IDELE website in the form of technical sheets ('*Cas-type*'). A '*cas-type*' consists of the technical and economic description of a model of farm based on the combination of real data from yearly surveys in the network '*Réseaux d'élevage*', and expertise. Information about grazing management, although retrieved in the survey, is not always displayed in the technical sheets, but is available on request.

Until recently, most farm data was only available to those who collected it, and was recorded on paper or sometimes in digital format. Thus, not only was the data invisible to other stakeholders, but it often lacked the necessary metadata required for an in-depth analysis. On the whole, data at the farm scale is limited by its availability, precision and interoperability (Bahlo et al., 2019). The increasing availability of software able to record and display farm data, combined with the deployment of smart sensors, has paved the way to a better availability, reliability and utilisation of farm data. However, the question of the property of such data raises much debate. Data collected with commercial tools and managed by private companies is usually displayed, after automated analysis, on the user interface of commercial software; raw data may be retrieved on request by the farmer.

In a variety of EU countries, the public sector has started to design and maintain data platforms where farmers can store and access their data, and eventually share it with advisers or researchers. PastureBase Ireland is an example of such tools; another, more recent, example is P@stor-all in France (see case study in Section 4.2). Such initiatives are recent, and to date, only a small proportion of farm data is available for large-scale and long-term analysis, due to legal and organisational limitations. Thus, further work is needed in order to use more and better of the large amounts of data collected in commercial farms.

2.2 Data collected within the framework of research programmes (experiments and observatories)

In the last century, research produced huge amounts of data about grasslands, both in experimental and in commercial farms (Dumont et al., 2019). The collection of grassland data by research was first oriented towards the understanding of the mechanisms that determine grass growth and quality and the improvement of forage production. With time, other issues such as adaptation to climate change and provision of non-agricultural services have been tackled in order to raise awareness among farmers and other strategic stakeholders and to support policy-making. Huge amounts of historical scientific data are stored in non-digitised archives. The effort required to make them widely available should be assessed against the interest of such data today, in a different context. Especially, in order to be reused, such data needs to be standardised, contextualised geographically and aggregated to provide a suitable level of information (Capalbo et al., 2017).

Data collected within the framework of scientific programmes is site-specific when collected at the plot or farm scale in experimental stations, or refers to a network of sites and/or farms when collected in an observatory, usually in collaboration with local stakeholders and/or technical services. In the first case, the upscaling and generalisation of the results might be difficult and may require mechanistic models. In the second case, efforts need to be made in order to normalise the type and amount of data collected across sites, to secure the quality and reliability of the subsequent analysis. For instance, huge amounts of data were collected in the framework of the AEOLE project (Carrère et al., 2022), which was a collaborative initiative mobilising a consortium of 80 experts (agronomy, animal science, ecology, botany, soil science, nutrition) and 15 stakeholders. The data originated from a network of 143 plots throughout the Massif Central area, monitored during 2 consecutive years at 3 key periods (early, middle and end of the first cycle of vegetation growth in spring). Overall, 729 agronomical samples were collected, 400 botanical surveys were performed, together with 138 soil analyses and 143 practice surveys. A huge amount of work was required to put together all this data and produce a multifunctional typology of grasslands[5]. Another example of a research-driven observatory is the Biodiversity Exploratories[6], which encompasses 300 grassland and forest plots distributed in three regions of Germany. Since 2006, data have been collected according to a common research design and can therefore be directly analysed and compared to investigate (a) how land use intensification affects biodiversity and species interactions and (b) how biodiversity and land use intensity affect ecosystem functioning and services (Fischer et al., 2010). Figure 4 shows the research design applied to the three interconnected exploratories (biodiversity, land use and ecosystem processes); the combination of various measurement intensities enables to monitor a large area with reasonable investment in time and funds, while the association of monitoring and experiments is useful to generalise the results obtained in controlled conditions and in a few sites. The interest of such an observatory has grown with the involvement of scientists from neighbouring countries and with the collaborative projects carried out with them. To date, a variety of grassland observatories are active worldwide; they are usually managed by research and implemented at regional to national levels, with a focus on the botanical composition, on climate change (carbon cycle), or on the plant and animal biodiversity (for semi-natural environments).

The main advantage of scientific data is that the methods and conditions for data collection are described in detail. Another interesting point is that, for a given site and experiment, a great number of indicators are measured or estimated: forage quality and production, soil characteristics, botanical composition, fluxes of greenhouse gases, biodiversity, etc. In recent

5 Available in digital version at: https://www.sidam-massifcentral.fr/developpement/aeole/
6 https://www.biodiversity-exploratories.de/en/

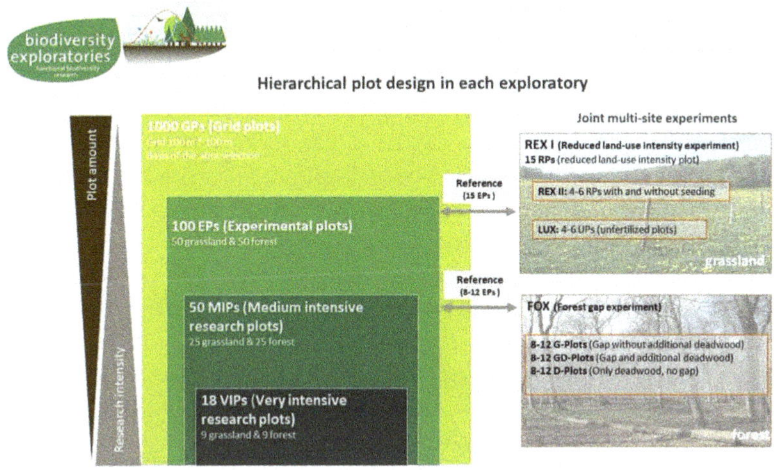

Figure 4 Schematic representation of the research design of Biodiversity Exploratories. Source: https://www.biodiversity-exploratories.de/en/about-us/research-design/. GP: grid plots; EP: experimental plots; MIP: plots of medium research intensity; VIP: plots of high research intensity; REX: reduced land use experiment; LUX: land use experiment.

decades, large efforts have been made to document the diversity of semi-natural grasslands and to acquire long-term data about grasslands and their management within the framework of observatories at regional to national scales (France: Michaud et al., 2016; Carrère et al., 2022; Germany: Morris et al., 2014; Vogt et al., 2019). The amount and availability of scientific data have benefited from the use of smart sensors (Wang et al., 2022), the development of information systems shared among laboratories or institutions such as the Sen2Grass cloud (Hardy et al., 2021) and the general movement towards open science. Such evolutions are favourable to the development of references and simulation models for a variety of environments.

2.3 Agricultural statistics

Agricultural statistics provide large-scale data about land use, farm types and structure, agricultural practices and the economic value of agricultural products (see Eurostat[7], USDA statistical services[8], OECD Agricultural statistics[9], FAO Statistic[10]). Such statistics may be based on administrative declarations or on specific in-depth agricultural surveys carried out at the national or international

7 https://ec.europa.eu/eurostat/web/agriculture
8 https://www.nass.usda.gov/
9 https://www.oecd-ilibrary.org/agriculture-and-food/data/oecd-agriculture-statistics_agr-data-en
10 https://www.fao.org/statistics/en

scale. In the EU, the common CAP framework implies yearly declarations by farmers and, therefore, the collection of large amounts of data. However, CAP regulations and indicators may vary among member states, contributing to uncertainty in the analysis at the international level. For example, in recent years in France, the conditions for declaring extensive pastures as agricultural areas have changed from a utilisation-based approach (the grasslands must be used every year, with a minimum stocking rate) to a condition-based approach (the grassland must be fit for grazing, and a pro-rata principle is applied based on the assumed productivity in order to calculate the CAP subsidies). As a consequence, the estimated area of permanent grassland has increased in the regions dominated by large-scale pastoral systems.

Despite their limitations, agricultural statistics have proven useful in order to monitor major land use and management changes over time, such as the distribution and long-term evolution of grassland types, mainly in terms of surface area of semi-natural grasslands and, for temporary grasslands, in terms of plant species sown. For example, Estel et al. (2018) combined satellite image time series and agricultural statistics to produce classes of grassland management intensity based on information about mowing frequency and livestock density, and assess their distribution within continental Europe. Generally, agricultural statistics are an interesting data source for measuring the impact of agricultural policies and shaping future national or international policies. Nevertheless, the precision and reliability of such data are difficult to estimate. The application of international standards for the collection of a few key indicators would be an asset for the large-scale utilisation of such data.

At an international level, other approaches are possible: the Land Use/Cover Area frame Survey (LUCAS[11]) is a harmonised *in situ* land cover and land use data collection exercise that extends over the EU's territory. The survey is conducted in two phases: (1) systematic sampling of approximately 1.1 million points spaced 2 km apart in the four cardinal directions, and photo-interpretation in order to assign them to a pre-defined land cover class; and (2) direct observation on a stratified sample of points by local field surveyors. Figure 5 shows the visualisation of the latest LUCAS land use survey.

2.4 Satellite imagery and unmanned aerial systems

In the last decade, the development of remote sensing based on satellites or unmanned aerial systems (commonly named 'drones') has produced more and more data, freely available and of increasing resolution. Satellite images contain an immense amount of information, collected *a priori* without consideration of its further analysis. The interest in grassland science and management lies in

11 https://ec.europa.eu/eurostat/web/lucas

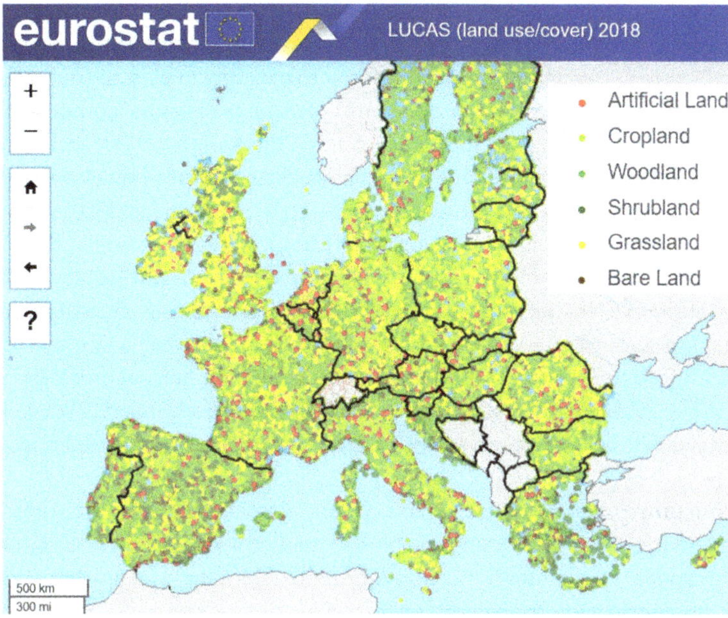

Figure 5 Map of land use in Europe produced on the basis of LUCAS data collected in 2018. Available at: https://ec.europa.eu/statistical-atlas/viewer.

the analysis and interpretation of this data. Various steps are required to obtain indicators concerning the grassland state: (1) choose the images with the least amount of missing information (little or no clouds); (2) calculate indicators correlated with useful information about grasslands (typically, Normalized Difference Vegetation Index or infrared spectral bands); and (3) develop and apply statistical models to estimate sward characteristics, for example forage mass and quality (Bareth, 2021). A variety of applications (mapping grassland types, monitoring of harvests, monitoring of grass growth, measurement of hedgerow lines, estimation of forage abundance and quality) are available from open sources or as a commercial product.

A strategic utilisation of satellite images is to determine land use classes. The CORINE Land Cover database[12] was first created in 1990 and has since provided essential information on European land cover/land use. The most recent update was made in 2018, based on satellite images from Sentinel-2 and Landsat-8 for gap filling, broken down into 44 thematic classes with a minimum mapping unit of 25 ha. Gallardo and Cocero (2023) analysed the uses of such a classification in the last decade and noted an increasing interest as a means of tracking deforestation, pressure on protected natural areas and soil

12 https://land.copernicus.eu/en/products/corine-land-cover

erosion risk. Although extremely valuable, the CORINE Land Cover database has a resolution grid that is too wide to address issues at farm or landscape scales. Besides, grasslands are distributed into a variety of classes (ten different classes, with only 'Permanent pastures' and 'Natural grasslands' composed only of grasslands), which makes it difficult to track the surface areas and effective uses of the diversity of grasslands. Alternative classifications have been tried in the EU at the local and national levels in order to more precisely assess strategic land uses connected with CAP subsidies (Campos-Taberner et al., 2023). The increasing availability of high-resolution free satellite data will certainly lead, in the near future, to the production of high-resolution land use maps that include information about agricultural and non-agricultural services. Two major issues will be (1) to determine and discriminate types of semi-natural grasslands that make sense in terms of management and services rendered; and (2) to validate such maps with *in situ* observations in order to ensure their reliability at local scales.

Unmanned aerial systems have been increasingly used in grassland ecosystem monitoring, focusing mainly on grassland vegetation and especially on the spatial distribution of aboveground biomass, on quality indexes based on chemical composition, on biodiversity indexes or on the presence of undesired species (see reviews of: Ali et al., 2016; Lyu et al., 2022). For vegetation quality assessment, which is to date the most common application of UAS in grasslands, the methods of data analysis are evolving from simple regression to machine learning algorithms. Despite continuous progress, remote sensing with UAS is still limited by technical performance (quality of the sensors, stability of the vehicle), cost (price of the sensors), reproducibility of monitoring and analysis methods across sites and practical applications for decision-making (see reviews of: Astor and Geipel, 2021; Lyu et al., 2022). A promising perspective is to combine unmanned aerial systems with ground and satellite data in order to provide reliable information about grasslands at large spatial scales.

3 Tools to analyse grassland data and support management decisions at different levels

Decision support tools may come in a variety of forms, but have a common function: to use data in order to represent the state of a system and to put it into perspective, thus contributing to identifying solutions that can be implemented through management decisions. Recently, the H2020 SuperG[13] project published a review of existing tools used by farmers, advisers, policy-makers and other stakeholders to support permanent grassland management

13 www.super-g.eu

within the EU (Sagoo et al., 2020). Most of the 127 tools identified had not been published in the scientific literature. Fifty-nine per cent of them were based on local data and thus specific to a particular country or region, and only a small proportion of the rest was available in more than one language. Most tools (56%) included at least one software element, and 'traditional' paper-based tools (usually available to download from the internet) were still common (44%). Software-based tools make it possible to save and store data and, consequently, to keep records as well. Such data might be pooled at a regional scale in order to feed simulation models and to add a prospective dimension. Almost all tools were intended for farmers and often their advisers as well; less than 15% included policy-makers in the target audience. Most tools (77%) were free or had a free version available, while much of the commercial farm management software involved paying a monthly or annual subscription. Thus, although very little information is available about the number of users, we may consider the availability of decision support tools for grassland management to be good.

The criteria differentiating the existing decision support systems for grassland management are summarised in Fig. 6. Historically, technical sheets based on reference data and aimed at operational decision-making by farmers

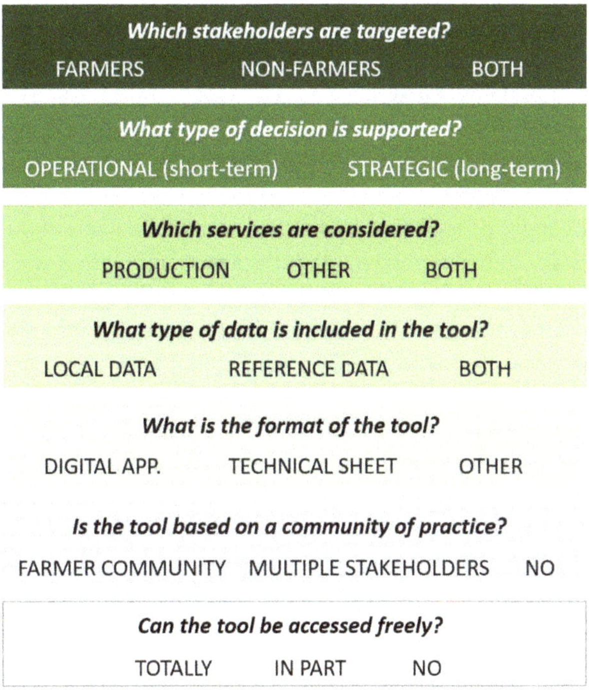

Figure 6 Main criteria that differentiate the various decision support tools for grasslands (based upon Sagoo et al., 2020 and expertise from the authors).

were most common. Since the 2010s, decision support tools based on farm data, including strategic decisions and other services other than grassland productivity, have been increasingly available. The most recent tools often rely on communities of practice and come in 'new' formats such as apps and serious games; they are usually based on data collected at a local scale by local stakeholders (farmers, technicians and even citizens), and are designed to meet at least one of the following objectives: (1) optimise the use of the forage resource in terms of animal production and self-sufficiency at the farm level; (2) document the variety of agricultural and non-agricultural services rendered by grasslands, at the farm and territorial scales. This second objective might interest a variety of stakeholders, including land managers. The availability of free broad-spectrum decision support tools is an asset to building a shared vision among stakeholders. In order to include extensive or marginal systems, which have a limited investment capacity, such tools should be free of charge.

3.1 Tools to optimise the use of grassland from an agricultural perspective

Many tools supporting technical decisions about grassland management are intended to optimise the use of the forage resource at the farm level. Optimisation may rely on grass rotations, on best dates for grazing/cutting/fertilising a paddock, or on the monitoring of the available forage stock at the paddock and farm scales. Optimisation tools are usually based on data about forage biomass, extrapolated from measurements of grass height and sometimes combined with simulation models of grass growth that take meteorological conditions and farm production objectives into account. In Ireland, a variety of tools are available: on the one hand, the free national grassland database (PastureBase Ireland[14]) associated with a web application and with a weekly report on television; on the other hand, commercial tools such as the 'GrasshopperTM' measurement system (a high precision plate metre) to be associated with the GrasslandTools[15] software at the farm scale, which provides an overview of the grassland state and management and of herd condition if the system is connected to an automated weighing device. Similar to PastureBase Ireland, the '*Pâtur-Plan*' French tool (Delaby and Bignon, 2015) collects grassland data at paddock and farm scale and provides the farmer with simulation results for possible management strategies. This may trigger discussions between the farmer and the adviser, and encourage management changes. In Portugal, a free smartphone application was recently released in order to help traditional farmers manage grazing, by providing means to

14 https://pasturebase.teagasc.ie/V2/login.aspx
15 https://www.grasslandtools.ie

record, calculate and visualise the amount and distribution of grazing pressure within the Montado (Sales-Baptista et al., 2018); a similar tool is also available in Argentina (PastorINTA[16]). Such software-based decision support tools are most efficiently disseminated and used when associated with a network of farms (O'Brien et al., 2015). If more than one service is considered, the optimisation logic gives way to a sub-optimal agricultural utilisation of grasslands, which improves adaptation to hazards and reconciles different services.

3.2 Tools to document and discuss services rendered by grasslands

Another important function of grassland decision support tools is to document the variety of services rendered by grasslands, at the farm and territorial levels. This second objective might interest a variety of stakeholders. A number of tools have been developed in order to assess and discuss the impact of management decisions on the provision of services by grasslands and grassland-based farming systems.

In the early 2000s, simulation models at the farm scale enabled research to explore possible compromises between agricultural production and biodiversity conservation, carbon stocks, nutrient losses, etc. For example, the SEBIEN model investigated trade-offs between agricultural production and biodiversity conservation in grass-based suckler cow farms in the medium mountains of Central France (grassland model: Jouven et al., 2006a, 2006b; global model: Jouven and Baumont, 2008). The utilisation of such a model with contrasted case studies showed that the effects of biodiversity-friendly management practices (in terms of biodiversity enhancement and animal production) varied widely depending on the local conditions and on the integration level considered (paddock or farm). The DairyMod and EcoMod simulation models developed for Australia and New Zealand (Johnson et al., 2008) simulated not only grassland production in response to management, but environmental impacts, such as leaching and greenhouse gases (GHG) emissions, as well. At the landscape level, Sabatier et al. (2010) explored the trade-offs between livestock grazing and the conservation of bird populations. Most of these models, although simulating management decisions or their consequences, were not intended to be used as decision support tools, but rather as research tools to produce knowledge; nevertheless, the fact that such models were designed and calibrated on the basis of local situations makes that knowledge meaningful to the local stakeholders and eventually useful for decision support.

16 See official communication in 2023: https://intainforma.inta.gob.ar/pastorinta-una-app-para-calcular-la-superficie-de-pastoreo/

In the same period of time, grassland typologies including information about non-agricultural services were published in a variety of countries, including Italy (Cavallero et al., 2007) and France (Galliot et al., 2020). Grassland typologies are highly effective tools for organising the diversity of grasslands found in a given area (region, country) and providing a concise, accessible description. Furthermore, when they draw connections between botanical diversity (taxonomic and functional), environmental conditions and management strategies, they make it possible to assess the functioning of the plant communities and to evaluate the potential ecosystem services they provide (Michaud et al. 2013).

Since the 2010s, serious games at farm and landscape levels have been designed for the purpose of educating a variety of stakeholders and/or providing information to stimulate discussions and support decisions in multi-stakeholder groups. Examples of such games are Vivagrass[17] for various regions within the EU or AEOLE in France (Carrère et al., 2022; Fig. 7). The latter was designed to raise awareness among a variety of local stakeholders about the diversity of grassland types and the importance of agreed individual decisions in the achievement of a common collective objective. It capitalises on grassland typologies to describe different types of grasslands, which may be associated with a farm to produce different levels of ecosystem services. AEOLE can be used at the farm or landscape level to test the impact of management practices on the services provided by grasslands and to assess the benefits of diversifying land parcels in response to random climatic events.

Only a small fraction of decision support tools are described in scientific or technical publications, thus it is difficult to assess their number. Based on the survey carried out in the SuperG project, a small proportion of decision support tools would take both the agricultural and non-agricultural services of grasslands into account. The vast majority of the decision support tools for grasslands studied in the SuperG project considered the agricultural services (provision of forage and other products); 15% or more documented non-agricultural services such as biodiversity and pollination, landscape and recreation, or water quality; and less than 8% considered flood and erosion control or carbon storage and GHG. Assessing multiple services implies considering different scale levels and requires the development of metrics suited to be scaled up from plot to landscape (Stokes et al., 2023); the priority is to produce integrated information from data that is easy to acquire or access. Taugourdeau et al. (2024), based on academic knowledge and expertise, identified and aggregated several vegetation criteria to construct indicators of the services provided by grasslands. Such indicators were calculated based on over 2000 botanical surveys distributed over a wide biogeographical

17 https://vivagrass.eu/integrated-planning-tool/

Figure 7 AEOLE, the game[1]. (a) Group of technical advisers and teachers playing together; (b) example of cards representing grassland types, with information about agricultural, environmental and socio-cultural services; (c) boards representing individual farms; up to five grassland types may be associated (upper right corner); farm scores may be calculated based on the scores for each grassland type (tables on the left and in the centre); the figure on the bottom right corner represents the evolutions paths between grassland types, and thus enables the players to identify which management changes need to be applied in order to diversify the available semi-natural grasslands.

1 https://www.sidam-massifcentral.fr/developpement/aeole/

gradient. Their results show that an aggregation method based on a large dataset of botanical surveys could be appropriate for studying the temporal dynamics of services and useful for managers to evaluate the impact of their practices.

3.3 Issues for grassland management decision support tools

To date, only part of the data collected about grasslands has been analysed, included in decision support tools, and effectively used for decision-making. The proportion of data used should increase with the development and intensification of farm networks and large-scale observatories, which make it possible to collect with similar methods, gather, normalise and analyse data at various spatiotemporal resolutions. Such networks and observatories also have the advantage of triggering collaborations between various stakeholders, with a shared tool and complementary objectives. Their major drawback remains their funding: who should pay for their development, and how to secure long-term financial resources for their maintenance.

Another issue will be to cross-analyse data from various sources, at various spatial resolutions. Presently, the upscaling of results obtained at the plot or farm scale might be difficult to implement due to the high impact of data collection methods and geographical location. Conversely, large-scale data are rarely applicable at local or farm scales due to the large impact of local conditions on the actual state of grasslands and on the provision of agricultural and non-agricultural services. In order to secure the consistency of grassland management decisions at farm to country scales, tools able to provide indicators at multiple spatiotemporal resolutions would be most useful. In order to achieve this, they should be able to cross-analyse satellite imagery, scientific data and farm technical data.

To be useful and used, decision support tools considering a variety of grassland services should combine indicators from academic research for credibility and precision and empirical results to allow the assessment of management impact. Stokes et al. (2023) point out three qualities that indicators of agricultural and non-agricultural services should have in order to be used for decision-making: salience, legitimacy and feasibility. Two main issues are raised by the authors: (1) the necessity of including both ecological and socio-economic indicators; and (2) the co-design of indicators with local stakeholders in order to increase their operability.

Besides, there is a need for tools able to bring together public and private data for decision support (Bahlo et al., 2019). The effective adoption of decision support tools by end-users is rarely documented. Tracking the number of users is possible for tools that require opening an account; even then, it can be difficult to tell if all accounts are equally active unless a special tracking system is implemented. The uptake of decision support tools is much easier and more efficient when they are associated with a network of stakeholders (usually at a local or national level). Commercial tools that do not count enough users will probably disappear soon enough. Non-commercial tools should be managed in close collaboration with a community of users in order to improve the existing ones and, if needed, propose complementary or alternative solutions.

4 Case study: semi-natural grasslands in the mountain and Mediterranean areas of France

Semi-natural grasslands are common in areas that have not undergone agricultural intensification due to unfavourable pedo-climatic conditions. They include wet areas, mountain and sub-alpine meadows and pastures, grazed steppes and dry pasture, grazed wooded areas and grazed/browsed shrubby zones (Peeters et al., 2014; Fig. 8). They are characterised by high plant diversity

Figure 8 Examples of semi-natural grasslands across France: (a) mountain pasture in the Pyrenean; (b) Causse rangeland in the calcareous plateaux of southern Massif Central; (c) wetland in the Camargue area; (d) *garrigue* rangeland in the Mediterranean area (photo credit: Magali Jouven).

unevenly distributed over space, including edible and non-edible plants and plant parts (Marriott and Carrère, 1998). Due to the harsh conditions, annual and non-grass species are abundant and imply strong seasonal variations in the amount and quality of forage; these dynamics are enhanced by the extensive management practices applied (Jouven et al., 2010; Carrère et al., 2018; Etienne et al., 2021). The consequences for data collection in view of decision-making are: (1) the sward cannot be represented by a single indicator; (2) the question of spatial resolution/spatial unit becomes crucial; and (3) metadata about the ecological, technical and seasonal context is absolutely necessary.

As mentioned earlier in this chapter, the intrinsic characteristics of such grasslands and the diversity of services they support call for sub-optimal management for agricultural utilisation (Bardgett et al., 2021). Moreover, the fact that the current state of semi-natural grasslands (both in terms of conservation and of forage resource) depends mainly on the medium to long-term utilisation by livestock implies adding a long-term objective (i.e. vegetation stability or targeted changes) to the specifications of grassland management. Decision-making about semi-natural grasslands thus aims at two types of compromise: between a variety of services and between short-term production and long-term conservation of the vegetation (Jouven et al., 2010).

4.1 Data acquisition in semi-natural grasslands still raises methodological issues

The first necessary step towards data acquisition in semi-natural grasslands is to describe the macro-heterogeneity of the vegetation. This can be performed by visually identifying *topo-facies*, i.e. areas with similar slope and vegetation cover and, thus, with similar ecological and pastoral issues. Ideally, each *topo-facies* should be characterised, whereas often, only certain areas are studied, making it necessary in this case to contextualise the results by considering the diversity within the management unit (paddock or landscape). Due to the major long-term impact of management on the dynamics of grassland vegetation, and the interdependency between the utilisation of different management units (especially when it comes to grazing), metadata about grassland management at paddock and farm scale is also required. A comprehensive method for the eco-pastoral characterisation of semi-natural grasslands from *topo-facies* to farm was designed in the framework of the Mil'Ouv project (Lepart et al., 2019), where technical advisers and engineers in agricultural and environmental fields worked together with a network of farmers. The Mil'Ouv method[18] is entirely based on visual assessments, and has been adapted for educational purposes and for application in other geographical contexts (Albania). Although time-consuming to implement, it has provided an interesting alternative to the traditional approaches based on utilisation or animal behaviour.

Increasingly, farmers invest in collars tracking the location of livestock with Global Positioning Systems (GPS), which provides information about the spatial behaviour of the herd that may be interpreted in terms of the attractiveness of different *topo-facies* (Llaria et al., 2024). Usually, the company selling the collars also provides an application in order to visualise herd behaviour on a map, and the raw data can be accessed on request. In the areas where agri-environmental measures are contracted, technical advisers need to record the state of the grassland and its utilisation (management practices). Unfortunately, such information is usually imprecise (as reported by the advisers themselves) due to the limited time available to implement the assessment, and in any case, the data is not in a format that could support further analysis. Data may also be collected through interviews with farmers, but without a shared and simple measurement tool (such as the plate metre for homogeneous grasslands), the information might be strongly dependent on the individual framework of reference.

In the field of research, temporary or permanent grasslands have mainly been investigated by agricultural science, while semi-natural grasslands have been mainly studied by plant ecologists. Fortunately, research about plant

18 https://idele.fr/detail-dossier/methode-milouv

functional traits has established bridges between these apparently contrasted approaches. For example, functional ecology approaches have shown that intra-specific diversity (Litrico and Violle, 2015) and interspecific diversity (Hector et al., 2010; de Mazancourt et al., 2013), through processes of niche or function complementarity (growth strategies, space occupation methods, temporality and seasonality of growth), are a source of stability in the functioning of plant communities. Studying semi-natural grasslands from an agricultural perspective still raises methodological issues for research (Fig. 9): developing experimental designs is difficult, given the diversity of plant communities, the uneven spatial distribution of plants and the high number of interacting factors. Thus, to date, most research has either focused on the long-term evolution of plant biodiversity (Louault et al., 2017) or on the direct observation of animal behaviour over seasons and years (Baraza et al., 2006; Azambuja et al., 2020). Given the number of plant species co-existing in semi-natural grasslands, a key issue in order to better document the forage availability is to organise plant diversity into categories which make sense for the services of interest. For forage services, such categories need to take animal behaviour into account, and especially the bite size allowed, the plant part ingested, or the overall chemical composition of the latter (Agreil et al., 2004; Bonnet et al., 2015; Mikicic et al., 2023). In any case, the research methods involve close, repeated or long-term monitoring and are highly time-consuming and do not

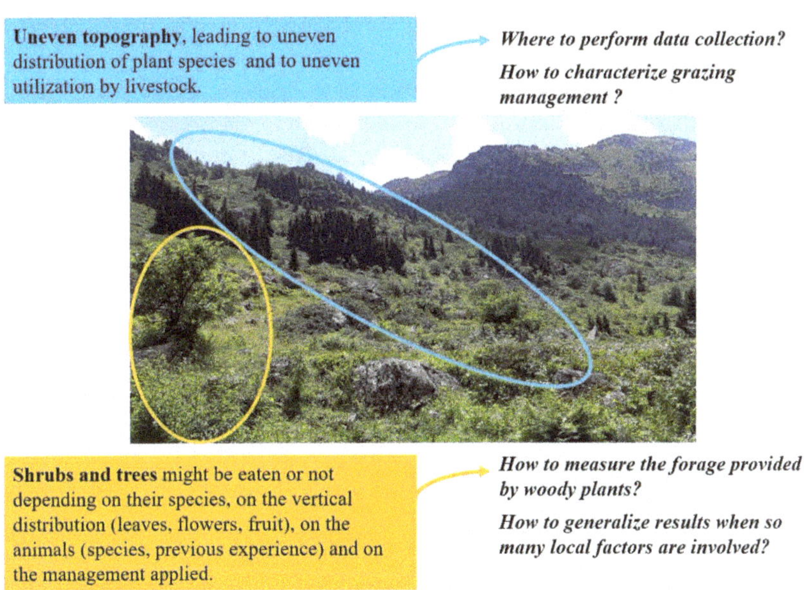

Figure 9 Key features of heterogeneous semi-natural grasslands which are most challenging for research (photo credit: Magali Jouven).

entirely solve the question of result generalisation, unless they are multiplied over many sites, in the framework of networks of farms or experimental sites.

At the landscape and country levels, the acquisition and interpretation of data about semi-natural grasslands is also difficult. In agricultural statistics, different types of semi-natural grasslands may or may not be included in the CAP declarations, depending on the country, on the declaration strategy of the farmer and on the presence or absence of a contract to access the land. Satellite data is a promising source of information, although a number of methodological complications associated with the presence of rocky outcrops and woody plants still need to be solved (Shaqura and Lasseur, 2021; Kalenga et al., 2022a). This is especially the case with pictures taken from the sky that represent only part of the herbaceous layer in shrubby or wooded areas: the woody plants either cover the sward or project their shadow on it, changing its colour.

4.2 Combining data sources and stakeholders in order to produce useful tools

Each source of data about semi-natural grasslands provides incomplete information: farmers and technical advisers collect site-specific data with little associated metadata, which is difficult to upscale and mainly serves the development of their own expert knowledge. Science documents multiple factors, but in a small range of conditions that are far from the variety of situations encountered in the field. Public data, such as agricultural statistics or satellite imagery, provide incomplete information due to the limitations of the underlying data acquisition methods. Nevertheless, a number of useful and powerful decision support tools are available for semi-natural grasslands. Such tools have the combination of technical data, scientific data and expert knowledge in common, as well as the support of a multi-stakeholder network. Three examples of tools recently developed in France are presented hereafter (Fig. 10).

Pastothèque (Dodier et al., 2023; Bonnet, 2025) is a book with a typology of mountain and Mediterranean grasslands that summarises the current knowledge about the diversity, vegetation dynamics and services of semi-natural grasslands. The latter are categorised based on pedo-climatic conditions and vegetation characteristics (especially the presence and type of woody plants). Each type of semi-natural grassland is extensively described, with a variety of indicators pertaining to agricultural and non-agricultural services. The main difference with other typologies is that seasonal and long-term dynamics in response to management and climatic hazards are included, and the evaluation of services is related to grassland management, especially in terms of animal species, season of utilisation and grazing intensity. Pastothèque is the result

Figure 10 Complementary functions (capital bold) of the three decision support tools for semi-natural grasslands in France presented in this chapter. K: contribution to knowledge enhancement; E: contribution to exchanges between stakeholders; DS: contribution to decision support.

of years of collaboration between technical and environmental services, with the contribution of research. Although the tool would benefit from a simplified digital version, it provides a valuable and functional classification of semi-natural grasslands, which is a necessary prerequisite for the acquisition and interpretation of data.

The Pâtur'Ajuste[19] network (Agreil et al., 2011) includes farmers, technical advisers, environment officers, teachers and a few researchers. It was created in the 2010s to trigger discussions and innovations about the technical aspects of pasture management, focusing on heterogeneous semi-natural pastures and rangelands. The network's activities include the organisation of technical exchange days and short training courses, the creation of technical sheets combining scientific and expert knowledge, and participation in education and applied research programmes. The results of such activities are available on a dedicated website. Ten years after its creation, the network is well known all over France, is recognised by a variety of institutions, and has helped many farmers to improve the utilisation of semi-natural grasslands. To date, it counts around 220 members distributed in 55 departments of continental France. The main

19 www.paturajuste.fr

reason for this success is probably the fact that it has provided a bottom-up national alternative to the traditional top-down regional advisory services.

P@stor-all[20] (Kalenga et al., 2022b) is an information system on a computer or smartphone devoted to the management of semi-natural grasslands. Co-designed with farmers, it gathers heterogeneous data: satellite images, GPS tracking of herds, links to technical sheets, popularised scientific results and, last but not least, feedback from farmers pertaining to their own experiences on the subject. The information system is designed mainly for farmers; advisers and education institutes interacting with them may also contribute. Farmers have access to: (1) a private area where they can upload GPS tracks and monitor various indicators of herd spatial behaviour; and (2) a public area where they can find expert knowledge, technical and scientific information about various subjects, and a forum. The information system is free and is based (similarly to PastureBase Ireland) on the exchange of services between stakeholders: farmers' data gives researchers insight into the diversity of situations, while scientists provide farmers with free analysis of farm data and access to a large pool of knowledge. The tool will be disseminated in France in 2024 and 2025. Its success will very much depend on the participation of a sufficient number of farmers and the interest of other stakeholders who may use it to more efficiently interact with the farmers, or to complement training courses.

The three tools presented previously have two objectives in common: (1) to increase the knowledge about the functioning of the grazed ecosystem and (2) to provide practical examples (described in a more or less sophisticated way) of grassland management options, with their consequences in terms of services provided and long-term evolution of the vegetation. The difficulty of providing general knowledge about semi-natural grasslands is overcome by the combination of a variety of case studies, general knowledge and local knowledge. Especially, local knowledge is valued and enhanced by the associated network. The three tools are also an asset for the interactions between stakeholders, for example between the farmer and his adviser, since they trigger reflexion, critical analysis, and raise operational questions while providing a common framework. In all cases, the user does not receive advice or data about the state of his grasslands, but rather has access to a bunch of information which might help to evaluate the services that could be provided by a given grassland, and identify management options in order to perform them best. The fact that these three tools are different in nature (book, network with website, application on computer or smartphone) but share a common approach to decision-making makes them highly compatible, and encourages a complementary use (Fig. 10).

20 https://pastorall.supagro.inrae.fr/

5 Contribution of research in decision-making in grasslands

Research is not directly involved in the process of decision-making about grasslands (researchers do not make decisions), but it has many key roles to play in this respect.

Regarding data collection, it provides reliable data with ample metadata for a range of local situations; the wider the range and the longer the period of time, the more interesting the dataset, because it may support the generalisation of scientific results. In this respect, observatories are powerful tools, especially if they involve a multi-stakeholder and possibly international network. The development and management of observatories is costly and time-consuming; thus, compromise needs to be made between the number of sites monitored and the density of data collected. An efficient solution is to design observatories associating highly monitored experimental sites with lightly monitored sites in real farms.

Research also has a key role to play in providing robust methods and tools for data collection and analysis. Co-designing such methods and tools with farmers, technical advisers or other stakeholders is an asset to build a common framework and to share the burden of data collection. Images from satellites and UAS need specific treatments, which involve a close collaboration between agronomists (in the broader sense) and IT scientists. Two key aspects are involved here: on the one hand, the pre-treatment of raw data in order to compensate for incomplete datasets (for e.g. clouds on satellite images) or interference (for e.g. shade of trees); on the other hand, the design and validation of different models predicting agricultural or non-agricultural services on the basis of the available information. Similar issues and procedures may also be applied to animal-borne sensors. To compensate for the limitations of each source of data, combination and cross-analysis of heterogeneous data can be performed; especially, expertise may be used in order to compensate for the lack of data or metadata.

Moreover, research may participate actively in the development of decision support tools. To ensure the relevance and dissemination of such tools, they are preferably developed in the framework of participative projects. The contribution of research to the design of robust tools consists of the provision of reference data and models for the design and validation of tools. Research may also contribute to the design and management of the participative process itself.

Finally, research contributions to data collection, analysis and to the design of decision support tools are more meaningful if integrated in the framework of a multi-stakeholder group. Building such groups and making them work effectively is not simple, which encourages to focus on a few local groups, scattered in the territory and possibly connected in the framework of large projects.

6 Future trends in research

Data about grasslands is already diverse and abundant, and should further increase in the future. Rather than collecting more data, the issue is now to organise it and make it available and interoperable, with observatories and information systems that gather long-term information about grasslands. Three critical characteristics of grassland data are: (1) its availability in digital format, (2) the methods of collection and (3) the associated metadata. Research and public institutions have a key role to play in order to propose and disseminate reliable data collection methods and tools which may be used by a variety of stakeholders, but also manage databases designed to gather large amounts of data and support a mutual benefit relation between research (who will cross-analyse data to increase grassland knowledge) and farmers (who will use the associated tools and indicators for decision support). The addition of local knowledge to the pool of data would be an asset, for example, to document services for which no other data acquisition method is currently available. Research should also progress in the analysis of satellite imagery in order to extract spatial information about grassland types and production, which may help to extrapolate on-farm data collected at a local scale, especially for semi-natural grasslands, which are more diverse and more difficult to characterise with simple methods.

Decision support tools already exist in most European countries. Such tools are often adapted to the regional/national agricultural context, and especially the dominant type of grassland and the main objective of grassland management. In this respect, there are two strategies that require different data and different types of decision support tools: (1) optimisation of grass utilisation, which can be based on a single indicator of grass height or biomass; (2) sub-optimal utilisation, which requires understanding and comparing different services provided by grasslands. Whatever the strategy, decision support tools may help the farmer to develop local knowledge about his grasslands and, if associated with simulation models, to compare the results of different management strategies. The main difference is that sub-optimal utilisation may involve other stakeholders, and thus require tools and networks shared between farmers, naturalist technicians, land managers, etc. For all types of decision support tools, the inclusion of uncertainty and hazards in the analysis and predictions is of utmost importance in the present context of climate change. Another critical aspect is to improve the methods for cross-analysis of grassland data in order to represent the multiple characteristics and services of grassland in the best possible way, at various levels of scale. This is especially true for semi-natural grasslands, which are at the crossroads of multiple agricultural and non-agricultural objectives.

7 Conclusion

Data about grasslands is abundant but diverse in nature, with limited interoperability between sources. Presently, we may assume that only a small amount of such data is effectively used for decision-making, although the decision-making process itself is difficult to document. There is certainly room for improvement in the availability of grassland data to decision-makers, and in this respect research has a key role to play in order to gather data in digital format, qualify it with suitable metadata and use it to fuel decision support tools. Interdisciplinary reseach projects associating a variety of stakeholders are needed to combine satellite imagery, agricultural statistics, measurements from observatories and local observations from farmers and shepherds. Such combination of data, and the role of expert data from farmers and advisers, is particularly important for the management of semi-natural grasslands, which diversity complicates the acquisition of scientific data and the interpretation of satellite images. The tools and approaches developed for semi-natural grasslands support a sub-optimal utilisation of the vegetation, which is a prerequisite to reconciling a variety of services. Similar considerations could benefit more intensively managed grasslands, in a context of climate change and of agricultural policies increasingly supporting ecological services.

8 Where to look for further information

Many review articles are cited along the text, listed here, and can be easily found with the information given in the 'References' section:

Ali et al. (2016); Astor and Geipel (2021); Bahlo et al. (2019); Bareth (2021); Bengtsson et al. (2019); Brown and Fagerholm (2015); Dumont et al., (2019); Jouven et al. (2010); Llaria et al. (2024); Lyu et al. (2022); Marriot and Carrère (1998); Murphy et al. (2021); Sagoo et al. (2020); Stokes et al. (2023); Wang et al (2022).

Generally, most research and innovation in grassland science is shared within the meetings of the EGF (European Grassland Federation, https://www.europeangrassland.org/) and the IGC (International Grassland Congress, https://www.internationalgrasslands.org/). The IRC (International Rangeland Congress, https://rangelandcongress.org/) focuses on semi-natural grasslands in harsh conditions across the world, with multidisciplinary approaches.

In the last 5 years, a few major projects have been carried out at a European level. Information about such projects is usually available, for a time, on a dedicated website. A list of projects (in alphabetical order) is provided:

- INNO4GRASS (www.inno4grass.eu) was a thematic network project involving stakeholders from eight European countries. The aim was to put forward farmers implementing innovative practices adapted to the current challenging context, and bridge the gap between practice and science communities to ensure the implementation of innovative systems on productive grasslands.
- SHOWCASE (www.showcase-project.eu) involves 23 organisations across 15 European countries, with the aim to facilitate the integration of biodiversity into farming practices. In order to deliver new insight and innovative tools facilitating the agricultural sector's transition towards more sustainable farming, leading scientists in the field of agro-ecology and socio-economy join forces with farmer and citizen science networks, science communication specialists and nature conservation NGOs.
- SUPER G (www.super-g.eu) involved a variety of stakeholders from 13 European countries. Using complementary approaches, the project enabled to better understand the importance and functioning of permanent grasslands, benchmark their performance across the EU, and co-develop sustainable grassland management options and also tools and policy mechanisms.
- TRANSECT (www.agroecology-transect.net) stands for Trans-disciplinary approaches for Systemic economic, Ecological and Climate change Transitions. Innovation hubs and on-farm trials often involving grasslands are carried out in ten European countries.

A few grassland observatories, at the national or international level, provide valuable data on the long-term:

- LUCAS database (Europe): https://ec.europa.eu/eurostat/web/lucas.
- EU Grassland watch (Europe): https://ec.europa.eu/eu-grassland-watch/.
- Biodiversity exploratories (Germany): https://www.biodiversity-exploratories.de/en.
- Sentinel mountain pastures program (France): https://portal.osug.fr/Sentinel-Mountain-Pastures-program and https://www.alpages-sentinelles.fr/.

Finally, a number of decision support tools have been cited in this chapter. We report here those which can be accessed – at least in part – with a free digital version, in alphabetical order. Obviously, a long list of tools could be established; these are only a few examples:

- AEOLE (typology, diagnosis tool, serious game): https://www.sidam-massifcentral.fr/developpement/aeole/.

- MIL'OUV (diagnosis and education tool): https://idele.fr/detail-dossier/methode-milouv.
- PARTUR'AJUSTE (network and technical sheets): https://www.paturajuste.fr.
- P@STOR-ALL (web application): https://pastorall.supagro.inrae.fr/.
- PASTURE BASE Ireland (web application): www.teagasc.ie/crops/grassland/pasturebase-ireland/.
- VIVAGRASS (serious game): https://vivagrass.eu/integrated-planning-tool/.

9 References

Agreil, C., et al. (2004). 'GRENOUILLE: une méthode pour gérer les ressources alimentaires pour des ovins sur milieux embroussaillés'. *Fourrages*, 180, 467–481.

Agreil, C., et al. (2011). 'La gestion pastorale des milieux naturels: mise en oeuvre des MAE-t et gestion adaptative avec la démarche PATUR'AJUSTE'. *Fourrages*, 208, 293–304.

Ali, I., et al. (2016). 'Satellite remote sensing of grasslands: from observation to management'. *Journal of Plant Ecology*, 9(6), 649–671.

Astor, T. and Geipel, J. (2021). 'Remote sensing for grassland quality assessment: status and Prospects'. *Grassland Science in Europe*, 26, 30–34.

Azambuja, J., et al. (2020). 'Functional classification of potential bites in Pampa grassland based on their NIR spectrum'. *Rangeland Ecology and Management*, 73, 358–367.

Bahlo, C., et al. (2019). 'The role of interoperable data standards in precision livestock farming in extensive livestock systems: a review'. *Computers and Electronics in Agriculture*, 156, 459–466.

Bareth, G. (2021). 'Towards an informed grassland farming – Sensors, platforms and algorithms'. *Grassland Science in Europe*, 26, 11–29.

Bardgett, R.D., et al. (2021). 'Combatting global grassland degradation'. *Nature reviews Earth and Environment*, 2, 720–735.

Baraza, E., et al. (2006). 'Conditional outcomes in plant–herbivore interactions: neighbours matter'. *Oikos*, 113(1), 148–156.

Beillouin, D., et al. (2023). 'A global meta-analysis of soil organic carbon in the Anthropocene'. *Nature Communications*, 14, 3700.

Bengtsson, J., et al. (2019). 'Grasslands– more important for ecosystem services than you might think'. *Ecosphere*, 10(2): e02582.

Bonnet, O., et al. (2015). 'Continuous bite monitoring: a method to assess the foraging dynamics of herbivores in natural grazing conditions'. *Animal Production Science*, 55(3), 339–349.

Bonnet, O. (2025). *La Pastothèque. Tome 2. Méditerranée*. Cardère, 464p.

Brown, G. and Fagerholm, N. (2015). 'Empirical PPGIS/PGIS mapping of ecosystem services: a review and evaluation'. *Ecosystem Services*, 13, 119–133.

Campos-Taberner, M., et al. (2023). 'Land use classification over smallholding areas in the European Common Agricultural Policy framework'. *Journal of Photogrammetry and Remote Sensing*, 197, 320–334.

Capalbo, S.M., et al. (2017). 'Next generation data systems and knowledge products to support agricultural producers and science-based policy decision making'. *Agricultural Systems*, 155, 191–199.

Carrère, P., et al. (2018). 'La diversité des prairies d'Auvergne'. In: P. Carrère, et al. (eds) *Revue d'Auvergne «L'agriculture entre plaine et montagne; d'hier à aujourd'hui»*, Alliance Universitaire d'Auvergne, N°627–628: 395–414.

Carrère, P., et al. (2022). 'AEOLE. A collaborative initiative which benefit both farmers and biodiversity'. *Grassland Science in Europe*, 27, 803–806.

Cavallero, A., et al. (2007). *I tipi pastorali delle Alpi piemontesi*. Oasi Alberto Perdisa, 470p.

Delaby, L. and Bignon, E. (2015). 'Pâtur'Plan, nouvel outil dynamique pour anticiper la gestion du pâturage'. *Fourrages*, 223, 249–252.

de Mazancourt,, C. et al. (2013). 'Predicting ecosystem stability from community composition and biodiversity'. *Ecology Letters*, 16(5), 617–625.

Dixon, A. P., et al. (2014). 'Distribution mapping of world grassland types'. *Journal of Biogeography*, 41, 2003–2019.

Dodier, H., et al. (2023). *La Pastothèque, Tome 1. Montagne*. Cardère, 512p.

Dumont, B., et al. (2019). 'Associations among goods, impacts and ecosystem services provided by livestock farming'. *Animal*, 13, 1773–1784.

Estel, S., et al. (2018). 'Combining satellite data and agricultural statistics to map grassland management intensity in Europe'. *Environmental Research Letters*, 13, 074020.

Etienne, L., et al. (2021). 'Des outils et méthodes pour enseigner et accompagner l'utilisation des parcours'. *Fourrages*, 245, 53–64.

Eurostat (2022). *Data Browser*. Available at: https://ec.europa.eu/eurostat/databrowser/explore/all/all_themes, accessed 20 January 2024

Eurostat (2023). *Glossary: Permanent Grassland*. Available at: https://ec.europa.eu/eurostat/statistics-explained/index.php?title=Glossary:Permanent_grassland, accessed 20 January 2024

FAO (2019). *Grasslands, Rangelands and Forage Crops*. Available at: https://www.fao.org/agriculture/crops/thematic-sitemap/theme/spi/grasslands-rangelands-and-forage-crops/en/, accessed 10 August 2024

Fischer, M., et al. (2010). 'Implementing large-scale and long-term functional biodiversity research: the Biodiversity Exploratories'. *Basic and Applied Ecology*, 11, 473–485.

Fraser, L.H., et al. (2013). 'Coordinated distributed experiments: an emerging tool for testing global hypotheses in ecology and environmental science'. *Frontiers in Ecology and the Environment*, 11, 147–155.

Gallardo, M. and Cocero, D. (2023). 'Using the European CORINE land cover database: a 2011–2021 specific review'. In: De Lázaro Torres, et al. (eds) *Sustainable Development Goals in Europe*. Key Challenges in Geography. Springer.

Galliot, J. N., et al. (2020). *Typologie multifonctionnelle des prairies du Massif central*. SIDAM-AEOLE, 284 p.

Gautier, G. (2017). 'Influence de la Politique agricole commune sur les pratiques pastorales des Causses et des Cévennes: Analyse et recommandations'. *Mémoire de thèse professionnelle pour le Mastère spécialisé PAPDD*, 95p. Available at: https://draaf.occitanie.agriculture.gouv.fr/etude-de-l-influence-de-la-politique-agricole-commune-sur-les-pratiques-a2045.html

Hanrahan, L., et al. (2017). 'PastureBase Ireland: a grassland decision support system and national database'. *Computers and Electronics in Agriculture*, 136, 193–201.

Hardy, T., et al. (2021). 'Sen2Grass: a cloud-based solution to generate field-specific grassland information derived from sentinel-2 imagery'. *AgriEngineering*, 3, 118–137.

Hector, A., et al. (2010). 'General stabilizing effects of plant diversity on grassland productivity through population asynchrony and overyielding'. *Ecology*, 91(8), 2213–2220.

Heyl, K., et al. (2022). 'Potentials and limitations of subsidies in sustainability governance: the example of agriculture'. *Sustainability*, 14, 15859.

Johnson, I. R., et al. (2008). 'DairyMod and EcoMod: biophysical pasture-simulation models for Australia and New Zealand'. *Australian Journal of Experimental Agriculture*, 48(5), 621–631.

Jouven, M. and Baumont R. (2008). 'Simulating the dynamic interactions between management, herd and grassland resources in beef suckler systems based on permanent pasture'. *Agricultural Systems*, 96, 260–272.

Jouven, M., et al. (2006a). 'Model predicting dynamics of biomass, structure and digestibility in managed permanent pastures: 1- Model description'. *Grass and Forage Science*, 61, 112–124.

Jouven, M., et al. (2006b). 'Model predicting dynamics of biomass, structure and digestibility in managed permanent pastures: 2- model evaluation'. *Grass and Forage Science*, 61, 125–133.

Jouven, M., et al. (2010). 'Rangeland utilization in Mediterranean farming systems'. *Animal*, 4, 1746–1757.

Kalenga Tshingomba, U., et al. (2022a). 'Cartographie des surfaces pastorales à l'aide des données Sentinel 2 L3A et des données ouvertes: promesses et réalités'. *Revue Internationale de Géomatique*, 30, 245–277.

Kalenga Tshingomba, U., et al. (2022b). 'A spatialised information system to support decisions regarding grazing management in mountainous and Mediterranean rangelands', *Computers and Electronics in Agriculture*, 198, 107100.

Lepart, J., et al. (2019). 'Enhancing coviability through an eco-pastoral approach, the European project LIFE + MIL'OUV'. In: Barrière, O., et al. (eds) *Coviability of Social and Ecological Systems: Reconnecting Mankind to the Biosphere in an Era of Global Change*. Springer.

Lindborg, R., et al. (2022). 'Ecosystem services provided by semi-natural and improved grasslands–synergies, trade-offs and bundles'. *Grassland Science in Europe*, 27, 55–63.

Litrico, I. and Violle, C. (2015). 'Diversity in plant breeding: a new conceptual framework'. *Trends in Plant Science*, 20(10), 604–612.

Llaria, A., et al. (2024). 'Embedded systems for small ruminants behaviour identification: application to pastoral farms in Mountain and Mediterranean Areas'. *IEEE Sensors Applications Symposium (SAS)*, Naples, Italy.

Louault, F., et al. (2015). 'Complex plant community responses to modifications of disturbance and nutrient availability in productive permanent grasslands'. *Journal of Vegetation Science*, 28(3), 538–549.

Lyu, X., et al. (2022). 'Unmanned Aerial Vehicle (UAV) remote sensing in grassland ecosystem monitoring: a systematic review'. *Remote Sensing*, 14, 1096.

Maes, J., et al. (2020). *Mapping and assessment of ecosystems and their services: an EU ecosystem assessment*. EUR 30161 EN, Publications Office of the European Union.

Manning, P., et al. (2018). Redefining ecosystem multifunctionality. *Nature Ecology and Evolution*, 2, 427–436.

Marriot, C. A. and Carrère, P. (1998). 'Structure and dynamics of grazed vegetation'. *Annales de Zootechnie*, 47, 359–369.

Matos, I. S., et al. (2020). 'Critical transitions in rainfall manipulation experiments on grasslands'. *Ecology and Evolution*, 10, 2695–2704.

MEA, Millenium Ecosystem Assesment (2005). *Ecosystems and Human Well-Being. Synthesis. A Report of the Millenium Ecosystem Assesment.* Island Press, 219 p.

Mehrabi, Z., et al. (2020). 'Livestock policy for sustainable development'. *Nature Food*, 1, 160–165.

Michaud, A., et al. (2013). 'Construire des typologies de prairies pour évaluer leur potentiel à rendre des services agro-environnementaux'. *Fourrages*, 213, 35–44.

Michaud, A., et al. (2016). 'A French classification of permanent grasslands at national level to evaluate their forage and environmental services'. *Grassland Science in Europe*, 21, 554–556.

Mikicic, E., et al. (2023). 'Contribution of woody plants to horses' diets in Mediterranean rangelands'. *Revue d'Elevage et de Médecine Vétérinaire des pays Tropicaux*, 76, 36956.

Morris, E. K., et al. (2014). 'Choosing and using diversity indices: insights for ecological applications from the German biodiversity exploratories'. *Ecology and Evolution*, 4, 3514–3524.

Murphy, D. J., et al. (2021). 'A review of precision technologies for optimising pasture measurement on Irish grassland'. *Agriculture*, 11, 600.

O'Brien, B., et al. (2015). 'Modelling precision grass measurements for a web-based decision platform to aid grassland management'. In: *Proceedings of the 9th European Conference on Precision Livestock Farming*, Ireland, 858–863.

Peeters, A., et al. (2014). 'Grassland term definitions and classifications adapted to the diversity of European grassland-based systems'. *Grassland Science in Europe*, 19, 743–747.

Ruelle, E., et al. (2021a). 'PastureBase Ireland, un outil national d'aide à la gestion du pâturage'. In: *Journées AFPF – Prairies et Fourrages 2.0*, France, 116–117.

Ruelle, E., et al. (2021b). 'La prévision de la croissance d'herbe en Irlande: une information attendue, de l'éleveur au gouvernement'. *Fourrages*, 247, 33–39.

Sabatier, R., et al. (2010). 'Modelling trade-offs between livestock grazing and wader conservation in a grassland agroecosystem'. *Ecological Modelling*, 221(9), 292–1300.

Sagoo, L., et al. (2020). 'Review of existing tools available for use on permanent grassland'. *EC Project No. 774124, SUPER-G Deliverable Report 5.1*, 277 p.

Sales-Baptista, E., et al. (2018). 'Pastor.i: a smartphone application to facilitate grazing management'. *IV Congreso Nacional de Sistemas Silvopastoriles: una oportunidad para el desarollo sustentable*. Villa la Angostura, Neuquén, Argentina.

Schils, R. L. M., et al. (2022). 'Permanent grasslands in Europe: land use change and intensification decrease their multifunctionality'. *Agricultural, Ecosystems and Environnement*, 330, 107891.

Shaqura, I. and Lasseur, J. (2021). 'Characterising pastoral land on a large scale: a way to assess better the resilience of Mediterranean livestock activities. Lessons learned from a study on the scale of the Provence Alpes Côte d'Azur region (France)'. *Options Méditerranéennes, Série A*, 125, 387–392.

Stokes, A., et al. (2023). 'Services provided by multifunctional agroecosystems: questions, obstacles and solutions'. *Ecological Engineering*, 191, 106949.

Taugourdeau, S., et al. (2024). 'A set of ecosystem service indicators for European grasslands based on botanical surveys'. *Grassland Research*, 3(1), 43–56.

Tibi, A. and Therond, O. (2017). *Evaluation des services écosystémiques rendus par les écosystèmes agricoles. Une contribution au programme EFESE.* Report synthesis, Inra (France), 118 p. Available at: https://www.inrae.fr/sites/default/files/pdf/efese-services-ecosystemiques-rendus-par-les-ecosystemes-agricoles-synthese-2.pdf

Vogt, J., et al. (2019). 'Eleven years' data of grassland management in Germany'. *Biodiversity Data Journal*, 7, e36387.

Wang, Z., et al. (2022). 'Review of remote sensing applications in grassland monitoring'. *Remote Sensing*, 14, 2903.

Chapter 9

Advances in managing grassland weeds

Maria-Teresa Sebastià, Forest Science and Technology Centre of Catalonia – University of Lleida (UdL-CTFC), Spain; Ana Fossas, Forest Science and Technology Centre of Catalonia (CTFC), Spain; and Cristina Chocarro, University of Lleida (UdL-Agrotecnio), Spain

1 Introduction

The term 'weed' is both relative and subjective because it is defined according to context and in relation to humans. The most generally accepted definition of weed is any plant growing where it is not wanted (Britannica, 2024). Weeds are usually fast-growing, opportunistic species that can cause economic problems and reduce yield by up to 100% in poorly managed cropping systems (Gu et al., 2021). The main effect of weeds in arable crop production relates to competition for space, light, water and nutrients between weeds and the crop (Riemens et al., 2022). Other threats include weeds as reservoirs of pests and diseases (Chauhan, 2020). Another form of negative crop–weed interaction is allelopathy where crops and/or weeds produce allelochemicals that interfere with crop and weed growth (Riemens et al., 2022). However, as we will see, not all allelopathic interactions are negative and competition is only one of the multiple biological interactions among species, including those within weed–crop systems. From this perspective, and considering grasslands are often multispecies systems, we could ask: what is a weed in a grassland?

Given the wide diversity of grassland ecosystems, the answer to that question might be dependent on the type of grassland. Peeters et al. (2014) recognize over 20 categories of grasslands in Europe. Firstly, grasslands are

http://dx.doi.org/10.19103/AS.2025.0154.15

areas devoted for production of forage for harvest, grazing or cutting, or for other purposes such as renewable energy production. They can be temporary or permanent. Among those, the three principal groups, the most extended, and economically and culturally important, are

Leys are species sown on arable land for forage production and harvesting for green material and include temporary grasslands. Grazed fallow lands are areas that are being extensively grazed and uncultivated after a cropping episode, between 1 and 4 years. Permanent grasslands refer to those grasslands that used to grow grasses or other forage species, which have not been completely renewed by ploughing or herbicides, for 10 years or more. Within permanent grasslands, we can find hay meadows and semi-natural and natural grasslands. Hay meadows are grasslands devoted to hay production and semi-natural and natural grasslands are permanent grasslands of low production that are dominated by grass communities that occur naturally. The difference between semi and natural grasslands is that the latter is not modified and thereby affected by anthropogenic disturbances. Because the separation between semi-natural and natural grasslands in Europe is very tenuous and has a complex cultural background and evolutionary history (Hejcman et al., 2013; Pärtel & Bruun, 2015; Schils et al., 2022), we will use the term natural grasslands henceforth.

However, depending on the grassland age or the fodder management, there are a lot of different established definitions. For example, we can distinguish two main grassland categories:

- meadows that are defined as grasslands that have been harvested mainly through mowing over the last 5 years or since the establishment of the sward, if it is less than 5 years old; and
- pastures, which refer to grasslands that have been mainly harvested by grazing.

Here we will discuss weeds in the context of different grassland types, defined according to plant diversity management and intensification, and complexity of the agroecosystem (Fig. 1). For leys, we will discuss:

- monocrops;
- intercrops; and
- polycrops.

Monocrops are referred to those crops that are cultivated alone in time and space. Polycrops are those that are sown with different crop species on the same field during most part of their growing periods (Gu et al., 2021). One type of polycrops are intercrops, which does not necessarily mean that they

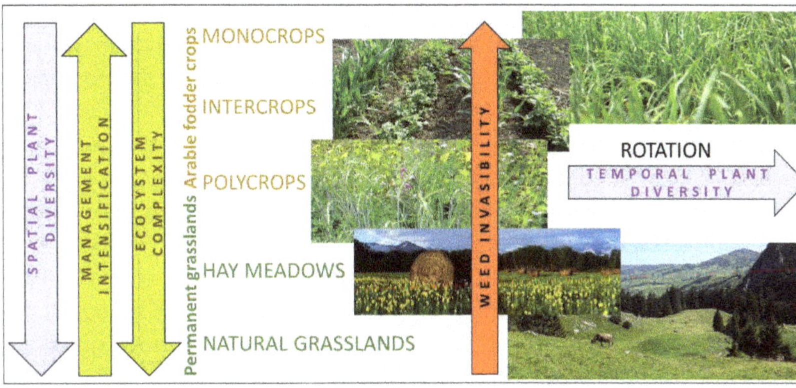

Figure 1 Five grassland types involving six broad management systems based on spatial changes in grassland species diversity, including number of species, species identity, and evenness. High spatial species diversity involves high ecosystem complexity usually accompanied by a reduction in management intensification, and the other way around. Rotation involves temporal changes in either sown species identity or species diversity and can include almost all combinations within and between grassland types. Increasing species diversity in grasslands is expected to reduce weed invasion. Bottom picture from Y. Piñuela.

share the same space, but they can be found in the same field. They can be in different rows and complement each other in some ways.

The five grassland types can be positioned across a gradient of increasing diversity of co-occurring plant species, linked to an increase in system complexity and a decrease in management intensification (Fig. 1). However, sown diversity can be also manipulated throughout time, involving another management system, rotation, which can be combined within and between most of the five grassland types in almost any order (Fig. 1). The changes in sown diversity throughout time can involve substitutions of either plant species identity (e.g. fodder cereal monocultures following legume lay monocultures) or plant diversity (e.g. a monoculture followed by a polycrop), or both. What generates a grassland composition is the combination of number (richness) and identity of sown species, together with their proportions (evenness). We can consider the species proportions in terms of sowing proportions in leys or in terms of realized biomass or any other abundance indicator (species frequency, species cover and so on) in the developed natural grasslands.

We would expect leys to function close to arable crops, particularly those including pure grass or pure legume and generally any fodder system, including monocultures. Intercropping and polycropping systems are closer to natural systems in the sense that they have higher sown diversity and thus are systems of higher complexity than monocultures (Fig. 1). Those would function in between fodder monocultures and permanent grasslands. Finally, permanent grasslands have the highest number of plant species and are expected to function as natural ecosystems, where the rules that control their functioning are less regulated by anthropogenic intervention than by natural ecological mechanisms, particularly in natural grasslands.

In this context, most weeds in European grasslands are considered native species, either unpalatable for livestock or toxic (Schaffner et al., 2022). Considering that grasslands in Europe have a long history of human use and intervention since the Neolithic Period (Poschlod et al., 2009; Feurdean et al., 2018), plant (and other) species that might have invaded the original pristine grasslands have had time to adapt and be part of the current community. This might be the destiny of exotic invader species colonizing those grasslands today but, as we will see, not all new arrivals are welcome. Furthermore, in the context of global change and globalization, the concept of grassland weed and grassland invader might be taking a completely new and much more complex dimension.

In this context of increasing complexity and multi-specificity from monocrops to natural grasslands, in the framework of current complex human societies driving extreme environmental changes, it is important to answer the following questions regarding weeds and weed invasion in grasslands:

1 Are weeds always a problem in grassland ecosystems, or are there occasions when they can be overlooked, or even when they contribute to solutions in some grassland ecosystems?
2 Is it possible to control weed invasion through nature-based solutions (NbSs)?
3 Is global change interfering with weed dynamics and invasion potential and, if so, how can we address this challenge to protect grasslands?

Our hypotheses are as follows:

1 The higher the intensification, the higher the agroecosystem simplification, and thus the higher the vulnerability of the grassland to weed invasion and the higher the need for external inputs.
2 The higher the plant diversity, the higher the complexity of the agroecosystem, and thus, the higher the capacity of the grassland to control weed invasion internally.

In the following sections, we will try to test our hypotheses and answer those questions. Furthermore, we will present a Case Study in Catalonia, in the Western Mediterranean, based on leys, and focusing on a very important fodder legume, alfalfa.

2 Weeds: a threat or an opportunity?

Reduction in crop yield is the most cited damage of weeds on crops, with estimated losses of 25% of crop production worldwide (Patterson et al., 1999) and associated financial losses in agricultural production (Baucom & Holt, 2009), which have been estimated above US$100 billion per year, plus the cost of herbicides, estimated in US$25 billion (Gonzalez-Andujar, 2023). Weeds have been estimated to produce the highest potential crop production loss (34%), followed by animal pests and pathogens with half the proportion (18% and 16%) (Oerke, 2006).

Weeds can negatively affect both the yield and the quality of forage. However, crop yield losses by weeds depend on factors including weed emergence time, weed density, type of weeds, type of crops, etc. (Chauhan, 2020). For instance, weed effects might be more significant in lays grown as monocultures for export, than in multispecies systems, including natural grasslands managed for extensive livestock farming. In addition, the proportion of weeds in the fodder can also make a difference, and a low weed proportion might be acceptable. The question is to determine from which proportion – or farmer's needs or market requirements – it is worth an aggressive intervention, and which is the threshold to let the weeds be. In addition, other considerations might play a role, for instance, if weeds are introducing toxicity in the feed, or if they harbor pathogens and insects that attack crop plants (Chauhan, 2020).

Thus, weeds are usually characterized by their negative effect on crops, rather than their benefits in the ecosystem (Cordeau, 2022; Boutagayout et al., 2023). Benefits of weeds under certain circumstances have also been cited (Frick & Johnson, 2002). A list of threats and opportunities by weeds in grasslands can be found in Table 1.

2.1 Threats

- Yield and feed quality: Weeds usually reduce growth in arable crop production by increasing competition for resources and light, estimating a crop production loss average value of 25% (Parker & Fryer, 1975; Patterson et al., 1999). Moreover, weeds might inhibit the growth of surrounding plants due to allelopathic processes (Zhang et al., 2020). Allelopathy is the process mediated by allelochemicals, which are secondary metabolites that are not needed for plant metabolism, but in turn affect other organisms. Allelochemicals can affect the crop system in many ways, such as changing weed species composition, affecting crop yield, feed and forage quality (Varanasi et al., 2016; Gu et al., 2021).
- Biodiversity and crop health: Biological invasions pose a threat to global biodiversity, altering ecosystem functioning (Levine et al., 2003;

Table 1 Threats and opportunities of weeds in grasslands, according to ecosystem functions, goods and services

Service	Threats	Opportunities
Yield and feed quality	Crop yield reduction by competition (Oerke & Dehne, 2004; Varanasi et al., 2016; Gu et al., 2021; Ribas et al., 2023). Crop yield reduction by allelopathy (Zhang et al., 2020; Boutagayout et al., 2023). Herbage quality reduction (Gu et al., 2021; Monteiro & Santos, 2022). Herbage palatability reduction Herbage protein reduction (Canevari et al., 2007).	Potential for synergistic growth (Fargione & Tilman, 2006; Ryel et al., 2010; Leng et al., 2013; Zhang et al., 2020) Crop yield losses can be mitigated by a diverse weed community (Adeux et al., 2019). Reduced competition if crops and weeds don't overlap in phenology, niches and/or functional traits (Fargione & Tilman, 2006; Ryel et al., 2010; Leng et al., 2013; Zhang et al., 2020). Increased nutrient diversity in feed Increased micronutrients in feed.
Biodiversity and crop health	Introduction of pests (Boydston et al., 2008; Varanasi et al., 2016). Introduction of pathogens (Zandstra & Motooka, 1978). Changes in community composition (Traveset & Richardson, 2006; Gassó et al., 2009). Increased mammal disturbance (Bueno et al., 2011). Weed extension through negative selection by livestock (Zhang et al., 2020). Disruption of mutualistic plant–animal interactions (Traveset & Richardson, 2006). Highly flammable which contributes to wildfire risks in drought areas (Pausas & Keeley, 2021).	Biodiversity conservation (Kiehl et al., 2010; Rene & Tracy 2013). Pollinators' attraction (Qasem & Foy, 2001; Boutagayout et al., 2023). Protection from pests by hosting natural enemies (Carnus et al., 2006). Protection from herbivore pressure by microenvironment formation around weeds (Sun et al., 2009; Zhang et al., 2020) Potential for ecological restoration of degraded grasslands (Zhang et al., 2020) Can be exploited for feed livestock (Ekwealor et al., 2019).
Soil functioning	Soil nutrient depletion (Parton et al., 2007; McLeod et al., 2016). Soil water depletion (Zimdahl, 2018). Accumulation of soil pathogens through time (Maron et al., 2014). Increased susceptibility to wind and rain erosion (Zhang et al., 2020). Changes in soil microbial communities (Larios & Suding, 2015; Canals et al., 2019) Changes in C:N ratio (Checinska Sielaff et al., 2018). Change in mineralization rates and a source of CO_2 (Checinska Sielaff et al., 2018).	Nutrient pumping from deep soil layers (Sun et al., 2009; Zhang et al., 2020) Water pumping from deep soil layers under drought conditions (Sun et al., 2009; Zhang et al., 2020). Increased soil infiltration (Leng et al., 2013; Zhang et al., 2020). Fertility islands formation (Sun et al., 2009; Zhang et al., 2020). Reduce soil erosion (Boutagayout et al., 2023). Increase phosphorus solubility (Boutagayout et al., 2023).
Other	Toxicity (Zhao et al., 2013; Wu et al., 2016; Zhang et al., 2020). Livestock poisoning (Zhang et al., 2020; Krieger et al., 2022).	Potential for medicines due to aromatic and medicinal properties (Qasem & Foy, 2001; Jayasundera et al., 2021). Potential for pest control through bioactive compounds (Qasem, 2012). Potential food crops (Rapoport et al., 1995; Molina et al., 2014).

MacDougall & Turkington, 2005; Traveset & Richardson, 2006). Weeds can alter quality and availability of resources and space (Levine et al., 2003), modify evolutionary ecosystem trajectories and animal–plant mutualisms, which natural communities depend on for their regeneration and for maintaining its diversity (Gurevitch & Padilla, 2004). Grasslands dominated by exotic species often have lower plant diversity and altered C3-C4 biomass proportions (Christian & Wilson, 1999; Martin et al., 2014; Xu et al., 2015; Wilsey et al., 2018; Checinska Sielaff et al., 2018). Furthermore, weeds can serve as hosts for crop pests and diseases (Boydston et al., 2008). Weeds can indeed alter disturbance regimes (Brooks et al., 2004), beyond the range of variation to which native species are adapted (D'Antonio et al., 1999), which results in ecosystem and community changes. In fact, there are weed species, usually grass invaders (Brooks et al., 2004) that can alter fire regime and frequency, to a point that natural plant populations aren't able to recover (Ditomaso, 2000), affecting animals that depend on these environments for forage. As animals avoid poisonous toxic weeds, overgrazing of grassland palatable plants might occur (Zhang et al., 2020). In addition, the presence of weeds with nutritious underground storage organs might attract mammals, including wild boar, which might produce extensive damage in both leys and permanent grasslands (Bueno et al., 2011). Abandonment of many rural areas and decreases in the human population and livestock farming activities has produced an increase of the populations of wild boars and other mammals, increasing their capacity for destruction (Bueno et al., 2011). Farming practices used to reduce abundance of weeds can also affect community structure such as food webs, species composition, species evenness and richness (Chapin III et al., 2000). For example, changing the proportion of functional groups, from dicotyledons to monocotyledons, putting in danger nature conservation values of grasslands (Krieger et al., 2022; Farmilo et al., 2023).

- Soil functioning: Weeds are often opportunists that outcompete neighboring plant species, and this can lead to soil nutrient and water depletion (Parton et al., 2007; McLeod et al., 2016; Zimdahl, 2018) as well as soil pathogens accumulation through time (Maron et al., 2014). Exotic species can alter aboveground conditions, which in turn can affect belowground nutrient cycling and soil microbial communities (Bardgett & van der Putten, 2014; Checinska Sielaff et al., 2018). For example, exotic plants could alter nitrogen-fixing microbes due to tissue type variations in plant communities (Checinska Sielaff et al., 2018); increase mineralization rates in systems dominated by C3 plant species; and constitute a source of CO_2 (Wedin & Tilman, 1993; Stark & Norton, 2015; Checinska Sielaff et al., 2018). In systems dominated by C4 or annual plant species, weeds

might increase the C:N and the lignin:N ratios, which in turn reduce mineralization rates (Evans et al., 2001).
- Other: Toxic weeds can affect livestock, humans and wild herbivores (Zimdahl, 2007; Zhao et al., 2013; Krieger et al., 2022), as they accumulate toxins to a level whose concentration can affect either the environment (Zhao et al., 2013; Wu et al., 2016; Zhang et al., 2020) and domestic animals if toxins are ingested or inhaled (Zhao et al., 2013).

2.2 Opportunities

- Yield and feed quality: Weeds might successfully coexist with other plant species (Zhao et al., 2013). For example, graminoids present horizontal roots whereas weed forbs present axial roots and can reach nutrients from deep layers compared to grass forages (Sun et al., 2009). This alleviates competition and allows their co-existence. A potential for synergistic growth has also been observed between crops and weeds (Fargione & Tilman, 2006; Ryel et al., 2010; Xin et al., 2013; Zhang et al., 2020), when they present niche differentiation (Zhang et al., 2020). Legume weeds can promote growth of certain crops through nitrogen fixation, increasing N soil pools. Also, chemicals released by weeds can be stimulatory by themselves or after being modified by soil microorganisms. Finally, weeds could increase nutrient diversity and micronutrients in the grassland forage, for instance, by introducing forbs into grass-dominated crops (Fisher et al., 1996; Belesky et al., 2001).
- Biodiversity and crop health: Weeds provide an important gene pool and, if they are forbs, their invasion increases the diversity of insects and invertebrates, which maintains biodiversity. Weeds attract pollinators because of their reproductive strategies, which involve floral traits with often brilliant colors (Qasem & Foy, 2001). The species surrounded by exotic weeds are protected from grazing in overgrazed grasslands because of the toxic smell and by altering their microenvironmental conditions (Zhang et al., 2020). Regarding degraded pastures, weeds can improve plant community structure and promote an excessive disturbance from livestock, a fact that can allow recovery in degraded grasslands (Tan & Zhou, 1995; Sun et al., 2009; Zhang et al., 2020). Eliminating weeds may promote ecosystem collapse, as grazing pressure on grasslands is found to be higher without the protection of toxic grasses (Holechek, 2002).
- Soil functioning: Weeds might present well-developed root systems, and thus can fix sand and capture nutrients and water from deeper soil layers than crops, and thus, supply nutrients to crops and increase soil infiltration (Zhang et al., 2020). Also, it has been found that weeds can indeed pump

water from deep soil layers under drought conditions (Sun et al., 2009; Zhang et al., 2020). Moreover, weeds can redistribute soil nutrients and create a cool environment that promotes soil moisture retention (Qasem, 2012). Weeds can also create a phenomenon known as fertile islands, as they present higher tissue nitrogen and less lignin compared to non-weeds, and thus, soil microbial biomass activity and enzymes are higher in weed patches than in areas between these patches. In turn, it stimulates the protection and improves soil conditions, which provides a superior material basis for plant growth and benefit the recovery of degraded grasslands (Sun et al., 2009; Zhang et al., 2020).

- Other: Weeds are traditionally known to have medicinal properties because of the presence of bioactive compounds, which are often secondary active metabolites. Four families of plants have high percentages of medicinally valuable weed species: Asteraceae, Solanaceae, Lamiaceae and Papilionaceae (Jayasundera et al., 2021). Bioactive compounds can either be used in human medicine or in pest control. When used for controlling pests, allelopathic water extracts are often used for organic parasitic weed management either as such or by incorporation of dead or living plants that release allelochemicals into the surrounding environment (Qasem, 2012). In addition, weeds can have nutraceutical values and constitute important food and medicinal assets (Ceccanti et al., 2018). It has been suggested that some weeds may have received dispersal assistance by humans linked to gastronomic habits in prehistoric times (Rapoport et al., 1995).

3 Nature-based solutions for weed management

NbS are defined as actions based on ecosystem functioning designed to protect, sustainably manage and restore ecosystems meant to address societal challenges effectively and adaptively, simultaneously providing benefits for human well-being and biodiversity (Hrabanski & Le Coq, 2022). NbS require a profound observation and understanding of nature in interaction with the social and cultural environments. Applied to agriculture, this means natural processes relying on ecosystem functioning to ensure food and livelihood security, healthier diets and more inclusive rural economies (Hrabanski & Le Coq, 2022). NbS is built on the former concept of agroecology, which focus on ecosystem processes in agricultural systems as a whole, including mineral cycles, energy transformations, biological processes and socioeconomic relationships (Altieri, 1999).

NbS applied to weed control include the use of ecosystem in-built regulatory mechanisms. Here we address NbS by considering three spatial scales (Fig. 2):

Figure 2 Left: Weeds can act as an entire ecosystems, attracting other organisms, including the shoot, root and soil microbiome, and producing a diversity of metabolites, including allelochemicals. Center: Weeds inside a grassland can have obnoxious and positive effects, including snapdragon, *Anthirrinum majus* L., a weed with the capacity to increase the attractiveness of the grassland. Right: a diversified landscape can bring in the advantages of biodiversity, for instance, by combining monocultures of different crops, or mono and polycrops. Pictures: left, Y. Piñuela; center, C. Chocarro, right, M.T. Sebastià.

- the weed as an ecosystem;
- weeds inside grassland ecosystems; and
- weeds in agricultural landscapes.

We also analyze the role of biodiversity as both an asset and part of the solution.

A plant can be considered an ecosystem in itself. Plants attract herbivores, pollinators, pathogens and other organisms. Weeds are not an exception (Martín-Forés et al., 2017) (Fig. 2). Furthermore, plants, similarly to animals, host a wide range of microorganisms, and those can modify the relationship of the plant with the rest of the world. In addition, plants and associated microbiota, above and belowground, as well as the microbiota in the soil where plants reside, generate a huge amount of metabolites, which interact with the environment and regulate organismic interactions, although we are just starting to perceive their magnitude and significance (Fig. 3). The capacity to generate a wide diversity of metabolites applies to both weeds and sowed crops – as in plants of interest – in grasslands, and some authors have started looking at metabolomics for weed control (Barrett et al., 2021). Metabolomics involves the identification and quantification of metabolites resulting from biochemical processes in the environment, the soil and/or the plants (Razzaq et al., 2012). Thus, metabolomics is a recent addition to the analysis of ecosystem functioning and species interactions, and the potential application in weed control still needs to be explored (Ngondya et al., 2016).

Nonetheless, allelochemicals have been known to exist for a long time. However, it has been recognized for a long time as a sustainable weed

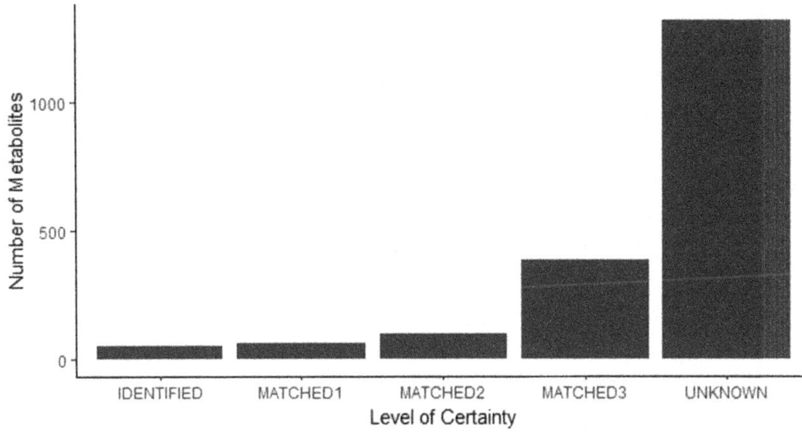

Figure 3 Number of metabolites found in soils of 60 mesic and xeric alpine grasslands in five National Parks in mountains of the Iberian Peninsula across a latitudinal gradient: Sierra Nevada, Sierra de Guadarrama, Ordesa y Monte Perdido, Aigüestortes i Estany de Sant Maurici, and Picos de Europa. From 1910 metabolites, only 53 were unequivocally identified. Matched 1 to Matched 3 indicate possible matches with a decreasing certainty. Around 45% of the total extracted metabolites were totally unknown. Source: Sebastià et al., unpublished.

management tool (Zimdahl, 2018; Li et al., 2019). Allelopathy can have stimulatory or inhibitory effects on other nearby plants (Farooq et al., 2019) and is an option for finding alternatives for conventional weed management, so the reliance on traditional herbicides is reduced (Inderjit & Keating, 1999). Also, allelopathic compounds are biodegradable and more sustainable than traditional herbicides because they are less polluting. On the other hand, we should consider the possible role of biochar as a tool to handle allelopathy of resistant weeds (Sujeeun & Thomas 2017). Thus, plants in grasslands, including weeds, are not isolated from other organisms and the influence of the environment but grow inside grassland ecosystem, and thus ecosystem functioning rules apply. In addition to allelopathy, competition is the most frequently cited antagonistic interaction between crops and weeds. Weeds usually reduce growth in arable crop production, caused by increased competition for light, water and nutrients (Oerke & Dehne, 2004; Varanasi et al., 2016; Gu et al., 2021; Krieger et al., 2022; Ribas et al., 2023). However, competition is not the only form of interaction between organisms, and synergistic plant interactions can also be relevant for weed management and control. Sown diversity has been found to play a role in weed suppression in poliphytic arable grasslands, which are grasslands that are built up on a complex variety of herbaceous plants that grow spontaneously, which depend upon different environmental factors, including irrigation, mowing and fertilization practices. For instance, the agrodiversity experiment found that the average weed biomass in mixture was 52% lower than that of the most suppressive monoculture (Connolly et al., 2018). The experiment compared a series of benefits of sown diversity in forage systems and found beneficial effects of mixtures compared to monocultures across a wide range of environmental conditions for all the studied functions, including yield (Kirwan et al., 2007), weed suppression (Connolly et al., 2018, Ribas et al., 2023), forage quality (Sturludóttir et al., 2014), nitrogen efficiency use (Suter et al., 2015), function stability and sward persistence (Finn et al., 2013). Later studies have found benefits of mixtures on leachate reduction and greenhouse gas regulation through an increase in net CO_2 absorbance (Ribas et al., 2015; Cummins et al., 2021). Increased plant diversity in grasslands has also been related to resistance to drought (Caldeira et al., 2001; Lüscher et al., 2022; but see Sebastià et al., 2023). In addition to polycrops, intercrops also share the advantages of poliphytic systems (Dong et al., 2024).

Thus, once again, we must discriminate between leys and permanent grasslands. A leys type of grassland starts the community assembly process anew every time the field is ploughed and sown. Newcomers need to make room for themselves into the ecosystem, and tilling will help them with this process by offering bare soil to germinate. On the other hand, native weeds, or long-term invaders, have had the opportunity to integrate into the community in permanent grasslands, particularly in natural grasslands. Long-established and healthy permanent grasslands are more difficult to invade. However, changes in management and climate might alter the ecosystem dynamics and favor invasions by native and non-native weeds (Canals et al., 2017). Finally, native plant species might be toxic for livestock, thus behaving as undesirable weeds (Wagner et al., 2023).

Another practice that contributes to increase biodiversity provides resilience and decreases weed invasion by natural methods is the maintenance of a diverse landscape including the combination of several ecosystem types and crops (Epanchin-Niell et al., 2010). Large extensions of the same monoculture need strong external inputs for support, including pesticides and herbicides. On the contrary, the spatial variation of crops in the landscape can contribute to the depletion of weeds in the soil seed bank, particularly in combination with rotation. Rotation is another practice contributing to reducing weeds, in this case using the advantages of sown diversity throughout time.

On the other hand, certain practices can contribute to weed transportation across landscapes. For instance, weed seeds can be dispersed long distances by water in grassland-irrigated systems (Ghersa & Roush, 1993). Animals can also transport weed seeds from one grassland to another, sometimes several hundred kilometers away, for instance, through transhumance. Transhumance involves the movement of livestock from the winter grasslands in the south to the summer grasslands in the north, for instance, from the dehesas in the southern Iberian Peninsula to alpine grasslands in the center and north of

Iberia, representing many hundreds of kilometers. Other livestock movements might be short distance, a few hundreds, for instance, from the Ebro valley to the Pyrenees, in the eastern Iberian Peninsula (Barrantes et al., 2016). Still, some movements involve very short distances, from the villages in the valley to the summits of the surrounding mountains (Rodríguez et al., 2018; Sebastià et al., 2008a).

Rodríguez et al. (2018) found that sheep-grazed mountain areas maintained a more diverse plant composition in isolated semi-natural grassland patches compared to isolated grasslands in cattle-grazed areas; shepherds let sheep move widely in landscapes, while cattle are mostly left on their own in the mountain grasslands, and they use only the more easily accessible grassland patches (Sebastià et al., 2008a). In the same natural mountain grasslands, cattle have been found to increase plant diversity within the grassland community compared to sheep, but many of those species were banal species, and several could be associated to weeds (Sebastià et al., 2008b).

4 Weeds and management within models of sustainable agriculture

With the world increasingly warming, extreme climatic events increasingly frequent, and trade and transportation increasingly long-distance, organisms are expected to move, within the native area and between distant regions, within and between ecosystems. Plus, our world is also experiencing numerous land use and management changes. The combination leads to unbalances in grassland ecosystems, and modifications in grassland compositions, with the consequence of new combinations being tested, new neighbors being found together.

Both leys and permanent grasslands are intrinsically disturbed habitats. Leys are regularly ploughed. Natural grasslands are regularly used in extensive livestock farming and are disturbed by grazing by wildlife and livestock (Sebastià et al., 2008a). This large-scale grazing disturbance opens small gaps and allows the entrance of several plant species, increasing the overall biodiversity (Sebastià et al., 2008a). Other biotic disturbances at smaller scales, including activity by voles, moles, ants, etc., are also frequent and contribute to the dynamics, functioning and biodiversity of the grassland (Canals & Sebastià, 2000; Sebastià & Puig, 2008). Those are usually long-term interactions and are part of the regular dynamics of permanent grasslands. Even regular climatic changes from year to year can be included as part of the regular functioning of the grassland (Schäppi & Körner, 1996). The problem emerges when the disturbances are too new, or too big, or the deviations are too wide (Sebastià, 2007), and the ecosystem cannot handle those disturbances.

Because feeding people is important and because industrial agriculture has caused many environmental problems, including threats to the basic capacity to feed people, a new agriculture is evolving. Overall, we can summarize current philosophies and attitudes in the agronomic field into two convergent types:

- sustainable agriculture and
- sustainable agricultural intensification.

Sustainable agriculture is a way of farming that meets current food needs without compromising the food security of the next generations. This model is often associated to organic agriculture but is not necessarily so. It is often included among the NbS. The second model relies on climate-smart agriculture, but smart agriculture is not limited to this model either. Nevertheless, determining the most effective solutions to facilitate sustainable intensification is inherently dependent on the specific context, influenced by the characteristics of existing production systems and their broader economic, institutional and socio-cultural frameworks. Additionally, diverse sustainability outcomes and associated trade-offs may arise beyond the farm scale. Consequently, a comprehensive understanding and evaluation of this farming practice benefits and trade-offs across various geographical regions and temporal scales are essential. In the next chapters, we will discuss in more detail the different approaches for weed management associated with those agricultural models.

4.1 General solutions

There is a set of transversal solutions available to approaches contributing to weed control with reduced emissions. For instance, planting arrangement, seed density and sowing date can be manipulated both as mitigation mechanisms and adaptation measures. Sustainable agriculture might sometimes rely on precision agriculture, where images by cameras including those flying in drones can contribute to the identification of particularly noxious weeds or weed patches and latter chirurgical removal (Esposito et al., 2021). Agricultural modeling is also considered to be a strong tool in agricultural sustainable intensification (Brown et al., 2023), but again this can be applied to all agricultural approaches.

4.2 Diversity-based solutions

Because crop diversification negatively affects weed seed germination and growth, it reduces weed density (Sharma et al., 2021). Crop diversification implies reduced pesticide use and crop association leads to associated crops are less susceptible to diseases and improve water flow. At the same time, agricultural intensification leads to decreased weed richness and diversity; crop diversification can prevent this, allowing the survival of more weed species at harmless densities, and thus favoring biodiversity while reducing weed damage (Gu et al., 2021).

For instance, intercropping and polycropping consist of cultivating multiple crop species in the same field, usually during an important part of their growing periods (Willey, 1990; Kirwan et al., 2007; Gu et al., 2021), blocking light which is essential for weed growth (high crops, fast-growing crops, allelochemicals, etc.). Intercropping is known to improve resilience to multiple stressors, either biotic or abiotic. Weeds can be suppressed mixing varied species using intercropping and polycropping (Fukai & Trenbath, 1993; Liebman & Dyck, 1993; Boudreau, 2013; Zhang et al., 2019; Gu et al., 2021, 2022). Moreover, those methods might also increase yield and lower production risks, land use efficiency, resource use efficiency and crop disease (Gu et al., 2021). The combination of increased grassland competition with inundative biological control by herbivory through targeted inoculation with root-boring *Pyropteron* spp had a significant synergistic effect on growth reduction of *Rumex obtusifolius* (Klötzli et al., 2023). Cover crops and companion crops would be other methods of weed control using increased sown diversity in space. On one hand, cover crops are those that are sown with the crop species of interest for erosion control, rather than for harvesting. They can be sown as well for weed and pest control and to increase soil fertility. On the other hand, companion crops are defined as crops sown together during the establishment phase to prevent soil erosion and suppress weeds (Canevari et al., 2007).

Introducing crop diversity throughout time is another important strategy for weed control, including rotation. There has been a decline in its use because of the increase in N synthetic fertilizers and pesticides, which removes the need for crop rotation. There are few studies that evaluate the effects of crop rotation on weed control. However, this technique is especially important for controlling perennial weeds. Crop rotation is usually practiced for 2–3 years to decrease disease pressure and disrupt weed and insect life cycles. Benefits happen particularly when legumes are included in the system (Selim, 2019).

5 Case study: weed management in alfalfa as a forage crop

Alfalfa (*Medicago sativa* L.) is classified as a typical perennial Mediterranean legume (Lloveras et al., 1998) and it is considered the most productive forage crop in temperate areas (Frame et al., 1998). It is contemplated to be one of the oldest and most important forage crops in the world (Michaud et al., 2015;

Mseddi et al., 2017). The interest of alfalfa cropping is due to its yield, nutritive value, low temperature tolerance and disease resistance (Pozo & Ibáñez, 1983), as well as its multiple benefits to the soil ecosystem, such as nitrogen fixation or organic matter input coming from its roots (Lloveras et al., 1998). In Catalonia, alfalfa is normally cultivated for either hay or seed production, followed by a grazing period during winter (Fig. 4; Lloveras et al., 2012).

In the Mediterranean region, alfalfa is usually grown under irrigated conditions in monoculture, although it can also be found combined with grasses (Delgado et al., 2020). Alfalfa production in irrigated Mediterranean regions is characterized by long growing seasons and intensive production systems (with 5–6 cuts per year) (Del Pozo & Ibáñez, 1984; Dovrat, 1993). Around 80% of the alfalfa cultivated in Spain takes place in the Ebro Valley, with 75% of the alfalfa produced under irrigation systems. The Ebro Valley is a large depression extending along the north-east of the Iberian Peninsula. It is a relatively flat area in the otherwise mountainous Iberian landscape, delimited by the Pyrenees, the Iberian Mountain range, and the Catalan pre-littoral and littoral ranges. From the Ebro Valley, alfalfa is dehydrated and sent to other regions (Lloveras, 2001). In Catalonia, after cereal, forages are the most cultivated product, of which one fourth corresponds to alfalfa in monoculture (Delgado et al., 2022).

Weed species that usually affect alfalfa are dycotiledons, anuals and pluriannual, such as *Amarathus retroflexus*, *Capsella bursa-pastoris*, *Malva sylvestris* or *Plantago lanceolata*. Cereals that usually affect alfalfa are *Lolium rigidum*, *Lolium multiflorum* and *Sorghum halepense*. Parasite plant species

Figure 4 Development of alfalfa cultivation. a: Germination, seedling stage. b: Vegetative development. c: Flowering of *Medicago sativa* var. Aragon. d: Harvest in a commercial field in the Ebro Valley (Spain) for forage production. e: Seed production. f: Alfalfa with composite species weeds. Pictures: C. Chocarro.

are usually *Cuscuta ephithymum* and *Cuscuta campestris*. These compete for resources required for alfalfa's growth and development, reducing its feed value and yield (Forney et al., 1985). Perennial weeds are the most dangerous because they are difficult to remove once established (Chocarro & Lloveras, 2015).

To ensure crop's longevity and to increase quality and productivity of alfalfa, its physiology and correct agricultural practices should be considered. It is constituted by leaves, stems, a crown, that unites the aerial part and the roots. In the last two organs, the crop accumulates the carbohydrate and protein reserve, which ensure a correct response to cutting and its persistence (Lloveras et al., 2020). The state of the crown root is crucial, as it is directly related to increased crop productivity and reduce weed presence (Lloveras et al., 1998). So, agricultural practices should ensure a successful establishment of alfalfa seedlings because they are vulnerable to weed competition, as they are not impetuous enough to compete with weeds (Lloveras et al., 2008). Therefore, to ensure the optimal development of alfalfa and to avoid weed proliferation, agricultural practices leading to weed control should be performed.

5.1 Conventional agricultural practices

The most commonly used conventional practices, mostly applied to alfalfa monocultures, are as follows:

- Herbicide control: It is the most used practice for weed management (Gu et al., 2021). However, it simplifies the weed community in the soil, consisting of one or several weeds highly adapted to the growing conditions continuously maintained for the crop (MacLaren et al., 2020; Sharma et al., 2021). However, herbicide control should be the last to consider, as there is no selective herbicide, and we must take advantage of alfalfa's competitive capacity once established (Canevari et al., 2007).
- Mechanical weeding: Herbicide application is usually followed by mechanical weeding, which consists of mowing tall weeds that shade the small seedling alfalfa (Lloveras et al., 2020).
- Soil flattening: Soil flattening should be taken into consideration to uniform germination and to promote the rapid growth of alfalfa seedlings. If the soil is not flattened, water coming from irrigation reaches the lowest part of the field, drowning the alfalfa and reducing its competitive capacity (Delgado et al., 2020).
- Soil tillage: Another technique is false sowing before the actual sowing of the alfalfa. It consists of preparing the soil for sowing (tillage), so the weeds emerge, and then they are erased mechanically or using a non-selective herbicide (Delgado et al., 2020). The critical period to eliminate weeds

is 40 days after sowing the alfalfa and on the first stage of development (Canevari et al., 2007). Soil tillage is an effective practice in controlling weeds, usually when it is performed before alfalfa sowing, to reduce competition. However, it is a fuel-intensive practice and a CO_2 emission source. Also, it exacerbates soil losses due to erosion (Paustian et al., 2000).

5.2 Nature-based solutions

- Intercropping: It is suggested that there is a resource complementarity between crops in these systems, which allows them a higher resource capture compared to monoculture, and thus, fewer resources are available for weeds (Gu et al., 2021). For alfalfa, when cultivated with olive trees, it has been found that competition for light is more important than competition for water (Mantino et al., 2021).
- Polycropping: Polycropping including alfalfa has been identified as a sound agricultural practice that reduces weed infestation, among other benefits (Ribas et al., 2023). The agrodiversity experimental location in Gósol, Eastern Pyrenees, showed an excellent performance of alfalfa in mixtures with two grasses and another legume, sainfoin (*Onobrychis viciifolia* Scop., Ribas et al., 2023), despite harsh climatic conditions. However, more recently, a similar combination of grasses and legumes showed how sainfoin outperformed alfalfa under the extreme drought and high temperatures experienced in Catalonia the last years (Sebastià et al., 2023). The polycropping system in this case functioned because of the portfolio effect, that is, at least one of the sown species showed some significant growth (Sebastià et al., 2023). However, another challenge for alfalfa in polycrops is the preference some buyers show for pure alfalfa. Under reduced water availability and considering the drought has been so severe that even irrigation systems were shut down, the future of alfalfa in the Ebro Valley must be reconsidered.
- Crop rotation: Crop rotation is efficient in lowering weed presence and makes their control more sustainable (Hunt et al., 2017). It is important to rotate with different types of crops, such as summer and winter crops, used for hay or seed, with different rooting depth (Lloveras et al., 1998). In alfalfa crops, rotation takes place usually between alfalfa and an annual crop, such as wheat or corn, which aggressively compete with perennial weeds (Canevari et al., 2007; Weisberger et al., 2019). Zhao et al. (2020) recently performed a meta-analysis on 45 studies and reported a 20% increment in crop yields due to crop rotation. Moreover, another meta-analysis on 54 studies showed that crop rotation leads to a 49% reduction in weed density (Weisberger et al., 2019). Thus, crop rotation helps to reduce weed pressure and increase crop yield.

- Companion crops: It is usually selected an alfalfa's companion crop that grows faster than alfalfa, because it competes with fast-growing weeds, suppressing the need to use herbicides and improving forage quality and yield. Companion crops contribute to enhanced weed control and forage yield while providing protection to developing seedlings during the establishment phase of alfalfa. Additionally, they facilitate improved alfalfa emergence on soils susceptible to crusting and offer protection to young seedlings from the impacts of wind-blown soils and soil erosion (Canevari et al., 2007). When utilizing a companion crop, all management practices should prioritize the alfalfa over the companion species. The companion crop serves as short-term aid, typically for a single cutting, to support the establishment of a long-term alfalfa stand intended for 20 or more cuttings. Management practices should focus on the fact that the predominant error associated with companion crops is seeding at excessively high rates. Pure oat stands for forage are generally sown at a rate of 80–100 pounds of seed per acre. When such high rates are applied in companion seeding, they can induce excessive competition, leading to suboptimal alfalfa stands. The critical task is to determine a seeding rate that optimizes forage yield and weed suppression while minimizing adverse effects on the alfalfa. Moreover, there are crops that have high allelochemical potential. Because of this, it has been suggested that those with a high allelochemical potential should be planted as companion crops to exert their effect on weeds (Zimdahl, 2018), such as rye and its residues (Putnam et al., 1983; Zimdahl, 2018).
- Grazing: Chocarro et al. (2005) found that one severe winter grazing by sheep in the Mediterranean areas of the Ebro Valley was beneficial, because it increased forage quality in spring, giving an additional economic input to the grower of alfalfa when they rent the fields to the sheep producers in winter outside the growing season. Grazing by the end of autumn and in winter can reduce the presence of weeds, reducing weed competition. However, it can negatively affect the root crown, which allows the entry of pathogens. Also, it can promote soil compaction, lowering water infiltration and crop yield (Chocarro et al., 2005).

5.3 Climate change and alfalfa in Catalonia and other Mediterranean areas

The Mediterranean region is a hotspot for climate change because it is warming above the world's annual temperature mean (Cramer et al., 2020), and precipitation is decreasing below the World's annual precipitation mean (Tuel et al., 2021). From 2021 to 2023, a catastrophic drought occurred in Catalonia

(Altava-Ortiz et al., 2023). This situation added up to the elevated temperature extremes, in a region traditionally hit by summer droughts. The deficit of water in Catalonia has been so severe that all the irrigation systems in the Ebro Valley were shut. Under those conditions, what will the future of alfalfa cultivation, and in general, forage cultivation in the region be?

The field experiment of the SUSFORAGE project in the Catalan Pyrenees had to be prematurely terminated because of those harsh climatic conditions (Sebastià et al., 2023). The experiment included mixtures and monocultures of two grasses: raygrass (*Lolium perenne* L.) and orchard grass (*Dactylis glomerata* L.) and two legumes, alfalfa and sainfoin (*Onobrychis viciifolia* Scop.). Despite the productivity reduction, surprisingly sainfoin performed better than alfalfa, and both performed better than any of the grasses (Sebastià et al., 2023). Having a portfolio of sown species was more important than the synergistic effect of mixtures to confront drought. Those results contrast with a previous experiment in a close-by area with higher precipitation, experiencing the 2003–2004 heatwave. In that particular experiment including the same four forage species, yield was increased in the forage mixtures compared to monocultures, with excellent performance of alfalfa in the mixtures (Ribas et al., 2023). In addition to enhanced yield, weed invasion was highly reduced in the mixtures (Ribas et al., 2023).

Thus, some important questions remain:

- Should the pressure be on developing better adapted varieties of alfalfa as well as a research line on sainfoin potential and varieties?
- Should the forage production in Catalonia move to Atlantic areas where precipitation is expected to be maintained or increase?
- Should the focus be put on more sustainable and water use efficient systems, including polycrops and intercrops?
- Should forage production be externalized to other, more productive areas outside Catalonia?

The socioeconomic impact of moving forage production to other areas, within or outside Catalonia, is important. Issues of food security emerge. The conversion towards more sustainable production systems where at least some of the species are expected to perform under challenging climatic conditions might crash against market needs and the requirements of countries currently buying alfalfa from Catalan producers. The biggest questions, however, remain: how far will we go before we hopefully revert the climatic crisis, and which areas will remain productive enough to feed an increasing human population under the path humanity started two centuries ago?

6 Future trends in weed management

The most promising venues at the moment seem to be the introduction of biodiversity to support agrifood and feed systems (Burgess et al., 2022), and those have been widely tested under extreme climates but also temperate conditions (Kirwan et al., 2007; Finn et al., 2013). In addition, grasslands include a highly diversity of ecosystems, and besides feed, they provide multiple additional values, including carbon sequestration (Ibañez et al., 2020; Rodríguez et al., 2020; Ibañez & Sebastià, 2022), biodiversity conservation and recreation (Marzetti et al., 2011).

Thus, our review on weeds in grasslands seems to support our hypotheses that introducing sown diversity and diversifying feed systems and landscapes should contribute to weed control as NbS, through ecosystem functioning and compatible with natural values, answering part of our questions. For a thorough review about Integrated Weed Management in grasslands, check the review by Schaffner et al. (2022).

Nonetheless, although sometimes weeds can be overlooked, it is difficult to determine if this can be applied to toxic native species or new invaders, where extreme measures might be necessary. Global change, and in particular climate change, is expected to produce shifts in grassland species distributions (Engler et al., 2011), producing unbalances and new species combinations, pushing ecosystems far away from their currently stable states and anticipated disturbance regimes.

7 Acknowledgments

We would like to thank Pepe Manjón-Córdoba and Yasmin Piñuela for comments, pictures and data. This work was funded by the Spanish Science Foundation through the project IMAGINE (CGL2017-85490-R); the Autonomous Organization of Spanish National Parks through the project IBERALP (PN2021-2820s); and the SUSFORAGE Project funded by the Partnership for Research and Innovation in the Mediterranean Area (PRIMA) and the National Funding Agencies (PRIMA Call 2020, Section 2/PCI 2021-121982).

8 References

Adeux, G., et al. (2019). Mitigating crop yield losses through weed diversity. *Nature Sustainability*, *2*(11), 1018–1026. https://doi.org/10.1038/s41893-019-0415-y

Altava-Ortiz, V., et al. (2023). Characterisation and temporal evolution (1915–2022) of drought and pluvial periods in Catalonia. *EMS Annual Meeting 2023, Bratislava, Slovakia, 20.*

Altieri, M. A. (1999). The ecological role of biodiversity in agroecosystems. In *Invertebrate Biodiversity as Bioindicators of Sustainable Landscapes* (pp. 19–31). Elsevier. https://doi.org/10.1016/b978-0-444-50019-9.50005-4

Bardgett, R. D. & van der Putten, W. H. (2014). Belowground biodiversity and ecosystem functioning. *Nature*, *515*(7528), 505–511. https://doi.org/10.1038/nature13855

Barrantes, O., et al. (2016). Transhumant GPS tracked sheep flocks from lowlands to highlands in Spain: grazing resources use and difficulties of walking/herding. *Options Méditerranéennes. Série A, Séminaires Méditerranéens*, *115*, 347–351.

Barrett, D. P., et al. (2021). Metabolomic analysis of host plant biochemistry could improve the effectiveness and safety of classical weed biocontrol. *Biological Control*, *160*, 104663. https://doi.org/10.1016/j.biocontrol.2021.104663

Baucom, R. S. & Holt, J. S. (2009). Weeds of agricultural importance: bridging the gap between evolutionary ecology and crop and weed science. *New Phytologist*, *184*(4), 741–743. https://doi.org/10.1111/j.1469-8137.2009.03077.x

Belesky, D. P., et al. (2001). Mineral composition of swards containing forage chicory. *Agronomy Journal*, *93*(2), 468–475. https://doi.org/10.2134/agronj2001.932468x

Boudreau, M. A. (2013). Diseases in intercropping systems. *Annual Review of Phytopathology*, *51*(1), 499–519. https://doi.org/10.1146/annurev-phyto-082712-102246

Boutagayout, A., et al. (2023). Agroecological practices for sustainable weed management in Mediterranean farming landscapes. *Environment, Development and Sustainability*. https://doi.org/10.1007/s10668-023-04286-7

Boydston, R. A., et al. (2008). Effect of hairy nightshade (solanum sarrachoides) presence on potato nematodes, diseases, and insect pests. *Weed Science*, *56*(1), 151–154. https://doi.org/10.1614/WS-07-035.1

Britannica, T. (2024). *Editors of Encyclopaedia. Weed.*

Brooks, M. L., et al. (2004). Effects of invasive alien plants on fire regimes. *Bioscience*, *54*, 677–688.

Brown, M. E., et al. (2023). Integrated modeling framework for sustainable agricultural intensification. *Frontiers in Sustainable Food Systems*, *6*. https://doi.org/10.3389/fsufs.2022.1039962

Bueno, C. G., et al. (2011). Assessment of wild boar rooting on ecological and pastoral values of alpine pyrenean grasslands. *Pirineos*, *166*(0), 51–67. https://doi.org/10.3989/pirineos.2011.166003

Burgess, A. J., et al. (2022). The deployment of intercropping and agroforestry as adaptation to climate change. *Crop and Environment*, *1*(2), 145–160. https://doi.org/10.1016/j.crope.2022.05.001

Caldeira, M. C., et al. (2001). Mechanisms of positive biodiversity–production relationships: insights provided by δ^{13}C analysis in experimental Mediterranean grassland plots. *Ecology Letters*, *4*(5), 439–443. https://doi.org/10.1046/j.1461-0248.2001.00238.x

Canals, R.-M. & Sebastià, M.-T. (2000). Analyzing mechanisms regulating diversity in rangelands through comparative studies: a case in the southwestern Pyrennees. *Biodiversity and Conservation*, *9*(7), 965–984. https://doi.org/10.1023/A:1008967903169

Canals, R. M., et al. (2017). Plant-herbivory feedbacks and selective allocation of a toxic metal are behind the stability of degraded covers dominated by Brachypodium pinnatum in acidic soils. *Plant and Soil*, *415*(1–2), 373–386. https://doi.org/10.1007/s11104-016-3153-1

Canals, R. M., et al. (2019). Soil bacterial functional diversity mirrors the loss of plant diversity by the expansion of a native tall-grass in high mountain grasslands. *Plant and Soil*, *445*(1–2), 243–257. https://doi.org/10.1007/s11104-019-04281-w

Canevari, M., et al. (2007). *Weed Management in Alfalfa*. http://alfalfa.ucdavis.edu/IrrigatedAlfalfa

Carnus, J.-M., et al. (2006). Planted forests and biodiversity. *J. Forestry*, *104*, 65–77. https://doi.org/10.1093/jof/ 104.2.65

Ceccanti, C., et al. (2018). Mediterranean wild edible plants: weeds or "new functional crops"? *Molecules*, *23*(9), 2299. https://doi.org/10.3390/molecules23092299

Chapin III, F. S., et al. (2000). Consequences of changing biodiversity. *Nature*, *405*(6783), 234–242. https://doi.org/10.1038/35012241

Chauhan, B. S. (2020). Grand challenges in weed management. *Frontiers in Agronomy*, *1*. https://doi.org/10.3389/fagro.2019.00003

Checinska Sielaff, A., et al. (2018). Microbial community structure and functions differ between native and novel (exotic-dominated) grassland ecosystems in an 8-year experiment. *Plant and Soil*, *432*(1–2), 359–372. https://doi.org/10.1007/s11104-018-3796-1

Chocarro, C. & Lloveras, J. (2015). The effect of row spacing on alfalfa seed and forage production under irrigated Mediterranean agricultural conditions. *Grass and Forage Science*, *70*(4), 651–660. https://doi.org/10.1111/gfs.12146

Chocarro, C., et al. (2005). Effects of winter grazing on spring production of lucerne under Mediterranean conditions. *Grass and Forage Science*, *60*(2), 146–150. https://doi.org/10.1111/j.1365-2494.2005.00461.x

Christian, J. M. & Wilson, S. D. (1999). Long-term ecosystem impacts of an introduced grass in the northern great plains. *Ecology*, *80*(7), 2397. https://doi.org/10.2307/176919

Connolly, J., et al. (2018). Weed suppression greatly increased by plant diversity in intensively managed grasslands: a continental-scale experiment. *Journal of Applied Ecology*, *55*(2), 852–862. https://doi.org/10.1111/1365-2664.12991

Cordeau, S. (2022). Conservation agriculture and agroecological weed management. *Agronomy*, *12*(4), 867. https://doi.org/10.3390/agronomy12040867

Cramer, W., et al. (2020). *Climate and Environmental Change in the Mediterranean Basin – Current Situation and Risks for the Future*. First Mediterranean Assessment Report.

Cummins, S., et al. (2021). Beneficial effects of multi-species mixtures on N2O emissions from intensively managed grassland swards. *Science of The Total Environment*, *792*, 148163. https://doi.org/10.1016/j.scitotenv.2021.148163

D'Antonio, C. M., et al. (1999). Disturbance and biological invasions: Direct effects and feedbacks. *Ecosystems of Disturbed Ground*.

Del Pozo, M. & Ibáñez, M. (1984). *La alfalfa. Su cultivo y aprovechamiento*. Ediciones Mundi-Prensa.

Delgado, I., et al. (2022). *Superfície agrícola. Per productes. Províncies*. Institut d'Estadística de Catalunya. https://www.idescat.cat/indicadors/?id=aec&n=15422

Ditomaso, J. (2000). Invasive weeds in rangelands: species, impacts, and management. *Weed Science Society of America*, *48*(2), 255–265.

Dong, B., et al. (2024). Competition for light and nitrogen with an earlier-sown species negatively affects leaf traits and leaf photosynthetic capacity of maize in relay intercropping. *European Journal of Agronomy*, *155*, 127119. https://doi.org/10.1016/j.eja.2024.127119

Dovrat, A. (1993). *Irrigated Forage Production*. Elsevier.

Ekwealor, K. U., et al. (2019). Economic importance of weeds: a review. *Asian Plant Research Journal*, 1–11. https://doi.org/10.9734/aprj/2019/v3i230063

Engler, R., et al. (2011). 21st century climate change threatens mountain flora unequally across Europe. *Global Change Biology*, *17*(7), 2330–2341. https://doi.org/10.1111/j.1365-2486.2010.02393.x

Epanchin-Niell, R. S., et al. (2010). Controlling invasive species in complex social landscapes. *Frontiers in Ecology and the Environment*, *8*(4), 210–216. https://doi.org/10.1890/090029

Esposito, M., et al. (2021). Drone and sensor technology for sustainable weed management: a review. *Chemical and Biological Technologies in Agriculture*, *8*(1), 18. https://doi.org/10.1186/s40538-021-00217-8

Evans, H. C., et al. (2001). Fungal biocontrol agents of weeds. In *Fungi as Biocontrol Agents: Progress, Problems and Potential* (pp. 169–192). CABI Publishing. https://doi.org/10.1079/9780851993560.0169

Fargione, J. & Tilman, D. (2006). Plant species traits and capacity for resource reduction predict yield and abundance under competition in nitrogen-limited grassland. *Functional Ecology*, *20*(3), 533–540. https://doi.org/10.1111/j.1365-2435.2006.01116.x

Farmilo, B., et al. (2023). Transient shifts in composition of degraded temperate native grassland following grazing exclusion. *Applied Vegetation Science*, *26*(2), 1–12. https://doi.org/10.1111/avsc.12731

Farooq, N., et al. (2019). *Allelopathy for Weed Management* (pp. 1–16). Springer. https://doi.org/10.1007/978-3-319-76887-8_16-1

Feurdean, A., et al. (2018). Biodiversity-rich European grasslands: ancient, forgotten ecosystems. *Biological Conservation*, *228*, 224–232. https://doi.org/10.1016/J.BIOCON.2018.09.022

Finn, J. A., et al. (2013). Ecosystem function enhanced by combining four functional types of plant species in intensively managed grassland mixtures: a 3-year continental-scale field experiment. *Journal of Applied Ecology*, *50*(2), 365–375. https://doi.org/10.1111/1365-2664.12041

Fisher, G. E. J., et al. (1996). Herbage production from swards containing a range of grass, forb and clover species and under extensive management. *Grass and Forage Science*, *51*(1), 58–72. https://doi.org/10.1111/j.1365-2494.1996.tb02038.x

Forney, D. R., et al. (1985). Weed suppression in No-Till Alfalfa (Medicago sativa) by prior cropping of summer-annual forage grasses. *Weed Science*, *33*(4), 490–497. https://doi.org/10.1017/S0043174500082710

Frame, J., et al. (1998). *Temperate Forage Legumes* (pp. 107–159). Cab International.

Frick, B. & Johnson, E. (2002). *Weeds on Organic Farms* (pp. 9–11). Minnesota Institute for Sustainable Agriculture.

Fukai, S. & Trenbath, B. R. (1993). Processes determining intercrop productivity and yields of component crops. *Field Crops Research*, *34*(3–4), 247–271. https://doi.org/10.1016/0378-4290(93)90117-6

Gassó, N., et al. (2009). Exploring species attributes and site characteristics to assess plant invasions in Spain. *Diversity and Distributions*, *15*(1), 50–58. https://doi.org/10.1111/j.1472-4642.2008.00501.x

Ghersa, C. M. & Roush, M. L. (1993). Searching for solutions to weed problems. *BioScience*, *43*(2), 104–109. https://doi.org/10.2307/1311971

Gonzalez-Andujar, J. L. (2023). Integrated weed management: a shift towards more sustainable and holistic practices. *Agronomy*, *13*(10), 2646. https://doi.org/10.3390/agronomy13102646

Gu, C., et al. (2021). Annual intercropping suppresses weeds: a meta-analysis. *Agriculture, Ecosystems and Environment*, *322*. https://doi.org/10.1016/j.agee.2021.107658

Gu, C., et al. (2022). A predictive model for weed biomass in annual intercropping. *Field Crops Research*, *277*, 108388. https://doi.org/10.1016/j.fcr.2021.108388

Gurevitch, J. & Padilla, D. (2004). Are invasive species a major cause of extinctions? *Trends in Ecology & Evolution*, *19*(9), 470–474. https://doi.org/10.1016/j.tree.2004.07.005

Hejcman, M., et al. (2013). Origin and history of grasslands in Central Europe – a review. *Grass and Forage Science*, *68*(3), 345–363. https://doi.org/10.1111/gfs.12066

Holechek, J. L. (2002). Do most livestock losses to poisonous plants result from "poor" range management? *Journal of Range Management*, *55*(3), 270. https://doi.org/10.2307/4003134

Hrabanski, M. & Le Coq, J. F. (2022). Climatisation of agricultural issues in the international agenda through three competing epistemic communities: climate-smart agriculture, agroecology, and nature-based solutions. *Environmental Science & Policy*, *127*, 311–320. https://doi.org/10.1016/j.envsci.2021.10.022

Hunt, N. D., et al. (2017). Reducing freshwater toxicity while maintaining weed control, profits, and productivity: effects of increased crop rotation diversity and reduced herbicide usage. *Environmental Science & Technology*, *51*(3), 1707–1717. https://doi.org/10.1021/acs.est.6b04086

Ibañez, M. & Sebastià, M. T. (2022). Net ecosystem CO2 exchange in mountain grasslands is seriously endangered by the temperature increase in the Eastern Pyrenees. *Atmosphere*, *13*(12), 1980. https://doi.org/10.3390/atmos13121980

Ibañez, M., et al. (2020). Phenology and plant functional type dominance drive CO2 exchange in seminatural grasslands in the Pyrenees. *The Journal of Agricultural Science*, *158*(1–2), 3–14. https://doi.org/10.1017/S0021859620000179

Inderjit, M. & Keating, K. I. (1999). *Allelopathy: Principles, Procedures, Processes, and Promises for Biological Control* (pp. 141–231). Kluwer Academic Publishers. https://doi.org/10.1016/S0065-2113(08)60515-5

Jayasundera, M., et al. (2021). Medicinal value of three agricultural weed species of the asteraceae family: a review. *Pharmacognosy Journal*, *13*(1), 264–277. https://doi.org/10.5530/pj.2021.13.36

Kiehl, K., et al. (2010). Species introduction in restoration projects–Evaluation of different techniques for the establishment of semi-natural grasslands in central and northwestern Europe. *Basic and Applied Ecology*, *11*, 285–299.

Kirwan, L., et al. (2007). Evenness drives consistent diversity effects in intensive grassland systems across 28 European sites. *Journal of Ecology*, *95*(3), 530–539. https://doi.org/10.1111/j.1365-2745.2007.01225.x

Klötzli, J., et al. (2023). Synergistic effects of grass competition and insect herbivory on the weed *Rumex obtusifolius* in an inundative biocontrol approach. *Scientific Reports*, *13*, 18508. https://doi.org/10.1038/s41598-023-45609-y

Krieger, M. T., et al. (2022). Controlling the abundance of a native invasive plant does not affect species richness or functional diversity of wet grasslands. *Applied Vegetation Science*, *25*(3), e12676. https://doi.org/10.1111/avsc.12676

Larios, L. & Suding, K. N. (2015). Competition and soil resource environment alter plant-soil feedbacks for native and exotic grasses. *AoB Plants*, *7*(1). https://doi.org/10.1093/aobpla/plu077

Leng, X., et al. (2013). Differential water uptake among plant species in humid alpine meadows. *Journal of Vegetation Science*, *24*(1), 138–147. https://doi.org/10.1111/j.1654-1103.2012.01439.x

Levine, J. M., et al. (2003). Mechanisms underlying the impacts of exotic plant invasions. *Biological Sciences*, *270*, 775–781.

Li, M., et al. (2019). A combination of rice cultivar mixed-cropping and duck co-culture suppressed weeds and pests in paddy fields. *Basic and Applied Ecology*, *40*, 67–77. https://doi.org/10.1016/j.baae.2019.09.003

Liebman, M. & Dyck, E. (1993). Crop rotation and intercropping strategies for weed management. *Ecological Applications*, *3*(1), 92–122. https://doi.org/10.2307/1941795

Lloveras, J. (2001). Alfalfa (Medicago sativa L.) management for irrigated Mediterranean conditions: the case of the Ebro Valley. *Quality in Lucerne and Medics for Animal Production*, 115–126.

Lloveras, J., et al. (1998). Harvest management effects on alfalfa (Medicago sativaL.) production and quality in Mediterranean areas. *Grass and Forage Science*, *53*(1), 88–92. https://doi.org/10.1046/j.1365-2494.1998.00100.x

Lloveras, J., et al. (2008). Yield, yield components, and forage nutritive value of alfalfa as affected by seeding rate under irrigated conditions. *Agronomy Journal*, *100*(1), 191–197. https://doi.org/10.2134/agronj2006.0333

Lloveras, J., et al. (2012). Alfalfa yield components and soil potassium depletion as affected by potassium fertilization. *Agronomy Journal*, *104*(3), 729–734. https://doi.org/10.2134/agronj2011.0293

Lloveras, J., et al. (2020). *La alfalfa: Agronomía y utilización* (Universitat de Lleida, Ed.).

Lüscher, A., et al. (2022). Using plant diversity to reduce vulnerability and increase drought resilience of permanent and sown productive grasslands. *Grass and Forage Science*, *77*(4), 235–246. https://doi.org/10.1111/gfs.12578

MacDougall, A. S. & Turkington, R. (2005). are invasive species the drivers or passengers of change in degraded ecosystems? *Ecology*, *86*(1), 42–55. https://doi.org/10.1890/04-0669

MacLaren, C., et al. (2020). An ecological future for weed science to sustain crop production and the environment: a review. *Agronomy for Sustainable Development*, *40*(4), 24. https://doi.org/10.1007/s13593-020-00631-6

Mantino, A., et al. (2021). Competition for light affects alfalfa biomass production more than its nutritive value in an olive-based alley-cropping system. *Forests*, *12*(2), 233. https://doi.org/10.3390/f12020233

Maron, J. L., et al. (2014). Invasive plants escape from suppressive soil biota at regional scales. *Journal of Ecology*, *102*(1), 19–27. https://doi.org/10.1111/1365-2745.12172

Martin, L. M., et al. (2014). Biodiversity, photosynthetic mode, and ecosystem services differ between native and novel ecosystems. *Oecologia*, *175*(2), 687–697. https://doi.org/10.1007/s00442-014-2911-0

Martín-Forés, I., et al. (2017). Weed abundance is positively correlated with native plant diversity in grasslands of southern Australia. *PLoS One*, *12*(6), e0178681. https://doi.org/10.1371/journal.pone.0178681

Marzetti, S., et al. (2011). Conservation and recreational values from semi-natural grasslands for visitors to two Italian parks. *Journal of Environmental Planning and Management*, *54*(2), 169–191. https://doi.org/10.1080/09640568.2010.505792

McLeod, M. L., et al. (2016). Exotic invasive plants increase productivity, abundance of ammonia-oxidizing bacteria and nitrogen availability in intermountain grasslands. *Journal of Ecology*, *104*(4), 994–1002. https://doi.org/10.1111/1365-2745.12584

Michaud, R., et al. (2015). *World Distribution and Historical Development* (pp. 25–91). Springer. https://doi.org/10.2134/agronmonogr29.c2

Molina, M., et al. (2014). Weeds and food diversity: natural yield assessment and future alternatives for traditionally consumed wild vegetables. *Journal of Ethnobiology*, *34*(1), 44–67. https://doi.org/10.2993/0278-0771-34.1.44

Monteiro, A. & Santos, S. (2022). Sustainable approach to weed management: the role of precision weed management. *Agronomy*, *12*(1), 118. https://doi.org/10.3390/agronomy12010118

Mseddi, K., et al. (2017). Screening of weeds and their effect on alfalfa (Medicago sativa). *Indian Journal of Agricultural Sciences*, *87*(11), 1565–1571. https://doi.org/10.56093/ijas.v87i11.75795

Ngondya, I. B., et al. (2016). A nature-based approach for managing the invasive weed species Gutenbergia cordifolia for sustainable rangeland management. *SpringerPlus*, *5*(1), 1787. https://doi.org/10.1186/s40064-016-3480-y

Oerke, E.-C. (2006). Crop losses to pests. *The Journal of Agricultural Science*, *144*(1), 31–43. https://doi.org/10.1017/S0021859605005708

Oerke, E.-C. & Dehne, H.-W. (2004). Safeguarding production–losses in major crops and the role of crop protection. *Crop Protection*, *23*(4), 275–285. https://doi.org/10.1016/j.cropro.2003.10.001

Parker, C. & Fryer, J. D. (1975). Weed control problems causing major reductions in world food supplies. *FAO Plant Protection Bulletin*, *23*, 83–95.

Pärtel, M. & Bruun, H. H. (2015). Biodiversity in temperate European grasslands: origin and conservation. *Grassland Science in Europe*, *10*, 1–14.

Parton, W., et al. (2007). Global-scale similarities in nitrogen release patterns during long-term decomposition. *Science*, *315*(5810), 361–364. https://doi.org/10.1126/science.1134853

Patterson, D. T., et al. (1999). Weeds, insects and diseases. *Climatic Change*, *43*(4), 711–727. https://doi.org/10.1023/A:1005549400875

Pausas, J. G. & Keeley, J. E. (2021). Wildfires and global change. *Frontiers in Ecology and the Environment*, *19*(7), 387–395. https://doi.org/10.1002/fee.2359

Paustian, K., et al. (2000). Management options for reducing CO2 emissions from agricultural soils. *Biogeochemistry*, *48*(1), 147–163. https://doi.org/10.1023/A:1006271331703

Peeters, A., et al. (2014). Grassland term definitions and classifications adapted to the diversity of European grassland-based systems. *25th EGF General Meeting on "EGF at 50: The Future of European Grasslands*, *19*, 743–750.

Poschlod, P., et al. (2009). Origin and development of grasslands in Central Europe. In *Grasslands in Europe* (pp. 15–25). KNNV Publishing. https://doi.org/10.1163/9789004278103_003

Pozo, M. & Ibáñez, M. (1983). *La Alfalfa: su cultivo y aprovechamiento* (3rd ed.). S. A. Mundi-Prensa Libros.

Putnam, A. R., et al. (1983). Exploitation of allelopathy for weed control in annual and perennial cropping systems. *Journal of Chemical Ecology*, *9*(8), 1001–1010. https://doi.org/10.1007/BF00982207

Qasem, J. R. (2012). Allelopathy in crop-weed interaction, field studies. *Allelopathy Journal*, *30*(2), 159–176.

Qasem, J. R. & Foy, C. L. (2001). Weed allelopathy, its ecological impacts and future prospects. *Journal of Crop Production*, *4*(2), 43–119. https://doi.org/10.1300/J144v04n02_02

Rapoport, E. H., et al. (1995). Edible weeds: a scarcely used resource. *Bulletin of the Ecological Society of America*, *76*(3), 163–166.

Razzaq, A., et al. (2012). Reduced herbicide doses used together with allelopathic sorghum and sunflower water extracts for weed control in wheat. *Journal of Plant Protection Research*, *52*(2), 281–285. https://doi.org/10.2478/v10045-012-0045-0

Renne, I. J., & Tracy, B. F. (2013). Disturbance intensity, timing and history interact to affect pasture weed invasion. *Basic and Applied Ecology*, *14*(1), 44–53. https://doi.org/10.1016/j.baae.2012.10.006

Ribas, A., et al. (2015). Plant identity and evenness affect yield and trace gas exchanges in forage mixtures. *Plant and Soil*, *391*(1–2), 93–108. https://doi.org/10.1007/s11104-015-2407-7

Ribas, À., et al. (2023). Sown diversity effects on yield and resistance to weed invasion: Clues to improve mixture design under climatic change in the Mediterranean. *Agriculture, Ecosystems and Environment*, *356*. https://doi.org/10.1016/j.agee.2023.108601

Riemens, M., et al. (2022). An integrated weed management framework: a pan-European perspective. *European Journal of Agronomy*, *133*, 126443. https://doi.org/10.1016/j.eja.2021.126443

Rodríguez, A., et al. (2018). Interactions between global change components drive plant species richness patterns within communities in mountain grasslands independently of topography. *Journal of Vegetation Science*, *29*(6), 1029–1039. https://doi.org/10.1111/jvs.12683

Rodríguez, A., et al. (2020). Interactions between biogeochemical and management factors explain soil organic carbon in Pyrenean grasslands. *Biogeosciences*, *17*(23), 6033–6050. https://doi.org/10.5194/bg-17-6033-2020

Ryel, R. J., et al. (2010). Functional differences in water-use patterns of contrasting life forms in great basin steppelands. *Vadose Zone Journal*, *9*(3), 548–560. https://doi.org/10.2136/vzj2010.0022

Schaffner, U., et al. (2022). *Integrated Weed Management in Grasslands* (pp. 339–360). Springer. https://doi.org/10.19103/as.2021.0098.15

Schäppi, B. & Körner, C. (1996). Growth responses of an alpine grassland to elevated CO2. *Oecologia*, *105*(1), 43–52. https://doi.org/10.1007/BF00328790

Schils, R. L. M., et al. (2022). Permanent grasslands in Europe: land use change and intensification decrease their multifunctionality. *Agriculture, Ecosystems & Environment*, *330*, 107891. https://doi.org/10.1016/j.agee.2022.107891

Sebastià, M. (2007). Plant guilds drive biomass response to global warming and water availability in subalpine grassland. *Journal of Applied Ecology*, *44*(1), 158–167. https://doi.org/10.1111/j.1365-2664.2006.01232.x

Sebastià, M., et al. (2008a). Grazing as a factor structuring grasslands in the Pyrenees. *Applied Vegetation Science*, *11*(2), 215–222. https://doi.org/10.3170/2008-7-18358
Sebastià, M., et al. (2008b). Strong shifts in plant diversity and vegetation composition in grassland shortly after climatic change. *Journal of Vegetation Science*, *19*(3), 299–306. https://doi.org/10.3170/2008-8-18356
Sebastià, M.-T., et al. (2023). Quality production of sainfoin swards challenged by global change in mountain areas in the Western Mediterranean. *Agronomy*, *14*(1), 6. https://doi.org/10.3390/agronomy14010006
Sebastià, M.-T. & Puig, L. (2008). Complex vegetation responses to soil disturbances in mountain grassland. *Plant Ecology*, *199*(1), 77–88. https://doi.org/10.1007/s11258-008-9413-2
Selim, M. (2019). A review of advantages, disadvantages and challenges of crop rotations. *Egyptian Journal of Agronomy*, *0*(0), 0–0. https://doi.org/10.21608/agro.2019.6606.1139
Sharma, G., et al. (2021). Crop diversification for improved weed management: a review. *Agriculture*, *11*(5), 461. https://doi.org/10.3390/agriculture11050461
Stark, J. M. & Norton, J. M. (2015). The invasive annual cheatgrass increases nitrogen availability in 24-year-old replicated field plots. *Oecologia*, *177*(3), 799–809. https://doi.org/10.1007/s00442-014-3093-5
Sturludóttir, E., et al. (2014). Benefits of mixing grasses and legumes for herbage yield and nutritive value in Northern Europe and Canada. *Grass and Forage Science*, *69*(2), 229–240. https://doi.org/10.1111/gfs.12037
Sun, G., et al. (2009). Stellera chamaejasme L. increases soil N availability, turnover rates and microbial biomass in an alpine meadow ecosystem on the eastern Tibetan Plateau of China. *Soil Biology and Biochemistry*, *41*(1), 86–91. https://doi.org/10.1016/j.soilbio.2008.09.022
Suter, M., et al. (2015). Nitrogen yield advantage from grass–legume mixtures is robust over a wide range of legume proportions and environmental conditions. *Global Change Biology*, *21*(6), 2424–2438. https://doi.org/10.1111/gcb.12880
Sujeeun, L. & Thomas, S. C. (2017). Potential of biochar to mitigate allelopathic effects in tropical island invasive plants: evidence from seed germination trials. *Tropical Conservation Science*, 10, 1–12. https://doi.org/10.1177/1940082917697
Tan, Z. M. & Zhou, S. C. (1995). Grassland degrading in Luolong County, Tibet. *Journal of Sichuan Grassland*, *13*(1), 25–28.
Traveset, A. & Richardson, D. (2006). Biological invasions as disruptors of plant reproductive mutualisms. *Trends in Ecology & Evolution*, *21*(4), 208–216. https://doi.org/10.1016/j.tree.2006.01.006
Tuel, A., et al. (2021). Understanding climate change over the Southwestern Mediterranean using high-resolution simulations. *Climate Dynamics*, *56*(3–4), 985–1001. https://doi.org/10.1007/s00382-020-05516-8
Varanasi, A., et al. (2016). Impact of climate change factors on weeds and herbicide efficacy. In *Advances in Agronomy* (Vol. 135, pp. 107–146). Academic Press Inc. https://doi.org/10.1016/bs.agron.2015.09.002
Wagner, T. C., et al. (2023). Summer rain and wet soil rather than management affect the distribution of a toxic plant in production grasslands. *Scientific Reports*, *13*(1), 13530. https://doi.org/10.1038/s41598-023-40646-z

Wedin, D. & Tilman, D. (1993). Competition among grasses along a nitrogen gradient: initial conditions and mechanisms of competition. *Ecological Monographs*, *63*(2), 199–229.

Weisberger, D., et al. (2019). Does diversifying crop rotations suppress weeds? A meta-analysis. *PLoS One*, *14*(7), e0219847. https://doi.org/10.1371/journal.pone.0219847

Willey, R. W. (1990). Resource use in intercropping systems. *Agricultural Water Management*, *17*(1–3), 215–231. https://doi.org/10.1016/0378-3774(90)90069-B

Wilsey, B. J., et al. (2018). Phenology differences between native and novel exotic-dominated grasslands rival the effects of climate change. *Journal of Applied Ecology*, *55*(2), 863–873. https://doi.org/10.1111/1365-2664.12971

Wu, C., et al. (2016). The toxicology mechanism of endophytic fungus and swainsonine in locoweed. *Environmental Toxicology and Pharmacology*, *47*, 38–46. https://doi.org/10.1016/j.etap.2016.08.018

Xin, L., et al. (2013). Differential water uptake among plant species in humid alpine meadows. *Journal of Vegetation Science*, *24*(1), 138–147. https://doi.org/10.1111/j.1654-1103.2012.01439.x

Xu, X., et al. (2015). Plant invasions differentially affeted by diversity and dominant species in native- and exotic-dominated grasslands. *Ecology and Evolution*, *5*, 5662–5670.

Zandstra, B. H. & Motooka, P. S. (1978). Beneficial effects of weeds in pest management – a review. *PANS*, *24*(3), 333–338. https://doi.org/10.1080/09670877809411634

Zhang, C., et al. (2019). Intercropping cereals with faba bean reduces plant disease incidence regardless of fertilizer input; a meta-analysis. *European Journal of Plant Pathology*, *154*(4), 931–942. https://doi.org/10.1007/s10658-019-01711-4

Zhang, Z., et al. (2020). Don't judge toxic weeds on whether they are native but on their ecological effects. In *Ecology and Evolution* (Vol. 10, Issue 17, pp. 9014–9025). John Wiley and Sons Ltd. https://doi.org/10.1002/ece3.6609

Zhao, J., et al. (2020). Does crop rotation yield more in China? A meta-analysis. *Field Crops Research*, *245*, 107659. https://doi.org/10.1016/j.fcr.2019.107659

Zhao, M., et al. (2013). A review of the most economically important poisonous plants to the livestock industry on temperate grasslands of China. *Journal of Applied Toxicology*, *33*(1), 9–17. https://doi.org/10.1002/jat.2789

Zimdahl, R. L. (2007). *Fundamentals of Weed Science* (3rd ed.). Elsevier.

Zimdahl, R. L. (2018). Allelopathy. In *Fundamentals of Weed Science* (pp. 253–270). Elsevier. https://doi.org/10.1016/B978-0-12-811143-7.00009-3

Part 3

Grasslands, grazing and animal nutrition: effects on health and product quality

Chapter 10

Quantifying forage quality to improve nutrient use efficiency in ruminant feeding

Martin Gierus, BOKU University, Austria

1 Introduction

Demands of society for environmentally friendly agricultural practices in Europe are increasing. Measures within the European Union (EU) to regulate N and P losses to the environment are intended to support the reduction of N and P inputs from mineral fertilization. Farmers are under pressure to reduce their surpluses (especially for N), to avoid greenhouse gas emissions at the farm level and to integrate aspects of ecosystem services in their farm operations. This applies to intensive dairy production, where, in dependence of the different European contrasting environments, permanent grassland and/or field forage production play an essential role in providing nutrient and energy supply. In particular, nitrate levels in groundwater have reached critical levels, exacerbated in areas that are prone to nitrate leaching. A decrease in mineral

http://dx.doi.org/10.19103/AS.2025.0154.16

N fertilization in forage production is expected to achieve high N use efficiency (low N surpluses at the farm gate). To meet the demand for higher production, on the one hand, field forage cultivation (or temporary grassland) will be used more, and on the other hand, grassland cultivation will be intensified, using also integrated crop-livestock-forestry systems. In this regard, temporary grassland is an important strategy in grassland management and forage production. Temporary grassland is characterized by a grass phase of one to three years. It has its origin in the positive response to residual effects of the grass phase in the subsequent arable crop, especially on soil fertility, reduced crop diseases and weeds. This kind of grass-arable system is important for some regions, to obtain adequate succeeding crop yields. The establishment of temporary grassland may be advantageous to overcome feed shortages observed in permanent grassland (Søegaard et al., 2007).

The increasing herd size in dairy farms in some regions, or the increasing performance of individual animals in smaller herds, and the variable forage quality of the offered fodder along the vegetation period are decisive limits for the use of grassland as pasture or conserved forage (hay or silage). Improved N efficiency is required at the same time. Reduced mineral N fertilizer use makes the cultivation of forage legumes and their mixtures with grasses as native protein sources interesting again. Dry matter (DM) yield and forage quality of legumes can vary widely between production systems and legume species. White clover is commonly found in pastures. In addition to white clover, red clover and alfalfa are important forage legumes for cut use (hay, silage). Birdsfoot trefoil is of lesser importance, but due to its condensed tannin content (see Section 10) it could be an interesting alternative forage for certain production systems.

The main objective of the present work is to integrate the knowledge about evaluation of feed quality with that of grassland science in order to optimize nutrient fluxes in the plant–animal system. In this way, it should be possible to obtain a better understanding of processes influencing the variation in forage quality and relate them to possible relevant aspects in animal nutrition. Several laboratory methods are available to estimate the protein availability in the rumen, the energy content of feeds and the organic matter digestibility (OMD) for ruminants.

2 Evaluation of the nutritive value of forages

Among forage crops, grasses are quantitatively the most important and most studied functional group in grassland. Grasses are also present in several different ecosystems worldwide. Depending on biotic and abiotic factors, grasses are characterized by a high yield potential with high-energy density, provided a corresponding management. The forage quality is determined by the energetic potential, the content of crude protein and carbohydrates, the digestibility as well as the feed intake, which determines the high nutritive value of the forage. The multitude of factors determining feed quality results in the necessity of a complex evaluation of the nutritional value. The basis for this is the adaptation to the supply recommendations. For practical reasons, there is necessary to ensure simple and straightforward application of suitable methods in feed evaluation. This means, although animal performance is the most accurate answer to define appropriate feed quality, *in vitro* laboratory methods (enzymatic, chemical or with rumen fluid) are often used to evaluate feed quality by feed composition, digestibility or degradability of defined groups of substances.

The common proximal analyses (Weender analysis) are used to characterize forage nutritive value, measuring crude protein, crude fat, crude fiber and non-nitrogen extracts from the organic matter. A sufficient characterization of forages with respect to ruminal protein and carbohydrate availability is not achieved via Weender analysis. A more differentiated view of carbohydrates is achieved using the detergent method. Here, the cell complexes are subdivided into the fractions cell wall, i.e. neutral detergent fiber (NDF), acid detergent fiber (ADF) and acid detergent lignin (ADL), and cell content, i.e. proteins, sugars and minerals. The degradation rate (Kd in %/h) measurements support the more precise characterization of the nutritive value of forages, as parameters of ruminal kinetics are involved. Ideally, achieving the optimization between protein and carbohydrates degradation in the rumen (synchronism) is of great importance with regard to the performance of animals but also for the environmental compatibility of grassland-based milk production.

2.1 Evaluation of the protein value of forages

In the evaluation of protein quality of forages, methods should not only quantify (crude) protein content but also include parameters of ruminal kinetics, e.g. protein degradation. In this case, the quality of the protein is not differentiated in the proximate analysis, nor is it differentiated in the detergent method. The relatively intensive protein degradation of grasses and legumes in the rumen or during the ensiling process in the silage is not captured. This disparity in the differentiation of proteins (and carbohydrates) in forages is overcome by the Cornell Net Carbohydrate and Protein System (CNCPS, Sniffen et al., 1992 and other more recent versions, e.g., Higgs et al., 2015), which models the conversions of cell wall and reserve carbohydrates and proteins in the forestomachs and postruminal digestive tract of ruminants. Using standard chemical analyses, the extent of biologically significant carbohydrate and protein fractions is estimated.

A limitation of the CNCPS model is that the CP or carbohydrate fractions should have a measurable rumen degradation rate. Together with the passage rate (Kp, %/h), the effective degradability in the rumen is determined, assuming a range of degradation rates that is similar within each fraction and among different feed types. Measuring the degradation rate of fractions B1, B2, B3 would be important to adapt different grass accessions (in breeding programs), grass silages (conservation effects on the nutritive value) and grass–legume mixtures (botanical composition) to their intrinsic contribution to the bypass protein.

3 Ruminal forage protein degradation

The definition of forage quality is closely related to the availability of nutrients and energy in feed to meet animal requirements. Nutrients and energy content to be supplied to ruminants have to be determined not only by its chemical constituents but also should involve knowledge about the digestive processes and physiological status of the animal to be fed.

The protein content of feeds for ruminants in particular undergoes two main transformations. The first one occurs in the forestomachs, where dietary protein is to a great extent degraded to peptides, amino acids and ammonia by microbial activity in the reticulo-rumen. In this way, microbes contribute extensively to the host converting intermediates of dietary protein degradation to valuable high-quality protein in form of microbial protein, which is available for absorption in the small intestine accounting for a major proportion of the absorbed amino acids.

The second transformation of dietary proteins in the ruminant occurs in the transfer of free amino acids and peptides from the gut lumen into the bloodstream. Energy expenditure for nutrient transport and replacement of cells from the gut wall, i.e. digestion work, are the main factors underlying the transformation (Gierus et al., 2001; Bannink et al., 2006). This means that the amino acids appearing in the bloodstream, liver and peripheric organs do not necessarily reflect the amino acid composition available for absorption or present as dietary protein. Microbial protein synthesis in the rumen and the contribution of post-absorptive amino acids and peptides are the final sources for net use.

Although the absorption of amino acid to meet animal requirements of maintenance and production is the main concern in animal nutrition, the determination of the protein value of a feed for ruminants is based on the amount of degradable crude protein, its rate of degradation in the rumen, the protein flow to the small intestine (bypass or escape protein) and the small intestine digestibility. The feed protein characteristics are assessed using the *in situ* nylon bag technique and *in vitro* techniques, whereas flows are measured *in vivo* (Tamminga and Chen, 2000; Huhtanen et al., 1997). The main constraint to these approaches is that bias in the analytical determination as well as bias in the metabolism and utilization of protein by the animal occurs (Martineau et al., 2023). For instance, bias due to analytical determination may be the case with the washout fraction of the nylon bag technique, which is considered completely and immediately degraded in the rumen (Gierus, 2005). Correction for losses of insoluble particles in the nylon bag was introduced to reduce the variation of estimates based on this technique (Hvelplund and Weisbjerg, 2000). Results suggest that soluble protein is not necessarily immediately and completely degraded, contributing with 10% on average of the escape protein of different feeds (Gierus, 2005).

Unlike *in vivo* methods, *in vitro* methods are not able to estimate the variation related to nutrient utilization efficiency for animal product conversion, such as the response to different levels of grinding of concentrate ingredients or the suppression of rumen fermentation due to the addition of excessive amounts of concentrates in the diet (Tyrrell and Moe, 1975; Kleinmanns and Potthast, 1984), resulting in partially incorrect measurements of the nutritive value. The effect of intake on digestibility is another example. As the level of intake increases, digestibility may decrease, depending on the feed type and quality (Reynolds et al., 1991). Energy expenditure for transport of nutrients, replacement of cells from the gut wall, transamination and interorgan amino acid exchange are the main reasons for a high protein turnover rate in the gut tissues (Webster et al., 1975; Reynolds et al., 1994; Reynolds, 1995). Amino acids contribute to the interchange of nutrients between tissues and in gut tissue maintenance. Therefore, in a deeper insight, the N composition of escaped and microbial protein available for absorption in the small intestines does not necessarily reflect the N compounds appearing post-absorption, i.e. in the portal vein (Tagari and Bergman, 1978; Lindsay, 1993).

In vivo methods comprise the more reliable information obtained for biological processes in the animal. *In vitro* and *in situ* methods are necessary to provide specific data on a routine basis. In addition, the degradation rate of nutrients is important for the determination of energy supply for the ruminant, because microbial activity and the host are involved. In contrast, *in vitro* methods are performed on standardized conditions, and several samples at once can be analyzed and their nutritive value is estimated. *In vitro* methods are important tools for the screening of forage plant accessions, analyses of similar feeds but variable charges and serve as basis for modelling purposes, where parameters of digestion kinetics are necessary inputs (Gierus et al., 2024).

Besides *in vitro* and *in vivo* approaches, the proportion of crude protein that occurs as degradable crude protein or undegraded crude protein (UDP) is influenced by the protein degradation rate (Kd) and the passage rate (Kp) of the solid and liquid phases of the rumen contents. Estimation of UDP is of great importance when protein flowing in the duodenum is to be quantified in order to avoid, if necessary, excesses of intraruminal degradable protein. With the determination of the UDP fraction and the amount of microbial protein synthesis, the amount of protein flowing in the small intestine is estimated, which occurs in connection with estimates of the OMD or the metabolizable energy (ME).

The supply of high-quality protein is the primary concern for forage crops. High protein quality (i.e. composition of amino acids, ruminal degradation rate compatible with the availability of fermentable carbohydrates) of the feed ration of high-yielding dairy cows is present at a level of about 30% of crude protein undegraded in the rumen (UDP). Rapid protein degradation is expected for grasses and forage legumes, although this can vary greatly between grass and legume species. Rapid degradation, in the presence of carbohydrate deficiency, limits N use efficiency in high-yielding dairy cows and inefficient utilization. The addition of UDP-rich concentrate in diet formulation becomes necessary to optimize animal performance.

As a rationale for the rapid degradation of plant protein, intact proteins of the photosynthetic apparatus are potentially rapidly degradable. Approximately 1/3 of the protein is in the form of Rubisco (CO_2-fixing enzyme ribulose-1,5-bisphosphate carboxylase/oxygenase), which is generally rapidly degradable (McNabb et al., 1994). The degradability of proteins in the rumen is dependent on the predisposition to rumen microbial activity; rapid degradability is known to result in lower fluxes of high-quality protein in the small intestine. Several papers demonstrate that under such conditions, amino acids absorption in the small intestine is limiting for lamb, lactating ewe and dairy cow performance (Tedeschi and Fox, 2017). Rapid degradation of the protein fraction usually leads to excess nitrogen in the rumen, which is correlated with the non-protein N content in milk (milk urea). Since about 90% of the protein in leaves consists of pure protein of photosynthesizing enzymes, a higher excretion of N in urine and of milk urea can be expected in case of excess nitrogen in the rumen. Systematic determination of protein breakdown of individual grass or forage legumes as a function of management (stage of development at utilization) is therefore a prerequisite for (1) deriving forage production strategies or grazing practices with higher protein quality and higher N use efficiency of the grazing animal, (2) modeling N fluxes in the soil–plant–animal system as a function of forage crop species and (3) to provide forage plant breeders with suitable selection targets.

The nylon bag method (*in situ* method) is used worldwide in the various protein evaluation systems for estimating the UDP proportion of crude protein, especially since the proportions of degradable and undegradable crude protein are also considered essential in most systems for estimating the protein quality of a feed for ruminants (Fig. 1). *In vitro* characterization of the UDP content

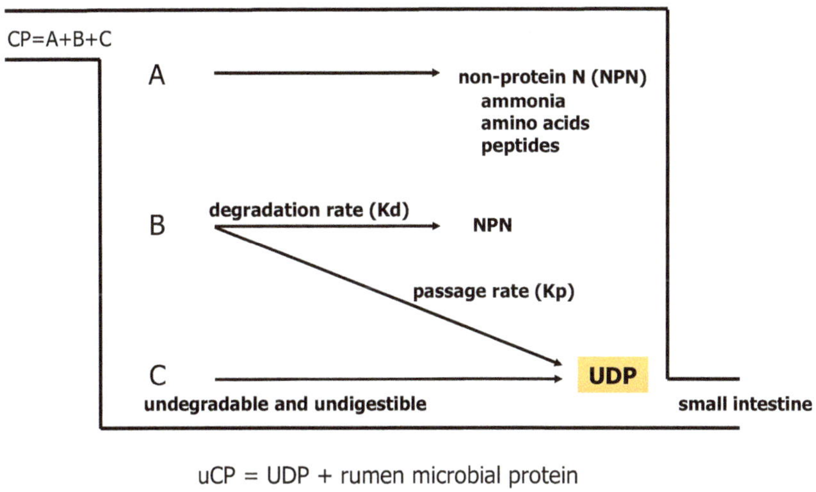

Figure 1 Schematic representation of the formation of undegraded crude protein (UDP) in the rumen (own representation); uCP: utilizable crude protein in small intestines.

in the different grass and legume species depends on their utilization regime (cutting frequency, fertilization rates, season). For the use as selection target in breeding programs for forage crops, the characterization of UDP has not yet been sufficiently validated with respect to efficient N utilization in animals.

3.1 Availability of undegradable protein in the small intestine

There are considerable differences in protein digestion between ruminants and non-ruminant animals. In ruminants, the supply of amino acids is carried out by the rumen microbes, and in this way, about 0.75 of the amino acid requirement is supplied. For this reason, the amino acid pattern in the duodenum of ruminants differs greatly from the amino acid pattern of the ingested food, as explained above. In ruminants, the feed proteins in the rumen are broken down into amino acids by protein-degrading microorganisms and deaminated (= cleavage of an ammonium ion or ammonia). The result of protein degradation in the rumen is ammonia (NH_3) and short-chain carboxylic acids. Ammonia is an important source of nitrogen, especially for bacteria that break down cellulose and hemicellulose. In addition, CO_2 and volatile fatty acids are also obtained by breaking down the carbon skeleton of the amino acids, especially enhanced when fermentable organic matter in the rumen is insufficient. The protein that cannot be broken down in the rumen flows on into the abomasum and small intestine. The indigestible protein proportion (not absorbed during transit in the small intestine) is represented by fraction C (see Fig. 2). The UDP and microbial protein provide ruminants with amino acids.

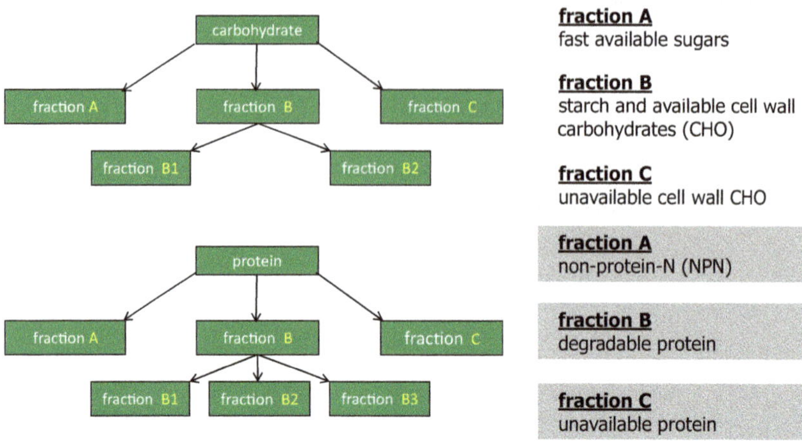

Figure 2 Breakdown of carbohydrate and protein fractions of feeds (after CNCPS, Sniffen et al., 1992, expanded; Licitra et al., 1996). For protein, fraction B is differentiated into 3 subfractions [B1 (rapidly degradable in the rumen), B2 (potentially completely degradable), B3 (slow, not necessarily completely degradable)]. For carbohydrates, fractions B1 and B2 are differentiated. B1 includes the slowly available starch and the soluble cell wall components. B2 comprises the insoluble but available cell wall components.

It is unclear if the higher intestinal digestibility of rumen UDP may have resulted from the lower protein degradability. Factors affecting the low intestinal digestibility of UDP may be related to processing steps (in concentrate feeds), heat damaged forages (hay, silage), but also due to high proportion of secondary plant compounds (e.g. polyphenols) – see Section 10.

4 Protein and carbohydrate fractionation to evaluate different forage crops

A quantification of the protein and carbohydrate fractions was performed for different forage crops. Potential genotype-related differences in the cultivar spectrum have not yet been systematically investigated, especially for the protein fractions. Genotypes with low proportions of protein fraction A are desirable. Potential N losses in the animal are less environmentally effective if they are excreted in the form of fraction C (N in the cell wall) via the feces, which, e.g., may show slower mineralization in grazing systems.

In ruminant nutrition, protein fractionation represents a chemical-analytical *in vitro* method for evaluating the protein quality of grassland growths (Licitra et al., 1996). Moreover, together with the ME content or the OMD, it allows an estimation of the amount of protein digestible in the intestines (undegraded forage protein and microbial protein). In Germany, since 2023 (GfE, 2023), the small intestine digestibility of proteins (sidCP) and amino acids (sidAA) is being considered for diet formulation, including the proportion of amino acids coming from the undegraded protein (UDP) and microbial protein predicted.

The CNCPS is a mathematical-dynamic model that estimates and predicts nutrient requirements and supply. The system uses breakdown and passage rates of carbohydrates and proteins to predict rumen fermentation, intestinal absorption and the content of metabolic energy and protein (Fox et al., 2004). Protein fractionation is an important component to ensure nutrient requirements and supply. There are various approaches to this. The focus is on protein fractionation according to Licitra et al. (1996), as depicted in Fig. 2. A significant difference between the two methods lies in the subdivision into the respective fractions. According to Licitra et al. (1996), the fractionation is as follows: A (non-protein), B1 (buffer-soluble pure protein), B2 (buffer-insoluble pure protein), B3 (cell-wall-bound soluble pure protein) and C (cell-wall-bound insoluble pure protein). Higgs et al. (2015), on the other hand, subdivides fraction A into A1 (ammonia) and A2 (buffer-soluble pure protein). There is a further deviation with B1 (insoluble pure protein) and B2 (fiber-bound protein). Fraction C with the indigestible protein, however, is identical. The separation into fractions A1 and A2 enables a more precise prediction of the metabolic protein from non-protein nitrogen, free amino acids and small peptides (Higgs et al., 2015). The free amino acids and small peptides were thus shifted to fraction A2 (Van Amburgh et al., 2015).

Over the years, the CNCPS model has become a tool to characterize the nutritive value of forages. In this regard, fraction A in forages, determined by this method, includes non-protein nitrogen (NPN) compounds such as nitrate, free amino acids, etc., which do not contribute to undegradable protein formation. In plants, NPN compounds accomplish important functions in N metabolism as storage and transport forms. The size of fraction A depends on the N metabolism of the forage crop (legume, non-legume) and the development stage at harvest time (see Tables 1, 2, 3). As a reference, grass silages of high nutritive value with respect to protein quality are considered as such, when at least 50% of crude protein is true protein (true protein in g N/kg N = crude protein content (g N/kg N) – fraction A (g N/kg N)). The systematic determination of protein fractions in forage crops during the growing season is a prerequisite for the derivation of forage production strategies or grazing practices with increased protein quality and N use efficiency for high-yielding animals.

In contrast to protein, a greater proportion of fraction A and a lower proportion of fraction C is desirable for the carbohydrate fractions. Selective breeding for water-soluble carbohydrates (fraction A) has been carried out in perennial ryegrass (*Lolium perenne* L.). This effort was aimed at synchronizing the supply of N and energy in the rumen for microbial protein synthesis (Humphreys, 1989). Increased efficiency of microbial protein synthesis has been demonstrated *in vitro* but not *in vivo*. Feeding trials demonstrate increased weight gains for lambs

Table 1 Classification of developmental stage for horst-forming grasses

Stage	Abbreviation	Description
1	S0	No extended leaf sheath
2	S1	1 extended leaf sheath
3	S2	2 elongated leaf sheaths
4	S3	3 or more elongated leaf sheaths
5	K1	1. culm thickening is palpable
6	K2	2. culm thickening is palpable
7	K3	3. culm thickening is palpable
8	B	Thickening of the uppermost leaf sheath
9	G1.1	Beginning of ear or panicle emergence
10	G1.5	50% of inflorescence is visible
11	G1.9	Spike or panicle neck becomes visible
12	OH	Top stem limb becomes visible
13	G2.1	Beginning of flowering, some stamens visible
14	G2.5	Full flowering, maximum pollen formation
15	G2.9	Flowering, only minor pollen formation visible
16	G3	Milk ripening
17	G4	Yellow ripening

Source: Adapted from: Salama et al. (2017)

Table 2 Classification of developmental stage for alfalfa

Stage	Stage name	Description
0	Early vegetative	Stem length <15 cm; no buds, flowers or seed bearers
1	Medium vegetative	stem length 16–30 cm; no buds, flowers or seed bearers
2	Late vegetative	stem length >31 cm; no buds, flowers or seed bearers
3	Early bud	1 to 2 nodes with buds, no flowers or seed bearers
4	Late bud	>3 nodes with buds, no flowers or seed bearers
5	Early flower	1 node with 1 open flower, no seed bearers
6	Late flowering	>2 nodes with open flowers, no seed bearers
7	Early seed formation	1 bi 3 nodes with green seed bearers
8	Late seed formation	>4 nodes with green seed bearers
9	Seed maturation	Occurrence of brown mature seeds

Source: Adapted from: Kalu and Fick (1983).

and increased milk yield and N retention in milk for dairy cows. Such sugar-rich ryegrass cultivars do not seem to have sufficient winter hardiness and resistance to early summer drought under appropriate climatic site conditions.

Table 4 shows results of a study with different forage crops. Despite the higher N content, forage legumes do not show higher ruminal degradation

Table 3 Classification of developmental stage for red clover

Stage	Description
1	Leaf stage
2	1st node
3	2nd node
4	3rd node
5	4th node
6	5th node
7	6th node
8	Bud palpable
9	Bud appearing
10	Beginning of flowering
11	Full flowering
12	Flowering – early seed formation
13	Late seed formation (milk and dough maturity)
14	Seed ripening (yellow or full ripening)

Source: Adapted from: Fagerberg (1988).

rates (Kd, in %/h). The degradation rates determined for the forage legumes agree with literature data. The *in vitro* method also allows the estimation the content of proteins digestible in intestines at different passage rates (4 and 8%/h) and of novel forage legumes (Caucasian clover and birdsfoot trefoil). A low protein degradability was detectable in birdsfoot trefoil, which was probably due to the effect of condensed tannins in the latter.

4.1 The protein quality of clover-grass mixtures under grazing

In concentrates (e.g. soybean extraction meal, rapeseed cake), the UDP content can be controlled by physical (heat) or chemical treatments. Feed preservation processes (ensiling, drying) can optimize the UDP content of forages based on proper conservation procedures. Several papers find a positive effect on the amount of non-ammonia nitrogen (NAN) flowing in the duodenum as the UDP content of the forage ration increases (e.g. Santos et al., 1998). As mentioned above, the optimal range for high-yielding dairy cows is approximately 30% of crude protein. Compared to forage conservation, in grazing the selective change of the UDP fraction in the forage is hardly controllable and variable because of the growth dynamics of the grazing plants. One possibility of control is the choice of grazing timing: within 1 week, the crude protein content of an alfalfa stand decreases by about 2% while the cell wall fraction increases by 3%. The variation in quality is dependent on the development stage.

Table 4 Nitrogen content, protein degradation rates (Kd), UDP contents and uCP estimates of different forage legume species, perennial ryegrass and silage maize

Forage	N, % DM	Degradation rate, %/h	UDP-4[1], %	UDP-8, %	uCP[2], g/kg TS		uCP[3], g/kg TS	uCP[4], g/kg TS
					4%	8%		
WK[5]	3.86^{a}	6.1^{a}	31.5^{b}	44.8^{bc}	194	217	191	172
RK	3.50^{a}	6.6^{a}	30.7^{bc}	44.0^{bc}	181	201	178	164
KU	3.56^{a}	6.2^{a}	33.2^{b}	46.5^{b}	193	213	186	–
LU	3.66^{a}	7.2^{a}	27.0^{bc}	38.8^{cd}	175	194	180	154
HO	3.72^{a}	2.6^{b}	53.2^{a}	64.8^{a}	223	242	185	–
DW	1.55^{bc}	8.1^{a}	27.7^{bc}	41.3^{bcd}	141	144	133	156
RK-G	1.91^{b}	8.0^{a}	29.1^{bc}	43.4^{bc}	138	146	133	156
Oldham	1.23^{cd}	6.2^{a}	28.7^{bc}	39.3^{cd}	122	122	114	128
Fuego	1.08^{d}	8.2^{a}	24.8^{c}	36.0^{d}	126	128	120	128
SE	*0.2*	*0.7*	*1.6*	*1.5*				

[1] UDP-4, UDP-8: calculated undegraded crude protein with a passage rate (Kp) of 4 and 8%/h, respectively. The portion of unavailable nitrogen was not subtracted.
[2] calculated with the formula: $uCP = [11.93 - (6.82 \times (UDP/XP))] \times ME + 1.03 \times UDP$.
[3] calculated with the formula: $uCP = 8.76 \times ME + 0.36 \times XP$.
[4] DLG feed value tables.
[5] WK: white clover, RK: red clover, KU: Caucasian clover – *Trifolium ambiguum*, LU: Alfalfa, HO: birdsfoot trefoil, DW: perennial ryegrass, RK-G: perennial ryegrass as companion grass to red clover; Oldham, Fuego: Silage maize cultivars.
Source: Excerpted from Gierus et al. (2007).
Means followed by different letter(s) within the same column are different at $p < 0.05$

In clover–grass mixtures under grazing, the size of the UDP fraction depends on the rate of degradation (%/h) of the individual plant species (see Table 4). In addition to the clover fraction in the stand, several factors influence the rate of degradation, such as the species-specific variability of legumes (e.g. content of secondary plant constituents such as condensed tannins), differences between grasses and legumes, and the assumed use regime (grazing/cut) (Gierus et al., 2012). The rate of protein degradation can be influenced by the leaf-weight ratio (ratio of leaf weight to tiller weight). The leaf-weight ratio is known to decrease with increasing developmental stage, but at a slower rate in legumes than in grasses. Legume species also differ among themselves: birdsfoot trefoil, alfalfa and red clover form stems with significantly increased cell wall contents as they develop. As the plants develop, degradation rates decrease in a species-specific manner. Grazing results in more frequent use of the above-ground biomass; consequently, younger growths are grazed (selectively). Furthermore, it can be assumed that a greater proportion of leaf mass is taken up compared to stem mass, provided there are sufficient selection possibilities. In this case, rapid protein degradation is caused by the higher uptake of leaf mass, which is accelerated by the easy accessibility of leaf protein to rumen microorganisms. Since rapid protein degradation is of great importance for the characterization of different legumes, the determination of the degradation rate in a short period of time (<4 hours) seems to be particularly suitable for this purpose (see also Section 9).

There are species-related differences between grasses and legumes with regard to protein degradation and the proportion of the UDP fraction. Dairy cows consume more alfalfa silage compared to grass silage and produce more milk (Hoffman et al., 1998). Despite comparable cell wall degradation rates of both silages (NDF, about 10%/h), the passage rate of the solid fraction of alfalfa silage increases. The higher milk yield when fed alfalfa silage is due to a faster formation of small feed particles in the rumen (Waghorn et al., 1989). Such effects can also be observed in feeding behavior on pasture. In addition to the higher palatability of alfalfa, the degradability of the cell wall, especially of leaves or petioles, plays an important role in the nutrient intake of grazing animals. The species-specific variation in forage quality requires a management strategy for grasslands that goes beyond the purely agronomic point of view.

5 Estimation of the energy value of forages

The estimation of the energy value of forages is becoming increasingly important. It is common to apply *in vitro* methods and equations to predict the ME content of forages. This is not only due to the great variability of forages but also due to the increasing importance of the accuracy of balancing appropriately the diet and its controllability aiming efficient N use. Various methods and prediction equations have been developed and tested to predict the energy value of forages, specifically the content of ME, which represents the central measurement of the energy supply for livestock (e.g. Gierus et al. 2016a). Estimates of OMD are crucial for the prediction of ME. In this regards, currently recognized methods are Tilley and Terry (1963) and De Boever et al. (1986), which estimate the OMD and ME content of the feed, and Steingass and Menke (1986), who uses the gas production *in vitro* based on rumen fluid. Very large differences are observed between the estimation methods for legume-rich grassland crops. The reason for this is the sensitivity and differentiation of the enzymes and microorganisms used. The latter vary between different agronomic traits like cutting times and botanical composition. The increasingly precise analysis options also require constant questioning of prediction equations. Most important, OMD seems to be the simple way to compare genetic progress in forage plants (Casler and Vogel, 1999). In addition, OMD is highly correlated to the ME content.

5.1 Estimation of organic matter digestibility

Digestibility plays a significant role in determining the energy value of forages. Measuring OMD on animals are the reference method for evaluating the energy content in any feed evaluation system. ME content estimation commonly uses measurements of OMD. OMD indicates the proportion of nutrients and energy available in the feed for the animal, which are not excreted by faeces. Various techniques are used to measure OMD using *in vivo* and *in vitro* methods (Kowalski et al., 2014). All currently used equations to predict the energy value of forages are based on different crude nutrients and are corrected by OMD measurements. While the cell content of the plant consists of carbohydrates, lipids, proteins and nitrogen-containing compounds, which have OMD of almost 100%, the cell wall is only degradable in the rumen to a lesser extent. Digestibility is largely determined by the fermentability of the cell wall (Forejtová et al., 2005).

The accuracy of prediction equations is validated against the *in vivo* determined OMD. In many cases, wethers are used to measure *in vivo* OMD. Briefly, wethers are fed according to their maintenance requirements. After an adaptation time of 14 days with the test feed, the total collection of faeces takes place. The *in vivo* digestibility is calculated from the difference of a nutrient between feeding and faecal excretion. The main disadvantages are the large amount of sample material required, the time needed to collect the faecal samples, feed and care for the animals, as well as the costly husbandry of the animals. This makes the routine analyses (e.g. accessions in breeding programs for forages, control analyses for farmers of their silage) on a regular basis almost impossible. Apart from these limitations, the *in vivo* OMD determination should serve as a reference method as far as possible in order to align the various *in vitro* methods with it.

5.2 In vitro *estimation of digestibility*

In vitro methods have been developed as they can be carried out much more quickly. Those methods based on the collection of rumen fluid may become limiting, as fistulated animals are necessary. In this regard, common *in vitro* methods using rumen fluid are the two-step method according to Tilley and Terry (1963) to estimate OMD, and the Hohenheim Feed Value test according to Menke and Steingass (1988), the latter determines the gas formation (GP) under controlled conditions. These tests results can either be entered directly into the prediction equation for the energy value created for this purpose or can be used to estimate digestibility.

Measurement using enzymes, like cellulases extracted from microorganisms, are getting important. The enzyme soluble organic matter (ESOM) according to De Boever (1986) is the one possibility to generate data for the prediction equation without the need of rumen fluid. Steingass and Leberl (2008) stated the following requirements on *in vitro* methods: the methods must be very precise and easily comparable with the reference method, easy to standardize, quick and easy to perform, easy to calibrate using NIR-Systems, and inexpensive.

A comparison of the three *in vitro* methods mentioned shows that the two-stage test according to Tilley and Terry and the Hohenheim feed value test have a clear disadvantage, due to fluctuations in the rumen fluid quality. The rumen fluid has a different effect in each laboratory run, which is why the repeatability of this method suffers. The concentration of rumen microbes is therefore subject to fluctuations, as the host of the donor animals cannot be standardized to such an extent that the same rumen fluid can be expected. The cellulase method according to De Boever is the most suitable, as this method shows a standardizable and favorable agreement with the *in vivo* digestibility using wethers. Iantcheva et al. (1999) compared the cellulase method and the Hohenheim feed value test with the *in vivo* method using a linear regression and also found a better correlation with the cellulase test. This was 0.85 for alfalfa hay and 0.88 for grass-rich hay. In contrast, the correlation for gas formation was only 0.54 and 0.75, respectively, compared to *in vivo* digestibility. Nevertheless, Kowalski et al. (2014) suggest correction equations, as the *in vitro* digestibility of enzyme-based methods is often lower than the *in vivo* digestibility.

When comparing the two-stage method with rumen fluid (Tilley and Terry 1963) with the pepsin cellulase method according to Iowerth et al. (1975), significant differences were also found by Barchiesi-Ferrari et al. (2011). The digestibility of different grass species was again significantly higher using rumen fluid than with the cellulase method.

5.3 Prediction equations for the estimation of ME content in forages

Grassland is gaining in ecological importance within the productions systems. Along with the current performance level of high yielding dairy cows, the demands on the quantity and quality of protein from forage are also increasing. In order to achieve an increased supply of utilizable crude protein (uCP) in the small intestines for dairy cows from grassland, appropriate levels of ME are required, which support microbial protein synthesis in the rumen. The uCP represents the sum of microbial protein synthesis in the rumen and the amount of bypass protein from feeds. Therefore, obtaining high energy yields from grassland, i.e. adjusting the balance between the right cutting time, an optimal crop composition and high DM yields is important.

Estimation of the energy value of harvested forage is crucial for the optimal nutrient composition of diets of dairy cows. Weissbach et al. (1999) proposed a model to predict the digestibility of organic matter from cellulase. This contains the crude ash and the IOM (insoluble organic matter) from the cellulase method according to De Boever et al. (1986). A validation revealed an estimation error of 4.2%, which is why the equation was classified by the GfE as generally valid and superior due to the low effort involved. In this regard, the ME content estimation can be computed applying the equation for legumes of Weissbach et al. (1996) on the values obtained with the *in vitro* procedure:

$$\mathrm{ME}\left(\mathrm{MJ/kg\,DM}\right) = 13.98 - 0.0147 \times \mathrm{CA} - 0.0137 \times \mathrm{IOM} + 0.00234 \times \mathrm{CP} \qquad (1a)$$

The equation for grasses of Weissbach et al. (1996) is as follows:

$$\mathrm{ME}\left(\mathrm{MJ/kg\,DM}\right) = 13.96 - 0.0147 \times \mathrm{CA} - 0.0108 \times \mathrm{IOM} + 0.00234 \times \mathrm{CP}, \qquad (1b)$$

where CA is the crude ash content (g/kg DM), IOM is enzymatically insoluble organic matter (g/kg DM) and CP is crude protein content (g/kg DM).

These equations were used until 2008, when new estimations of ME content in forages were available, including all types of plant species and forage conservation procedures (hay, silage, fresh). The current equation is as follows (GfE, 2008):

$$\begin{aligned}\mathrm{ME}\left(\mathrm{MJ/kg\,DM}\right) = {} & 5.51 + 0.00828 \times \mathrm{ESOM} - 0.00511 \\ & \times \mathrm{CA} + 0.02507 \times \mathrm{CL} - 0.00392 \times \mathrm{ADForg}\end{aligned} \qquad (2)$$

The equations 1a, 1b and 2 above use cellulase as determinant for the estimation of ME. For the *in vitro* procedure using rumen fluid for determination of gas production, the equation applies:

$$\begin{aligned} ME\,(MJ/kg\,DM) = 7.81 + 0.07559 \times GP - 0.00384 \\ \times CA + 0.00565 \times CP + 0.01898 \\ \times CL - 0.00831 \times ADForg, \end{aligned} \quad (3)$$

where CA is crude ash, ESOM is enzymatically soluble organic matter, CL is crude lipid, GP is gas production and ADForg is ash-free ADF content. All parameters are expressed as g/kg DM, except GP, which is expressed as mL/200 mg DM.

Under the climatic conditions of the grassland regions of Central Europe, a high utilization frequency of 4 and more cuts per year is required to achieve an energy density of the silage of 6.4 MJ NEL/kg DM, a value that is targeted in the feeding of high-performance animals.

6 Factors determining the nutritive value of forages

The breeding progress improved the digestibility of forages changing their chemical composition with higher nutritive value. As a result, methods were developed that considered forage OMD and the resulting prediction equations to calculate the energy value (ME). This was necessary, as farmers increased their demands on the accuracy aiming precise diet formulation and feeding techniques to account for the increased need for efficient production. This led to new equations to improve the prediction of animal performance.

High milk yields are only possible and justifiable if animals can be supplied with energy and nutrients to meet their performance requirements, the largest proportion coming from forage intake. This requires the use of high-quality forage and improved grassland management strategies. Dairy cows need a supply of high structure forage, especially during peak lactation. As the feed intake capacity is limited, there is a risk of energy deficiency at high performance, resulting in the mobilization of body reserves, especially just after calving. A prolonged energy deficit can lead to metabolic and fertility problems. While improvements in the supply of protein and energy in terms of their availability to the rumen microbiota have been achieved through diet formulation (rumen-protected protein, rumen-protected fat and rumen-protected starch), targeted modification of forage plants, e.g. through breeding, has been largely ignored.

In the following, special attention will be paid to variables influencing the forage quality of forage plants. Agronomic characteristics and biochemical processes, such as the effect of plant proteases and secondary plant constituents, will be discussed.

7 Agronomic traits

An important agronomic characteristic of forage production is land use efficiency (yield of milk or meat per unit of area). Cropland and grassland differ in land use efficiency. As only cell-wall-rich biomass (forage) is produced on grassland, special attention is paid to the high area efficiency of grassland as a source of energy and protein (Fig. 3). The comparison of potentials of milk yield in kilogram per hectare or in kilogram per ton DM grassland is an important tool to compare suitable cultivars in combining preferences of high forage quality with restrictions on yield. Or, in opposite directions, high yield amount with restrictions in forage quality may be the preference for some cultivars. The decision is dependent on the suitability of the selected cultivars for the growing conditions in the forage-based production system.

Several factors are responsible for the high efficiency of land use. In addition to soil and climate, fertilization is of great importance for yield and forage quality. It influences species composition and provides nutrients for the metabolic processes of grassland plants. It is important not only to achieve a satisfactory dry matter yield from the forage area but also to ensure high forage digestibility. Grassland stands are usually multi-flora in composition. Botanical composition has a major influence on forage quality. It varies both in the natural plant communities of permanent grassland, depending on the dominant species, and in more species-poor plant communities, such as in temporary

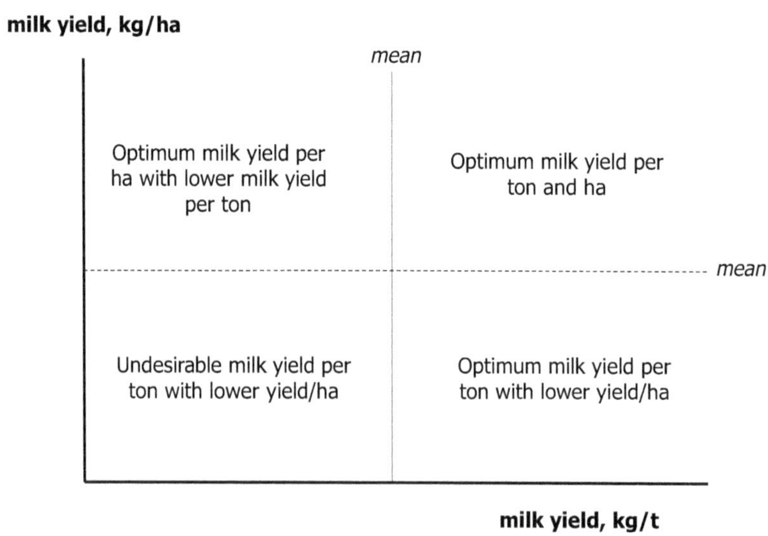

Figure 3 Recording genotype-related differences in forage crops with effect on milk yield potential. Cultivars that achieve optimal milk yield per ton and ha (top right) are desirable (own illustration).

grassland production, which mostly consist of pure or simple mixed clover or grass stands.

Well-managed permanent grassland often exceeds the forage quality of pure field forage stands. In the case of multi-cut stands, the frequency of use is decisive for forage quality and may achieve high ME contents per cut. In general, there is a negative correlation between yield and quality. The energy content per kilogram DM decreases with increasing yield. Leaf mass and the proportion of leaf mass in the total harvested crop (leaf weight ratio) are indicators of forage quality. As the crop develops, yield increases and the leaf weight ratio decreases. The different characteristics of species and cultivars should be considered when choosing a cultivar adapted to the site (Fig. 3).

8 Developmental stage at harvest

As mentioned previously, the development stage at harvest is determinant for the nutritive value of the forage. This is important to observe if comparison between cultivars is carried out. It is known that the forage quality changes in composition depending on the development stage, i.e. the concentration of protein and non-structural carbohydrates decreases, and the cell wall (NDF, ADF) and lignin contents increase at later maturity stages. Protein and carbohydrate fractionation can be used to detect these changes (see Fig. 2). If the effect of different cultivars or utilization frequencies on forage quality is to be investigated, the MSC (mean stage by count) value is suitable for characterizing the development stage. The MSC represents the mean of the developmental stage of 30–50 representative shoots of the dominant plant species of the stand and is calculated as follows:

$$(\mathrm{MSC}) = \sum_{0}^{17} \frac{S_i N}{C}$$

where, for grasses, S_i = developmental code (0–17, see Table 1, or Table 2 for alfalfa and Table 3 for red clover); N = number of tillers in S_i; and C = total number of tillers.

Tables 1 through Table 3 provide suggestions for determining the stage of development of dominant plant community species in grasslands. Fig. 4 illustrates with the example of perennial ryegrass (*Lolium perenne* L.) that the MSC value reflects OMD with a comparatively high degree of precision (Gierus et al., 2024).

In addition to MSC, which is calculated based on the number of tillers, the mean stage by weight (MSW) uses a similar approach. The difference is that the weight of the tillers is determined and then the tillers are sorted in a similar manner to calculate the MSW. The formula is as follows:

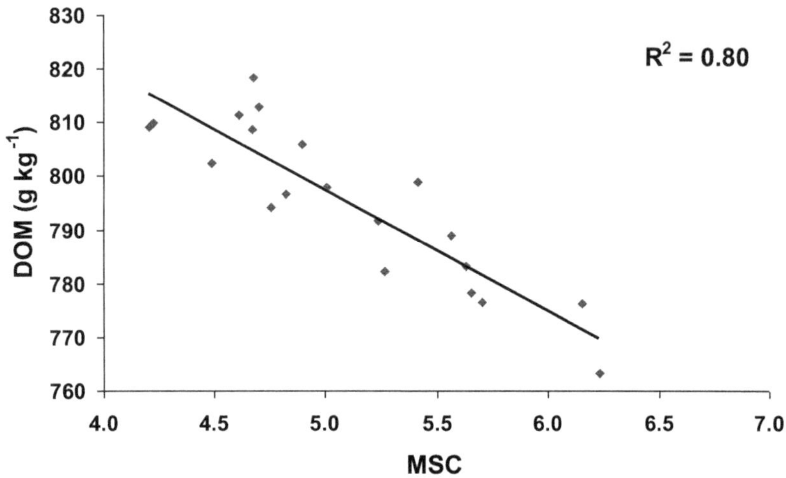

Figure 4 Relationship between developmental stage, measured as mean stage by count (MSC) and *in vitro* organic matter digestibility (DOM) of perennial ryegrass.

$$(\text{MSW}) = \sum_{0}^{17} \frac{S_i D}{W}$$

where, for grasses, S_i = developmental code (Tables 1–3); D = number of tillers in S_i; and W = total dry weight (60°C until constant weight) of tillers.

MSW tends to be a more homogeneous representation of the developmental stage of harvested forage. MSC is preferred because it is easier to determine, but MSW is more closely related to the nutritional value of the forage, as MSC does not represent the plasticity of stem growth, e.g., caused in botanically diverse plant communities, i.e. competition for light, water or morphological plasticity due to cutting frequency (Van Soest, 1994; Gastal and Lemaire, 2015).

9 Plant proteases and protein degradation

Protein degradation begins at harvest and is limited by increasing loss of moisture (hay/wilt silage). Although it is generally assumed that degradation of plant protein in the rumen is mediated by microbial enzymes, studies demonstrate the importance of plant proteases in protein degradation after harvest, as shown by Gierus et al. (2016b) and Seebeck et al. (2011) in Table 5. In pastures, after fresh plant biomass reaches the rumen and before more extensive degradation of the plant cell wall can begin, a period of intense proteolysis is often observed, which has been suggested to be caused mainly by plant enzymes (Kingston-Smith and Theodorou, 2000; Beha et al., 2002). This enzymatic activity is likely to be of paramount importance not only

Table 5 Effect of specific activity of polyphenol oxidase (PPO) in red clover on protein levels in leaf incubated under rumen-like conditions (39°C, anaerobic, dark)

	Red clover			White clover
First cut, Incubation	Cultivar 1	Cultivar 2	Cultivar 3	Cultivar 1
0 hour, mg/g DM	1.54	1.55	1.50	1.51
6 hours, mg/g DM	0.25	0.20	0.21	0.27
24 hours, mg/g DM	0.18	0.15	0.15	0.09
Degradation rate, %/h	**9.2[b]**	**10.0[b]**	**9.8[b]**	**12.2[a]**
spec. PPO-activity (IU/µg prot./g DM)	1.80	1.15	1.05	0.06

Source: After Seebeck et al. (2011).
Means followed by different letter(s) are different at $p < 0.05$

with respect to ruminant N efficiency during grazing but also for proteolytic processes during wilting and ensiling. Considerable variability in proteolytic activity among forage crop species supports the hypothesis that selective breeding for delayed proteolytic activity could increase N use efficiency on dairy/forage farms. Protease activity can degrade over 50–60% of protein from its original state in the plant to NPN when temperatures, humidity, and pH are favorable (Kingston-Smith et al., 2005).

Protease-mediated crude protein degradation can be understood in two levels: (1) protein degradation is very pronounced in the silo, which can be explained by the relationship with the residence time of the green material in the silo and with the pretreatment (including degree of wilting, use of silage additives); (2) immediately after the intake of pasture forage, the own proteases of plants exert a marked influence on proteolysis (Beha et al., 2002; Gierus et al, 2016b). In both cases, however, it is not clear what causes the triggering of proteolytic activity when plant tissues are still intact. In the case of grazing, assuming that intact cells enter the rumen, the cells face a wide range of stressors, including darkness, lack of oxygen, elevated temperature of 39°C, pH ranging between 5.5 and 6.8, and active rumen microbial population (Fig. 5). Under such conditions, plant cells may initiate premature senescence. Plant metabolism initially remains active in the rumen after ingestion of fresh forages. Proteolysis in the first hours after feed intake from grazed plants is therefore of crucial importance for the protein quality of grassland growths.

The biochemical transformation in plant cell post-ingestion (grazing) or after cutting (field phase) is crucial to estimate the nutritive value of forages. Delays in the field phase may support an intensive proteolysis in cutting systems, reducing the protein quality (proportion of true protein in the crude protein). Similarly, post-ingestive biochemical transformation in plant cells may trigger proteolysis decreasing the protein quality. Such effects are decisive to understand variation in nutritive value of forages and should be considered in breeding programs

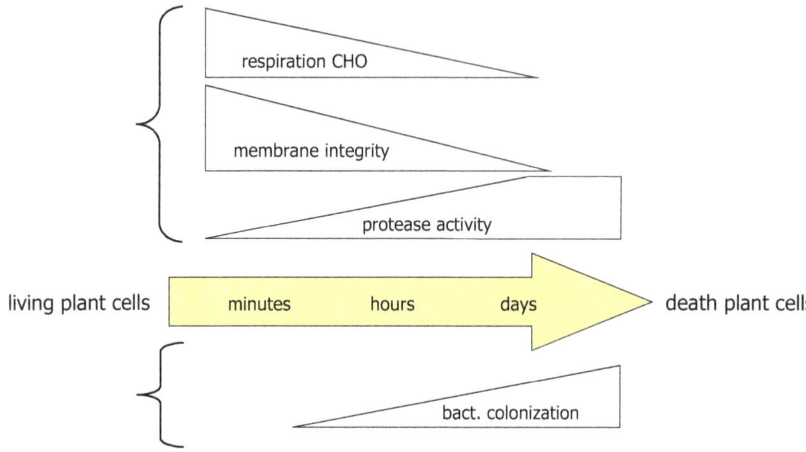

Figure 5 Schematic representation of biochemical reactions in plants after cutting/biting, where CHO = carbohydrates (own representation, modified from Kingston-Smith et al., 2008).

of forage plants. Selection of more resistant proteins to degradation processes by plant proteases should help to overcome this problem. Investigations in this regard developed the low-initial-rate-of-degradation (LIRD) cultivars of alfalfa (Coulman et al., 2000; Berg et al., 2000). The limitation of this approach was the increase in cell wall content in the LIRD-cultivars, with poor digestibility. This can be explained by the degradation rate of cell wall proteins being lower, and so their important contribution to the by-pass protein content using the current methods of UDP determination (*in situ* method). New approaches are necessary which consider the fast rate of degradation of proteins in forage plants, especially in the first hours after cutting for the field phase in silage/hay production, or the first hours post-ingestion under grazing conditions. For forage legumes, secondary plant compounds like polyphenols are interesting.

10 Secondary plant compounds in forage plants

Botanically, secondary plant constituents are the result of biosynthesis that follows primary metabolism. The active natural substances can be roughly divided into the substance classes of terpenoids, phenol derivatives and alkaloids (nitrogen-containing compounds). In the plant, and also in forage plants, they perform functions as signaling, recognition, defense, inhibitory or toxic substances. As a result, the content in the plant varies in the individual organs, tissues and cells and changes depending on the stage of development and the course of the year. Factors such as nutrient availability and stressors that occur can also induce or enhance the content of secondary plant compounds.

In forage plants, these compounds receive attention because they can have different effects on the animal depending on the substance or their concentration. The range extends from components classified as nutritionally negative or harmful, such as alkaloids or glucosinolates, which are sometimes deliberately bred out for forage and crop plants, to ingredients with a positive effect on the metabolism of animals, in which there is increasing interest by the feed industry to produce phytogenic feed additives (plant-based feed additives, mainly with origin from secondary plant metabolism). Table 6 shows examples of the most important forage plants and the content of condensed tannins (polyphenols).

The content of secondary plant constituents of some forage legumes such as birdsfoot trefoil and red clover is associated with reduced protein degradation rates or proteolysis in the ensiled material (Krawutschke et al., 2013). Secondary plant constituents such as condensed tannins can form complexes with proteins as well as affect the activity of microorganisms and enzymes in the forestomachs of ruminants, which has a positive effect on the efficient N use of the animal (Table 7). Plant proteases can also be inhibited in their action by condensed tannins. An overview of the effect of tannins in the nutrient cycle of a dairy cattle feedlot is presented in Fig. 6. Both the use of tannin-containing forage plants and the use of extracts from tannin-containing plants are considered.

The rapid degradation of proteins in the rumen can be inhibited and the availability of amino acids in the small intestine increased at a condensed tannin content of about 50 g/kg DM. Other studies cite a content of 20–40 g/kg DM as beneficial in terms of N utilization in the animal (Aerts et al., 1999; Min et al., 2018). This value is controversial in literature, which is due to the properties of tannins, i.e. the potential for complexation with proteins, molecular weight, and composition of tannins, in addition to the content itself. The effect of condensed

Table 6 Average contents of condensed tannins in forage plants from various literature sources

Species		Condensed tannins (g/kg DM)
Lotus pedunculatus	Greater Birdsfoot trefoil	77.0
Hedysarum coronarium	Sulla	45.0
Coronilla varia	Crown vetch	30.0
Ornithopus sativus	Serradella	24.0
Onobrychis viciifolia	Sainfoin	22.0
Lotus corniculatus	Birdsfoot trefoil	21.0
Chicorium intybus	Chicory	4.2
Trifolium pratense	Red clover	1.7
Lolium perenne	Ryegrass	1.1
Medicago sativa	Lucerne	0.5

Source: Adapted from: Barry and McNabb (1999); Aerts et al. (1999); Kardel et al. (2013).

Table 7 Potential benefits of forage crops with condensed tannins in ruminant nutrition

Change	Process	Area
Reduction	Proteolysis during fermentation in the silo	Plant
Inhibition	Bacterial growth in the silo	Silage
Increase	Absorption of essential amino acids, with subsequent increase in animal performance (meat, milk, wool)	Animal
Shift	Urinary N excretion → Fecal N excretion	Animal
Reduction	Methane emission from ruminants	Animal
Reduction	Flatulence (in legumes)	Animal
Reduction	Mineralization rate of C and N in soil	Soil

Source: Adapted from: Waghorn et al. (1987); Wang et al. (1996); Misselbrook et al. (2005); Mutabaruka et al. (2007); Bueno et al. (2020).

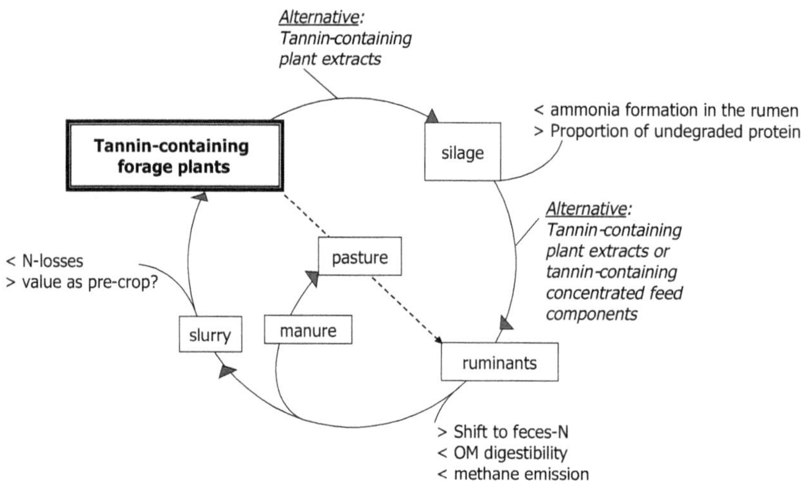

Figure 6 Importance of condensed tannins in the nutrient cycle. Positive effects (>) consist in the shift from urine N to fecal N excretion. The reduction (<) of ammonia formation in the rumen (proteolysis inhibition) as well as the increase in the proportion of undegraded protein are positive effects of tannin inclusion. Overall, this can lead to a reduction in N losses in the nutrient cycle. Negative effects of feeding tannins are the reduction of digestibility of organic matter if the tannin content in the feed is exceeded (own presentation).

tannins can be observed, among other things, in the lower urinary N excretion as a % of N uptake and higher N retention in red clover and birdsfoot trefoil silage compared to alfalfa silage (Hymes-Fecht et al., 2013).

The tannin contents of forage legumes show considerable variations (see Table 6). High concentrations in forage-relevant species are found especially in birdsfoot trefoil. In addition to the legume species itself, the plant organ, age, nutrient supply of the plant and the weather also have a modifying effect on the tannin content. The influence of tannins on the rapid degradation rate of crude protein in the rumen is not yet clearly understood. Identification of factors controlling the effect of tannins on N utilization as well as changes in tannin content over the growing season of pastures is of great importance. Moderate tannin content could be beneficial under grazing conditions with respect to the N loss potential of legume/grass mixtures.

The example of condensed tannins shows that there are regulatory mechanisms in the genetic diversity of forage plants that have a direct influence on forage quality. Precise knowledge can contribute to sustainable positive effects at various levels if applied in a targeted manner (Fig. 6). The effects on N utilization in ruminants presented here result first of all in improved N use efficiency in the animal, and consequently in higher animal performance and simultaneously reduced environmental pollution by lower N outputs. At the same time, however, the current limitations become clear here, because for sensible forage production, the plants must show suitability for the usual forage production systems and the climatic requirements of the respective region with corresponding yield reliability. Birdsfoot trefoil or sainfoin as forage legumes have high contents of tannins but can be classified as weaker in competition with other species in the plant community in this respect due to possible difficulties in establishing the crop and moderate biomass yields. Targeted breeding of appropriately adapted cultivars could achieve a higher importance in agriculture.

11 Conclusion

The definition of forage quality is closely related to the availability of nutrients and energy in forages to meet ruminant requirements. There is need to ensure simple and straightforward application of suitable methods in feed evaluation to evaluate forage quality and include this knowledge in plant breeding programs to improve the nutritive value for forage plants based on ruminant needs.

Nutrients and energy content to be supplied to ruminants have to be determined not only by its chemical constituents but also should involve knowledge about the digestive processes and physiological status of the animal to be fed. Although animal performance is the most accurate answer to define appropriate forage quality, *in vitro* laboratory methods (enzymatic, chemical, or with rumen fluid) are validated with *in vivo* measurements and often used to evaluate forage quality based on botanical composition, digestibility, or degradability of defined groups of substances.

12 References

Aerts, R. J., et al. (1999): Condensed tannins from *Lotus corniculatus* and *Lotus pedunculatus* exert different effects on the *in vitro* rumen degradation of ribulose-1,

5-bisphosphate carboxylase/oxygenase (Rubisco) protein. *J. Sci. Food Agric.*, 79(1), 79–85.

Bannink, A., et al. (2006): Physiology, regulation and multifunctional activity of the gut wall: a rationale for multicompartmental modelling. *Nutr. Res. Rev.*, 19, 227–253.

Barchiesi-Ferrari C., et al. (2011): Pepsin-cellulase digestibility of pasture silages: effects of pasture type, maturity stage, and variations in the enzymatic method. *Chil. J. Agric. Res.*, 71(2), 249.

Barry, T. N. and McNabb, W. C. (1999): The implications of condensed tannins on the nutritive value of temperate forages fed to ruminants. *Br. J. Nutr.*, 81(4), 263–272.

Beha, E. M., et al. (2002): Grass cells ingested by ruminants undergo autolysis which differs from senescence: implications for grass breeding targets and livestock production. *Plant, Cell Environm.*, 25(10), 1299–1312.

Berg, B. P., et al. (2000): Bloat in cattle grazing alfalfa cultivars selected for a low initial rate of digestion: a review. *Can. J. Plant Sci.*, 80(3), 493–502.

Bueno, A. V. I., et al. (2020): Fermentation profile and hygienic quality of rehydrated corn grains treated with condensed tannins from quebracho plant extract. *Anim Feed Sci. Technol.*, 267, 114559.

Casler, M. D. and Vogel, K. P. (1999): Accomplishments and impact from breeding for increased forage nutritional value. *Crop Sci.*, 39, 12–20.

Coulman, B., et al. (2000): A review of the development of a bloat-reduced alfalfa cultivar. *Can. J. Plant Sci.*, 80(3), 487–491.

De Boever J. L., et al. (1986): The use of an enzymatic technique to predict digestibility, metabolizable and net energy of compound feedstuffs for ruminants. *Anim Feed Sci. Technol.*, 14(3–4), 203–214.

Fagerberg, B. (1988): The change in nutritive value in timothy, red clover and lucerne in relation to phenological stage, cutting time and weather condition. *Acta Agric. Scand.*, 38, 347–362.

Forejtová, J., et al. (2005): Comparison of organic matter digestibility determined by *in vivo* and *in vitro* methods. *Cz. J. Anim. Sci.*, 50(2), 47–53.

Fox, D. G., et al. (2004): The cornell net carbohydrate and protein system model for evaluating herd nutrition and nutrient excretion. *Anim. Feed Sci. Technol.*, 112(1–4), 29–78.

Gastal, F. and Lemaire, G. (2015): Defoliation, shoot plasticity, sward structure and herbage utilization in pasture: review of the underlying ecophysiological processes. *Agriculture*, 5(4), 1146–1171.

GfE (2008): New equations for predicting metabolizable energy of grass and maize products for ruminants. *Proc. Soc Nutr. Physiol.*, 17, 191–197.

GfE (2023): *Energie- und Nährstoffbedarf landwirtschaftlicher Nutztiere. Nr. 12. Empfehlungen zur Energie- und Nährstoffversorgung von Milchkühen.* DLG-Verlag, Frankfurt a.M. 287p.

Gierus, M., et al. (2024): Protein and carbohydrate fractionation to evaluate perennial ryegrass (*Lolium perenne* L.) accessions. *Agronomy*, 14(1),168.

Gierus, M., et al. (2005): Physico-chemical characteristics and degradation rate of soluble protein obtained from the washout fraction of feeds. *Liv. Prod. Sci.*, 97, 219–229.

Gierus, M., et al. (2016a): Using methods to estimate metabolizable energy content of five forage legumes harvested under different defoliation systems/Einsatz von Methoden zur Schätzung der umsetzbaren Energie in fünf Futterleguminosen

aus unterschiedlichen Nutzungssystemen. *Die Bodenkultur: J. Land Manag., Food Environm.*, 67(3), 121–131.
Gierus, M., et al. (2007): Untersuchungen *in vitro* zum Abbau von Rohprotein verschiedener Futterleguminosen, Deutsches Weidelgras und Silomais. *Züchtungskde*, 79, 466–475.
Gierus, M., et al. (2012): Forage legume species determine the nutritional quality of binary mixtures with perennial ryegrass in the first production year. *Anim. Feed Sci. Technol.*, 172(3–4), 150–161.
Gierus, M., et al. (2016b): Protein degradation rate as affected by plant proteases among fresh samples of perennial ryegrass cultivars (*Lolium perenne* L). *Die Bodenkultur, J. Land Manag., Food Environm.*, 67(2), 61–68.
Gierus, M., et al. (2001): Appearance of aminogenic nutrients in the portal vein as a tool for feed protein evaluation in the DVE/OEB System. *Proc. Soc. Nutr. Physiol.*, 10, 82.
Higgs, R., et al. (2015): Updating the cornell net carbohydrate and protein system feed library and analyzing model sensitivity to feed inputs. *J. Dairy. Sci.*, 98, 6340–6360.
Hoffman, P. C., et al. (1998): Performance of lactating dairy cows fed alfalfa silage or perennial ryegrass silage. *J. Dairy. Sci.*, 81(1), 162–168.
Huhtanen, P., et al. (1997): Omasal sampling technique for assessing fermentative digestion in the forestomach of dairy cows. *J. Anim. Sci.*, 75(5), 1380–1392.
Humphreys, M. O. (1989): Water-soluble carbohydrates in perennial ryegrass breeding. II. Cultivar and hybrid progeny performance in cut plots. *Grass Forage Sci.*, 44, 237–244.
Hvelplund, T. and Weisbjerg, M. R. (2000): *In situ* techniques for the estimation of protein degradability and postrumen availability. In: Givens, D.I., et al. (eds) *Forage evaluation in ruminant nutrition*. CABI Publishing, New York, NY, pp. 233–258.
Hymes-Fecht, U. C., et al. (2013): Replacing alfalfa or red clover silage with birdsfoot trefoil silage in total mixed rations increases production of lactating dairy cows. *J. Dairy. Sci.*, 96(1), 460–469.
Iantcheva, N., et al. (1999): A comparison of *in vitro* rumen fluid and enzymatic methods to predict digestibility and energy value of grass and alfalfa hay. *Anim. Feed Sci. Technol.*, 81(3), 333–344.
Iowerth, D., et al. (1975): The effect of pepsin pretreatment of herbage on the prediction of dry matter digestibility from solubility in fungal cellulase solutions. *J. Sci. Food Agric.*, 26(5), 711–718.
Kalu, B. A. and Fick, G. W. (1983): Morphological stage of development as a predictor of alfalfa herbage quality. *Crop Sci.*, 23, 1167–1172.
Kardel, M., et al. (2013): Different approaches to evaluate tannin content and structure of selected plant extracts-review and new aspects. *J. Appl. Bot. Food. Qual.*, 86(1), 154–166.
Kingston-Smith, A. H., et al. (2008): From plants to animals; the role of plant cell death in ruminant herbivores. *J. Exp. Bot.*, 59(3), 521–532.
Kingston-Smith, A. H., et al. (2005): Evidence in support of a role for plant-mediated proteolysis in rumens of grazing animals. *Br. J. Nutr.*, 93, 73–79.
Kingston-Smith, A. H. and Theodorou, M. K. (2000): Post-ingestion metabolism of fresh forage. *New Phytol.*, 148, 37–55.
Kleinmanns, J. and Potthast, V. (1984): Zur "Verdrängung" von Grundfutter durch Kraftfutter in der Milchviehfütterung. Übers. *Tierernährg.*, 12, 187–214.
Kowalski Z. M., et al. (2014): The use of cellulase and filter bag technique to predict digestibility of forages. *Anim. Feed Sci. Technol.*, 198, 49–56.

Krawutschke, M., et al. (2013): Changes in crude protein fractions of forage legumes during the spring growth and summer regrowth period. *J. Agric. Sci.*, 151(1), 72–90.

Licitra, G., et al. (1996): Standardization of procedures for nitrogen fractionation of ruminant feeds. *Anim. Feed Sci. Technol.*, 57(4), 347–358.

Lindsay, D. B. (1993): Metabolism of the portal-drained viscera. In: Forbes, J.M. and France, J. (eds) *Quantitative aspects of ruminant digestion and metabolism*. CAB International, London, UK, pp. 267–289.

Martineau, R., et al. (2023): Ability of three dairy feed programs to predict post-rumen outflows of nitrogenous compounds in dairy cows: a meta-analysis. *J. Dairy Sci.*, 106(12), P8583–P8610.

McNabb, W. C., et al. (1994): *In-vitro* rates of rumen proteolysis of ribulose -1,5-bisphosphate carboxylase (Rubisco) from lucerne leaves, and of ovalbumin, vicilin and sunflower albumin 8 storage proteins. *J. Sci Food Agric.*, 64, 53–61.

Menke, K. H. and Steingass, H. (1988): Estimation of energetic feed value obtained from chemical analysis and *in vitro* gas production using rumen fluid. *Anim. Res. Dev.*, 28, 7–12.

Min, B. R. and Solaiman, S. (2018): Comparative aspects of plant tannins on digestive physiology, nutrition and microbial community changes in sheep and goats: a review. *J. Anim. Physiol. Anim. Nutr.*, 102(5), 1181–1193.

Misselbrook, T. H., et al. (2005): Dietary manipulation in dairy cattle: laboratory experiments to assess the influence on ammonia emissions. *J. Dairy Sci.*, 88(5), 1765–1777.

Mutabaruka, R., et al. (2007): Microbial degradation of hydrolysable and condensed tannin polyphenol–protein complexes in soils from different land-use histories. *Soil Biol. Biochem.*, 39(7), 1479–1492.

Reynolds, C. K. (1995): Quantitative aspects of liver metabolism in ruminants. In: Tsuda, T., et al. (eds) *Ruminant physiology: digestion, metabolism, growth and reproduction. International Symposium on Ruminant Physiology*, Academic Press, New York, NY, pp. 351–371.

Reynolds, C. K., et al. (1994): Absorption and delivery of nutrients for milk protein synthesis by portal-drained viscera. *J. Dairy Sci.*, 77, 2787–2808.

Reynolds, C. K., et al. (1991): Effects of forage-to-concentrate ratio and intake on energy metabolism in growing beef heifers: whole body energy and nitrogen balance and visceral heat production. *J. Nutr.*, 121, 994–1003.

Salama, H., et al. (2017): A simplified maturity index to quantify the development stage of perennial ryegrass (*Lolium perenne* L.) and its relationship with yield and nutritive value. *Die Bodenkultur, J. Land Manag., Food Environm.*, 68, 89–101.

Santos, F. A. P., et al. (1998): Effect of rumen-undegradable protein on dairy cow performance: a 12-year literature review. *J. Dairy Sci.*, 81, 3182–3213.

Seebeck, M., et al. (2011): Plant-mediated proteolysis among red clover (*Trifolium pratense* L.) cultivars and the relationship with the polyphenol oxidase activity. In: *Eight International Symposium on the Nutrition of Herbivores (ISNH8)*. September 6th–9th, Aberystwyth, Wales.

Sniffen, C. J., et al. (1992): A net carbohydrate and protein system for evaluating cattle diets: II. Carbohydrate and protein availability. *J. Anim. Sci.*, 70(11), 3562–3577.

Søegaard, K., et al. (2007): Temporary grassland - challenges in the future. *Grassl Sci. Europ.*, 12, 27–38.

Steingass, H. and Leberl, P. (2008): *In vitro* Verfahren: eine notwendige Ergänzung zur Nährstoffanalytik bei Futtermitteln. *Übers. Tierernährg.*, 36, 31–46.

Steingass, H. and Menke, K.-H. (1986): Schätzung des energetischen Futterwertes aus der *in vitro* mit Pansensaft bestimmten Gasbildung und der chemischen Analyse. I: Untersuchungen zur Methode. *Übers. Tierernährg.*, 14, 251–270.

Tagari, H. and Bergman, E. N. (1978): Intestinal disappearance and portal blood appearance of amino acids in sheep. *J. Nutr.*, 108, 790–803.

Tamminga, S. and Chen, X. B. (2000): Animal-based techniques for the estimation of protein value of forages. In: Givens, D.I., et al. (eds) *Forage evaluation in ruminant nutrition*. CABI Publishing, New Yord, NY, pp. 215–232.

Tedeschi, L. O. and Fox, D. G. (2017): *The ruminant nutrition system: an applied model for predicting nutrient requirements and feed utilization in ruminants (No. Ed. 2)*. XanEdu Publishing, Inc., Acton/MA, USA, 616 pg.

Tilley, J. M. A. and Terry, R. A. (1963): A two-stage technique for the *in vitro* digestion of forage crops. *Grass Forage Sci.*, 18(2), 104–111.

Tyrrell, H. F. and Moe, P. W. (1975): Effect of intake on digestive efficiency. *J. Dairy Sci.*, 58, 1152–1163.

Van Amburgh, M. E., et al. (2015): The cornell net carbohydrate and protein system: updates to the model and evaluation of version 6.5. *J. Dairy Sci.*, 98(9), 6361–6380.

Van Soest, P. J. (1994): *Nutritional ecology of the ruminant*. Cornell University Press, Ithaca, 476 p.

Waghorn, G. C., et al. (1989): Particle breakdown and rumen digestion of fresh ryegrass (*Lolium perenne* L.) and lucerne (*Medicago sativa* L.) fed to cows during a restricted feeding period. *Br. J. Nutr.*, 61, 409–423.

Waghorn, G. C., et al. (1987): The effect of condensed tannins on the site of digestion of amino acids and other nutrients in sheep fed on *Lotus corniculatus* L. *Br. J. Nutr.*, 57(1), 115–126.

Wang, Y., et al. (1996): Effect of condensed tannins upon the performance of lambs grazing *Lotus corniculatus* and lucerne (*Medicago sativa*). *J. Agric. Sci.*, 126(1), 87–98.

Webster, A. J. F., et al. (1975): The influence of food intake on portal blood flow and heat production in the digestive tract of sheep. *Br. J. Nutr.*, 34, 125–139.

Weissbach, F., et al. (1999): Schätzung der Verdaulichkeit und der umsetzbaren Energie von Gras und Grasprodukten. *Proc. Soc. Nutr. Physiol.*, 8, 72.

Weissbach, F., et al. (1996): Schätzung der umsetzbaren Energie von Grundfutter mittels einer Cellulase-Methode. *Proc. Soc. Nutr. Physiol.*, 5, 115.

Chapter 11

The effects of pasture grazing on milk quality[1]

Anjo Elgersma, Independent Scientist, The Netherlands

1 Introduction

Milk contains numerous important nutrients such as fat, protein, lactose, mineral salts, and vitamins. Milk composition is economically important to milk producers and processors and nutritionally important to consumers. Variations in milk composition arise from differences in relative rates of synthesis and secretion of milk components by the mammary gland. The processes involved for lactose, protein, and fat synthesis and secretion are independent but regulated through nutrient or substrate availability and hormonal control of nutrient utilization. Thus, genetics, which mediates hormonal effects, and

1 The chapter is based on an article (Elgersma, 2015) published in the *European Journal of Lipid Science and Technology* and on a review article (Elgersma et al., 2006a) published in Animal Feed Science and Technology.

http://dx.doi.org/10.19103/AS.2025.0154.17

diet, which regulates nutrient availability, are the major factors affecting milk composition.

Milk protein percentage is best changed through genetics. Dietary protein percentage has a minimal effect on milk protein percentage when it is within practical feeding ranges. Levels of other nutritional components of milk (lactose, vitamins, and minerals) are rather constant and not subject to large changes through genetic or nutritional manipulation (NRC, 1988); for further reading of recent work in this area, the reader is referred to a review article of Alothman et al. (2019). This chapter therefore focuses on milk fat composition.

Animal milk composition is affected by several factors (animal breed, health status, stage of lactation, feeding regimen, and seasons). This chapter focuses on cow milk; the milk fatty acid (FA) profile in other animal species was reviewed by Mollica et al. (2021). Feeding regime and animal diet can be changed at the farm level. Milk fat content and FA composition and concentration mainly depend on the animals' diet, e.g. Kalač and Samkova (2010). Nutrition, therefore, constitutes a natural and economical way for farmers to markedly and rapidly modulate the milk FA composition.

As dairy products are a substantial source of FA in the human diet, milk FA composition is of great interest. Because the FA profile in dairy product fats is comparable to that of the milk fat from which these products are obtained (Dhiman et al., 1999; Lavillonnière et al., 1998), farm milk composition is important. This is relevant not only from a nutritional perspective for dairy products but also in view of alteration of functional traits (processability of the milk) and sensory characteristics of dairy products (Alothman et al., 2019).

The animal diet, amount, and quality of herbage consumed and the concentration and fate of FA in feed, digestive tract, and mammary gland (Fig. 1) determine the FA profile in cow milk, farm milk, and dairy products. What can farmers do? Animals, plants, and soils are inextricably linked in grazed systems, so achieving successful outcomes at the farm level requires recognition of these interactions.

Several factors dictate the feeding system used on dairy farms for milk production in different regions of the world, including weather, land availability, and the dairy cows' energy requirements. Grazed grass is the most economical means of feeding dairy cows (e.g. Van den Pol-van Dasselaar et al., 2020). Pasture-based feeding systems prevail in regions with fertile soils, a temperate climate and abundant rainfall, thus environments that favour the growth of grass (e.g. Dillon et al., 2005), e.g. Ireland and New Zealand. However, also economic factors (production costs, labour) and social factors (intrinsic motivation, farmer mindset) play a role (Van den Pol-van Dasselaar et al., 2020).

Milk from grazed ruminants has higher levels of polyunsaturated fatty acids (PUFAs) than from ruminants that are kept indoors (e.g. Benbrook et al., 2018;

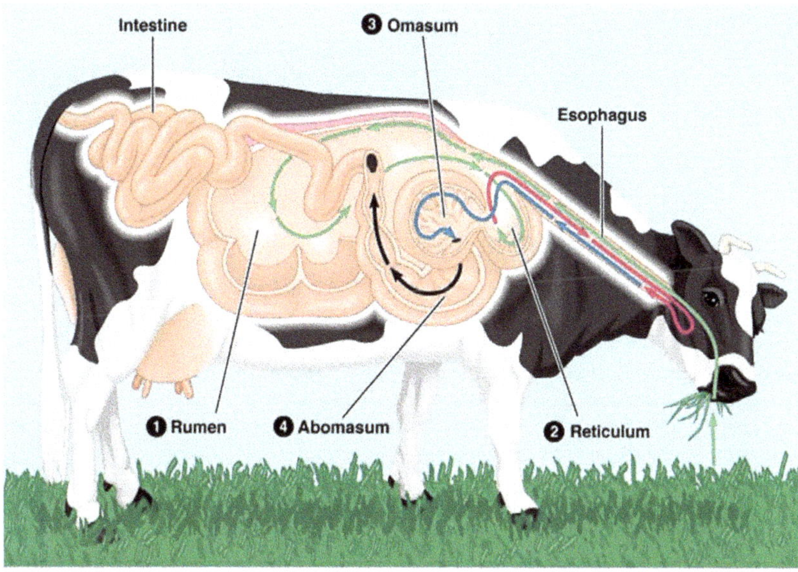

Figure 1 Schematic picture of the digestive system of a grazing cow. Source: Addison Wesley Longman, Inc.

Elgersma, 2015; Precht and Molkentin, 1995; Timlin et al., 2023). Herbage intake affects the milk FA profile and grazing cows collect fresh herbage while grazing. High herbage allowance enhanced the PUFA concentration in milk of grazing cows (Elgersma et al., 2006a, 2006b).

The FA profile of milk is an important factor in determining its nutritional value (Jensen, 2002; Hirahatake et al., 2020), with special attention paid to essential FA, namely linoleic acid (LA; C18:2n-6) and α-linolenic acid (ALA; C18:3n-3), which contribute to the human daily intake of PUFAs. These FA cannot be synthesized in humans (Hu et al., 2001).

Green herbage as consumed by grazing animals is especially rich in ALA, which is extensively biohydrogenated in the rumen to vaccenic acid (VA; C18:1 *trans*-11) and finally to stearic acid (C18:0). VA is desaturated in the mammary gland to yield the naturally predominant isomer of CLA: rumenic acid (RA; C18:2 *cis*-9, *trans*-11). Stearic acid is desaturated into oleic acid (C18:1 *cis*-9) (Fig. 2)

Data suggest that milk fat composition with respect to human health can be optimized by exploiting grazing in the diet of dairy cows (e.g. Larsen et al., 2010). Benbrook et al. (2018) mentioned that in the U.S., consumer demand for milk and meat from grass-fed cattle is growing, driven mostly by perceived health benefits and concerns about animal welfare.

It is interesting to unravel effects of grazed herbage as opposed to offered (cut) herbage. Offered cut herbage can be distinguished into freshly cut,

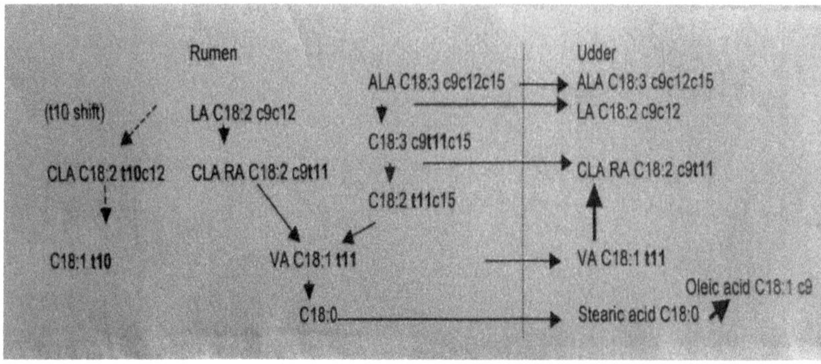

Figure 2 Simplified scheme of biohydrogenation and desaturation pathways of C18 fatty acids in rumen and udder of dairy cows. ALA, α-linolenic acid; LA, linoleic acid; CLA, conjugated linoleic acid; RA, rumenic acid; VA, vaccenic acid. *Trans* positions and desaturation pathways in the mammary gland are in bold.

stall-fed forage (cut and carry), or conserved forage (ensiled or dried as hay). Capuano et al. (2014) found that milk from cows with daily rations of fresh grass could be successfully distinguished from milk from cows with no fresh grass in their diet. Milk from cows at pasture could easily be distinguished from milk from cows kept indoors without fresh grass in the diet, as has been known since long, but the correct prediction of milk from non-grazing cows that were fed freshly cut grass indoors proved difficult. The difference between grazed pasture and cut fresh forage lies in animal behavior, selection opportunities for the animal, and freshness and composition of the cut forage when ingested, which are affected by processes during transport from field to barn and the time since cutting till the moment of ingestion. The grass used for grazing is generally less mature than the grass cut for stall-feeding or silage making, which further decreases the levels of PUFA compared with fresh grass, especially that of ALA. The difference with conserved forage also lies in changes in herbage dry matter content and composition, and in physical changes that occurred post-harvest i.e. after cutting, transport from the field, during conservation and the feed-out phase of the conserved product. Besides the above-mentioned aspects, the proportion of grazed, fresh or conserved forage in the total animal diet may be relevant, as well as the forage-to-concentrate (F:C) ratio. The botanical composition and the developmental stage of the pasture plants during grazing or cutting, defoliation height, animal selection options and intake are relevant.

Extensive reviews have been published on FA profiles of milk fat, e.g. (Chilliard et al., 2007; Jensen, 2002; Precht and Molkentin, 1995; 2000; Schwendel et al., 2015), and to avoid repetition, the reader is referred to their tables. Although milk from grazed ruminants has higher PUFA levels then milk from diets with conserved forage, grazing-based production systems, including certified organic systems (Halfa et al., 2013; Schwendel et al., 2015) do not have exclusive claims to greater levels of PUFA in milk. It is possible to increase such compounds in confinement feeding through the use of supplemental feeds such as oilseeds or fish or algae sources, although this may have undesirable effects on the unsaturated FA composition of ruminant-derived milk and meat (Aldai et al., 2013; Alves et al., 2021). This review on the effects of grazing on the FA composition of milk aims to describe various aspects of grazing and to compare and discuss observed differences in FA in herbage and in cow milk. First, however, an overview of the importance in view of reported health effects of FA in milk is presented.

2 Milk fat composition in relation to human health

2.1 Saturated fatty acids

Concerns have been raised about the role of fatty acids in human health. In this respect, ruminant products have been criticized for their relatively high levels of medium-chain saturated fatty acids (SFA) (mainly C12:0, C14:0, and C16:0) and low levels of PUFA.

Milk fat is comprised of approximately 400 different FA of various chain length, making its composition very complex (Table 1 shows the main FA). Approximately 50–70% of these FAs are long-chain FA ($C \geq 16$); 30–50% are short ($4 \leq C \leq 8$); and medium-length chain FAs ($10 \leq C \leq 14$) (Alothman et al., 2019). Palmitic acid (PA; C16:0) is considered a long-chain FA by some authors and a medium-chain FA by others.

In milk fat, most FA are SFA, with a lesser amount of mono-unsaturated FA (MUFA) and only 2–5% PUFA.

In 2006, the state of the art was summarized as follows: 'High-fat human diets, especially those rich in saturated fats, are often claimed to have detrimental effects on cardiovascular disease risk factors such as blood low-density lipoprotein (LDL) cholesterol (Ascherio et al., 1996; Williams, 2000). Dairy products contribute 15–20% of human intake of total fat, 25–33% of saturated fat, and about 15% of dietary cholesterol in the U.S.(Havel, 1997). At present, about 2% of total FA in milk are PUFA and about 70% are saturated, but less than 40% of SFA are in categories considered to be less healthy. These values can be modified by changing the animal diets (Tamminga, 2001; Kennelly, 2001; Chilliard et al., 2001), and it is thought that lauric acid (C12:0), but especially myristic (C14:0) and palmitic (C16:0) acid, raise total and LDL cholesterol, whereas stearic acid (C18:0) is neutral relative to the MUFA oleic acid (C18:1) in animal fats (Grundy and Vega, 1998). Cardiovascular risk might be reduced by lowering intake of undesirable SFA and/or by making alterations

Table 1 Yearly (March until November) average (g/100 g of milk fat ± SD), statistical significance (*P*-value) of fatty acid compositional results from raw milk fats derived in 2000 in Ireland from spring-calving cows that were kept outdoors on (grazed) grass (GRS), indoors (on a total mixed ration (TMR) of grass silage, maize silage, and concentrates) or had a partial mixed ration (PMR) as they were allowed daytime grazing and were kept indoors at night with access to TMR feed

	GRS	TMR	PMR
C4:0 Butyric acid	4.46 ± 0.59a[1]	4.76 ± 0.67a	4.64 ± 0.56a
C6:0 Caproic acid	2.52 ± 0.26b	2.80 ± 0.28a	2.68 ± 0.24a
C8:0 Caprylic acid	1.52 ± 0.19c	1.74 ± 0.17a	1.65 ± 0.17b
C10:0 Capric acid	3.40 ± 0.61c	4.07 ± 0.41a	3.79 ± 0.51b
C12:0 Lauric acid	3.90 ± 0.76c	4.74 ± 0.53a	4.40 ± 0.70b
C14:0 Myristic acid	11.93 ± 1.44b	12.65 ± 0.90a	12.68 ± 1.34a
C16:0 Palmitic acid (PA)	28.88 ± 2.77b	30.97 ± 3.15a	31.60 ± 3.13a
C18:0 Stearic acid	10.64 ± 1.76a	9.48 ± 1.80b	8.79 ± 1.89b
C18:1n-9 *cis* oleic acid	22.12 ± 4.02a	18.99 ± 3.02b	19.35 ± 3.88b
C18:1n-9 *trans* elaidic acid	0.34 ± 0.08b	0.40 ± 0.12ab	0.43 ± 0.20a
C18:2n-6 *cis* linoleic acid (LA)	1.12 ± 0.15b	2.28 ± 0.55a	2.08 ± 0.38a
C18:2n-6 *trans* linolelaidic acid	1.66 ± 0.37a	1.06 ± 0.30c	1.31 ± 0.29b
C18:2 *cis*-9, *trans*-11 rumenic acid (CLA)	1.47 ± 0.44a	0.61 ± 0.15c	0.84 ± 0.21b
C18:3n-3 α-linolenic acid (ALA)	0.73 ± 0.12a	0.40 ± 0.08c	0.54 ± 0.10b
SFA	69.51 ± 4.63b	73.37 ± 4.17a	72.29 ± 4.37a
MUFA	25.51 ± 4.17a	22.29 ± 3.47b	22.94 ± 3.97b
PUFA	4.98 ± 0.72a	4.34 ± 0.88b	4.77 ± 0.63a
n-3	0.73 ± 0.12a	0.40 ± 0.08c	0.54 ± 0.10b
n-6	2.78 ± 0.47b	3.33 ± 0.74a	3.40 ± 0.51a
LA/ALA	1.56 ± 0.16c	5.81 ± 1.42a	3.99 ± 1.06b
HH ratio[2]	0.62 ± 0.17a	0.49 ± 0.12b	0.51 ± 0.15b
Atherogenic index[3]	2.73 ± 0.65b	3.34 ± 0.68a	3.23 ± 0.70a
Desaturase index[4]	0.34 ± 0.05a	0.30 ± 0.05b	0.30 ± 0.05b
Thrombogenic index[5]	3.19 ± 0.65b	3.87 ± 0.81a	3.66 ± 0.72a
Health-promoting index[6]	0.39 ± 0.10a	0.31 ± 0.07b	0.33 ± 0.09b
Unsaturation index[7]	36.20 ± 5.23a	31.37 ± 4.99b	33.02 ± 4.89b

1 a–c Mean values in the same row with different superscripts differ ($P < 0.05$) between feeding systems.
2 Hypochlosterolemic/hypercholesterolemic (HH) ratio = (C18:1 *cis* + ΣPUFA)/(12:0 + C14:0 + C16:0).
3 Atherogenic index = [C12:0 + (4 × C14:0) + C16:0]/ΣUFA.
4 Desaturase index = (C14:1 + C16:1 + C18:1n-9 *cis* + CLA *cis*-9, *trans*-11)/(C14:0 + C14:1 + C16:0 + C16:1 + C18:0 + ΣC18:1 + CLA *cis*-9, *trans*-11).
5 Thrombogenic index = (C14:0 + C16:0 + C18:0)/[(0.5 × ΣMUFA) + (0.5 × Σn-6 PUFA) + (3 × n-3 PUFA) + (n-3/n-6)].
6 Health-promoting index = ΣUFA/[C12:0 + (4 × C14:0) + C16:0].
7 Unsaturation index = (1 × % monoenics) + (2 × % dienoics) + (3 × % trienoics).
Source: Timlin et al., 2023; only a selection of FA is shown.

in the FA profile of the fat consumed. Increasing the concentration of desired FA in ruminant products is receiving current attention.' (Elgersma et al., 2006). New insights, e.g. regarding *trans* FA, are decribed hereafter.

2.2 Mono- and polyunsaturated fatty acids, n-3 and n-6

There is growing interest among scientists and consumers in altering diets to prevent or slow the progression of metabolic, cardiovascular (CVD), and other chronic diseases (e.g. Benbrook et al., 2018). Potential diet alterations include reducing intakes of omega-6 (ω-6 or n-6) FA and increasing intakes of omega-3 (ω-3 or n-3) FA, thus decreasing dietary n-6/n-3 ratios. These ratios have become historically high in Western diets during the last century, reaching about 15, compared to estimated evolutionary ratios near 1 (Hibbeln et al., 2006; Simopoulos, 2006). These large changes are due to both increased intakes of n-6 FAs and decreased intakes of n-3 FAs. Modern grain feeding of farm animals has contributed to these n-6 increases and n-3 decreases in Western diets.

The main n-3 FA in milk is α-linolenic acid (C18:3 *cis*-9, *cis*-12, *cis*-15). Another category of PUFA is conjugated linoleic acid (CLA), the main CLA isomer in milk being rumenic acid (RA; C18:2 *cis*-9, *trans*-11). The MUFA in milk consist mainly of C18:1 *cis*-9 (oleic acid) and also C18:1 *trans*-11 (vaccenic acid; VA).

The n-6:n-3 ratio in cow milk essentially describes the concentrations of LA versus ALA as they represent the most abundant n-6 and n-3 FA. A lower n-6:n-3 ratio is therefore indicative of a forage-based diet. The composition of conjugated FA in the fat of forage-fed ruminants is different from that of animals raised under intensive feeding practices. High intake of PUFA from fresh herbage affects rumen fermentation pattern and gives milk fat with less SFA and more VA, RA, and ALA compared with indoor feeding (Elgersma et al., 2006b; 2015; Lock and Garnsworthy, 2003; Wiking et al., 2010). Strangely, VA was not measured by Timlin et al. (2023) (Table 1), but it was reported by Timlin et al. (2024).

Elgersma et al. (2006a) wrote 'Rumenic acid has been associated with anticarcenogenic properties in rats and possibly in humans. Other potential beneficial effects of CLA for human health were mentioned, but more work is needed to elucidate the safety and efficacy of isomers and their required doses.' Whereas the effects of C12:0, C14:0, and C16:0 on consumer cholesterol are well-established, debate has been ongoing on the impact of dairy PUFA on healthy diets in view of their low concentrations in milk relative to other food sources (e.g. Benbrook et al., 2018).

2.3 Trans *fatty acids*

The terms *cis* and *trans* relate to positioning of components in a molecule, with a *cis* double bond they are on the same side of some plane, whereas with a *trans* bond they are in opposite (transverse) directions as illustrated in Fig. 3.

Trans FA has a bad reputation because of evidence that sources from partially hydrogenated vegetable oils strongly increase LDL cholesterol, decrease HDL cholesterol, and have multiple other metabolic effects associated with CVD. The industrial process results in the formation of predominantly mono-unsaturated *trans* FA, of which elaidic acid (C18:1 *trans*-9) is a major component; *trans* bond distribution occurs from the 6th- to 16th carbon atom (Stender & Dyerberg, 2003). *Trans* FA from partially hydrogenated vegetable oils (i.e. industrial *trans*) have been phased out of foods in many countries due to their promotion of CVD.

The bacterial process in the rumen results mainly in the formation of VA (C18:1 *trans*-11), and conjugated isomers of LA, the CLAs such as RA (C18:2 *cis*-9, *trans*-11). Humans are also able to convert some VA in milk to RA (Lock and Bauman, 2004; Turpeinen et al., 2002; Tyburczy et al., 2008). As a result of the presence of one double bond in the *trans* configuration (Fig. 3), CLAs are technically considered to be *trans* fatty acids. Ruminant fats now represent the main dietary source of naturally occurring *trans* FAs for humans. In forage-fed ruminants, these FA peak strongly at the 11th carbon atom, with only small

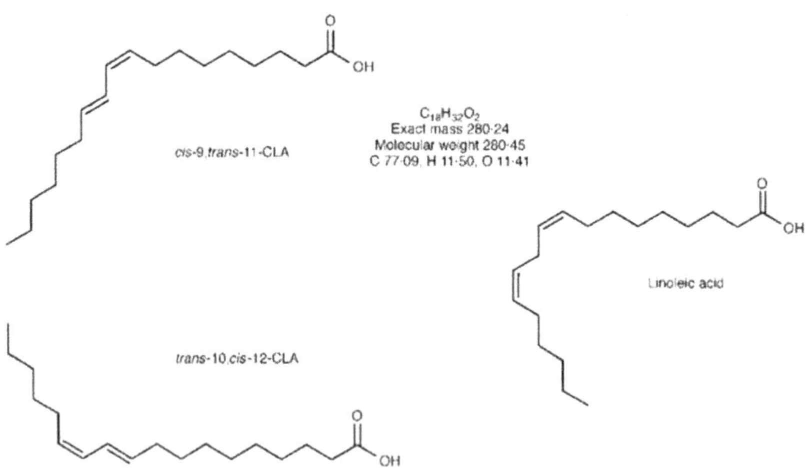

Figure 3 The structure of rumenic acid (C18:2 *cis*-9, *trans*-11 (conjugated linoleic acid; CLA)), C18:2 *trans*-10, *cis*-12-CLA and linoleic acid (C18:2 *cis*-9 *cis*-12). Counting from the methyl end, the double bond in rumenic acid is at position n-7 and C18:2 *trans*-10, *cis*-12-CLA and linoleic acid the double bond is at position n-6.

amounts at other positions. The two major *trans*-11 FA (i.e. position n-7) in forage-fed ruminants are VA and RA. However, CLA is currently not classified as a *trans*-fat by the United States Food and Drug Administration and was given a 'Generally Regarded as Safe (GRAS)' designation in 2008, exempting it from classification as a *trans*-fat on nutrition labels (den Hartigh, 2019). Also EU regulations made an exception for *trans* fat naturally occurring in fat of animal origin.

VA has little or no adverse effect on risk factors for CVD (Lacroix et al., 2012). Despite having a *trans* double bond, RA has proven benefits in animals, especially anticarcinogenic activity against diverse cancer types (Lock & Bauman, 2004). In humans, there is suggestive support for activity against colon cancer from a large, epidemiological study in Sweden (Larsson et al., 2005) and possibly against breast cancer (Dilzer & Park, 2012). Gayet-Boyer et al. (2014) found in intervention trials that used dairy products as the primary source of *trans* FA, in amounts as high as 4.2% of energy, that these levels of *trans* FA have no harmful effects on HDL cholesterol, LDL cholesterol, or their ratio. Hence, there is a clear need to distinguish between natural and industrial sources of *trans* FA, but this will take time and careful reporting, because of long-standing assumptions to the contrary (Benbrook et al., 2018).

The composition of conjugated FA changes, however, when ruminants are raised under intensive feeding practices. The feeding of high concentrate rations leads to a marked C18:1 *trans*-10 enrichment plus other *trans*-FA isomers. A number of negative health effects were found associated with the CLA C18:2 *trans*-10, *cis*-12 isomer. However, this so-called *t*11 to *t*10 shift is not uncommon in high-input systems (Aldai et al., 2013; Alves et al., 2021).

3 Lipids in feed and their fate in the digestive tract: conversions in the rumen and mammary gland

3.1 Lipids from feed components

It is well documented that diet, management regime, and extent of biohydrogenation in the rumen are critical in determining the composition of FA in the milk of dairy cows. Changes in FA composition can be accomplished by altering the diet to produce changes in fermentation patterns or the composition of fat absorbed from the digestive tract. The milk fatty acids are derived almost equally from two sources, the feed and the microbial activity in the rumen of the cow (Månsson, 2008; Parodi, 2004).

Dairy cow diets are usually composed of a mix of forages (i.e. mainly leaf blades and also stems) that can be grazed or stall-fed as fresh herbage, conserved forages (dried as hay or ensiled e.g. grass or maize silage), and concentrates (i.e. plant seeds). Diet composition of dairy cattle differs greatly among seasons and regions of the world. The lipid fraction in leaves of herbs and grasses ranges from 30 g/kg DM to 100 g/kg DM, much of which is contributed by chloroplast lipids (Bauchart et al., 1984). Sources of variation in lipid concentration are plant species, growth stage, temperature, and light intensity. There are five major FA in grasses and approximately 95% consists of linolenic, linoleic, and palmitic acids (Hawke, 1973). Fresh grass contains a high proportion (0.50–0.75) of total FA content as ALA.

Lipids in fodder crops and in concentrate ingredients have a different FA profile. Lucerne (alfalfa) (*Medicago sativa*) is usually harvested as a mixture of leaves and stems, and a maize crop (*Zea mays*) for silage contains about 50% seeds that have a high starch content at the time of harvest. Fatty acids in maize (corn) silage originate from membrane lipids in the leaves and stems (mainly ALA) and from storage lipids in the kernels (oleic acid and LA), resulting in high levels of the latter in mature maize silage (Khan et al., 2011). Whereas forage is rich in ALA, cereals (e.g. barley, maize, oats (*Avena sativa*)) are rich in oleic acid and LA. Other concentrate ingredients (e.g. soybean) generally contain high amounts of LA (Khiaosa-ard et al., 2010). The FA composition in concentrates will differ depending on the raw material composition. In concentrates, FA are either included as part of the original seed or added to it as oil. Supplemental fat sources can be used in dairy diets to increase energy intake and to change milk FA profile. These supplemental fat sources can be rich in either PA (palm oil sources; *Elaeis guineensis*), C18:0 (animal fat sources), oleic acid (rapeseed sources; *Brassica napus*), LA (soybean and sunflower sources), ALA (linseed sources; *Linum usitatissimum*), C20:5n-3 (fish oil sources), or C22:6n-3 (marine algae, fish oil sources). In addition, many by-products of the food industry, highly variable in quality and FA composition, may be included in the diet of dairy cows (Palmquist et al., 2005).

3.2 Rumen processes: fatty acid in the digestive tract

Ruminants are highly dependent on ruminal microbiota that degrade carbohydrates to short-chain FA, transform the majority of dietary nitrogen to microbial protein, biohydrogenate unsaturated FA, and synthesize vitamins. The rumen microbial population performs two major processes: lipolysis and biohydrogenaton (Jenkins et al., 2008). Lipolysis is the metabolic process through which triacylglycerols break down via hydrolysis into their constituent molecules: glycerol and free FA. Biohydrogenation is a process that occurs in the rumen in which bacteria convert unsaturated FA to SFA.

FA in cow milk originate partly from FA that are ingested and either reach the mammary gland unmodified or have been modified in the rumen or in the animal's metabolism, partly from *de novo* synthesis (only FA with carbon chain length of C16 and less) and partly from mobilization of body reserves (Palmquist and Jenkins, 1980). C18 acids in milk fat result mainly from dietary intake of ALA and LA from green pasture, abundant in the grass growing season (Hutton et al., 1969; Hall, 1970).

Palmitic acid (C16:0) in milk fat is both synthesized *de novo* in mammary tissue and obtained from blood triacylglycerols. Lower milk fat proportions of PA may be explained by inhibited *de novo* synthesis due to negative energy balance, leading to mobilization of FA from the adipose tissue.

Acetate and butyric acid are generated in the rumen by fermentation of feed components. Acetate and butyrate were significantly higher in samples derived from a pasture-based diet, whereas sugar-related metabolites were higher in concentrate-based samples (Connolly et al., 2024). Even-chain SFA C4:0–C14:0 and part of PA are synthesized *de novo* in the mammary gland from these short-chain FA formed by microbial digestion of carbohydrates in the rumen (Palmquist, 2006). Those short-chain FA synthesized in the mammary gland (C4:0 to C14:0) and are largely influenced by animal genetics (Knutsen et al., 2018). The FA sourced from blood are predominantly of diet origin, with lipolysis and the mobilization of body fat accounting for 5% in a well-fed animal to over 20% of milk FA in early lactation when cows are in a negative energy balance (Lock and Bauman, 2004) or of microbial origin. The content and composition of FA of microbial origin varies markedly and typically represent the odd-chain and branched FA (Vlaeminck et al., 2006).

Milk FA derived from the diet is also variable and represents long-chain PUFA. Diet FA, particularly PUFA, are extensively biohydrogenated in the rumen, which reduces their concentration in milk (Jenkins et al., 2008). Plant factors can influence this process, providing opportunities to manipulate FA proportions in the rumen and thereby in the milk (Toral et al., 2018). Following feed intake, dietary FA are extensively altered in the rumen, resulting in marked differences between FA intake (mostly PUFA) and FA outflow (mostly SFA). Rumen microbiota metabolize most dietary PUFA by complex processes (Figure 2). VA in the cow's udder is partially converted to RA, the major CLA in milk (75–90%) (Lock and Bauman, 2004; Tyburczy et al., 2008). The fate of dietary lipids in the rumen depends on their release from the feed, their adherence to or incorporation in bacterial cells, their biohydrogenation and their passage to the lower gut. Both lipolysis and biohydrogenation are affected by the lipid source (amount and composition of FA, technological form) and the characteristics of the basal diet (forage type, F:C ratio, fiber content, and starch content). Especially simultaneous changes in these dietary characteristics have a major impact on ruminal FA metabolism (Palmquist et al., 2005). Lipolysis is affected by the type of fat, pH, and feed intake level. A reduced degree of lipolysis is predicted when the retention time of lipid in the rumen decreases.

3.3 Biohydrogenation

The majority of RA is produced through the Δ^9-desaturase of VA supplied to the mammary tissue as a result of the biohydrogenation of PUFA in the rumen. Due to the higher proportions of PUFA in forage at pasture compared with maize, pasture-derived milk fats, and milk fat products can possess twice the proportion of RA compared with that of TMR. Increases in total CLA are associated with the high ALA content of forage grasses (Elgersma, 2015), so forage feeding increases not only CLA and n-3 FA levels in milk (Slots et al., 2009; Molkentin, 2009) but also *trans*-11 FAs. Increased levels of CLA and n-3 FAs in milk have been associated with health benefits (Larsson et al., 2005; Smit et al., 2010; Stender and Dyerberg, 2003; Vargas-Bello-Pérez and Garnsworthy, 2013).

LA has been shown to be biohydrogenated to a smaller extent than ALA; VA is an important product of biohydrogenation of both LA and ALA. Because grazing cows have a higher intake of ALA, this also results in higher amounts of post-ruminal ALA as measured in their milk. In cows' milk, about 70% of RA originates from conversion of the precursor VA by Δ^9-desaturase in the mammary gland and other tissues (Bauman and Griinari, 2001). A varying proportion of PUFA, ranging from less than 2% to over 20%, is recovered in milk as VA. There is a strong relationship between the concentrations of RA and VA in cow milk, with the concentration of VA being 2–2.5 times that of RA (e.g. Elgersma et al., 2006a; 2006b). Higher proportions of VA and RA, partly derived from the rumen biohydrogenation intermediate VA, indicate an inhibition of the terminal biohydrogenation step to stearic acid. An accumulation of biohydrogenation intermediates has been observed by including fish oil in the diet (Lee et al., 2008; Shingfield et al., 2012) or by feeding botanically diverse forages (Lourenço et al., 2008). The rate of biohydrogenation can be described as a function of the FA pool size, the ruminal retention time, and the hydrogenation capacity of the microbial population (Harvatine and Allen, 2006). The hydrogenation capacity depends on the species and concentration of the microbial population in combination with the rumen environment. Due to the changes in fiber and starch concentrations in the diets, changes in the microbial population in the rumen will likely have appeared. These changes in microbial fermentation can be characterized by a decreased rumen pH and shifts in the rumen pattern of VFA produced, i.e. lower acetate and higher propionate proportions (Fuentes et al., 2009). It was speculated that an increased starch content in the diet decreases the activity/number of *Butyrivibrio fibrisolvens* in the rumen which may alter biohydrogenation routes (Klieve et al., 2003; Nielsen et al., 2006).

There are a number of ways to increase the concentration of PUFA in cows' milk, such as by increasing the amount (i.e. intake) or concentration of their precursors in the diet, by reducing the extent of biohydrogenation in the rumen (Lashkari et al., 2019) and/or by enhancing activity of the Δ^9-desaturase enzyme that converts VA into RA in the udder. For example, diets high in readily fermentable carbohydrates are known to increase total VFA and reduce pH during ruminal fermentation, limiting lipolysis, and thus, biohydrogenation (Chouinard et al., 1999; Kolver and De Veth, 2002). Chilliard et al. (2000) reported a 35–50% decrease in ruminal biohydrogenation of PUFA, such as LA and ALA when concentrates formed more than 70% of the diet, a result of reduced biohydrogenation at lower rumen pH. However, this may have disadvantages for animal health and is not applicable in a forage-based diet.

4 Grazing as opposed to stall-feeding

Wild ruminants are always outdoors and consume forages by grazing (herbaceous plants) or browsing (trees or shrubs). Domesticated ruminants are often (partially) housed and/or fed feedstuffs other than forage. When discussing the effect of grazing it is, therefore, important to consider effects of diet composition, eating process, and housing system. The quantity and quality of milk is related to the amount and quality of ingested herbage.

Pasture is typically supplemented with grain or pelleted concentrates fed in the milking parlor. High input farming systems are characterized by the use of high-yielding plant and animal breeds, generally associated with an increased level of purchased feed. Grazing is reduced and replaced by feeding a more constant total mixed ration (TMR) during the year in confinement.

Grazing is perceived as more natural and more sustainable than indoor feeding due to the ability of animals to express their normal behavior in their natural environment (Charlton and Rutter, 2017). Because the labor demand is reduced, costs are lower than in more intensive systems, although this depends on farm and herd size. There are interactions between herd size, diet composition, and grazing. A larger proportion of farms with large herds has an automated milking system, housed animals, or restricted stocking, cows with a high milk production, more concentrate feed, maize silage, and grass silage in the diet, and less fresh grass than farms with smaller herds.

In organic systems, cows must be offered a minimum proportion of forage of total DMI and have access to pasture for grazing when conditions allow (Halfa et al., 2013; Schwendel et al., 2015). Marketers of organic products emphasize the naturalness of grazing and added value of the resulting animal-derived products.

There are interactions between animal breed and feeding system. Cows that have been selected for high milk production in confinement (e.g. Holstein) are not best suited to grazing systems where they must be able to graze effectively, survive fluctuations in feed supply and walk long distances. Breeds differ for milk yield, milk composition, and FA profile. This hampers comparison

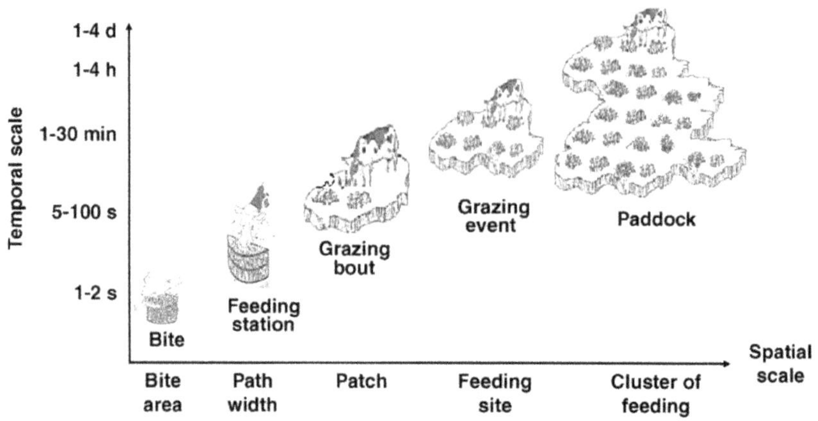

Figure 4 Attributes of different extents of scale in the foraging of a large herbivore. Source: Gregorini et al. (2006).

among experiments, e.g. in conventional and organic systems because in the latter often other breeds are used.

In grazing systems whether intensive or not, the plant–animal interface is a central feature (Gibb et al., 1997; McGilloway et al., 1999). Animals at pasture walk outdoors and can harvest fresh herbage from a three-dimensional canopy bite by bite, while grazing (Fig. 4).

Grazing animals encounter a vertical gradient while grazing down the sward (Fig. 5). DMI is the major determinant of animal production (Forbes, 1995) and basis for management decisions. Although humans regard utilization as a problem to be solved on a paddock-by-paddock basis, grazing animals are constrained to meeting their dietary requirements bite by bite (Parsons and Chapman, 2000). The bite is the functional link between the spatial and morphological properties of the vegetation and the ingestive apparatus of the animal (Laca et al., 1992; 1994). During grazing, intake rate is calculated as the product of bite mass and bite rate.

Fig. 6 illustrates the components of ingestive behavior. Grazing periods are followed by rumination periods. Ruminants have a diurnal behavioral pattern with three main daily grazing events (bouts) during which they graze continuously for more than 1 h: morning, afternoon, and dusk time, the latter being the longest (Phillips and Leaver, 1986; Taweel et al., 2006). Management decisions might be taken on a daily or larger time scale, but the underlying processes take place continuously (digestion) or in discrete bouts (ingestion) within a day. To integrate intake on a daily basis, DMI under grazing is commonly expressed as the product of the time spent eating and intake rate. At the level of the pasture/animal interface, the weight of herbage consumed in each bite is constrained by the morphology of the sward. Sward state can be described in terms of tiller density

Figure 5 Illustrations (Hodgson (1990) and own photo) of sward structure and grass tiller morphology with leaf blades in upper horizons and pseudostems near the ground surface showing a vertical decline in organic matter digestibility with sward surface height.

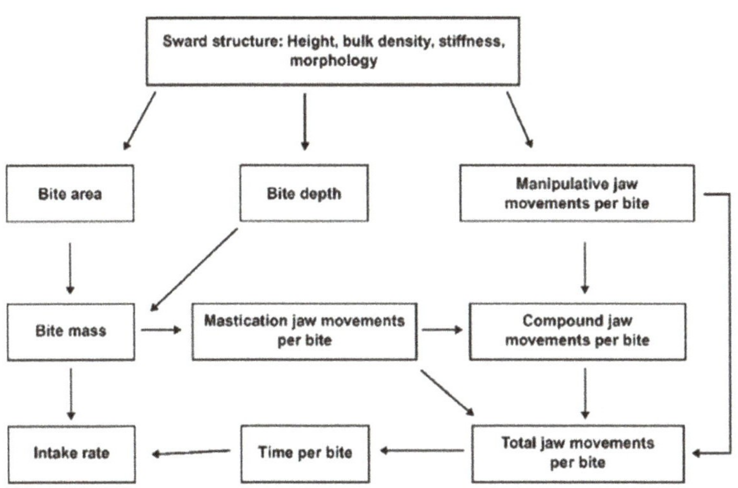

Figure 6 The components of ingestive behavior, which mediate between sward structure and short-term intake rate. Source: Gregorini et al. (2006).

and sward surface height (SSH); the mass of plant material shows a gradient from the soil surface to the top of the three-dimensional canopy. A grass tiller consists of leaf blades and leaf sheaths, the latter form the so-called pseudostem that is normally not grazed and forms the major part of biomass in the stubble which contains sugar reserves for regrowth after defoliation. The leaf/stem ratio declines from the canopy surface to the stubble, as does the concentration of protein, while concentrations of fiber compounds and lignin increase. Grazed grasslands contain predominantly leaves, grasslands cut for silage can have high stem proportions during parts of the year. Green leaf mass is an important parameter because leaf blades have a higher nutritional quality than stems. Grazing animals can select between species and plant parts. Therefore, their ingested diet is usually of higher quality than that of a random sample of the sward (e.g. Fraser et al., 2022; Grings, 2020). Farmers can maximize opportunities for diet selection by offering grazing animals a high herbage allowance.

To ensure an adequate supply of energy and nutrients, different systems of grazing and feeding management have been developed: (1) unrestricted stocking (day and night at pasture); (2) restricted stocking, usually only during the day; (3) zero grazing (no access to pasture) with cut fresh forage offered indoors (green chop); and (4) zero grazing with only conserved forage. The amount of cow traffic (walking) differs in the various systems and affects the proportion of energy needed for locomotion or producing milk. Grazing management can be continuous variable stocking management where animals have access to a large paddock over a longer period. In this system, a balance is maintained between the production of new herbage and consumption by the grazing animals, with little of the herbage becoming senescent. A more common system nowadays is rotational grazing where animals have access to a paddock for hours (strip grazing) or days (e.g. 4 days) before they move on. Under intermittent defoliation management, forage is allowed to accumulate between stocking of the pasture and is grazed down over a relatively short period during which the morphology of the sward is greatly modified.

Low herbage DMI is considered a major limiting factor for energy intake and hence milk production of high producing cows under grazing systems. Herbage allowance, SSH, tiller density, leaf strength, eating time, time allocated for rumination, and chewing all influence the capacity of an animal to consume herbage (Gibb et al., 1997; McGilloway et al., 1999). A better understanding and predictability of DMI under grazing conditions should result in better management and utilization of the available resources.

There are trade-offs between herbage utilization and herbage quality. Animal selection is both horizontal (among patches, plants, and species; Fig. 4) and vertical (Fig. 5) as they start grazing from the top of the canopy in so-called horizons. The mass of herbage within the grazed horizon has a more profound effect on bite mass than SSH per se (Orr et al., 2001). It is, therefore, important to maximize the green leaf mass by providing dense, leafy swards to enable the grazing animal to achieve large bite masses. Appropriate choice of grassland species and cultivars and correct grassland and grazing management can help to achieve this.

5 Effect of feeding system

5.1 Diet selection and effects on rumen passage rate

A fundamental difference between grazing and stall-feeding is that grazing animals eat fresh living herbage while cut forage is either living (green chop, with freshness depending on duration and conditions from cutting until eating) or dead (hay, silage). Furthermore, during stall-feeding selection possibilities, i.e. options to reject plant parts of detached material are minimal; animals can pick up a detached plant part with their mouth but are not able to tear off the most palatable portion, unlike grazing animals which can use their tongue and the force of their head. Apart from the more limited option of biting off material once it has been picked up, detached tillers can either be rejected as a whole or have to be ingested as a whole. The eating and chewing pattern are different for ruminants that have to bite or tear off forage during grazing compared to animals that eat cut forage (A. Elgersma, personal observation). With indoor feeding, detached plant parts are often longer than in grazed bites as forage is usually cut at a stubble height of 4 cm or 5 cm. After cutting, wilting begins and plants lose water, crispness, and rigor. This physical change can affect the chewing pattern and fragmentation of particles of ingested feed boli that are swallowed. It can affect the rumination process and fermentation rate. Since bacteria change the *cis*-11 of LA to the *trans*-11 of RA in the rumen, concentrations depend not only on levels in the feed but also on the retention time in the rumen. As intake increases, the main retention time of feed in the rumen decreases. Feed moves out of the rumen more rapidly as pasture intake increases and may reduce the extent of biohydrogenation, thereby resulting in higher concentrations of RA. The large variation in RA concentration of milk fat seen between individual animals eating a similar diet is related to differences in the cows' ability to produce RA from VA through Δ^9-desaturase.

Offer (2002) compared milk from cows grazing pasture with milk from cows offered cut material from the same field. Zero-grazing led to a 0.24 % reduction in milk ALA proportion, but a much larger reduction in RA and VA (0.52%). This effect may be due to a greater potential by cattle to select high-quality pasture in the field. This may explain the reduction in ALA, but clearly the effect on biohydrogenation intermediates was higher and some other mechanism is probably involved. Elgersma (unpublished results) disentangled the housing/cow traffic effect when comparing grazing and feeding cut herbage. Eight groups of two cows were formed. In a 1 day rotational system with ample allowance one cow could graze while the other got identical herbage that had been cut just prior to the moment that cows entered the new paddocks (zero-grazing outside). After 1 week, treatments were changed. Cows that had grazed during the first week showed a 23% decline in CLA concentration in their milk at the end of the second week when they were fed cut grass compared to the first week, whereas cows with the opposite treatment showed a 23% increase. However, there was a greater residual herbage mass in the grazed plots. Thus grazing cows would have eaten higher sward layers, with a higher leaf blade proportion compared to cows offered the cut grass, which is in line with Offer (2002). Visual observations (A. Elgersma, unpublished results) showed different behavior: after accessing a fresh paddock the cows in the grazing treatments immediately started grazing, at a high bite rate, whereas the other cows were less eager to start eating the freshly cut grass. Jaw movements of the grazing cows were faster.

5.2 Herbage allowance and grazing management strategies

Herbage allowance can be manipulated by increasing the area of the offered pasture (the size of the paddock) per animal or increasing the yield per hectare. Care must be taken to distinguish increased pasture allowance achieved by altering land area at a given herbage mass versus by altering herbage mass per unit area, which probably means altering plant developmental stage. For a given paddock, farmers could increase herbage allowance by reducing stocking rate (animal units/ha) or the time animals stay at a certain paddock. Forage lipids are the most economic source of n-3 FA and an important strategy would be to increase the delivery of FA from plants to animal products. Palladino et al. (2009) investigated the effect of different grazing strategies, based on the quality and quantity of herbage offered, on the milk FA of dairy cows. Pre-grazing herbage mass and daily herbage allowance were compared. Cows grazing a low SSH had a higher intake of total FA, LA, and ALA, which was related to the high-herbage concentration of these FA. This is explained because low SSH swards had had a shorter regrowth period and thus consisted of younger herbage (Elgersma et al., 2003a; Dewhurst et al., 2001). Variation in FA intake across the season was more a function of differences in FA composition of the pasture than differences in DMI, in line with Elgersma et al. (2003a). Differences in FA intake among treatments were not reflected in differences in milk for most of the FA (Palladino et al., 2009). Dewhurst et al. (2006) in a review also found no effect of daily herbage allowance on milk FA. However, high variation among individual animals has been observed. Higher variation in Δ^9-desaturase activity was observed under grazing conditions than with TMR. This grazing effect on the enzyme activity could potentially mask differences in experiments carried out on grazing systems. Cows at pasture grazing lush green grass (Dhiman et al., 1999; Khanal and Ohlson, 2004) at a high herbage allowance (Elgersma et al., 2004) produce milk with the highest concentrations of PUFA. Data on effects of grazing system on ingested FA are scarce, especially in relation to FA in the milk. In rotational grazing systems, amount and quality of herbage on offer differ greatly between the first and later days of a grazing cycle.

Elgersma et al. (2008) determined FA concentrations in herbage in ungrazed and pre-grazed swards in relation to FA intake and CLA concentration in milk of grazing dairy cows. Twelve cows with on average 22 g CLA/k milk fat were allocated to two treatments, i.e. starters, offered daily ungrazed swards and followers, offered 1-day pre-grazed grass that still enabled *ad libitum* intake; treatments changed after a week. The vertical gradient in FA concentration in grass canopies was related to leaf blade proportion. Cows with access to upper sward horizons had the highest CLA concentrations in their milk. After the first week, CLA concentrations in milk had increased in the starters to on average 120% of their initial level and had declined in the followers to 65% of their initial level. After the second week, milk CLA concentrations of the first group (now followers) declined to 80% of their initial level, whereas that in the second group increased to 125%. A positive relation between the CLA concentration of the milk and the ALA concentration of the grass was found. Grazing management strategies could thus modify milk FA composition.

6 Effect of forage species on milk: grasses, legumes, and botanically diverse pastures

6.1 Plant species have different effects on rumen processes and milk composition

Positive correlations were reported (Collomb et al., 2002) between several dicotyledoneous species (forbs) in pastures in the Alps and milk fat proportions of PUFA, total CLA, and VA. The higher PUFA concentration in milk and meat of animals grazing species-rich relative to grass swards may relate to inhibited or modified FA metabolism in the rumen. The changes may be caused by secondary plant metabolites which are associated with the numerous forbs common in species-rich grassland. However, the presence of specific metabolites in forbs of botanically diverse forage which might modify rumen FA metabolism, or transfer efficiency of ALA from the duodenum to the mammary gland still needs to be tested *in vivo*.

No or inverse effects on proportions of ALA were reported when cows grazed red clover (*Trifolium pratense*) – rich pastures compared to white clover (*Trifolium repens*) – rich pastures (Wiking et al., 2010; Larsen et al., 2012). Increasing milk fat proportions of RA were found (Soder et al. 2006) for cows grazing more botanically diverse forage mixtures; however, ALA was not affected. The effect of white clover on ALA is most likely related to increased rumen passage rate (Dewhurst et al., 2003).

Feeding of grass-red clover silage versus grass silage or grass-white clover silage has in most cases decreased rumen biohydrogenation, leading to higher apparent recovery and higher milk fat proportions of ALA (Dewhurst et al., 2003).

Less extensive biohydrogenation for white clover-rich pastures (Adler et al., 2013) may be explained by higher rumen passage rate caused by white clover (Dewhurst et al., 2003). Likewise, in the study of Larsen et al. (2012), white clover increased milk fat proportions of ALA, despite lower ALA intake.

An accumulation of biohydrogenation intermediates has been observed by feeding botanically diverse forages (Lourenço et al., 2008). Van Ranst et al. (2009) postulated that PUFA could be protected against biohydrogenation through encapsulation in protein–phenol complexes. Specific secondary plant metabolites, such as condensed tannins or saponins may inhibit lipase activity. Phenols are present in various forb species, e.g. salad burnet (*Sanguisorba minor*), chicory (*Cichorium intybus*), and ribwort plantain (*Plantago lanceolata*) (e.g. Bryant et al., 2019).

The high vitamin concentrations of some forbs (e.g. Elgersma et al., 2013) also offer perspectives for naturally improved milk vitamin composition. In a pilot study comparing the transfer efficiency and concentration in milk of n-3 and n-6 FA and vitamins of cows fed TMR, fresh grass-clover forage or a mixture of forb species, FA increased with forbs due to an increased transfer rate from feed to milk. However, apart from a higher retinol concentration with forbs, no significant differences were observed in the vitamin concentration of the various milks (Petersen et al., 2011). There were, however, increased concentrations of n-3 and n-6 FA in milk of forb-fed cows that could be due to an increase in transfer efficiency from feed to milk for n-3 FA compared to both grass-clover and TMR diets. Petersen and Jensen (2014) compared rumen fluid collected from cows fed either TMR, species-rich silage, or grass silage. Rumen fluid from cows on the TMR diet had the highest concentration of LA. LA is more toxic than C18:1 isomers to bacteria cells responsible for conversion of LA to VA (Choi et al., 2009). The slow and low apparent appearance of VA and fast formation of oleic acid from the TMR diet indicated such possible effects of the TMR diet on the rumen bacterial cultures compared to forb-rich and grass silage diets, as the apparent appearance of VA was similar in the latter. The supplies of energy and fermentable carbohydrates for microbes were also highest on the TMR diet and possibly provided better bacteria growth. Combined with a numerically lower pH value and thus more hydrogen available for biohydrogenation, this could contribute to optimized conditions for biohydrogenation in the rumen and might explain the higher biohydrogenation rate in the TMR diet. Lucerne and salad burnet had the lowest biohydrogenation of ALA and LA, whereas the fastest biohydrogenation occurred in birdsfoot trefoil (*Lotus corniculatus*) (Petersen and Jensen, 2014). Lucerne contains no tannins but flavonoids (Prati et al., 2007), which could have a specific impact on rumen biohydrogenation (Bacells et al., 2012). The often found positive effect on milk concentration of ALA with a lucerne diet (Chilliard et al., 2007) warrants further studies.

6.2 Fatty acids in forages

6.2.1 Lipid gradient in canopy height and relevance to intake by grazing animals

The bulk of nutrients utilized by ruminants are of plant origin. The chemical and morphological characterization of plants and plant parts (leaves: leaf blades, leaf sheaths and inflorescences: stems, flowers, seeds) is thus very relevant. Development of stem formation in grasses of temperate climate zones is regulated by temperature and day length. After winter, primary growth is leafy but in the course of spring, generative stem-producing tillers are formed. After stem removal (by cutting or grazing), vegetative leafy tillers are formed in later re-growths. Lipids are mainly found in plant membranes. The lipid fraction in leaves of grasses and forbs ranges from 30 g/kg DM to 100 g/kg DM, much of which is contributed by lipids in the chloroplasts (Bauchart et al., 1984). About 95% of FA consists of ALA, LA, and PA chloroplasts (Hawke, 1973), C16:1 *cis*-9 (palmitoleic acid) and C18:1 *cis*-9 (oleic acid) are the main 'other' FA. The synthesis of PUFA plays an important role in the adaptation and survival of plants in cold temperatures as it contributes to the maintenance of membrane fluidity in the chloroplast. Therefore, PUFA are present in larger amounts in plants in temperate climates (Hugly and Somerville, 1992). Global warming may cause declines in PUFA concentrations.

The upper part of a canopy has the highest proportion of leaf blades (Fig. 5) and in rotational grazing systems, animals that remain in a fenced paddock will consume lower layers with older leaf blades and more pseudostems when grazing down a sward. Elgersma et al. (2008) determined FA concentrations in herbage in ungrazed perennial ryegrass (*Lolium perenne*) swards prior to grazing by cows, and the next day in the grazed swards and found that the FA concentrations of total FA, ALA, and C16:0 in herbage were lower ($P < 0.001$) in the grazed canopy while the concentration of LA was higher ($P < 0.001$). Thus the FA profile changed as the proportion of ALA declined ($P < 0.001$) from 80% in ungrazed herbage to 75% in lightly grazed herbage, while the proportion of LA increased ($P < 0.001$) from 8% to 11%. Although the PA concentration declined ($P < 0.01$), its proportion increased from 10.4% to 11.6%. Thus the PUFA concentration was lower in the herbage that remained after 1-day grazing compared to that in the ungrazed canopy (Elgersma et al., 2008).

6.2.2 Variation in lipid concentrations and fatty acid composition within and among plant species

FA concentrations in fresh forage vary among plant species. FA profiles were distinctive to species when grasses (e.g. Dewhurst et al., 2001; Wyss et al., 2010) and forbs (e.g. Elgersma et al., 2013; Wyss et al., 2010; Clapham et al., 2005; Warner et al., 2010) received the same management (i.e. measured at the same cutting date), confirming a strong genetic basis. However, interactions with seasonal development could be further explored by cutting at the same developmental and phenological stage. Boufaïed et al. (2003) harvested species at a similar phenological stage. Grasses had higher ALA concentrations than legumes, the latter had higher concentrations of total FA, LA, and PA. However, large variation was observed among grasses and legumes. Whetsell et al. (2022) determined the FA content in pasture grasses, legumes, and non-leguminous forbs in northeast West Virginia. Forbs had the highest LA content followed by legume and grass species. Grasses and forbs had the highest ALA content. Forbs had the highest total FA content.

Within a species, the concentration varies with plant and environmental factors such as variety/cultivar, regrowth stage, stage of maturity (leaf/stem ratio), season, light intensity, growing conditions during the preceding regrowth period, and N availability. Leaves have higher FA concentrations than stems. Younger plants have higher FA concentrations than older plants. Pasture management thus plays an important role in forage FA composition. Elgersma (2015) reviewed total FA concentrations and proportions of the major FA, i.e. ALA, LA, and PA, reported in the literature in various species and also illustrated the range of FA concentrations and associated N concentrations in perennial ryegrass herbage from the same sward, as a function of N fertilizer application rate, regrowth stage (harvest date), and season (Witkowska et al., 2008). N application resulted in higher concentrations of FA in grasses (Boufaïed et al., 2003; Witkowska et al., 2008) and in a higher proportion of ALA. A longer regrowth period significantly decreased the FA concentration and the proportion of ALA in grasses (e.g. Dewhurst et al., 2001; Witkowska et al., 2008). A strong positive overall linear relation was found between ALA and N concentrations in the herbage (Fig. 7).

Concentrations were higher in autumn (September/October) than in May. Harvest date is in general confounded with weather conditions, (re-)growth stage, and phenological development of the plants. This illustrates limitations to generalize and compare FA concentrations among experiments where in addition to local factors, different soil type and growing conditions (altitude, climate) occur. In vegetative herbage as found in early spring and in autumn, FA concentrations are generally higher than in reproductive (stemmy) herbage

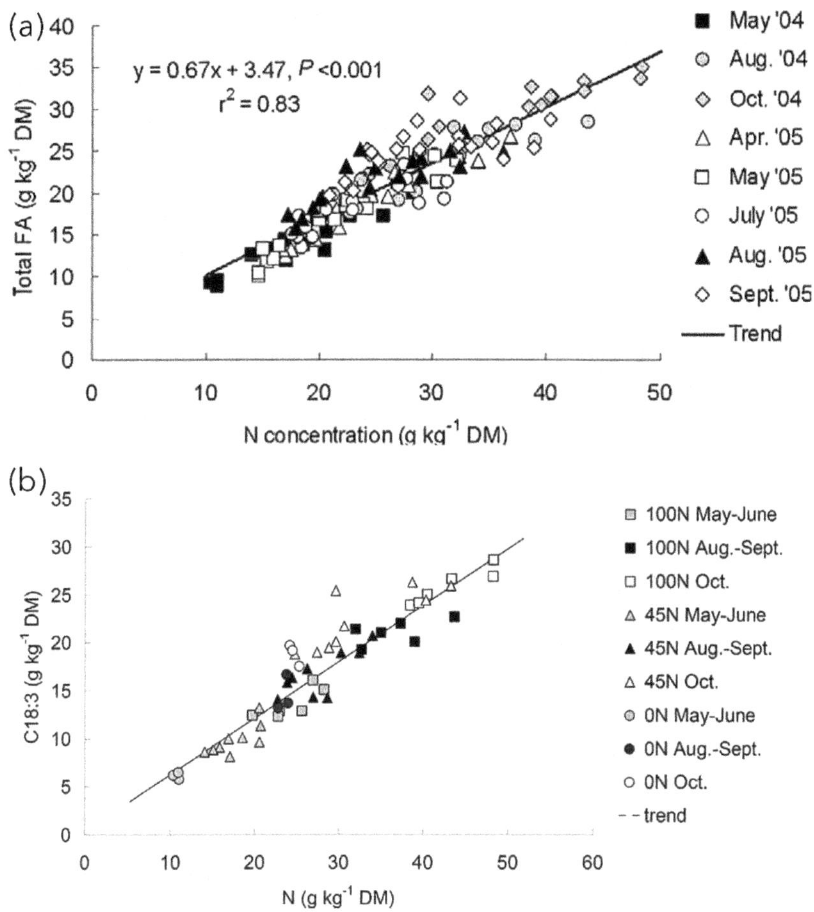

Figure 7 The relationship between concentrations of (a) total fatty acids (FAs) and nitrogen (N) in perennial ryegrass during eight experimental periods in 2004 and 2005 (Witkowska et al., 2008), and (b) α-linolenic acid (ALA) (C18:3) and N in 2004 for annual N application levels of 0 kg N/ha, 45 kg N/ha, and 100 kg N/ha; C18:3 concentration = 0.49 + 0.59 * N; R2 = 0.87. Source: Modified from data of Witkowska et al. (2008).

as found in late spring and early summer (Bauchart et al., 1984; Dewhurst et al., 2001; Elgersma et al., 2013).

Clapham et al. (2005) found a decline in FA concentration with increasing growth interval, which was stronger in chicory than in perennial ryegrass and white clover. Wyss et al. (2010) harvested first-cut herbage of cocksfoot (*Dactylis glomerata*), lucerne, and dandelion (*Taraxacum officinale*) at different dates; in older herbage, FA and crude protein concentrations declined. The proportions of ALA declined and of LA increased in older herbage. Stems and leaves were also analyzed separately. Stems of dandelion had a different FA profile with a high LA concentration. Cocksfoot stems had very low FA concentrations compared with leaves.

Because grazed grass is often younger than grass cut for stall-feeding and certainly younger than grass cut for silage-making, and because grazing animals prefer leaf blades over stems, these findings imply that intake of FA and proportional intake of ALA under grazing is higher than in indoor-fed cows. Plants in the vegetative stage have a higher concentration of FA than in the generative (reproductive) stage because leaf blades are richer in FA then stems, as most FA are localized in the chloroplast (Hawke, 1973). This can explain in part effects of cutting date on the FA concentration and composition. Interactions between species or cultivars and cutting date can also be due to differences in generative period and/or generative growing stage.

Dewhurst et al. (2001) compared ryegrass species, cut on three or five occasions during the year. All species had high concentrations of FA and high ALA proportions during vegetative growth in late April, while there was a marked declined in both of parameters in all species after this date. After the flush of reproductive stem growth during May and June, leaf proportions increased to the end of the season. Total FA concentrations began to increase again during July and August. Differences between cultivars of herbage species were mostly not significant or of minor importance (Elgersma et al., 2003b; 2015; Van Ranst et al., 2009); the higher FA concentration in late than in intermediate heading perennial ryegrass cultivars reported by Palladino et al. (2009) was predominantly due to one diploid late-heading cultivar. Boufaïed et al. (2003), however, did report variation among cultivars within species when cut at the same phenological stage. Also in Elgersma et al. (2006a), a difference in FA concentration was reported between two perennial ryegrass cultivars that had the same heading date and thus were cut at the same phenological stage. Significant species by cut interaction effects indicate that management factors will be as important as plant breeding in manipulating FA levels.

Choice of seed mixture when resowing is a rapid way to modify the botanical composition of a sward. Pasture management including fertilization and cutting/grazing regime has a major effect on FA concentrations in animal feed. Forbs such as chicory, that had a high mean FA concentration but markedly greater variability, may present special challenges when defining pasture species combinations to achieve modified FA profiles in milk. Variability in FA concentration with plant development could affect the utility of forages in grazing systems. Rodríguez et al. (2020) found that a plantain/chicory pasture contained higher LA concentrations compared to a grass-based permanent pasture, with no difference in ALA concentrations detected. Also Mangwe et al. (2020) found higher LA concentrations in chicory and reported that allocation of mature chicory herbage to dairy cows at 50% of their ration modified rumen fermentation and improved both milk yield and the FA profile of the milk.

Chicory inclusion increased the concentration of PUFA in the milk compared to the ryegrass-white clover regime. Kearns et al. (2025) showed that the FA, tocopherol and carotenoid concentrations in grass, grass - white clover and multispecies sward types can differ, depending on the plant species included and the season.

6.2.3 Post-harvest changes and analytical constraints

Lipids in plants are not static entities and lipases are also normally present. Significant modifications of the lipid fraction may occur during senescence (Thomas, 1986), immediately after detachment and during storage after cutting. Following cutting, pre-wilting of herbage prior to ensiling is a widely used swath management practice. During field wilting of grass prior to ensiling or hay making, oxidative losses of the PUFA occur (Dewhurst et al., 2006) and lead to substantially lower ALA concentrations in conserved grasses compared with fresh pasture (Dewhurst and King, 1998; Doreau and Poncet, 2000; Elgersma et al., 2003a). In perennial ryegrass – red clover herbage there was a significant decline in FA during 29 h wilting and the ALA proportion showed a small but significant decline compared with fresh herbage (Elgersma et al., 2015). Further losses of unsaturated FA can occur during silage fermentation, storage, and the feedout phase of silage or hay.

Lipids in forage are part of a complex matrix and their release in the rumen is different than that of oils and fats added to rations that can adversely affect rumen function when fed in high concentrations (Elgersma et al., 2013). Analytical constraints of measurements were discussed by Elgersma (2015). Standard feed testing procedures are typically performed on forage samples which have been conserved by oven-drying or freeze-drying, making them easy to store without degradation. However, using dried ground tissue to predict rumen availability of compounds from plant cells of fresh forage may not adequately reflect the rumen interactions of freshly grazed forages (Bryant et al., 2019). In ingested fresh herbage, compounds are part of a living complex matrix that is subject to chewing, exposure to saliva, rumen microbes, regurgitation, etcetera, which hampers replication of these processes in the laboratory as the rumen interactions of freshly grazed forages may not adequately be reflected. Measured quantities in dried samples may not equal those in living herbage (Wyss, 2012).

7 Seasonal variation in fatty acid composition in milk from cows in pasture-based systems

Animal diet is especially relevant when comparing FA in milk from concentrate-fed and pasture-based systems. Milk FA composition in pasture-based systems is, additionally, subject to seasonal variations that influence the quantity and quality of available forages. However, also stage of lactation plays a role and these factors are confounded in countries like Ireland. In seasonal systems, cows are traditionally calved in a compact calving period in spring designed to maximize pasture feeding in accordance with the nutritive demands of dairy cows throughout the lactation period. Cows will graze during most of the year in these systems. When grass growth slows down in late autumn and winter, the cows are brought back indoors and fed pasture-based silage as their primary feed, which was cut and ensiled from surplus pasture earlier in the year. In such systems, there is an interaction with calving date of the herd that is not found in systems with year-round calving. From calving up to 8-week postpartum, cows can be in a state of negative energy balance, resulting in reduced synthesis of short- and medium-chain FA and a greater mobilization of tissue FA into milk (Palmquist et al., 1993).

Seasonal variation in milk FA composition has long been recognized (Mattsson, 1949; Stadhouders and Mulder, 1955; Riel, 1963) and is caused by differences in diet between summer and winter. Fresh herbage is only available in the growing season. Traditionally, grasslands (mixed vegetation of various grass species and forbs) were grazed during the growing season and surplus forage was made into hay that was fed in winter or during dry periods. Leguminous fodder crops such as lucerne and red clover were used as protein source, often fed as hay. Mattsson (1949) reported in 1949 concentrations of conjugated dienoic acids (representing total CLA) of 6–24 g/kg milk fat, and lower levels for stall-fed cows in Sweden. In the Netherlands, a threefold difference was reported in 1955 (Stadhouders and Mulder, 1955) in conjugated dienoic acid concentrations in milk of cows fed in stalls versus pastured (i.e. 7 g/kg vs. 25 g/kg milk fat). In systems maximizing grazed forage intake, as still practiced in countries with a long grass growing season, e.g. Ireland and New Zealand, seasonal calving is desired at the start of the growing season which implies high fertility. However, many pasture-based production systems have changed with the pattern of milk supply to the manufacturer becoming less seasonal, as illustrated for southern Australia. Although grazed herbage remains a predominant feed of dairy cows on the majority of farms, there has been increased use of supplementary feeds and a shift from seasonal calving in late winter to early spring toward split-calving (Downey and Doyle, 2007). The authors attributed this in part not only to seasonal price incentives but also to the failure of cows to conceive, improved knowledge of feeding and use of supplements, and a sequence of dry winters. As stage of lactation tends to become less synchronized in more intensive systems, the confounding effect of herd calving pattern with the grass growing season diminishes. For comparison of studies on changes in milk FA profiles at a farm or regional level as cows move from summer to winter diets and *vice versa* at turnout (Agenas et al., 2002; Khanal et al., 2008), herd information is, therefore, relevant.

Seasonal variation in the CLA concentration of milk fat was reported and highest concentrations were found in grazing cows, i.e. 12.7 g/kg fat in New Zealand and 17.0 g/kg fat in the UK (Lock and Garnsworthy, 2003); lowest concentrations occurred in winter and amounted to 6.0 g/kg milk fat in both countries. There was no relationship between milk fat level and CLA proportion of milk fat. Seasonal fluctuations in milk CLA concentrations in the Netherlands ranged from 3.7 g/kg to 9.2 g/kg fat in winter and summer, respectively (Elgersma et al. 2006a).

Nałęcz-Tarwacka (cited in Elgersma et al., 2006a) studied 429 cows of 23 farms in central Poland during 2 years. The proportion of unsaturated FA was 344 g/kg fat in summer when all cows had green grass, and 308 g/kg fat in winter when they were fed maize silage, grass silage, and brewers' grain. The omega-3 FA declined from 9 g/kg to 8 g/kg fat, and the CLA concentration from 7.2 g/kg to 5.4 g/kg fat. In another study, CLA concentrations in milk of black-and-white cows in Poland (Reklewska et al., 2003) were higher in summer (8.4 g/kg fat) and autumn (8.9 g/kg fat) than in winter (6.3 g/kg fat); CLA values in milk from cows on a TMR and on pasture were 6.1 g/kg and 11.7 g/kg fat, respectively. There was a quick response of the total CLA concentration in milk when cows changed from unrestricted stocking to a 1:1 grass silage – maize silage diet (Elgersma et al., 2004) or to a maize silage diet (Elgersma et al., 2006a), maintaining 3 kg/day concentrates feeding. After transition to maize silage, the average CLA concentration decreased from 21 g/kg to 11 g/kg fat within 2 days and stabilized at 5 g/kg fat (Elgersma et al., 2004), which is in line with the statement (Chilliard et al., 2001) that the maximum effect is reached within about 5 days. Increasing the consumption of grazed herbage increases the concentration of PUFA, notably RA and ALA in milk. The CLA concentrations in milk of pasture-fed cows can be 2 to 5 times higher than in cows given TMR (Kelly et al, 1998: 1.09 vs. 0.46 g/100 g fat; Dhiman et al, 1999: 2.21 vs. 0.39 g/kg fat). Although effects of fresh forage on increasing ALA and RA are consistent among a number of studies, there are no consistent effects on other FA – perhaps because of differences in the level and type of concentrates offered.

A number of reviews have compared milk FA profiles from grazing cows to milk of cows fed hay or silage-based diets (Dewhurst et al., 2006; Elgersma et al., 2006a; Schwendel et al., 2015). Underlying studies range from surveys comparing milk produced from pasture during the grazing season with milk produced from conserved forages with housed feeding to more controlled comparisons, although it is obviously not possible to compare fresh forage with material conserved from the same forage at the same time. In fact, comparisons are further complicated by differences in the F:C ratio for cows fed silage-based rations versus pasture. As mentioned by Dewhurst et al. (2006), it is not possible to calculate F:C ratios for early studies, while pastured cows received proportionately less concentrate than those fed TMR. In Kelly et al. (1998), forage comprised 47% and 100% of TMR and pasture diets and in White et al. (2001), forage comprised 59% and approximately 75% of TMR and pasture diets, respectively. In Offer (2002) and studies of Elgersma et al. (2003a, 2004, 2008), however, concentrate feeding level was low (3 kg concentrates/day) and a relatively fixed proportion of DM intake. Schroeder et al. (2004) compared concentrations of RA in milk of cows fed pasture-based diets to that of cows fed TMR diets from seven studies and reported an increase of 134% compared to the control, with considerable variation in the responses ranging from 15% to 396%. Differences in herbage allowance (Bargo et al., 2006; Stanton et al., 1997; Stockdale et al., 2003; Timlin et al., 2023), herbage mass (Stanton et al., 1997; Stockdale et al., 2003), time at pasture (Khanal et al., 2008), grass species and cultivar (Dewhurst et al., 2006), phenological stage (Griinari et al., 1998), and N fertilization rate (Boufaïed et al., 2003; Witkowska et al., 2008) might explain these large variations.

A review of on-farm feeding practices to enhance the quality of grassland-based ruminant dairy and meat products showed that the differences between feeding management practices observed on farm were smaller than those observed under controlled trials (Cabiddu et al. 2022). As fertilization and herbage allowance can be managed easily by farmers, it might represent an interesting and feasible way to obtain higher concentrations of FA in herbage. Also decisions on the timing of cutting or grazing, i.e. the regrowth stage and phenological development at which herbage is harvested or cows are allowed to enter the pasture, can be a practical means of influencing herbage quality and milk FA profile.

8 Trends in grazing and dairy farming systems and implications for milk fatty acid composition

In view of expected increased demand for animal products, whereas land scarcity, limited resources, and restrictions, to emissions must be addressed, the dairy industry focuses on increased efficiency of the use of resources while attempting to reduce the environmental footprint. High throughput phenotyping and 'omics' technologies accelerate breeding of cows with ever-increasing genetic potential for milk production. Care should be taken that search for alternative feed resources as a way to increase efficiency of feed chains does not lead to undesirable side effects on milk FA composition. For example, starch-rich diets can lower methane losses but, as shown in this review, cause a shift toward *trans*-10 FA in milk. Developments in breeding, feeding, housing, and milking mentioned in the introduction have led to sharp declines in grazing and in the proportion of fresh grass in cows' diets. Feed composition rather than farming system per se (organic vs. conventional) influences concentration of compounds in milk. However, because organic dairy production standards require a reliance on forage, organic dairy products contain significantly higher protein, ALA, total n-3 PUFA, RA, and a lower n-6 to n-3 ratio compared to dairy products derived from non-organic systems (Palupi et al., 2012). Also greater concentrations of fat-soluble antioxidants have been found in organic milk compared to milk from silage and concentrates-based TMR systems (Butler et al., 2008; 2011).

High fresh grass proportions in the diet are not exclusive to organic systems; milk from conventional farms with grazing cows has similar, high concentrations of PUFA. The latter, however, is diluted in retail milk as raw milk from grass-based conventional farms is pooled prior to processing with milk from TMR-based farms. Whether these represent much larger volumes, depends on the region. Generally, variability in milk fat quality across systems tends to mirror the diversity in diet composition with the concentrations of RA, ALA, and/or antioxidants appearing to be elevated by the inclusion of fresh herbage grazing.

Nevertheless, there are region-specific, farm-specific and farmer-specific constraints for farmers to use a grazing system (Van den Pol-van Dasselaar, 2020). Rewarding farmers for grazing can be divided into financial incentives/initiatives within the market, common agreements to stimulate grazing, and policies/legislation. Following the first initiatives of a small cheese-producing dairy (Elgersma et al., 2006a), all Dutch dairy cooperatives made a switch in favour of promoting and rewarding grazing and pay their farmers who let their cows graze at least 120 days per year for minimal 6 h per day a premium since 2012 at the expense of a milk price reduction for all farms. This was launched to stimulate farmers to give their cows access to pasture. Access to pasture, however, is no guarantee for high fresh herbage intake: herbage allowance should be adequate, and actual grazing time of cows may be reduced when high levels of supplementary feeds are provided. Following abolishment of milk quota in the EU in April 2015, milk prices went up and the use of rumen-protected fat with high levels of SFA increased in the Netherlands. Such so-called bypass fat, mainly from palm oil sources, enhances the energy density of the ration as it resists lipolysis and biohydration in the rumen, gets digested in lower digestive tract, and leads to higher milk production. From 2014 onward, the PA concentration in Dutch summer milk increased with 1% each year and also winter milk levels went up such that cheese producers noticed that the texture of Gouda-type cheeses became too hard. On November 1, 2017, Dutch dairy and animal feed industries made a joint decision to stop the use of fractionated fats in dairy cow rations for the winter 2017–2018. The increase rate of PA content in milk then slowed down. In Ireland, a Grass Fed Standard was launched for Irish dairy products; products sold with this logo are from a feeding system with at least 90% grass on a fresh matter basis annually (Timlin et al., 2023).

To increase the competitiveness of mountain farms (Lombardi et al., 2014), interests in human health and animal welfare led in 2015 to the marketing in Italy, and later Mexico, of 'Latte Nobile' (Noble Milk), produced by cows fed primarily grass and hay. The protocol requires at least 150 days per year of grazing and 70% DMI from species-diverse pasture and hay throughout the year, with up to 30% DMI from grains and concentrates allowed. The milk fat must contain at least 0.25% of total CLA and 0.5% of total n-3 FA, and the LA/ALA ratio must be lower than 4-year round. However, the project was not further developed commercially. Consumer perception and willingness to pay for sustainable and traceable milk production models was explored (Peira et al., 2020). A new initiative 'Filierba' is being developed (G. Lombardi, pers. comm.). In France, there are various initiatives beyond protected designation of origin (PDO) and protected geographical indication (PGI) that enable farmers to get extra payment for their milk (M. Jouvin, pers. comm.).

The Wisconsin-based cooperative CROPP, a leading U.S. supplier of organic milk, launched a new, whole milk, organic product called Grassmilk in 2011. This milk comes from cows fed a nearly 100% forage-based diet. Besides prohibiting grain in cow rations, standards for pasture access, supplemental feeds, and animal care were set. Farmers in the Grassmilk program receive a price premium of ~15% compared to the organic milk price. CROPP closely monitors the FA content of raw Grassmilk to assure compliance with its minimum requirements of (1) 39–41 mg total n-3 FA/100 g of milk, depending on geographical region, (2) 26.6–32.8 mg total CLA/100 g of milk, and (3) an n-6/n-3 ratio ≤ 1.2. Benbrook et al. (2018) compared milk samples from cows fed a nearly 100% forage-based diet (Grassmilk) and compared it to profiles from milk from cows under conventional (TMR), and organic management where at least 30% of DMI must come from grazing during a minimum of 120 days per year. The n-6/n-3 rations were, respectively, 0.95, 5.77, and 2.28. C.M. Benbrook (pers. comm. 5 Oct. 2025) mentioned that the grassmilk market has taken off in the U.S.. These are some examples to illustrate the effects of pasture grazing on milk quality in combination with rewards for farmers. In the long run, increased use of pasture can be economically, environmentally, and socially sustainable if forage-use efficiency is high and consumers continue to pay a premium for grass-fed dairy products.

9 Conclusion

This review showed consistent effects on milk FA of feeding system, grazing system, herbage allowance, grassland management, and herbage species. The fate of FA in the digestive tract was discussed and effects of plant species on biohydrogenation and transfer efficiency of PUFA from feed to milk were illustrated. Seasonal variation in herbage quality and FA composition in plants, feed compounds, and milk was illustrated. Scope for management options at the farm level aiming to increase the PUFA concentration in milk was provided. Processes at the plant–animal interface during grazing and implications for PUFA intake were indicated. Insights in the various aspects of grazing are useful for ongoing discussions in society and in the dairy industry: there are trends in animal genetics and increased herd size, TMR feeding, and automated milking systems. These developments will enhance further reduction of grazing in Europe and elsewhere. Reduction of grazing and of fresh herbage intake by dairy cattle can affect milk FA profiles toward less PUFA and more *trans*-10 FA and reduce naturally occurring fat-soluble vitamins and antioxidants in dairy products.

10 Abbreviations

ALA, α-linolenic acid; CLA, conjugated linoleic acid; CVD, cardiovascular disease; DM, dry matter; DMI, dry matter intake; FA, fatty acid; F:C, forage:concentrate; HDL, high-density lipoprotein; LA, linoleic acid; LDL, low-density lipoprotein; MUFA, mono-unsaturated fatty acid; PA, palmitic acid; PUFA, poly-unsaturated fatty acid; RA, rumenic acid; SFA, saturated fatty acid; SSH, sward surface height; TMR, total mixed ration; VA, vaccenic acid.

11 References

Adler, S. A., et al., Fatty acid composition, fat-soluble vitamin concentrations and oxidative stability in bovine milk produced on two pastures with different botanical composition. *Livestock Sci.* 2013, 154, 93–102.

Agenas, S., et al., Effects of turnout to pasture and dietary fat supplementation on milk fat composition and conjugated linoleic acid in dairy cows. *Anim. Sci.* 2002, 52, 25–33.

Aldai, N., et al., What are the *trans* fatty acids issues in foods after discontinuation of industrially produced *trans* fats? Ruminant products, vegetable oils, and synthetic supplements. *Eur. J. Lipid Sci. Technol.* 2013, 115, 1378–1401.

Alothman, M., et al., The "grass-fed" milk story: understanding the impact of pasture feeding on the composition and quality of bovine milk. *Foods* 2019, 8, 350.

Alves, S. P., et al., *Trans*-10 18:1 in ruminant meats: a review. *Lipids* 2021, 56(6), 539–562. https://doi.org/10.1002/lipd.12324

Ascherio, A., et al., Dietary fat and risk of coronary heart disease in men: cohort follow up study in the United States. *Brit. Med. J.* 1996, 313, 84–90.

Balcells, J., et al., Effect of an extract of plant flavonoids (Bioflavex) on rumen fermentation and performance in heifers fed high-concentrate diets. *J. Anim. Sci.* 2012, 90, 4975–4984.

Bargo, F., et al., Supplementing total mixed rations with pasture increase the content of conjugated linoleic acid in milk. *An. Feed Sci. Tech.* 2006, 131, 226–240.

Bauchart, D., et al., Long-chain fatty acid digestion in lactating cows fed fresh grass from spring to autumn. *Can. J. An. Sci.* 1984, Suppl., 330–331.

Bauman, D. E. and Griinari, J. M., Regulation and nutritional manipulation of milk fat: low-fat milk syndrome. *Livest. Prod. Sci.* 2001, 70, 15–29.

Benbrook, C. M., et al. Enhancing the fatty acid profile of milk through forage-based rations, with nutrition modeling of diet outcomes. *Food Sci Nutr.* 2018; 6: 681–700. https://doi.org/10.1002/fsn3.610

Butler, G., et al., Fatty acid and fat-soluble antioxidant concentrations in milk from high- and low-input conventional and organic systems: Seasonal variation. *J. Sci. Food Agric.* 2008, 88, 1431–1441.

Butler, G., et al., Fat composition of organic and conventional retail milk in northeast England. *J. Dairy Sci.* 2011, 94, 24–36.

Boufaïed, H., et al.., Fatty acids in forages. I. Factors affecting concentrations. *Can. J. Anim. Sci.* 2003, 83, 501–511.

Bryant, R. H., et al., Can alternative forages substantially reduce N leaching? findings from a review and associated modelling. *N.Z. J. Agri. Res.* 2019, 63(1), 3–28. https://doi.org/10.1080/00288233.2019.1680395

Cabiddu. A, et al. A quantitative review of on-farm feeding practices to enhance the quality of grassland-based ruminant dairy and meat products. Animal 2022, 16, 100375. https://doi.org/10.1016/j.animal.2021.100375

Capuano E., et al., Verification of fresh grass feeding, pasture grazing and organic farming by cows farm milk fatty acid profile. *Food Chem.* 2014, 164, 234–241.

Charlton, G. L. and Rutter S. M. 2017. The behaviour of housed dairy cattle with and without pasture access: a review. *Appl. Anim. Behav. Sci.* 192, 2–9.

Chilliard, Y., et al., Effect of different types of forages, animal fat or marine oils in cow's diet on milk fat secretion and composition, especially conjugated linoleic acid (CLA) and polyunsaturated fatty acids. *Livest Prod. Sci.* 2001, 70, 31–48.

Chilliard Y., et al., Ruminant milk fat plasticity: nutritional control of saturated, polyunsaturated, trans and conjugated fatty acids. *Ann. Zootech.* 2000;49:181–205. https://doi.org/10.1051/animres:2000117

Chilliard, Y., et al., rumen biohydrogenation and nutritional quality of cow and goat milk fat. *Eur. J. Lipid Sci. Tech.* 2007, 109, 828–855.

Choi, N., et al.., Characterizations of environmental factors in conjugated linoleic acid production by mixed rumen bacteria. *J. Agric. Food Chem.* 2009, 57, 9263–9267.

Chouinard, P. Y., et al.. Conjugated linoleic acids alter milk fatty acid composition and inhibit milk fat secretion in dairy cows. *J. Nutr.* 1999;129:1579–1584. https://doi.org/10.1093/jn/129.8.1579

Clapham, W. M., et al., Fatty acid composition of traditional and novel forage species. *J. Agric. Food Chem.* 2005, 53, 10068–10073.

Collomb, M., et al., Correlation between fatty acids in cows' milk fat produced in the lowlands, mountains and highlands of Switzerland and botanical composition of the fodder. *Int. Dairy J.* 2002, 12, 661–666.

Connolly, C., et al., The impact of varying pasture levels on the metabolomic profile of bovine ruminal fluid. *Metabolites* 2024, 14, 476. https://doi.org/10.3390/metabo14090476

Dewhurst, R. J. and King, P. J., 1998. Effects of extended wilting, shading and chemical additives on the fatty acids in laboratory grass silages. *Grass Forage Sci.* 1998, 53, 219–224.

Dewhurst, R. J., et al., Influence of species, cutting date and cutting interval on the fatty acid composition of grasses. *Grass Forage Sci.* 2001, 56, 68–74.

Dewhurst, R. J., et al., Comparison of grass and legume silages for milk production. 1. Production responses with different levels of concentrate. *J. Dairy Sci.* 2003, 86(8), 2598–2611.

Dewhurst, R. J., et al., Increasing the concentrations of beneficial polyunsaturated fatty acids in milk produced by dairy cows in high-forage systems. *Anim. Feed Sci. Tech.* 2006, 131, 168–206.

Dhiman, T. R., et al., Conjugated linoleic acid content of milk from cows fed different diets. *J. Dairy Sci.* 1999, 82, 2146–2156.

Dillon, P., et al., Optimising financial return from grazing in temperate pastures. In *Utilisation of Grazed Grass in Temperate Animal Systems*; Murphy, J.J., Ed.; Wageningen Academic Publishers: Wageningen, The Netherlands, 2005; pp. 131–147.

Dilzer, A. and Park, Y. Implication of conjugated linoleic acid (CLA) in human health. *Crit. Rev. Food Sci. Nutr.* 2012, 52, 488–513. https://doi.org/10.1080/10408398.2010.501409

Doreau, M. and Poncet, C., Ruminal biohydrogenation of fatty acids originating from fresh or preserved grass. *Repr. Nutr. Dev.* 2000, 40, 201.

Downey, L. and Doyle, P. T., Cow nutrition and dairy product manufacture – implications of seasonal pasture-based milk production systems. *Austr. J. Dairy Technol.* 2007, 62, 3–11.

Elgersma, A., 2015. Grazing increases the unsaturated fatty acid concentration of milk from grass-fed cows: a review of the contributing factors, challenges and future perspectives. *Eur. J. Lipid Sci. Technol.* 2015, 117, 1345–1369.

Elgersma, A., et al., 2006a. Modifying milk composition through forage – a review. *Animal Feed Sci. Tech.* 131, 207–225.

Elgersma, A., et al., 2006b. Lipids in herbage. In: *Fresh Herbage for Dairy Cattle*. Eds A. Elgersma, et al.. Springer: Wageningen, The Netherlands, 2006; pp. 175–194.

Elgersma, A., et al., Influence of cultivar and cutting date on the fatty acid composition of perennial ryegrass. *Grass and Forage Sci.* 2003a, 58, 323–331.

Elgersma, A., et al., Comparison of the fatty acid composition of fresh and ensiled perennial ryegrass (*Lolium perenne* L.), affected by cultivar and regrowth interval. *Anim. Feed Sci. Technol.* 2003b, 108, 191–205.

Elgersma, A., et al., Quick changes in milk fat composition after transition from fresh grass to a silage diet. *Anim. Feed Sci. Techn.* 2004, 117, 13–27.

Elgersma, A., et al., Effects of the grazed horizon in perennial ryegrass swards on the conjugated linoleic acid concentration in milk of dairy cows. *Grassl. Sci. Europe* 2008, 13, 388–390.

Elgersma, A., et al., Fatty acids, α-tocopherol, β-carotene and lutein contents in forage legumes, forbs and a grass – clover mixture. *J. Agric. Food Chem.* 2013, 61, 11913–11920.

Elgersma, A., et al., Changes in vitamin and fatty acid contents in grass – red clover herbage after cutting. *Grassl. Sci. Eur.* 2015, 20, 212–214.

Forbes, J. M., *Voluntary Feed Intake and Diet Selection in Farm Animals*. CAB International, Wallingford, UK. Fulkerson, W. J. and Donaghy, D. J. 1995.

Fraser, M. D., et al., Animal board invited review: grassland-based livestock farming and biodiversity. *Animal*. 2022, 16(12), 100671. https://doi.org/10.1016/j.animal.2022.100671

Fuentes, M. C., et al., Effect of pH and level of concentrate in the diet on the production of biohydrogenation intermediates in a dual-flow continuous culture. *J. Dairy Sci.* 2009, 92, 4456–4466.

Gayet-Boyer, C., et al., Is there a linear relationship between the dose of ruminant trans-fatty acids and cardiovascular risk markers in healthy subjects: results from a systematic review and meta-regression of randomised clinical trials. *Br. J. Nutr.* 2014, 114, 1914–1922. https://doi.org/10.1017/S0007114514002578

Gibb, M. J., et al., Effect of sward surface height on intake and grazing behaviour by lactating Holstein Friesian cows. *Grass Forage Sci.* 1997, 52, 309–321.

Gregorini, P., et al., Review: behavior and daily grazing patterns of cattle. *Prof. Anim. Sci.* 2006, 22, 201–209.

Griinari, J. M., et al., Effect of dietary sunflower oil and pasture forage maturity on conjugated linoleic acid (CLA) content in milk fat from lactating dairy cows. *J. Dairy Sci.* 1998, 81, 300.

Grings, E. E. 2020. Cattle Grazing Behavior and Diet Selection – SDSU Extension Chapter 49. Extension.sds.edu, 7 pp.

Grundy, S. M. and Vega, G. L., 1998. Plasma cholesterol responsiveness to saturated fatty acids. *Am. J. Clin. Nutr.* 47, 822–824.

Halfa, A. N., et al., Sustainability of US organic beef and dairy production systems: soil, plant and cattle interactions. Review. *Sustainability* 2013, 5, 3009–3034.

Hall, A. J., Seasonal and regional variation in the fatty acid composition of milk fat. *Dairy Ind.* 1970, 35, 20–24.

den Hartigh, L. J. 2019. Conjugated linoleic acid effects on cancer, obesity, and atherosclerosis: a review of pre-clinical and human trials with current perspectives. *Nutrients* 11, 370. https://doi.org/10.3390/nu11020370

Harvatine, K. J. and Allen, M. S., Fat supplements affect fractional rates of ruminal fatty acid biohydrogenation and passage in dairy cows. *J. Nutr.* 2006, 136, 677–685.

Havel, R. J., *Milk Fat Consumption and Human Health: Recent NIH and Other American Governmental Recommendations*. Welch, R. A. S., et al. Eds., Milk Composition, Production And Biotechnology. CAB International, Wallingford, UK, 1997, pp. 13–22.

Hawke, J. C., Lipids. In *Chemistry and Biochemistry of Herbage*. G. W. Butler, R. W. Bailey, Eds. London: Academic Press, 1973, pp. 213–263.

Hibbeln, J. R., et al., Healthy intakes of ω-3 and ω-6 fatty acids: estimations considering worldwide diversity. *Am. J. Clin. Nut.* 2006, 83, 1483S–1493S.

Hirahatake, K. M., et al., Potential cardiometabolic health benefits of full-fat dairy: the evidence base. *Adv. Nutr.* 2020, 11, 533–547.

Hodgson, J., *Grazing Management: Science into Practice*. Longman Handbooks in Agriculture, Longman, Harlow (Essex, UK), 1990, p. 203.

Hu, F. B., et al., Types of dietary fat and risk of coronary heart disease: a critical review. *J. Am. Coll. Nutr.* 2001, 20, 5–19.

Hugly, S. and Somerville, C., A role for membrane lipid polyunsaturation in chloroplast biogenesis at low temperature. *Plant Physiol.* 1992, 99, 197–202.

Hutton, K., et al., The variation throughout the year in the fatty acid composition of milk fat from 2 dairy herds. *J. Dairy Res.* 1969, 36, 103–113.

Jenkins, T. C., et al., Board-invited review: recent advances in biohydrogenation of unsaturated fatty acids within the rumen microbial ecosystem. *J. Anim. Sci.* 2008, 86, 397–412.

Jensen, R. G., The composition of bovine milk lipids: January 1995 to December 2000. *J. Dairy Sci.*, 2002, 85, 295–350.

Kalač, P. and Samkova, E., The effects of feeding various forages on fatty acid composition of bovine milk: a review. *Czech J. Anim. Sci.* 2010, 55, 521–537.

Kelly, M. L, et al., Effect of intake of pasture on concentrations of conjugated linoleic acid in milk of lactating cows. *J. Dairy Sci.* 1998, 81, 1630–1636.

Kennelly, J. J., The fatty acid composition of milk fat as influenced by feeding oilseeds. *Bull. Int. Dairy Fed.* 2001, 366, 28–36.

Kearns, M., et al., Fatty acid, tocopherol and carotenoid content of botanically diverse pastures rotationally grazed by ruminant livestock in Ireland. J. Agric. *Food Research. 2025*, 22, 102050. https://doi.org/10.1016/j.jafr.2025.102050

Khan, N. A., et al., 2011. Changes in fatty acid content and composition in silage maize during grain filling. *J. Sci. Food Agric.* 2011, 91, 1041–1049.

Khanal, R. C., et al., Changes in fatty acid composition of milk from lactating dairy cows during transition to and from pasture. *Livest. Sci.* 2008, 114, 164–175.

Khanal, R. C. and Ohlson, K. C., Factors affecting conjugated linoleic acid (CLA) content in milk, meat and egg: a review. *Pakistan J. Nutr.* 2004, 3, 82–98.

Khiaosa-ard, R., et al., Transfer of linoleic and linolenic acid from feed to milk in cows fed isoenergetic diets differing in proportion and origin of concentrates and roughages. *J. Dairy Res.* 2010, 77, 331–336.

Klieve, A. V., et al., Establishing populations of *Megasphaera elsdenii* YE 34 and *Butyrivibrio fibrisolvens* YE 44 in the rumen of cattle fed high grain diets. *J. Appl. Microbiol.* 2003, 95, 621–630.

Knutsen, T. M., et al. Unravelling genetic variation underlying de novo-synthesis of bovine milk fatty acids. *Sci. Rep.* 2018, 8, 2179. https://doi.org/10.1038/s41598-018-20476-0

Kolver, E. S. and De Veth, M. J. Prediction of ruminal pH from pasture-based diets. *J. Dairy Sci.* 2002, 85, 1255–1266. https://doi.org/10.3168/jds.S0022-0302(02)74190-8

Laca, E. A., et al., Mechanisms of handling time and intake rate of a large mammalian grazer. *Appl. Anim. Behav. Sci.* 1994, 39, 3–19.

Laca, E. A., et al., Effects of sward height and bulk density on bite dimensions of cattle grazing homogeneous swards. *Grass Forage Sci.*, 1992, 47, 91–102.

Lacroix E, et al., (2012) Randomized controlled study of the effect of a butter naturally enriched in trans fatty acids on blood lipids in healthy women. *Am J Clin Nutr*, 95, pp. 318-325

Lashkari, S., et al., (2019). Milk fatty acid profile of dairy cows is affected by forage species, parity, and milking time. *70th Annual Meeting of the European Federation of Animal Science*, Ghent, Belgium. Wageningen Academic Publishers, 713.

Larsson, S. C., et al. High-fat dairy food and conjugated linoleic acid intakes in relation to colorectal cancer incidence in the Swedish Mammography Cohort. *Am. J. Clin. Nut.* 2005, 82, 894–900.

Larsen, M. K., et al. Fatty acid, tocopherol and carotenoid content in herbage and milk affected by sward composition and season of grazing. *J. Sci. Food Agric.* 2012, 92(14), 2891–2898.

Larsen, M. K., et al., Milk quality as affected by feeding regimens in a country with climatic variation. *J. Dairy Sci.* 2010, 93, 2863–2873.

Lavillonnière, F., et al., Analysis of conjugated linoleic acid isomers and content in French cheeses. *J. Am. Oil Chem. Soc*. 1998, 75, 343–352.

Lee, M. R. F., et al., Effect of fish oil on ruminal biohydrogenation of C18 unsaturated fatty acids in steers fed grass or red clover silages. Animal 2008, 2, 1859–1869.

Lock, A. L. and Bauman, D. E., Modifying milk fat composition of dairy cows to enhance fatty acids beneficial to human health. *Lipids*. 2004, 39, 1197–1206. https://doi.org/10.1007/s11745-004-1348-6

Lock, A. L. and Garnsworthy, P. C., Seasonal variation in milk conjugated linoleic acid and Δ^9 – desaturase activity in dairy cows. *Livestock Prod. Sci*. 2003, 79, 47–59.

Lombardi, G., et al. The Piedmont Noble Milk as a tool to improve the competitiveness of mountain farms. *J. Nutr. Ecol. Food Res*. 2014, 2, 233–236. https://doi.org/10.1166/jnef.2014.1081

Lourenço, M., et al., Influence of different dietary forages on the fatty acid composition of rumen digesta as well as ruminant meat and milk. *Anim. Feed Sci. Technol*. 2008, 145, 418–437.

Mangwe, M., et al. Rumen fermentation and fatty acid composition of milk of mid-lactating dairy cows grazing chicory and ryegrass. *Animals* 2020 10, 169. https://doi.org/10.3390/ani10010169

Månsson, H. L., 2008. Fatty acids in bovine milk fat. *Food Nutr. Res*. 2008. https://doi.org/10.3402/fnr.v52i0.1821

Mattsson, S., Poly-unsaturated fatty acids in butter and their influence on the oxidation of butter. Methods of analysis and preliminary results. In: *Proceedings of the XIIth International Dairy Congress*, Section II. Stockholm, 1949, pp. 308–324.

McGilloway, D. A., et al., The relationship between level of sward height reduction in a rotationally grazed sward and short-term intake rates of dairy cows. *Grass Forage Sci*. 1999, 54, 116–126.

Molkentin, J., Authentication of organic milk using δ13C and the α-linolenic acid content of milk fat. *J. Agric. Food Chem*. 2009, 57, 785–790. https://pubs.acs.org/doi/abs/10.1021/jf8022029

Mollica, P. M., et al., 2021. Milk fatty acid profiles in different animal species: focus on the potential effect of selected PUFAs on metabolism and brain functions. *Nutrients* 2021, 13(4), 1111. https://doi.org/10.3390/nu13041111

National Research Council (US) Committee on Technological Options to Improve the Nutritional Attributes of Animal Products. *Designing Foods: Animal Product Options in the Marketplace*. National Academies Press (US); 1988. Factors Affecting the Composition of Milk from Dairy Cows, Washington, DC. https://www.ncbi.nlm.nih.gov/books/NBK218193/

Nielsen, T. S., et al., Effect of silage type and concentrate level on conjugated linoleic acids, *trans*-C18:1 isomers and fat content in milk from dairy cows. *Reprod. Nutr. Dev*. 2006, 46, 699–712.

Offer, N. W., Effects of cutting and ensiling grass on levels of CLA in bovine milk. In: *Proceedings of the XIIIth International Silage Conference*. Gechie, L.M. and Thomas, C. Eds. Auchincruive, Scotland. Scottish Agricultural College, UK, 2002, pp. 16–17.

Orr, R. J., et al., Matching grass supply to grazing patterns for dairy cows. *Grass Forage Sci*. 2001, 56, 352–361.

Palladino, R. A., et al., Fatty acid composition and nutritive value of twelve cultivars of perennial ryegrass. *Grass Forage Sci*. 2009, 64, 219–226.

Palmquist, D. L., Milk fat: origin of fatty acids and influence of nutritional factors thereon. In: *Advanced Dairy Chemistry*, Vol. 2. Fox, P. F. and McSweeney, P. L. H Eds.. Springer Science, New York, 2006, pp. 43–92.

Palmquist, D. L., et al., Feed and animal factors influencing milk fat composition. *J. Dairy Sci.* 1993, 76, 1753–1771.

Palmquist, D. L. and Jenkins, T. C., Fat in lactation rations. Review. *J. Dairy Sci.* 1980, 63, 1–14.

Palmquist, D. L., et al., Biosynthesis of conjugated linoleic acid in ruminants and humans. *Adv. Food Nutr. Res.* 2005, 50, 179–217.

Palupi, E., et al., Comparison of nutritional quality between conventional and organic dairy products: a meta-analysis. *J. Sci. Food Agric.* 2012, 92, 2774–2781.

Parodi, P. Milk fat in human nutrition. *Aus. J Dairy Tech.* 2004, 59, 3–59.

Parsons, A. J. and Chapman, D. F., The principles of pasture growth and utilisation. In: *Grass, Its Production and Utilisation*. Hopkins, A. Ed. Blackwell Science, Oxford, UK, 2000, pp. 31–89.

Peira, G., et al., Grass-Fed Milk Perception: Profiling Italian Consumer. *Sustainability 2020*, 12, 10348. https://doi.org/10.3390/su122410348

Petersen, M. B., et al., Herb feeding increases n-3 and n-6 fatty acids in cow milk. *Livest. Sci.* 2011, 141, 90–94.

Petersen, M., B. and Jensen, S. K., Bohydrogenation of fatty acids is dependent on plant species and feeding regimen of dairy cows. *J. Agric. Food Chem.* 2014, 62, 3570–3576.

Phillips, C. J. C. and Leaver, J. D., *Seasonal and Diurnal Variation in Grazing Behaviour of Dairy Cows*. Grazing, Hurley. 1986, pp. 98–104.

Prati, S., et al., Composition and content of seen flavonoids in forages and grain legume crops. *J. Sep. Sci.* 2007, 30, 491–501.

Precht, D. and Molkentin, J., *Trans* fatty acids: implications for health, analytical methods, incidence in edible fats and intake. *Die Nahrung* 1995, 39, 343–374.

Precht, D. and Molkentin, J., Frequency distributions of conjugated linoleic acid and *trans* fatty acid contents in European bovine milk fats. *Michwissenschaft* 2000, 55, 687–691.

Ranst, G., et al, Influence of herbage species, cultivar and cutting date on fatty acid composition of herbage and lipid metabolism during ensiling. *Grass Forage Sci.* 2009, 64, 169–207.

Reklewska, B., et al., [Concentration of milk functional components in Black-and-White cows, depending on the season and feeding system] in Polish, *Zeszyty Naukowe Przeglądu Hodowlanego* 2003, 68, 85–98.

Riel, R. R., Physico-chemical characteristics of Canadian milk fat. Unsaturated fatty acids. *J. Dairy Sci.* 1963, 46, 102–106.

Rodríguez, R., et al. Milk and meat fatty acids from sheep fed a plantain-chicory mixture or a grass-based permanent sward. *Anim. Int. J. Anim. Biosci.* 2020. https://doi.org/10.1017/S1751731119002611

Schroeder, G. F., et al., Effects of fat supplementation on milk production and composition by dairy cows on pasture: a review. *Livest. Prod. Sci.* 2004, 86, 1–18.

Schwendel, B. H., et al., Invited review: organic and conventionally produced milk - An evaluation of factors influencing milk composition. *J. Dairy Sci.* 2015, 98, 721–746.

Shingfield, K. J., et al., Dietary fish oil supplements modify ruminal biohydrogenation, alter the flow of fatty acids at the omasum, and induce changes in the ruminal *Butyrivibrio* population in lactating cows. *J. Nutr.* 2012, 142, 1437–1448.

Simopoulos, A. P., Evolutionary aspects of diet, the omega-6/omega-3 ratio and genetic variation: Nutritional implications for chronic diseases. *Biomedicine and Pharmacotherapy*, 2006, 60, 502–507. https://doi.org/10.1016/j.biopha.2006.07.080

Slots, T., et al., Potentials to differentiate milk composition by different feeding strategies. *J. Dairy Sci.* 2009, 92, 2057–2066. https://doi.org/10.3168/jds.2008-1392

Smit, L. A., et al. Conjugated linoleic acid in adipose tissue and risk of myocardial infarction. *Am. J. Clin. Nut.* 2010, 92, 34–40. https://doi.org/10.3945/ajcn.2010.29524

Soder, K. J., et al., Intake and performance of lactating cows grazing diverse forage mixtures. *J. Dairy Sci.* 2006, 89, 2158–2167.

Średnicka-Tober, D., et al. Higher PUFA and n-3 PUFA, conjugated linoleic acid, α-tocopherol and iron, but lower iodine and selenium concentrations in organic milk: a systematic literature review and meta- and redundancy analyses. *Br. J. Nutr.* 2016, 115, 1043–1060. https://doi.org/10.1017/s0007114516000349

Stadhouders, J. -and Mulder, H., The composition of Dutch butterfat. 1. Seasonal variations in the unsaturated fatty acid composition of butter fat. *Neth. Milk Dairy J.* 1955, 9, 182–193.

Stanton, C., et al., Dietary influences on bovine milk *cis*-9, *trans*-11-conjugated linoleic acid content. *J. Food Sci.* 1997, 62, 1083–1086.

Stender, S. and Dyerberg, J., The influence of trans fatty acids on health. *Ann. Nutr. Metab.* 2004, 48, 61–66. https://doi.org/10.1159/000075591

Stockdale, C. R., et al., Influence of pasture and concentrates in the diet of grazing dairy cows on the fatty acid composition of milk. *J. Dairy Res.* 2003, 70, 267–276.

Tamminga, S., Effects of feeds, feed composition and feed strategy on fat content and fatty acid composition in milk. *Bull. Int. Dairy Fed.* 2001, 366, 15–24.

Taweel, H. Z., et al., Grazing behavior, intake, rumen function and milk production of dairy cows offered *Lolium perenne* containing different levels of water-soluble carbohydrates. *Livest. Sci.* 2006, 102, 33–41.

Thomas, H., The role of polyunsaturated fatty acids in senescence. *J. Plant Physiol.* 1986, 123, 97–105.

Timlin, M., et al., Impact of varying levels of pasture allowance on the nutritional quality and functionality of milk throughout lactation. *J. Dairy Sci.* 2023, 106(10), 6597–6622. https://doi.org/10.3168/jds.2022-22921

Timlin, M., et al. (2024). The impact of seasonal calving systems with varying pasture allowance on Cheddar cheese composition, nutritional quality and ripening characteristics. Journal Dairy Science. https://doi.org/10.3168/jds.2024-24745

Toral, P. G., et al., Review: modulating ruminal lipid metabolism to improve the fatty acid composition of meat and milk. Challenges and opportunities. *Animal.* 2018, 12, S272–S281. https://doi.org/10.1017/S1751731118001994

Turpeinen, A. M., et al. Bioconversion of vaccenic acid to conjugated linoleic acid in humans. *Am. J. Clin. Nut.* 2002, 76, 504–510. https://doi.org/10.1093/ajcn/76.3.504

Tyburczy, C., et al. Uptake and utilization of *Trans* octadecenoic acids in lactating dairy cows. *J. Dairy Sci.* 2008, 91, 3850–3861. https://doi.org/10.3168/jds.2007-0893

Van den Pol-van Dasselaar, A., et al. Grazing of dairy cows in Europe—an in-depth analysis based on the perception of grassland experts. *Sustainability.* 2020, 14, 1102–1109. https://doi.org/10.3390/su12031098

Vargas-Bello-Pérez, E. and Garnsworthy, P. C. *Trans* fatty acids and their role in milk of dairy cows. *Ciencia e Investigación Agraria*, 2013, 40, 449–473.

Vlaeminck, B., et al. Factors affecting odd- and branched-chain fatty acids in milk: A review. *Anim. Feed Sci. Technol.* 2006, 131, 389–417. https://doi.org/10.1016/j.anifeedsci.2006.06.017

Warner, D., et al., Fatty acid composition of forage herb species. *Grassl. Sci. Eur.* 2010, 15, 491–493.

Whetsell, M. and Rayburn, E. Variation in fatty acids concentration in grasses, legumes, and forbs in the Allegheny plateau. *Agronomy* 2022, 12, 1693. https://doi.org/10.3390/agronomy12071693

White, S. L., et al., Comparison of fatty acid content of milk from Jersey and Holstein cows consuming pasture or a total mixed ration. *J. Dairy Sci.* 2001, 84, 2295–2301.

Wiking, L., et al., Effect of grazing fresh legumes or feeding silage on fatty acids and enzymes involved in the synthesis of milk fat In dairy cows. *J. Dairy Res.* 2010, 77, 337–342.

Williams, C. M., Dietary fatty acids and human health. *Ann. Zootech*. 2000, 49, 165–180.

Witkowska, I. M., et al., Effects of nitrogen rate and regrowth interval on perennial ryegrass fatty acid content during the growing season. *Agron. J.* 2008, 100, 1371 – 1379.

Wyss, U., Influence of conservation method on fatty acid composition of herbages of a permanent grassland meadow. *Grassl. Sci. Eur.* 2012, 17, 323–325.

Wyss, U. and Collomb, M., Fatty acid composition of different grassland species. *Grassland Sci. Eur*. 2010, 15, 631–633.

Chapter 12

Grasslands and One Health: the human health component

Frédéric Leroy, Vrije Universiteit Brussel, Belgium; Peter J. Ballerstedt, Grass Based Health LLC, USA; and Stefaan De Smet, Ghent University, Belgium

1 Introduction

During a complex evolutionary niche transition that occurred over a period of millions of years, the human lineage emerged from the African grasslands, within an idiosyncratic ecological and dietary context mostly unknown to its forest-inhabiting ancestors (Ulijaszek et al., 2012; Mann, 2018; Pontzer and Wood, 2021; Leroy et al., 2023). The exact timeline and the specific hominid species involved in this transition have not been fully elucidated, but it is known that early hominins already displayed a flexible mix of arboreal and bipedal adaptations. Approximately 2–3 million years ago, the *Homo* genus made a decisive move into the savanna landscape, acquiring the physical capacity to cover extensive distances. Within this ecosystem, tool usage became a lot more advanced, whilst enabling some degree of manipulation of the environment and more efficient access to nutritional resources. The resulting dietary transition, away from fibrous plant materials toward foods with a higher nutritional density that was derived from the grassland niche, translated into encephalization (i.e. an increase of brain size) and a shrinkage of the digestive

http://dx.doi.org/10.19103/AS.2025.0154.19

system, some 1–2 million years ago (Aiello and Wheeler, 1995). This led to enhanced cognitive capacity, which became a defining feature of *Homo* and contributed to the development of sophisticated social structures and various adaptations to dynamic environments.

When *Homo erectus* began to migrate out of Africa, a higher level of ecological flexibility developed. This evolution towards niche diversification reached its peak with the anatomically modern *Homo sapiens*, around 200 000–300 000 years ago. As *H. sapiens* conquered the globe, they encountered a variety of environments beyond the savanna homeland, adapting their diets accordingly. The most drastic dietary changes, leaving behind the ancestral hunter-gatherer model that defined *Homo* for a few million years, occurred with the adoption of agriculture during the Neolithic era, the transition of early agrarian societies to more structured urban environments, the industrialization of the food chain in the nineteenth century, and the transition to the current globalized diets dominated by commodity foods, often highly processed in nature. With these fundamental changes in dietary composition, the status of human health also went through various phases (Ulijaszek et al., 2012).

In this chapter, an admittedly sketchy overview of the main shifts in diet–health interactions during human evolution and (pre)history will be provided, with a particular focus on the role of grassland-derived foods. Whereas these foods were paramount for *Homo* during most of its existence on earth as hunter-gatherers, and perceived as such, their current role in the human diet is heavily debated, generating confusion where there should be none.

2 A species-adapted diet for humans: the evolutionary argument

Even if there is no singular true human diet (Pontzer et al., 2018), it is nonetheless plausible that dietary solutions that are in line with evolutionary requirements are more likely to result in a robust health status than dietary options that provide an evolutionary mismatch (Ulijaszek et al., 2012). Although it is theoretically possible to design a *de novo* diet that would still result in beneficial health outcomes, such approaches tend to be nutrient-centric and reductionist, ignoring the intricacies of physiological optimization obtained over long-stretched periods of finetuning between diverse food and dietary matrices and evolving human bodies. In other words, there is no good reason to claim that humans deviate from the general rule that a species-adapted diet is the best physiological option for animal health (Leroy and Cofnas, 2020). The question thus becomes: can we define such a species-adapted diet for humans with enough confidence, even if only with broad strokes or by identifying some basic dietary and health boundaries?

What is established beyond doubt is that *H. sapiens* evolved as a habitual rather than facultative eater of meat and animal fat (Henneberg et al., 1998; Leroy et al., 2023). Even if some still argue that modern humans are naturally herbivorous based on a phylogenetic relationship with other apes, such arguments overlook the existence of a major evolutionary split between both groups (Barnard and Leroy, 2020). Hominin ancestors adapted to derive nutrients not only from plants but also substantial amounts from animal-source foods. Whereas some members of *H. sapiens* may sustain health on plant-based diets for extended periods, it may not be generalisable at species level owing to interindividual differences in nutritional preferences and needs, as well as genetic variations (e.g. linked to lipid metabolism and brain function; Yaseen et al., 2023). Moreover, strict vegetarianism lacks an evolutionary precedent (O'Hearn et al., 2022) and hunter-gatherers are known to obtain at least 1/4 of their calories from animals (Cordain et al., 2000).

Multiple lines of evidence suggest that a significant intake of animal-source foods throughout most of human evolution has played a pivotal role in various morphological and physiological changes (Ben-Dor et al., 2021). This includes cranial-dental modifications, adaptation of erect posture, reproductive characteristics, extended lifespan, a lower stomach pH, a lower *in vivo* conversion rate for precursors of nutrients that are naturally present in animal source foods (e.g. choline, taurine, and essential long-chain fatty acids), a preferential absorption of haem iron over ionic forms, the inability to absorb sufficient vitamin B12, a reduced fermentative capacity in the gut, and the expansion of brain size (Ulijaszek et al., 2012; Mann, 2018; Leroy et al., 2023). These adaptations are further supported by anatomical and genetic changes indicative of ecological adaptations related to the pursuit and hunting of animals, endurance running, heat regulation, enhanced vision, respiratory efficiency, throwing, and tool use.

Based on these substantial effects of evolutionary diets on human bodies and functions, ancestral dietary models of the hunter-gatherer type can at least offer clues of what is to be considered as 'suitable nutrition'. We grant that over a shorter timeline, of about 10 000 years, other valuable dietary options have been developed based on agricultural outputs, which to some degree have also introduced adaptations (e.g. lactase persistence, the ability to digest lactose into adulthood, as well as increased salivary alpha-amylase, an enzyme produced by the salivary glands which plays a role in the initial stages of carbohydrate digestion in the mouth). However, such agricultural diets may sometimes cause protein and micronutrient deficiencies, especially when heavily based on cereal staples, because of the low content and/or bioavailability of some important nutrients (e.g. lysine and iron) and the presence of anti-nutritional factors (e.g. phytates). Another point of attention is that diets that are evolutionary tuned to optimal metabolic health during the reproductive years are not *necessarily* optimal for longevity. That said, the average modal lifespan for hunter-gatherers has been estimated at 68–78 years, i.e. when abstracting causes of early death like infant mortality (Gurven and Kaplan, 2007). Another reason why the hypothesis of evolutionary relaxation of optimal metabolic health at old age may not hold up in the case of humans is that fitness of ancestral human communities has likely been heavily dependent on contributions from elder clan members, for instance for providing alloparental care (contributions to childrearing provided by individuals other than parents).

3 The grassland-derived diet of hunter-gatherers

Before entering into the specifics of what evolutionary grasslands diet may or may not have looked like, it is paramount to point out that there exists a diverse range of hunter-gatherer diets with large regional and seasonal variations. Also, these diets are spread out over a vast period, ranging from the Palaeolithic era to the present day, so large fluctuations over time – within dramatically evolving ecological and climatic contexts – are to be expected.

For early humans, meat and animal fat, both 'outside-bone nutrients' in muscle and organs and 'inside-bone nutrients' in marrow and brains (Thompson et al., 2019) may initially only have served as modest sources of nutrition, but their significance increased as access improved through aggressive scavenging and predation, with the latter likely ending up as the most important form of sourcing (Domínguez-Rodrigo et al., 2021). The consumption of large animals is well-documented with humans becoming the only living primates that regularly kill and consume animals of similar or larger body size (Butynski, 1982) to the point that some of the early human populations may even have been hypercarnivorous, deriving >70% of their diet from meat (Ben-Dor et al., 2021). Yet, the disappearance of mega-herbivores prompted humans to adapt by hunting smaller and faster animals, resulting in the selection of more agile and cognitively developed hominins. The development of long-range weapons, such as spearthrowers, contributed to increased dietary diversity and flexibility of prey choice. In addition to meat, organs, and marrow, the gathering of eggs, fish, seafood, insects, honey, and a wide variety of plants played a crucial role in food security, especially in regions where these resources were abundant (Pontzer and Wood, 2021). Carbohydrate-rich foods, including honey and plant carbohydrates, likely functioned as preferred foods when available, leading to protein-sparing effects, especially when meat would have been very lean. In contrast, data from modern hunter-gatherer communities suggest that some plants like tubers and bulbs must have served as fall-back and supplementary foods due to their relatively low energetic rewards (Marlowe and Berbesque, 2009).

Attempts to reconstruct the nature of Palaeolithic diets have largely been made drawing on modern hunter-gatherer communities. Such communities derive 2/3 of their energetic intake from hunting and fishing, on average, or about 1/2 in the warmer climate zones (Cordain et al., 2000; Pontzer and Wood, 2021). For about 3/4 of worldwide hunter-gatherer societies, animal-source foods contribute to more than half of the caloric intake (Cordain et al., 2000). Meat consumption, specifically, varies widely, ranging from 5% to 90% of caloric intake, and even within a single community, intake levels fluctuate considerably throughout the year (Pontzer and Wood, 2021). Yearly per capita meat intake estimates mostly not only span from 40 kg to 220 kg but also reach up to 350–650 kg for certain groups, including Arctic populations (Pontzer and Wood, 2021). Even if all these numbers are imperfect estimates for a Pleistocene grassland context, due to disparities in large prey availability, a modelling study has similarly proposed that at least 30% and up to 70% of the total caloric intake of East African Palaeolithic diets may have been of animal origin (Kuipers et al., 2010). With respect to macronutrient composition, variable proportions of carbohydrates (20–50% of total kcal), fat (20–70%), and protein (10–35%) seem plausible for Paleolithic diets (Kuipers et al., 2010). For present-day Hadza populations (a hunter-gatherer community in Tanzania), these levels are estimated at 20–70%, 13–36%, and 11–43%, respectively (Pontzer and Wood, 2021). Saturated fat intake in ancestral diets is estimated to be within a range of 7–19% of total kcal or 20–60 g/day, with significant variability among populations (Kuipers et al., 2010). Such pronounced flexibility in food and nutrient compositions indicates that a single dietary solution to the problem of human health optimization would be illusionary. Taken together, some populations may have followed low-carbohydrate and meat-heavy diets, while others have been more reliant on starches or sugars like honey (Pontzer et al., 2018; Pontzer and Wood, 2021).

The important observation at this point is that, despite what is now seen as unwise dietary choices (substantial amounts of red meat and saturated fat, relatively low amounts of carbohydrates and cereals, and sometimes high protein levels), hunter-gatherer communities that follow these diets are not only known to be generally in good physical health but also almost free of the range of cardiometabolic diseases that are now common in global populations (Cordain et al., 2002).

4 The impact of the development of farming on the human diet

During the Neolithic era (the 'New Stone Age', following the Paleolithic era and starting around 12 000 years ago), dietary reliance underwent a significant shift induced by the domestication of plants, as crops, and animals, as livestock.

The spread of farming from Anatolia, initially into the south-eastern European regions around 6000–7000 BCE, eventually led to the onset of the Copper Age (from about 5000–3200 BCE). This period was characterized by intercultural interactions, especially between farmers and nomadic herders from the Pontic Steppes in the Western Black Sea region around 4500 BCE. These interactions spurred innovations in herd management, food processing, and dairying practices. This, in turn, led to the expansion of fully developed pastoralist groups during the third millennium BCE, contributing to the complex cultural and genetic mosaic of European ancestries derived from hunter-gatherers, farmers, and pastoralists, and – therefore – a patchwork of dietary solutions based on local resources and know-how.

Although the original motives for animal husbandry are still debated, and probably multiple, livestock soon became essential to the food and nutrient security of settled communities. Domesticated animals were of utmost importance to such communities for multiple reasons, not only including the provision of food, traction, and manure, but also for their major religious, social, and economic significance. Even though animals were still central to the food system, the consumption of meat by farming communities dropped considerably compared to the hunter-gatherer communities or to those groups that evolved into pastoralist lifestyles. This decline was compensated for by the newly discovered option of dairy as well as crops. However, the limited variety of crops often failed to compensate for nutritional deficiencies, essentially leading to a diet characterized by low meat and high cereal intake. This novel dietary pattern correlated on average with a shift from good health to disease and with higher reproduction rates, leading to population booms and busts (Williams and Hill, 2017). The direct consequences of this Neolithic food system included malnutrition and infectious diseases, reflected in reduced stature, poor bone health, nutritional deficiencies, and dental caries (Ulijaszek et al., 2012; Marciniak et al., 2022). Reduced stature is indicative of deprivation and low dietary quality, resulting from a combination of the low protein intake, high glycaemic load, and micronutrient deficiencies of at least some early farmer diets (Marciniak et al., 2022).

With an increasing supply throughout the Bronze Age (3300 BCE–1200 BCE) and beyond (Münster et al., 2018), the consumption of meat and animal source foods may have reached very substantial levels in some regions and eras, occasionally exceeding levels seen in current Western diets. In the Middle Ages, certain areas of Western Europe enjoyed high supplies of mutton, beef, pork, cheese, and butter (Banegaz López, 2010; Dunne et al., 2019). Nonetheless, malnutrition persisted among the poorer segments of agricultural populations. Even today, in the context of abundant animal-source foods in industrialized countries and the increasing share of foods of animal origin in diets in some low- and middle-income countries with rising economic status, an unbalanced reliance on cereal-heavy diets and poor availability of micronutrient-dense foods persists in many regions of the Global South. In contrast, pastoralist communities in low- and middle-income countries, both ancestral and modern, have traditionally secured levels of animal-source foods derived from grasslands that are usually higher than what is obtained in agricultural communities. Examples include the Turkana people, a nomadic pastoralist community in Kenya, who still derive 62% of their calories from dairy and 12% from meat, fat, and blood. It is only those Turkana who adopt Western diets that are high in refined carbohydrates and ultra-processed foods, while moving into urban centers, who tend to manifest clear symptoms of poor cardiometabolic health (Lea et al., 2020). This observation can be extrapolated to most human communities that abandon their ancestral lifestyles while adopting Western-type diets.

In summary, the above-mentioned information shows that diets that are rich in grassland-derived foods lead to nutritional adequacy and robust health in those ancestral food systems that correspond with either the human evolutionary prehistory of hunter-gatherer lifestyles or, more recently, pastoralist communities. The food system transformation towards agriculture offered new possibilities to feed populations, albeit often in a less robust manner because of a high dependency on a limited variety of crops. In such systems, sufficient access to animal-source foods acts as a game changer, facilitating the quest for nutritional adequacy. However, if such access is also coupled with the Westernization of diets, additional problems emerge related to cardiometabolic diseases, an issue which will be discussed below. Some of the major global food system challenges we are currently facing stem at least partially from a legacy of diets dominated by domesticated plant staples and reduced dietary quality (Larse, 2023). Nonetheless, there is more to the story, as these problems are compounded by the emergence of zoonoses and epidemics, which are related to animal husbandry, as well as by the problem of the 'Western diet' and its highly processed convenience foods. The current ultra-processed food environment drives overconsumption and metabolic maladaptation and, by doing so, probably lies at the basis of the steady rise in non-communicable diseases. Such diseases are generally linked to a physiological state of hyperinsulinemia and the symptoms of the metabolic syndrome (excess abdominal fat, low high-density lipoprotein cholesterol, high blood pressure, glucose, and triglycerides).

5 Industrialisation and the human diet

In the eighteenth- and nineteenth-century industrializing West, animal-source foods became widely available because of technological advancements, including cooling technologies, improved logistics, and a surge in urban demand for meat (Leroy and Degreef, 2015). Notwithstanding initial gains in nutritional status and an increase in stature that hinted at diets better suited to human biological needs, this by no means equated to a return to Pleistocene-type dietary patterns, neither in terms of quality nor quantity. Over time, the diets of these countries were increasingly marked by the incorporation of large amounts of heavily processed foods. At the same time, it is a common misconception that current Western diets are typified by historically excessive and unnaturally high consumption of meat and other animal-source foods, at least within an evolutionary context. The current contribution of animal-source foods to the caloric intake of Western diets (approximately 30% in the US and EU) is lower than that of most pre-agricultural diets, which averaged around 65%, albeit somewhat lower at 40–50% in plant-abundant regions (see above). This deviation is exacerbated by the large inclusion of refined grains, added sugars, plant-derived oils, and ultra-processed foods, magnifying the mismatch between today's eating patterns and the evolutionary diets that the human body has metabolically adapted to over millions of years. Contrary to assumptions, red meat consumption levels are not systematically rising in high-income nations. While there is a global increase in meat consumption due to economic growth and urbanization in low- and middle-income countries, high-income nations have experienced a decrease in red meat intake over the last four decades, partly offset by increased consumption of poultry (https://www.fao.org/faostat/en/#data/FBS). There is also a decline in nose-to-tail eating; offal meats (e.g. liver, kidney, heart, tongue, and brain) are still moderately consumed in some Western countries but have become almost negligible in others.

In this light, it is remarkable that grassland-derived foods, red meat, offal meats, whole-fat dairy, and animal fat are now often seen as culprits for the current public health crisis centred around obesity and cardiometabolic disease. The EAT-Lancet commission, for instance, specifically designated red meat as an unhealthy food choice, while promoting whole grains and plant oils as staple foods (Willett et al., 2019). EAT-Lancet's own Planetary Health Diet, however, in some aspects similar to Neolithic diets or the current cereal-dominated diets of the Global South, is deficient in certain key nutrients that are most bioavailable in animal-source foods (e.g. vitamin B12, iron, and zinc). To overcome these micronutrient shortfalls, the proposed share of animal-source foods in the Planetary Health Diet would have to double from 14% to 27% of the caloric intake (Beal et al., 2023). Interestingly, and perhaps not by coincidence, the latter is similar to the lower limit of hunter-gatherer diets worldwide (Cordain et al., 2000). It also matches the threshold observed in national diets below which micronutrient deficiencies start to surge (Nordhagen et al., 2020). In other words, a caloric contribution of animal-source foods of about 30% roughly corresponds with (1) the lower intake level of most ancestral-type hunter-gatherer diets, (2) the threshold of increased micronutrient risk in national diets worldwide, and (3) the current share of animal-source foods in Western diets (for the example of the USA, see Rehkamp, 2016). We do not wish to set this 30% value in stone; a lot will of course depend on the quality of the foods involved, at both sides of the animal/plant divide. Clearly, the current Western diet is problematic, even if the consumption of animal-source foods is in line with said threshold, both at the level of cardiometabolic health (O'Hearn et al., 2022) and micronutrient deficiencies, especially in populations with elevated needs (Devarshi et al., 2021; Bird et al., 2023).

The preceding paragraph makes a case for a *minimum* amount of animal-source foods needed to increase the nutritional robustness of diets. Yet, a similar question may be asked concerning a potential *maximum* amount of animal-source foods for optimal health. In any case, and without ruling out possible detrimental health effects on at least some individuals, the current evidence does not seem indicative of strong evidence when it comes to the causal link between cardiometabolic and all-cause mortality risk, on the one hand, and the consumption of unprocessed red meat or whole-fat dairy products, on the other hand. Whereas the effects of dairy tend to be mostly protective, to begin with (Feng et al., 2022; Mente et al., 2023), the case for unprocessed red meat has been more contested. Some individuals, for instance, those genetically or metabolically predisposed to haemochromatosis (a disorder which causes the body to absorb more iron from food than usual), may have to be cautious when exceeding a certain intake level of meat high in haem iron. Also, some preparation or processing methods, leading to the accumulation of toxic chemicals (e.g. polycyclic aromatic hydrocarbons or heterocyclic amines), may have harmful effects on consumer health. Finally, and with respect to the evolutionary rationale followed in this chapter, the argument may also be made that the composition of current meat obtained from livestock agriculture is different than that of meat coming from wild animals, for instance concerning fat composition and the presence of phytochemicals in feedlot-finished cattle. Although potentially meaningful, it is unclear if those differences are substantial enough to translate into actual health outcomes. Moreover, the adoption of certain animal genetics and grazing strategies in contemporary animal husbandry may approximate the more ancestral variants.

Regardless of the above-mentioned considerations, and based on the totality of data, the certainty of evidence to designate unprocessed red meat as a causal factor in the development of Western-type diseases has been characterized as low to very low (Johnston et al., 2019, 2023; Hill et al., 2020, 2024; Lescinsky et al., 2022). When evaluating the data globally, using the transnational PURE (Prospective Urban and Rural Epidemiological) study's cohort, the associations with cardiovascular disease or mortality even appear as non-existent, with a tendency for higher intake levels to display protective rather than harmful associations (Iqbal et al., 2021; Mente et al., 2023). Meanwhile, and despite dietary guidelines urging further restrictions on red meat intake, the steady decrease in red meat consumption in the West is contrasting oddly with the rise of cardiometabolic disease, prompting a need to reconsider the current dietary paradigm and the often excessive vilification of grassland-derived foods.

6 Case study: reclaiming health with grassland-derived foods

Animal foods derived from grasslands include milk, meat, and edible offal from herbivorous species, predominantly ruminants such as cattle, sheep, goats, deer, buffalo, etc. Additionally, several monogastric species with large hindgut fermentation, such as horses, geese, and rabbits, can be productive with a diet primarily composed of grass and forage. In a broader sense, foraging represents the natural eating behaviour of farm animals, exhibiting diversity – cattle grazing, goats preferentially browsing, and chickens pecking invertebrates, grains, and seeds. However, animal feeding and husbandry practices have evolved, particularly with the intensification of livestock production in many parts of the world, starting from the 1960s and 1970s. Presently, intensive pig and chicken production systems are predominantly based on feeding concentrates largely composed of grains and (proteinaceous) by-products. Nevertheless, even today, more than 70% of globally used feedstuffs for animals consist of grass, leaves, fodder crops, and crop residues (Mottet et al., 2017).

Grasslands exhibit natural diversity in botanical composition, leading to variations in nutrient content and the presence of a wide array of plant secondary compounds. These differences are influenced by soil characteristics, climate conditions, and grazing management. Grasslands within extensive and semi-intensive livestock systems tend to be more diverse (i.e. comprised of grasses, legumes, herbs, shrubs, etc.) and are enriched in phytochemicals compared to those in intensive simple mixture or monoculture systems (Van Vliet et al., 2021). This prompts the question of whether the variation in grass and forage composition translates into biologically relevant differences in the composition and nutritive value of animal-source foods. It is essential to consider that in intensive systems – whether grass or forage-based ruminant systems or concentrate-based monogastric systems – animal diets are now optimized to meet maximal productivity by carefully selecting major ingredients and incorporating various additives. A straightforward answer to this question is thus challenging, as other factors, including genetics, also play a role and impact the composition of animal-source foods.

The most evident and well-documented distinctions in animal-derived foods from various grassland-based production systems pertain to the polyunsaturated fatty acid (PUFA) content and composition, as illustrated by a comparison between conventional and forage-only farming systems. Generally, there is no effect on the gross composition. However, meta-analyses have indicated a higher PUFA content, particularly n-3 PUFAs (i.e. linolenic acid and/or long-chain n-3 PUFA), in organic and/or grass-fed milk and meat (Elgersma, 2015; Prache et al., 2022). This is nutritionally advantageous given the deficit in n-3 PUFA intake in many populations with low fish consumption. Similarly, multiple studies highlight that feeding leguminous and biodiverse forages, compared to intensively managed ryegrass, results in increased n-3 PUFA and CLA c9t11 (conjugated linoleic acid) contents in milk and intramuscular fat of cattle and sheep due to altered rumen metabolism via several potential mechanisms, especially through shifts in microbial populations and their associated activities, in particular biohydrogenation (Lourenço et al., 2008; Provenza et al., 2019). It is important to note that the effects of biodiverse grasslands, forages, and associated phytochemical richness on bioactive compounds and nutrients other than PUFA in animal-source foods are less explored and less apparent. While the notion of enriching animal tissues and products with bioactive compounds from the intake of secondary plant compounds is appealing from a One Health perspective, the properties, digestion, and metabolism of these compounds complicate their deposition in lean meat. Fat tissue, eggs, and particularly milk demonstrate much higher plasticity in this regard. Several studies reviewed by Van Vliet et al. (2021) show increased levels of terpenoids, polyphenols, carotenoids, and tocopherols in dairy products and fat from (plant species-rich) grass-fed cattle compared to conventional cattle, with concurrent improved oxidative stability of these products.

A well-known example of premium pork derived from an agroforestry system is the long-ripened *Iberico* ham manufactured from rustic pigs foraging in the *Dehesa* on grass and acorns during specific times of the year; characterized by exclusive sensory properties and compositional differences (Tejerina et al., 2012). Although challenging to disentangle due to interfering factors, further research is warranted to understand the relationship between animal diet richness and the wholesomeness and complexity of the resultant meat and dairy *matrix* (i.e. the complex composition and structure of a food and its various interacting components, which ultimately will affect the digestion, absorption, and metabolism in the human body).

The final question is whether the consumption of foods derived from those livestock and grassland-based systems that aim at linking animal, human, and environmental health might have a different impact on human health compared to foods from conventional systems. While the literature on this topic is currently limited, a small number of studies suggest potential anti-inflammatory effects and improved lipoprotein profiles when individuals consume pasture-raised meat and dairy (Van Vliet et al., 2021) or meat and meat products enriched in n-3 PUFA (Vossen and De Smet, 2016). These effects appear to be modest but could be relevant in the long term. Given the growing consumer interest in animal-source foods from more 'natural' systems, there is a clear need for clinical nutrition studies that explore the potential of these foods to modulate human health. It must be noted that active debates within the medical and human nutrition disciplines are not well known outside of those communities and that the 'conventional wisdom' coming from one side of these current controversies influences the efforts of other disciplines (e.g. grassland sciences). The *Energy Balance Model* (in its most reductive setup, known as *Calories In Calories Out*) is an example of the conventional wisdom of obesity while the *Carbohydrate Insulin Model* is an opposing theory within a larger unresolved debate (Ludwig et al., 2021; Hall et al., 2022). Another, the *Diet-Health Hypothesis*, ascendent since the mid-1950s, promoted low-fat, carbohydrate-rich diets for heart health (Taubes, 2007). An alternative theory is that hyperinsulinemia (chronically elevated insulin) with insulin resistance is a potentially unifying theory of chronic disease (Crofts, 2015; Reaven, 1992). The resolution of these debates will have profound implications for the biological significance of compositional differences of meat and milk.

7 Conclusion and future trends

Given the current triple burden of malnutrition, consisting of the coexistence of overnutrition alongside undernutrition and micronutrient deficiencies, and its massive impact on public health and sustainability, the question needs to be asked if dietary solutions can help mitigate the problem that was created by diets in the first place? And, more importantly, if we can identify what those dietary solutions need to be? A common-sense approach we would like to propose is to drastically restrict highly processed foods that typify the Western diet and to reintroduce dietary patterns that are not necessarily replicating but mimicking those ancestral-type diets that can be considered evolutionary appropriate and nutritionally adequate. Such diets do not need to be narrowed down to a single formulation; well on the contrary, there are numerous manners to combine micronutrient-rich and minimally or lightly processed animal and plant foods in wholesome diets.

8 Where to look for further information

8.1 Further reading

One of the most comprehensive books on human evolutionary diets is as follows:

Ulijaszek, S., et al. (2012). *Evolving Human Nutrition: Implications for Public Health,* Cambridge Studies in Biological and Evolutionary Anthropology, Cambridge University Press, Cambridge, UK.

An evolving overview of the scientific literature on the role of animal source foods in human and planetary health can be found on the 'dynamic' website https://aleph2020.org

Two Special Issues on animal source foods in the journal Animal Frontiers:

- Foods of Animal Origin: A Prescription for Global Health. 2019. Animal Frontiers. Vol 9, No. 4. https://academic.oup.com/af/issue/9/4.
- The Societal Role of Meat. 2023. Animal Frontiers. Vol 13, No. 2. https://academic.oup.com/af/issue/13/2.

9 Declaration of interest statement

FL is a non-remunerated board member of various academic non-profit organizations including the Belgian Association for Meat Science and Technology (president), the Belgian Society for Food Microbiology (president), and the Belgian Nutrition Society. On a non-remunerated basis, he also serves on the Scientific Board of World Farmers' Organization (WFO) and FAO/COAG Sub-Committee on Livestock. The views the authors express in this commentary are theirs alone and not necessarily those of the aforementioned organizations. PB is a non-remunerated board member of the American Forage and Grassland Council (past-president). SDS serves as vice-president (non-remunerated) to the Belgian Association for Meat Science and Technology (academic non-profit organization) and is a member of the Flemish Strategic Advisory Council for Agriculture and Fisheries. FL acknowledges financial support from the Research Council of the Vrije Universiteit Brussel, under grants SRP71, IOF3017, and IRP21.

10 References

Aiello, L.C. and Wheeler, P. (1995). The expensive-tissue hypothesis: the brain and the digestive system in human and primate evolution, *Curr. Anthropol.* 36: 199–221.

Banegas López, R.A. (2010). Consumption of meat in Western European cities during the Late Middle Ages: A comparative study, *Food Hist.* 8: 63–86.

Barnard, N.D. and Leroy, F. (2020). Children and adults should avoid consuming animal products to reduce the risk for chronic disease: debate consensus, *Am. J. Clin. Nutr.* 112: 937–940.

Beal, T., et al. (2023). Estimated micronutrient shortfalls of the EAT–Lancet planetary health diet, *Lancet Planet. Health* 7: E233–E237.

Ben-Dor, M., et al. (2021). The evolution of the human trophic level during the Pleistocene, *Yearb. Phys. Anthropol.* 175: 27–56.

Bird, J.K., et al. (2023). Micronutrient intakes in the Dutch diet: foods, fortified foods and supplements in a cross sectional study, *Eur. J. Nutr.* 62: 3161–3179.

Butynski, T.M. (1982). Vertebrate predation by primates: a review of hunting patterns and prey, *J. Hum. Evol.* 11: 421–430.

Cordain, L., et al. (2000). Plant-animal subsistence ratios and macronutrient energy estimations in worldwide hunter-gatherer diets, *Am. J. Clin. Nutr.* 71: 682–692.
Cordain, L., et al. (2002). The paradoxical nature of hunter-gatherer diets: meat-based, yet non-atherogenic, *Eur. J. Clin. Nutr.* 56: S42–S52.
Crofts, C.A.P. (2015). Hyperinsulinemia: a unifying theory of chronic disease?, *Diabesity* 1: 34.
De Smet, S. and Vossen, E. (2016). Meat: the balance between nutrition and health. A review. *Meat Science, Meat for Global Sustainability: 62nd International Congress of Meat Science and Technology (62nd ICoMST)*, August 14–19, 2016, Bangkok, Thailand, 120: 145–156.
Devarshi, P.P., et al. (2021). Total estimated usual nutrient intake and nutrient status biomarkers in women of childbearing age and women of menopausal age, *Am. J. Clin. Nutr.* 113: 1042–1052.
Domínguez-Rodrigo, M., et al. (2021). Early Pleistocene faunivorous hominins were not kleptoparasitic, and this impacted the evolution of human anatomy and socio-ecology, *Sci. Rep.* 11: 16135.
Dunne, J., et al. (2019). Reconciling organic residue analysis, faunal, archaeobotanical and historical records: diet and the medieval peasant at West Cotton, Raunds, Northamptonshire, *J. Archaeol. Sci.* 107: 58–70.
Elgersma, A. (2015). Grazing increases the unsaturated fatty acid concentration of milk from grass-fed cows: a review of the contributing factors, challenges and future perspectives, *Eur. J. Lipid Sci. Technol.* 117: 1345–1369.
Feng, Y., et al. (2022). Consumption of dairy products and the risk of overweight or obesity, hypertension, and type 2 diabetes mellitus: a dose-response meta-analysis and systematic review of cohort studies, *Adv. Nutr.* 13: 2165–2179.
Gurven, M. and Kaplan, H. (2007). Longevity among hunter-gatherers: a cross-cultural examination, *Popul. Dev. Rev.* 33: 321–365.
Hall, K.D., et al. (2022). The energy balance model of obesity: beyond calories in, calories out, *Am. J. Clin. Nutr.* 115: 1243–1254.
Henneberg, M., et al. (1998). Human adaptations to meat eating, *Hum. Evol.* 13: 229–234.
Hill, E.R., et al. (2020). Red meat intake and cardiometabolic disease risk: an assessment of causality using the Bradford Hill criteria, *Curr. Dev. Nutr.* 4: 31.
Hill, E.R., et al. (2024). Red and processed meat intakes and cardiovascular disease and type 2 diabetes mellitus: an umbrella systematic review and assessment of causal relations using Bradford Hill's criteria, *Crit. Rev. Food Sci. Nutr.*: 2423–2440.
Iqbal, R., et al. (2021). Associations of unprocessed and processed meat intake with mortality and cardiovascular disease in 21 countries [Prospective Urban Rural Epidemiology (PURE) Study]: a prospective cohort study, *Am. J. Clin. Nutr.* 114: 1049–1058.
Johnston, B.C., et al. (2019). Unprocessed red meat and processed meat consumption: dietary guideline recommendations from the Nutritional Recommendations (NutriRECS) Consortium, *Ann Intern Med.* 171: 756–764.
Johnston, B.C., et al. (2023). Non-communicable disease risk associated with red and processed meat consumption-magnitude, certainty, and contextuality of risk? *Anim. Front.* 13: 19–27.
Kuipers, R., et al. (2010). Estimated macronutrient and fatty acid intakes from an East African Paleolithic diet, *Br. J. Nutr.* 104(11): 1666–1687.

Larse, C.S. (2023). The past 12,000 years of behavior, adaptation, population, and evolution shaped who we are today, *PNAS* 120: e2209613120.
Lea, A.J., et al. (2020). Urbanization and market integration have strong, nonlinear effects on cardiometabolic health in the Turkana, *Sci. Adv.* 6: eabb1430.
Leroy, F. and Cofnas, N. (2020). Should dietary guidelines recommend low red meat intake?, *Crit. Rev. Food Sci. Nutr.* 60: 2763–2772.
Leroy, F. and Degreef, F. (2015). Convenient meat and meat products. Societal and technological issues, *Appetite* 94: 40–46.
Leroy, F., et al. (2023). The role of meat in the human diet: evolutionary aspects and nutritional value, *Animal Front.* 13: 11–18.
Lescinsky, H., et al. (2022). Health effects associated with consumption of unprocessed red meat: a Burden of Proof study, *Nat Med.* 28: 2075–2082.
Lourenço, M., et al. (2008). Influence of different dietary forages on the fatty acid composition of rumen digesta as well as ruminant meat and milk, *Anim. Feed Sci. Technol.* 145: 418–437.
Ludwig, D.S., et al. (2021). The carbohydrate-insulin model: a physiological perspective on the obesity pandemic, *Am. J. Clin. Nutr.* 114: 1873–1885.
Mann, N.J. (2018). A brief history of meat in the human diet and current health implications, *Meat Sci.* 144: 169–179.
Marciniak, S., et al. (2022). An integrative skeletal and paleogenomic analysis of stature variation suggests relatively reduced health for early European farmers, *PNAS* 119: e2106743119.
Marlowe, F.W. and Berbesque, J.C. (2009). Tubers as fallback foods and their impact on Hadza hunter-gatherers, *Am. J. Phys. Anthropol.* 140: 751–758.
Mente, A., et al. (2023). Diet, cardiovascular disease, and mortality in 80 countries, *Eur. Heart J.* 44: 2560–2579.
Mottet, A., et al. (2017). Livestock: on our plates or eating at our table? A new analysis of the feed/food debate, *Glob. Food. Sec.* 14: 1–8. https://doi.org/10.1016/j.gfs.2017.01.001
Münster, A., et al. (2018). 4000 years of human dietary evolution in central Germany, from the first farmers to the first elites, *PLoS One* 13: e0194862.
Nordhagen, S., et al. (2020). The role of animal-source foods in healthy, sustainable, and equitable food systems. Global Alliance for Improved Nutrition (GAIN). Discussion Paper Series #5. Geneva, Switzerland.
O'Hearn, M., et al. (2022). Trends and disparities in cardiometabolic health among US adults, 1999–2018, *J. Am. Coll. Cardiol.* 80: 138–151.
Pontzer, H., et al. (2018). Hunter-gatherers as models in public health, *Obesity Rev.* 19: 2435.
Pontzer, H. and Wood, B.M. (2021). Effects of evolution, ecology, and economy on human diet: insights from hunter-gatherers and other small-scale societies, *Ann. Rev. Nutr.* 41: 363–385.
Prache, S., et al. (2022). Review: quality and authentication of organic animal products in Europe, *Animal* 16: 100405.
Provenza, F.D., et al. (2019). Is grassfed meat and dairy better for human and environmental health?, *Front. Nutrition* 6: 26.
Reaven, G.M. (1992). The role of insulin resistance and hyperinsulinemia in coronary heart disease, *Metabolism* 41: 16–19.

Rehkamp, S. (2016). A look at calorie sources in the American diet. Economic Research Service, US Department of Agriculture. https://www.ers.usda.gov/amber-waves/2016/december/a-look-at-calorie-sources-in-the-american-diet

Taubes, G. (2007). *Good Calories, Bad Calories: Challenging the Conventional Wisdom on Diet, Weight Control, and Disease*, Knopf, New York, NY.

Tejerina, D., et al. (2012). Effect of production system on physical–chemical, antioxidant and fatty acids composition of *Longissimus dorsi* and *Serratus ventralis* muscles from Iberian pig, *Food Chem.* 133: 293–299.

Thompson, J.C., et al. (2019). Origins of the human predatory pattern: the transition to large-animal exploitation by early hominins, *Curr. Anthropol.* 60: 1–23.

Ulijaszek, S., et al. (2012). *Evolving Human Nutrition: Implications for Public Health*, Cambridge Studies in Biological and Evolutionary Anthropology, Cambridge University Press, Cambridge, UK.

van Vliet, S., et al. (2021). Health-promoting phytonutrients are higher in grass-fed meat and milk, *Front. Sust. Food Syst.* 4: 555426.

Willett, W., et al. (2019). Food in the anthropocene: the EAT–Lancet Commission on healthy diets from sustainable food systems, *Lancet* 393: 447–492.

Williams, A.C. and Hill, L.J. (2017). Meat and nicotinamide: a causal role in human evolution, history, and demographics, *Int. J. Tryptophan Res.* 10: 1178646917704661.

Yaseen, N.R., et al. (2023). Genetics of vegetarianism: a genome-wide association study, *PLoS One* 18: e0291305.

Part 4

Developments in botanical composition of grasslands

Chapter 13

Improving biodiversity in agricultural grassland systems

Irina Herzon, University of Helsinki, Finland; James M. Bullock, UK Centre for Ecology and Hydrology, UK; Jean-Yves Humbert and Raphaël Arlettaz, University of Bern, Switzerland; Péter Török, University of Debrecen, Hungary; Karin Stein-Bachinger, Leibniz Centre for Agricultural Landscape Research, Germany

1 Introduction

Agricultural grasslands include some of the most biodiverse habitats in the world, but also some of the most species-poor and inhospitable environments to non-farmland species. In attempts to make grassland-based agricultural production more sustainable, one must ensure that the biodiversity of grasslands and the multiple ecosystem services supported by above- and below-ground biodiversity are maintained or reinstated. In their global review, Zhao (2023) identified biodiversity conservation and restoration as among the emerging research challenges in the domain of sustainable grassland management, with a steadily growing curve of scientific contributions. In many parts of the world, natural grasslands are the ecosystems most extensively transformed by human activities, mostly for food and biofuel production (Carbutt, Henwood and Gilfedder, 2017 for temperate grassland biomes). Grasslands are also among ecosystems that are projected to undergo the largest biodiversity losses in the

http://dx.doi.org/10.19103/AS.2025.0154.20

future due to their sensitivity to the main drivers of biodiversity erosion, notably land use changes (Sala et al., 2000). Apart from the biodiversity occurring within the grasslands themselves, it is also important to consider that grassland-based production directly or indirectly affects a large fraction of the biodiversity in other habitats than grasslands. Such biodiversity includes, e.g. forests and wetlands that are converted into agricultural grasslands and aquatic species affected by pollutants of agricultural origin. In most instances, the expansion of agricultural land, including grassland, into the native natural ecosystems has impoverished if not annihilated their native biodiversity (Newbold et al., 2016; Ramankutty et al., 2018). The increasing human population and dietary shifts towards animal-sources foods accompanied by the rapid globalisation of modern food production systems also mean that indirect impacts on biodiversity extend far beyond the boundaries of countries and continents, via for instance the widespread and increasing reliance on mineral fertilisers and feed, and massive production of feed, i.e. agricultural processes that increase greenhouse gas emissions.

Natural grasslands are found in areas where forests cannot grow due to climate constraints (e.g. in steppe or alpine zones) and where hydrology, soil conditions, relief and/or natural disturbances such as grazing of woody vegetation by large wild herbivores or forest fires prevent tree growth (Dengler et al., 2014). For long grazed by wild ungulates, pre-existing open landscapes have seen their native herbivore community progressively replaced by domesticated animals, involving controlled pasturing and hay mowing. This also occurred in predominantly forested regions, including most of Europe, where widespread grazing by large native herbivores had created semi-open woodlands, i.e. an intricated mosaic of forest stands and grassy clearings (Pearce et al., 2023) (Fig. 1a). It is the species adapted to such semi-open conditions that later colonised expanding agricultural landscapes. The potential for agricultural grasslands to support high levels of biodiversity is considerable, providing that the intensity of management is not too high. All depends on how many species of different groups – plants, invertebrates, birds – are tolerated, maintained or encouraged during production on grasslands. The diversity of species typically inhabiting grasslands will thus primarily depend on both landscape structural diversity and grassland management practices (e.g. Vickery & Arlettaz, 2012).

A significant and increasing fraction of modern agricultural grasslands across the world are secondary grasslands: they result from the expansion of agricultural land into natural ecosystems such as forests and wetlands, and their mosaics (e.g. review in Porqueddu et al., 2016). Some of these grasslands are among the most floristically diverse habitats in the world (Wilson et al., 2012) (Fig. 1b, Fig. 2a,b,c). They are most common in regions where traditional forms of agricultural use predominate, as it is a case for High Nature Value farmland in the context of Europe (Keenleyside et al., 2014), or in rangelands

Figure 1 General relationship between biodiversity and biomass production by agriculture across main grassland types: (a) pre-agricultural ecosystem with wild herbivores maintaining native grasslands and creating mosaics of semi-open grassy patches in woodland; (b) semi-natural grassland grazed by domesticated herbivores or used for traditional haymaking; (c) and (d) agronomically improved permanent grassland under different levels of management intensity for pasturing and forage production; (e) rotational grasslands cultivated as part of arable field rotations. Image by Lola Fernández Multigner.

with native vegetation. Such grasslands are often referred to as 'semi-natural'. Because they are managed with local (not imported) fertiliser inputs and are not reseeded, their species composition is largely determined by natural processes, albeit mediated by grazing or haying (Peeters et al., 2014). These grasslands, or rangelands, depend on livestock grazing and/or mowing to prevent succession to scrub and woodland, and are common in major regions dominated by ruminant production (e.g. Brazil (Leidinger et al., 2017), South Africa (Bengtsson et al., 2019), North America (Bartolome et al., 2014) and Mediterranean climate zones around the world (Porqueddu et al., 2016)). Biodiversity in such grasslands is threatened by intensification of production in

Figure 2 Overview of secondary grasslands under different management and land use intensities: (a) traditional agroforestry system with extensive pasture areas (Image by HNV-Link); (b) extensively managed mountainous pasture (Image of Wikipedia, https://commons.wikimedia.org/wiki/File:Pojejena,_Romania_-_panoramio.jpg); (c) extensively managed permanent subalpine hay meadow; (d) intensively managed permanent grassland; (e) cultivated grass-legume ley; (f) intensively managed cultivated pasture for dairy production (Aino Pietikäinen). All other images by Jean-Yves Humbert.

many regions (Dengler et al., 2014; Porqueddu et al., 2016) but also large-scale abandonment in other regions (Dengler et al., 2014). In the boreal region of Europe, e.g. the current extent of semi-natural grasslands is only a few percent of what it was 100 years ago due to both processes (Herzon et al., 2022). Other impacts come from pesticide drift (Bullock et al., 2024; Newton et al., 2012), invasive species and nutrient deposition from agricultural and industrial sources (Maskell et al., 2010).

Published by Burleigh Dodds Science Publishing Limited, 2025.

By improving conditions for production – through drainage, irrigation, liming, reseeding or additional seeding of existing swards, addition of organic and/or mineral fertilisers and other agrochemicals, increased mowing frequency and/or stocking rates – the biodiversity of natural and semi-natural grasslands has been progressively reduced, resulting in highly productive but species-poor grasslands (Habel et al., 2013, Fig. 1c, Fig. 2d). There is a wealth of research on the biodiversity of permanent grasslands across a range of productivity levels (reviewed in Schils et al., 2022). The lowest levels of biodiversity tend to be on permanent grasslands with particularly high fertiliser inputs, grazing pressure, frequent mowing operations and temporary grasslands (also called leys or artificial grasslands), which are regularly re-seeded with one or more species of grass or grass-legume mixtures, either on the same plot or in crop rotations (Fig. 1d, Fig. 2e,f). The latter are particularly common in dairy production and are part of organic arable farming. Relatively few studies have focused on the biodiversity of grasslands under intensive use for high productivity (reviews in Plantureux et al., 2005; Tiainen et al., 2020), apart from research on the diversification of seed mixes for sown crops (Nyfeler et al., 2009).

In general, permanent agricultural grasslands are known to support higher levels of biodiversity than arable land, at least for some taxa (reviewed in Schils et al., 2022). For example, species richness of earthworms, collembolans and oribatid mites is lower in arable systems than in intensive grasslands across Europe (Rutgers et al., 2016; Tsiafouli et al., 2015). A global meta-analysis showed that intensive grasslands have higher arthropod species richness, especially in the decomposer guild, than arable systems, but no such difference was found for predator species richness (Attwood et al., 2008). On the other hand, farmland with flowering crops, but not with livestock systems, increased wild bee species richness (four European countries; Le Féon et al., 2010). A bioenergy crop of *Miscanthus* increased earthworm diversity compared to intensive grassland (McCalmont et al., 2017). Also, there is some evidence that the biodiversity of high-intensity ley grasslands does not exceed that of arable fields (Toivonen et al., 2013), though comparative studies are few.

The wide range of land use histories, soil conditions and production intensities of grasslands makes generalisations on biodiversity patterns in this type of farmland difficult. Yet, one pattern is consistent across reviews: the intensification of grassland use decreases its multifunctionality, including biodiversity (Lindborg et al., 2023; Porqueddu et al., 2016; Ribeiro et al., 2023; Schils et al., 2022). After analysing a dataset of over 4,000 species from 12 trophic groups, Gossner et al. (2016) showed how even moderate increases in local land-use intensity lead to biotic homogenisation, i.e. loss of the fairly rare specialist species, across the microbes, plants and animals typical of grasslands, and for both above and below ground communities.

There are well-documented approaches to increasing biodiversity in all types of grassland, this for many different taxa. In this chapter, we review recent research on the effects of agricultural management of grasslands in temperate regions on different groups of organisms and synthesize the available scientific evidence on the effectiveness of different management options for enhancing their biodiversity. We focus primarily on the biodiversity typically inhabiting grasslands but also consider the key role of grasslands for biodiversity in both the overall agri-food system and wider landscape. Though the largest body of reviewed evidence comes from European grasslands, many of the findings and management implications raised here are relevant to all grasslands used for agricultural production globally. However, the development of locally specific measures to enhance grassland biodiversity must account for differences in agricultural contexts, notably ecosystem and farming histories and amenable agricultural practices.

2 Grassland management practices and biodiversity

Key features beneficial for biodiversity on grassland are the permanence of the vegetation compared to annual cropping, low levels of disturbance that allow a diversity of species to colonise and persist, low use of pesticides, grazing at moderate-to-low stocking rates that create a diverse vegetation structure within a parcel, or combination of high-stocking grazing with long rest periods that creates a mosaic of vegetation within the grassland matrix at the landscape scale. Due to a long history of evolutionary co-adaptation between native herbivores and other species, grazing by domesticated livestock is generally associated with higher pastoral biodiversity and is widely used also as a conservation tool. Increasing management pressure along any of the above aspects is likely to adversely affect biodiversity. The levels of biodiversity in grasslands are strongly influenced also by the availability of natural and semi-natural habitats at farm and landscape levels (see Section 5).

Few plant species, including non-native cultivars, can tolerate the high soil nutrients and defoliation rates under grazing or mowing in intensively used agricultural grasslands. Low plant diversity leads to low arthropod diversity due to shrinkage of ecological space (Attwood et al., 2008; Simons et al., 2015). This is due to a reduction in floristic diversity per se, but also because the swards become structurally homogeneous and lack flowers or seed heads to support other taxa (Bendel et al., 2018; Humbert et al., 2010b, 2021; Woodcock et al., 2007b). For example, silage grassland in Germany – usually managed for high productivity – had an 86% lower level of the butterfly conservation value indicator than traditionally managed hay grassland (Hannappel & Fischer, 2020).

Frequent cutting, rendered possible by regular mineral fertiliser inputs, further reduces arthropod abundance and diversity (Humbert et al., 2010b), especially among specialist species (Gossner et al., 2016). Declines in farmland birds across Europe are associated with reduced seed and arthropod prey availability, frequent mowing or high grazing pressure (Donald et al., 2002; Strebel et al., 2015; Vickery et al., 2001, Vickery & Arlettaz, 2012). In particular, early-season mechanised mowing for silage, which destroys nests and kills chicks, has been shown to be a major cause of population decline of several formerly common species – such as the critically endangered corncrake (*Crex crex*), skylark (*Alauda arvensis*) and whinchat (*Saxicola rubetra*) – from grasslands across Europe (Buckingham et al., 2015; Donald et al., 2002; Fay et al., 2021; Green & Gibbons, 2000). Intensive grazing has been shown to exert negative effects on biodiversity and habitat conservation in Mediterranean landscapes (review in Ribeiro et al., 2023). In arid climates, bare ground cover created by the excessing grazing can reduce seed germination for species that rely on standing biomass for shelter from aridity, leading to lower diversity and/or native cover (Lunt et al., 2007), although bare ground may benefit ground-foraging insectivorous vertebrates (Schaub *et al.*, 2010). Overgrazing is particularly detrimental to above and below-ground biodiversity, also in the long term (Donovan & Monaghan, 2021; Eldridge et al., 2013).

On rotational grassland, high fertiliser inputs as well as repeated defoliation for silage or bioenergy impede species colonisation and completion of their breeding cycle after the establishment. For example, quasi-experimental multi-taxa surveys across field types in Finland found few differences in plant and arthropod multi-taxa diversity between cultivated pasture, grass ley field and several types of arable crops (Toivonen et al., 2022).

3 Enhancing biodiversity on grasslands

3.1 Introducing plant species

There is increasing interest in establishing more biodiverse vegetation on agronomically improved but species-poor grasslands, such as herbal leys, to enhance forage production, minimise inputs and provide additional resources for wild species, such as pollen and nectar for pollinators (Finn et al., 2013). For example, the increasingly species-rich swards (from five grass species to up to 18 species of grasses, legumes and native forbs) also supported increasing diversity of phytophagous beetles (Woodcock et al., 2012), predatory beetles (Woodcock et al., 2013) and bees, butterflies and hoverflies (Woodcock et al., 2014). The addition of forbs was particularly important for maintaining biodiversity gains. In these and other studies, the use of multi-species seed mixtures in grasslands also led to increased productivity (reviewed in Ashworth et al., 2018). However, such studies mostly manipulate low levels of diversity (below 10 species), and production gains with an increase of diversity from one to several sown species appear to stop at low levels (Ashworth et al., 2018). Such diversity levels of vegetation are well below those achievable in semi-natural grasslands.

3.2 Leaving uncut refuges

Harvesting of grassland biomass, from mowing to baling, kills a large proportion of invertebrates: up to 65–85% of orthopteran populations in some studies (Humbert et al., 2010a, 2010b). Mortality of small mammals such as European hare, roe deer, amphibians, reptiles and ground-nesting birds is also substantial (reviewed in Oppermann et al., 2000; Broyer, 2003; Saumure, Herman & Titman, 2007; Humbert, Ghazoul & Walter, 2009; Grendelmeier, 2011; Grüebler et al., 2012). There is evidence that intensification of grassland production has turned formerly biodiversity-friendly grasslands into so-called sink habitats or even ecological traps (Müller et al., 2005). This means that organisms, such as birds, start breeding on grasslands, but are unable to reproduce due to intensified agricultural practices, thus becoming ecologically trapped in their choice of breeding habitat (Britschgi et al., 2006). As no biomass harvesting method is damage-free (see Section 3.4), leaving uncut grass strips or patches as refugia between each harvest could to some extent mitigate some of the negative impacts (Humbert et al., 2018; Schwarz et al., 2023; Fig. 3).

A long-term experimental field study in Switzerland showed that, compared to controls without a refuge, grasslands with an uncut grass refuge had: (1) higher species richness of specialist butterflies (Bruppacher et al., 2016; see also Konvicka et al., 2008); (2) twice the density and 23% higher species richness of orthopterans (Buri, Arlettaz and Humbert, 2013; see also Humbert *et al.*, 2012; Kaláb, Šipoš & Kočárek, 2020; Schwarz et al., 2023); (3) higher abundance and species richness of wild bees and hoverflies, this due to

Figure 3 Examples of a mowing refuge on a grassland. Images by Jean-Yves Humbert.

an enhanced and prolonged availability of pollen and nectar (Buri et al., 2014; Meyer et al., 2017); and (4) higher abundance of ground beetles (van Klink et al., 2019). Yet, no such effects were found for rove beetles, moths, plant- and leafhoppers and spiders (Buri et al., 2016; van Klink et al., 2019). Other studies have shown that the spider community in wet litter meadows benefits from annual uncut grass strips that remain in the same place for at least 1 year, providing overwintering habitat (Cattin et al., 2003; Frenzel et al., 2022; Schmidt et al., 2008).

Sward architecture is a primary factor to promote phytophagous and predatory grassland arthropod species and biomass in general (Andrey et al., 2014; Woodcock et al., 2009). Therefore, leaving some uncut areas when harvesting biomass is an important measure for ensuring the continuity of complex vegetation during the growing season. This is especially relevant in landscapes where most of the grasslands are mown within a narrow time window. To avoid changes in the plant community, the location of the uncut refuge(s) needs to change at each mowing operation (Rossier et al., 2023; van Klink et al., 2017). Leaving unmown refuges has proven or is highly likely to be beneficial also for mammals, amphibians, reptiles and ground-nesting birds (e.g. Broyer, 2003). Intensive research on an endangered bird, the Corncrake (*Crex crex*), has demonstrated that 10-m-wide strips should be considered as the absolute minimum width for refuge during mowing operations (Arbeiter et al., 2017).

Leaving a grassland parcel fallow for one to several years, or following an arable field with a seed mix of grasses and forbs, is an efficient way of creating biodiversity refugia at the landscape scale. Such biodiversity features have been successfully tested and are now financially supported in several European countries (including Hungary, Finland). Together with long-term set-asides, such low- or non-productive grasslands have been shown to be beneficial for numerous plant taxa, pollinators and other invertebrates, and birds (Hertzog et al., 2023; Herzon et al., 2011; Revaz et al. 2008; Toivonen et al., 2013). In some cases, the use of such fields for extensive production (single late mowing for forage, or low-intensity grazing) may be compatible with biodiversity (Toivonen et al., 2013). A similar approach can work on also pastures, including those managed under rotational grazing: refuges without grazing are left during the most sensitive times of the year (see Section 3.6).

3.3 Maintaining non-productive margins

Similar to the use of field margins on arable land, non-productive margins on intensively managed grassland can be used to enhance biodiversity. This is achieved by renouncing agrochemicals, fencing off livestock from the margins for the duration of the breeding season and/or late single mowing, and

Figure 4 Partially fenced-out margin along a stream: a combination of the grazed and fenced-out parts creates a diversity of conditions that benefit numerous organisms. Image by I. Herzon.

sowing diversified seed mixtures of grass and herb species (Fig. 4). Benefits of establishing 10-m-wide margins on intensive livestock farms, managed so as to increase sward architectural complexity and promote floral species richness, enhanced abundance and species richness of beetles (Woodcock et al., 2007a), bumblebees and butterflies (Potts et al., 2009). Fencing was required to exclude livestock during flowering to prevent negative impacts on a wide range of arthropods (Anderson et al., 2013; Cole et al., 2012). Also, 1.5–3.5 m wide fenced margins of permanent, intensively grazed pastures harboured more springtails (Collembola: Anthropleona) and spiders (Araneae) than control margins (Sheridan et al., 2008). Margins that were rotavated and allowed to regenerate naturally also had more flies (Diptera) and plant species than control plots (ibidem). Birds may also benefit due to increased abundance of large aerial insects (Wiggers et al., 2016), although such a positive effect could not be demonstrated in other studies on birds (Davey et al., 2010) and hares (Zellweger-Fischer et al., 2011).

Margins occupy a relatively small part of the field, have a fairly limited impact on overall production (Pywell et al., 2015), and act as refuges for some species (see Section 3.2). These are especially important for amphibians when placed around ponds in sensitive areas (Schuler et al., 2013). Such margins along watercourses also operate as buffer zones for water protection.

3.4 Delaying and modifying mowing patterns

Delaying first mowing and modifying mowing practices can potentially benefit grassland biodiversity. A review of 24 studies showed that delaying the first mowing date in European grasslands has either positive or neutral effects on biodiversity (Dicks et al., 2020). Benefits included: higher plant species richness, higher densities of two rare arable weeds, more insect species and individuals visiting flowers, higher abundances of some spiders and ground beetles, and higher breeding densities of wading birds. On the other hand, delaying the first cut to the end of the growing season (September in central Europe) can reduce plant species richness (Humbert et al., 2012b). Reduced mowing frequency, omission of mowing in parts of the grassland (refuges), or extensive grazing instead of mowing have been shown to be of the greatest benefit for arthropod populations (Berger et al. 2024).

Mowing from the centre of the field outwards has also been recommended to avoid concentrating individuals of highly mobile species such as birds, mammals and insects in the centre of the field and to facilitate their escape (Fig. 5; Green et al., 1997; Tyler, Green & Casey, 1998; Prochnow & Meierhöfer, 2003; Saumure, Herman & Titman, 2007). Some studies recommend the use of cutter bar mowers over rotary and flail mowers to reduce mortality of mobile species, but the relative reductions in mortality are still poorly understood for most growing conditions, especially for less mobile species (reviewed in Flade et al., 2006; Humbert et al., 2009). Some studies have reported that higher cutting heights are less damaging to field-dwelling animals (reviewed in Humbert et al., 2009). Also abstaining from some operations, such as conditioning, rolling and levelling at critical times for sensitive species is regarded as beneficial for many taxa (Gottwald & Stein-Bachinger, 2016; Humbert et al., 2010b). However, there is little research on the overall mortality caused by the various steps of the harvesting process, including the removal of baled grass. For example, most of the common skylark (*Alauda arvensis*) nest failures are caused by being run over by machinery during the entire silage collection process (Peach et al., 2007), which is particularly difficult to mitigate in practice for small passerine species. A large study in the UK testing ways to improve the reproductive output of the species by manipulating silage harvesting at the field level had limited success: changes in the timing and method of harvesting, or cutting height, proved insufficient to increase breeding success above the rate required for population stability (Buckingham et al., 2015).

In several countries in Europe, there are schemes, in which volunteers carry to search for nests of wader birds, such as Black-tailed Godwit (*Limosa*

Figure 5 Adapted mowing for mobile species, which allows species to escape the mowing equipment towards outer edges of the mown grassland. Image by Lola Fernández Multigner.

limosa), and mark them, so that farmers can avoid nest destruction during the operations (Herzon et al., 2018). Despite nest protection locally, there is somewhat inconsistent evidence about the success of such interventions on the species populations in The Netherlands (e.g. Verhulst et al., 2007). A potential reason is that approaching and designating the nests and leaving uncut areas during mowing can increase predation rates, so that nest success may not be increased, or that predation rate in production grasslands is too high for meadows birds regardless of nest protection (Grüebler et al., 2012). On the other hand, such collaborative partnership between a farmer network and an NGO has been shown to contribute to farmers' knowledge about the conservation of birds on their grasslands and to recognition of the conservation role of farmers among citizens, decision-makers and agri-food companies (Runhaar & Polman, 2018).

The effectiveness of different methods to scare mammals and birds away from mowers has been researched, but the evidence is still limited (Green, 2007 for ducks in the USA). Loud noises, scarecrows, dogs or extension arm bars were found to be inefficient in reducing mortality of fawns (Fackelmeier & Biebl, 2009; Lehner et al., 2015). This is partly because young mammals try to hide by lying down when frightened. However, scaring female deer with offspring away from grassland to be mown in consequent days seems to work (Jarnemo, 2002). Mechanical or recently developed electronic acoustic and visual scaring devices can be used (Leune et al., 2022).

In summary, modern biomass harvesting on agricultural grassland managed for maximum attainable productivity, whether for feed or bioenergy, poses a serious challenge to biodiversity conservation on grassland. Nevertheless, there is modest evidence for the benefits of modifying harvesting practices for some species. The use of new technologies may also help to resolve the conflict between grassland production and biodiversity. The importance of creating landscape-level variation in management across grasslands in collaboration among producers as well as NGOs is also needed.

3.5 Reducing agrochemical inputs

Some research has shown that reducing fertiliser or pesticide inputs can benefit plants, invertebrates and birds, while other studies have shown no or little effect (reviewed in Dicks et al., 2020). One reason may be that many improved grasslands receive such substantial amounts of organic and/or mineral fertiliser that they have accumulated soil phosphorus reserves in the long run. Reducing or even temporarily stopping inputs therefore makes relatively little difference to plant communities on intensively managed grasslands, as indirectly confirmed by an EU-wide survey (Kleijn et al., 2008). The benefits are likely to be indirect through a concomitant reduction in stocking density, and a delay in the first mowing or grazing date, i.e. via an overall extensification of management (see Section 3.7).

The type of fertiliser is likely to be important. Using farmyard manure was shown to lead to both production benefits and biodiversity gains (plant and insect communities), while the addition of mineral fertilizer at the highest tested level (of about 100 N t kg^{-1} $year^{-1}$) had the most negative impact on biodiversity (Villa-Galaviz et al., 2023). Production benefits aside, additional fertilisation of grassland decreases plant diversity irrespective of the source of nutrients, as compared to a complete lack of fertilizer (Villa-Galaviz et al., 2023). According to Onrust et al. (2019), replacing slurry and slit-injection-based management with the use of farmyard manure increases surface soil moisture to benefit earthworms and the birds that feed on them.

3.6 Grazing management

Grazing is generally considered to be more benign for farmland biodiversity than mowing forage (Metera et al., 2010; Tälle et al., 2016). However, it has declined in many regions in favour of indoor housing of ruminants, especially in the dairy sector (van den Pol-van Dasselaar et al., 2020). It is therefore important to maintain grazing, at least for part of the herd, if biodiversity maintenance within grassland-based systems is desired. Three factors are key in designing livestock grazing on grasslands to benefit biodiversity: (i) livestock type, (ii) grazing intensity and (iii) grassland type such as semi-natural or sown, dry or mesic, etc. (Peco et al., 2012; Tóth et al., 2018). Due to the high variability of grazing regimes, the most appropriate management for biodiversity needs to be adapted to the region, climate, soil type, land-use history and evolutionary co-existence of species with herbivores. It should be further fine-tuned at parcel level and adjusted between years and within a year, ideally following biomass production, vegetation structure and plant or bird indicator species (review in Herzon et al., 2018, and see Section 5).

Livestock identity has been shown to have a strong influence on the effects of grazing on grassland biodiversity (Jerrentrup et al., 2015; Tóth et al., 2018). Most research has focused on cattle and sheep with evidence for other livestock species or wild grazers being limited (Metera et al., 2010). The grazing strategy of livestock is influenced by grassland type such as semi-natural (e.g. Mládek et al., 2013), and pasture management (e.g. rotational or mob grazing forces grazers to mass foraging), and may change over the years (Zhang et al., 2023). Because of the evolutionary adaptation to the past conditions, continuity in the use of the livestock type best adapted to the biophysical conditions (soil, terrain, climate) is also likely to be best for species communities adapted to specific grazing regimes. This includes also less common herbivores, such as pigs (e.g. Iberian agroforestry systems), and mixed grazing (e.g. sheep and cattle or cattle and horses). For example, the fact that most pre-agricultural grassland communities in Europe were formed by and adapted to wild bovine and equine herbivores, may explain why cattle grazing has been shown to be more beneficial for maintaining grassland biodiversity than sheep grazing (Bullock et al., 2001), especially those characterised by high forb diversity, such as in loess steppe grasslands or other forb-rich grasslands (Török et al., 2018). In the global context, however, the benefits of cattle grazing for grassland butterflies revealed only slightly more positive effects than negative, with many inconclusive results due to variability in management, environmental and evolutionary factors (Bussan, 2022). Even the replacement of a wild bovine species (such as bison) by a related one (cattle) may lead to changes in biodiversity due to differences in their behavior and management (Lueders et al., 2006). In some parts of the world, such as Australia, grazing by cattle and sheep has been transplanted on the original open habitats with no history of such type of herbivory (Lunt et al., 2007). A review on the use of domestic herbivores for ecosystem management in Mediterranean landscapes demonstrated highly varied effects on biodiversity and habitat conservation (Ribeiro et al., 2023).

As shown above, maintaining a diverse vegetation structure and plant composition is a basis for overall biodiversity in grasslands (see Section 2). Therefore, grazing regimes should ideally vary in both space and time, also reflecting small-scale spatial and temporal variation in primary productivity. For example, in a semi-arid rangeland, grazing affected the diversity of annual plant communities differently along the gradient of community productivity (Osem et al., 2002). In each grazing year, some pastures could be kept at relatively low grazing levels to provide a diversity of flowering resources and safe breeding of ground-nesting birds. Creating grazing exclusion zones or having sufficiently long rest periods in paddocks under rotational grazing, can also enhance overall biodiversity especially species sensitive to trampling and defoliation. It works similarly to refuges left during mowing (see Section 3.2).

Using native prairies as rangelands, in which domesticated bovines and sheep, mimic the ecological functions of native herbivores, is often regarded among sustainable agricultural systems because it allows production in original grassland ecosystems, preserving their biodiversity. Research from North America indicates that such multiple benefits arise at comparatively low stocking rates and grazing intensities on multispecies permanent grasslands with low to no additional inputs (Lusk & Koper, 2013). For example, maintaining a gradient of grazing regimes on such farms maximizes the availability of suitable nest sites for various species in mixed-grass prairies in production (Pintaric et al., 2019). There is also evidence that the ecological effects of cattle grazing differ in numerous ways from historical effects of bison grazing, due to different selections of grass vs. forbs, and use of the landscape (Lueders et al., 2006).

On low-productive grasslands grazed for biodiversity conservation, research often focuses on year-round free grazing (e.g. Köhler et al., 2016) or grazing with shepherds (Török et al., 2016a) (see Section 3.8). In some landscapes, such as extensive grasslands on alkaline soils, the effects on species and functional grassland biodiversity were similar under both year-round free and low-intensity seasonal grazing with a shepherd (Török et al., 2016b). Rotational or mob grazing has been promoted recently for its potential production and environmental benefits. However, research to date shows that this grazing regime does not universally promote biodiversity due to heavy trampling and intensive defoliation (reviewed in Morris, 2021). Additional measures are needed to manage also for biodiversity under this regime. For example, leaving wildlife refuges on central paddocks that are excluded from grazing during the breeding season is promoted as beneficial for some typical grassland bird species in the USA (Fig. 6; Undersander et al., 2000), but evidence for improved nest survival is still scarce.

When planning fine-tuned management at the pasture and landscape level, traditional knowledge of experienced herders (Molnár et al., 2020), extensive research on modern pastures and innovative technological tools for monitoring pasture development and indicator species can be integrated in a complementary way to achieve the multiple benefits of pasture-based production. The importance of the landscape structure, into which pastures are embedded – the presence of trees, bushes, water courses, buffer zones, woodland etc. – will heavily influence biodiversity. Retention or planting of trees and shrubs on pastures also improves carbon sequestration without compromising overall pasture quality (Tölgyesi et al., 2023), and provides shade for grazing animals and other benefits to production (England et al., 2020).

Finally, biodiversity includes species that challenge pasture-based production: apex predators such as wolves, wolverines and eagles. For many pastoralists worldwide, the conflict between wildlife and grazing animals is seen as a severe problem (e.g. Bisi et al., 2010; The World Bank, 2023). Several studies have demonstrated the effectiveness of various methods of resolving such conflicts by the use of shepherd dogs and specially designed fences (Bruns et al., 2020). Some of these also minimise the risk of disease transmission from wildlife to livestock (Jori et al., 2021). In some cases, the conflict is managed through payments to herders linked to the confirmed

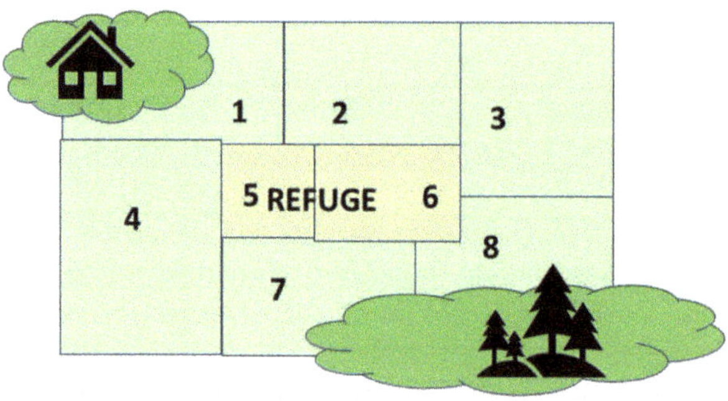

Figure 6 Refuge for ground nesting meadows birds recommended for grassland managed under rotational, or mob grazing. It is recommended to place it in central paddocks away from trees or farmsteads and exclude it from grazing during the breeding season of birds species sensitive to trampling. Re-drawn from Undersander et al. (2000).

Published by Burleigh Dodds Science Publishing Limited, 2025.

breeding cases of predators of conservation importance (review by Herzon et al., 2018). There is an emerging understanding of the importance of bringing stakeholders together in conflict resolution processes and developing national policies for human-wildlife conflict (e.g. Salvatori et al., 2020; The World Bank, 2023).

In summary, grazing is an important and perhaps most appropriate form of management on grasslands due to the evolutionary adaptation to herbivory by large mammals on native grasslands. As a rule of thumb, low to moderate grazing intensities on multi-species grasslands without high additional inputs of agrochemicals would support the highest multi-taxa diversity. Due to the high variability of grazing regimes, the most appropriate management for biodiversity needs to be adapted to the region, climate, soil type, land-use history and evolutionary coexistence of species with herbivores. As with other forms of grassland management, there are trade-offs between production and conservation in pastoral systems. One type of conflict specific to pastoral systems is that between native apex predators and domesticated grazers. Conflict resolution processes for human-wildlife conflicts should also apply to this conflict context.

3.7 Extensifying management

As shown above, reducing input and land use intensity on grasslands has been confirmed to have biodiversity benefits, at least for some taxa. Many studies involve changing two or more management practices at the same time. For example, a replicated controlled study in the UK (Peach et al., 2007) found that plant seed heads and invertebrates, including those important as food for birds, were more abundant on plots with reduced overall management intensity (lighter grazing, seasonal removal of livestock, reduced fertiliser application). Noxious weed cover remained low on all plots throughout the study. Similarly, the index of the extensity of grassland management, as indicated by the levels of fertilisation, mowing frequency and grazing, positively affected species richness of several insect taxa, especially pollinators (Panassiti et al., 2023). Focusing on an overall reduction in the intensity of grassland use, rather than on individual practices, may allow producers to choose options that best suit their production type and achieve benefits for biodiversity while optimising overall productivity and resource use. This is exemplified in the Farming for Biodiversity project (see Case Study). Implementing extensification of agricultural management for biodiversity and other ecosystem services is disproportionately more costly in intensively managed areas than in extensively managed ones (e.g. Kleijn et al., 2008), and for some livestock systems than others (e.g. dairy is more dependent on high-energy forage than suckler cow production). Solutions and support systems need to be developed in a context-specific approach (see Section 5).

3.8 Restoring and managing semi-natural grasslands

Due to the exceptional biodiversity and currently high levels of endangerment of many types of semi-natural grasslands, research on the restoration of such grasslands is prolific in Europe. However, similar research exists in other regions with native grasslands used for livestock production under a range of intensities (e.g. Bartolome et al., 2014 for North America). Active restoration of semi-natural grasslands can be separated into two main categories: (1) the re-creation of grasslands; and (2) the restoration of existing but degraded or intensively managed, species-poor grasslands. Both are possible but may take time (Pywell et al., 2002; Török et al., 2011). Re-creation from arable land appears to be easier than restoration on intensive grassland, where the intact vegetation provides few opportunities for introduced species to establish and persist (Kiehl et al., 2010). In addition, high soil fertility in intensive grasslands may limit the success of plant introductions (Janssens et al., 1998). To restore a comparable plant community as in the past, grassland restoration often requires active measures, mainly the addition of seeds after soil disturbance such as ploughing, which facilitates seed germination (Klaus et al., 2017; Poschlod & Biewer, 2005; Pywell et al., 2007). Sowing a seed mixture and direct seeding by transferring hay from a freshly mown biodiversity-rich donor grassland are the two most common methods of seed addition used in semi-natural grassland restoration (Hedberg & Kotowski, 2010; Klimkowska et al., 2010). A recent meta-analysis of the respective effectiveness of these methods in restoring temperate grasslands provides quantitative evidence that such operations are generally effective (Slodowicz et al., 2023), supporting previous narrative reviews on the same topic (Hedberg & Kotowski, 2010; Kiehl et al., 2010). Restoration success of plant species richness is best achieved by combining seeds from a species-rich donor grassland with commercial seeds or solely using the seeds of a species-rich donor grassland.

The use of selected local breeds has been shown to work best on semi-natural grasslands (Köhler et al., 2016) due to their adaptation to local conditions or specific grassland types, and their ability to exploit more diverse and possibly less nutritious vegetation than on intensive grasslands. All countries with a history of traditional pastoralism have such breeds, and their conservation is supported by national governments. Livestock that move across large areas of grassland or whole landscapes, as in transhumance, can also spread seeds of native species (Ozinga et al., 2009), contributing to active restoration. Grazing management can thus be designed in a way that supports the restoration of diversity in degraded or enhanced grasslands. A solution is to start grazing in high-diversity grasslands and then move the herd towards species-poor grasslands so as to facilitate seed dispersal without any further human intervention.

Other forms of management, such as controlled burning and removal of the surface layer of litter and soil, can also be effective for such restoration (Valkó et al., 2014), especially in regions with a long history of fire regimes, such as the Mediterranean climates (Porqueddu et al., 2016) or Australia (Lunt et al., 2007).

4 Grassland biodiversity at farm and landscape levels

The diversity of the whole agro-landscape has been emphasised as the best approach to maintain biodiversity (Benton et al., 2003). A modern regional separation of arable and grassland-based production has undermined such diversity. Incorporating some arable crops into grassland-dominated landscapes is known to benefit mobile species that can exploit resources across landscapes and through time (Robinson et al., 2001). For example, some birds can nest on arable crops while moving with their chicks to grasslands to feed (Fig. 7) while some insectivores foraging on the ground require patches of bare ground to access their invertebrate prey base that dwells in the swards (e.g. Arlettaz et al., 2012; Schaub et al., 2010). The diversity of bird species overwintering in intensive grassland landscapes was similarly enhanced by mosaics of fields managed as short-term leys and permanent pastures with low-intensity cattle grazing during autumn and winter (Perkins et al., 2000). The systematic

Figure 7 Resource use by mobile species like birds benefit from a spatio-temporal juxtaposition of different land uses across the cultivated landscape, providing feeding grounds, shelter and even nesting opportunities, that is, optimal conditions for any stage of the life cycle. Image by Lola Fernández Multigner.

Published by Burleigh Dodds Science Publishing Limited, 2025.

re-integration of grasslands into arable production systems would reinstate the many biodiversity benefits of traditional land management (Liu et al., 2013; Porqueddu et al., 2016). This happens on organic arable farms, which are reliant on grass-legume leys for nutrient management (Stein-Bachinger et al., 2021). The maintenance and restoration of integrated production systems associating grasslands with other types of production generally boost biodiversity (e.g. reviewed for agroforestry in Castle et al., 2022, or traditional systems of China, Liu et al., 2013).

The presence and diversity of various non-cropped areas and landscape features – trees, bushes, water bodies, woodland – in and around agricultural grasslands, which are themselves managed with varied defoliation regimes, is likely to benefit all those elements of biodiversity that have undergone a long co-evolution under the pressure of natural herbivory in woodland-grassland landscapes. Numerous research on the potential benefits of organic production (as a case for more extensive production in terms of some agricultural inputs) for biodiversity have further confirmed that the farm and landscape structure has a considerably higher influence on farmland biodiversity than the very management regime of the fields *per se* (e.g. Schneider et al., 2014; Vickery & Arlettaz, 2012). In order to enhance biodiversity, as well as the ecological functions and ecosystem services it provides, future agriculture should reintegrate natural features into landscapes that are deprived of them following intensification.

Re-creating a range of grazing stages within a shifting mosaic of disturbance intensities in both space and time (from intensive short-period grazing to long rest periods of many months) and allowing diversities of productivities of natural vegetation on heterogeneous soil and terrain conditions – i.e. mimicking natural herbivory regimes – is likely to maintain biodiversity on originally open steppe-like landscapes. The so-called holistic grazing management claims to aim at this, but its implementation remains varied in its ambition and delivery of biodiversity (Morris, 2021).

There is a need for research and development projects that support farms and neighbouring farm clusters too, on one hand, optimise land use management for maximum benefits at the lowest costs to farm enterprise (see Section 5) and, on the other hand, improve public benefits in rural areas, while balancing local and global impacts so as to opt for the most suitable modes of agricultural production. Any such approach should (1) prevent the conversion of ancient permanent grassland to arable or temporary grassland, (2) extensify grassland production to minimise environmental impacts such as erosion and water pollution (Leidinger et al., 2017; Schils et al., 2022), and (3) maintain and restore semi-natural grasslands. Keeping or even increasing management intensity on some grasslands should in any case be compensated for by an extensification on some other grasslands in the same landscape (Qi et al.,

2018). Paludiculture, that is often understood as farming systems in areas with high water tables, represents an interesting management option that yields specialised agricultural products (rice, buffalo) and boosts the biodiversity typically encountered in wetlands (Martens et al., 2023). Finally, restoration of pre-agricultural woodland-pasture mosaics, wetlands and forests from some grassland areas, as well as native grassland ecosystems such as prairies, can contribute to the restoration of natural ecosystems, or rewilding, in regions that underwent extensive modification by humans (Garrido et al., 2019; Boyce et al., 2022).

5 Case study

The following example illustrates how farmers involved in both dairy and suckler cow production can increase the diversity of flora and fauna at the farm level, and how emerging conflicts between production and biodiversity can be resolved. Though the approach has been developed and implemented with organic farmers and for organic production systems, the general principles and specific measures are applicable, and indeed are in use, also in other contexts.

The 'Farming for Biodiversity' approach results from one of the first practice-orientated research of this kind, running since 2012 in Germany in organic farming systems (Gottwald & Stein-Bachinger, 2018). Operating in close cooperation with farmers, and drawing on expert knowledge, evidence-based information and their own scientific outputs, researchers have developed, tested and implemented a set of nature management options (overall about 150 options), accounting for the site conditions and farming practices (Stein-Bachinger et al., 2010; Stein-Bachinger & Fuchs, 2012; Gottwald & Stein-Bachinger, 2016). The overall precondition to join the scheme is following the standards of organic farming. To receive the certificate, farmers need to reach 120 points per 100 hectares from measures of their choice. The measures can be applied to arable land, grassland, landscape elements and farmsteads covering a broad range of options (see Table 1 for a selection of measures on grassland and grass-legume leys). Each measure is associated with credit points reflecting its ecological performance and area of implementation (Gottwald & Stein-Bachinger, 2018). The occurrence of special species (e.g. red list species) and habitats (e.g. wetlands) are evaluated as well (result-oriented modules). Since most of these nature conservation options entail additional costs or effort from farmers, or compromised yield and/or quality of agricultural products, every aspect was carefully quantified (Rühs & Stein-Bachinger, 2024), also using bio-economic modelling, which includes cost and benefit calculations alongside ecological gains (Schuler et al., 2013).

Involved land managers could get a premium price for selected products by the wholesaler EDEKA (e.g. meat, potatoes) if they reach a threshold of

Table 1 Selection of measures on grassland and grass-legume leys offered to organic farmers in 'Farming for Biodiversity' approach in Germany.

Measure	Description
Basic measure	A moderately extensive grassland management (fertilisation according to the standards of organic farming; max. 2 livestock units/ha of the stocking rate, grazing regime adapted to the pasture growth, prevention of overgrazing; leaving grazing residues where possible; max. 2 grass cuts and subsequent grazing, management of problem weed species). There are additional measures available for selected parcels with one or more options.
Extensive grazing on dry grassland or coastal grassland	On sites with very low productivity; max. 0.8 LU/ha, no fertilisation
Reduction of rolling and levelling	For periods when species are sensitive to disturbance (depending on the region and species or avoiding it completely).
8–10 weeks without farming operations during the breeding period of birds or late first utilization for mowing or grazing beginning/middle of July	Ground breeding birds could complete their nesting. Delayed use must be adapted to the vegetation, as in productive and early maturing stands the vegetation often becomes too dense with permanent late utilization.
Leaving refuge (unmown or grazing-free areas within grassland)	About 3–20 % of refuges, also possible on pastures. Especially recommended at flower-rich areas, least productive spots, waterlogged depressions, strips along ditches and fences, etc.
Perennial strips	Strips at least 3 m wide that are not used or disturbed from summer to mid-July of the following year. Available also for legume-grass leys.
High cut on small or large scale	Mowing height at least 12 cm. Due to the lower proportion of stalks, the fodder quality will be better, which reduces the losses for the farmer. Available also for legume-grass leys.
Individual measures	A variety of site-specific measures for biodiversity-rich biotopes (such as semi-natural grasslands), and endangered species (such as corncrake (*Crex crex*)) can be individually added with the help of the advisor.

ecological quality based on a whole farm assessment. Consultants individually advise the farmers on how to implement nature conservation measures and help to identify target species. For example, moderate extensive grazing still allows good productivity for the farm. More biodiversity-friendly options on suitable grassland parcels (e.g. delayed cutting, leaving strips unmown) are available additionally, and qualified for higher credit points and reaching the threshold more easily. Besides a premium price when having the 'Farming for Biodiversity' certificate, farmers could claim public financial support through agri-environmental schemes. Further non-monetary support to participating producers is provided by allowing consumers to get to know about biodiversity achievements of the farms through a tracking code (QR code) on the farm products leading to the project homepage (www.landwirtschaft-artenvielfalt.de).

The efficiency of the special measures for promoting biodiversity was analysed, confirming that, if implemented in the right way, combinations of these targeted measures benefit a wide range of taxa among plants, amphibians and birds (Gottwald & Stein-Bachinger, 2018). For example, measures implemented on legume–grass leys, such as 8–10 weeks rest period without farming operations or leaving refuge (Table 1), contributed to the reduction in the direct nest losses caused by mowing operations for four species of birds (Stein-Bachinger & Fuchs, 2012).

Research also confirmed the importance of implementing the most effective measures (i.e. those that lead to most biodiversity gain), even when run at small scales. Such are, for example, leaving small strips or patches of grassland unused for a prolonged time as refuges for birds (Fig. 8), insects and other animals (Gottwald et al., 2017), or the clearing of woody plants on the south-facing sections of kettle holes for amphibians (Stein-Bachinger et al. 2021). For example, the breeding success of an emblematic meadow bird whinchat (*Saxicola rubetra*) was nearly double on grasslands with measures than on plots without them (Fig. 9). Although some of such measures are expensive to implement (e.g. leaving out unmown strips or postponing mowing of hay or silage grass on dairy farms), the fact that they can be implemented at a small scale might render them economically more acceptable, while they provide advantages for biodiversity as long as there is good connectivity to other suitable habitats (Kruse et al., 2016). Yet, the options most favoured by farmers

Figure 8 A late mown refuge along the field edge, in which Whinchats (*Saxicola rubetra*) predominantly build their nests (Gottwald et al. 2017).

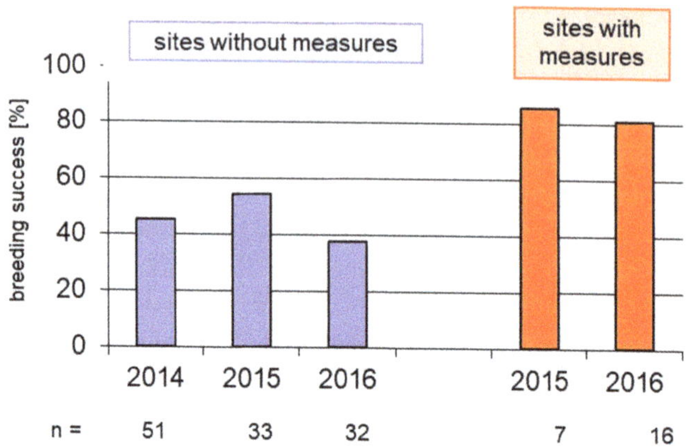

Figure 9 Breeding success of whinchat (*Saxicola rubetra*) (ratio of nests with at least one fledgling) for nest sites with and without measures (Gottwald et al. 2017).

on grassland were "reduction of rolling and levelling" and 'limitation in fertiliser application', which are relatively easy to implement on large areas (Gottwald & Stein-Bachinger, 2018).

This approach offers to farmers a high degree of flexibility for acquiring the nature conservation certificate. This is important because the feasibility, effectiveness and efficiency of the different measures are highly dependent on individual farm conditions, such as soil quality and weather circumstances. Because the measures are chosen mostly according to habitat type (as specified in the scheme) and characteristic species (proposed by advisors with knowledge of species), the approach offers a strategic and pragmatic way for improving biodiversity in any farm context. However, since this scheme is entirely voluntary, the decision to join it is ultimately based on the personal preferences and experiences of the farmers. The scoring scheme can be used by farmers also for a self-evaluation of their ecological performance, with the objective to ameliorate it (Birrer et al., 2014). This approach consisting of a scoring strategy and a series of optional measures for biodiversity promotion could be readily adapted to other geographic contexts and biomes.

Finally, the success of this approach relies largely on a long-term commitment of all actors: producers, advisors and regional agencies as shown in 'Farming for Biodiversity'. Over time, farmers' experience grows while advisors' competence improves, paving the way for consolidating efficient biodiversity-friendly and profitable farming systems into the future. The role of advisors is to help farmers identify target species and select suitable sites, taking both agricultural needs and species' ecological requirements into consideration. This may sound laborious but it is the price to pay for achieving tangible results for biodiversity while minimizing losses for the farmers. Compared to the mostly non-selective distribution of subsidies characterizing conventional input-based incentives, output-based strategies are economically and ecologically rational. Investing in expert advisory services specialised in biodiversity and agriculture is thus no waste of money. To conclude, farmers' readiness to cultivate land with nature objectives in mind increases if they can (1) choose between different measures from a sufficient set of options; (2) properly quantify inherent costs, which may be key for operating measures selection; (3) receive adequate remuneration given the effort and potential loss in production and (4) finally, assess their biodiversity achievements, preferably by themselves.

6 Conclusion

Biodiversity conservation is a rapidly advancing research area in sustainable grassland management. The biodiversity value of grasslands varies widely, primarily influenced by production-driven management intensity. There is considerable evidence by now that intensification of both grassland and arable systems has led to dramatic losses of farmland biodiversity, while the expansion of agricultural production, including of the livestock industry, has caused damages to the biosphere beyond agro-ecosystems. Among all habitat types, grasslands have one of the greatest potentials for supporting biodiversity within human-managed landscapes. Recognizing the impact of expanding grassland-based production on biodiversity is essential for realising this potential.

The well-established mechanisms through which management affects grassland biodiversity are known for many taxa, particularly in European grasslands, but also increasingly in North and South America, Africa, Australia and Asia. Due to ample research on the relationships between biodiversity and grassland management practices, practitioners can now choose from a wide array of approaches to promote and restore farmland flora and fauna, and deliver multiple ecosystem services. Optimal trade-offs between productivity, on one hand, and biodiversity and ecosystem services, on the other, may be difficult to achieve under all circumstances, especially on intensive permanent and temporary ley grasslands. Spatial and temporal optimisation of land use for multiple production and other purposes remains the crux. The underlying idea is to create a spatio-temporal mosaic of highly productive areas, extensively managed parcels, and native vegetation. Specific measures can then be adapted to the prevailing geographic, climatic, agricultural and socio-economic contexts, including the availability of public financial incentives. Sufficient societal and financial support to farmers, as our case study illustrates, are necessities for enhancing sustainability of grassland-based production.

The still prevailing mainstream focus on the volumes of agricultural yields negatively affects not only biodiversity but also the multifunctionality of the whole grassland-based production system. There is a growing recognition of the need to improve the sustainability of land use across wider rural landscapes to optimise overall benefits in the future, such as improving landscape hydrology, enhancing carbon storage and restoring nature, in which grasslands can play a crucial role. However, these benefits remain unrealised in most modern, intensively cultivated systems where grasslands occur. If the approaches outlined here are likely to deliver multiple benefits, including for biodiversity, they are likely to cause reductions in yields. Inherent trade-offs between achieving various benefits, including biodiversity, and high yields necessitate systemic changes, such as dietary adjustments on the food system level. Along with the adjustment of management, a shift towards more plant-based human diets is thus a prerequisite for achieving such a change of paradigm in agriculture and society. Only less extravagant patterns of food consumption, especially of animal proteins, in modern industrial societies will allow humanity to achieve multiple sustainability targets in food systems (Röös et al., 2022).

7 Future trends in research

We identify four key avenues of research that may in the future contribute to promote biodiversity and the provision of ecosystem services in cultivated grasslands. First, a rapidly progressing topic concerns the diversification of grassland vegetation through the addition of seed mixtures. Second, research is urgently needed on new modes of grassland management that allow other species to co-exist with production plants. Third, studies should focus on active restoration of biodiversity across grassland landscapes for improving whole-farm economic performance, while at the same time enhancing biodiversity and boosting key ecosystem services (Soliveres et al., 2016). Fourth, research should deal with minimising the indirect impacts of grassland-based production on biodiversity globally, including climatic impacts that fall outside of this chapter's focus.

Traditionally, research and education in the fields of grassland-based production and grassland conservation, respectively, tended to be separated by disciplinary lines (e.g. Herzon et al., 2021), but this is rapidly changing (Zhao, 2023). We need to seek adaptive solutions that include perspectives of ensuring adequate production and biodiversity conservation through flexible and goal-oriented management, and that are developed in close collaboration between scholars of grassland and ruminant production, grassland ecologists and producers (Durant et al., 2008). Such research should expand into regions with relatively little or no history of grazing by domesticated ruminants, such as Australia, so that the costs and benefits of grazing can be compared with the pros and cons of other sources of ecosystem disturbance aiming at maintaining grassy habitats, such as fire (Lunt et al., 2007).

Further research on ecosystem services of grasslands from a landscape-centered multi-scale perspective is essential (Zhao et al., 2020). Biodiversity must be measured across the whole range of present and potentially restorable farmed ecosystems devoted to grasslands-based production so that trade-offs can be grasped and balanced. Solutions should be developed across sectoral policies and their impacts modelled to minimise both direct and indirect impacts of grassland-based production on biodiversity at regional and global scales. The need to quantify the multifunctionality of grasslands has also been highlighted in view of monetary compensations to producers for efforts that go beyond production (e.g. EIP-AGRI, 2018; Lyons et al., 2023).

Research is also needed to improve and refine guidelines for successful implementation of specific biodiversity measures in grasslands, especially since today available methods are considered to be less efficient than on arable crops (e.g. Staggenborg & Anthes, 2022). Several gaps in knowledge have been highlighted in this chapter. In particular, we need controlled experiments that can tease apart which grazing and mowing modes are most beneficial to flora and fauna. In this respect, the impact of different types of grazers, combinations thereof and the influence of different grazing intensities across a variety of environmental conditions deliver precious information for sustainable management. However, given that processes at multiple levels are hardly amenable to controlled experimental trials, one must use novel methodological approaches based on modelling and simulations in order to integrate spatial and temporal processes at the relevant scale (e.g. Durant et al., 2008).

Research into belowground grassland biodiversity has made huge advances (Zhao, 2023), which is of prime importance considering the vital contribution of such biodiversity to soil fertility, plant nutrient uptake, disease suppression and carbon sequestration, as well as above-ground biodiversity (Lyons et al., 2023).

There is a need to streamline proper definitions of grassland types (Török & Dengler, 2021). This will ease the collection of statistics and the combination of remote-sensed data of different origins, and the development of common indicators. It will also help alerting about the grassland state and assist in decision-making for optimal management strategies. For now, there is a lack of studies comparing the functioning of permanent and temporary grasslands across the whole spectrum of ecosystem services as concluded in the review of Schils et al. (2022).

A further research agenda should be launched for combatting the degradation and promoting the restoration of grasslands for biodiversity, ecological functions and ecosystem services, while in the same time accounting for the long-term success of different restoration tactics from a socio-economic perspective (Bardgett et al., 2021).

Innovative technology development and deployment in all areas of environmental sustainability of grassland production has been particularly fast in recent times (Zhao, 2023). Potential negative impacts on biodiversity must be assessed critically while synergetic opportunities must be sought. The domain of alternative methods of grassland-based protein production that can be more compatible with wildlife conservation is an interesting avenue (Jori et al., 2021).

8 Acknowledgements

The authors are grateful to Lea Opitz for skillful technical support and Lola Fernández Multigner for illustrations. IH thanks Rowan Eisner for fruitful discussions on biodiversity and grasslands in Australian and tropical contexts. At a time of work on the chapter, the authors were financially supported: RA and JYH by the Swiss National Science Foundation (310030-215272), JMB by NERC (NE/W005050/1 AgZero+), PT by the National Research, Development and Innovation Office, Hungary (NKFIH; KKP 144068 and K 137573), IH by Research Council of Finland (350649).

9 Where to look for further information

The following articles and books provide a good overview of the subject:

Articles:

- Bengtsson, J. et al. (2019) 'Grasslands–more important for ecosystem services than you might think', *Ecosphere*, 10(2), p. e02582. Available at: https://doi.org/10.1002/ECS2.2582.
- Bullock, J.M. et al. (2020) 'Biodiversity in intensive grasslands: is a compromise possible?', *Grassland Science in Europe*, 25, pp. 384–393. Available at: https://www.europeangrassland.org/fileadmin/documents/Infos/Printed_Matter/Proceedings/EGF2020.pdf.
- Lindborg, R. et al. (2023) 'Ecosystem services provided by semi-natural and intensified grasslands: Synergies, trade-offs and linkages to plant traits and functional richness', *Applied Vegetation Science*, 26(2), pp. 1–13. Available at: https://doi.org/10.1111/avsc.12729.
- Plantureux, S., Peeters, A. and McCracken, D. (2005) 'Biodiversity in intensive grasslands: Effect of management, improvement and challenges', *Agronomy Research*, 3(2), pp. 153–164.
- Schils, R.L.M. et al. (2022) 'Permanent grasslands in Europe: Land use change and intensification decrease their multifunctionality', *Agriculture, Ecosystems and Environment*, 330(January). Available at: https://doi.org/10.1016/j.agee.2022.107891.

Books:

- Squires, V. R. Dengler, J., Hua, L., Feng, H. (eds.) (2018), *Grasslands of the World: Diversity, Management and Conservation*. CRC Press, ISBN 9780367780937 pp. 426.
- Grasslands and Shrublands - Sea of Plants. 2020. In: Goldstein, M. I. and DellaSala, D.A. (eds) *Encyclopedia of the World's Biomes*. Volume 3. Section 6. Elsevier. Available at: https://www.sciencedirect.com/referencework/9780128160978/encyclopedia-of-the-worlds-biomes: the most recent overview of grassland biomes of the world.
- Petermann J. S. and O. Y. Buzhdygan. 2021. Grassland biodiversity. *Current Biology* 31, 1195–1201, https://doi.org/10.1016/j.cub.2021.06.060.

Key research and research-based resources in this area can be found at the following organisations:

- Conservation Evidence (https://www.conservationevidence.com/): a free, authoritative information resource designed to support decisions about how to maintain and restore biodiversity; includes several summaries of evidence from the scientific literature about the effects of conservation actions on grasslands.
- Result based payment (RBP) network: https://www.rbpnetwork.eu/: a meeting point for sharing experiences on result based payments in Europe, of which many target grasslands.
- EIP-AGRI Focus Group. 2012. Wildlife and agricultural production. Available from https://ec.europa.eu/eip/agriculture/en/publications/eip-agri-focus-group-wildlife-and-agricultural-0.html
- HNV-Link (http://hnvlink.eu/): a networking project focusing on High Nature Value farmland and farming systems, and innovative solutions to increasing the socio-economic viability of such biodiverse systems.
- Biodiversity exploratories (https://www.biodiversity-exploratories.de/en/): a German large-scale project with research articles on grasslands ecology and biodiversity
- The Eurasian Dry Grassland Group (EDGG, https://edgg.org/): a European network of researchers and conservationists interested in natural and semi-natural grasslands.
- (https://www.biodivers.ch/): a webpage with practical guidelines and references for the management and restoration of divers habitat, including grasslands (in French and German).
- Farming for biodiversity. https://www.landwirtschaft-artenvielfalt.de/ (In German).

10 References

Anderson, A., Carnus, T., Helden, A.J., et al. 2013. The influence of conservation field margins in intensively managed grazing land on communities of five arthropod trophic groups. *Insect Conserv. Divers.* 6, 201–211. https://doi.org/10.1111/J.1752-4598.2012.00203.X

Andrey, A., Humbert, J.Y., Pernollet, C. & Arlettaz, R., 2014. Experimental evidence for the immediate impact of fertilization and irrigation upon the plant and invertebrate communities of mountain grasslands. *Ecol. Evol.* 4, 2610–2623. https://doi.org/10.1002/ECE3.1118

Arbeiter, S., Helmecke, A. & Bellebaum, J., 2017. Do Corncrakes Crex crex benefit from unmown refuge strips? *Bird Conserv. Int.* 27, 560–567. https://doi.org/10.1017/S0959270916000447

Arlettaz, R., Maurer, M.L., Mosimann-Kampe, P., et al. 2012. New vineyard cultivation practices create patchy ground vegetation, favouring Woodlarks. *J. Ornithol.* 153, 229-238.

Ashworth, A.J., Toler, H.D., Allen, F.L. & Auge, R.M., 2018. Global meta-analysis reveals agro-grassland productivity varies based on species diversity over time. *PLoS One* 13, e0200274. https://doi.org/10.1371/JOURNAL.PONE.0200274

Attwood, S.J., Maron, M., House, A.P.N. & Zammit, C., 2008. Do arthropod assemblages display globally consistent responses to intensified agricultural land use and management? *Glob. Ecol. Biogeogr.* 17, 585–599. https://doi.org/10.1111/J.1466-8238.2008.00399.X

Bardgett, R.D., Bullock, J.M., Lavorel, S., et al. 2021. Combatting global grassland degradation. *Nat. Rev. Earth Environ.* 210(2), 720–735. https://doi.org/10.1038/s43017-021-00207-2

Bartolome, J.W., Allen-Diaz, B.H., Barry, S. et al. 2014. Grazing for Biodiversity in Californian Mediterranean Grasslands. *Rangelands* 36, 36–43. https://doi.org/10.2111/Rangelands-D-14-00024.1

Bendel, C.R., Hovick, T.J., Limb, R.F. & Harmon, J.P., 2018. Variation in grazing management practices supports diverse butterfly communities across grassland working landscapes. *J. Insect Conserv.* 22, 99–111. https://doi.org/10.1007/S10841-017-0041-9/TABLES/4

Bengtsson, J., Bullock, J.M., Egoh, B. et al. 2019. Grasslands–more important for ecosystem services than you might think. *Ecosphere* 10, e02582. https://doi.org/10.1002/ECS2.2582

Benton, T.G., Vickery, J.A. & Wilson, J.D., 2003. Farmland biodiversity: Is habitat heterogeneity the key? *Trends Ecol. Evol.* 18, 182–188. https://doi.org/10.1016/S0169-5347(03)00011-9

Berger, J.L., Staab, M., Hartlieb, M., Simons, N.K., Wells, K., Gossner, M.M., Vogt, J., Achury, R., Seibold, S., Hemp, A. and Weisser, W.W., 2024. The day after mowing: Time and type of mowing influence grassland arthropods. *Ecological Applications*, p.e3022. https://doi.org/10.1002/eap.3022

Birrer, S., Zellweger-Fischer, J., Stoeckli, S. et al. 2014. Biodiversity at the farm scale: a novel credit point system. *Agric. Ecosyst. Environ.* 197, 195–203. https://doi.org/10.1016/J.AGEE.2014.08.008

Bisi, J., Liukkonen, T., Mykrä, S., et al. 2010. The good bad wolf–wolf evaluation reveals the roots of the Finnish wolf conflict. *Eur. J. Wildl. Res.* 56, 771–779. https://doi.org/10.1007/s10344-010-0374-0

Boyce, A.J., Shamon, H. & McShea, W.J. 2022. Bison reintroduction to mixed-grass prairie is associated with increases in bird diversity and cervid occupancy in riparian areas. *Front. Ecol. Evol.* 10, 821822. https://doi.org/10.3389/fevo.2022.821822

Britschgi, A., R. Spaar & R. Arlettaz. 2006. Impact of grassland farming intensification on the breeding ecology of an indicator insectivorous passerine, the Whinchat *Saxicola rubetra*: Lessons for overall Alpine meadowland management. *Biol. Conserv.* 130, 193–205. https://doi.org/10.1016/j.biocon.2005.12.013

Broyer, J., 2003. Unmown refuge areas and their influence on the survival of grassland birds in the Saône valley (France). *Biodivers. Conserv.* 12, 1219–1237. https://doi.org/10.1023/A:1023099901308/METRICS

Bruns, A., Waltert, M. & Khorozyan, I., 2020. The effectiveness of livestock protection measures against wolves (Canis lupus) and implications for their co-existence with humans. *Glob. Ecol. Conserv.* 21, e00868. https://doi.org/10.1016/J.GECCO.2019.E00868

Bruppacher, L., Pellet, J., Arlettaz, R. & Humbert, J.Y., 2016. Simple modifications of mowing regime promote butterflies in extensively managed meadows: evidence from field-scale experiments. *Biol. Conserv.* 196, 196–202. https://doi.org/10.1016/J.BIOCON.2016.02.018

Buckingham, D.L., Giovannini, P. & Peach, W.J., 2015. Manipulating grass silage management to boost reproductive output of a ground-nesting farmland bird. *Agric. Ecosyst. Environ.* 208, 21–28. https://doi.org/10.1016/J.AGEE.2015.04.018

Bullock, J.M., Franklin, J., Stevenson, M.J., et al. 2001. A plant trait analysis of responses to grazing in a long-term experiment. *J. Appl. Ecol.* 38, 253–267. https://doi.org/10.1046/J.1365-2664.2001.00599.X

Bullock, J.M., Jarvis, S.G., Fincham, W.N.W., et al. 2024. Mapping the ratio of agricultural inputs to yields reveals areas with potentially less sustainable farming. *Sci. Total Environ.* 909, 168491. https://doi.org/10.1016/J.SCITOTENV.2023.168491

Buri, P., Arlettaz, R. & Humbert, J.Y., 2013. Delaying mowing and leaving uncut refuges boosts orthopterans in extensively managed meadows: evidence drawn from field-scale experimentation. *Agric. Ecosyst. Environ.* 181, 22–30. https://doi.org/10.1016/J.AGEE.2013.09.003

Buri, P., Humbert, J.Y., Arlettaz, R., 2014. Promoting pollinating insects in intensive agricultural matrices: field-scale experimental manipulation of hay-meadow mowing regimes and its effects on bees. *PLoS One* 9, e85635. https://doi.org/10.1371/JOURNAL.PONE.0085635

Buri, P., Humbert, J.Y., Stańska, M., et al. 2016. Delayed mowing promotes planthoppers, leafhoppers and spiders in extensively managed meadows. *Insect Conserv. Divers.* 9, 536–545. https://doi.org/10.1111/ICAD.12186

Bussan, S.K., 2022. Can cattle grazing benefit grassland butterflies? *J. Insect Conserv.* 26, 359–374. https://doi.org/10.1007/S10841-022-00373-8/TABLES/3

Carbutt, C., Henwood, W.D. & Gilfedder, L.A., 2017. Global plight of native temperate grasslands: going, going, gone? *Biodivers. Conserv.* 26, 2911–2932. https://doi.org/10.1007/S10531-017-1398-5/TABLES/3

Castle, S.E., Miller, D.C., Merten, N., Ordonez, P.J. & Baylis, K., 2022. Evidence for the impacts of agroforestry on ecosystem services and human well-being in high-income countries: a systematic map. *Environ. Evid.* 11, 1–27. https://doi.org/10.1186/S13750-022-00260-4/FIGURES/10

Cattin, M.F., Blandenier, G., Banašek-Richter, C., Bersier, L.F., 2003. The impact of mowing as a management strategy for wet meadows on spider (Araneae) communities. *Biol. Conserv.* 113, 179–188. https://doi.org/10.1016/S0006-3207(02)00297-5

Cole, L.J., Brocklehurst, S., Mccracken, D.I., Harrison, W. & Robertson, D., 2012. Riparian field margins: their potential to enhance biodiversity in intensively managed Grasslands. *Insect Conserv. Divers.* 5, 86–94. https://doi.org/10.1111/J.1752-4598.2011.00147.X

Davey, C.M., Vickery, J.A., Boatman, N.D., et al. 2010. Assessing the impact of entry level stewardship on lowland farmland birds in England. *Ibis (Lond. 1859).* 152, 459–474. https://doi.org/10.1111/J.1474-919X.2009.01001.X

Dengler, J., Janišová, M., Török, P. & Wellstein, C., 2014. Biodiversity of Palaearctic grasslands: a synthesis. *Agric. Ecosyst. Environ.* 182, 1–14. https://doi.org/10.1016/j.agee.2013.12.015

Dicks, L.V., Ashpole, J.E., Dänhardt, J., et al. (Eds.). 2015. *What Works in Conservation*. Open Book Publishers, Cambridge, UK, pp. 283–321.

Donald, P.F., Pisano, G., Rayment, M.D. & Pain, D.J., 2002. The Common Agricultural Policy, EU enlargement and the conservation of Europe's farmland birds. *Agric. Ecosyst. Environ.* 89, 167–182. https://doi.org/10.1016/S0167-8809(01)00244-4

Donovan, M. & Monaghan, R., 2021. Impacts of grazing on ground cover, soil physical properties and soil loss via surface erosion: a novel geospatial modelling approach. *J. Environ. Manage.* 287, 112206. https://doi.org/10.1016/J.JENVMAN.2021.112206

Dorresteijn, I., Milcu, A.I., Leventon, J., Hanspach, J. & Fischer, J., 2016. Social factors mediating human–carnivore coexistence: understanding thematic strands influencing coexistence in Central Romania. *Ambio* 45, 490–500. https://doi.org/10.1007/s13280-015-0760-7

Durant, D., Tichit, M., Kernéïs, E. & Fritz, H., 2008. Management of agricultural wet grasslands for breeding waders: integrating ecological and livestock system perspectives - A review. *Biodivers. Conserv.* 17, 2275–2295. https://doi.org/10.1007/S10531-007-9310-3/FIGURES/3

EIP-AGRI, 2018. EIP-AGRI Focus Group Grazing for Carbon. *Eur. Comm.* 32 p.

Eldridge, D.J., Soliveres, S., Bowker, M.A. & Val, J., 2013. Grazing dampens the positive effects of shrub encroachment on ecosystem functions in a semi-arid woodland. *J. Appl. Ecol.* 50, 1028–1038. https://doi.org/10.1111/1365-2664.12105

England, J.R., O'Grady, A.P., Fleming, A., Marais, Z. & Mendham, D., 2020. Trees on farms to support natural capital: an evidence-based review for grazed dairy systems. *Sci. Total Environ.* 704, 135345. https://doi.org/10.1016/J.SCITOTENV.2019.135345

Fackelmeier, A. & Biebl, E.M., 2009. *A multistatic radar array for detecting wild animals during pasture mowing*. In: 2009 European Radar Conference (EuRAD). pp. 477–480.

Fay, R., Schaub, M., Banik, M. V., et al. 2021. Whinchat survival estimates across Europe: can excessive adult mortality explain population declines? *Anim. Conserv.* 24, 15–25. https://doi.org/10.1111/ACV.12594

Finn, J.A., Kirwan, L., Connolly, J., et al. 2013. Ecosystem function enhanced by combining four functional types of plant species in intensively managed grassland mixtures: a 3-year continental-scale field experiment. *J. Appl. Ecol.* 50, 365–375. https://doi.org/10.1111/1365-2664.12041

Flade, M., Plachter, H., Schmidt, R. & Werner, A., 2006. *Nature conservation in agricultural ecosystems: results of the Schorfheide-Chorin Research Project*. Quelle & Meyer, Wiebelsheim.

Frenzel, T., Rischen, T. & Fischer, K., 2022. Humid grassland fallows promote spider diversity in a traditionally managed landscape. *Basic Appl. Ecol.* 63, 59–70. https://doi.org/10.1016/J.BAAE.2022.05.007

Garrido, P., Mårell, A., Öckinger, E., Skarin, A., Jansson, A. & Thulin, C.G., 2019. Experimental rewilding enhances grassland functional composition and pollinator habitat use. *J. Appl. Ecol.* 56, 946–955. https://doi.org/10.1111/1365-2664.13338

Gossner, M.M., Lewinsohn, T.M., Kahl, T., et al. 2016. Land-use intensification causes multitrophic homogenization of grassland communities. *Nature* 540, 266–269. https://doi.org/10.1038/nature20575

Gottwald, F., Matthews, A., Matthews, A. & Stein-Bachinger, K., 2017. Enhancing the breeding success of whinchats–first results with small-scale measures on organic farms in north-eastern Germany. *WhinCHAT* 1, 42–52. https://www.zobodat.at/pdf/WhinCHAT_1_0042-0052.pdf

Gottwald, F. & Stein-Bachinger, K., 2018. 'Farming for biodiversity'–a new model for integrating nature conservation achievements on organic farms in north-eastern Germany. *Org. Agric.* 8, 79–86. https://doi.org/10.1007/S13165-017-0198-2/TABLES/2

Gottwald, F. & Stein-Bachinger, K., 2016. Landwirtschaft für Artenvielfalt. Ein Naturschutzmodul für ökologisch bewirtschaftete Betriebe. 2. Auflage, 208 p. Available www.landwirtschaft-artenvielfalt.de

Green, C., 2007. Reducing Mortality of Grassland Wildlife During Haying and Wheat-Harvesting Operations. *Div. Agric. Sci. Nat. Resour. Oklahoma State Univ. OSU Ext. Wildl. For.* 1–4.

Green, R.E. & Gibbons, D.W., 2000. The status of the Corncrake Crex crex in Britain in 1998. *Bird Study* 47, 129–137. https://doi.org/10.1080/00063650009461168

Green, R.E., Tyler, G.A., Stowe, T.J. & Newton, A. V., 1997. A simulation model of the effect of mowing of agricultural grassland on the breeding success of the corncrake (Crex crex). *J. Zool.* 243, 81–115. https://doi.org/10.1111/J.1469-7998.1997.TB05758.X

Grendelmeier, B., 2011. *Entwicklung einer junghasenschonenden Mähmethode*. Bachelorarb., Zürcher Hochsch. f. Angew. Wissensch.

Grüebler, M.U., Schuler, H., Horch, P. & Spaar, R., 2012. The effectiveness of conservation measures to enhance nest survival in a meadow bird suffering from anthropogenic nest loss. *Biol. Conserv.* 146, 197–203. https://doi.org/10.1016/J.BIOCON.2011.12.019

Habel, J.C., Dengler, J., Janišová, M., Török, P., Wellstein, C. & Wiezik, M., 2013. European grassland ecosystems: threatened hotspots of biodiversity. *Biodivers. Conserv.* 22, 2131–2138. https://doi.org/10.1007/S10531-013-0537-X/METRICS

Hannappel, I. & Fischer, K., 2020. Grassland intensification strongly reduces butterfly diversity in the Westerwald mountain range, Germany. *J. Insect Conserv.* 24, 279–285. https://doi.org/10.1007/S10841-019-00195-1/TABLES/1

Hedberg, P. & Kotowski, W., 2010. New nature by sowing? The current state of species introduction in grassland restoration, and the road ahead. *J. Nat. Conserv.* 18, 304–308. https://doi.org/10.1016/J.JNC.2010.01.003

Hertzog, L.R., Klimek, S., Röder, N., et al. 2023. Associations between farmland birds and fallow area at large scales: consistently positive over three periods of the EU Common Agricultural Policy but moderated by landscape complexity. *J. Appl. Ecol.* 60, 1077–1088. https://doi.org/10.1111/1365-2664.14400

Herzon, I., Birge, T., Allen, B., et al. 2018. Time to look for evidence: Results-based approach to biodiversity conservation on farmland in Europe. *Land Use Policy* 71, 347–354. https://doi.org/10.1016/J.LANDUSEPOL.2017.12.011

Herzon, I., Ekroos, J., Rintala, J., et al. 2011. Importance of set-aside for breeding birds of open farmland in Finland. *Agric. Ecosyst. Environ.* 143, 3–7. https://doi.org/10.1016/J.AGEE.2011.05.006

Herzon, I., Raatikainen, K.J., Helm, A., et al. 2022. Semi-natural habitats in the European boreal region: caught in the socio-ecological extinction vortex? *Ambio* 51, 1753–1763. https://doi.org/10.1007/S13280-022-01705-3/FIGURES/1

Herzon, I., Raatikainen, K.J., Wehn, S., et al. 2021. Semi-natural habitats in boreal Europe: a rise of a social-ecological research agenda. *Ecol. Soc. Publ.* https://doi.org/10.5751/ES-12313-260213

Humbert, J.Y., Buri, P., Unternährer, D. & Arlettaz, R., 2018. Alternative mowing regimes to favour meadow biodiversity. *Rech. Agron. Suisse* 9, 314–321.

Humbert, J.Y., Delley, S. & Arlettaz, R., 2021. Grassland intensification dramatically impacts grasshoppers: experimental evidence for direct and indirect effects of fertilisation and irrigation. *Agric. Ecosyst. Environ.* 314, 107412. https://doi.org/10.1016/J.AGEE.2021.107412

Humbert, J.Y., Ghazoul, J., Richner, N. & Walter, T., 2012a. Uncut grass refuges mitigate the impact of mechanical meadow harvesting on orthopterans. *Biol. Conserv.* 152, 96–101. https://doi.org/10.1016/J.BIOCON.2012.03.015

Humbert, J.Y., Pellet, J., Buri, P. & Arlettaz, R., 2012b. Does delaying the first mowing date benefit biodiversity in meadowland? *Environ. Evid.* 1, 1–13. https://doi.org/10.1186/2047-2382-1-9/FIGURES/5

Humbert, J.Y., Ghazoul, J., Richner, N. & Walter, T., 2010a. Hay harvesting causes high orthopteran mortality. *Agric. Ecosyst. Environ.* 139, 522–527. https://doi.org/10.1016/J.AGEE.2010.09.012

Humbert, J.Y., Ghazoul, J., Sauter, G.J. & Walter, T., 2010b. Impact of different meadow mowing techniques on field invertebrates. *J. Appl. Entomol.* 134, 592–599. https://doi.org/10.1111/J.1439-0418.2009.01503.X

Humbert, J.Y., Ghazoul, J. & Walter, T., 2009. Meadow harvesting techniques and their impacts on field fauna. *Agric. Ecosyst. Environ.* 130, 1–8. https://doi.org/10.1016/J.AGEE.2008.11.014

Janssens, F., Peeters, A., Tallowin, J.R.B., et al. 1998. Relationship between soil chemical factors and grassland diversity. *Plant Soil* 202, 69–78. https://doi.org/10.1023/A:1004389614865/METRICS

Jarnemo, A., 2002. Roe deer Capreolus capreolus fawns and mowing - mortality rates and countermeasures. *Wildlife Biol.* 8, 211–218. https://doi.org/10.2981/WLB.2002.035

Jerrentrup, J.S., Seither, M., Petersen, U. & Isselstein, J., 2015. Little grazer species effect on the vegetation in a rotational grazing system. *Agric. Ecosyst. Environ.* 202, 243–250. https://doi.org/10.1016/J.AGEE.2015.01.007

Jori, F., Hernandez-Jover, M., Magouras, I., Dürr, S. & Brookes, V.J., 2021. Wildlife–livestock interactions in animal production systems: what are the biosecurity and health implications? *Anim. Front.* 11, 8–19. https://doi.org/10.1093/AF/VFAB045

Kaláb, O., Šipoš, J. & Kočárek, P., 2020. Leaving uncut refuges during meadow harvesting increases the functional diversity of Orthoptera. *Entomol. Sci.* 23, 95–104. https://doi.org/10.1111/ENS.12404

Keenleyside, C., Beaufoy, G., Tucker, G. & Jones, G., 2014. High nature value farming throughout EU-27 and its financial support under the CAP – Final report. *Inst. Eur. Environ. Policy, London* 10. https://doi.org/doi/10.2779/91086

Kiehl, K., Kirmer, A., Donath, T.W., Rasran, L. & Hölzel, N., 2010. Species introduction in restoration projects – evaluation of different techniques for the establishment of semi-natural grasslands in Central and Northwestern Europe. *Basic Appl. Ecol.* 11, 285 299. https://doi.org/10.1016/J.BAAE.2009.12.004

Klaus, V.H., Schäfer, D., Kleinebecker, T., et al. 2017. Enriching plant diversity in grasslands by large-scale experimental sward disturbance and seed addition along gradients of land-use intensity. *J. Plant Ecol.* 10, 581–591. https://doi.org/10.1093/JPE/RTW062

Kleijn, D., Kohler, F., Báldi, A., Batáry, P., et al. 2008. On the relationship between farmland biodiversity and land-use intensity in Europe. *Proc. R. Soc. B Biol. Sci.* 276, 903–909. https://doi.org/10.1098/RSPB.2008.1509

Klimkowska, A., Kotowski, W., Van Diggelen, R., et al. 2010. Vegetation re-development after fen meadow restoration by topsoil removal and hay transfer. *Restor. Ecol.* 18, 924–933. https://doi.org/10.1111/J.1526-100X.2009.00554.X

Köhler, M., Hiller, G., Tischew, S., 2016. Year-round horse grazing supports typical vascular plant species, orchids and rare bird communities in a dry calcareous grassland. *Agric. Ecosyst. Environ.* 234, 48–57. https://doi.org/10.1016/j.agee.2016.03.020

Konvicka, M., Benes, J., Cizek, O., Kopecek, F., et al. 2008. How too much care kills species: Grassland reserves, agri-environmental schemes and extinction of Colias myrmidone (Lepidoptera: Pieridae) from its former stronghold. *J. Insect Conserv.* 12, 519–525. https://doi.org/10.1007/S10841-007-9092-7/FIGURES/4

Kruse, M., Stein-Bachinger, K., Gottwald, F., et al. 2016. Influence of grassland management on the biodiversity of plants and butterflies on organic suckler cow farms. *Tuexenia* 36, 97–119.

Le Féon, V., Schermann-Legionnet, A., Delettre, Y., et al. 2010. Intensification of agriculture, landscape composition and wild bee communities: a large scale study in four European countries. *Agric. Ecosyst. Environ.* 137, 143–150. https://doi.org/10.1016/J.AGEE.2010.01.015

Lehner, M., Eberhardt, M., Ascher, A. & Biebl, E., 2015. An active UHF RFID transponder for fawn saving during pasture mowing. *2015 Ger. Microw. Conf. GeMiC* 2015 303–306. https://doi.org/10.1109/GEMIC.2015.7107814

Leidinger, J.L.G., Gossner, M.M., Weisser, W.W., et al. 2017. Historical and recent land use affects ecosystem functions in subtropical grasslands in Brazil. *Ecosphere* 8, e02032. https://doi.org/10.1002/ECS2.2032

Leune, T., Koch, C., Fabry, S. & Struck, S., 2022. IoT Device for Reduction of Roe Deer Fawn Mortality During Haymaking. *Lect. Notes Comput. Sci. (including Subser. Lect. Notes Artif. Intell. Lect. Notes Bioinformatics)* 13533 LNCS, 375–384. https://doi.org/10.1007/978-3-031-20936-9_30/FIGURES/4

Lindborg, R., Hartel, T., Helm, A., et al. 2023. Ecosystem services provided by semi-natural and intensified grasslands: synergies, trade-offs and linkages to plant traits and functional richness. *Appl. Veg.* Sci. 26, 1–13. https://doi.org/10.1111/avsc.12729

Liu, J., Feng, C., Wang, D., Wang, L., Wilsey, B.J. & Zhong, Z., 2015. Impacts of grazing by different large herbivores in grassland depend on plant species diversity. *J. Appl. Ecol.* 52, 1053–1062. https://doi.org/10.1111/1365-2664.12456

Liu, Y., Duan, M. & Yu, Z., 2013. Agricultural landscapes and biodiversity in China. *Agric. Ecosyst. Environ.* 166, 46–54. https://doi.org/10.1016/J.AGEE.2011.05.009

Lueders, A.S., Kennedy, P.L. & Johnson, D.H. 2006. Influences of management regimes on breeding bird densities and habitat in mixed-grass prairie: an example from North Dakota. *J. Wildlife Manage.* 600–606. https://www.jstor.org/stable/3803710

Lunt, I.D., Eldridge, D.J., Morgan, J.W. & Witt, G.B., 2007. A framework to predict the effects of livestock grazing and grazing exclusion on conservation values in natural ecosystems in Australia. *Aust. J. Bot.* 55, 401–415. https://doi.org/10.1071/BT06178

Lusk, J.S. & Koper, N. 2013. Grazing and songbird nest survival in southwestern Saskatchewan. *Rangel. Ecol. Manag.* 66, 401–409. https://doi.org/10.2111/REM-D-11-00203.1

Lyons, K.G., Török, P., Hermann, J.M., et al. 2023. Challenges and opportunities for grassland restoration: a global perspective of best practices in the era of climate change. *Glob. Ecol. Conserv.* 46. https://doi.org/10.1016/j.gecco.2023.e02612

Martens, H.R., Laage, K., Eickmanns, M., et al. 2023. Paludiculture can support biodiversity conservation in rewetted fen peatlands. *Sci. Rep.* 131(13), 1–10. https://doi.org/10.1038/s41598-023-44481-0

Maskell, L.C., Smart, S.M., Bullock, J.M., Thompson, K. & Stevens, C.J., 2010. Nitrogen deposition causes widespread loss of species richness in British habitats. *Glob. Chang. Biol.* 16, 671–679. https://doi.org/10.1111/J.1365-2486.2009.02022.X

McCalmont, J.P., Hastings, A., McNamara, N.P., et al. 2017. Environmental costs and benefits of growing Miscanthus for bioenergy in the UK. *GCB Bioenergy* 9, 489–507. https://doi.org/10.1111/GCBB.12294

Metera, E., Sakowski, T., Słoniewski, K. & Romanowicz, B., 2010. Grazing as a tool to maintain biodiversity of grassland - a review. *Anim. Sci. Pap. Rep.* 28, 315–334.

Meyer, S., Unternährer, D., Arlettaz, R., et al. 2017. Promoting diverse communities of wild bees and hoverflies requires a landscape approach to managing meadows. *Agric. Ecosyst. Environ.* 239, 376–384. https://doi.org/10.1016/J.AGEE.2017.01.037

Mládek, J., Mládková, P., Hejcmanová, P., et al. 2013. Plant trait assembly affects superiority of grazer's foraging strategies in species-rich grasslands. *PLoS One* 8, e69800. https://doi.org/10.1371/JOURNAL.PONE.0069800

Molnár, Z., Kelemen, A., Kun, R., et al. 2020. Knowledge co-production with traditional herders on cattle grazing behaviour for better management of species-rich grasslands. *J. Appl. Ecol.* 57, 1677–1687. https://doi.org/10.1111/1365-2664.13664

Morris, C.D., 2021. How biodiversity-friendly is regenerative grazing? *Front. Ecol. Evol.* 9, 816374. https://doi.org/10.3389/FEVO.2021.816374/BIBTEX

Müller, M., Spaar, R., Schifferli, L. & Jenni, L., 2005. Effects of changes in farming of subalpine meadows on a grassland bird, the whinchat (Saxicola rubetra). *J. Ornithol.* 146, 14–23. https://doi.org/10.1007/S10336-004-0059-0/TABLES/4

Newbold, T., Hudson, L.N., Arnell, A.P., et al. 2016. Has land use pushed terrestrial biodiversity beyond the planetary boundary? A global assessment. *Science (80-.).* 353, 291–288. https://doi.org/10.1126/science.aaf2201

Newton, A.C., Walls, R.M., Golicher, D., Keith, S.A., Diaz, A. & Bullock, J.M., 2012. Structure, composition and dynamics of a calcareous grassland metacommunity over a 70-year interval. *J. Ecol.* 100, 196–209. https://doi.org/10.1111/J.1365-2745.2011.01923.X

Nyfeler, D., Huguenin-Elie, O., Suter, M., Frossard, E., Connolly, J., Lüscher, A., 2009. Strong mixture effects among four species in fertilized agricultural grassland led to persistent and consistent transgressive overyielding. *J. Appl. Ecol.* 46, 683–691. https://doi.org/10.1111/J.1365-2664.2009.01653.X

Onrust, J., Wymenga, E., Piersma, T. & Olff, H., 2019. Earthworm activity and availability for meadow birds is restricted in intensively managed grasslands. *J. Appl. Ecol.* 56, 1333–1342. https://doi.org/10.1111/1365-2664.13356

Oppermann, R., Handwerk, J., Holsten, M. & Krismann, A., 2000. *Naturverträgliche Mähtechnik für das Feuchtgrünland, voruntersuchung für das E \& E-Vorhaben*. ILN Singen, Bonn 236.

Osem, Y., Perevolotsky, A. & Kigel, J., 2002. Grazing effect on diversity of annual plant communities in a semi-arid rangeland: interactions with small-scale spatial and temporal variation in primary productivity. *J. Ecol.* 90, 936–946. https://doi.org/10.1046/J.1365-2745.2002.00730.X

Ozinga, W.A., Römermann, C., Bekker, R.M., et al. 2009. Dispersal failure contributes to plant losses in NW Europe. *Ecol. Lett.* 12, 66–74. https://doi.org/10.1111/J.1461-0248.2008.01261.X

Panassiti, B., Wolfrum, S., Birnbeck, S., Burmeister, J., Freibauer, A., Morinière, J. & Walter, R., 2023. Insects benefit from agri-environmental schemes aiming at grassland extensification. *Agric. Ecosyst. Environ.* 356, 108613. https://doi.org/10.1016/J.AGEE.2023.108613

Peach, W., Buckingham, D., Watts, P., et al. 2007. *Management of agricultural grassland to promote in-field structural heterogeneity, invertebrates and birds in pastoral landscapes., High value grassland: providing biodiversity, a clean environment and premium products*. Proceedings of the BGS/BES/BSAS Conference held at Keele University, Staffordshire, UK, 17–19 April, 2007. British Grassland Society (BGS).

Pearce, E.A., Mazier, F., Normand, S., et al. 2023. Substantial light woodland and open vegetation characterized the temperate forest biome before Homo sapiens. *Sci. Adv.* 9. https://doi.org/10.1126/SCIADV.ADI9135/SUPPL_FILE/SCIADV.ADI9135_DATA_S1_AND_S2.ZIP

Peco, B., Carmona, C.P., de Pablos, I. & Azcárate, F.M., 2012. Effects of grazing abandonment on functional and taxonomic diversity of Mediterranean grasslands. *Agric. Ecosyst. Environ.* 152, 27–32. https://doi.org/10.1016/J.AGEE.2012.02.009

Peeters, A., Beaufoy, G., Canals, R.M., et al. 2014. Grassland term definitions and classifications adapted to the diversity of European grassland-based systems. *Grassl. Sci. Eur.* 19, 743–750.

Perkins, A.J., Whittingham, M.J., Bradbury, R.B., et al. 2000. Habitat characteristics affecting use of lowland agricultural grassland by birds in winter. *Biol. Conserv.* 95, 279–294. https://doi.org/10.1016/S0006-3207(00)00042-2

Pintaric, A.L., Reid, R. & Nol, E. 2019. Variation in surrogate breeding habitat quality between continuously grazed rangelands and late-cut hayfields for a threatened grassland birds. *Range. Ecol. Manag.* 72, 474–483. https://doi.org/10.1016/j.rama.2019.01.001

Plantureux, S., Peeters, A. & McCracken, D., 2005. Biodiversity in intensive grasslands: effect of management, improvement and challenges. *Agron. Res.* 3, 153–164.

Porqueddu, C., Ates, S., Louhaichi, M., Kyriazopoulos, A.P., et al. 2016. Grasslands in 'Old World' and 'New World' Mediterranean-climate zones: past trends, current status and future research priorities. *Grass Forage Sci.* 71, 1–35. https://doi.org/10.1111/GFS.12212

Poschlod, P. & Biewer, H., 2005. Diaspore and gap availability are limiting species richness in wet meadows. *Folia Geobot.* 40, 13–34. https://doi.org/10.1007/BF02803041/METRICS

Potts, S.G., Woodcock, B.A., Roberts, S.P.M., et al. 2009. Enhancing pollinator biodiversity in intensive grasslands. *J. Appl. Ecol.* 46, 369–379. https://doi.org/10.1111/J.1365-2664.2009.01609.X

Prochnow, A. & Meierhöfer, J., 2003. Befahrmuster bei der Grünlandmahd: Faunaschonung und Aufwendungen. *Grartechnische Forsch.* 9, 36–43.

Pywell, R.F., Bullock, J.M., Hopkins, A., et al. 2002. Restoration of species-rich grassland on arable land: assessing the limiting processes using a multi-site experiment. *J. Appl. Ecol.* 39, 294–309. https://doi.org/10.1046/J.1365-2664.2002.00718.X

Pywell, R.F., Bullock, J.M., Tallowin, J.B., et al. 2007. Enhancing diversity of species-poor grasslands: an experimental assessment of multiple constraints. *J. Appl. Ecol.* 44, 81–94. https://doi.org/10.1111/J.1365-2664.2006.01260.X

Pywell, R.F., Heard, M.S., Woodcock, B.A., et al. 2015. Wildlife-friendly farming increases crop yield: evidence for ecological intensification. *Proc. R. Soc. B Biol. Sci.* 282. https://doi.org/10.1098/RSPB.2015.1740

Qi, A., Holland, R.A., Taylor, G. & Richter, G.M., 2018. Grassland futures in Great Britain – Productivity assessment and scenarios for land use change opportunities. *Sci. Total Environ.* 634, 1108–1118. https://doi.org/10.1016/J.SCITOTENV.2018.03.395

Ramankutty, N., Mehrabi, Z., Waha, K., et al. 2018. *Trends in global agricultural land use: implications for environmental health and food security.* https://doi.org/10.1146/annurev-arplant-042817-040256 69, 789–815. https://doi.org/10.1146/ANNUREV-ARPLANT-042817-040256

Revaz, E., Schaub, M. & Arlettaz, R. 2008. Foraging ecology and reproductive biology of the Stonechat *Saxicola torquata*: comparison between a revitalized, intensively cultivated and a historical, traditionally cultivated agro-ecosystem. *J. Ornithol.* 149, 301–312.

Ribeiro, I., Domingos, T., McCracken, D. & Proença, V., 2023. The use of domestic herbivores for ecosystem management in Mediterranean landscapes. *Glob. Ecol. Conserv.* 46, e02577. https://doi.org/10.1016/J.GECCO.2023.E02577

Robinson, R.A., Wilson, J.D. & Crick, H.Q.P., 2001. The importance of arable habitat for farmland birds in grassland landscapes. *J. Appl. Ecol.* 38, 1059–1069. https://doi.org/10.1046/J.1365-2664.2001.00654.X

Röös, E., Mayer, A., Muller, A., et al. 2022. Agroecological practices in combination with healthy diets can help meet EU food system policy targets. *Sci. Total Environ.* 847, 157612. https://doi.org/10.1016/J.SCITOTENV.2022.157612

Rossier, L.C.P., Auberson, C., Arlettaz, R. & Humbert, J.Y., 2023. Effects of uncut grass refuges on the plant community of extensively managed hay meadows. *Basic Appl. Ecol.* 72, 38–44. https://doi.org/10.1016/J.BAAE.2023.07.003

Rühs, M. & Stein-Bachinger, K., 2024. *Honorierung von Naturschutzleistungen – Grundlagen und Beispiele für ökologisch bewirtschaftete Betriebe*, 3rd Edition. www.landwirtschaft-artenvielfalt.de

Published by Burleigh Dodds Science Publishing Limited, 2025.

Runhaar, H. & Polman, N., 2018. Partnering for nature conservation: NGO-farmer collaboration for meadow bird protection in the Netherlands. *Land Use Policy* 73, 11–19. https://doi.org/10.1016/j.landusepol.2018.01.033

Rutgers, M., Orgiazzi, A., Gardi, C., et al. 2016. Mapping earthworm communities in Europe. *Appl. Soil Ecol.* 97, 98–111. https://doi.org/10.1016/J.APSOIL.2015.08.015

Sala, O.E., Chapin, F.S., Armesto, J.J., et al. 2000. Global biodiversity scenarios for the year 2100. *Science (80-.).* 287, 1770–1774. https://doi.org/10.1126/science.287.5459.1770

Salvatori, V., Balian, E., Blanco, J.C., et al., 2020. Applying participatory processes to address conflicts over the conservation of large carnivores: understanding conditions for successful management. *Front. Ecol. Evol.* 8, 182. https://doi.org/10.3389/fevo.2020.00182

Saumure, R.A., Herman, T.B. & Titman, R.D., 2007. Effects of haying and agricultural practices on a declining species: The North American wood turtle, Glyptemys insculpta. *Biol. Conserv.* 135, 565–575. https://doi.org/10.1016/J.BIOCON.2006.11.003

Schaub, M., Martinez, N., Tagmann-Ioset, A., et al. 2010. Patches of bare ground as a staple commodity for declining ground-foraging insectivorous farmland birds. *PLoS One* 5, e13115.

Schils, R.L.M., Bufe, C., Rhymer, C.M., et al. 2022. Permanent grasslands in Europe: land use change and intensification decrease their multifunctionality. *Agric. Ecosyst. Environ.* 330. https://doi.org/10.1016/j.agee.2022.107891

Schmidt, M.H., Rocker, S., Hanafi, J. & Gigon, A., 2008. Rotational fallows as overwintering habitat for grassland arthropods: the case of spiders in fen meadows. *Biodivers. Conserv.* 17, 3003–3012. https://doi.org/10.1007/S10531-008-9412-6/TABLES/2

Schneider, M.K., Lüscher, G., Jeanneret, P., et al. 2014. Gains to species diversity in organically farmed fields are not propagated at the farm level. *Nat. Commun.* 51(5), 1–9. https://doi.org/10.1038/ncomms5151

Schuler, J., Sattler, C., Helmecke, A., Zander, P., Uthes, S., Bachinger, J. & Stein-Bachinger, K., 2013. The economic efficiency of conservation measures for amphibians in organic farming – Results from bio-economic modelling. *J. Environ. Manage.* 114, 404–413. https://doi.org/10.1016/J.JENVMAN.2012.10.037

Schwarz, C., Fumy, F., Drung, M. & Fartmann, T., 2023. Insect-friendly harvest in hay meadows – Uncut refuges are of vital importance for conservation management. *Glob. Ecol. Conserv.* 48, e02731. https://doi.org/10.1016/J.GECCO.2023.E02731

Sheridan, H., Finn, J.A., Culleton, N. & O'Donovan, G., 2008. Plant and invertebrate diversity in grassland field margins. *Agric. Ecosyst. Environ.* 123, 225–232. https://doi.org/10.1016/J.AGEE.2007.07.001

Simons, N.K., Gossner, M.M., Lewinsohn, T.M., Lange, M., Türke, M. & Weisser, W.W., 2015. Effects of land-use intensity on arthropod species abundance distributions in grasslands. *J. Anim. Ecol.* 84, 143–154. https://doi.org/10.1111/1365-2656.12278

Slodowicz, D., Durbecq, A., Ladouceur, E., et al. 2023. The relative effectiveness of different grassland restoration methods: a systematic literature search and meta-analysis. *Ecol. Solut. Evid.* 4, 1–13. https://doi.org/10.1002/2688-8319.12221

Soliveres, S., Van Der Plas, F., Manning, P., et al. 2016. Biodiversity at multiple trophic levels is needed for ecosystem multifunctionality. *Nat.* 536, 456–459. https://doi.org/10.1038/nature19092

Staggenborg, J. & Anthes, N., 2022. Long-term fallows rate best among agri-environment scheme effects on farmland birds–a meta-analysis. *Conserv. Lett.* 15, e12904. https://doi.org/10.1111/CONL.12904

Stein-Bachinger, K & Fuchs, S., 2012. Protection strategies for farmland birds in legume-grass leys as trade-offs between nature conservation and farmers' needs. *Org. Agric.* 2, 145–162. https://doi.org/10.1007/S13165-012-0029-4/TABLES/10

Stein-Bachinger, K., Fuchs, S., Gottwald, F., et al. 2010. *Naturschutzfachliche Optimierung des Ökologischen Landbaus "Naturschutzhof Brodowin."* BfN-Schriftenvertrieb im Landwirtschaftsverlag, Münster.

Stein-Bachinger, K., Schoenbrodt, T., Schmidt, E., Dissanayake, M. & Gottwald, F., 2021. Improving kettle holes as habitat and reproduction areas for amphibians – a case study in organic farms in north-eastern Germany. *Amphibia-Reptilia* 43, 51–62. https://doi.org/10.1163/15685381-BJA10079

Stein-Bachinger, K., Gottwald, F., Haub, A., et al. 2021. To what extent does organic farming promote species richness and abundance in temperate climates? A review. *Org. Agr.* 11, 1–12. https://doi.org/10.1007/s13165-020-00279-2

Strebel, G., Jacot, A., Horch, P. & Spaar, R., 2015. Effects of grassland intensification on Whinchats Saxicola rubetra and implications for conservation in upland habitats. *Ibis (Lond. 1859).* 157, 250–259. https://doi.org/10.1111/IBI.12250

Tälle, M., Deák, B., Poschlod, P., Valkó, O., Westerberg, L. & Milberg, P., 2016. Grazing vs. mowing: a meta-analysis of biodiversity benefits for grassland management. *Agric. Ecosyst. Environ.* 222, 200–212. https://doi.org/10.1016/J.AGEE.2016.02.008

Tiainen, J., Hyvönen, T., Hagner, M., et al. 2020. Biodiversity in intensive and extensive grasslands in finland: The impacts of spatial and temporal changes of agricultural land use. *Agric. Food Sci.* 29, 68–97. https://doi.org/10.23986/afsci.86811

The World Bank. 2023. Human-Wildlife Conflict: Global Policy and Perception Insights. Brief 09.11.2023. https://www.worldbank.org/en/programs/global-wildlife-program/brief/human-wildlife-conflict-global-policy-and-perception-insights. Last accessed 11.03.2024.

Toivonen, M., Herzon, I. & Helenius, J., 2013. Environmental fallows as a new policy tool to safeguard farmland biodiversity in Finland. *Biol. Conserv.* 159, 355–366. https://doi.org/10.1016/J.BIOCON.2012.11.016

Toivonen, M., Huusela, E., Hyvönen, T., et al. 2022. Effects of crop type and production method on arable biodiversity in boreal farmland. *Agric. Ecosyst. Environ.* 337. https://doi.org/10.1016/j.agee.2022.108061

Tölgyesi, C., Kelemen, A., Bátori, Z., et al. 2023. Maintaining scattered trees to boost carbon stock in temperate pastures does not compromise overall pasture quality for the livestock. *Agric. Ecosyst. Environ.* 351, 108477. https://doi.org/10.1016/J.AGEE.2023.108477

Török, P. & Dengler, J., 2021. Palaearctic grasslands in transition: overarching patterns and future prospects. *Grasslands of the World* 29–40. https://doi.org/10.1201/9781315156125-9

Török, P., Hölzel, N., van Diggelen, R. & Tischew, S., 2016a. Grazing in European open landscapes: how to reconcile sustainable land management and biodiversity conservation? *Agric. Ecosyst. Environ.* 234, 1–4. https://doi.org/10.1016/J.AGEE.2016.06.012

Török, P., Valkó, O., Deák, B., Kelemen, A., et al. 2016b. Managing for species composition or diversity? Pastoral and free grazing systems in alkali steppes. *Agric. Ecosyst. Environ.* 234, 23–30. https://doi.org/10.1016/J.AGEE.2016.01.010

Török, P., Penksza, K., Tóth, E., et al. 2018. Vegetation type and grazing intensity jointly shape grazing effects on grassland biodiversity. *Ecol. Evol.* 8, 10326–10335. https://doi.org/10.1002/ece3.4508

Török, P., Vida, E., Deák, B., Lengyel, S. & Tóthmérész, B., 2011. Grassland restoration on former croplands in Europe: an assessment of applicability of techniques and costs. *Biodivers. Conserv.* 20, 2311–2332. https://doi.org/10.1007/S10531-011-9992-4/TABLES/3

Tóth, E., Deák, B., Valkó, O., et al. 2018. Livestock type is more crucial than grazing intensity: traditional cattle and sheep grazing in short-grass Steppes. *L. Degrad. Dev.* 29, 231–239. https://doi.org/10.1002/ldr.2514

Tsiafouli, M.A., Thébault, E., Sgardelis, S.P., et al. 2015. Intensive agriculture reduces soil biodiversity across Europe. *Glob. Chang. Biol.* 21, 973–985. https://doi.org/10.1111/GCB.12752

Tyler, G.A., Green, R.E. & Casey, C., 1998. Survival and behaviour of Corncrake Crex crex chicks during the mowing of agricultural grassland. *Bird Study* 45, 35–50. https://doi.org/10.1080/00063659809461076

Undersander, D., Temple, S., Bartlet, J., Sample, D. & Paine, L., 2000. *Grassland birds: fostering habitats using rotational grazing.* University of Wisconsin Extension, Madison, Wisconsin, USA.

Valkó, O., Török, P., Deák, B. & Tóthmérész, B., 2014. Review: prospects and limitations of prescribed burning as a management tool in European grasslands. *Basic Appl. Ecol.* 15, 26–33. https://doi.org/10.1016/J.BAAE.2013.11.002

van den Pol-van Dasselaar, A., Hennessy, D., Isselstein, J., 2020. Grazing of dairy cows in Europe–an in-depth analysis based on the perception of grassland experts. *Sustain.* 12, 1098. https://doi.org/10.3390/SU12031098

van Klink, R., Boch, S., Buri, P., et al. 2017. No detrimental effects of delayed mowing or uncut grass refuges on plant and bryophyte community structure and phytomass production in low-intensity hay meadows. *Basic Appl. Ecol.* 20, 1–9. https://doi.org/10.1016/J.BAAE.2017.02.003

van Klink, R., Menz, M.H.M., Baur, H., et al. 2019. Larval and phenological traits predict insect community response to mowing regime manipulations. *Ecol. Appl.* 29, e01900. https://doi.org/10.1002/EAP.1900

Verhulst, J., Kleijn, D. & Berendse, F., 2007. Direct and indirect effects of the most widely implemented Dutch agri-environment schemes on breeding waders. *J. Appl. Ecol.* 44, 70–80. https://doi.org/10.1111/j.1365-2664.2006.01238.x

Vickery, J.A., Tallowin, J.R., Feber, R.E., et al. 2001. The management of lowland neutral grasslands in Britain: effects of agricultural practices on birds and their food resources. *J. Appl. Ecol.* 38, 647–664.

Vickery, J. & Arlettaz, R. 2012. The importance of habitat heterogeneity at multiple scales for birds in European agricultural landscapes. In: *Birds and Habitat: Relationships in Changing Landscapes* (ed. R.J. Fuller), pp. 177–204. Cambridge University Press, Cambridge. https://doi.org/10.1046/J.1365-2664.2001.00626.X

Villa-Galaviz, E., Smart, S.M., Ward, S.E., Fraser, M.D. & Memmott, J., 2023. Fertilization using manure minimizes the trade-offs between biodiversity and forage production in agri-environment scheme grasslands. *PLoS One* 18, e0290843. https://doi.org/10.1371/JOURNAL.PONE.0290843

Wiggers, J.M.R.H., van Ruijven, J., Berendse, F. & de Snoo, G.R., 2016. Effects of grass field margin management on food availability for Black-tailed Godwit chicks. *J. Nat. Conserv.* 29, 45–50. https://doi.org/10.1016/J.JNC.2015.11.001

Wilson, J.B., Peet, R.K., Dengler, J. & Pärtel, M., 2012. Plant species richness: the world records. *J. Veg. Sci.* 23, 796–802. https://doi.org/10.1111/J.1654-1103.2012.01400.X

Woodcock, B.A., Bullock, J.M., Nowakowski, M., Orr, R., Tallowin, J.R.B. & Pywell, R.F., 2012. Enhancing floral diversity to increase the robustness of grassland beetle assemblages to environmental change. *Conserv. Lett.* 5, 459–469. https://doi.org/10.1111/J.1755-263X.2012.00262.X

Woodcock, B.A., Potts, S.G., Pilgrim, E., et al. 2007a. The potential of grass field margin management for enhancing beetle diversity in intensive livestock farms. *J. Appl. Ecol.* 44, 60–69. https://doi.org/10.1111/J.1365-2664.2006.01258.X

Woodcock, B.A., Potts, S.G., Westbury, D.B., et al. 2007b. The importance of sward architectural complexity in structuring predatory and phytophagous invertebrate assemblages. *Ecol. Entomol.* 32, 302–311. https://doi.org/10.1111/J.1365-2311.2007.00869.X

Woodcock, B.A., Potts, S.G., Tscheulin, T., et al. 2009. Responses of invertebrate trophic level, feeding guild and body size to the management of improved grassland field margins. *J. Appl. Ecol.* 46, 920–929. https://doi.org/10.1111/J.1365-2664.2009.01675.X

Woodcock, B.A., Savage, J., Bullock, J.M., et al. 2014. Enhancing floral resources for pollinators in productive agricultural grasslands. *Biol. Conserv.* 171, 44–51. https://doi.org/10.1016/J.BIOCON.2014.01.023

Woodcock, B.A., Savage, J., Bullock, J.M., et al. 2013. Enhancing beetle and spider communities in agricultural grasslands: The roles of seed addition and habitat management. *Agric. Ecosyst. Environ.* 167, 79–85. https://doi.org/10.1016/J.AGEE.2013.01.009

Zellweger-Fischer, J., Kéry, M. & Pasinelli, G., 2011. Population trends of brown hares in Switzerland: the role of land-use and ecological compensation areas. *Biol. Conserv.* 144, 1364–1373. https://doi.org/10.1016/J.BIOCON.2010.11.021

Zhang, M., Delgado-Baquerizo, M., Li, G., et al. 2023. Experimental impacts of grazing on grassland biodiversity and function are explained by aridity. *Nat. Commun.* 14, 1–8. https://doi.org/10.1038/s41467-023-40809-6

Zhao, G., 2023. Trends in grassland science: based on the shift analysis of research themes since the early 1900s. *Fundam. Res.* 3, 201–208. https://doi.org/10.1016/J.FMRE.2022.05.008

Zhao, Y., Liu, Z. & Wu, J., 2020. Grassland ecosystem services: a systematic review of research advances and future directions. *Landsc. Ecol.* 35, 793–814. https://doi.org/10.1007/S10980-020-00980-3/FIGURES/6

Chapter 14

Designing and developing multispecies swards for future grassland systems

Shona Baker and Thomas Moloney, Teagasc and DLF Ireland, Ireland

1 Introduction: grasslands in agriculture

Temperate grasslands that are actively managed by grazing or cutting span roughly one-quarter of the Earth's land surface. They are fundamental to livestock agriculture, supplying nearly half of all the feed consumed by farmed animals. Their crucial role in food security is underscored by their contribution to human nutrition, supplying approximately 15% of global energy and 25% of protein consumption (FAO, 2009; Herrero et al., 2013). Despite their vital importance, traditional grass-based livestock systems face mounting environmental challenges. In regions where grasslands dominate the agricultural landscape, such as Ireland (75% of Utilised Agricultural Area; Eurostat, 2023) and New Zealand (64% of Utilised Agricultural Area; Pearson, 2020), management practices significantly influence environmental outcomes and the ability to meet sustainability targets. These systems account for 12–19% of anthropogenic greenhouse gas (GHG) emissions through multiple pathways: enteric methane from livestock, nitrous oxide released from fertilisers and manure, and carbon dioxide from land-use changes (Vermeulen et al., 2012). Climate change further compounds these challenges, with rising temperatures and increasingly frequent

http://dx.doi.org/10.19103/AS.2025.0154.23

extreme weather events threatening system resilience (Durack et al., 2012). This convergence of environmental impacts and climatic vulnerabilities underscores the urgent need for a comprehensive approach to grassland management – one that enhances sustainability, improves efficiency, and builds resilience for the future.

2 Historical context and developments in grassland management

In the early to mid-twentieth century, agricultural researchers began developing grass–legume seed mixtures to boost productivity and durability in temperate grasslands (Elliot, 1948; Turner, 1955). However, subsequent decades saw a shift toward monoculture systems dominated by perennial ryegrass (*Lolium perenne* L.) due to its high forage yield, adaptability, resilience after defoliation, and reliable germination (Lee et al., 2010; McDonagh et al., 2016). The availability of inexpensive synthetic fertilisers further promoted these high-input systems. Although effective for yield, monoculture systems that rely heavily on nitrogen (N) hungry species can be costly both economically and environmentally (Baker et al., 2023a). This may present an opportunity for the use of more diverse pasture mixtures that could improve resource-use efficiency and enhance ecosystem health.

3 Renewed interest in grass-legume systems

Given the urgent need for farmers to reduce reliance on N fertilisers while maintaining sward productivity, incorporating legumes into grazing systems to promote biological N fixation presents an effective solution. Grass–legume mixtures have regained popularity for their capacity to enhance productivity and minimise synthetic fertiliser use. Legumes fix atmospheric N, supporting forage yields with reduced dependence on inorganic fertilisers (Muir et al., 2011). Countries such as France, New Zealand, and the United States increasingly integrate these species into grazing systems, finding combinations like white clover and perennial ryegrass particularly effective (Bouton, 2006; Peyraud et al., 2009; Moot, 2012). Advances in plant breeding have further improved the resilience and productivity of legumes, making them a viable option even in intensive systems (Rochon et al., 2004;Sanderson et al., 2004). Grass-legume systems enhance ecosystem function by promoting greater root diversity, which improves water and nutrient acquisition – an essential advantage during drought conditions (Sleugh et al., 2000; Cardinale et al., 2007). Additionally, legumes elevate forage quality by increasing crude protein levels and digestibility (Jensen et al., 2011; Lüscher et al., 2014).

4 Policy-driven shifts in grassland management

Current policy frameworks further encourage sustainable grassland management. For example, the EU's Nitrates Directive and European Green Deal aim to reduce N fertiliser use and achieve carbon neutrality by 2050 (Brouwer and Hellegers, 1996; European Commission, 2019). Similarly, the United States has set a goal of halving agricultural GHG emissions, and New Zealand is adopting strategies to decrease water pollution from nitrates and phosphorus (Scarsbrook and Melland, 2015; Wardropper et al., 2023; Favero et al., 2024). Countries in Asia, including China and India, have also committed to reducing emissions and are integrating sustainable agricultural practices as a path toward their climate targets (Das et al., 2023; Zhan et al., 2024).

5 Emergence and potential of multispecies swards

Building on the success of grass–legume systems, multispecies swards (MSSs) introduce additional biodiversity, incorporating not only grasses and legumes but also deep-rooted herb species. These diverse mixtures promote resilience and nutrient cycling, offering agronomic benefits and a range of ecosystem services (Grace et al., 2018; Sollenberger et al., 2019; Cummins et al., 2021; Baker et al., 2023a, 2023b, 2023c). Multispecies swards are intentionally designed plant mixes that combine grasses, legumes, and forage herbs to optimise grassland production and environmental resilience (Komainda et al., 2020). While traditional pastures focus on single species like perennial ryegrass, MSS use various plants to enhance ecological functions and decrease reliance on synthetic inputs (Baker et al., 2023a, 2023b; see Table 1 for a summary of species commonly sown as part of a MSS and their associated benefits, including their suitability based on soil type and moisture conditions.).

6 Design principles for multispecies swards

To maximise the potential of MSS in delivering both agronomic performance and ecosystem services, careful attention must be paid to the composition, richness, and management of species within the sward. The following design principles, grounded in recent research, offer a framework for creating productive, resilient grasslands that balance high forage yield with environmental benefits.

6.1 Optimising species composition over richness

While increasing species richness can enhance ecosystem functions, research consistently shows that it is the composition and functional traits of species within a mixture that often matter more than the sheer number of species (Finn et al., 2013; Lüscher et al., 2022). Targeting species with complementary traits – such as grasses that efficiently acquire nutrients and legumes that fix N – can result in synergistic interactions that boost yields and reduce the need for inorganic inputs (Nyfeler et al., 2011). Mixtures with two grasses and two legumes have been found to consistently outperform species-poor systems, providing a robust balance of productivity and ecosystem services (Connolly et al., 2018).

Furthermore, species identity plays a critical role in determining overall sward performance. For instance, selecting species that are high yielding under local conditions can significantly improve results, even in low-diversity mixtures (Grange et al., 2021). This suggests that MSS should prioritise species functionality over aiming for maximum species numbers, especially since additional species beyond a certain threshold – typically around four to six – may provide diminishing returns in terms of yield or ecosystem service delivery (Suter et al., 2021; Lüscher et al., 2022).

6.2 Functional diversity for yield stability

Incorporating temporal and spatial diversity into grassland design can significantly enhance productivity and resilience. Temporal diversity involves selecting species that grow and thrive at different times of the year, ensuring a continuous supply of high-quality forage across seasons. Spatial diversity, on the other hand, creates heterogeneity within the sward at any one time, further stabilizing productivity. A core advantage of MSS is their ability to maintain yields despite environmental variability, such as drought or extreme weather events. This resilience largely stems from functional diversity, which relies on the variety of ecological roles played by different species in the sward (Lüscher et al., 2022). For example, combining deep-rooted herbs with shallower-rooted grasses and legumes allows the sward to access water at different soil depths, reducing the impact of drought (Grange et al., 2021).

However, designing and managing this functional diversity requires more advanced skills and expertise than traditional monoculture systems. Research shows that community composition, rather than species richness alone, drives yield stability under stress conditions. Functionally diverse mixtures are often more resilient and productive by using niche complementarity – the efficient use of resources like light, water, and nutrients across species (Hofer et al., 2016; Komainda et al., 2020). Yet, managing this diversity to maintain optimal species interactions and balance can pose a significant challenge for farmers who may not have specific experience with MSS or access to resources for ongoing monitoring and adjustment. For these farmers, establishing and maintaining such diversity can be a barrier to adoption, as it demands a greater understanding of plant interactions and responsive management practices.

6.3 Balancing multifunctionality and agronomic performance

Designing MSS to simultaneously deliver multiple ecosystem services – such as carbon sequestration, nutrient cycling, and biodiversity – alongside high forage yields requires an understanding of the trade-offs between these functions. While it is possible to achieve multifunctionality with a limited number of species, it is crucial to select species that contribute to the desired services in a complementary way (Suter et al., 2021). For example, in mixtures with grasses, legumes, and herbs, complementary interactions between functional groups can enhance overall sward performance, particularly in terms of yield stability, weed suppression, and N efficiency (Connolly et al., 2018; Moloney et al., 2020; Grange et al., 2021; Baker et al., 2023a, 2023b; Shackleton et al., 2024). Increased plant diversity can also create habitats for various wildlife, supporting insects, earthworms, and microorganisms (Cong et al., 2020; Shnel et al., 2021; Ikoyi et al., 2023). However, rather than maximising species richness, MSS should aim to include species that synergise well and contribute to multiple ecosystem functions without sacrificing agronomic performance.

6.4 Adapting to local environmental conditions

Multispecies swards must be designed with site-specific conditions in mind, as the effectiveness of different species mixtures can vary widely depending on soil type, climatic conditions, and management practices (Wróbel et al., 2025). This principle holds particular significance in light of the growing challenges posed by climate change. For instance, mixtures that perform well under drought conditions may not be as effective in wetter environments or where soil fertility is higher. Similarly, the inclusion of deep-rooted species like chicory (*Cichorium intybus* L.) or plantain (*Plantago lanceolata* L.) can improve water infiltration and soil structure in some soils but may be less beneficial in others (Berdeni et al., 2021).

In rain-fed North-Western Europe, seasonal summer dry periods coupled with heat can cause feed scarcity due to reduced grassland productivity. For example, the drought in summer 2018 led to an average 38% decline in grassland biomass production across Europe (Fu et al., 2022). The frequency of such weather extremes is expected to increase in the future, negatively impacting grass-based livestock husbandry (Ergon et al., 2018). Traditional approaches to mitigate this, such as adjusting stocking rates, can limit farm profitability (Grinnell et al., 2023).

Comprehensive research initiatives, such as the international LegacyNet project, are essential for developing evidence-based recommendations to help adapt MSS to local conditions while maintaining their functional benefits. LegacyNet is implementing standardised experiments across 32 sites worldwide to investigate how grassland diversity affects both sward performance and subsequent crop productivity (O'Malley et al., 2023). By using systematic combinations of grasses, legumes, and herbs across diverse environments, this coordinated research initiative addresses the limitations of single-site experiments. It captures the complex interactions between soil, plants, management, and climatic conditions.

Such comprehensive research efforts are crucial for informing the design of flexible and adaptable sward compositions. The findings will help tailor MSS to specific environmental contexts while maximising both productivity and ecosystem services. Rather than a one-size-fits-all approach, this research will enable the development of site-specific recommendations that account for the nuances of local conditions. This, in turn, will enhance the long-term sustainability and resilience of MSS in the face of growing challenges posed by climate change.

6.5 Long-term stability and management

In addition to selecting the right species, it is crucial to ensure that species remain stable within the community over time. Even the most well-designed MSS can lose functionality if key species decline in abundance or disappear due to competition, poor adaptability, or mismanagement (Grace et al., 2018; Baker et al., 2023a, 2023b). As such, ongoing management practices, such as proper grazing or cutting regimes, can help maintain the desired species composition and prevent the dominance of less desirable species (Hamacher et al., 2021) (Table 1).

Long-term system-scale animal grazing experiments are critical to understanding the dynamic interactions that occur in real-world farming conditions. While small-scale or short-term studies provide valuable initial insights, they often fail to capture the full range of ecological and performance-related outcomes that emerge over time. System-scale experiments, on the other hand, allow researchers to observe how MSS respond to variables such as grazing intensity, stocking rates, seasonal changes, and species persistence under natural stressors. These experiments are essential for assessing the true longevity of species compositions, especially how well MSS maintain productivity and functional diversity when subjected to continuous animal grazing. Without this long-term perspective, it is difficult to predict which species will thrive or decline, and how such shifts might affect forage quality, animal performance, and environmental benefits over time.

Table 1 Overview of selected grass, legume, and herb species: selection of benefits, soil requirements, and soil moisture tolerance

Species	Selection of benefits	Soil requirements	Soil moisture tolerance	References
Grass				
Perennial ryegrass (*Lolium perenne* L.)	High digestibility; high-water-soluble sugar content; high yielding	Most soil types	Moderate moisture	Harris, 2001
Italian ryegrass (*Lolium multiflorum* L.)	Rapid establishment and growth; high quality; suitable for winter feed	Most soil types	Moderate moisture	Frame & Laidlaw, 2011; Frame et al., 1989
Timothy (*Phleum pratense* L.)	Winter-hardy; high yield; provides good-quality forage early in the season	Heavier soils	Wet conditions	Thorvaldsson & Andersson, 1986; Jørgensen et al., 2010
Cocksfoot (*Dactylis glomerata* L.)	Highly digestible when young; drought tolerant; deep roots for water extraction	Lighter soils	Drought tolerant	Čop et al., 2009; Volaire, 1995
Tall fescue (*Festuca arundinacea* Schreb.)	High dry matter yields; drought resistant; good for forage in dry regions	Most soil types	Drought tolerant	Frame et al., 1989; Frame & Laidlaw, 2011
Meadow fescue (*Festuca pratensis* L.)	Tolerant to wet conditions; high forage quality	Heavier soils	Wet conditions	Frame et al., 1989
Red fescue (*Festuca rubra* L.)	Suitable for diverse applications; good for lawns and forage	Lighter-to-heavier soils	Moderate moisture	Nesom, 2006
Legume				
White clover (*Trifolium repens* L.)	High protein content; N fixation; enhances forage quality	Most soil types	Moderate moisture	Dewhurst et al., 2009; Baker et al., 2023c
Red clover (*Trifolium pratense* L.)	High protein content; N fixation; enhances forage quality	Heavier soils	Moderate moisture	Carlsson & Huss-Danell, 2003; Harris et al., 1987
Lucerne (*Medicago sativa* L.)	High yield; deep roots; N fixation	Lighter soils	Drought tolerant	Elgersma & Schlepers, 1997
Sainfoin (*Onobrychis viciifolia* Scop.)	Deep rooting; drought tolerant; improves soil structure; N fixation	Lighter soils	Drought tolerant	Carlsson & Huss-Danell, 2003

(*Continued*)

Table 1 (*Continued*)

Species	Selection of benefits	Soil requirements	Soil moisture tolerance	References
Alsike clover (*Trifolium hybridum* L.)	Tolerant-to-wet conditions; high protein content; N fixation	Heavier soils	Wet conditions	Elgersma et al., 2014
Vetches (*Vicia* spp.)	High N fixation; improves soil structure	Heavier soils	Wet conditions	Ledgard & Steele, 1992
Herb				
Chicory (*Cicorium intybus* L.)	High yield and nutritive value; drought tolerant; high mineral content	Most soil types	Drought tolerant	Li & Kemp, 2005; Belesky et al., 2001
Plantain (*Plantago lanceolata* L.)	High palatability; high mineral content; good forage quality	Most soil types	Drought tolerant	Fraser & Rowarth, 1996; Wilman & Riley, 1993
Yarrow (*Achillea millefolium* L.)	Drought tolerant; improves soil structure; contributes to biodiversity	Lighter soils	Drought tolerant	Stewart, 1996
Salad burnet (*Sanguisorba minor* Scop.)	High nutritive value and mineral content; contributes to sward diversity	Lighter soils	Moderate moisture	Pirhofer-Walzl et al., 2011; Tilman et al., 1997
Caraway (*Carum carvi* L.)	Comparable digestibility to grasses; high nutrient value	Lighter soils	Moderate moisture	Elgersma et al., 2014

Moreover, while it is possible to use MSS in low-input systems, where N fixation by legumes reduces the need for synthetic fertilisers, it is important to ensure that management practices align with the species' nutrient needs and functional roles (Nyfeler et al., 2009). For example, N-fixing legumes can reduce the need for external N inputs, but the system must be managed in a way that supports the long-term health and productivity of both legumes and their companion species (Baker et al., 2023a). Well-designed system-scale studies also help refine these management practices, ensuring that long-term productivity goals are met without sacrificing the ecological benefits MSS are designed to provide.

6.6 Optimising species selection for animal performance

Incorporating legumes into MSS offers significant advantages for livestock nutrition by boosting crude protein levels and forage intake. This enhancement is vital for maintaining milk production in dairy systems and promoting optimal growth rates in beef and sheep production (McAuliffe, 2020; Beaucarne et al., 2021; McCarthy et al., 2023). Additionally, the inclusion of herbs in grass–clover mixtures contributes positively to animal production by improving both the palatability and digestibility of forage. These improvements lead to more consistent live weight gains and reduced finishing times in beef systems (Bryant et al., 2017; Grace et al., 2019; Boland et al., 2022).

Legumes and herbs in MSS contain condensed tannins and secondary metabolites like saponins and terpenes, which contribute to both environmental sustainability and animal health. Condensed tannins, in particular, have been shown to reduce methane emissions per unit of milk by up to 18% (Arndt et al., 2022), while also reducing nitrogen excretion and ammonia volatilisation in dairy systems (Beauchemin et al., 2020).

Secondary metabolites also play a notable role in managing livestock health by providing natural anthelmintic properties. In sheep systems, for instance, grazing on MSS high in secondary metabolites reduces gastrointestinal parasite burdens, thereby decreasing the need for synthetic anthelmintics and addressing concerns over resistance (Peña-Espinoza et al., 2018). Similarly, dairy and beef systems benefit from MSS in terms of rumen stability and reduced risk of digestive disturbances when compared to monoculture pastures (Enriquez-Hidalgo et al., 2014; Costes-Thiré et al., 2019). Moreover, livestock often exhibit selective grazing behaviour, instinctively choosing plants with specific secondary metabolites to self-regulate their health, a behaviour that highlights the importance of species diversity within MSS (Peña-Espinoza et al., 2018).

7 Practical aspects of establishing and managing multispecies swards

Establishing and managing MSS involves several practical considerations that can enhance their resilience and productivity over time. Although MSS may require a higher initial investment, these costs can often be offset by reduced synthetic input needs and the improved resilience of these systems.

7.1 Establishment methods and site preparation

Effective site preparation is crucial for MSS success. Soil testing before sowing guides adjustments in lime and fertiliser application, ensuring that optimal pH and nutrient levels support strong establishment (Wall and Plunkett, 2021). Innovative techniques such as broadcasting (oversowing) and direct drilling (undersowing), which introduce new species into existing swards, offer opportunities to enhance diversity with minimal soil disturbance/damage to the existing sward (Skinner and Dell, 2010; Raedts and Langworthy, 2018). These methods minimise soil disturbance and can reduce establishment costs by eliminating the need for ploughing and other intensive preparation practices.

7.2 Adaptive management strategies

Long-term success in MSS requires adaptive management to maintain species balance and prevent dominance by any single species. Regular monitoring of species composition, forage quality, and soil health enables precise adjustments in grazing frequency and nutrient management. This adaptive approach helps optimise productivity and extend sward longevity, particularly through rotational grazing, which promotes root development and reduces soil erosion (Pilon et al., 2017).

7.3 Silage production from multispecies swards

Multispecies swards generally produce silage with a lower dry matter content than perennial ryegrass-only swards, particularly when herb and legume content is high. Consequently, MSS silage often requires an extended wilting period to reach optimal DM levels for storage. Despite this, MSS silage tends to have a higher crude protein content, increased ash, and lower neutral detergent fibre, which can improve its nutritional profile relative to traditional perennial ryegrass silage (Moloney et al., 2020).

7.4 Economic implications

The multifunctional nature of MSS introduces both initial and ongoing costs. These include investments in diverse seed mixtures and increased labour for monitoring. However, reduced reliance on synthetic fertilisers and herbicides, coupled with the natural anthelmintic effects of certain herbs like chicory, can offset these expenses. Improved livestock productivity and resilience to environmental stressors further contribute to the economic viability of MSS. That said, the initial costs and maintenance needs requires careful planning, particularly for farmers working within tighter budgets or managing smaller operations.

8 Conclusion

The adoption of MSS represents a transformative approach to sustainable livestock systems, integrating grasses, legumes, and herbs into diverse sward mixtures that capitalise on functional diversity to improve nutrient cycling, pasture resilience, and forage quality. These benefits contribute to livestock performance, enhancing productivity in dairy, beef, and sheep systems and reducing reliance on synthetic inputs. However, it is essential to consider certain limitations of MSS. The successful MSS establishment and management require not only an initial investment but also ongoing expertise to ensure that species remain well-balanced over time. Farmers may encounter challenges in maintaining the desired diversity, especially if they lack specific knowledge in managing complex mixtures or face resource limitations for monitoring and adaptation. Additionally, the cost of seeds, equipment, and sometimes more labour-intensive management practices can be prohibitive for some farms, particularly smaller-scale or resource-limited operations. Despite these challenges, MSS swards contribute significantly to farm-level sustainability by promoting soil health, enhancing carbon sequestration, and supporting reductions in GHG emissions. They offer a viable strategy for farmers to adapt to climate change, improve soil and water quality, and align with regulatory demands to lower environmental impact. Looking forward, continued research and on-farm trials will be essential to optimise MSS, refine management practices, and address economic feasibility. By addressing the challenges and leveraging the advantages, MSS holds promise as a cornerstone of resilient, productive, and environmentally aligned agriculture.

9 References

Arndt, C., et al. (2022). Full adoption of the most effective strategies to mitigate methane emissions by ruminants can help meet the 1.5 °C target by 2030 but not 2050. *Proceedings of the National Academy of Sciences*, *119*, e2111294119.

Baker, S., et al. (2023a). Multispecies swards outperform perennial ryegrass under intensive beef grazing. *Agriculture, Ecosystems & Environment, 345*, 108335.

Baker, S., et al. (2023b). Dry-matter production and botanical composition of multispecies and perennial ryegrass swards under varying defoliation management. *Grass and Forage Science, 78*, 338–352.

Baker, S., et al. (2023c). A comparison of the nutritive value of perennial ryegrass, perennial ryegrass and white clover and multispecies-based farmlet systems. *Grassland Research, 2*, 129–139.

Beaucarne, G., et al. (2021). The effect of sward type on the performance of co-grazed cattle and sheep. *Journal of Animal Science, 99*, 96.

Beauchemin, K. A., et al. (2020). Fifty years of research on rumen methanogenesis: lessons learned and future challenges for mitigation. *Animal, 14*, s2–s16.

Belesky, D. P., et al. (2001). Mineral composition of swards containing forage chicory. *Agronomy Journal, 93*, 468–475.

Berdeni, D., et al. (2021). Soil quality regeneration by grass-clover leys in arable rotations compared to permanent grassland: effects on wheat yield and resilience to drought and flooding. *Soil and Tillage Research, 212*, 105037.

Boland, T. M., et al. (2022). The impact of sward type on animal performance and the carbon footprint of beef production. In *Proceedings of the 8th International Greenhouse Gas and Animal Agriculture Conference*, Orlando, Florida, 72.

Bouton, J. (2006). The economic benefits of forage improvement in the United States. *Euphytica, 154*, 263–270.

Brouwer, F. & Hellegers, P. (1996). The nitrate directive and farming practice in the European Union. *European Environment, 6*, 204–209.

Bryant, R. H., et al. (2017). Milk yield and nitrogen excretion of dairy cows grazing binary and multispecies pastures. *Grass and Forage Science, 72*, 806–817.

Cardinale, B. J., et al. (2007). Impacts of plant diversity on biomass production increase through time because of species complementarity. *Proceedings of the National Academy of Sciences, 104*, 18123–18128.

Carlsson, G. & Huss-Danell, K. (2003). Nitrogen fixation in perennial forage legumes in the field. *Plant and Soil, 253*, 353–372.

Cong, W., et al. (2020). Optimizing yield and flower resources for pollinators in intensively managed multi-species grasslands. *Agriculture, Ecosystems & Environment, 302*, 107062.

Connolly, J., et al. (2018). Weed suppression greatly increased by plant diversity in intensively managed grasslands: a continental-scale experiment. *Journal of Applied Ecology, 55*, 852–862.

Čop, J., et al. (2009). Influence of cutting regime and fertilizer application on the botanical composition, yield and nutritive value of herbage of wet grasslands in Central Europe. *Grass and Forage Science, 64*, 454–465.

Costes-Thiré, M., et al. (2019). Diet selection and trade-offs between condensed tannins and nutrients in parasitized sheep. *Veterinary Parasitology, 271*, 14–21.

Cummins, S., et al. (2021). Beneficial effects of multi-species mixtures on N_2O emissions from intensively managed grassland swards. *Science of the Total Environment, 792*, 148163.

Das, A., et al. (2023). Pathways to net zero emissions for the Indian power sector. *Energy Strategy Reviews, 45*, 101042.

Dewhurst, R. J., et al. (2009). Nutritive value of forage legumes used for grazing and silage. *Irish Journal of Agricultural and Food Research*, *48*, 167–187.

Durack, P. J., et al. (2012). Ocean salinities reveal strong global water cycle intensification during 1950 to 2000. *Science*, *336*, 455–458.

Elgersma, A. & Schlepers, H. (1997). Performance of white clover–perennial ryegrass mixtures under cutting. *Grass and Forage Science*, *52*, 134–146.

Elgersma, A., et al. (2014). Forage quality in legumes and non-leguminous forbs. *Proceedings of the European Grassland Federation*, *19*, 600–602.

Elliot, R. H. (1948). *The Clifton Park System of Farming and Laying Down Land to Grass. A Guide to Landlords, Tenants and Land Legislators* (5th ed., 6th impression). London: Faber and Faber.

Enriquez-Hidalgo, D., et al. (2014). Effect of rotationally grazing perennial ryegrass white clover or perennial ryegrass only swards on dairy cow feeding behaviour, rumen characteristics and sward depletion patterns. *Livestock Science*, *169*, 48–62.

Ergon, Å., et al. (2018). How can forage production in Nordic and Mediterranean Europe adapt to the challenges and opportunities arising from climate change? *European Journal of Agronomy*, *92*, 97–106.

European Commission. (2019). The European Green Deal. Retrieved from https://ec.europa.eu/info/strategy/priorities-2019-2024/european-green-deal_en

Eurostat. (2023). Agri-environmental indicator: cropping patterns. % Share of Utilised Agricultural Area, EU Countries 2020. Retrieved from https://ec.europa.eu/eurostat/statistics-explained/index.php?title=Agrienvironmental_indicator_-_cropping_patterns&oldid=587530#Analysis_of_cropping_patterns_at_country_level

FAO. (2009). *The State of Food and Agriculture: Livestock in the Balance*. Rome: FAO.

Favero, A., et al. (2024). US land sector mitigation investments and emissions implications. *Nature Communications*, *15*, 9625.

Finn, J. A., et al. (2013). Ecosystem function enhanced by combining four functional types of plant species in intensively managed grassland mixtures: a 3-year continental-scale field experiment. *Journal of Applied Ecology*, *50*, 365–375.

Frame, J. (1989). Herbage productivity of a range of grass species under a silage cutting regime with high fertilizer nitrogen application. *Grass and Forage Science*, *44*, 267–276.

Frame, J. & Laidlaw, S. (2011). Improved Grassland Management (2nd ed.). Marlborough, UK: Crowood Press.

Fraser, T. J., et al. (1996). Pasture species effects on carcass and meat quality. *Proceedings of the New Zealand Grassland Association*, *58*, 63–66.

Fu, Z., et al. (2022). Contribution of climate change to interannual variations in global terrestrial net ecosystem productivity. *Nature Communications*, *13*, 1–10.

Grace, C., et al. (2018). The effect of increasing pasture species on herbage production, chemical composition and utilization under intensive sheep grazing. *Grass and Forage Science*, *73*, 852–864.

Grace, C., et al. (2019). Grazing multispecies swards improves ewe and lamb performance. *Animal*, *13*, 1721–1729.

Grange, G., et al. (2021). Plant diversity enhanced yield and mitigated drought impacts in intensively managed grassland communities. *Journal of Applied Ecology*, *58*, 1864–1875.

Grinnell, N. A., et al. (2023). Long-term effects of extensive grazing on pasture productivity. *Animal Production Science*, *63*, 1236–1247.

Hamacher, M., et al. (2021). Evaluation of yield formation and nutritive value of forage legumes and herbs with potential for diverse grasslands due to their concentration in plant specialized metabolites. *European Journal of Agronomy*, *128*, 126307.

Harris, W. (1987). Population dynamics and competition. In M. J. Baker & W. M. Williams (Eds.), *White Clover* (pp. 203–297). Wallingford, UK: CAB International.

Harris, P. S. (2001). *Grassland Resource Assessment for Pastoral Systems*. FAO Plant Production and Protection Paper No. 162. Food and Agriculture Organization of the United Nations.

Herrero, M., et al. (2013). The roles of livestock in developing countries. *Animal*, *7*, 3–18.

Hofer, D., et al. (2016). Yield of temperate forage grassland species is either largely resistant or resilient to experimental summer drought. *Journal of Applied Ecology*, *53*, 1023–1034.

Ikoyi, I., et al. (2023). Plant diversity enhanced nematode-based soil quality indices and changed soil nematode community structure in intensively-managed agricultural grasslands. *European Journal of Soil Biology*, *118*, 103542.

Jensen, E. S., et al. (2011). Legumes for mitigation of climate change and feedstock in a bio-based economy – a review. *Agronomy for Sustainable Development*, *32*, 329–364.

Jørgensen, M., et al. (2010). De-hardening in contrasting cultivars of timothy and perennial ryegrass during winter and spring. *Grass and Forage Science*, *65*, 38–48.

Komainda, M., et al. (2020). Drought tolerance is determined by species identity and functional group diversity rather than by species diversity within multi-species swards. *European Journal of Agronomy*, *119*, 126116.

Ledgard, S. F. & Steele, K. W. (1992). Biological nitrogen fixation in mixed legume/grass pastures. *Plant and Soil*, *141*, 137–153.

Lee, J. M., et al. (2010). Plants modify biological processes to ensure survival following carbon depletion: a Lolium perenne model. *PLOS One*, *5*, e12306.

Lee, M. R. F. & Hodge, I. D. (2017). The role of livestock in climate change mitigation: a review of the evidence. *Agricultural Systems*, *156*, 3–12.

Li, G. D. & Kemp, P. D. (2005). Forage chicory (Cichorium intybus L.): a review of its agronomy and animal production. *Advances in Agronomy*, *88*, 187–222.

Lüscher, A., et al. (2014). Potential of legume-based grassland–livestock systems in Europe: a review. *Grass and Forage Science*, *69*, 206–228.

Lüscher, A., et al. (2022). Using plant diversity to reduce vulnerability and increase drought resilience of permanent and sown productive grasslands. *Grass and Forage Science*, *77*, 235–246.

McAuliffe, G. A., et al. (2020). Applications of nutritional functional units in commodity-level life cycle assessment (LCA) of agri-food systems. *The International Journal of Life Cycle Assessment*, *25*, 208–221.

McCarthy, K., et al. (2023). The effect of a zero-grazed perennial ryegrass, perennial ryegrass and white clover, or multispecies forage on the dry matter intake, milk production and nitrogen utilization of dairy cows in mid-late lactation. *Livestock Science*, *272*, 105234.

McDonagh, J., et al. (2016). Genetic gain in perennial ryegrass (*Lolium perenne*) varieties 1973 to 2013. *Euphytica*, *212*, 187–199.

Moloney, T., et al. (2020). Yield of binary-and multi-species swards relative to single-species swards in intensive silage systems. *Irish Journal of Agricultural and Food Research*, *59*, 12–26.

Moot, D. J. (2012). An overview of dryland legume research in New Zealand. *Crop and Pasture Science*, *63*, 726–733.
Muir, J. P., et al. (2011). Sustainable, low-input, warm-season, grass–legume grassland mixtures: mission (nearly) impossible? *Grass and Forage Science*, *66*, 301–315.
Nesom, G. L. (2006). Erigeron. In *Flora of North America North of Mexico* (Vol. 20, pp. 256–348). New York: Oxford University Press.
Nyfeler, D., et al. (2009). Strong mixture effects among four species in fertilized agricultural grassland led to persistent and consistent transgressive overyielding. *Journal of Applied Ecology*, *46*, 683–691.
Nyfeler, D., et al. (2011). Grass–legume mixtures can yield more nitrogen than legume pure stands due to mutual stimulation of nitrogen uptake from symbiotic and non-symbiotic sources. *Agriculture, Ecosystems & Environment*, *140*, 155–163.
O'Malley, J., et al. (2023). LegacyNet: introducing an international multi-site experiment investigating potential benefits of increasing the species diversity of grassland leys within crop rotations. *Grassland Science in Europe*, *28*, 249–251.
Pearson, D. (2020). Key roles for landscape ecology in transformative agriculture using Aotearoa–New Zealand as a case example. *Land*, *9*, 146.
Peña-Espinoza, M., et al. (2018). Antiparasitic activity of chicory (Cichorium intybus) and its natural bioactive compounds in livestock: a review. *Parasites & Vectors*, *11*, 475.
Peyraud, J. L., et al. (2009). Potential food production from forage legume-based systems in Europe: An overview. *Irish Journal of Agricultural and Food Research*, *48*, 115–135.
Pilon, C., et al. (2017). Long-term effects of grazing management and buffer strips on soil erosion from pastures. *Journal of Environmental Quality*, *46*, 364–372.
Pirhofer-Walzl, K., et al. (2011). Forage herbs improve mineral composition of grassland herbage. *Grass and Forage Science*, *66*, 415–423.
Raedts, P. & Langworthy, A. (2018). Establishing plantain in spring in existing perennial ryegrass pastures in northern Tasmania. *Animal Production Science*, *60*, 114–117.
Rochon, J. J., et al. (2004). Grazing legumes in Europe: a review of their status, management, benefits, research needs and future prospects. *Grass and Forage Science*, *59*, 197–214.
Sanderson, M. A., et al. (2004). Plant species diversity and management of temperate forage and grazing land ecosystems. *Crop Science*, *44*, 1132–1144.
Scarsbrook, M. R. & Melland, A. R. (2015). Dairying and water-quality issues in Australia and New Zealand. *Animal Production Science*, *55*, 856–868.
Shackleton, J., et al. (2024). Annual and seasonal dry matter production, botanical species composition, and nutritive value of multispecies, permanent pasture, and perennial ryegrass swards managed under grazing. *Grass and Forage Science*, *79*, 630–650.
Shnel, A., et al. (2021). The importance of forage legume inclusion in agricultural swards to enhance earthworm activity and water infiltration rates. In *Virtual Joint XXIV International Grassland and XI International Rangeland Congress, Kenya 2021, Congress Proceedings* (Vol. 1, pp. 341–345). Nairobi: Kenya Agricultural and Livestock Research Organization (KALRO).
Skinner, R. H. & Dell, C. J. (2010). Reestablishing chicory into multi-species perennial pastures. *Forage & Grazinglands*, *8*, 1–6.
Sleugh, B., et al. (2000). Binary legume–grass mixtures improve forage yield, quality, and seasonal distribution. *Agronomy Journal*, *92*, 24–29.
Sollenberger, L. E., et al. (2019). Grassland management affects delivery of regulating and supporting ecosystem services. *Crop Science*, *59*, 441–459.

Stewart, A. V. (1996). Plantain (Plantago lanceolata): a potential pasture species. *Proceedings of the New Zealand Grassland Association*, *58*, 77–86.

Suter, M., et al. (2021). Multispecies for multifunctions: combining four complementary species enhances multifunctionality of sown grassland. *Scientific Reports*, *11*, 3835.

Teagasc. (2020). Greenhouse gas emissions from agriculture in Ireland. Retrieved from https://www.teagasc.ie/publications/2020/greenhouse-gas-emissions-from-agriculture-in-ireland.php

Teagasc. (2021). Sustainable food systems: the role of grassland management. Retrieved from https://www.teagasc.ie/publications/2021/sustainable-food-systems-the-role-of-grassland-management.php

Teagasc. (2022). Grazing management: strategies for improving pasture quality and livestock performance. Retrieved from https://www.teagasc.ie/publications/2022/grazing-management-strategies-for-improving-pasture-quality-and-livestock-performance.php

Thorvaldsson, G. & Andersson, S. (1986). Variations in timothy dry matter yield and nutritional value as affected by harvest date, nitrogen fertilization, year and location in Northern Sweden. *Acta Agriculturae Scandinavica*, *36*, 367–385.

Tilman, D., Lehman, C. L., & Thomson, K. T. (1997). Plant diversity and ecosystem productivity: theoretical considerations. *Proceedings of the National Academy of Sciences of the United States of America*, *94*, 1857–1861.

Turner, F. (1955). *Fertility Pastures*. London: Faber and Faber Limited.

Vermeulen, S. J., et al. (2012). Climate change and food systems. *Annual Review of Environment and Resources*, *37*, 195–222.

Volaire, F. (1995). Growth, carbohydrate reserves and drought survival strategies of contrasting Dactylis glomerata populations in a Mediterranean environment. *Journal of Applied Ecology*, *32*, 56–66.

Wall, D. & Plunkett, M. (2021). Major and micro nutrient advice for productive agricultural crops. Retrieved from https://www.teagasc.ie/publications/2020/major--micro-nutrient-advice-for-productive-agricultural-crops.php

Wardropper, C. B., et al. (2023). Policy process and problem framing for state Nutrient Reduction Strategies in the US Upper Mississippi River Basin. *Journal of Soil and Water Conservation*, *78*, 70–81.

Wróbel, B., et al. (2025). Improving forage quality from permanent grasslands to enhance ruminant productivity. *Agriculture*, *15*, 1438.

Zhan, X., et al. (2024). The shape of reactive nitrogen losses from intensive farmland in China. *Science of the Total Environment*, *915*, 170014.

Chapter 15

Plantain (*Plantago lanceolata*): from weed to multi-functional component in sustainable dairy-farm systems

Ina (J.B.) Pinxterhuis, DairyNZ, New Zealand; and Mike (M.B.) Dodd, AgResearch, New Zealand

1 Introduction

Narrow-leaved plantain (ribgrass, *Plantago lanceolata* L.) has long been a common adventive species in managed grasslands globally. Its value as a livestock forage was enhanced with the development of productive and erect cultivars from the mid-1990s. Another major development has been the observation of its effect in reducing livestock urinary-nitrogen concentration. Low tissue dry matter content has been verified as the main contributor to this, but studies also point to effects of plant characteristics on livestock nitrogen partitioning and reduced drainage and rate of nitrification in soils. This functionality provides a way of mitigating nitrogen losses to air and water in intensive, pasture-based livestock systems that is relatively simple to implement.

This chapter summarises the broad characteristics and uses of modern forage plantain and identifies scope for research relating to management

http://dx.doi.org/10.19103/AS.2025.0154.24

for persistence, effects on animal health and milk composition and a better understanding of the biochemical mechanisms behind nitrogen cycle effects.

2 Nitrogen loss from grazed pastures

Nitrogen (N) losses to the environment, notably nitrate leaching (contaminating groundwater and surface water) and nitrous oxide emission (a potent greenhouse gas), remain important unwanted impacts of productive, intensively grazed pastures. Over time, increased use of synthetic N fertiliser has increased pasture production and animal N intake per hectare, increased animal production and increased associated N losses to the environment (Chapman and Parsons, 2017; Pinxterhuis et al., 2015; Van der Meer and Van Uum-Van Lohuyzen, 1986). Additionally, this system inefficiency has economic consequences for farms and is also associated with indirect greenhouse gas emissions through energy use for production and transport of synthetic N fertilisers. Examples for farm-scale N use efficiency (NUE, the ratio of N in product: N in inputs) and N surplus (the difference between N in inputs and N in product) of well-managed dairy farms in New Zealand, Ireland and The Netherlands were given by Pinxterhuis et al. (2015), with NUE varying between 32% and 50% and N surplus between 73 kg N/ha and 257 kg N/ha, including N inputs from fertiliser, imported supplementary feeds, and N_2 fixation via legumes, rainfall, irrigation and deposition. This leaves large amounts of N at risk of loss to the environment.

Sown pastures comprised of perennial ryegrass (*Lolium perenne* L.) and N-fixing white clover (*Trifolium repens* L.) are often N deficient for maximum herbage yield, if no N fertiliser is used (Gray, 2023). However, these pastures have shown high herbage yields relative to grass-only swards with no or low N fertiliser application rates. For example, this unfertilised two-species pasture attained herbage dry matter yields equivalent to perennial ryegrass monoculture fertilised at 300–600 kg N/ha/year in New Zealand (Ball and Field, 1985). At three sites in Europe, perennial ryegrass/white clover/red clover (*Trifolium pratense* L.) pastures with N fertiliser at 70–150 kg N/ha/year produced equal or more herbage dry matter (19–29 t DM/ha/year) than perennial ryegrass only with approximately double the N fertiliser inputs of 165–350 kg N/ha/year (17–30 t DM/ha/year; Collins et al., 2014). High herbage yields and grazing of perennial ryegrass/white clover or N-fertiliser grass swards result in high levels of N returned to the soil via death and decay of roots and stubble and via animal excreta. Whitehead (1986) estimated these pathways amount to 200–350 kg N/ha/year for death and decay of roots and stubble and 150–300 kg N/ha/year for animal excreta at harvested yields of 10–12 t DM/ha/year. Soil biological processes determine the proportion of this cycling N that is lost, either to the atmosphere or via drainage below the reach of plants.

The urine excreted by grazing stock has long been identified as the main source of N loss from grazed pastures due to the uneven distribution of relatively high N loads (Ball et al., 1979; Corré et al., 2014; Di and Cameron, 2002; Garwood and Ryden, 1986). The benefits of N leaching of reducing urine patches by standing animals off pasture have been well documented (Beukes et al., 2017; Wachendorf et al., 2004). However, this requires capital investment and maintenance of infrastructure and feeding systems, which increases costs of production and reduces profitability, even if the potential benefit of less pasture damage in wet conditions is considered (Beukes et al., 2013; Laurenson et al., 2016; Laurenson et al., 2017).

Alternative forage options present another means of reducing nitrate leaching and nitrous oxide emission from grazed highly productive pasture. This may be at lower cost and higher profit than systems with off-pasture infrastructure or systems with substantially lower overall N inputs and consequently lower production, such as reported by Ledgard et al. (1997, 2006), Shepherd et al. (2017) and Beukes et al. (2024). Various options have been proposed, such as utilising plants that can maintain high production with lower synthetic N inputs and lower N content (Chapman et al., 2014), plants that reduce urinary N excretion or concentration (Bryant et al., 2019b), or plants that reduce soil nitrification rate (Judson et al., 2018a).

The standard pastures of New Zealand dairy farms consist of perennial ryegrass and white clover mixtures, supplemented with synthetic N fertiliser at rates up to 190 kg N/ha/year, offered to ruminants at the 2–3.5 leaf stage of perennial ryegrass. These pastures frequently have crude protein concentrations of 15–25%, well above animal requirements (Pacheco and Waghorn, 2008). Bryant et al. (2019b) illustrated this with a comparison of crude protein supply versus requirement in pasture-only diets of dairy cows producing 340 kg or 400 kg milksolids per year, which were averages for the Waikato and Canterbury regions of New Zealand, respectively. Only in November (at peak lactation of a spring-calving dairy cow) were demand and supply equal and in other months crude protein concentration could be up to 10 percentage points above demand. Most of the excess N is excreted in urine, resulting in a wide range of reported urinary N concentrations (0.9–20.5 g/L) and N loads in urine patches (200–2000 kg N/ha for cattle; Selbie et al., 2015).

New Zealand research in the period 2005–2015 investigated if the addition of multiple herb species to perennial ryegrass/white clover swards would be beneficial for herbage production, animal production and NUE. Woodward et al. (2012) and Totty et al. (2013) reported similar yields and milk production from multispecies perennial pastures containing ryegrass, white clover and the herbs chicory (*Cichorium intybus* L.) and narrow-leaved plantain (*Plantago lanceolata*; also ribwort plantain, ribgrass, lamb's tongue, forage plantain – plantain hereafter). A potentially beneficial environmental factor was the reported lower urinary N concentration and urinary N excretion. Beukes et al. (2014) modelled the effects of the observed differences in herbage growth rates and feed composition in the Woodward et al. (2012, 2013) and Totty et al. (2013) studies and indicated N leaching could be reduced by 19% if half the farm area consisted of such grass–clover–herb mixture.

Animal feeding trials provided evidence that adding plantain to perennial ryegrass/white clover pasture lowered urinary N concentration (Cheng et al., 2017). Thus, the focus shifted from diverse grass–clover–herb mixtures to grass–clover–plantain mixtures. Judson et al. (2018a) suggested that plantain was a key species to achieve N leaching reductions via several mechanisms. These included reduced urinary N concentration, reduced partitioning of excreted N to urine and a soil nitrification inhibiting effect of plantain. It was suggested that this could occur both indirectly from urine patches of animals fed plantain and directly from plantain plant material and root exudates. These aspects of the role of plantain in the N cycle have been topics of ongoing research since.

The following sections provide a summary of current knowledge on plantain and are for a large part based on two review papers (Dodd et al., 2025; Pinxterhuis et al., 2024). Recent review papers also summarise current knowledge and confirm the potential of plantain in mitigating N losses to the environment from pasture-based grazed systems (Chesney et al., 2025; Eme and Roche, 2025).

3 Introduction to narrow-leaved plantain (*Plantago lanceolata*)

Plantain has long been seen as a weed species or minor contributor to pasture production, though valued for its animal health benefits (Burke, 2008; Pol et al., 2021; Stewart, 1996). Before research focussed on the effects of plantain on N loss to the environment, varieties of plantain were developed for agronomic attributes. These have been used in grazed pastures in New Zealand since the 1990s when two cultivars were released: Grasslands Lancelot, a prostrate cultivar best suited to close sheep grazing and Ceres Tonic, a more erect cultivar better suited to rotational cattle grazing. Their production and quality in summer-dry environments complemented perennial ryegrass/white clover pasture (Stewart and Judson, 2019) and they were mainly used in the drystock sector in mixtures with clover for finishing young livestock. However, the lack of persistence beyond 2–4 years of improved plantain cultivars as a monoculture forage crop or as a significant contributor in grass-dominant pastures has hampered their widespread use. The widespread use of dicot-specific herbicides in grass–clover pastures has also been a barrier to successful integration.

3.1 Growth habit

Wild-type plantain can be found throughout the world in a wide range of environments and is largely a winter-dormant prostrate perennial dicotyledon herb, forming rosettes with growing points emerging from axillary meristems on a crown below the soil surface. It has high phenotypic plasticity and will become more erect to avoid shading in higher vegetation. Commercial selection of genotypes resulted in upright cultivars with improved cool-season growth (Figure 1).

Flowering is initiated in spring when daylight lengthens and seed heads emerge more readily from the second year onwards. Fresh seeds are largely dormant and need after-ripening and cold stratification to germinate. Germination increases with higher temperatures, more light and greater availability of soil N. Plantain establishes relatively slowly in a multispecies sward and is susceptible to competition, particularly in environments conducive to strong grass growth, such as high N fertility, mild temperatures and sufficient soil moisture. Compared with perennial ryegrass, plantain tolerates higher temperatures and moisture stress, lower soil macro-nutrient fertility and lower soil pH.

Figure 1 Modern genotypes of forage plantain (*Plantago lanceolata* L.) are more productive and upright than wild types. Left wild type, right cultivar AgriTonic.

The root system of plantain is a combination of fibrous roots at the surface and courser roots that can reach >1 m deep in the soil, making the species relatively drought resistant. However, water use efficiency of plantain is lower than that of ryegrass (Neal et al., 2011).

Troelstra et al. (1992) showed that plantain herbage yield increased with increasing N supply, with plants allocating more growth to leaf rather than root material. This partitioning was reversed under moisture stress and low N supply, when a greater proportion of biomass growth was invested in roots (Skinner and Comas, 2010).

3.2 Herbage yield

Even though several plantain cultivars are currently commercially available (Dodd et al., 2025), most of the agronomic and N management research has been conducted with Ceres Tonic and more recently its successor, Grasslands Agritonic. This is largely due to the longevity of market availability of Tonic in New Zealand (c. 30 years) and the substantial research investment by its breeder, PGG Wrightson Seeds. Tonic was selected from Mediterranean germplasm for improved cool-season growth, a more upright growth habit (i.e. more harvestable material) and improved grazing tolerance. Grasslands Agritonic has been selected from Tonic germplasm for improved competitiveness in multispecies swards, more resilience when herbicides are used and traits that have been associated with improved NUE (Box et al., 2023b).

Plantain is used in a range of pasture types (for references, see Dodd et al., 2025):

- Monoculture – sown at 10–12 kg/ha, persisting for ~2 years. Perennial or Italian ryegrass (*Lolium multiflorum*) and white clover can be drilled into this monoculture in autumn of the first or second year. Weed invasion can occur in these monocultures. Reported yields in New Zealand varied between 7.5 t DM/ha/year and 20.6 t DM/ha/year.
- Herb–legume mixture – plantain sown at 6–8 kg/ha with clovers and sometimes chicory. High-quality summer-active crop, mainly used in summer-dry non-irrigated systems. Can persist well in these circumstances, without grass competition, but weeds can become a problem. Yields varying from 9.3 t DM/ha/year to 20.8 t DM/ha/year have been reported from New Zealand studies.
- Grass–clover–plantain mixture – perennial ryegrass and white clover sown with plantain at 1–4 kg/ha. Can be sown as a new pasture, or plantain seed can be drilled or over-sown into existing pasture. Provides higher quality summer/autumn forage than perennial ryegrass/white clover swards in summer-dry non-irrigated systems. Reported yields in

New Zealand were 8.9–18.1 t DM/ha/year, which was often higher than that of perennial ryegrass/white clover. This mixture (Figure 2) has been the focus of research in the New Zealand Plantain Potency and Practice programme.

- Multispecies mixture – plantain sown at 1–4 kg/ha with a range of other pasture species. An increasingly used mixture consists of perennial ryegrass, white and red clover, plantain and chicory. Depending on region (climate) and farmer preference, other grasses are added or substitute perennial ryegrass, such as tall fescue (*Festuca arundinacea*), Italian ryegrass, timothy (*Phleum pratense*), cocksfoot (*Dactylis glomerata*), kikuyu (*Pennisetum clandestinum*) and/or prairie grass (*Bromus willdenowii*). Other legumes that may be added to these mixtures are lucerne (*Medicago sativa*) and bird's-foot-trefoil species (*Lotus pedunculatus* and *Lotus corniculatus*). Yields of 11.6–17.9 t DM/ha/year for dairy pastures in Waikato, New Zealand, were reported for multispecies pastures by Woodward et al. (2013), which were similar to perennial ryegrass/white clover. However, yields in summer and autumn were higher than for perennial ryegrass/white clover. Daly et al. (1996) reported yields for multispecies drystock pasture in Canterbury, New Zealand, of 4–11.5 t DM/ha/year, being 10–30% higher than perennial ryegrass/white clover yields. In this study, the benefit was due to higher yields in summer.

Figure 2 Perennial ryegrass/white clover/plantain pasture in New Zealand.

As with any pasture, large differences in yield exist between regions, years, management treatments and sward types. In general, plantain in grass-based multispecies swards (grass–clover–plantain mixture and multispecies mixture pasture types described above) does not reduce annual average pasture or milk production, compared with perennial ryegrass/white clover pastures at similar N fertiliser rates or perennial ryegrass pasture with higher N fertiliser rates. However, plantain tends to grow less in winter and early spring than perennial ryegrass, but this is compensated for with greater summer and autumn growth, especially in summer-dry conditions (Dodd et al., 2025). Therefore, seasonality in the contribution of plantain to standing herbage in mixtures is often observed, with a greater proportion in late summer/autumn.

The development of a wider range of cultivars by various breeders in the last decade tended to focus on variation in winter growth activity and flowering date. In New Zealand, this started with the release of two cultivars in the 1990s: Ceres Tonic (erect, winter-active with large leaves) and Grasslands Lancelot (semi-erect, low winter growth, medium-to-large leaves; Box et al., 2023a). There have been surprisingly few inter-cultivar comparisons in the published literature. Differences between cultivars in annual and seasonal yield have been shown in South Africa (Ammann et al., 2021) and this is currently also under investigation in New Zealand in the Plantain Potency and Practice Program (DairyNZ: https://www.dairynz.co.nz/research/science-projects/plantain-potency-and-practice/).

3.3 Forage quality

A meta-analysis of herbage quality data showed that compared with perennial ryegrass, plantain has similar digestibility, lower percent dry matter, similar total N concentration in dry matter, greater non-structural fibre concentration, and lower concentrations of structural fibre, soluble and degradable N (Minneé et al., 2019; Table 1). Similar results were found in an Irish study (Minogue et al., 2025), though they did not find differences in soluble carbohydrate content.

The concentration of crude protein declines more in plantain in late summer compared with other species such as chicory and ryegrass (Dodd et al., 2022; Kemp et al., 2010). Plantain also has greater concentrations of trace elements than perennial ryegrass, when these are sufficiently available in the soil (Moorhead et al., 2002). However, these herbage characteristics of plantain will have a minor effect on multispecies pasture quality if plantain does not comprise a substantial proportion of the pasture. Plantain/clover mixtures have consistently shown better feed quality than perennial ryegrass/white clover swards in drystock systems, particularly in summer (Cranston et al., 2015a, 2015b; Somasiri et al., 2015), but this may be due to clover making up a higher proportion of the sward in these two-species swards.

Table 1 Herbage nutritive quality of perennial ryegrass and plantain monocultures. Shown are least square means and upper and lower 95% confidence limits, resulting from a meta-analysis of predominantly New Zealand data

	Perennial ryegrass	Plantain	*P*
Dry matter (%)	21.1 (19.2–23.0)	14.8 (12.9–16.7)	<0.001
Digestibility (%)	77.8 (n.p.)	77.2 (n.p.)	0.339
Total nitrogen (% of DM)	3.1 (2.7–3.4)	3.2 (2.8–3.5)	0.5207
Soluble N (% of total N)	38.4 (31.8–45.0)	12.0 (5.3–18.6)	0.0012
Degradable N (% of total N)	69.2 (65.9–72.5)	56.0 (52.7–59.3)	0.0011
Neutral detergent fibre (% of DM)	43.1 (40.0–46.2)	28.7 (25.5–31.8)	<0.001
Acid detergent fibre (% of DM)	24.6 (22.7–26.6)	20.8 (18.8–22.7)	<0.001
Non-structural carbohydrate (% of DM)	21.1 (19.6–22.6)	29.9 (28.4–31.4)	<0.001
Ash (% of DM)	10.2 (n.p.)	12.8 (n.p.)	<0.001

Source: Minneé et al. (2019).

Longer grazing intervals generally lead to lower crude protein concentration and higher neutral detergent fibre concentration of plantain (Labreveaux et al., 2004, Lee et al., 2015a, Navarrete, 2015). Box et al. (2017) reported no effect of morning versus afternoon harvesting on nutritive value of plantain, compared with higher soluble carbohydrate concentration in the afternoon for perennial ryegrass.

Low palatability of plantain is sometimes cited by dairy farmers observing livestock reluctant to graze it, but this has yet to be experimentally verified. Other farmers report no palatability issues with plantain, especially when livestock have been exposed to it for some time or on a regular basis. There is certainly the potential for old leaf tissue and hardened seed heads to be less palatable, whether as a result of long grazing intervals or the persistence of previously ungrazed material in residual herbage.

3.4 Management of grazed pastures

New Zealand pastoral systems are generally based on year-round in-situ rotational grazing of perennial ryegrass/white clover or other multispecies pastures. Forage crops are utilised where pastures do not provide sufficient feed in dry summers or cold winters. In summer-dry regions, plantain monocultures or plantain/clover mixtures have been shown to have greater herbage accumulation in summer and autumn than perennial ryegrass-based swards (Dodd et al., 2022). This marginal gain in forage supply can help fill moisture stress-induced feed gaps.

Optimum grazing management for high yield and improved persistence of plantain is still under investigation. In a field experiment with monocultures,

Lee et al. (2015a; Waikato, New Zealand) reported higher seasonal and annual yields with longer grazing intervals (21.3 t DM/ha grown over 18 months after sowing for defoliation at 150 mm extended leaf height vs. 29.8t DM/ha for 450 mm extended leaf height). However, stem-to-leaf ratio increased and in-vitro digestibility decreased, indicating a trade-off between yield and feed quality. The effect of residual heights was smaller (26.1 t DM/ha and 24.6 t DM/ha over 16 months for 30–50 mm and 60–80 mm sward height, respectively).

In chicory/plantain/red and white clover mixtures, Navarrete (2015; Manawatū, New Zealand) observed no significant herbage yield difference in two consecutive growing seasons between 2-week and 4-week grazing intervals with dairy cows. The proportion of plantain in this mixture was unaffected by grazing interval but decreased in the second year. In a cutting experiment with herb/clover mixtures, plantain yield was greatest at either a cutting interval of 7 days and a residual height of 75 mm or a cutting interval of 42 days and a residual height of 25 mm; taproot diameter was greater with longer intervals but not impacted by residual height (Cranston et al., 2021).

Cranston et al. (2015b) showed that under a sheep grazing regime of chicory/plantain/red and white clover swards with 3–5-week rotations and low (~4 cm) or high (~8 cm) residuals, the size of plantain's taproot and its water-soluble carbohydrate concentration remained stable over time and between grazing treatments, suggesting that these regrowth intervals were sufficient to maintain root mass and carbohydrate reserves irrespective of post-grazing residuals. In another field study, using monocultures, plantain root non-soluble carbohydrates declined for 14 days post-defoliation in summer and took another 21 days to reach pre-defoliation levels (Lee et al., 2015b). This meant dry matter production was reduced at defoliation intervals less than 35 days. Defoliation height was less influential, indicating that root reserves are more important for regrowth than remaining leaf area.

In a glasshouse study with constructed perennial ryegrass/white clover/plantain swards, increasing defoliation intensity from no cutting (average plant height approximately 40 cm) to cutting to 35 cm, 25 cm, 15 cm and 10 cm residual height above ground level every 2 weeks during spring resulted in the greatest proportion of plantain in the shoot mass in the 25 cm residual height treatment and the greatest plantain shoot production in the 15 cm residual height treatment (Mikola et al., 2001).

In ryegrass-dominant rotationally grazed dairy pastures, managed to standard industry guidelines for ryegrass/clover swards, the plantain proportion of herbage dry matter generally declines within a few years (Dodd et al., 2025). The results of the grazing or cutting experiments summarised above suggest that the optimal rotation length for mixed-species swards containing plantain may be longer than standard practise for ryegrass/clover swards to enable plantain to persist. However, the seasonality of such adjustments is important, as Wilson et al. (2024) suggested shortening rotation lengths in spring to avoid suppression by the more competitive ryegrass at this time of year.

Modern plantain cultivars establish well in newly sown pastures, but the success of establishment into existing pastures via under-sowing or broadcasting is mixed. Bryant et al. (2019a) showed greater plantain plant survival following an early grazing (21 days cf. 42 days after direct-drilling plantain into an existing ryegrass/clover pasture), when plantain seedlings were small enough to avoid defoliation and shading by existing pasture was reduced. Generally, plantain proportion increases over the first year after sowing into existing grass-dominant swards (Bryant et al., 2019a; Dodd et al., 2017; Hearn et al., 2024b) but rarely contributes more than 15% to herbage mass 2–3 years after sowing. To maintain a moderate proportion of plantain in existing swards that makes a meaningful contribution to seasonal herbage yield and animal intake (e.g. >20%), some New Zealand farmers add plantain seed to fertiliser applications on an annual or biennial basis. In this situation, coated seed reportedly spreads more evenly due to being closer in size and density to the fertiliser granules, i.e. there is less segregation of seed and fertiliser granules during transport or handling (resulting in smaller seed settling beneath larger fertiliser granules), and the seed is thrown as far as the fertiliser granules.

Poor persistence in grass-based swards was also reported for rotationally grazed sheep systems and rotationally grazed and cut beef systems in Ireland (Baker et al., 2023; Grace et al., 2018). Various reasons have been suggested, depending on environmental conditions, such as strong competition from grasses, insufficient replenishment of root reserves, weed ingress, pests and disease pressure, or plant damage from treading under wet conditions (Wilson et al., 2024).

The variable success of plantain establishment and relatively poor persistence in grass-based swards results in large differences in plantain content between paddocks within a farm, as illustrated in Fig. 3. Here, results are shown from five Plantain Potency and Practice Partner farms, where annual over-sowing of plantain across the whole farm is practised (Fransen et al., 2025). This variability makes it impractical to adjust management to plantain content per paddock, and farmers generally manage all paddocks the same.

Weed management when introducing multispecies pastures is challenging, given that most herbicides used in dairy systems target herbaceous dicotyledon weeds. For example, Gawn et al. (2012) showed that flumetsulam controlled broadleaved weeds in herb–clover pastures but severely supressed plantain initially. It is therefore important to control weeds prior to establishing plantain. Some farmers also control weeds on a 3-year cycle, followed by re-introducing plantain via over- or under-sowing. The plantain cultivar Agritonic was bred to better persist when herbicides are used (Box et al., 2023a) and this has been

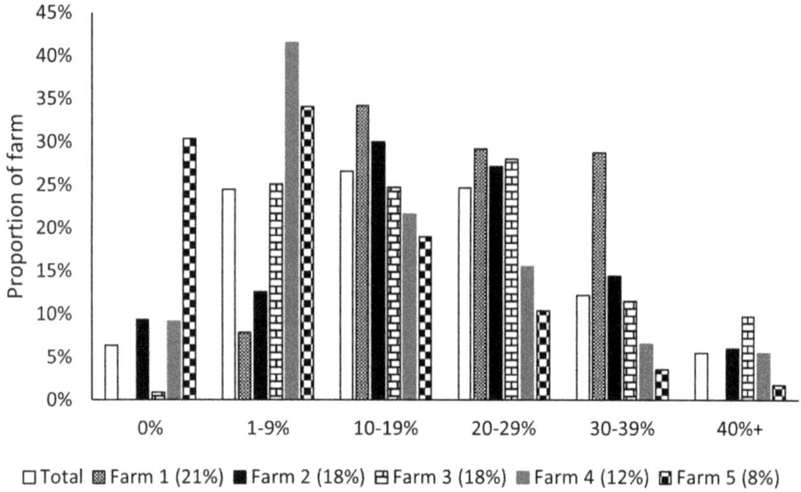

Figure 3 Distribution of Rotorua Lake Partner Farms' farm area across bins of visually assessed plantain content (% of total herbage DM), for autumn 2025. Average plantain content per farm is shown between brackets. Source: Fransen et al. (2025).

confirmed in dose–response experiments with phenoxy herbicides (Shrivastav et al., 2023).

Pests can also be a reason for declining proportion of plantain in pasture. Common pests in New Zealand are plantain moth (*Scopula rubraria* and *Epyaxa rosearia*), the native New Zealand grass grub (*Costelytra zealandica*), porina (*Wiseana* spp.) and slugs (Gerard et al., 2017). Cultivar selection procedures can include pest resistance (Gerard et al., 2018) and targeted insecticides are available (Dodd et al., 2025).

In terms of soil N fertility management, while the wild-type plantain prefers a lower soil N status (Berendse, 1982), modern cultivars are very responsive to added N fertiliser. Using the Ceres Tonic cultivar, Martin et al. (2017a) found plantain had a greater response to N fertiliser than cocksfoot, chicory, perennial ryegrass, white clover and red clover, measured over a 1–4-week regrowth interval of these monocultures during autumn in Canterbury, New Zealand. Measured over a full year at 0–450 kg N/ha, response of plantain was similar to perennial ryegrass (19.8 kg DM/kg vs. 21.1 kg DM/kg N; Martin et al., 2017b). Over three full years in an Irish experiment with ten different pasture mixtures (Hearn et al., 2024a), herbage DM yields were similar for perennial ryegrass/white clover and perennial ryegrass/white clover/plantain pastures. Increasing N fertiliser rates increased annual herbage DM production irrespective of sward mixture (e.g. 9.4 t DM/ha vs. 10.0 t DM/ha for perennial ryegrass/white clover and perennial ryegrass/white clover/plantain, respectively, at 0 kg N/ha/year, and 12.4 t DM/ha vs.

12.3 t DM/ha, respectively, at 200 kg N/ha). No significant N fertiliser rate x pasture mixture, nor N fertiliser rate x pasture mixture x year interactions were apparent for herbage DM yield. Additionally, no significant effects of N fertiliser rate nor second- or third-order interactions between N fertiliser rate, pasture mixture and year were apparent for the proportion of plantain (Hearn et al., 2024a).

4 Effect of plantain on nitrogen leaching

A large body of research on the effect of plantain on N losses from grazed pastures has emerged in the last decade, e.g. through major New Zealand research programmes and more recent research in Ireland. In New Zealand, the benefits for reductions in N leaching were first explored in the Forages for Reduced Nitrate Leaching programme (2013–2019) and the Greener Pastures programme (2014–2018) and are currently under investigation in the Plantain Potency and Practice programme (2021–2027).

Significant reductions in N leaching from simulated urine patches have been measured in several New Zealand lysimeter studies comparing perennial ryegrass/white clover swards and plantain monoculture or mixed-species swards, including plantain (summarised in Pinxterhuis et al., 2024). A recent Irish study added to these results using lysimeters with varying soils, from free- to poorly draining (Egan et al., 2025).

Lysimeter studies have inherent limitations associated with pasture and soil confinement. Leaching measurements from longer-term farmlet or paddock scale studies, where cows graze a specific pasture treatment, provide an environment where these limitations are obviated. A recent trial in New Zealand's summer-moist North Island (no irrigation) measured leaching from perennial ryegrass/white clover/plantain pastures compared with perennial ryegrass/white clover pastures in grazed, hydrologically isolated large plots (800 m^2). During a relatively low N drainage year, nitrate leaching was reduced by 60% in the first year. The reduction was 20–46% in the second year under higher drainage conditions (Navarrete et al., 2023). Plantain ranged from 30% to 50% of the sward DM offered and from 21% to 39% in the diet of the grazing dairy cows, lower than offered in pasture due to supplement feeding. In another New Zealand farmlet trial in a summer-dry environment with irrigation (Lincoln, Canterbury, South Island), total oxidisable N leaching (i.e. nitrate-N and nitrite-N) was reduced in the first 2 years of the trial by on average 18% in pastures with 20–30% plantain, compared with perennial ryegrass/white clover pasture (Fransen et al., 2024).

A further feature of these system trials is that they incorporate all of plantain's mechanisms so far purported to have a role in modifying the N cycle of grazed systems. These mechanisms include the following:

Effects on animal urinary N excretion:

1 Reduced urinary N concentration, associated with increased daily urine volume of the grazing animals and
2 Reduced total urinary N excretion, associated with a greater proportion of N intake partitioned to faeces and milk.

Effects on soil N processes:

1 Reduced rate of soil nitrification, associated with secondary compounds found in urine excreted by the grazing animals and from root exudates and herbage litter and
2 Reduced drainage, associated with greater water use by swards that include plantain.

These effects are briefly discussed in the following sections.

4.1 Animal urinary nitrogen excretion

To date, the most consistently demonstrated effect of including plantain in the dairy cow's diet is the effect on daily urination volume (associated with more urinations per day) and hence N concentration in urine. This was also observed when N intake was greater with increasing proportion of plantain in the grazed sward (Box et al., 2023a). This is thought to be mainly the result of plantain's relatively low dry matter percent (Minneé et al., 2019), i.e. animals ingesting more water with their feed, even if they compensate for this by drinking less from the water trough (Cheng et al., 2017; Minneé et al., 2020). A higher mineral concentration of plantain compared with grass (Minneé et al., 2019) may also contribute to greater overall water intake. Additionally, plantain in the diet can result in less N partitioned to urine and more to faeces and/or milk (Minneé et al., 2020).

Nguyen et al. (2022a) conducted a meta-analysis of animal experiments and reported the following relationships between plantain proportion in the animal diet and excreta characteristics:

1 Relative change of urinary N concentration (%) = −0.0645 × plantain proportion (g kg^{-1} DM) + 0.61 ($R^2 = 0.69$, $P < 0.001$).
2 Relative change of daily urine volume (%) = 0.0659 × plantain proportion (g kg^{-1} DM) − 12.73 ($R^2 = 0.64$, $P < 0.001$).
3 Relative change of total daily urinary N excretion (%) = −0.0297 × plantain proportion (g kg^{-1} DM) − 5.86 ($R^2 = 0.22$, $P = 0.036$).

Apart from less urinary N excretion per cow per day when plantain was fed to dairy cows, the analysis of Nguyen et al. (2022a) also indicated higher total milk and milk protein yield in late lactation (autumn), i.e. greater partitioning of N to milk. These authors did not report partitioning of N to faeces; however, some of the results used in this meta-analysis also showed that this was elevated when plantain was fed (Minneé et al., 2020; Marshall et al., 2021).

Plantain's effect on urinary N concentration and urine volume excreted per day have also been reported for red deer (Beck et al., 2020) and greater urine volumes have been reported for sheep (Judson et al., 2018b; O'Connell et al., 2016).

The main plant-based drivers of N partitioning in the animal are plant crude protein concentration (i.e. N concentration × 6.25, consisting of true protein and other nitrogenous compounds), protein degradability and water-soluble carbohydrate/crude protein ratio (Castillo et al., 2000; Bryant et al., 2019b; Minneé et al., 2020). Crude protein concentration of pasture in intensive forage-based livestock systems often exceeds animal requirements. For lactating dairy cows, requirements equate to approximately 3% N of dry matter (Pacheco and Waghorn, 2008) and lactating dairy cows utilise only 13–31% of N consumed (Castillo et al., 2000). The remaining N is excreted, mainly in urine (Kebreab et al., 2001).

Much of the protein in forages is degradable in the rumen (75–90%; Waghorn and Clark, 2004). Degradable protein consists of soluble protein which is rapidly degraded and insoluble degradable protein which is more slowly degraded in the rumen. The rumen degradable and soluble protein concentrations of forage influence the rate of ammonia accumulation in the rumen and in turn, urinary N excretion (depending on energy supply). Rumen undegradable protein is largely associated with plant cell walls and is difficult for rumen microbes to access. This protein fraction is either digested in the small intestine (digestible rumen undegradable protein) or is excreted in faeces. Minneé et al.'s (2019) meta-analysis showed that on average, plantain herbage contains less soluble protein than perennial ryegrass (12% vs. 38% of total N) and more rumen undegradable protein (64% vs. 31% of total N). In accordance, a greater plantain proportion in the diet of lactating dairy cows increased partitioning of N to faeces in the experiments of Marshall et al. (2021) and Minneé et al. (2020).

Soluble carbohydrates (i.e. sugars) are a key source of energy for rumen microbiota. With sufficient supply of soluble carbohydrates in the rumen, dietary N is synthesised in microbial protein, which is then available for use by the animal for production or growth. If energy is limiting, compared with available N, N accumulates in the rumen as ammonia, which is absorbed into the bloodstream and recycled or processed into urea in the liver and excreted in urine. Temperate pastures generally contain high concentrations of N and low concentrations of soluble carbohydrates. Edwards et al. (2007) showed that increasing soluble carbohydrate concentration in perennial ryegrass herbage reduced urinary N excretion. The same may occur when plantain is added to the diet, since plantain herbage contains greater soluble carbohydrate concentration than ryegrass (on average 299 g kg^{-1} vs. 211 g kg^{-1} DM) with similar total N concentration (Minneé et al., 2019).

Plant secondary metabolites in forages, such as terpenoids and tannins, can also impact rumen N processes (Bryant et al., 2019b), e.g. through their antimicrobial properties or functioning as an alternative energy source. Terpenoids known to be present in plantain are aucubin, acteoside (also known as verbascoside) and catalpol (Box et al., 2019). Navarrete et al. (2016) found ammonia production was reduced by 40% after 24 h of incubation when plantain was used as a substrate in an in-vitro study, compared with chicory as the substrate. In that case, the chicory forage did not have detectable levels of aucubin and acteoside, which were present in the plantain forage used. Addition of aucubin or acteoside to the chicory, at higher rates than were found in the plantain herbage used, reduced 24 h ammonia production by 5–20%. Minneé et al. (2017) measured lower rumen ammonia concentrations when late-lactation cows fed perennial ryegrass/white clover were supplemented with plantain or chicory, but no difference between the supplementation with chicory or plantain. In this study, secondary metabolites were not measured.

The concentrations of secondary compounds aucubin, acteoside and catalpol in plantain are highly variable both spatially and temporally and also vary between cultivars (e.g. Dietz et al., 2013). This presents a substantial challenge to identify potential effects of these compounds in feeding trials with plantain where the levels of secondary compounds are difficult to control.

4.2 Soil nitrogen processes

Plantain has been shown to impact several soil N processes, notably (1) nitrification rate, either directly via plant root and litter exudates or indirectly via compounds excreted in the urine of animals consuming herbage containing plantain, (2) leaching rate indirectly through influencing drainage. Inhibiting or reducing nitrification in the soil retains N in the form of ammonium for longer, which is less susceptible to leaching than nitrate. Reduced drainage will also retain mineral N in the rooted zone for longer, extending the time for plant N uptake.

Several studies have explored the impact of plantain on soil nitrification. Carlton et al. (2019) used micro-plots to assess changes in soil mineral N concentration following the application of standardised dairy cow urine. Nitrification appeared to be delayed under perennial ryegrass/white clover/ plantain swards (with 20–30% plantain) compared with perennial ryegrass/

white clover alone. This was shown by a lower soil ammonium-N concentration under perennial ryegrass/white clover up until day 30 after urine application. Also, soil nitrate-N concentration peaked earlier under perennial ryegrass/white clover (151 mg N/kg soil, day 14) compared with perennial ryegrass/white clover/plantain (140 mg N/kg soil, day 30). This was accompanied by higher growth rates of ammonia oxidising bacteria under perennial ryegrass/white clover.

Using Dutch peat soils and applying ammonium sulphate, rather than cow urine, Pijlman et al. (2020) found 39% lower potential nitrification in mesocosms with a plantain monoculture compared with a perennial ryegrass monoculture. They did not detect any significant differences between treatments in soil nitrate and ammonium concentration, but there were significant negative correlations between soil potential nitrification and soil nitrate concentration.

Peterson et al. (2023) collected root exudates from plantain grown in hydroponics (Fig. 4) and showed these exudates reduced rates of nitrification in a bioassay with a nitrifying bacterium. They also showed differences

Figure 4 Peterson et al. (2023) collected root exudates from plantain grown in hydroponics; shown are plants from the AgriTonic cultivar. These exudates reduced rates of nitrification in a bioassay with a nitrifying bacterium. However, results were variable between exudates from different plantain cultivars and within cultivars.

between exudates from different plantain cultivars and within cultivars and correlations with root exudate chemistry. Dietz et al. (2013) also showed a delay in nitrification rate in soil incubation studies, with no nitrate detected for the first 28 days, when soil samples were incubated with plantain leaf material, aqueous extract of plantain leaves or aucubin. However, they also showed varying levels of aucubin, catalpol and verbascoside in plantain leaf material over time, differences between two plantain cultivars, and variation between leaves from plants in vegetative or generative state. This presents a challenge in quantitatively linking sward plantain proportions with secondary metabolite abundance and their effects on soil N processes in the field.

Other work has shown indirect biological nitrification inhibition effects of plantain, i.e. a nitrification reduction effect of the urine of animals consuming plantain compared with perennial ryegrass/white clover (Judson et al., 2018a; Judson et al., 2018b; Judson et al., 2019; Peterson et al., 2022, 2023). This lower rate of nitrification was attributed to the organic fraction of urine and breakdown molecules of plantain's secondary compounds. The extent of the "urine effect" depended on the length of time the animals were grazing swards with plantain, the plantain cultivar and the amount of plantain in the diet. Additionally, seasonal differences may be expected due to temporal variability in secondary compounds in plantain (Box and Judson, 2018) and differences between soil types may be expected too (Clark et al., 2021). As yet, it is not clear which compounds are responsible and how they affect nitrification rate. Peterson et al. (2023) suggested an inhibiting effect on the enzymes that convert ammonium to nitrite.

The magnitude of the "plant–soil effect" of plantain on nitrate leaching was shown by Carlton et al. (2019), Welten et al. (2019), Talbot et al. (2021) and Egan et al. (2025). Carlton et al. (2019) applied standardised fresh dairy cow urine at 700 kg N/ha to lysimeters and measured on average 82% less nitrate leaching from a perennial ryegrass/white clover/plantain sward than from a perennial ryegrass/white clover sward when urine was applied in December (early summer) and 74% less when urine was applied in February (late summer). Drainage in these mesocosms was reduced by on average 21% and 28% for the December and February treatments, respectively, by inclusion of plantain. These results occurred despite the total herbage dry matter yield and N uptake being lower for perennial ryegrass/white clover/plantain than for perennial ryegrass/white clover in this study (12.2–14.2 t DM/ha vs. 13.3–14.9 t DM/ha and 355–441 kg N/ha vs. 408–466 kg N/ha, respectively).

Welten et al. (2019) applied urine collected from dairy cows grazing a perennial ryegrass/white clover pasture to lysimeters with either ryegrass or plantain monocultures. They observed 15% less N leaching from plantain swards than from ryegrass swards when urine was applied in summer and 50% less N leached when the urine was deposited in winter.

Talbot et al. (2021) also applied urine collected from dairy cows grazing a perennial ryegrass/white clover pasture to lysimeters with swards containing none or 25–35% plantain and found reductions in N leached of 14–24% with plantain, irrespective of the seasonal timing of urine application. These authors suggested the smaller reductions they found could have been due to the shallow stony soils in their study, i.e. more freely draining than the deeper loam soils used by Carlton et al. (2019) and Welten et al. (2019).

Egan et al. (2025) applied a mixture of urine, collected from cows grazing either perennial ryegrass only, perennial ryegrass/white clover or a multispecies sward including plantain to lysimeters, over two years and five soil types. Hence also here, the "animal effects" of plantain in the cows' diet were excluded as a factor. In their study, reductions in N leaching from perennial ryegrass/white clover/plantain mixtures compared with perennial ryegrass/white clover were much greater in the second drainage season after establishment than in the first drainage season (reductions of on average 97% in year two, compared with 46% in year one). In this study, there was limited additional benefit from the treatment aiming at 50% plantain in the sward DM, compared with the 30% plantain sward.

In a context where soil type and rainfall resulted in high drainage following urine application, no reductions in N leaching associated with plantain were observed (Graham et al., 2024; Healy et al., 2024). The variable results across studies imply that soil type and climate may be critical determinants of the magnitude of plantain impact on nitrification rates and leaching losses. This is an important consideration to understand the efficacy of plantain as a tool to reduce N losses to sensitive water bodies under a wide range of soil and climate combinations. Current research in the New Zealand "Plantain Potency and Practice" programme attempts to unravel these sources of variation by examining N dynamics in four different regions with different soils and climate.

5 Case study: use of plantain to reduce nitrogen leaching at catchment scale

In some New Zealand regions, N discharge into waterbodies is regulated. This case study illustrates the environmental and economic impact of implementing plantain in pasture to achieve the N reduction targets in the Rotorua Lake catchment, Bay of Plenty. Details of this case study are provided in Farrell et al. (2025) and Fransen et al. (2025).

This case study utilised two DairyNZ models: the Dairy Sector Pathways (DSP) model and the Catchment Accounting Framework (CAF). The Dairy Sector Pathways model simulates the production and economic changes in a population of dairy farms over time as they adopt farm system and management changes to achieve N leaching reductions. The CAF combines land use and geographical data with farm system data to account for N losses within a catchment for baseline scenarios and when farm systems or management changes are implemented.

The Plantain Potency and Practice Programme partnered with five dairy farmers in the Rotorua Lake catchment to investigate the implementation of plantain in pasture. Annual monitoring of plantain content (% of herbage DM) indicated that plantain could feasibly amount to 20% of DM in pasture (see Fig. 3), so this level was assumed in the modelling.

OverseerFM (https://overseer.org.nz/overseerfm/) and Farmax (https://www.farmax.co.nz/) models were used to estimate the impact of plantain and other N leaching mitigative actions on farm-level N leaching and operating profit for three of the partner farms. Other mitigative actions were reductions in imported supplementary feeds and/or N fertiliser, removal of cropping, utilising supplements with reduced N content, stand-off facilities to remove cows from pasture at critical times of the year and retiring land to unharvested pine forestry. Results of this modelling were used to validate and inform the DSP and CAF models. Of note is that OverseerFM currently only reflects the effects of plantain on urinary N excretion and dilution, and not the potential soil effects. Therefore, it is possible that N leaching reductions are underestimated.

Based on dairy statistics, DSP generated a population of 26 dairy farms in the Rotorua Lake catchment. On average, these farms were required to reduce N leaching by 31% relative to their baseline of 2017/18 to achieve their 2032/33 targets, as set by the Bay of Plenty Regional Council. For all farms, scenarios with or without plantain were modelled to assess the future value of the implementation of plantain at catchment scale. For individual farms, N leaching reduction targets ranged from 7% to 47%, hence the combination of generated mitigative actions varied widely between farms. In general, reductions in supplementary feed imports and N fertiliser use were necessary, which resulted in fewer cows and reduced milk production and operating profit, relative to the baseline.

Table 2 shows the impact at farm and catchment scale of the required mitigations with or without plantain. At catchment scale, implementing plantain meant that 151 ha of dairy land and 429 more dairy cows could be retained, producing 202 tonnes more milk solids per year and returning $0.6 million more operating profit per year. The total reduction in N leaching achieved from the dairy farms amounted to 35 t N/year from 20% plantain alone, which was 36% of the total required reduction.

The results of this catchment scale scenario modelling indicate a significant value of plantain as a mitigation option. Clearly, results will differ depending on variances in, e.g., milk price, costs of inputs, plantain % of DM achieved,

Table 2 Predicted average performance for 26 dairy farms in the Rotorua Lake catchment for the 2017/18 season (baseline) and two scenarios for the 2032/33 season when 31% nitrogen leaching reduction, compared with the baseline, needs to be met

	Baseline	P0	P20
Farm-scale parameters*	2017/18	2032/33	2032/33
Milksolids production (kg/ha)	1115	897	943
Gross farm revenue ($/ha)	7927	6466	6778
Operating expenses ($/ha)	5380	4783	4920
Operating profit ($/ha)	2547	1683	1858
Carbon sequestration revenue ($/ha)	0	38	6
Catchment-scale parameters			
Total dairy farm area (ha)	4498	4229	4380
Cow numbers (head)	12 327	10 782	11 211
Milksolids produced (tonnes)	4759	3752	3954
Operating profit ($ millions)	11.3	7.4	8.0

*Per hectare of land effectively used for milk production.
P0 is a scenario without plantain; P20 is a scenario with on average 20% plantain in pasture dry matter across the farm.

N fertiliser response rates, and effect of plantain on N leaching and herbage production. Some examples of the impact of varying assumptions are provided by Chikazhe et al. (2024). Their (limited) variations resulted in abatement costs varying from $11/kg N to $45/kg N leaching reduction, illustrative of the large uncertainty farmers face when deciding on steps to achieve future N leaching targets.

6 Effect of plantain on nitrous oxide emissions

To date, studies looking at both the "animal-urine effect" and the "plant-soil effect" have shown the inclusion of plantain can also reduce nitrous oxide emissions.

To investigate "animal-urine effects", Gardiner et al. (2017) added either plantain leaf extract or aucubin to the urine of dairy cows grazing perennial ryegrass/white clover swards and applied this urine to a perennial ryegrass/white clover sward. This resulted in 50% and 70% lower urinary nitrous oxide emissions, respectively, than from the same urine without additives.

Rodriguez et al. (2023) also investigated the effects on nitrous oxide emission and nitrate leaching of plantain, aucubin and urine derived from cows fed plantain monocultures or perennial ryegrass/white clover swards, with the latter diluted to the same N concentration as the plantain urine. These authors did not find an effect of plantain urine on nitrous oxide emission or nitrate leaching. Instead, nitrous oxide emission was reduced from perennial ryegrass/

white clover swards when aucubin was applied and from plantain swards compared with perennial ryegrass/white clover swards. However, in the latter case, they did not find aucubin in the plantain root exudates; rather, catalpol was the main compound detected.

In an investigation of "plant-soil effects", Luo et al. (2018) collected urine from dairy cows grazing perennial ryegrass/white clover sward and adjusted the N concentration to about 6.22 g N/L by adding urea or diluting the urine, before applying this urine to lysimeters. They showed a 39% and 35% reduction in cumulative nitrous oxide emissions (total over twelve weeks) from plantain monoculture lysimeters compared with ryegrass monoculture, when applying urine in March or May (early and late autumn), respectively. When urine was applied in February (late summer), emission from plantain was not significantly different from ryegrass. Luo et al. (2018) also measured nitrous oxide emission from field plots growing monocultures and reported a 22% lower emission from plantain than perennial ryegrass without urine applied and 74% lower emission with standardised urine applied in June (early winter).

Chibuike et al. (2024) also reports significant nitrous oxide emission reductions of on average 39% from soil columns with plantain, compared with perennial ryegrass, when synthetic urine was applied. In the field study of Pijlman et al. (2020), the presence of plantain resulted in 39% lower cumulative nitrous oxide emission compared with the monoculture of perennial ryegrass.

Simon et al. (2019) showed cumulative nitrous oxide emissions over a 15-week period decreased linearly with increasing proportions of plantain in both the diet and sward simultaneously. They also showed a linear decrease in nitrous oxide emissions with increasing plantain proportion in the sward when urine from dairy cows grazing perennial ryegrass/white clover swards was applied or when no urine was applied at all. These results imply an impact of plantain on the emission factor (ratio of N emitted to N applied), an important metric for assessing the efficacy of greenhouse gas mitigations. However, only with the standard urine treatment was the decline in the emission factor with increasing plantain proportion in the sward significant. Simon et al. (2019) also showed lower total nitrous oxide emissions when urine from cows grazing 45% plantain swards was applied to perennial ryegrass/white clover swards, compared with urine from cows grazing ryegrass/clover swards. There were no significant differences in emission factors between these treatments, nor was there a difference between the "45% plantain urine" treatment and a treatment with urine from cows grazing perennial ryegrass/white clover swards that was diluted to the same N concentration as the plantain urine. This implied that there were no measurable effects of plant secondary compounds via urine, rather only an effect of the lower N concentration in the urine.

Published by Burleigh Dodds Science Publishing Limited, 2025.

7 Future research trends

Despite a burgeoning body of research on the use of plantain in intensive pastoral systems, many questions remain. These span microbial to catchment scales and cover the domains of biochemistry, biology, soil chemistry, agronomy, hydrology, animal physiology, animal production and product quality. Because of the diffuse nature of the environmental impact of N within agricultural systems, future research will need to address some pertinent questions farmers and other stakeholders have regarding implementing plantain at scale and putting its use into the context of other novel mitigation options and systems configurations. For example, a modelling exercise indicated that the effectiveness of reduced urinary N concentration on N leaching and nitrous oxide emission can be less with higher stocking rates (Snow et al., 2025).

Specific areas for future research pursuit are indicated below.

7.1 Farm-scale impact of plantain on production and profit

To date, limited information is available on the productivity and profitability of farm systems with plantain in mixed-species pastures, relative to farm systems based on perennial ryegrass or perennial ryegrass/white clover pastures. At the paddock scale, various studies have shown either no significant or a beneficial impact of adding plantain in perennial ryegrass/white clover pasture on herbage yield (Dodd et al., 2023; Hearn et al., 2024a; Herath et al., 2023; Minneé et al., 2013; Monigue et al., 2025; Nguyen et al., 2022b). However, it is important to understand the implications for farm-scale milk production and profitability, if plantain is to be recommended to dairy farmers as a cost-effective means of reducing N loss to the environment.

The meta-analysis of Nguyen et al. (2022a) showed pastures with plantain on average increased milk yield by 0.07 kg MS/cow/day across a season, and 1.4 kg MS/cow/day in late lactation, i.e. autumn, in New Zealand conditions. However, a higher milk yield per cow does not necessarily mean higher milk production per hectare. A one-lactation Irish study showed similar annual milk yield per cow for perennial ryegrass/white clover pastures with or without plantain under the same stocking rate (Minogue et al., 2025). The key difference was 6% lower daily milksolids yield in early lactation and 9% higher milksolids yield in late lactation for pastures with plantain. In a replicated farm system study, comparing farmlets with perennial ryegrass/white clover swards and farmlets with perennial ryegrass/white clover/plantain swards, Herath et al. (2023) showed similar pasture production (13.0 t DM/ha), milksolids production (1356 kg MS/ha) and profitability (NZ$4347/ha) for the first year of the study. These limited results suggest that the addition of plantain to pasture mixtures would be a low-cost and revenue-neutral means of reducing N loss from grazed pasture on dairy farms, compared with other mitigations involving reducing stocking rate, reducing N inputs and implementing off-paddock infrastructure.

7.2 Plantain persistence in mixed-species pasture

Challenges persist in establishing and maintaining relevant proportions of plantain in New Zealand grass-based mixed-species pastures (Dodd et al., 2025; Eme and Roche, 2025), especially where conditions are conducive to high growth rates for perennial ryegrass. By stark contrast, in Ireland, Hearn et al. (2024a) observed the highest plantain proportion of on average 37% plantain in the third year after sowing across a range of species mixtures and fertiliser rates (0–200 kg N/ha/year). None of the possible reasons for poor persistence in New Zealand (suggestions include grazing management, pests, treading damage etc.) have yet been conclusively identified. To date, the only methods shown to improve the proportion of plantain in existing swards, without spraying and tillage, have been over-sowing or direct drilling, with best results if competition from the existing sward is managed during establishment, e.g. by hard grazing before over-sowing (Hearn et al., 2024b) and/or a light grazing after over-sowing (Bryant et al., 2019a). For successful application on commercial farms, it will be crucial to understand how the proportion of plantain in pasture can be maintained or improved after establishment, since frequent pasture renewal is undesirable from cost and environmental perspectives. An example of the detrimental environmental impact of pasture renewal was shown by Graham et al. (2024), with significantly higher N leaching after pasture renewal.

Further research could investigate the potential effects on plant longevity and/or recruitment of new plants of differing grazing rotation lengths, e.g. shorter intervals to avoid flowering and potentially extend plant longevity, or longer intervals during the flowering stage to support natural reseeding. There is also a need to explore the use of highly diverse species mixtures or grass species less competitive than perennial ryegrass. Silvia Cid et al. (2011) reported greater contribution of plantain to annual yield when sown with tall fescue and cocksfoot (compared with annual ryegrass), in a seasonally dry climate in Argentina. However, Dodd et al. (2017) observed no significant difference in the proportion of plantain in perennial ryegrass or tall fescue-based pastures in summer-dry Waikato in New Zealand.

7.3 Weed and pest management

Another challenging aspect of maintaining sward productivity and quality that farmers consistently identify is the management of weeds and pests. In pastures where white clover and plantain are desirable species, the ability to chemically control weeds is limited, and other non-chemical preventative management practices warrant further research. Some pasture species control weeds better than others (Connolly et al., 2018). Biocontrol of some weeds has been shown to be effective in New Zealand (Hayes et al., 2013), presenting an unexplored approach to minimising the detrimental effects of broad-action herbicides on plantain.

For most relevant invertebrate pests, pesticides are available and some opportunities for non-chemical management have been identified, such as the potential biopesticide bacterium *Yersinia entomophaga* (Jones et al., 2015) and potential for genetic selection within plantain to improve resistance to pests (Gerard et al., 2018).

The weed and pest challenges for plantain represent an example of a core challenge for research: they are a barrier to widespread uptake of a novel forage species, which in turn hinders the justification for research investment in relevant solutions.

7.4 Cow health and milk composition

To date, there has been limited research on minor milk components and fatty acid composition (Pinxterhuis et al., 2024). The inclusion of plantain in cow diets led to small reductions in milk fat concentrations (Nguyen et al., 2022a), greater concentration of poly-unsaturated fatty acids (Mangwe et al., 2018; Marshall et al., 2022) and differing concentrations of phytochemicals (Marshall et al., 2022). These results indicate that the potential human nutritional benefits of including plantain in the cow's diet warrant further research. To date, no risks to human health nor processability of milk has been found from milk originating from cows grazing pastures with plantain (unpublished results from the Plantain Potency and Practice programme; https://www.mpi.govt.nz/dmsdocument/70348-March-to-May-2025-Plantain-Potency-and-Practice-progress-report/).

A potential risk to animal and human health is the higher uptake of cadmium from the soil by plantain, compared with perennial ryegrass and white clover (Stafford et al., 2016). Where soil cadmium levels are high, e.g. through the repeated use of cadmium-containing phosphorus fertiliser, cadmium can accumulate in plantain leaves and via feed intake can lead to elevated concentrations in the animal's liver and kidney.

No immediate animal health issues have been reported to date, when plantain was included in the pasture or fed as a monoculture crop. However, some potential issues may occur as results to date have been derived from short-term research results and there may be merit in further investigation in longer-term experiments or on commercial farms where plantain is a structural component of the animals' diet. The short-term studies showed reduced water intake from the trough with increasing proportions of plantain in the dairy cows' diet (Minneé et al., 2020), which has implications for dosing minerals or animal health products via the water trough (e.g. bloat-prevention products).

The generally higher dietary sodium and calcium concentration and lower potassium concentration associated with plantain (Minneé et al., 2020) may impact dietary cation anion difference (DCAD) status and calcium absorption, potentially forming a risk or a benefit for the health of lactating dairy cows.

7.5 Methane and ammonium emissions

Enteric methane emission from grazing livestock systems is an important contributor to global greenhouse gas emissions (Hristov et al., 2013). While ammonia is not considered an environmental issue in New Zealand, it is regulated in Europe. *In vitro* studies indicate the potential for plantain to reduce both methane and ammonia emissions relative to other herb and brassica species (Durmic et al., 2016) and to perennial ryegrass/white clover (Sivanandarajah et al., 2025a). Further *in vitro* compound-specific work showed less methane and net ammonia production from plantain substrates with varying content of aucubin and acteoside, compared with perennial ryegrass, which does not contain these secondary compounds (Sivanandarajah et al., 2025b). Incubations of perennial ryegrass enriched with commercially available aucubin, acteoside and catalpol showed no reduction in methane production rates but a decrease in ammonia production rates. Li et al. (2025) investigated a range of binary pasture mixtures and showed a negative correlation between *in vitro* methane yields and saponins and condensed tannins; plantain had elevated concentrations of these compounds in their study.

Very few publications to date investigate the effect of plantain on methane emission from dairy cattle, and what is published has shown variable effects of multi-species pastures on methane emission from dairy cattle compared with perennial ryegrass/white clover pastures. Della Rosa et al. (2022) measured 15% and 28% less methane emission/kg DM intake from non-lactating cows when fed plantain in vegetative and reproductive growth stage, respectively, compared with perennial ryegrass, and attributed this to lower herbage digestibility. Koning et al. (2024) measured higher methane production (g/cow/day) and yield (g/kg DM intake) from lactating dairy cows grazing perennial ryegrass/plantain mixtures (maximum of 25% plantain in DM) compared with perennial ryegrass monocultures. They also suggest an effect of a lower herbage digestibility of the swards with plantain, due to higher acid detergent fibre (ADF) and acid detergent lignin (ADL) content than perennial ryegrass swards in their study.

Published by Burleigh Dodds Science Publishing Limited, 2025.

Clearly, the mechanisms of the effects of herbage composition, and specifically plantain, on methane emission requires further research, and the overall effect of incorporating forage plantain on the various gaseous emissions from livestock systems has yet to be clearly quantified.

7.6 Effects of secondary compounds on rumen and soil microbiome

The mechanisms of the observed impacts of plantain and its secondary compounds or their derivates on microbial processes in the rumen and in soil need to be better understood, to ensure optimal use of plantain to reduce environmental impact of grazed, productive dairy pastures and to guide future plant breeding efforts and research in other potentially successful forage options. Effects of secondary compounds and their derivates, either directly from the plantain plant or via excreted urine from animals consuming plantain (Peterson et al., 2023), are apparent but variable. It is not clear which compounds matter most for which processes, and under which circumstances (e.g. potential interactions with animal breed, overall feed composition, soil type, climate). It is also not clear what their impact is relative to other traits of plantain that reduce N losses, and if there are any unintended consequences of elevating secondary compounds and associated changes in soil microbial communities. No risks to human health and processability of milk derived from swards with plantain have been identified to date.

Additionally, concentration of some secondary compounds has been shown to vary over time in plantain (e.g. Box and Judson, 2018), so further understanding of these variations and whether the concentration can be managed or bred for will be needed to support consistent positive environmental effects over time.

8 Conclusion

Including plantain in perennial ryegrass/white clover pasture appears to be a low-cost option to reduce N leaching and nitrous oxide emission, without any substantive negative effects on pasture production, animal production, milk quality and animal health. On the contrary, under certain situations various positive effects of plantain on all these factors have been observed.

Some gaps in current knowledge have been identified that require further investigation to optimise the implementation of plantain on farm and to achieve maximum environmental benefits. Notable areas for deeper study include management of mixed-species swards to maintain the proportion of plantain in the sward and understanding the mechanisms in which plant secondary compounds and their derivatives impact rumen and soil processes and their relative importance for reducing N loss to the environment.

9 Acknowledgements

Time for writing this chapter was supported by the Plantain Potency and Practice programme, a 7-year Aotearoa New Zealand-wide collaborative research and development initiative. The funding partners are the Ministry for Primary Industries through Sustainable Food and Fibre Futures, New Zealand dairy farmers through DairyNZ, PGG Wrightson Seeds Ltd. and Fonterra. Delivery partners include DairyNZ, Fonterra, Lincoln University, Massey University, Lincoln Agritech, Agricom, Manaaki-Whenua Landcare Research, AgResearch and The New Zealand Institute for Plant and Food Research Limited.

10 Where to look for further information

Plantain webpage DairyNZ is regularly updated with new research results from the Plantain Potency and Practice programme. It includes links to case studies on commercial farms and management guidelines. https://www.dairynz.co.nz/feed/crops/plantain-overview/.

The Forages for Reduced Nitrate Leaching research programme (2013–2019); https://www.dairynz.co.nz/frnl. includes a link to the FRNL New Knowledge Register, which summarises research results and provides links to published scientific manuscripts.

Reviews:

- Chesney, L.E., et al. (2025), Plantain (*Plantago lanceolata* L.) as an alternative forage to build resilience and reduce the environmental footprint of grazing dairy systems in temperate Northern climates: a review, *Sustainability*, 17, 3131. https://doi.org/10.3390/su17073131.
- Dodd. M.B., et al. (2025), Narrow-leaved plantain (*Plantago lanceolata* L.): a review of research on forage management within temperate grazing systems, *Grassland Research*, 2025, 1–16. https://doi.org/10.1002/glr2.12107.
- Eme, P.E. and Roche, J.R. (2025), Invited review: Cows grazing pastures containing narrow-leaved plantain have lower urine N concentrations: implications for nitrate leaching, *Journal of Dairy Science*, 108, 7876–7895. https://doi.org/10.3168/jds.2024-26169.
- Pinxterhuis, J.B., et al. (2024), Implementing plantain (*Plantago lanceolata*) to mitigate the impact of grazing ruminants on nitrogen losses to the environment: a review, *Grass and Forage Science*, 79, 144–157. https://doi.org/10.1111/gfs.12649.

11 References

Ammann, S., et al. (2021), *Forage herbs (Chicory and plantain) cultivar evaluation seasonal yield results for October 2018 to February 2021*. Information Day 2001 milk

Production from planted pastures, Plant and Animal Sciences Directorates, Western Cape Department of Agriculture, South Africa. https://www.elsenburg.com/wp-content/uploads/2022/08/Information-Day-2021-Booklet.pdf#page=70

Baker, S., et al. (2023), Multispecies swards outperform perennial ryegrass under intensive beef grazing, *Agriculture, Ecosystems and Environment*, 345, 108355. https://doi.org/10.1016/j.agee.2022.108335

Ball, P.R. and Field, T.R.O. (1985), Productivity and economics of legume-based pastures and grass swards receiving fertilizer nitrogen in New Zealand. In Barnes, R.F., et al. (Eds), *Forage legumes for energy-efficient animal production*, NTIS, Springfield, USA, pp. 47–55.

Ball, P.R., et al. (1979), Nitrogen balance in urine-affected areas of a New Zealand pasture, *Agronomy Journal*, 71, 309–314. https://doi.org/10.2134/agronj1979.00021962007100020022x

Beck, M.R., et al. (2020), Plantain (*Plantago lanceolata*) reduces the environmental impact of farmed red deer (*Cervus elaphus*), *Translational Animal Science*, 4(3), txaa160. https://doi.org/10.1093/tas/txaa160

Berendse, F. (1982), Competition between plant populations with different rooting depths. III. Field experiments, *Oecologia*, 53, 50–55. https://doi.org/10.1007/BF00377135

Beukes, P.C., et al. (2014), The potential of diverse pastures to reduce nitrogen leaching on New Zealand dairy farms, *Animal Production Science*, 54, 1971–1979. https://doi.org/10.1071/AN14563

Beukes, P.C., et al. (2013), Evaluating the benefits of standing cows off pasture to avoid soil pugging damage in two dairy farming regions of New Zealand, *New Zealand Journal of Agricultural Research*, 56(3), 224–238. https://doi.org/10.1080/00288233.2013.822002

Beukes, P.C., et al. (2017), The performance of an efficient dairy system using a combination of nitrogen leaching mitigation strategies in a variable climate, *Science of The Total Environment*, 599–600, 1791–1801. https://doi.org/10.1016/j.scitotenv.2017.05.10

Beukes, P.C., et al. (2024), Stacking nitrogen mitigation strategies for future pasture-based dairy farms: impacts on leaching and profit, *Frontiers in Animal Science*, 5, 1277131. https://doi.org/10.3389/fanim.2024.1277131

Box, L.A. and Judson, H.G. (2018), The concentration of bioactive compounds in *Plantago lanceolata* is genotype specific, *Journal of New Zealand Grasslands*, 80, 113–118. https://doi.org/10.33584/jnzg.2018.80.334

Box, L.A., et al. (2017), Milk production and urinary nitrogen excretion of dairy cows grazing plantain in early and late lactation, *New Zealand Journal of Agricultural Research*, 60(4), 470–482. https://doi.org/10.1080/00288233.2017.1366924.

Box, L.A., et al. (2019), Seasonal and diurnal changes in aucubin, catalpol and acteoside concentration of plantain herbage grown at high and low N fertiliser inputs, *New Zealand Journal of Agricultural Research*, 62(3), 343–353. https://doi.org/10.1080/00288233.2018.1505641

Box, G.M., et al. (2023a), The development of environmentally functional plantains, *Proceedings of the 25th International Grassland Congress*, 1834–1837. https://doi.org/10.52202/071171-0450

Box, L.A., et al. (2023b), Varying plantain content in temperate ryegrass-white clover pastures affects urinary-nitrogen excretion of non-lactating dairy cows, *Science of the Total Environment*, 862, 160847. http://dx.doi.org/10.1016/j.scitotenv.2022.160847

Bryant, R.H., et al. (2019a), Effectiveness of strategies used to establish plantain into existing pastures, *Journal of New Zealand Grasslands*, 81, 131–138. https://doi.org/10.33584/jnzg.2019.81.406

Bryant, R., et al. (2019b), Can alternative forages substantially reduce N leaching? findings from a review and associated modelling, *New Zealand Journal of Agricultural Research*, 63, 3–28. https://doi.org/10.1080/00288233.2019.1680395

Burke, G. (2008), *Research review: the role and management of herbal pastures for animal health, productivity and product quality*, Institute of Organic Training and Advice, United Kingdom.

Carlton, A.J., et al. (2019), Nitrate leaching losses are lower from ryegrass/white clover forages containing plantain than from ryegrass/white clover forages under different irrigation, *New Zealand Journal of Agricultural Research*, 62, 150–172. https://doi.org/10.1080/00288233.2018.1461659

Castillo, A.R., et al. (2000), A review of efficiency of nitrogen utilisation in lactating dairy cows and its relationship with environmental pollution, *Journal of Animal and Feed Sciences*, 9, 1–32. https://doi.org/10.22358/jafs/68025/2000

Chapman, D. and Parsons T. (2017), Doing more for the environment with lower inputs in pasture-based livestock systems: does this always lead to lower outputs? *Proceedings of the 18th Australian Society of Agronomy Conference*, 24–28 September, Ballarat, Australia. http://www.agronomyaustraliaproceedings.org/

Chapman, D.F., et al. (2014), Plants for dairy grazing systems operating under nitrate leaching limits, *Proceedings of the New Zealand Society of Animal Production*, 74, 102–107.

Cheng, L., et al. (2017), Live weight gain, urinary nitrogen excretion and urination behaviour of dairy heifers grazing pasture, chicory and plantain, *Journal of Agricultural Science*, 155, 669–678. https://doi.org/10.1017/S0021859616001076

Chesney, L.E., et al. (2025), Plantain (*Plantago lanceolata* L.) as an alternative forage to build resilience and reduce the environmental footprint of grazing dairy systems in temperate Northern climates: a review, *Sustainability*, 17, 3131. https://doi.org/10.3390/su17073131

Chibuike, G., et al. (2024), Does plantain reduce N_2O emissions from a pasture soil previously exposed to elevated atmospheric CO2? In Christensen, C.L., et al. (Eds), *Opportunities for improved farm and catchment outcomes*, Occasional Report No. 36, Farmed Landscapes Research Centre, Massey University, Palmerston North, New Zealand.

Chikazhe, T., et al. (2024), Exploring the robustness of nitrogen leaching estimates for stacked mitigations, through targeted sensitivity and uncertainty analyses. *Proceedings of the Australasian Dairy Science Symposium*, 178–180.

Clark, I.M., et al. (2021), Metagenomic approaches reveal differences in genetic diversity and relative abundance of nitrifying bacteria and archaea in contrasting soils, *Scientific Reports*, 11(1), 15905. https://doi.org/10.1038/s41598-021-95100-9

Collins, R.P., et al. (2014), Biomass production in multispecies and grass monoculture swards under cutting and rotational grazing, *Grassland Science in Europe*, 19, 719–721.

Connolly, J., et al. (2018), Weed suppression greatly increased by plant diversity in intensively managed grasslands: a continental-scale experiment, *Journal of Applied Ecology* 55: 852–862. https://doi.org/10.1111/1365-2664.12991

Corré, W.J., et al. (2014), Nitrate leaching and apparent recovery of urine-N in grassland on sandy soils in the Netherlands, *NJAS - Wageningen Journal of Life Sciences*, 70/71, 25–32.

Cranston, L.M., et al. (2015a), A review of the use of chicory, plantain, red clover and white clover in a sward mix for increased sheep and beef production, *Journal of New Zealand Grasslands*, 77, 89–94. https://doi.org/10.33584/jnzg.2015.77.475

Cranston, L.M., et al. (2015b), Effect of post-grazing height on the productivity, population and morphology of a herb and legume mix, *New Zealand Journal of Agricultural Research*, 58(4), 397–411. https://doi.org/10.1080/00288233.2015.1044014

Cranston, L.M., et al. (2021), The effect of defoliation frequency and height on plantain and chicory-based swards, *Agronomy New Zealand*, 51, 1–12. https://www.agronomysociety.org.nz/files/ASNZ_2021_01._Defoliation_effect_plantain_chicory.pdf

Daly, M.J., et al. (1996), A comparison of multi-species pasture with ryegrass-white clover pasture under dryland conditions, *Proceedings of the New Zealand Grassland Association*, 58, 53–58.

Della Rosa, M., et al. (2022), Effect of feeding fresh forage plantain (*Plantago lanceolata*) or ryegrass-based pasture on methane emissions, total-tract digestibility, and rumen fermentation of nonlactating dairy cows, *Journal of Dairy Science*, 105, 6628–6638. https://doi.org/10.3168/jds.2021-21757

Di, H.J. and Cameron, K.C. (2002), Nitrate leaching in temperate agroecosystems: sources, factors and mitigating strategies, *Nutrient Cycling in Agroecosystems*, 64, 237–256. https://doi.org/10.1023/A:1021471531188

Dietz, M., et al. (2013), Inhibitory effects of *Plantago lanceolata* L. on soil N mineralization, *Plant and Soil*, 368, 445–458. https://doi.org/10.1007/s11104-012-1524-9

Dodd, M.B., et al. (2017), Establishment year productivity, botanical composition and nutritive value of grass/lucerne/plantain dairy pasture mixtures, *Journal of New Zealand Grasslands*, 79, 229–235. https://doi.org/10.33584/jnzg.2017.79.534

Dodd, M.B., et al. (2019), A paddock survey of on-farm plantain use, *Journal of New Zealand Grasslands*, 81, 125–130. https://doi.org/10.33584/jnzg.2019.81.408

Dodd, M.B., et al. (2022), Performance of plantain pastures on Tararua dairy farms, *Journal of New Zealand Grasslands*, 84, 67–72. https://doi.org/10.33584/jnzg.2022.84.3573

Dodd, M.B., et al. (2023), Adoption of plantain within New Zealand farming systems, *Proceedings of the 25th International Grassland Congress*, 1111–1114. https://doi.org/10.52202/071171-0269

Dodd, M.B., et al. (2025), Narrow-leaved plantain (*Plantago lanceolata* L.): a review of research on forage management within temperate grazing systems, *Grassland Research*, 2025, 1–16. https://doi.org/10.1002/glr2.12107

Durmic, Z., et al. (2016), In vitro fermentability and methane production of some alternative forages in Australia, *Animal Production Science*, 56, 641–645. https://doi.org/10.1071/ AN15486

Edwards, G.R., et al. (2007), High sugar ryegrasses for livestock systems in New Zealand, *Proceedings of the New Zealand Grassland Association*, 69, 161–171. https://doi.org/10.33584/jnzg.2007.69.2674

Egan, A., et al. (2025), Ribwort plantain inclusion reduces nitrate leaching from grass-clover swards; a multi-year five soils study, *Agriculture, Ecosystems and Environment*, 380, 109376. https://doi.org/10.1016/j.agee.2024.109376

Eme, P.E. and Roche, J.R. (2025), Invited review: Cows grazing pastures containing narrow-leaved plantain have lower urine N concentrations: implications for nitrate leaching, *Journal of Dairy Science*, 108, 7876–7895. https://doi.org/10.3168/jds.2024-26169

Farrell, L., et al. (2025), Catchment level modelling of dairy farm adaptation to meet nitrogen loss targets and the value of including plantain as a mitigation, *Journal of New Zealand Grasslands*, 87, XXX-XXX. (in press)

Fransen, K.E., et al. (2024), Comment on 'an examination of the ability of plantain (*Plantago lanceolata* L.) to mitigate N leaching from pasture systems', *New Zealand Journal of Agricultural Research*, 68, 158–170. https://doi.org/10.1080/00288233.2024.2398149.

Fransen, K.E., et al. (2025), Using plantain (Plantago lanceolata) with other on-farm strategies to meet N loss reduction targets while maintaining profit in Rotorua Lake catchment, *Journal of New Zealand Grasslands*, 87, XXX-XXX. (in press)

Gardiner, C.A., et al. (2017), Potential inhibition of urine patch nitrous oxide emissions by *Plantago lanceolata* and its metabolite aucubin, *New Zealand Journal of Agricultural Research*, 61(4), 495–50. https://doi.org/10.1080/00288233.2017.1411953

Garwood, E.A. and Ryden, J.C. (1986), Nitrate loss through leaching and surface runoff from grassland: effects of water supply, soil type and management. In Van der Meer, H.G., et al. (Eds), *Nitrogen fluxes in intensive grassland systems*, Martinus Nijhoff Publishers, Dordrecht, The Netherlands, pp. 99–113.

Gawn, T., et al. (2012), Weed control in establishing mixed swards of clover plantain and chicory, *New Zealand Plant Protection*, 65, 59–63. https://doi.org/10.30843/nzpp.2012.65.5425.

Gerard, P.J., et al. (2017), *Annual Report on SFF Project 404890: tools for managing plantain weeds and pests*. AgResearch Client Report 6662, AgResearch, Hamilton.

Gerard, P., et al. (2018), Effect of narrow-leaved plantain cultivar on development of two geometrid pests, *Scopula rubraria* and *Epyaxa rosearia*, *New Zealand Journal of Agricultural Research*, 61(4), 403–413. https://doi.org/10.1080/00288233.2017.1398763

Grace, C., et al. (2018), The effect of increasing pasture species on herbage production, chemical composition and utilization under intensive sheep grazing, *Grass and Forage Science*, 73, 852–864. https://doi.org/10.1111/gfs.12379

Graham, S.L., et al. (2024), Assessing the potential of diverse pastures for reducing nitrogen leaching, *Frontiers in Environmental Science*, 12, 1445212. https://doi.org/10.3389/fenvs.2024.1445212

Gray, C. (2023), Nitrogen fertiliser use in grazed pasture-based systems in New Zealand: a summary, *New Zealand Journal of Agricultural Research*, 67(6), 670–721. https://doi.org/10.1080/00288233.2023.2198719

Hayes, L., et al. (2013), Biocontrol of weeds: achievements to date and future outlook. In Dymond, J.R. (Ed), *Ecosystem services in New Zealand – conditions and trends*, Manaaki Whenua Press, Lincoln, New Zealand.

Healy, M., et al. (2024), The effect of sward species on nitrate leaching: a lysimeter study, *Grassland Science in Europe*, 29, 645–647.

Hearn, C., et al. (2024a), Can the inclusion of ribwort plantain or chicory increase the seasonal and annual dry matter production of intensive dairy grazing swards? *European Journal of Agronomy*, 152, 127020. https://doi.org/10.1016/j.eja.2023.127020

Hearn, C., et al. (2024b), Does sowing rate affect establishment of ribwort plantain when over-sown in grass-clover swards? *Grassland Science in Europe*, 29, 612–614.

Herath, G., et al. (2023), Incorporating plantain into ryegrass-white clover mixed sward for an economically and environmentally sustainable dairy system: Year one of a farm system study, *Journal of New Zealand Grasslands*, 85, 321–331. https://doi.org/10.33584/jnzg.2023.85.3635

Hristov, A.N., et al. (2013), *Mitigation of greenhouse gas emissions in livestock production – A review of technical options for non-CO2 emissions*. Gerber, P.J., et al. (Eds). FAO Animal Production and Health Paper No. 177, FAO, Rome, Italy.

Jones, S., et al. (2015), Assessing the potential of *Yersinia entomophaga* to control plantain moth in a laboratory assay, *New Zealand Plant Protection*, 68, 146–150. https://doi.org/10.30843/nzpp.2015.68.5833

Judson, H.G., et al. (2018a), Reducing nitrate leaching from the urine patch – a plant-based approach, *Proceedings of the 7th Australasian Dairy Science Symposium*, 46–53. https://australasiandairyscience.com/wp-content/uploads/2020/03/2018-ADSS-proceedings.pdf

Judson, H.G., et al. (2018b), Specific genotypes of plantain (*Plantago lanceolata*) vary in their impact on sheep urine volume and nitrification in the urine patch, *Journal of New Zealand Grasslands*, 80, 125–128. https://doi.org/10.33584/jnzg.2018.80.335

Judson, H.G., et al. (2019), Nitrification inhibition by urine from cattle consuming *Plantago lanceolata*, *Journal of New Zealand Grasslands*, 81, 111–116. https://doi.org/10.33584/jnzg.2019.81.413

Kebreab, E., et al. (2001), Nitrogen pollution by dairy cows and its mitigation by dietary manipulation, *Nutrient Cycling in Agroecosystems*, 60, 275–285. https://doi.org/10.1023/A:1012668109662

Kemp, P.D., et al. (2010), The use of legume and herb forage species to create high performance pastures for sheep and cattle grazing systems, *Revista Brasileira de Zootecnia*, 39, 169–174. https://doi.org/10.1590/S1516-35982010001300019

Koning, L., et al. (2024), The effect of ribwort plantain (Plantago lanceolata) on enteric methane emission during grazing, *Grassland Science in Europe*, 29, 214–216.

Labreveux, M., et al. (2004), Productivity of chicory and plantain cultivars under grazing, *Agronomy Journal*, 96(3), 710–716. https://doi.org/10.2134/agronj2004.0710

Laurenson, S., et al. (2016), Assessing the production and economic benefits from preventing cows grazing on wet soils in New Zealand, *Journal of the Science of Food and Agriculture*, 96, 4584–4593. https://doi.org/10.1002/jsfa.7676

Laurenson, S., et al. (2017), Evaluating the economic and production benefit of removing dairy cows from pastures in response to wet soil conditions, *New Zealand Journal of Agricultural Research*, 60, 223–244. https://doi.org/10.1080/00288233.2017.1298630

Ledgard, S.F., et al. (1997), Nitrogen balances and losses on intensive dairy farms, *Proceedings of the New Zealand Grasslands Association*, 59, 49–53. https://doi.org/10.33584/jnzg.1997.59.2265

Ledgard, S.F., et al. (2006), Nitrogen leaching as affected by dairy intensification and mitigation practices in the Resource Efficient Dairying (RED) trial. In Currie, L.D. and Hanly, J.A. (Eds), *Implementing sustainable nutrient management strategies in agriculture*, Occasional Report No. 19, Fertilizer and Lime Research Centre, Massey University, Palmerston North, New Zealand, pp. 263–268. https://assets.ctfassets.net

/bo1h2c9cbxaf/44hYXIlacAXH18zJRK7Szv/91ec599c634f5753f1289e8321ef52c8/Nitrogen_leaching_as_affected_by_dairy.pdf

Lee, J.M., et al. (2015a), Management strategies for chicory (*Cichorium intybus*) and plantain (*Plantago lanceolata*): impact on dry matter yield, nutritive characteristics and plant density, *Crop and Pasture Science*, 66, 168–183. https://doi.org/10.1071/CP14181

Lee, J.M., et al. (2015b), Patterns in non-structural carbohydrate and nitrogen reserves in chicory (*Cichorium intybus L.*) and plantain (*Plantago lanceolata* L.) during regrowth in summer, *Crop and Pasture Science*, 66, 1071–1078. http://dx.doi.org/10.1071/CP15122

Li, G.D., et al. (2025), Agronomic performance, herbage quality, methane yield and methane emission potential of pasture mixtures, *Crop & Pasture Science*, 76, CP24356. https://doi.org/10.1071/CP24356

Luo, J., et al. (2018), Using alternative forage species to reduce emissions of the greenhouse gas nitrous oxide from cattle urine deposited onto soil, *Science of the Total Environment*, 610/611, 1271–1280. https://doi.org/10.1016/j.scitotenv.2017.08.186

Mangwe, M.C., et al. (2018), Grazed chicory, plantain or ryegrass–white clover alters milk yield and fatty acid composition of late-lactating dairy cows, *Animal Production Science*, 60, 107–113. https://doi.org/10.1071/AN18537

Marshall, C.J., et al. (2021), Nitrogen balance of dairy cows divergent for milk urea nitrogen breeding values consuming either plantain or perennial ryegrass, *Animals*, 11, 2464. https://doi.org/10.3390/ani11082464

Marshall, C.J., et al. (2022), Dietary and animal strategies to reduce the environmental impact of pastoral dairy systems result in altered nutraceutical profiles in milk, *Animals*, 12(21), 2994. https://doi.org/10.3390/ani12212994

Martin, K.E., et al. (2017a), Effect of autumn regrowth interval and nitrogen fertiliser on dry matter yield and plant characteristics of six forage species, *Journal of New Zealand Grasslands*, 79, 67–72. https://doi.org/10.33584/jnzg.2017.79.560

Martin, K.E., et al. (2017b), Herbage dry-matter yield and nitrogen concentration of grass, legume and herb species grown at different nitrogen-fertiliser rates under irrigation, *Animal Production Science*, 57, 1283–1288. https://doi.org/10.1071/AN16455

Mikola, J., et al. (2001), Effects of defoliation intensity on soil food-web properties in an experimental grassland community, *Oikos*, 92, 333–343. https://doi.org/10.1034/j.1600-0706.2001.920216.x

Minneé, E.M.K., et al. (2013), Herbage production from five grazeable forages, *Proceedings of the New Zealand Grassland Association*, 75, 245–250. https://doi.org/10.33584/jnzg.2013.75.2906

Minneé, E.M.K., et al. (2017), Including chicory or plantain in a perennial ryegrass/white clover-based diet of dairy cattle in late lactation: feed intake, milk production and rumen digestion, *Animal Feed Science and Technology*, 227, 52–61. https://doi.org/10.1016/j.anifeedsci.2017.03.008

Minneé, E.M.K., et al. (2019), Meta-analyses comparing the nutritional composition of perennial ryegrass (*Lolium perenne*) and plantain (*Plantago lanceolata*) pastures, *Journal of New Zealand Grasslands*, 81, 117–124. https://doi.org/10.33584/jnzg.2019.81.402

Minneé, E.M.K., et al. (2020), Substituting a pasture-based diet with plantain (*Plantago lanceolata*) reduces nitrogen excreted in urine from dairy cows in late lactation, *Livestock Science*, 239, 104093. https://doi.org/10.1016/j.livsci.2020.104093

Minogue, C.T. et al. 2025. Effect of plantain in perennial ryegrass-white clover pasture on the ingestive behaviour, rumen fermentation parameters, and enteric methane emissions of spring-calving dairy cows. Animal Feed Science and Technology 329: article 116500.

Moorhead, A.J.E., et al. (2002), Liveweight gain of lambs grazing Ceres Tonic plantain (*Plantago lanceolata*) or perennial ryegrass (*Lolium perenne*), *Proceedings of the New Zealand Society of Animal Production*, 62, 171–173. http://www.nzsap.org/proceedings/2002/liveweight-gain-lambs-grazing-ceres-tonic-plantain-plantago-lanceolata-or-perennial

Navarrete, S. (2015), *Evaluation of herb pastures for New Zealand dairy systems*. PhD Thesis, Massey University. https://mro.massey.ac.nz/handle/10179/10121

Navarrete, S., et al. (2016), Bioactive compounds, aucubin and acteoside, in plantain (*Plantago lanceolata* L.) and their effect on in vitro rumen fermentation, *Animal Feed Science and Technology*, 222, 158–167. https://doi.org/10.1016/j.anifeedsci2016.10.008

Navarrete, S., et al. (2023), Evidence of environmental benefits of plantain pastures, *Proceedings of the 25th International Grassland Congress*, 1846–1879. https://doi.org/10.52202/071171-0453

Neal, J., et al. (2011), Differences in water-use efficiency among perennial forages used by the dairy industry under optimum and deficit irrigation, *Irrigation Science*, 29, 213–232. https://doi.org/10.1007/s00271-010-0229-1

Nguyen, T.T., et al. (2022a), Forage plantain (*Plantago lanceolata* L.): meta-analysis quantifying the decrease in nitrogen excretion, the increase in milk production and the changes in milk composition of dairy cows grazing pastures containing plantain, *Animal Feed Science and Technology*, 285, 115244. https://doi.org/10.1016/j.anifeedsci.2022.115244

Nguyen, T.T., et al. (2022b), Incorporating plantain with perennial ryegrass-white clover in a dairy grazing system: dry matter yield, botanical composition and nutritive value response to sowing rate, plantain content and season, *Agronomy*, 12, 2789. https://doi.org/10.3390/agronomy12112789

O'Connell, C.A., et al. (2016), Sustained diuretic effect of plantain when ingested by sheep, *Proceedings of the New Zealand Society of Animal Production*, 76, 14–17. https://www.nzsap.org/proceedings/sustained-diuretic-effect-plantain-when-ingested-sheep

Pacheco, D. and Waghorn, G. (2008), Dietary nitrogen – definitions, digestion, excretion and consequences of excess for grazing ruminants, *Proceedings of the New Zealand Grassland Association*, 70, 107–116. https://doi.org/10.33584/jnzg.2008.70.2738

Peterson, M., et al. (2022), Soil nitrification inhibition by urine of sheep consuming plantain (*Plantago lanceolata*), *Biology and Fertility of Soils*, 58, 265–276. https://doi.org/10.1007/s00374-021-01573-1.

Peterson, M., et al. (2023), Soil nitrification inhibition with plantain (*Plantago lanceolata*), *Proceedings of the 25th International Grassland Congress*, 1741–1744. https://doi.org/10.52202/071171-0428

Pijlman, J., et al. (2020), Can the presence of plantain (*Plantago lanceolata* L.) improve nitrogen cycling of dairy grassland systems on peat soils?, *New Zealand Journal of Agricultural Research*, 63(1), 106–122. https://doi.org/10.1080/00288233.2019.1698620

Pinxterhuis, J.B., et al. (2015), Eco-efficient pasture based dairy farm systems: a comparison of New Zealand, the Netherlands and Ireland, *Grassland Science in Europe*, 20, 349–366. https://europeangrassland.org/fileadmin/documents/Infos/Printed_Matter/Proceedings/EGF2015.pdf#page=356

Pinxterhuis, J.B., et al. (2024), Implementing plantain (*Plantago lanceolata*) to mitigate the impact of grazing ruminants on nitrogen losses to the environment: a review, *Grass and Forage Science*, 79(2), 144–157. https://doi.org/10.1111/gfs.12649

Pol, M., et al. (2021), *Plantago lanceolata* – An overview of its agronomically and healing valuable features, *Open Agriculture*, 6, 479–488. https://doi.org/10.1515/opag-2021-0035

Rodriguez, M., et al. (2023), Can secondary metabolites of plantain reduce N losses from urine patches? *New Zealand Journal of Agricultural Research*, 66(1), 83–100. https://doi.org/10.1080/00288233.2021.1991388

Selbie, D., et al. (2015), The challenge of the urine patch for managing nitrogen in grazed pasture systems, *Advances in Agronomy*, 129, 229–292. https://doi.org/10. 1016/bs.agron.2014.09.004

Shepherd, M., et al. (2017), A summary of key messages arising from the Pastoral 21 research programme. In Currie, L.D. and Hedley, M.J. (Eds), *Farming's future: minimising footprints and maximising margins*, Occasional Report No. 23, Fertilizer and Lime Research Centre, Massey University, Palmerston North, New Zealand, pp. 174–181. https://flrc.massey.ac.nz/workshops/17/Manuscripts/Paper_Shepherd_2017.pdf

Shrivastav, N., et al. (2023), An assessment of resistance of two plantain (*Plantago lanceolata*) cultivars to phenoxy herbicides, *New Zealand Journal of Agricultural Research*, 1–12. https://doi.org/10.1080/00288233.2023.2260774

Silvia Cid, M., et al. (2011), Seedling establishment and forage accumulation of 'Ceres Tonic' plantain in pure stands and in grass mixtures, *Chilean Journal of Agricultural Research*, 71(3), 370–375. https://oes.chileanjar.cl/files/V71_I3_2011_ENG_MSilviaCid.pdf

Simon, P.L., et al. (2019), The efficacy of *Plantago lanceolata* for mitigating nitrous oxide emissions from cattle urine patches, *Science of the Total Environment*, 691, 430–441. https://doi.org/10.1016/j.scitotenv.2019.07.141

Sivanandarajah, K., et al. (2025a), Plantain-mixed pasture collected in different climatic seasons produced less methane and ammonia than ryegrass–white clover pasture *in vitro*, *Animal Production Science*, 65, AN24287. https://doi.org/10.1071/AN24287

Sivanandarajah, K., et al. (2025b), Effects of plantain (*Plantago lanceolata* L.) metabolites aucubin, acteoside, and catalpol on methane emissions *in vitro*, *Journal of Agriculture and Food Chemistry*, 73, 12132–12145. https://doi.org/10.1021/acs.jafc.4c12140

Skinner, R.H. and Comas, L.H. (2010), Root distribution of temperate forage species subjected to water and nitrogen stress, *Crop Science*, 50, 2178–2185. https://doi.org/10.2135/cropsci2009.08.0461

Snow, V.O., et al. (2025), Scaling from patch to paddock: are diuretic-based mitigations of leaching and nitrous oxide emissions from urine patches effective at all scales and stocking rates? *Agricultural Systems*, 230, 104467. https://doi.org/10.1016/j.agsy.2025.104467

Somasiri, S.C., et al. (2015), Growth performance and carcass characteristics of lambs grazing forage mixes inclusive of plantain (*Plantago lanceolata* L.) and chicory (*Cichorium intybus* L.), *Small Ruminant Research*, 127, 20–27. https://doi.org/10.1016/j.smallrumres.2015.04.005

Stafford, A.D., et al. (2016), Cadmium accumulation by forage species used in New Zealand livestock grazing systems, *Geoderma Regional*, 7, 11–18. https://doi.org/10.1016/j.geodrs.2015.11.003

Stewart, A.V. (1996), Plantain (*Plantago lanceolata*) - a potential pasture species, *Proceedings of the New Zealand Grassland Association*, 58, 77–86. https://doi.org/https://doi.org/10.33584/jnzg.1996.58.2221

Stewart, A.V. and Judson, H.G. (2019), Developments in the use of plantain (*Plantago lanceolata*) cultivars in New Zealand, *Proceedings of the Joint 20th Symposium of the European Grassland Federation and the 33rd Meeting of the EUCARPIA*, 422. https://uknowledge.uky.edu/cgi/viewcontent.cgi?article=5803&context=igc

Talbot, W.D., et al. (2021), Effect of timing of cattle urine deposition and pasture composition on nitrogen leaching losses, *Soil Use and Management*, 37(4), 723–735. https://doi.org/10.1111/sum.12652

Totty, V.K., et al. (2013), Nitrogen partitioning and milk production of dairy cows grazing simple and diverse pastures, *Journal of Dairy Science*, 96, 141–149. https://doi.org/10.3168/jds.2012-5504

Troelstra, S.R., et al. (1992), Mineral nutrient concentrations in the soil and in the plant. In Kuiper P.J.C. and Bos, M. (Eds), *Plantago: a multidisciplinary study (Chapter 5.2)*, Ecological Studies Analysis and Synthesis, Vol. 89, Springer-Verlag, Berlin, Germany.

Van der Meer, H.G. and Van Uum-Van Lohuyzen, M.G. (1986), The relationship between inputs and outputs of nitrogen in intensive grassland systems. In Van der Meer, H.G., et al. (Eds), *Nitrogen fluxes in intensive grassland systems*, Martinus Nijhoff Publishers, Dordrecht, The Netherlands, pp. 1–18.

Wachendorf, N., et al. (2004), Performance and environmental effects of forage production on sandy soils. II. Impact of defoliation system and nitrogen input on nitrate leaching losses, *Grass and Forage Science*, 59, 56–68. https://doi.org/10.1111/j.1365-2494.2004.00401.x

Waghorn, G.C. and Clark, D.A. (2004), Feeding value of pastures for ruminants, *New Zealand Veterinary Journal*, 52, 320–331. https://doi.org/10.1080/00480169.2004.36448

Welten, B.G., et al. (2019), Efficacy of different temperate pasture species to reduce nitrogen leaching from cattle urine applied in different seasons: a soil lysimeter study, *Soil Use and Management*, 35, 653–663. https://doi.org/10.1111/sum.12512

Whitehead, D.C. (1986), Sources and transformations of organic nitrogen in intensively managed grassland soils. In Van der Meer, H.G., et al. (Eds), *Nitrogen fluxes in intensive grassland systems,* Martinus Nijhoff Publishers, Dordrecht, The Netherlands, pp. 46–58.

Wilson, S., et al. (2024), Investigating the impact of treading damage on the plantain (*Plantago lanceolata* L.) content and performance of a plantain/ perennial ryegrass (*Lolium perenne* L.) pasture over two years, *Journal of New Zealand Grasslands*, 86, 97–107. https://doi.org/10.33584/jnzg.2024.86.3682

Woodward, S.L., et al. (2012), Can diverse pasture mixtures reduce nitrogen losses? *Proceedings of the 5th Australasian Dairy Science Symposium*, 463–464. https://australasiandairyscience.com/wp-content/uploads/2022/11/Proceedings-of-the-Australasian-Dairy-Science-Symposium-2012.pdf

Woodward, S.L., et al. (2013), Are diverse species mixtures better pastures for dairy farming? *Proceedings of the New Zealand Grassland Association*, 75, 82–86. https://doi.org/10.33584/jnzg.2013.75.2926

Chapter 16

Advances in *Epichloë* endophyte plant secondary metabolites in grasslands

David E. Hume and Sarah C. Finch, AgResearch Limited, New Zealand; Alan V. Stewart, PGG Wrightson Seeds Limited, New Zealand; and John R. Caradus, Grasslanz Technology Limited, New Zealand

1 Introduction

Two of the most widely cultivated cool-season forage grasses in mesic, temperate, perennial grasslands are tall fescue (*Festuca arundinacea* syn. *Lolium arundinaceum*) and perennial ryegrass (*Lolium perenne*) (Easton *et al.*, 1994; Jung *et al.*, 1996; Fribourg *et al.*, 2009; Ball *et al.*, 2019). These grasses

http://dx.doi.org/10.19103/AS.2025.0154.25

are adapted to a range of climatic conditions and various soils, tolerate a wide range of grazing managements and cutting regimes, combine well with legumes and herbs in sown mixed-species pastures and in general provide nutritious feed for livestock such as dairy cows, beef cattle, sheep and horses. They became of particular importance in large areas of the *New World*, including countries of North and South America, and Australia and New Zealand, as European settlers began to farm livestock globally. However, in these regions over time, these grasses became associated with poor animal performance and health, in some seasons of the year (Hume *et al.*, 2016; Ball *et al.*, 2019). These detrimental effects on farmed animals were seldom apparent in the regions of Europe, North Africa and Asia where these grasses originated.

In the late 1970s and early 1980s, breakthrough discoveries in the USA and New Zealand found that the toxicity to livestock was due to asexual species of the *Epichloë* fungi that endophytically colonised tall fescue and perennial ryegrass (Bacon *et al.*, 1977; Fletcher and Harvey, 1981). The naturalised, well-adapted local grass populations were highly infected with asymptomatic mammalian toxic strains of *Epichloë*, having been unknowingly introduced from the centres of origin by the European farmers (Johnson *et al.*, 2013). Of equal importance was the discovery that *Epichloë* were also providing protective properties to their grass hosts, which in some cases were the key determinant of their persistence in these countries (Mortimer *et al.*, 1982; Read and Camp, 1986). For tall fescue, this manifested as a greater tolerance to drought and pests such as nematodes, and for ryegrass this was due primarily to less damage caused by several invertebrate pests (Hume *et al.*, 2016; Ball *et al.*, 2019; Caradus *et al.*, 2021b; Caradus, 2023a).

The anti-quality effects for livestock, and many of the bioprotective properties of the grass, can be attributed to bioactive secondary metabolites produced by the endophyte when in association with the grass host. The bioactive metabolites are produced by both the asexual and sexual species of *Epichloë*, which collectively infect 29 genera of temperate grasses spread across all six farmed continents of the world (Schardl *et al.*, 2013a; Schardl *et al.*, 2013b; Leuchtmann *et al.*, 2014; Caradus *et al.*, 2021a; Anon., 2023b).

In this chapter we describe the biology of the *Epichloë*-grass association, the secondary metabolites that are linked to the bioactivity, control of invertebrate pests, the occurrence of mammalian toxicity globally, toxic responses in terms of animal productivity, health and welfare, utilisation of *Epichloë* strain variation in commerce, along with flow-on effects to animal product quality and food safety. There is a particular focus on the most well-studied, mutualistic tall fescue and perennial ryegrass endophyte associations due to their significant economic impacts and prospects to advance future sustainable grassland farming.

Published by Burleigh Dodds Science Publishing Limited, 2025.

2 *Epichloë* biology

Epichloë is a filamentous fungal genus within the tribe Balansieae (family Clavicipitaceae). Over many millennia, *Epichloë* have co-evolved with their temperate (cool season) grass hosts within the sub-family Pooideae (Schardl *et al.*, 2008; Schardl, 2010). Over this period of co-evolution, *Epichloë* have developed a high degree of host specificity which in nature generally restricts each *Epichloë* species to a single or closely related grass genera, although there are notable exceptions. *Epichloë* have been documented in over 100 grass species from a number of the tribes within the Pooideae sub-family (Caradus *et al.*, 2021a).

As an obligate symbiont, *Epichloë* can only live within a grass plant and seed from that plant. In vegetative plants, the growth of *Epichloë* is fully synchronised and highly regulated by the host grass (Christensen *et al.*, 2002; Christensen and Voisey, 2007b; Christensen *et al.*, 2008; Voisey, 2010). *Epichloë* hyphal strands colonise the apical meristem of the vegetative grass, through extensive hyphal branching between the grass cells. This enables the fungus to colonise the meristems of new tillers and inflorescence primordia as they develop. Fungal tip growth occurs in meristematic tissue, while a unique process of intercalary hyphal extension occurs in the leaf. Intercalary growth allows hyphae to grow with the leaf when it is in the phase of rapid elongation. As the leaf matures, hyphae branch through tip growth, where it occupies the intercellular spaces of the leaf sheath and to some extent leaf blade. In most cases, hyphal mass decreases with increasing distance from the crown of the plant. While *Epichloë* are systemic in the shoot, there have only been rare observations in roots, and they are not present in pollen (Siegel *et al.*, 1984; Schardl *et al.*, 2004; Christensen and Voisey, 2007a).

When a grass host becomes reproductive, hyphal growth varies between asexual and sexual species of *Epichloë*, affecting different paths of transmission for the fungus (Tadych *et al.*, 2014; Johnson *et al.*, 2021). Asexual species colonise the developing seed, particularly the seed embryo which is critical for the endophyte to continue its mutualistic, asymptomatic symbiosis when the seed germinates (Card *et al.*, 2011; Liu *et al.*, 2017; Zhang *et al.*, 2017). Some sexual species of *Epichloë* also colonise the seed and form new associations with the germinating seed. Transmission of *Epichloë* via the host seed is termed vertical transmission.

On the culms of reproductive tillers, sexual *Epichloë* may also form epiphytic reproductive structures (stroma) termed 'choke', as part of the process of sexual recombination and spread of *Epichloë* by ascospores. This is termed horizontal transmission. Grass associations with horizontal transmission are pathogenic for the host, as 'choke' prevents mature seed formation therefore preventing sexual propagation of the grass. Asexual *Epichloë* grass associations are always asymptomatic, while the presence of choke is a clear symptom of a sexual *Epichloë*-grass association.

On the basis of the form of transmission and reproduction, *Epichloë*-grass associations can be classified into three types (White, 1988; Leuchtmann and Clay, 1997) (Fig. 1). For type I associations, 'choke' occurs on all seed heads which completely prevents the grass from producing seed. In these associations, there are restrictions on reproduction for both symbionts which are likely to contribute, along with other factors, to type I grass-endophyte populations remaining small. For type II associations, not all individuals in a grass population, and not all reproductive tillers on a plant, are affected by choke. In this case, the endophyte can propagate not only via sexual reproduction but also vegetative reproduction, as it infects the grass seeds on tillers without choke. In type III associations, the asexual endophytes do not form choke in the seedheads. This allows for the full expression of the host's sexual reproductive capacity. However, the endophyte is clonal and needs to adapt to the new genetics of the grass progeny, which may not be so favourable for the endophyte.

In an agricultural context, the negative impact of sexual endophytes on seed yields is a disadvantage (particularly type I associations) of what is otherwise a mutualistic symbiosis. While each type of grass-endophyte association has varying pros and cons during this reproductive phase, for both organisms, of the 48 *Epichloë* described as of 2022, the majority are asexual (33 type III

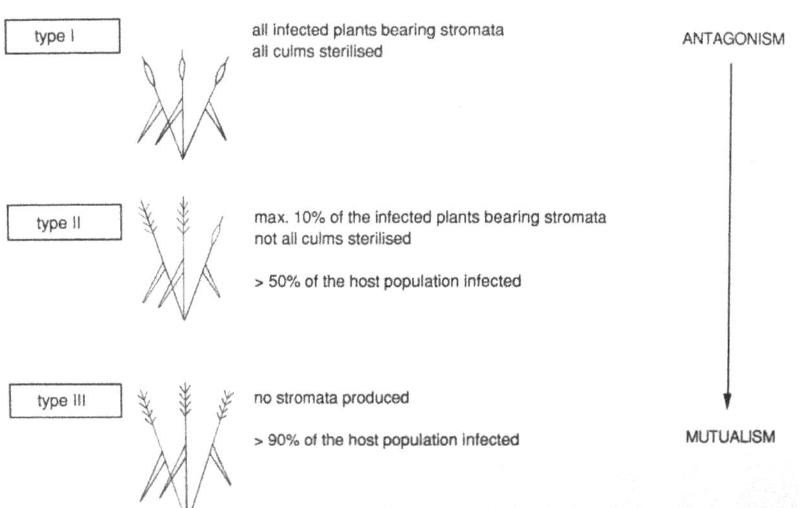

Figure 1 Types of endophyte-grass associations formed by sexual (types I and II) and asexual (type III) forms of *Epichloë*. A continuum of interactions from antagonism (pathogenic) to mutualism occurring in the reproductive grass tillers. (Prepared by Erika Bucheli; provided courtesy of Adrian Leuchtmann, ETH, Zurich, Switzerland).

associations) (Anon, 2023b). Of the 15 sexual *Epichloë*, just 4 could be classified as type I associations, the remainder being type II associations. The asexual associations have been of most interest to agriculture, possibly due to their high expression of mammalian and invertebrate toxic secondary metabolites (Leuchtmann *et al.*, 2000). This high secondary metabolite expression is consistent with the defence mutualism theory (Clay, 1988; Saikkonen *et al.*, 2016). The asexual *Epichloë* needs to provide the host with as many protective properties as possible (e.g. via secondary metabolite production), as the endophyte is totally reliant on the survival of the host and spread via the host seed. Sexual endophytes can achieve spread via horizontal transmission, and with less secondary metabolite production this lowers any metabolic cost for the host.

3 Benefits for the host grass

Across the various *Epichloë*-grass associations, endophyte has been shown to improve the fitness of the plant under a range of abiotic and biotic stresses. As an obligate symbiont, *Epichloë* must improve host performance in some manner to ensure that infected plants are as or are more competitive than endophyte-free equivalents, and/or those of competing plant species, thus overcoming any metabolic cost for hosting the endophyte (Rodriguez *et al.*, 2009). In some cases, endophyte-infected plants may be inferior for some traits, which may be more so in natural than cultivated grass species (Faeth and Saari, 2012).

There are multiple ways by which the endophyte-driven increase in host fitness is achieved. For abiotic stresses, *Epichloë* has been shown to improve tolerance to drought, flooding, heavy metals, aluminium, nutrient deficiency, salinity, heat stress and winter cold (Malinowski and Belesky, 2000; Song *et al.*, 2015; Caradus and Johnson, 2020; Hewitt *et al.*, 2021; Caradus and Hume, 2023). For biotic stresses, plant pathogens may be reduced and of greatest importance is a reduction in herbivory by invertebrate pests and farmed animals (Xia *et al.*, 2018; Caradus and Johnson, 2020; Pérez *et al.*, 2020; Caradus *et al.*, 2021a; Card *et al.*, 2021). Various mechanisms are involved which include changes to plant morphology, physiology, levels of primary metabolites, interactions with other microorganisms and, in particular, bioactive secondary metabolites that are not produced in endophyte-free plants (Malinowski and Belesky, 2000; Caradus *et al.*, 2021a; Johnson *et al.*, 2021; Bastías *et al.*, 2024).

In an agricultural context, reduced herbivory is favourable in terms of invertebrate pests but not for grazing livestock as it manifests in poor livestock performance and serious health disorders. In addition, results from indoor experiments, or highly controlled field studies, may differ in the farm environment where multiple stresses may be active and multi-trophic

Published by Burleigh Dodds Science Publishing Limited, 2025.

interactions occur (Fuchs *et al.*, 2020; Hewitt *et al.*, 2021). It may therefore be difficult to predict the endophyte-driven outcomes, particularly under a rapidly changing global environment (Bastías *et al.*, 2023).

The benefits of endophyte infection for the host grass reported in one grass species cannot be assumed to occur in another as each endophyte–grass association has unique characteristics. Even within endophyte species different strains may vary significantly in their secondary metabolite profile due to mutations. The discovery and development of endophyte strains lacking certain secondary metabolites have been utilised in commerce to achieve optimal agronomic performance with minimal or no negative effects on livestock (Caradus *et al.*, 2021a).

4 Secondary metabolites produced by endophyte-grass associations

Endophyte–grass associations produce a myriad of secondary metabolites, with those associated with known bioactivity falling into four classes. These include indole-diterpenes (lolitrem B and epoxyjanthitrems), ergot alkaloids (ergovaline), pyrrolopyrazine alkaloids (peramine) and saturated 1-aminopyrrolizidine alkaloids (lolines) (Fig. 2). Endophyte-infected tall fescue is associated with ergot alkaloids (ergovaline is the major compound) (Lyons *et al.*, 1986), peramine (Rowan and Gaynor, 1986; Rowan, 1993) and loline alkaloids (N-formyl loline is the major compound) (Bush *et al.*, 1982). Naturally occurring perennial ryegrass is associated with lolitrems (lolitrem B is the major compound) (Miles *et al.*, 1994), ergot alkaloids (ergovaline is the major compound) (Caradus *et al.*, 2020) and peramine. However, to maximise the benefits of endophytes in commerce, novel selected endophyte–grass associations have been created for both tall fescue and perennial ryegrass (see section 10.4). This has led to the expression of further indole-diterpenes in perennial ryegrass, the epoxyjanthitrems (five compounds, epoxyjanthitrem I is dominant) (Fig. 2) (Finch *et al.*, 2020).

Rather than a single pathway end product, most of the secondary metabolite biosynthetic pathways also express intermediate compounds (Johnson *et al.*, 2021; Bastías *et al.*, 2023). While this appears to be inefficient, some of these intermediates have bioactivity that contributes to the diversity of defensive secondary metabolites in the plant (Panaccione, 2005; Schardl *et al.*, 2013b; Caradus and Johnson, 2020).

4.1 Factors affecting concentrations

Concentrations of secondary metabolites in plant material are influenced by multiple factors including plant host, plant parts, season and other

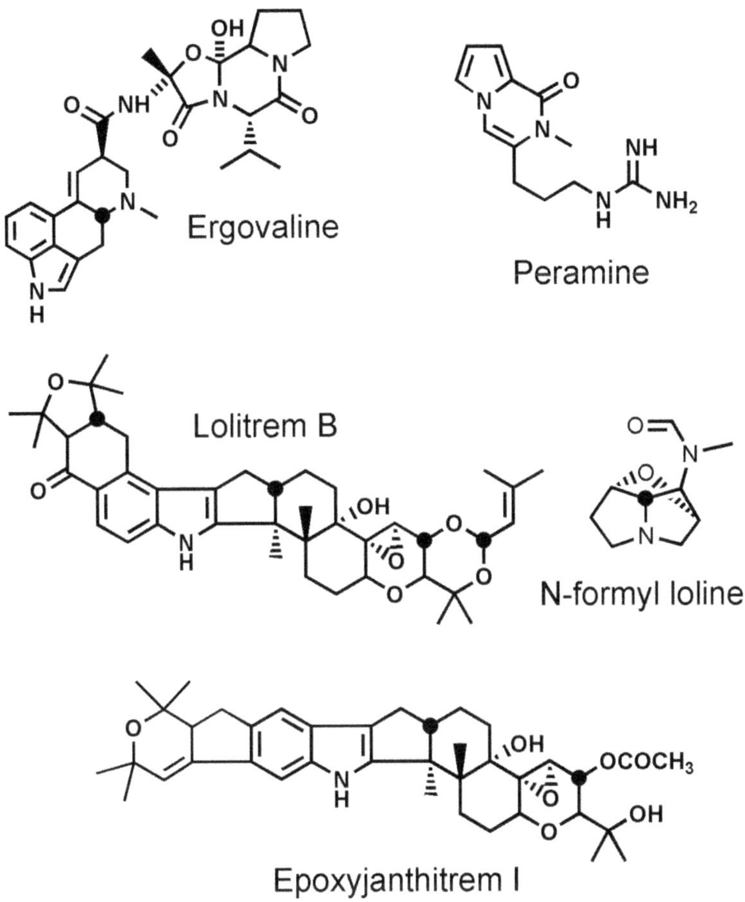

Figure 2 Major secondary metabolites produced by perennial ryegrass and tall fescue infected by *Epichloë* endophytes.

environmental factors (Agee and Hill, 1994; Easton *et al.*, 2002; Brosi *et al.*, 2011; Hennessy *et al.*, 2016; Fuchs *et al.*, 2017; Caradus *et al.*, 2020). In endophyte-infected ryegrass both lolitrem B and ergovaline are highest in mature seeds. In vegetative tissue, the endophyte is highest in the basal material corresponding to high lolitrem B and ergovaline concentrations. However, unlike ergovaline, lolitrem B is also high in old leaves and dead material (Repussard *et al.*, 2014). In contrast, peramine is more evenly distributed throughout the plant (Ball *et al.*, 1991). Although at a much higher concentration, ergovaline in endophyte-infected tall fescue shows the same distribution as that observed in ryegrass (Caradus *et al.*, 2020). Lolines are highest in seed heads and pseudostems compared to leaf blades (Justus *et al.*, 1997) and lolines are also found in root tissue (Patchett *et al.*, 2008b).

During winter, lolitrem B and ergovaline are at their lowest concentrations in endophyte-infected ryegrass, with both peaking in summer/autumn, but peramine concentrations remain relatively constant throughout the year (Ball *et al.*, 1991). In endophyte-infected tall fescue, both ergovaline and lolines are low in winter although concentrations peak at different times with ergovaline concentrations highest in late spring and lolines in summer (Bush *et al.*, 1993; Agee and Hill, 1994).

Temperature is a major factor influencing secondary metabolite expression with Hennessy *et al.* (2016) showing that endophyte-infected ryegrass grown at 20°C had much higher epoxyjanthitrem concentrations than plants grown at 7°C. Lolines in endophyte-infected tall fescue have also been shown to be affected by temperature (Huizing *et al.*, 1991) although this trend is less clear for ergovaline (Brosi *et al.*, 2011). Ryegrass grown at ambient CO_2 with added nitrogen showed an increase in peramine and ergovaline whereas the combination of elevated CO_2 and added nitrogen induced no effect on secondary metabolite production (Hunt *et al.*, 2005). In contrast, for tall fescue grown at elevated CO_2, lolines and ergovaline decreased but added nitrogen increased ergovaline (Arechavaleta *et al.*, 1992). Although ergovaline in tall fescue is increased by high nitrogen, this effect was not seen with lolitrem B, peramine or ergovaline in endophyte-infected perennial ryegrass (Rasmussen *et al.*, 2007). Plant damage is another factor to consider since endophyte secondary metabolites are increased with plant wounding (Salminen and Grewal, 2002; Bultman *et al.*, 2004).

Additionally, water deficit has been shown to influence secondary metabolite expression, with lolines increasing significantly when tall fescue plants were under severe water deficit (Kennedy and Bush, 1983) although the evidence of an effect on ergovaline concentrations is conflicting (Belesky *et al.*, 1989; Arechavaleta *et al.*, 1992). The effect of water deficit on the secondary metabolites produced by perennial ryegrass is also uncertain with studies published in the literature reporting opposite effects. In one study, water stress was shown to increase the concentrations of lolitrem B and ergovaline but with no effect on peramine (Eerens *et al.*, 1998) whereas in another study drought stress was shown to increase peramine but not lolitrem B (Lin *et al.*, 2022).

5 Toxicity to invertebrate pests

Endophytes in grass associations provide a critical role in the protection of pasture from invertebrate pests (mainly insects) due to their production of secondary metabolites. This effect is vast, with 45 insect species from 10 different families affected (Malinowski and Belesky, 2019). The most important known secondary metabolites for pest protection are peramine and epoxyjanthitrems in ryegrass and lolines in tall fescue. These compounds are 'beneficial' to the farming system because loline and peramine have no adverse effects on animals and the insect bioactivity of epoxyjanthitrems outweighs the negative ones on animals (see Section 6). In contrast, lolitrem B and ergot alkaloids are 'detrimental' to the farming system since they have severe effects on animals which outweigh their effects on insects.

The range of insects affected by the classes of secondary metabolites, as discussed in Section 4, are detailed in Table 1. The bioactivity listed in this table includes only that known to be induced by a specific metabolite, whereas the list of effects induced by endophyte-grass associations is far more extensive (Caradus and Johnson, 2020). For example, the AR37 strain of *Epichloë* that infects perennial ryegrass shows excellent bioactivity against a variety of insect pests including Argentine stem weevil (*Listronotis bonariensis*) (Popay and Wyatt, 1995), African black beetle (*Heteronychus arator*) (Popay and Thom, 2009), pasture mealybug (*Balanococcus poae*) (Pennell *et al.*, 2005), porina larvae (*Wiseana cervinata*) (Jensen and Popay, 2004) and a root aphid (*Aploneura lentisci*) (Popay and Cox, 2016). Epoxyjanthitrems are the only known secondary metabolites produced by this association and to date the correlation between this compound and insect bioactivity has been confirmed for only pasture mealybug and porina. The bioactivity of the AR37 endophyte against the other three key insect pests may be due to epoxyjanthitrems or it may be due to an unknown secondary metabolite.

6 Toxicity to livestock for major cultivated grasses

For cultivated grasses, the toxicity of endophytes to livestock are of greatest importance for tall fescue and perennial ryegrass where many, but not all, of the commonly occurring *Epichloë* strains of endophyte in these grasses produce indole diterpenes and/or ergot alkaloids.

6.1 Tall fescue toxicity

The toxicity of endophyte (*Epichloë coenophiala*)-infected tall fescue to livestock has been studied most in the USA due to the magnitude of the toxicity and its widespread occurrence, leading to significant economic impact. Ergot alkaloids, primarily ergovaline, cause three toxicity syndromes: fescue foot, bovine fat necrosis and fescue toxicosis, with fescue toxicosis having the greatest economic impact (Schmidt and Osborn, 1993; Strickland *et al.*, 2009; Waller, 2009).

Table 1 The bioactive effects of secondary metabolites against invertebrate pests

Metabolite and insect	Effect	References
Lolitrem B		
Listronotis bonariensis – Argentine stem weevil larvae	Deterrent and toxic	Dymock *et al.*, 1989
Epoxyjanthitrems		
Balanococcus poae – Pasture mealybug	Reduced population	Pennell *et al.*, 2005
Wiseana cervinata – Porina	Antifeedant	Finch *et al.*, 2020
Ergot alkaloids		
Agrostis ipsilon – Black cutworm	Deterrent and toxic	Potter *et al.*, 2008; Baldauf *et al.*, 2011
Heteronychus arator – African black beetle adults	Antifeedant	Ball *et al.*, 1997b
Spodoptero frugiperda – Fall armyworm	Deterrent	Clay and Cheplick, 1989
Pratylenchus scribneri – Lesion nematode	Toxic	Bacetty *et al.*, 2009
Listronotis bonariensis – Argentine stem weevil larvae	Deterrent	Popay and Wyatt, 1995
Popilla japonica – Japanese beetle larvae	Deterrent	Patterson *et al.*, 1991
Oncopeltus fasciatus – Large milkweed bug	Toxic	Yates *et al.*, 1989
Peramine		
Listronotis bonariensis – Argentine stem weevil larvae and adults	Deterrent and toxic	Dymock *et al.*, 1989; Rowan *et al.*, 1990; Popay *et al.*, 2003
Schizaphis graminus – Aphid	Deterrent and toxic	Siegel *et al.*, 1990
Balanococcus poae – Pasture mealybug	Reduced population	Pennell *et al.*, 2005
Agrostis ipsilon – Black cutworm	Deterrent and toxic	Baldauf *et al.*, 2011
Lolines		
Costelytra zealandica – Grass grub	Deterrent	Patchett *et al.*, 2011a
Heteronychus arator – African black beetle adults and larvae	Deterrent	Barker *et al.*, 2015a
Listronotis bonariensis – Argentine stem weevil	Deterrent and toxic	Patchett *et al.*, 2008a; Popay *et al.*, 2009
Aploneura lentisci – Root aphid	Reduced population	Schmidt, 1993
Agrostis ipsilon – Black cutworm	Toxic	Baldauf *et al.*, 2011
Oncopeltus fasciatus – Large milkweed bug	Deterrent and toxic	Johnson *et al.*, 1985; Bush *et al.*, 1993
Rhopalosiphum padi – Aphid	Deterrent and toxic	Johnson *et al.*, 1985; Siegel *et al.*, 1990; Bush *et al.*, 1993

(Continued)

Table 1 (*Continued*)

Metabolite and insect	Effect	References
Adoryphorus coulonii – Red-headed cockchafer	Deterrent	Bryant *et al.*, 2010
Lepidogryllus spp. – Mottled field cricket	Deterrent	Barker *et al.*, 2015b
Schizaphis graminum – Aphid	Toxic	Johnson *et al.*, 1985; Siegel *et al.*, 1990
Teleogryllus commodus – Black field cricket	Deterrent	Barker *et al.*, 2015b
Spodoptero frugiperda – Fall armyworm	Deterrent and toxic	Riedell *et al.*, 1991
Trigonotylus caelestialium – Rice leaf bug	Resistance	Shiba and Sugawara, 2009
Ostrinia nubilalis – European corn borer larvae	Toxic	Riedell *et al.*, 1991
Wiseana cervinata – Porina	Deterrent	Popay and Lane, 2000
Popilla japonica – Japanese beetle larvae	Toxic	Patterson *et al.*, 1991
Pratylenchus scribneri – Lesion nematode	Toxic	Bacetty *et al.*, 2009

6.1.1 Fescue foot

The restriction in peripheral blood in the extremities caused by ergovaline can result in lameness, and at its worst, necrosis and sloughing of hooves and tips of ears and tails (Garner and Cornell, 1978; Waller, 2009). The lameness and loss of hooves is a significant animal welfare issue. While most prominent in cattle, it may also occur in sheep and horses. It occurs in the cold temperatures of winter, a time of the year when it is typical for blood supply to the extremities to be reduced in response to the cold.

6.1.2 Fat necrosis

Cattle that have consumed toxic endophyte-infected tall fescue can build up a mass of necrotic fat (liptomatosis) in the abdominal cavity. This mass can become so large that it restricts internal organs causing digestive problems and reduced reproductive capacity. This phenomenon, commonly termed as bovine fat necrosis (Waller, 2009), can only be determined through post-mortem examination but contributes to the overall symptoms of fescue toxicosis (Waller, 2009). It has also been reported to be associated with fescue toxicosis in goats and deer (Wolfe *et al.*, 1998; Smith *et al.*, 2004).

6.1.3 Fescue toxicosis

Fescue toxicosis is generally a failure of livestock grazing endophyte-infected tall fescue to thrive, resulting in production losses. The ill thrift is particularly evident in cattle grazing in hot humid summer conditions in the USA, where it was commonly termed summer slump or summer syndrome. Cattle (and sheep) are intolerant of heat, with elevated body temperature (hyperthermia) and increased respiration rate. Compared with animals grazing forages free of ergot alkaloids, affected animals spend less time grazing and more time in shade and cooling off in dams and forming wallows. This inability to cope with heat can be attributed to the vaso-constrictive effects of ergovaline, disrupting thermo-regulatory functions in the animal. Ergovaline is also well known for its feeding deterrence (Caradus *et al.*, 2020). Additional signs of fescue toxicosis in cattle are excessive salivation, a rough hair coat and a general unthrifty appearance. This disease occurs most commonly in late spring, summer and autumn, but can also occur in winter.

Productivity losses are evident in poor weight gains, milk production, conception rates and birth weights. In US studies that have compared endophyte treatments within the same cultivar, growth rates of steers and lambs have been 30–100% greater in non-ergot alkaloid treatments (endophyte-free and selected endophytes) than toxic endophyte (Stuedemann and Hoveland, 1998; Bouton *et al.*, 2002; Parish *et al.*, 2003a, 2003b). For tall fescue cultivar Kentucky-31, dairy cows grazing endophyte-free grass have been shown to produce 22–31% more milk than those grazing toxic endophyte (Strahan *et al.*, 1987).

For horses, fescue toxicosis causes a similar reduction in growth rate as it does in cattle (Aiken *et al.*, 1993). However, effects on reproductive performance are much greater for horses and are a significant animal welfare issue (Cross, 2009). Effects are numerous, including low pregnancy rates, increased gestation lengths, birthing difficulty (dystocia), increased foal and mare deaths, thick and retained placentas, little or no milk production (agalactia), and foals may be weak and deformed.

6.2 Ryegrass toxicity

The clinical disorder referred to as ryegrass staggers occurs when livestock graze perennial and long-term hybrid ryegrasses (*Lolium boucheanum* syn. *L. hybridum*) infected with what is typically referred to as common-toxic, standard or wild-type strains of *E. festucae* var. *lolii*. Ryegrasses with common-toxic endophytes are widely naturalised in Australia and New Zealand, where ryegrass staggers have gained so much prominence. The disorder is neuromuscular and is characterised by tremors, staggering gait, convulsions and collapse, all of which are worsened by exercise (di Menna *et al.*, 2012). It has been recorded to occur in both ruminants and monogastrics, including horses, sheep, cattle, donkeys, deer, goats, llamas, camels, alpacas and rhinoceri, in countries from all around the world. Of the tremorgenic indole-diterpenes produced *in-planta*, lolitrem B is the primary causative secondary metabolite of ryegrass staggers (Gallagher *et al.*, 1984). When livestock are removed from toxic pastures, recovery is rapid (di Menna *et al.*, 2012).

The common-toxic strains of ryegrass endophyte also produce ergot alkaloids, particularly ergovaline. Ergovaline in tall fescue causes fescue toxicosis/summer slump. A similar effect is seen in ryegrass, where the common-toxic endophyte is considered to be a significant contributor to 'summer/autumn ill thrift' (Fletcher, 1999).

Concentrations of lolitrem B and ergovaline are highly seasonally dependent, being greatest in the warmer months of the year (Watson *et al.*, 1999; Moate *et al.*, 2012; Thom *et al.*, 2013). They are closely associated with the mass of *Epichloë* mycelia, so are at greatest concentrations in the base of the pasture and in the reproductive tiller (Realini *et al.*, 2024). Concentrations become toxic to livestock from late spring as reproductive tillers rapidly elongate, and staggers can occur. In summer and autumn, pasture secondary metabolite concentrations are at their greatest and consequently most toxic, particularly in dry conditions and when closely grazed.

Along with ryegrass staggers, losses occur through reduced milk production in dairy cows and lower liveweight gain is achieved in young stock, relative to livestock grazing ryegrass that is endophyte free or infected with a selected endophyte (Fletcher, 1999; Bluett *et al.*, 2005). Sheep have faeces with higher moisture contents which leads to greater faecal soiling in the breech area (termed 'dags'), and as a consequence, a higher incidence of myiasis (flystrike). Reproductive performance in sheep can be compromised, plasma prolactin levels reduced, body temperatures and respiration rates are elevated under warm humid conditions, and there is greater mortality in sheep and cattle (Foot *et al.*, 1988; Fletcher, 1999; Fletcher *et al.*, 1999; Reed *et al.*, 2005).

6.3 Secondary metabolite thresholds to guide use in commerce

For clinical toxicity, secondary metabolite thresholds have been established through dosing experiments and/or through a series of reported cases on farms (Table 2). Results can vary between studies, so ranges rather than specific threshold values are referred to. This may reflect the influence of environment, a range in susceptibility of individual animals and/or breed differences, height of grazing and grazing compared with indoor feeding studies. The values for camels, alpaca and white rhinoceri are based on a single study or observation and indicate these species are more sensitive than cattle and sheep, at least

Table 2 For clinical toxicity, threshold concentration values of endophyte secondary metabolites produced by endophyte-infected tall fescue (ergovaline) and perennial ryegrass (ergovaline and lolitrem B) when fed to horses, cattle, sheep, camels, rhinoceri and alpaca

	Threshold concentration range in forage (ppm)	
Animal species	Ergovaline	Lolitrem B
Horse	0.30–0.50 [a]	0.8–1.2
Cattle	0.40–0.75	1.8–2.0 [b]
Sheep	0.50–0.80	1.8–2.5
Camel	nd [c]	≤1.1
White rhinoceros	nd	<1.0
Alpaca	nd	0.9

Compiled from di Menna *et al.* (1992), Tor-Agbidye *et al.* (2001), Aldrich-Markham *et al.* (2003), Fink-Gremmels (2005), Alabdouli *et al.* (2014), Reed *et al.* (2010) and Bluett *et al.* (2004).
[a]Threshold is zero for mares in late pregnancy.
[b]Threshold toxicity could be lower for lolitrem B when fed over an extended period (Duringer *et al.*, 2021).
[c]Not determined.

for lolitrem B. It is generally accepted that deer as a species are sensitive to endophyte in ryegrass, particularly Wapiti deer (*Cervus elaphus manitobensis*) (Prestidge, 1993). For endophyte-infected perennial ryegrass, there is no evidence that the production of two mammalian toxins (ergovaline and lolitrem B) interact and worsen toxicity, at least for sheep (Finch *et al.*, 2018).

While not absolute, the thresholds of Table 2 have been deployed in commerce. They have been particularly useful in determining if conserved endophytic hay and straw can be 'safely' used to feed livestock. This has been applied in Europe to hay and in Oregon, USA, for seed production straw exported to Asia (Fink-Gremmels, 2005; Craig, 2009).

While much focus has been put on thresholds for clinical toxicity, research has shown that in general the subclinical impacts increase in a linear manner as secondary metabolite concentrations increase, with no threshold (Crawford Jr. *et al.*, 1989; Schmidt and Osborn, 1993; Fletcher *et al.*, 1999; Layton *et al.*, 2004). For each 10 percentage units increase in endophyte infection in pastures, 45 g/day/animal in cattle and up to 10 g/day/animal in sheep of potential liveweight gain are lost in the USA for tall fescue and in New Zealand for ryegrass, respectively. Economic thresholds have not been defined but can be calculated on animal production losses and costs of upgrading to a safe pasture.

7 Tall fescue endophyte toxicity across the globe

The toxicity of endophyte-infected tall fescue has been reported across the globe, with its greatest impact being in North and South America (Table 3) (Young *et al.*, 2013).

Table 3 *Epichloë* endophyte toxicities reported for livestock on farms globally for tall fescue

Country	Toxicity syndrome	Impacts	Livestock affected	Annual costs	Comments	Reference(s)
USA	Fescue toxicosis, fescue foot, bovine fat necrosis	Lower liveweight gain and milk production, loss of extremities, heat stress, reproductive losses	Beef cattle, dairy cows, horses	US$1 B	Fescue toxicosis has by far the greatest economic impact	Aiken and Strickland (2013), Young *et al.* (2013), Ball *et al.* (2019)
Chile, Argentina and Uruguay	Fescue toxicosis, fescue foot	Lower weight gain, excitement, hyperesthesia, elevated body temperature, lameness, loss of distal parts of limbs	Dairy cows, cattle	US$54 M in Argentina	Economic impact only assessed for Argentina	Sepulveda *et al.* (1996), De Battista *et al.* (1997), De Battista (2005), Petigrosso *et al.* (2013)
Australia and New Zealand	Fescue toxicosis, fescue foot	Heat stress, weight loss, lameness, loss of distal parts of limbs	Cattle	-[1]		Cunningham (1948), Pulsford (1950), Easton *et al.* (1994), Hume and Sewell (2014), Reed (2014)
South Africa	Fescue foot	Lameness, necrosis of the lower tail	Cattle	-[1]	Along with fescue foot, suspected case of fescue toxicosis in dairy cows; may be exacerbated by ergot (*Claviceps purpurea*) consumption	Botha *et al.* (2004)
Japan	Fescue toxicosis, fescue foot	Weight loss, lameness, abortions	Cattle	-[1]	In cattle-fed straw imported from USA	Tor-Agbidye *et al.* (2001), Craig (2009), Young III and Silberstein (2012), Craig *et al.* (2014).
France	Fescue toxicosis, fescue foot	Reduced milk production, necrosis of tails and feet	Cattle	-[1]	Cases reported in France, with unconfirmed cases in Spain	Raynal (1991), Le Bars and Le Bars (1996), Bony and Delatour (2001), Bony *et al.* (2001)

[1]No estimates made of losses but are now likely to be very low given the very low prevalence of toxic tall fescues in pastures or conserved forage, and the use of selected animal-safe endophyte-infected cultivars in some markets.

7.1 USA

Tall fescue endophyte toxicity is most well-studied, quantified and of greatest impact in the USA. There are approximately 14 million ha of tall fescue in the USA, predominately in the east where it is referred to as the 'fescue belt' due to its dominance in grasslands, turf and roadsides (Young *et al.*, 2013; Ball *et al.*, 2019). Endophyte is needed for tall fescue pasture persistence in the southern half of the fescue belt and it is mainly livestock-toxic endemic endophytes from the widely sown cultivar Kentucky 31 and similar cultivars developed from it.

Due to the widespread occurrence of toxic fescue, and the large effects on livestock, estimates made over a decade ago place the total costs of toxic endophyte to the USA forage-based livestock industry at over US$1 billion/year (Aiken and Strickland, 2013). It is not surprising that tall fescue endophyte toxicity has been considered the most important grass-induced toxicity in the country (Cheeke, 1995). Much of this loss occurs in the beef cattle industry, through reduced calf numbers and growth rates, with fescue toxicosis having the greatest economic impact of the three forms of toxicity (Table 3). It is generally accepted that a potential liveweight gain of 45 g/day (0.1 lb/day) is lost with each 10% increase in toxic endophyte infection in the pasture. Additional costs to the cattle industry occur through cases of fescue foot and fat necrosis. This cost has been difficult to determine, due to cattle being affected by long-term exposure, with cases not well documented or unreported (Aiken and Strickland, 2013).

In the equine sector, where ~700 000 horses are raised on endophyte-infected tall fescue, the reduced reproductive performance of mares is by far the greatest economic impact and a major welfare concern (Cross, 2009). Milk production in dairy cows is also impacted, with one study reporting a 20% loss in production for cows grazing toxic endophyte-infected tall fescue compared with endophyte-free tall fescue (Strahan *et al.*, 1987).

7.2 South America

Clinical and sub-clinical toxicity of tall fescue endophytes has also been documented in southern South America. Endophyte infection is widespread in tall fescue pastures and seed lots in Chile, Argentina and Uruguay. Fescue toxicosis and fescue foot have been reported for cattle and have been of greatest concern in Argentina and Uruguay due to large areas sown and/or high reliance on tall fescue grasslands. In Argentina, estimated annual losses in 1995 were US$54 million due to lower weight gains in cattle.

Old pastures in South America are of the greatest concern, often having been sown with cultivar Kentucky 31 sourced from the USA. The frequency of toxicity has varied over the decades and was greatest in the 1980s. Compared to the USA, the lower prevalence and impact of the tall fescue endophyte on livestock production systems has been attributed to several factors. These include greater farmer awareness of the problem, greater contents of legumes that dilute the intake of endophyte toxins, a faster rate of pasture renewal, regulation of endophyte levels in certified seed in Argentina and the use of locally bred cultivars that are now predominately endophyte free.

7.3 Beyond the Americas

Outside of the Americas, reports of endophyte toxicity are few and now of minor significance for livestock production systems.

In Australia and New Zealand, the widely naturalised populations of tall fescue are highly infected with toxic *Epichloë* in the temperate 'high rainfall' pasture zones. Its toxicity was renowned, which in some cases when it invaded pastures, earnt it the reputation as a weed rather than a valuable pasture grass. At least in New Zealand, local naturalised 'roadside' tall fescue has been recorded with at least double the ergovaline concentrations of Kentucky 31 infected with toxic endophyte, therefore increasing the severity of the toxicosis (Easton *et al.*, 1994; Christensen *et al.*, 1998). As in South America, some paddocks in Australia had been sown in the past to cultivar Kentucky 31 which would also have contributed to toxicity.

Like Australia/New Zealand, tall fescue in South Africa is widely naturalised; however its endophyte content has not been assessed. The single case reported of endophyte toxicity was for a sown pasture and the source of the seed was not described.

Cases of endophyte toxicity in cattle in Japan have been the result of importing large volumes of endophytic seed production straw from Oregon, USA. Straw has also been sent to other southeast Asian countries but no endophyte toxicities have been reported. Through establishing secondary metabolite toxicity thresholds and secondary metabolite testing of tall fescue (and ryegrass) straw that is destined for export, the occurrence of clinical cases of toxicity in Japan has greatly diminished.

Despite being a centre of origin for tall fescue and the source of *New World* endophyte toxicity, reports of tall fescue endophyte toxicity in commerce in Europe are rare. Various studies have shown ergovaline to be at or above toxic thresholds concentrations on occasions (Bony and Delatour, 2001; Bony *et al.*, 2001; Vázquez-de-Aldana *et al.*, 2001; Oliveira *et al.*, 2003; Vázquez-de-Aldana *et al.*, 2003; Zabalgogeazcoa and Bony, 2005), but even in experimental conditions, it has been difficult at times to demonstrate clear effects (Emile *et al.*, 2000). Endophyte infection in cultivars is highly variable (Latch *et al.*, 1987; Vázquez-de-Aldana *et al.*, 2001). The species diversity of old pastures, and the lack of endophyte in modern cultivars, are likely to contribute to the low toxicity. In the UK, cultivars on the recommended National Variety List must have ≤5% endophyte, and seed from New Zealand, Australia, South America and the USA is routinely tested for endophyte (Anon., 2023a).

8 Ryegrass endophyte toxicity across the globe

Reports of toxicity of endophyte-infected perennial ryegrass (Table 4) occur in the same regions of the world as do tall fescue endophyte toxicities (Table 3). However, as these grasses are best adapted to different climatic zones, and endophyte infection is variable across the world, the prevalence of toxicity is not equivalent for both species in all regions (Young *et al.*, 2013).

8.1 New Zealand and Australia

Globally, the impacts of perennial ryegrass endophyte toxicity have been greatest in New Zealand and the 'high rainfall' (i.e. >650 mm/year) zone of south-eastern Australia (Reed *et al.*, 2005; Johnson *et al.*, 2013). Perennial ryegrass is well adapted to these climates and soils, being a commonly sown pasture grass and widely naturalised in the landscape. Old ryegrass pastures are infected at high frequencies with toxic *Epichloë*, and locally-adapted cultivars were generally infected with toxic endophyte prior to the use, some 30 years ago, of selected endophytes (Reed *et al.*, 2000; Reed, 2014; Caradus *et al.*, 2023). In many districts, endophyte-infected seed needs to be sown to ensure pasture productivity and persistence.

In these countries, a range of toxic endophyte effects have been identified and well-characterised. The first pivotal study was in New Zealand, where the link was made between ryegrass staggers in sheep and endophyte-infected ryegrass (Fletcher and Harvey, 1981). Ryegrass staggers had been a significant animal disorder that researchers had been investigating for many decades and had failed to identify the cause (Hume *et al.*, 2020). In further research, it became abundantly clear that the toxic endophytes were responsible for a range of negative impacts on livestock, beyond clinical cases of ryegrass staggers. Some of these effects mirrored those of the ergovaline-driven fescue toxicosis seen in the USA for tall fescue, so ergovaline became a key component of the research. In New Zealand, toxic endophyte is considered to be a significant contributor to 'summer/autumn ill thrift', as it is in Australia, where the combined impacts are termed 'perennial ryegrass toxicosis (PRGT)'.

As described above, reduced reproductive performance, dags, flystrike, heat stress and deaths are greater for sheep grazing toxic endophyte ryegrass pastures than ryegrass free of endophyte or infected with a selected endophyte. The major economic impact is subclinical effects on production, with reduced weight gain in livestock (~17% for sheep) and milk production

Table 4 *Epichloë* endophyte toxicities reported for livestock on farms globally for perennial ryegrass

Country	Toxicity syndrome	Impacts	Livestock affected	Annual costs	Comments	Reference(s)
USA	Ryegrass staggers	Staggers, deaths	Cattle, sheep	-[1]		Galey *et al.* (1991), Galey *et al.* (1993), Tor-Agbidye *et al.* (2001), Fisher *et al.* (2004)
Chile, Argentina	Ryegrass staggers	Staggers	Dairy cows, cattle	-[1]	In addition to staggers, indirect evidence for grazing avoidance, reduction in forage intake and drop in milk production in dairy cows	Odriozola *et al.* (1993), Butendieck *et al.* (1994), De Battista *et al.* (1997)
Australia and New Zealand	Ryegrass staggers, sub-clinical toxicoses	Staggers, reduced milk and meat production, flystrike, faecal soiling, deaths, heat stress, reproductive losses	Dairy cows, cattle, sheep, horses, deer, alpaca	Australia: AU$100 M. NZ: 16–18% $ losses for sheep farms; 9% less dairy cow milk solids production	As well as ryegrass staggers in farmed animals, white rhinoceri in a New Zealand zoo have suffered staggers when fed hay	Valentine *et al.* (1993), Fletcher (1999), Fletcher *et al.* (1999), Watson *et al.* (1999), Lean (2001), Bluett *et al.* (2004), Bluett *et al.* (2005), Reed *et al.* (2005), Reed *et al.* (2010), Reed *et al.* (2011)
South Africa	Ryegrass staggers	Staggers	Cattle, sheep	-[1]		Kellerman *et al.* (1993)
Japan	Ryegrass staggers	Staggers	Cattle, horses	-[1]		Miyazaki *et al.* (2001), Miyazaki (2003), Miyazaki *et al.* (2004), Craig (2009), Young III and Silberstein (2012), Craig *et al.* (2014)
UK and Europe	Ryegrass staggers	Staggers	Cattle, sheep, horses, alpaca	-[1]		Lewis (1997), Holmes *et al.* (1999), Bony and Delatour (2001), Zabalgogeazcoa and Bony (2005), di Menna *et al.* (2012)

[1]No estimates made of losses but are likely to be very low given the generally low prevalence of toxic ryegrass in pastures or conserved forage, and the use of selected animal-safe endophyte-infected cultivars in some markets.

in dairy cows (9%). However, in Australia when PRGT is at national epidemic levels, deaths of sheep in particular can be significant, exceeding 100,000 in some years.

8.2 The Americas, South Africa and Japan

Ryegrass staggers are the only ryegrass endophyte toxicity reported in the Americas, South Africa and Japan, and is infrequent.

In the USA, most pasture ryegrass cultivars are endophyte free and despite being adapted to large areas in the north, perennial ryegrass use is low, particularly relative to that of tall fescue (Young *et al.*, 2013). Coastal northern California is the only region where ryegrass staggers have been reported in livestock grazing pastures, on several properties and years. One case has been reported for sheep on pasture in Oregon. High volumes of seed production straw from Oregon of endophyte-infected turf ryegrass cultivars do represent a significant risk of toxicoses when fed to livestock within the USA or when exported (Alabdouli *et al.*, 2014; Craig *et al.*, 2014).

Of the countries in South America, only Chile and Argentina have significant areas of perennial ryegrass. In these two countries, endophyte infection is common but variable in frequency of infection in pastures and seed lots. Occurrence of ryegrass staggers is infrequent and while not proven, subclinical toxicity is likely. In South Africa just a single case of staggers has been reported on a perennial ryegrass pasture. In both South America and South Africa, some cases have been linked to seeds imported from New Zealand.

Ryegrass staggers have been reported in Japan for straw imported from the seed production fields of Oregon, USA, where the sensitivity of Japanese Black cattle may be greater than other breeds of cattle (Miyazaki *et al.*, 2001). As is the case for tall fescue straw, the occurrence of clinical cases of toxicity diminished greatly once secondary metabolite toxicity thresholds were determined and secondary metabolite testing of exported ryegrass straw commenced.

8.3 UK and Europe

Endophyte has been detected in wild perennial ryegrass populations from many but not all countries in UK/Europe and at varying rates (1–100%) (Lewis and Clements, 1986; Latch *et al.*, 1987; Ribeiro *et al.*, 1996; Lewis *et al.*, 1997; Leyronas and Raynal, 2001; Jensen and Roulund, 2004; König *et al.*, 2017). While ryegrass staggers are not considered a major problem, it has been widely reported over many decades to occur in a number of countries in UK/Europe (di Menna *et al.*, 2012). This is in keeping with UK/European endophyte-infected ryegrass being the source of endophyte toxicity in the *New World*. As occurs for ergovaline in endophytic tall fescue, lolitrem B in various European studies has shown at times to be at or above toxic threshold concentrations (Bony and Delatour, 2001; Bony *et al.*, 2001).

For the UK, as an example, a reminder about ryegrass staggers was published for veterinarians in 1995 (Pritchard and Lewis, 1995). This listed 11 counties in England and southern Wales, where ryegrass staggers had been reported, and described a surge of cases in drought years. This is despite ryegrass cultivars being largely free of endophyte and the introduction of regulations for endophyte in the UK National List agronomic testing. These regulations limit the endophyte content of new ryegrass cultivars to a maximum of 5% (Anon, 2023a). In addition, there is specific endophyte testing for cultivars originating from the Americas, Australia and New Zealand. Ryegrass is however commonly endophyte-infected in old pastures, and while being only a minor component of the sward, outbreaks of ryegrass staggers occur in drought years (Lewis and Clements, 1986; Pritchard and Lewis, 1995).

In France, endophyte frequency in pastures is highly correlated with the level of summer drought, with highly infected populations being mostly in the Mediterranean regions (Lewis *et al.*, 1997; Leyronas and Raynal, 2001). Ryegrass staggers have been reported in The Netherlands, Germany, Belgium and France (see reviews by Bony and Delatour (2001), di Menna *et al.* (2012), Lewis (1997) and Zabalgogeazcoa and Bony (2005)). As seen in the UK, cases were associated with drought years, feeding of hay and the use of New Zealand-sourced seed. Highly endophyte-infected straw from turf cultivars was involved in many cases in The Netherlands and France.

The relatively low level of reported toxicity is influenced by the use of cultivars that are largely free of endophyte, the climatic conditions that influence secondary metabolite production being mostly moderate, generally low-to-moderate rates of infection in old pastures and ryegrass often being a minor component in botanically diverse permanent pastures. In addition, the *Epichloë* strain diversity that has been utilised for selected endophytes in New Zealand and Australia, with their low mammalian-toxic secondary metabolite profiles, have been derived from these European pastures (see Oliveira *et al.*, (2003); Vikuk *et al.*, (2019) for examples of lolitrem B-free or ergovaline-free endophyte-infected ryegrasses). The factors that determine toxicity might change under the rapidly changing climate associated with global warming, resulting in greater risks of toxicity in the future (McCulley *et al.*, 2014; König *et al.*, 2017; Vikuk *et al.*, 2019).

9 Livestock toxicity for other *Epichloë*-infected grasses

Epichloë-driven toxicity in livestock is not confined to ryegrass and tall fescue, with reports of toxicity for grass-endophyte associations native to southern Africa, Australia, China/Mongolia, USA and Argentina (Table 5) (Hume *et al.*,

Table 5 Intoxication and/or narcosis of animals consuming asexual *Epichloë*-infected grasses other than ryegrass and tall fescue

Local name for toxicity/grass or English translation (country)	Grass – *Epichloë*	Livestock affected	Toxicosis	Endophyte chemistry of relevance to livestock toxicity
Intoxication				
Dronkgras (South Africa and Lesotho)[1]	*Melica decumbens* – *E. melicicola*	Cattle, horses, donkeys, sheep	Intoxicated, narcotic	Indole diterpenes
Staggers-like (Australia)	*Echingopogon ovatus* – *E. aotearoae*, *E. australiensis*	Cattle, sheep, goats, turkeys	Staggers, weight loss, deaths	Indole diterpenes[2]
Huecú or tembaldera (Argentina)[3]	*Festuca* and *Poa* spp. – *Epichloë tembladerae*	Cattle, horses, sheep	Drunk, tremble	Indole-diterpenes. Ergot alkaloids
Narcosis				
Sleepy grass (SW USA)[4]	*Achnatherum robustum* – *E. funkii* and another *Epichloë* species	Cattle, horses, sheep	Narcosis, elevated body temperature	Ergot alkaloids (including lysergic acid amide). Indole diterpenes
Intoxication and narcosis				
Drunken horse grass (NW China and Mongolia)[5]	*Achnatherum inebrians* – *E. gansuensis*, *E. inebrians*	Cattle, horses, sheep, donkeys, mules, goats, rabbits	Intoxication, narcosis, panting, loss of tails and hooves	Indole diterpenes. Ergot alkaloids (including lysergic acid amide). Stipatoxin

[1] Gibbs Russell and Ellis (1982), Miles *et al.* (1995a) and Hoare (2014).
[2] Only *E. australiensis* has been shown to produce indole diterpenes (Seddon and Carne, 1926; Miles *et al.*, 1998; Moon *et al.*, 2002).
[3] Secondary metabolite profile varies with grass species. Pomilio *et al.* (1989), Miles *et al.* (1995b), Cabral *et al.* (1999), Iannone *et al.* (2011) and Iannone *et al.* (2012).
[4] Petroski *et al.* (1992), Jones *et al.* (2000), Faeth *et al.* (2006), Moon *et al.* (2007) and Shymanovich *et al.* (2015).
[5] *E. gansuensis* produces indole diterpenes and *E. inebrians* produces ergot alkaloids. Dang *et al.* (1992), Miles *et al.* (1996), Schardl *et al.* (2013a), Chen *et al.* (2015) and Liang *et al.* (2017).

Published by Burleigh Dodds Science Publishing Limited, 2025.

2016). Despite a diversity of grass host and *Epichloë* species, and irrespective of location in the world, several of these grass-endophyte associations cause symptoms of intoxication (drunken behaviour). Symptoms are similar to the staggers that can occur in *Epichloë*-infected ryegrass, and isolated tremorgens are implicated. In addition, a narcosis effect is reported for some grass-endophyte associations, and these are associated with the presence of ergot alkaloids. For some associations, both intoxication and narcosis are reported. Variable symptoms or occurrence of toxicity may be a result of more than one *Epichloë* species being involved or strain variation within a species.

For dronkgras in Africa and drunken horse grass in Asia, overgrazing of grasslands has been cited as a reason, amongst others, for their increased prevalence in the landscape. For drunken horse grass, this has resulted in large areas (>0.5 million hectares) that are dominated by toxic, endophyte-infected plants, being up to 100% cover in some regions. However, toxicoses are infrequent, as livestock in these areas generally show a strong avoidance of these infected grasses, and this is also the case for sleepy grass in the USA. However, the grazing avoidance results in economic impacts on the overall livestock production system, particularly so in China and to some extent in the USA.

The livestock toxicities reported to date occur only for grasses that are infected with asexual *Epichloë*, with none reported for sexual *Epichloë*. This may be somewhat surprising, as sexual *Epichloë* produce the same classes of secondary metabolites which include the mammalian toxic indole diterpenes and ergot alkaloids (Leuchtmann *et al.*, 2000; Schardl *et al.*, 2013b). However, asexual *Epichloë*–grass associations in general have more secondary metabolites and in higher concentrations (Leuchtmann *et al.*, 2000; Schardl *et al.*, 2012). This is consistent with the defence mutualism theory, where asexual *Epichloë* need to provide as many protective properties to the host as possible (e.g. via secondary metabolite production), as the endophyte is totally reliant on the survival of the host and spread via the host seed. Sexual endophytes can achieve spread via horizontal transmission, with less secondary metabolite production lowering any metabolic cost for the host.

As an example, for a range of sexual and asexual *Epichloë*-infected grasses in Spanish dehesa grasslands, ergovaline was detected in four *Festuca* species and *Holcus lanatus* and was absent in another four grass species (Vázquez-de-Aldana *et al.*, 2003). The fine fescues *F. ovina* and *F. rubra* infected with sexual *E. festucae* had a five-fold lower concentration of ergovaline than tall fescue infected with asexual *E. coenophiala*. There is an absence of clinical signs of ergot alkaloid toxicity in cattle grazing these grasslands. This is despite the high potential for toxicity, as *F. rubra* is a common species and highly infected with endophyte. The lack of toxicity is attributed to the high plant species diversity of these grasslands, diluting any mammalian toxic endophyte secondary metabolites in the animal's diet (Zabalgogeazcoa *et al.*, 1999; Vázquez-de-Aldana *et al.*, 2003). Grazing avoidance due to deterrence caused by ergovaline may also play a role (Bazely *et al.*, 1997).

10 Reducing toxicity to livestock

In regions of the world where toxicity of *Epichloë* to livestock is a significant concern, several options exist for farmers to minimise or eliminate adverse effects on livestock for toxic tall fescue and ryegrass pastures. These have been well reviewed, mostly for tall fescue endophyte toxicity, and include the novel use of selected endophyte strains for tall fescue and ryegrass (Ball *et al.*, 1993; Roberts and Andrae, 2004; Aiken and Strickland, 2013; Young *et al.*, 2013; Caradus *et al.*, 2021a; Poudel *et al.*, 2023; Bastías *et al.*, 2024). The options range from total avoidance of *Epichloë* toxins to dilution of toxins in the diet, pasture/grazing management to limit toxin intake and the use of selected strains of *Epichloë*. Each has advantages and disadvantages. In Australia, New Zealand and USA, where endophyte underpins pasture productivity, trade-offs may occur when maximising the pasture performance benefits of *Epichloë* while minimising or eliminating the detrimental effects on livestock.

10.1 Forages free of Epichloë *mammalian toxins*

As the production of mammalian toxins is a trait of the *Epichloë* endophyte, endophyte-free cultivars of tall fescue and ryegrass can be utilised on farms with confidence that livestock will thrive. These cultivars can be used in regions where endophyte is only occasionally needed to boost pasture performance, although this may be hard to predict, particularly in the face of climate change and the occurrence of new pasture pests through biosecurity breaches. In addition, endophyte-free cultivars may be utilised in farming systems where pastures are planned to be short-term e.g. in a mixed livestock/cropping systems and/or where practices such as irrigation may lessen stresses on pastures.

There is a range of non-*Epichloë* grass species and other forage species (legumes, herbs and brassicas) that can be utilised as alternative forages or as conserved feed, particularly during periods of the year when *Epichloë*-infected pastures are most toxic. Each of these species has their own set of limitations which may include restricted area of adaption, risk of other livestock disorders and difficulties in establishment and grazing management, all of which limit their application and widespread adoption.

In Argentina, Iannone *et al.* (2011) also highlighted the prospects of using grass species that are infected with asexual *Epichloë* where no toxic effects for cattle are known. In this region, some of the endophytic *Bromus* and *Poa* species are known for their forage quality. In addition, some strains of *Epichloë* infecting tall fescue and ryegrass are livestock safe and these are discussed in the section 'Selected Endophytes'.

10.2 Dilution of endophyte toxins

Various species can be used in mixed species pastures to dilute the concentration of *Epichloë* mammalian toxins in the forage that is consumed by livestock while grazing. This may include non-endophytic grasses, legumes and herbs. Of these, the use of clovers (*Trifolium* spp.) in mixed pastures has received the greatest attention, as they are highly valued for their ability to boost animal performance through high forage quality and for the addition of atmospheric nitrogen (N) to the soil through N-fixation in clover roots.

In the USA, the addition of perennial clovers (white and red; *T. repens and T. pratense*, respectively) to toxic endophyte tall fescue pastures has been widely promoted as a tool to dilute and so alleviate the toxicosis (Roberts and Andrae, 2004). In southern South America, the widespread use of legumes in pastures is considered one of the factors that has reduced the economic impact of tall fescue endophyte toxicity in the region (De Battista *et al.*, 1997). In Australia and New Zealand, ryegrass swards commonly have subterranean (*T. subterraneum*) and/or white clovers, with limited use of red clover. Cosgrove *et al.* (1996) found that under severe ryegrass-endophyte challenge, white clover reduced the severity of ryegrass staggers in cattle in central New Zealand and advocated for high clover contents in swards, along with other management factors. However, it can be difficult to maintain sufficiently high contents of white clover in endophytic ryegrass swards, as endophyte-infected perennial ryegrass is known to suppress clovers (Foot *et al.*, 1988; Sutherland *et al.*, 1999). In addition, the dry hot conditions in summer/autumn that increase the risks of toxicity are unfavourable for white clover, and as an annual species, subterranean clover is not present. Recently, Poudel *et al.* (2023) have postulated that condensed tannins produced by some legumes could be usefully deployed to reduce tall fescue toxicity.

Another form of dilution is to breed for low secondary metabolite producing grass–endophyte associations. This has been progressed by selecting lower ergovaline concentrations within mammalian-toxic endophyte-infected 'Jesup' tall fescue. Selection proved highly effective with high heritability (Adcock *et al.*, 1997). However, tall fescue persistence of the low ergovaline selection was reduced relative to unselected mammalian-toxic endophyte-infected tall fescue, and weight gains of lambs were also lower than endophyte free (Bouton, 2000; Bouton and Hopkins, 2003).

10.3 On-farm management

Grazing management has received the greatest attention of the various strategies that farmers can deploy to lessen the concentrations of secondary metabolites in the diets of livestock. For ryegrass in New Zealand, rapid rotational grazing of sheep, avoiding grazing to low residuals, can be highly effective in reducing ryegrass staggers and most likely results in less subclinical toxicity. Under this management, livestock primarily consume leaf blades in the upper portion of the sward, the horizon in the pasture where toxins are of lowest concentration. It is likely that this is why toxicity on-farm is less in dairy cows than that seen in sheep (Prestidge, 1993). The same principle applies to tall fescue.

The toxicity profile of the sward changes in late spring as rapid elongation of reproductive tillers elevates seed heads, that have high concentrations of alkaloids, into the grazing horizon, therefore increasing the risk of toxicity (Rottinghaus *et al.*, 1991; Ball *et al.*, 1997a; Roberts and Andrae, 2004). This is particularly so if seedheads are selectively grazed by livestock (Aiken and Strickland, 2013). This risk can be reduced by mowing ('topping') reproductive tillers, or close grazing or silage making in the very early stage of elongation, being aware that lolitrem B and ergovaline are present in silage (and hay) and this may take some time to decline to safe levels or require ammoniation treatment (Roberts and Andrae, 2004; Fink-Gremmels, 2005; Fletcher, 2005). Research in the USA has shown that tall fescue seedheads can be suppressed chemically, and severity of toxicity in cattle reduced, through application of plant growth regulators or herbicides (Roberts and Andrae, 2004; Aiken and Strickland, 2013). In addition, these chemicals can improve forage quality, such as digestibility, but they also pose risks to tall fescue and non-toxic companion species by hindering growth leading to decreased forage availability.

10.4 Selected endophytes

In nature, it is common to find a dominant endophyte strain within an *Epichloë* species. This is not surprising when certain secondary metabolite profiles provide a selective advantage to the grass–endophyte association (Schardl *et al.*, 2012). However, where researchers have collected many endophytes, in particular, from asexual *Epichloë* species, considerable genetic and secondary metabolite diversity has been found (van Zijll de Jong *et al.*, 2008; Ekanayake *et al.*, 2012; Takach and Young, 2014; Kaur *et al.*, 2015). In many situations, strains have been discovered that lack some of the metabolites that are problematic for animal health but express metabolites which provide insect pest tolerance. These secondary metabolite profiles are confirmed by extensive chemical analysis of grass–endophyte associations, along with gene studies of

Epichloë to determine the absence/presence and/or mutation of genes in the biosynthetic pathways (Young *et al.*, 2013).

This diversity has been capitalised on by selecting *Epichloë* strains with favourable secondary metabolite profiles and transferring them, usually by artificial inoculation in the laboratory, to elite grass cultivars to form novel host grass–endophyte associations for use on farms (Johnson *et al.*, 2013; Caradus and Johnson, 2020). This has been critical for Australia and New Zealand for perennial ryegrass, and the USA for tall fescue. In these countries, the widely naturalised populations and locally adapted cultivars were highly infected with mammalian-toxic *Epichloë* strains, with no useful strain variation identified (Simpson *et al.*, 2012; Reed, 2014; Young *et al.*, 2014; Ball *et al.*, 2019; Caradus *et al.*, 2023).

Selected endophytes are the first patented technology to be placed in grass cultivars (Bouton and Hopkins, 2003). The first release of a selected endophyte was in the early 1990s for use in perennial ryegrass (Johnson *et al.*, 2013). Since then, a number of selected strains have been deployed in tall fescue, several ryegrass species, Festulolium and meadow fescue (Table 6).

Some selected strains from tall fescue and ryegrass lack the mammalian toxic secondary metabolites ergovaline, lolitrem B and epoxyjanthitrems. While this ensures that the endophytes are benign for livestock, it does lead to reduced pest protection of the host grass, the possibility of overgrazing by livestock, and so poorer agronomic performance relative to cultivars infected with naturalised highly toxic endophytes. Relative to endophyte-free counterparts, selected endophytes improve the agronomic performance of cultivars (Caradus *et al.*, 2021a).

An approach to improve the agronomic efficacy of novel grass–endophyte associations for tall fescue and ryegrass is the utilisation of strains that retain the expression of lolitrem B and/or ergovaline, as these metabolites provide some level of pest protection (Caradus and Johnson, 2020). To reduce the risk of toxicity to livestock, these metabolites are at medium-to-low levels, as achieved through a combination of using certain endophyte strains and/or selecting grass–endophyte combinations for reduced metabolite concentrations as their expression is under strong plant genetic control (Easton, 2007; Faville *et al.*, 2015). This strategy does run the risk of livestock toxicity in some seasons or in adverse environments, which is predicted to occur more frequently due to climate change (Caradus and Johnson, 2020).

The discovery of strain AR37 (*Epichloë* sp. LpTG-3) for use in ryegrasses has illustrated that naturally occurring strains exist that can exceed the pest protection imparted by the common mammalian toxic strains and yet have reduced toxicity to livestock (Caradus and Johnson, 2020; Caradus *et al.*, 2021b). In New Zealand, the use of this strain has led to significant environmental and economic benefits (Caradus, 2023a).

10.4.1 Interspecies transfer

To improve the pest protection profile of ryegrass without compromising livestock health and performance, non-ergovaline-producing endophytes have been transferred from tall fescue and meadow fescue to enable loline expression in perennial ryegrass and Italian ryegrass (Matsukura *et al.*, 2012; Johnson *et al.*, 2019). While this interspecies transfer of *Epichloë* has been achievable, commercial application has been hindered by poor loline expression, sub-optimal transmission to seed and longevity in stored seed, highlighting the strong host specificity of *Epichloë* endophytes (Johnson *et al.*, 2019; Caradus and Johnson, 2020).

The host plant of novel endophyte–grass associations can have a major impact on the secondary metabolites produced and the biological effects induced. For example, the selected endophyte AR542 (*E. coenophiala*) combined with Mediterranean-type tall fescue, a tall fescue morphotype genetically distinct from Continental and rhizomatous types (Hand *et al.*, 2010) induced a previously unknown disease, equine fescue oedema. The same endophyte combined with Continental-type tall fescue did not (Munday *et al.*, 2017). To add to the complexity, equine fescue oedema was only observed in horses, with sheep, cows and rabbits unaffected. Although a loline analogue was initially implicated (Bourke *et al.*, 2009) this was later disproved, and the cause of equine fescue oedema is still unknown (Finch et al 2017).

10.4.2 Persistence

The persistence of elite cultivars with selected endophytes, in comparison to tall fescue and ryegrass with their mammalian toxic endophytes, has been questioned. This is particularly so for the range of selected endophytes in perennial ryegrass, where persistence is correlated with which secondary metabolites are produced, and in some cases the concentration of those metabolites, as this largely determines the range and effectiveness of insect pest control (Tables 1 and 6).

Overgrazing of cultivars with selected endophytes, due to no or reduced expression of secondary metabolites associated with feed deterrence, can lead to reduced persistence and can be a concern in stressful conditions in the warmer months of the year (Waller, 2009; Fletcher, 2012; Aiken and Strickland, 2013; Young *et al.*, 2013; Ball *et al.*, 2019). Farmers fail to recognise overgrazing is occurring, particularly so when previous farmer experience is with the mammalian toxic wild-type endophyte-infected tall fescue or ryegrass which can self-regulate grazing intensity. The ergovaline produced by wild-type endophytes has a strong deterrent effect on livestock when they graze close to the base of the pasture where concentrations are at their highest (Caradus

Table 6 Major end pathway secondary metabolite profiles for selected endophytes used on farms since the early 1990s for cultivars of tall fescue, ryegrasses, Festulolium and meadow fescue, in comparison to the highly mammalian toxic strains in tall fescue and perennial ryegrass. For further details see (Caradus and Johnson, 2019; Caradus and Johnson, 2020; Eady, 2021; Caradus and Hume, 2023)

Major end pathway secondary metabolites[1]					
Peramine	Ergovaline	Lolitrem B	Epoxyjanthitrems	Lolines	Strain name[2]
Original host: tall fescue[3]					
+	+	–	–	+	Toxic
+	–	–	–	+	AR542 (MaxQ, MaxP), AR584 (MaxQ II, MaxP), Strain 4 (ArkPlus, ArkShield), E647 (Protek)
+	+ (low)	–	–	+	BE9301A (E34)
Original host: perennial ryegrass[4]					
+	+	+	–	–	Common toxic, wild-type, standard
+	+	–	–	–	AR5 (Endosafe, Endo5), AR6 (Endosafe)
+ (low)	+ (low)	+ (low)	–	–	nea2 (NEA)[5]
+	+	–	–	–	nea3, nea6[5]
+	–	–	–	–	AR1, Edge
–	–	–	+	–	AR37, CM142, NEA12 and RGT18
Original host: meadow fescue[6]					
–	–	–	–	+	U2, Happe, AR1017 (MaxR)

[1]'+' = present, '–' = absent.
[2]Strain name, with tradename given in parentheses if different to strain name.
[3]Utilised in Continental-type tall fescue.
[4]Utilised in perennial, long and short-term hybrid, and Italian ryegrasses (*L. perenne*, *L. boucheanum* syn. *L. hybridum* and *L. multiflorum*, respectively).
[5]May be sold in mixtures as NEA2 (nea2 + nea6) and NEA4 (nea2 + nea3).
[6]Utilised in perennial ryegrass, Festulolium or meadow fescue.

et al., 2020). Selected endophytes either lack or have low concentrations of ergovaline (Caradus and Johnson, 2020).

10.4.3 Adoption

The development and use in commerce of selected endophytes is not surprising, for as biological control agents (BCAs), they are cost-effective, environmentally sound, enduring and highly effective at reducing biotic stresses, along with abiotic stresses in some cases. They are considered ideal BCAs, due to fungal plasticity, effective mechanism of action, secondary metabolite and genetic diversity, propagation via seed and obligate nature (Card *et al.*, 2016; Bastías *et al.*, 2024).

On-farm, there are no special techniques, equipment or new management procedures required, as the selected endophytes are already incorporated into the seed of elite grass cultivars by the plant breeding/seed industry. To minimise reinvasion of toxic endophytes, farmers need greater adherence to best management preparation of seedbeds and take simple steps to reduce reintroduction of seed with toxic endophytes as these may be more competitive than selected endophytes (Hume and Barker, 2005). In addition, greater attention to grazing residuals are needed to prevent overgrazing (as discussed earlier).

The traits that make *Epichloë* an effective BCA have also enabled the plant breeding and seed industries to achieve cost-effective product development, intellectual property protection, marketing and distribution of selected endophyte-infected grass seed products (Bouton and Hopkins, 2003; Card *et al.*, 2016; Caradus *et al.*, 2021b). This has involved the discovery and characterisation of endophytes, transfer from wild host grasses to elite grass genetics, and breeding for compatible, stable associations that provide effective bioprotective properties, as well as transmitting and storing well in seed at commercial scale (Johnson and Caradus, 2019). This has created extra complexity and costs to breeding programmes (Eady, 2021), and, in particular, the need to thoroughly test the pest protection profile and livestock safety of new grass-endophyte cultivars (Bouton and Hopkins, 2003; Thom *et al.*, 2012; Caradus *et al.*, 2021b).

While selected endophytes have some potential disadvantages, Ball *et al.* (2007) considered that for tall fescue in the USA their use is the only long-term solution for eliminating toxicity to livestock while still having the agronomic benefits of endophyte. It is a similar situation for perennial ryegrass in New Zealand and Australia (Heeswijck and McDonald, 1992; Prestidge, 1993), but greater compromises may be needed between livestock performance/health and agronomic performance. Despite this, the uptake of selected endophytes has been rapid and high in New Zealand where it is estimated that 90% of the proprietary perennial ryegrasses are sold with selected endophytes (Caradus *et al.*, 2013, 2021b).

In contrast to New Zealand, there has been poor uptake in the USA of selected endophytes in tall fescue, despite early significant sowing of selected endophytes and numerous grazing experiments and on-farm demonstrations (Ball *et al.*, 2007). Reasons for poor uptake are numerous, including a lack of awareness of endophyte toxicity, uncertainty regarding the process of killing old and planting new pasture, high seed costs and unclear financial benefits (Roberts and Andrae, 2005; Poore *et al.*, 2019). This has driven the formation of the Alliance for Grassland Renewal (https://grasslandrenewal.org/), a concerted collaborative multi-organisation, multi-state effort to enhance farmer understanding and adoption of 'Novel endophyte tall fescue' (Poore *et al.*, 2019; Roberts and Andrae, 2023).

10.4.4 Extending use of grasses through selected endophytes

The value of using selected endophytes goes beyond replacing the use of mammalian toxic endophytic grasses. Selected endophytes that have no mammalian toxins can be utilised to enhance agronomic performance of pastures in regions where *Epichloë* are not currently used in specific grass species. This may become of increasing importance as climate change increases biotic and abiotic stresses on pastures (Caradus and Hume, 2023).

In Australia and New Zealand, the use of tall fescue has been very limited despite being well adapted to harsh soil and climatic conditions where perennial ryegrass struggles (Easton *et al.*, 1994). There is little occurrence of toxic endophyte-infected tall fescue in pasture and it has been well-documented that mammalian-safe endophytes can greatly improve agronomic performance of tall fescue in these regions, with no negative effects on livestock (Easton *et al.*, 1994; Hume *et al.*, 2009; Fletcher, 2012; Hume and Sewell, 2014). The greatest opportunity is in Australia, where currently it is only used in 7% of its potential adaptive area of 16.8 M ha (Young *et al.*, 2013). Opportunities also exist in other regions of the world, such as Chile, Argentina and Uruguay (De Battista *et al.*, 1997; De Battista, 2005), and possibly southern Europe.

A similar but less certain situation exists for utilising selected endophytes to extend the use of perennial ryegrass in the USA, UK/Europe and southern South America, where currently pasture ryegrass cultivars are largely endophyte free. In the USA, endophyte has been shown to reduce invertebrate pests (Young *et al.*, 2013) but clearly demonstrated agronomic efficacy in the field is lacking with only limited (reported) experimentation (Anon, 2024). In contrast, endophyte is known to enhance the performance of turf ryegrasses in the USA (Brilman, 2005; Young III and Silberstein, 2012; Stewart *et al.*, 2022). In the UK/Europe, a region where perennial ryegrass is used widely, currently there is no clear, compelling evidence for agronomic benefit and/or control of important insect pests, based on limited experimentation. This is despite endophyte being common in wild populations and high in old pastures and drier environments implying that endophyte is advantageous for plant performance (Lewis, 2001; Zabalgogeazcoa and Bony, 2005). In southern South America, based on limited data, there appears to be no agronomic benefits of endophyte infection in perennial ryegrass (De Battista *et al.*, 1997; De Battista, 2005). However, given the agronomic advantages of endophyte infection in tall fescue in this region, there may be untapped opportunities to utilise selected endophytes to enhance perennial ryegrass pastures.

11 Mammalian safe *Epichloë* of annual ryegrasses and meadow fescue

Not all associations of *Epichloë* endophytes with their natural grass host have been found to be toxic to grazing animals. In particular, the two agricultural species Italian ryegrass (*L. multiflorum*) and meadow fescue (*F. pratensis*) may contain asexual endophytes which while producing some secondary metabolites are benign on livestock.

11.1 Italian ryegrass

E. occultans is an asexual seed-transmitted species found in most of the annual *Lolium* species: *L. canariense, L. multiflorum, L. persicum, L. remotum, L. rigidum, L. temulentum* and *L. subulatum* (Moon *et al.*, 2000; Moore *et al.*, 2015; Hume *et al.*, 2020). It is a natural allopolyploid of *E. baconii* and *E. bromicola*. Hyphae colonise meristematic tissue, but in leaves its hyphae are only found as a dense mycelial mass located at the very base of the leaf sheath (Christensen *et al.*, 2002). This contrasts with most other *Epichloë* species where hyphae are found throughout the leaf sheath, aligned to the leaf axis (Christensen *et al.*, 2002).

Lolines are produced by *E. occultans* when infecting *L. multiflorum* and several other annual *Lolium* species (Sugawara *et al.*, 2006; Moore *et al.*, 2015; Bastías *et al.*, 2019). The occurrence of lolines in *L. temulentum* was discovered many years before it was known to be associated with endophyte infection (Hofmeister, 1892). In a comparative review, Realini *et al.* (2024) found that loline concentrations in *E. occultans*-infected vegetative tissue of *L. multiflorum* and *L. rigidum* were similar to that found in both tall and meadow fescue but were only 7–13% of the concentrations in reproductive tissues of the fescues. The reduced secondary metabolites may reflect the unique hyphal distribution in the *E. occultans*-infected plants.

Loline concentrations in *E. occultans*-infected seed can be considerable, with Moore *et al.* (2015) reporting a maximum value of 2220 ppm in a *L.*

persicum accession. Peramine has been reported in *E. occultans*-infected inflorescences and/or seed for *L. multiflorum*, *L. persicum*, *L. rigidum* and *L. temulentum* (Sugawara *et al.*, 2006; Moore *et al.*, 2015). It is likely that lolines and peramine reduce seed predation over the crucial summer period.

In New Zealand, early seedling establishment of annual/Italian ryegrass is often enhanced by *E. occultans* infection through increased protection against Argentine stem weevil (Stewart, 1987; Hume *et al.*, 1993). Similarly, Cooper *et al.* (2007) in northern New Zealand reported that Italian ryegrass infected with *E. occultans* had advantages in plant survival and productivity of mature plants, which was attributed to reduced damage from African black beetle. In other studies, *E. occultans*-infected *L. multiflorum* has been shown to have reduced survival of rice leaf bug in Japan (Shiba and Sugawara, 2009; Shiba *et al.*, 2011) and reduced feeding damage by black cutworm in Argentina (Gundel *et al.*, 2020). In a further Argentinian experiment, *E. occultans* also reduced the impact of two out of four soil-borne pathogens at seedling establishment for *L. multiflorum* (Pérez *et al.*, 2013). Regardless of these favourable Argentinian pest and disease results, agronomic benefits of *E. occultans* are not evident in limited trialling in pastures in southern South America. This is despite the high rates of *E. occultans* infection in naturalised annual ryegrasses in this region (De Battista *et al.*, 1997; De Battista, 2005).

11.2 Meadow fescue

F. pratensis ecotypes are commonly infected with the endophyte *E. uncinata*, a hybrid of *E. bromicola* and *E. typhina*. In addition, a single occurrence has been reported in *F. pratensis* of a second endophyte *E. siegelii*, a hybrid of *E. amarillans* and *E. elymi* (Craven *et al.*, 2001). In northern Europe and the European Alps, endophyte-infected ecotype populations can be found in many countries and in cultivars at varying levels of infection (Saari *et al.*, 2009; Cagnano *et al.*, 2019). Surprisingly, the only *Epichloë*-infected meadow fescue ecotype found in the southern hemisphere does not occur in a cold temperate region but instead occurs in the subtropical north of New Zealand where insect pressure can often be severe (Cooper, 1996).

Both *E. uncinata* and *E. siegelii* produce high concentrations of loline alkaloids, usually more N-formylloline than N-acetylloline and N-acetylnorloline (Patchett *et al.*, 2011b; Cagnano *et al.*, 2019). These alkaloids are not known to cause any livestock problems but are well known for providing resistance to many insects (Fletcher *et al.*, 2001; Schardl *et al.*, 2007; Caradus and Johnson, 2020). Neither *E. siegelii* nor *E. uncinata* produce peramine due to mutations in the peramine gene (Berry *et al.*, 2015).

The use of meadow fescue and meadow fescue types (Festuloliums), with their *E. uncinata* endophyte, have found a small commercial role on farms in

New Zealand due to its effects on a wide number of insect pests (Fletcher *et al.*, 2001; Bryant *et al.*, 2010; Caradus and Johnson, 2020; Caradus *et al.*, 2021b; Caradus and Hume, 2023). For *E. uncinata*-meadow fescue, it is often sown in mixtures with tall fescue with a selected endophyte. As well as insect protection, *E. uncinata* endophyte infection may increase the drought tolerance and so the competitive ability of meadow fescue in the field (Malinowski *et al.*, 1997a; Malinowski *et al.*, 1997b).

12 Endophyte effects on food quality and safety

As well as animal toxicity, consideration also needs to be given to the risk of endophyte-expressed compounds to human health (Miyazaki *et al.*, 2004; Shimada *et al.*, 2013). Since animals are consuming these metabolites, it is possible that they go up the food chain to be consumed by humans eating animal products. To determine the risk to human health, toxicity to monogastric animals must be determined as well as the concentrations of metabolites in food. Once this information is available, a risk assessment can be performed.

12.1 *Toxicity of endophyte metabolites to monogastric animals*

Different routes of administration can be used for the determination of toxicity using non-ruminant animals. Using intravenous (i.v.) and intraperitoneal (i.p.) administration, the compounds become immediately bioavailable, whereas compounds dosed orally must be first absorbed by the gastrointestinal tract before going through the pre-system 'first pass' of the liver. Oral administration is obviously the most relevant to human consumption, with both i.v. and i.p. dosing giving an overestimation of toxicity. Different oral dosing regimes can also be used, with acute toxicity being a one-off dose and chronic toxicity being repeated dosing.

The toxicity of ergot alkaloids has been determined. For ergotamine, an acute oral LD_{50} for mice, rats and rabbits has been determined to be 3200, 1300 and 550 mg/kg body weight (BW), respectively (Griffith *et al.*, 1978). Chronic dosing of ergotamine, ergometrine or α-ergrocryptine to rats has demonstrated a no observable adverse effect level (NOAEL) of 0.22–0.60 mg/kg/day. As ergot alkaloids have been used for the treatment of headaches since the 1800s, toxicokinetic and bioactivity data are available. These studies have shown that humans have poor gastrointestinal absorption of ergot alkaloids, and following absorption, >90% is eliminated by 'first pass' metabolism. The combination of these factors results in an oral bioavailability of <1% for ergot alkaloids (Little *et al.*, 1982; Saper and Silberstein, 2006).

Published by Burleigh Dodds Science Publishing Limited, 2025.

Using mice, the acute oral LD_{50} of chanoclavine exceeded the limit dose specified by OECD guideline 425 (2000 mg/kg) (OECD, 2022). Chanoclavine fed as part of mouse diet, yielding a dose rate of 123 mg/kg/day, induced no adverse effects (Finch *et al.*, 2019).

The toxicity of indole-diterpenoids has also been investigated (lolitrems and epoxyjanthitrems). The lolitrem class of compounds has been assessed by incorporating pure lolitrem B into the diet of mice. The mice were fed for 90 days at an average dose rate of 0.15 mg/kg lolitrem B and showed no adverse effects (Finch *et al.*, 2022a). Additionally, a further trial demonstrated that a diet containing 1.06 ppm lolitrem B fed to two full generations of mice had no effect on gestation period, number of pups born or growth of pups (Finch *et al.*, 2022a).

Epoxyjanthitrems have been tested by incorporating endophyte-infected AR37 seed into the diet of mice. A dose rate of 3.17 mg/kg/day had no effect on mice over 90 days of feeding and a diet containing 20.1 ppm epoxyjanthitrems had no effect on mice in a two-generation reproduction study (Finch *et al.*, 2022a).

For loline alkaloids, a toxicological assessment was conducted by incorporating endophyte-infected meadow fescue seed into the diet of mice. For 3 weeks, mice ate lolines at a daily rate of 415 mg/kg with no adverse effects (Finch *et al.*, 2016). In a further trial, mice were fed seed containing high concentrations of lolines and peramine with no adverse effects (Finch *et al.*, 2022b).

12.2 Endophyte expressed metabolites in food

For ergot alkaloids, dosing experiments with livestock suggest a quick clearance. The serum of steers dosed with ergolines, ergotamine or ergine by i.v. injection showed an immediate drop in toxin concentration with elimination complete within 2 hours (Moubarak *et al.*, 1996). Similarly, the i.v. administration of ergovaline to sheep showed rapid metabolism/elimination with the toxin being undetectable within 1 hour (Jaussaud *et al.*, 1998). These observations suggest that residues of ergot alkaloids in animal products should be low. Consistent with this hypothesis, Schumann *et al.* (2009) were unable to detect ergot alkaloids in the milk or blood of cows fed ergot at 4.1–16.3 μg/kg BW and similarly dairy cows fed 3 μg/kg BW ergot had no ergot alkaloids in milk (Wolff *et al.*, 1995). Although below the limit of quantification, ergovaline was detectable in milk of lactating ewes consuming endophyte-infected ryegrass hay (Zbib *et al.*, 2015). In the same study, ergovaline could be detected in the liver, fat and kidney of the sheep but not in muscle or brain (Zbib *et al.*, 2015). Realini *et al.* (2005) showed ergovaline residues (2.81 ppb) in the subcutaneous fat of steers grazing endophyte-infected tall fescue.

The presence of indole-diterpenoids (lolitrem B and epoxyjanthitrems) in livestock consuming endophyte-infected ryegrass has also been investigated. In Japan, Miyazaki *et al.* (2004) showed that Japanese Black cows consuming endophyte infected ryegrass straw, and exhibiting ryegrass staggers, had lolitrem B residues in perirenal fat (210 ppb) but no residues in muscle, liver, kidney or brain. A further study conducted in the US also detected lolitrem B in the fat of Angus steers fed a diet of endophyte-infected ryegrass straw, where concentrations reached a maximum of 40 ppb (Duringer *et al.*, 2021). Similarly, monthly fat biopsies in New Zealand of sheep grazing ryegrass pastures infected with common toxic (containing lolitrem B) or AR37 (containing epoxyjanthitrems) endophytes showed maximum concentrations of 61.8 and 1032 ppb lolitrem B and epoxyjanthitrems, respectively. Toxins did not accumulate in fat, instead reflecting the concentrations being consumed at the time (Finch *et al.*, 2012). In the French study by Zbib *et al.* (2015), trace amounts of lolitrem B were detected in the milk of lactating ewes fed endophyte-infected hay. The greatest concentration of lolitrem B was found in abdominal fat (2.39 ppm) with detectable amounts in liver, kidneys and muscle, although undetectable in brain tissue. In a further New Zealand experiment, dairy cows fed endophyte infected common toxic and AR37 ryegrass pastures had toxin carryover into milk with maximum concentrations of 5 and 109 ppb for lolitrem B and epoxyjanthitrems, respectively. Concentrations in milk responded quickly to what the cows were eating and within 8 days of feeding toxin-free pastures the toxins could no longer be detected (Finch *et al.*, 2013).

12.3 Risk assessment

Ergot alkaloid residues which have been observed in fat and milk are low, and this, coupled with the very low bioavailability in humans, appears to make this class of alkaloids a very low risk to human health. Since ergots contaminate grain and humans eat grain this presents a genuine safety risk. For this reason, the European Food Safety Authority (EFSA) convened committees in both 2012 and 2017 to provide scientific opinions on ergot alkaloids (EFSA, 2012, 2017). An acute reference dose (ARfD), defined as the amount that can be eaten safely within a 24-hour period, was set at 1 μg ergot alkaloids/kg bodyweight. To reach this ARfD, a 70 kg human would have to eat 24 900 kg of fat contaminated at the concentration detected by Realini *et al.* (2005) (2.81 ppb), which is physically impossible.

Unlike the ergot alkaloids neither lolitrems nor epoxyjanthitrems are associated with grain so there is no risk assessment data available. However, using the animal data and applying uncertainty factors, a human risk assessment can be completed. A safety factor of 10 is applied to account for the species difference (mouse and human) and a further 10 is applied to account for possible variation in the human population. In the mouse toxicological assessment of lolitrem B and epoxyjanthitrems, dose rates of 0.15 and 3.17 mg/kg/day were shown to have no effect. Applying the combined 100-fold safety factors and considering a 70 kg human, a total of 0.105 and 2.22 mg lolitrem B and epoxyjanthitrems could be safely eaten in 24 hours (ARfD). Taking the highest concentrations measured in milk (Finch *et al.*, 2013) this human would have to drink 20 L of milk to reach the ARfD. Considering the highest concentrations detected in fat (Finch *et al.*, 2012), a 70 kg human would have to consume 1.7 kg and 2.2 kg of fat contaminated with lolitrem B or epoxyjanthitrems, respectively, to reach the ARfD. Assuming that meat is 20% fat then this would equate to 8.5 kg and 11 kg of meat.

12.4 Conclusion on impacts of endophyte metabolites on human health

The available evidence suggests that it is highly unlikely that the endophyte metabolites lolitrem B, ergovaline and epoxyjanthitrems are a risk to human health through the consumption of animal-derived food products. Due to the presence of sclerotia (ergot) in grain they are regulated and ergot alkaloids are regulated in animal feed. In contrast, there are no regulations for indole-diterpenoids, but as new selected endophytes are commercialised in elite grass cultivars, new endophyte expressed metabolites could be introduced into human food. Without regulation, the onus is on the developers of these new endophytes to undertake toxicological evaluations to ensure food safety.

13 Beyond pasture grasses

The discoveries of the effects of endophyte in ryegrass and tall fescue in pastures, and the benefits of seeking and utilising endophyte strain diversity, have spurred research into application of endophytes in turf grasses and cereals.

13.1 Application for turf

As occurred with pasture grasses, in particular regions of the world such as the USA and New Zealand, breeders had been unwittingly selecting for endophyte in turf grasses before the role of *Epichloë* in pastures was discovered in the 1970s/1980s (Funk *et al.*, 1994; Stewart, 2005). In the USA, the significance of endophytes in the breeding of turf grasses was first recognised some 40 years ago (Funk *et al.*, 1983) and by the early 1990s Funk *et al.* (1993) reported 'turf-grasses have frequently shown dramatic enhancement of resistance to many

Published by Burleigh Dodds Science Publishing Limited, 2025.

foliage-feeding insect pests'. To date the application of *Epichloë* endophyte to turf grasses has been mostly in perennial ryegrass and tall fescue. Opportunities exist for other turf species such as fine fescues (*Festuca* species) but there are significant barriers for application such as epiphytic growth by sexual species of *Epichloë* causing 'choke' in seed heads and therefore limiting seed yields (Funk *et al.*, 1993, 1994). It is likely that research on and application of *Epichloë* endophytes for turf grasses will increase, as one of the avenues to address reductions and bans on pesticide use (Wiewióra and Żurek, 2023).

Research on endophyte strain variation to enhance the agronomic performance of the host turf cultivar has been limited, possibly due to little need to consider the detrimental impacts on livestock. Selections based on elevated ergovaline alkaloid production of specific endophyte strains, and endophyte-grass associations, have led to endophyte-infected cultivars being developed and deployed for their wildlife deterrent properties (Stewart *et al.*, 2022). Endophyte-infected tall fescue and ryegrass that express a wide range of secondary metabolites have been utilised at airports, sports fields and recreational areas.

While the final use of endophyte-infected turf cultivars does not involve livestock, the production of seed of these cultivars does intersect with livestock production systems. The endophyte strains in these turf grasses produce endophyte alkaloids such as lolitrem and ergovaline that are known to cause endophyte toxicosis in farmed livestock. Risks of toxicity for livestock occur in mixed farming systems such as used in Australia and New Zealand, where livestock may graze seed production stands prior to and/or post-seed harvest. In addition, endophytic straw from the seed harvest may be used as feed for livestock. The toxicity of straw has been of greatest concern in the large seed production systems of the Willamette Valley in Oregon, USA, where large volumes of straw are shipped to Asia for its use in beef production (Craig, 2009). This came to prominence late in the twentieth century when straw burning reduced, and the number of endophyte-infected turf cultivars rose rapidly. In the future, turf breeders may incorporate selected endophyte strains that have no or reduced toxicity to livestock, to enhance pest resistance and possibly other traits, such as the epoxyjanthitrem-producing AR37 strain (Stewart, 2005; Popay *et al.*, 2008).

13.2 Application for cereals

Cereal grasses (tribe Hordeeae, formerly Triticeae) have for millennia been critical as food for humans and as a feed for livestock (Feuillet and Muehlbauer, 2009). Some wild *Elymus* and *Hordeum* relatives of the domesticated modern cereals harbour a range of *Epichloë* (Card *et al.*, 2014) and at least some are bioactive (Clement *et al.*, 2005). These *Epichloë* have secondary metabolites that are known for their anti-insect activity in pasture grasses, offering the potential for utilizing them in crops such as barley (*Hordeum vulgare* L.), wheat (*Triticum* spp.) and rye (*Secale cereale* L.) (Simpson *et al.*, 2014; Li *et al.*, 2021; Wang *et al.*, 2022; Popay *et al.*, 2023). Once the endophyte is transferred to a modern cereal, *Epichloë* has proved effective in reducing at least some insect pests and diseases and promoting growth (Hume *et al.*, 2018; Li *et al.*, 2021; Popay *et al.*, 2023).

As with pasture grasses, compatibility with the cereal host genetics has proved critical in the formation of a symbiosis and, in particular, a stable mutualistic *Epichloë* association with the new cereal host. This is not unexpected given the relatively large genetic distance between the original host of the *Epichloë* (wild cereal relative) and the newly inoculated host (domesticated cereal). Where a symbiosis has been established, the most immediate sign of incompatibility is a stunted morphological phenotype of the infected host (Simpson *et al.*, 2014). While this can also occur in pasture grasses (Simpson and Mace, 2012; Simpson *et al.*, 2012), the frequency and magnitude of the responses in phenotype in the cereals have been substantial (Simpson *et al.*, 2014). In outbreeding rye, selection has been possible from the range of expressed phenotypes, while phenotype diversity in wheat has required the use of specific genetics from wheat-alien addition/substitution lines (Johnson *et al.*, 2018).

Given the increasing limitations placed on the use of synthetic chemicals, and the promotion of environmentally friendly agriculture, further development of mammalian-safe *Epichloë* for cereals is likely to contribute to the environmental and sustainability goals of regions such as Europe (Barzman *et al.*, 2015; Lamichhane, 2017) and other areas of the world.

14 Genetic manipulation

Finding new chemistry or variation in naturally occurring endophyte strains, to improve on the current suite of selected strains used commercially in ryegrass and tall fescue, is becoming increasingly difficult (Caradus *et al.*, 2021b). Given the increasing breadth of knowledge of the major secondary metabolites and their known bioactivity and gene control, genetic manipulation to form an ideal endophyte for commerce has been investigated. This has opened up new possibilities to design endophyte strains with endophyte secondary metabolite profiles that maximise invertebrate pest protection to the host grass, without harming livestock, a combination of traits that have not been found to date, or do not exist, for strains in nature.

An early example of investigations using genetic manipulation was for the perennial ryegrass Lp1 (AR6) hybrid strain (*E. hybrida*) (Campbell *et al.*, 2017). This was the first selected endophyte strain to be released for use in ryegrass (branded Endosafe). It had limited success due to concerns over high levels of ergovaline, the ergot alkaloid of most concern for its detrimental effects on livestock (Johnson *et al.*, 2013). Using homologous recombination, gene knock outs were achieved in Lp1 that prevented the production of ergovaline but did not disrupt the compatibility of this strain with the ryegrass host (Panaccione *et al.*, 2001). In a follow-up bioactivity study with modified strains, it was found that the ergot alkaloids were not responsible for the suppression of plant parasitic nematodes (*Pratylenchus scribneri*) (Panaccione *et al.*, 2006b). In addition, the importance of intermediates in the ergot alkaloid pathway was revealed for effects on herbivore responses in a model animal species (Panaccione *et al.*, 2006a).

Application of the CRISPR (Clustered Regularly Interspaced Short Palindromic Repeats)-Cas (CRISPR Associated protein) gene editing system (Shi *et al.*, 2017) to *Epichloë* has been a step change for researchers. This technique has enabled precise gene manipulation of *Epichloë* alkaloids, without off-target effects or introduction of foreign DNA (footprint-less). This newer and safer technology may be more readily accepted by the public, particularly for gene disruption or deletion edits (SDN1 (site-directed nuclease)), leading to faster and more widespread commercial adoption of endophytes developed using CRISPR or other gene-editing techniques (Johnson *et al.*, 2021; Caradus, 2023b).

CRISPR-Cas9 has been successfully used to manipulate ergot alkaloid production in the toxic e19 tall fescue endophyte (*E. coenophiala*) strain. This has overcome the considerable difficulties encountered using older genetic technologies on e19 due to the slow growth of this strain in the laboratory and its triple interspecific hybrid ancestry (Florea *et al.*, 2009; Florea *et al.*, 2021). Florea *et al.* (2021) also successfully applied CRISPR-Cas9 to the ergot alkaloid pathway of the Lp1 ryegrass endophyte. In another perennial ryegrass endophyte, CRISPR-Cas9 has been used to genetically modify the AR37 strain, a strain that had previously been intractable to genetic manipulation (Miller *et al.*, 2022). For AR37, manipulation of the epoxyjanthitrem pathway has enabled its use to further study the bioactivity of pathway intermediates, genes involved in the pathway, and the possibility of revealing bioactivity of other secondary metabolite pathways (Miller *et al.*, 2022).

The use of gene editing in *Epichloë* to improve environmental outcomes, such as reduction in methane emissions, has also been proposed (Caradus, 2023a). An advantage of delivering new traits of value through genetic modification of asexual *Epichloë* is its life cycle which means it can only be maternally inherited and is not transferred to other germplasm through pollen spread (Mirlohi *et al.*, 2004).

15 Conclusion

Epichloë fungal endophytes, present in widely used temperate pasture grasses such as tall fescue and perennial ryegrass, provide protection against invertebrate pests as well as causing livestock toxicoses. Both the positive and negative effects are due to endophyte-expressed secondary metabolites. Optimising the secondary metabolite profile is key to achieving the desired bioactivity while minimising the unwanted toxicity to livestock. Undoubtedly changes to primary metabolites and the overall functioning of the plant, and interactions with the surrounding biota, are also important in enabling *Epichloë* to impart a range of biotic and abiotic protective properties to host grasses (Hume *et al.*, 2016; Caradus *et al.*, 2021a; Johnson *et al.*, 2021). Beyond pasture grasses, *Epichloë* endophytes have been shown to provide benefits for wildlife deterrence and improving the tolerance of cereals to biotic challenges.

In regions of the world where *Epichloë* is key to the productivity and persistence of the grass host, the trade off in terms of negative impacts on livestock can be addressed in various ways. This includes reducing the intake of mammalian toxins through the use of non-*Epichloë* pasture species, along with changes in pasture and grazing management. In areas where pasture renewal is economically viable, selected strains of endophyte in combination with elite cultivars can be used to replace the previous toxic endophytic pastures. These selected endophytes have proved highly effective, and despite some limitations, they have been rapidly adopted in some markets. There are opportunities to use mammalian safe endophyte strains to extend the use of tall fescue and ryegrass globally without risk of toxicity to livestock.

Enhancing endophytes through genetic manipulation and particularly the new application of targeted gene-editing technology offer new possibilities to elucidate secondary metabolite pathways and their bioactivity and enable science to design endophytes which could benefit livestock farmers and reduce the environmental impacts of livestock farming.

16 Future trends

In some regions of the world, *Epichloë* infection is a critical component of the tall fescue and ryegrass pastures that underpin large ruminant livestock production systems. The importance of *Epichloë* in these regions is only likely to grow, as global warming drives climate change, resulting in even greater biotic and abiotic stresses and increased toxicity of old pastures. Given the high reliance on *Epichloë* for pasture performance in these regions, a greater scientific understanding of the grass-endophyte symbiosis could be utilised to enhance *Epichloë*-driven properties in addition to invertebrate pest protection and drought tolerance. These properties include improving resistance to plant pathogens, synergistic interactions with surrounding biota, and greater tolerance to temperature extremes, flooding, low soil nutrients, salinity, heavy metals and aluminium. In addition, given the effectiveness of *Epichloë* in alleviating environmental stresses for tall fescue and ryegrass, consideration should be given to utilisation of other grass species known to host *Epichloë* to enhance overall pasture supply and resilience of livestock production systems (Caradus and Hume, 2023).

In the regions where *Epichloë* is critical for pastoral production systems, it is likely that research and innovation will continue to investigate and deliver naturally occurring strains that improve the bioprotective properties of grasses while minimising or eliminating the negative impacts on livestock. However, there is a diminishing rate of return from discovery of naturally occurring strains that have novel properties, and due to climate change, there will be a loss of germplasm in the Mediterranean basin, a key centre of diversity for tall fescue and ryegrass and their associated *Epichloë* endophytes. As an avenue to address this, genetic manipulation of endophytes will continue to grow in importance. The CRISPR-Cas9 gene-editing system has already proven to be an effective genetic tool for understanding key endophyte secondary metabolite pathways of *Epichloë*. This technology may drive a new generation of *Epichloë* strains with greater benefits for pastoral-based livestock production systems. Its application to enable *Epichloë* to act as a carrier of new traits to be expressed by the grass-endophyte association for better environmental outcomes will test the extent by which science can understand and manipulate this mutualistic association.

The opportunity to broaden the application of the bioprotective properties of *Epichloë* seen in a particular grass species-endophyte association, by cross species transfer of endophytes to other grasses, is severely hampered by endophyte-grass host incompatibility. While considerable progress has been made in understanding this fundamental aspect of the biology (Johnson *et al.*, 2021), further work is needed to enable the scientific knowledge to deliver problem-solving outcomes.

Beyond the purpose of addressing livestock toxicity, greater global application of mammalian safe selected endophytes will be driven by global climate change, increasing limitations on the use of synthetic chemicals, and the overall need for natural processes to underpin environmentally friendly production of safe food from livestock. Given the rapid pace of change, researchers need to be proactive in understanding the biology of the grass-*Epichloë* association and its effects on the ecosystem thus allowing its effective application in the field.

There are good prospects for the application of *Epichloë* metabolites beyond livestock production systems (Caradus, 2009). These include use as natural pesticide products and possible use in human medicine. Bulk extracts from some endophyte-infected seed have already been shown to be active in deterring insect pests of lettuce and roses. The application to medicine is still to be pursued but given there are still a number of classes of *Epichloë* secondary metabolites yet to be fully investigated (Johnson *et al.*, 2007; Johnson *et al.*, 2015), the prospects are favourable.

17 Where to look for further information

In addition to the numerous references provided throughout this chapter as detailed above, the following will be of value.

For the USA, the following books and websites provide a range of useful online overviews, historic perspective and practical resources on tall fescue-endophyte for pastures and turf:

- Understanding the tall fescue endophyte https://forages.ca.uky.edu/files/understanding_the_tall_fescue_endophyte.pdf.
- Tall fescue for the twenty-first century (Fribourg *et al.* 2009).
- The wonder grass: the story of tall fescue in the United States (Ball *et al.*, 2019).
- The Alliance for Grassland Renewal. https://grasslandrenewal.org.

For an overview of the importance of endophytes in ryegrass and tall fescue for New Zealand and Australia, along with the development of selected endophytes for these countries and other regions of the world, see the following: Caradus (2023a); Caradus and Johnson (2020); Card *et al.* (2024); Hume and Sewell (2014); Hume *et al.* (2020); Johnson *et al.* (2013); and Reed *et al.* (2011).

For the UK and Europe, see Kauppinen *et al.* (2016); for China, see Song and Nan (2015) and for South America, see Iannone *et al.* (2021).

Since 1990, the international science community have met every 3–4 years to share the latest findings on *Epichloë*. At these symposia, a wide range of topics are presented, ranging from genetics, ecology, mechanisms and trophic interactions, through to the practical application of endophytes for plant protection and production and impacts on livestock. Over recent symposia, the breadth of grass endophytes has been broadened beyond *Epichloë*, as evidence mounts on the critical role of microbes in plant growth. Abstracts from the most recent (10th) Symposium held in Salamanca Spain 2018 can be found here: https://digital.csic.es/bitstream/10261/167629/1/Book%20of%20abstracts%20ISFEG10%20Salamanca2018.pdf.

For the taxonomy of *Epichloë*, see Leuchtmann *et al.* (2014).

18 References

Adcock, R. A., Hill, N. S., Bouton, J. H., et al. (1997) 'Symbiont regulation and reducing ergot alkaloid concentration by breeding endophyte-infected tall fescue', *J. Chem. Ecol.*, 23(3), 691–704. https://doi.org/10.1023/B:JOEC.0000006404.33191.60.

Agee, C. S. and Hill, N. S. (1994) 'Ergovaline variability in *Acremonium*-infected tall fescue due to environment and plant genotype', *Crop Sci.*, 34(1), 221–226. https://doi.org/10.2135/cropsci1994.0011183X003400010040x.

Aiken, G. E., Bransby, D. I. and McCall, C. A. (1993) 'Growth of yearling horses compared to steers on high-and low-endophyte infected tall fescue', *J. Equine Vet. Sci.*, 13(1), 26–28. https://doi.org/10.1016/S0737-0806(07)80013-X.

Aiken, G. E. and Strickland, J. R. (2013) 'Forages and pastures symposium: managing the tall fescue-fungal endophyte symbiosis for optimum forage-animal production', *J. Anim. Sci.*, 91(5), 2369–2378. https://doi.org/10.2527/jas.2012-5948.

Alabdouli, K. O., Blythe, L. L., Duringer, J. M., et al. (2014) 'Physiological effects of endophyte-infected perennial ryegrass straw on female camels in the Middle East', *Emirates J. Food & Agric.*, 26(1), 82–92. https://doi.org/10.9755/ejfa.v26i1.16473.

Aldrich-Markham, S., Pirelli, G. and Craig, A. M. (2003) *Endophyte toxins in grass seed fields and straw effects on livestock.* EM 8598, Oregon State University Extension Service, USA.

Anon (2023a) *United Kingdom variety list trials: Trial procedures for official examination of value for cultivation and use (VCU) harvest 2024. Perennial, Italian and hybrid ryegrass, timothy, Festulolium, cocksfoot, tall and meadow fescue.* Animal and Plant Health Agency, Department for Environment, Food & Rural Affairs, United Kingdom. Available at: https://assets.publishing.service.gov.uk/media/653fe4b846532b000d67f59b/grasses-vcu-procedures-2024.pdf.

Anon (2023b) *Wikipedia: Epichloë.* Available at: https://en.wikipedia.org/wiki/Epichlo%C3%AB (Accessed: 2 February 2024).

Anon (2024) *Forage variety trials. Kentucky Cooperative Extension Service.* Available at: https://forages.ca.uky.edu/variety_trials (Accessed: 27 February 2024).

Arechavaleta, M., Bacon, C. W., Plattner, R. D., et al. (1992) 'Accumulation of ergopeptide alkaloids in symbiotic tall fescue grown under deficits of soil water and nitrogen fertilizer', *Appl. Environ. Microbiol.*, 58(3), 857–861. https://doi.org/10.1128/aem.58.3.857-861.1992.

Bacetty, A. A., Snook, M. E., Glenn, A. E., et al. (2009) 'Toxicity of endophyte-infected tall fescue alkaloids and grass metabolites on *Pratylenchus scribneri*', *Phytopath.*, 99(12), 1336–1345. https://doi.org/10.1094/PHYTO-99-12-1336.

Bacon, C. W., Porter, J. K., Robbins, J. D. and Luttrell, E. S. (1977) '*Epichloë typhina* from toxic tall fescue grasses', *Appl. Environ. Microbiol.*, 34(5), 576–581. https://doi.org/10.1128/aem.34.5.576-581.1977.

Baldauf, M. W., Mace, W. J. and Richmond, D. S. (2011) 'Endophyte-mediated resistance to black cutworm as a function of plant cultivar and endophyte strain in tall fescue', *Environ. Entomol.*, 40(3), 639–647. https://doi.org/10.1603/en09227.

Ball, D. M., Lacefield, G. D., Agee, C. S. and Hoveland, C. S. (2007) 'Introduction and acceptance of novel endophyte tall fescue in the USA'. In Popay, A. J. and Thom, E. R. (Eds), *Proceedings of the 6th International Symposium on Fungal Endophytes of Grasses. Grassland Research and Practice Series No. 13,* New Zealand Grassland

Association, Dunedin, New Zealand, pp. 249–251. https://doi.org/10.33584/rps.13.2006.3070.

Ball, D. M., Lacefield, G. D. and Hoveland, C. S. (2019) *The wonder grass: the story of tall fescue in the United States*. Oregon Tall Fescue Commission, Salem, Oregon, USA. https://aurora.auburn.edu/handle/11200/49449.

Ball, D. M., Pederson, J. F. and Lacefield, G. D. (1993) 'The tall-fescue endophyte', *Am. Sci.*, 81(4), 370–379. https://www.jstor.org/stable/29774973.

Ball, O. J.-P., Barker, G. M., Prestidge, R. A. and Sprosen, J. M. (1997a) 'Distribution and accumulation of the mycotoxin lolitrem B in *Neotyphodium* lolii-infected perennial ryegrass', *J. Chem. Ecol.*, 23(5), 1435–1449. https://doi.org/10.1023/B:JOEC.0000006474.44100.17.

Ball, O. J.-P., Miles, C. O. and Prestidge, R. A. (1997b) 'Ergopeptine alkaloids and *Neotyphodium lolii*-mediated resistance in perennial ryegrass against adult *Heteronychus arator* (Coleoptera: Scarabaeidae)', *J. Econ. Entomol.*, 90(5), 1382–1391. https://doi.org/10.1093/jee/90.5.1382.

Ball, O. J.-P., Prestidge, R. A., Sprosen, J. M. and Lauren, D. R. (1991) 'Seasonal levels of peramine and lolitrem B in *Acremonium lolii*-infected perennial ryegrass', *Forty Fourth New Zealand Weed and Pest Control Conference*, 176–180. https://doi.org/10.30843/nzpp.1991.44.10827.

Barker, G. M., Patchett, B. J. and Cameron, N. E. (2015a) '*Epichloë uncinata* infection and loline content afford *Festulolium* grasses protection from black beetle (*Heteronychus arator*)', *N. Z. J. Agric. Res.*, 58(1), 35–56. https://doi.org/10.1080/00288233.2014.978480.

Barker, G. M., Patchett, B. J. and Cameron, N. E. (2015b) '*Epichloë uncinata* infection and loline content protect *Festulolium* grasses from crickets (Orthoptera: Gryllidae)', *J. Econ. Entomol.*, 108(2), 789–797. https://doi.org/10.1093/jee/tou058.

Barzman, M., Bàrberi, P., Birch, A. N. E., et al. (2015) 'Eight principles of integrated pest management', *Agron. Sust. Develop.*, 35(4), 1199–1215. https://doi.org/10.1007/s13593-015-0327-9.

Bastías, D. A., Applegate, E. R., Gundel, P. E., et al. (2024) 'After air, light, and water, the next most important thing is grass: An introduction to the *Epichloë*–grass symbiosis'. In Hsueh, Y.-P. and Blackwell, M. (Eds), *Fungal associations, The Mycota 9,* Springer Nature, Switzerland, pp. 195–230. https://doi.org/10.1007/978-3-031-41648-4_8.

Bastías, D. A., Martínez-Ghersa, M. A., Newman, J. A., et al. (2019) '*Sipha maydis* sensitivity to defences of *Lolium multiflorum* and its endophytic fungus *Epichloë occultans*', *PeerJ*, 2019(12). https://doi.org/10.7717/peerj.8257.

Bastías, D. A., Ueno, A. C. and Gundel, P. E. (2023) 'Global change factors influence plant-*Epichloë* associations', *J. Fungi*, 9(4), 446. https://doi.org/10.3390/jof9040446.

Bazely, D. R., Vicari, M., Emmerich, S., et al. (1997) 'Interactions between herbivores and endophyte-infected *Festuca rubra* from the Scottish islands of St. Kilda, Benbecula and Rum', *J. Appl. Ecol.*, 34(4), 847–860. https://doi.org/10.2307/2405276.

Belesky, D. P., Stringer, W. C. and Plattner, R. D. (1989) 'Influence of endophyte and water regime upon tall fescue accessions. II. Pyrrolizidine and ergopeptine alkaloids', *Ann. Bot.*, 64(3), 343–349. https://doi.org/10.1093/oxfordjournals.aob.a087850.

Berry, D., Takach, J. E., Schardl, C. L., et al. (2015) 'Disparate independent genetic events disrupt the secondary metabolism gene perA in certain symbiotic *Epichloë* species', *Appl. Environ. Microbiol.*, 81(8), 2797–2807. https://doi.org/10.1128/aem.03721-14.

Bluett, S. J., Stephens, S. L., Hume, D. E. and Tapper, B. A. (2004) 'Incidence of ryegrass staggers in white rhinoceros (*Ceratotherium simum*) in Auckland Zoo', *N. Z. Vet. J.*, 52(1), 48. https://doi.org/10.1080/00480169.2004.36392.

Bluett, S. J., Thom, E. R., Clark, D. A., et al. (2005) 'Effects of perennial ryegrass infected with either AR1 or wild endophyte on dairy production in the Waikato', *N. Z. J. Agric. Res.*, 48(2), 197–212. https://doi.org/10.1080/00288233.2005.9513650.

Bony, S. and Delatour, P. (2001) 'Relevance and impact of grass endophyte toxins in Europe', *Fourth International Neotyphodium/ Grass Interactions Symposium*, Soest, Germany, 207–218.

Bony, S., Emile, J. C., Durix, A., et al. (2001) *8emes Rencontres autour des Recherches sur les Ruminants*. Institut National de la Recherche Agronomique, Paris, France, pp. 149–152.

Botha, C. J., Naude, T. W., Moroe, M. L. and Rottinghaus, G. E. (2004) 'Gangrenous ergotism in cattle grazing fescue (*Festuca elatior* L.) in South Africa', *J. South African Vet. Assoc.*, 75(1), 45–48. https://doi.org/10.4102/jsava.v75i1.449.

Bourke, C. A., Hunt, E. and Watson, R. (2009) 'Fescue-associated oedema of horses grazing on endophyte-inoculated tall fescue grass (*Festuca arundinacea*) pastures', *Aust. Vet. J.*, 87(12), 492–498. https://doi.org/10.1111/j.1751-0813.2009.00519.x.

Bouton, J. H. (2000) 'The use of endophytic fungi for pasture improvement in the USA', *Fourth International Neotyphodium/ Grass Interactions Symposium*, Soest, Germany, 163–168.

Bouton, J. H. and Hopkins, A. A. (2003) 'Commercial applications of endophytic fungi'. In White, J. F., Bacon, C. W., Hywel-Jones, N. L. and Spatafora, J. W. (Eds), *Clavicipitalean fungi: Evolutionary biology, chemistry, biocontrol and cultural impacts*, Marcel Dekker, New York, USA, pp. 495–516.

Bouton, J. H., Latch, G. C. M., Hill, N. S., et al. (2002) 'Reinfection of tall fescue cultivars with non-ergot alkaloid producing endophytes', *Agron. J.*, 94(3), 567–574. https://doi.org/10.2134/agronj2002.5670.

Brilman, L. A. (2005) 'Endophytes in turfgrass cultivars'. In Roberts, C., West, C. P. and Spiers, D. (Eds), *Neotyphodium in cool-season grasses*. Blackwell Publishing, Ames, IA, 341–349. https://doi.org/10.1002/9780470384916.ch16.

Brosi, G. B., McCulley, R. L., Bush, L. P., et al. (2011) 'Effects of multiple climate change factors on the tall fescue-fungal endophyte symbiosis: infection frequency and tissue chemistry', *New Phytol.*, 189(3), 797–805. https://doi.org/10.1111/j.1469-8137.2010.03532.x.

Bryant, R. H., Cameron, N. E. and Edwards, G. R. (2010) 'Response of black beetle and red-headed pasture cockchafer larvae to loline alkaloids in meadow fescue roots', *N. Z. Plant Prot.*, 63, 219–223. https://doi.org/10.30843/nzpp.2010.63.6571.

Bultman, T. L., Bell, G. and Martin, W. D. (2004) 'A fungal endophyte mediates reversal of wound-induced resistance and constrains tolerance in a grass', *Ecol.*, 85(3), 679–685. https://doi.org/10.1890/03-0073.

Bush, L. P., Cornelius, P. L., Buckner, R. C., et al. (1982) 'Association of N-acetyl loline and N-formyl loline with *Epichloe typhina* in tall fescue', *Crop Sci.*, 22, 941–943. https://doi.org/10.2135/cropsci1982.0011183X002200050010x.

Bush, L. P., Fannin, F. F., Siegel, M. R., et al. (1993) 'Chemistry, occurrence and biological effects of saturated pyrrolizidine alkaloids associated with endophyte-grass interactions', *Agric., Eco. & Envir.*, 44(1–4), 81–102. https://doi.org/10.1016/0167-8809(93)90040-V.

Butendieck, B. N., Romero Y, O., Hazard, T. S., et al. (1994) 'Caida del consumo y produccion de leche en vacas alimentadas con *Lolium perenne* infectada con *Acremonium lolii* [Drop in intake and milk production of cows fed with *Lolium perenne* infected with *Acremonium lolii*]', *Agricultura Tecnica (Santiago)*, 54(1), 1–6. https://doi.org/20.500.14001/35203.

Cabral, D., Cafaro, M. J., Saidman, B., et al. (1999) 'Evidence supporting the occurrence of a new species of endophyte in some South American grasses', *Mycologia*, 91(2), 315–325. https://doi.org/10.1080/00275514.1999.12061021.

Cagnano, G., Roulund, N., Jensen, C. S., et al. (2019) 'Large scale screening of *Epichloë* endophytes infecting *Schedonorus pratensis* and other forage grasses reveals a relation between microsatellite-based haplotypes and loline alkaloid levels', *Frontiers in Plant Sci.*, 10. https://doi.org/10.3389/fpls.2019.00765.

Campbell, M. A., Tapper, B. A., Simpson, W. R., et al. (2017) '*Epichloë hybrida*, sp. nov., an emerging model system for investigating fungal allopolyploidy', *Mycologia*, 109(5), 715–729. https://doi.org/10.1080/00275514.2017.1406174.

Caradus, J. R. (2009) 'Endophyte futures'. In Emerson, A. and Rowarth, J. (Eds), *Future food farming. New Zealand Inc. meeting tomorrow's markets*, NZX Limited, Wellington, New Zealand, pp. 40–41.

Caradus, J. R. (2023a) '*Epichloë* fungal endophytes – a vital component for perennial ryegrass survival in New Zealand', *N. Z. J. Agric. Res.*, 1–18. https://doi.org/10.1080/00288233.2023.2170426.

Caradus, J. R. (2023b) 'Processes for regulating genetically modified and gene edited plants', *GM Crops & Food*, 1–41. https://doi.org/10.1080/21645698.2023.2252947.

Caradus, J. R., Card, S. D., Finch, S. C., et al. (2020) 'Ergot alkaloids in New Zealand pastures and their impact', *N. Z. J. Agric. Res.*, 63, 1–41. https://doi.org/10.1080/00288233.2020.1785514.

Caradus, J. R., Card, S. D., Hewitt, K. G., et al. (2021a) 'Asexual *Epichloë* Fungi–Obligate Mutualists', *Encyclopedia*, 1(4), 1084–1100. https://doi.org/10.3390/encyclopedia1040083.

Caradus, J. R., Chapman, D. F., Cookson, T., et al. (2021b) '*Epichloë* endophytes – new perspectives on a key ingredient for resilient perennial grass pastures'. In Douglas, G. (Eds), *Resilient Pastures – Grassland research and practice series No. 17*. New Zealand Grassland Association, pp. 57–70. https://doi.org/10.33584/rps.17.2021.3435.

Caradus, J. R., Duder, R., Kerr, G. A., et al. (2023) 'Mangere to mainstream – the story of a perennial ryegrass ecotype, a determined farmer, and a progressive seed company', *N. Z. J. Agric. Res.*, 1–17. https://doi.org/10.1080/00288233.2023.2238674.

Caradus, J. R. and Hume, D. E. (2023) 'Future role of *Epichloë* endophytes in temperate pasture environments', *XXV International Grassland Congress*, Covington, Kentucky, USA, 1074–1077. https://doi.org/10.52202/071171-0259.

Caradus, J. R. and Johnson, L. J. (2019) 'Improved adaptation of temperate grasses through mutualism with fungal endophytes'. In Schouten, A. (Ed.), *Endophyte biotechnology: promise for agriculture and pharmacology*, CAB International, pp. 85–108. https://doi.org/10.1079/9781786399427.0085.

Caradus, J. R. and Johnson, L. J. (2020) '*Epichloë* fungal endophytes–from a biological curiosity in wild grasses to an essential component of resilient high performing ryegrass and fescue pastures', *J. Fungi*, 6(4) https://doi.org/10.3390/jof6040322.

Caradus, J. R., Lovatt, S. and Belgrave, B. (2013) 'Adoption of forage technologies', *New Zealand Grassland Association Conference*, Tauranga, New Zealand, 39–44. https://doi.org/10.33584/jnzg.2013.75.2917.

Card, S. D., Bastías, D. A. and Caradus, J. R. (2021) 'Antagonism to plant pathogens by *Epichloë* fungal endophytes–a review', *Plants*, 10(10), 1997. https://doi.org/10.3390/plants10101997.

Card, S. D., Faville, M. J., Simpson, W. R., et al. (2014) 'Mutualistic fungal endophytes in the *Triticeae* – survey and description', *FEMS Micro. Eco.*, 88, 94–106. https://doi.org/10.1111/1574-6941.12273.

Card, S. D., Johnson, L., Teasdale, S. and Caradus, J. (2016) 'Deciphering endophyte behaviour: The link between endophyte biology and efficacious biological control agents', *FEMS Microbiol. Ecol.*, 92(8), fiw114. https://doi.org/10.1093/femsec/fiw114.

Card, S. D., Rolston, M. P., Park, Z., et al. (2011) 'Fungal endophyte detection in pasture grass seed utilising the infection layer and comparison to other detection techniques', *Seed Sci. & Tech.*, 39(3), 581–592. https://doi.org/10.15258/sst.2011.39.3.05.

Card, S. D., Bastías, D. A., Zhang, W., Hume, D. E., and Caradus, J. R. (2024) 'Epichloë – a key element of New Zealand's agricultural landscape', *N. Z. J. Bot.*, 62, 519–547. https://doi.org/10.1080/0028825X.2024.2331570.

Cheeke, P. R. (1995) 'Endogenous toxins and mycotoxins in forage grasses and their effects on livestock', *J. Anim. Sci.*, 73(3), 909–918. https://doi.org/10.2527/1995.733909x.

Chen, L., Li, X., Li, C., et al. (2015) 'Two distinct *Epichloë* species symbiotic with *Achnatherum inebrians*, drunken horse grass', *Mycologia*, 107(4), 863–873. https://doi.org/10.3852/15-019.

Christensen, M. J., Bennett, R. J., Ansari, H. A., et al. (2008) '*Epichloë* endophytes grow by intercalary hyphal extension in elongating grass leaves', *Fungal Genet. Biol.*, 45, 84–93. https://doi.org/10.1016/j.fgb.2007.07.013.

Christensen, M. J., Bennett, R. J. and Schmid, J. (2002) 'Growth of *Epichloë/Neotyphodium* and p-endophytes in leaves of *Lolium* and *Festuca* grasses', *Mycol. Res.*, 106(1), 93–106. https://doi.org/10.1017/S095375620100510X.

Christensen, M. J., Easton, H. S., Simpson, W. R. and Tapper, B. A. (1998) 'Occurrence of the fungal endophyte *Neotyphodium coenophialum* in leaf blades of tall fescue and implications for stock health', *N. Z. J. Agric. Res.*, 41, 595–602. https://doi.org/10.1080/00288233.1998.9513343.

Christensen, M. J. and Voisey, C. R. (2007a) 'The biology of the endophyte /grass partnership'. In Popay, A. J. and Thom, E. R. (Eds), *Proceedings of the 6th International Symposium on Fungal Endophytes of Grasses. Grassland Research and Practice Series No. 13,* New Zealand Grassland Association, Dunedin, New Zealand, pp. 123–134. https://doi.org/10.33584/rps.13.2006.3078.

Christensen, M. J. and Voisey, C. R. (2007b) 'The biology of the endophyte/grass partnership'. In Popay, A. J. and Thom, E. R. (Eds), *Proceedings of the 6th International Symposium on Fungal Endophytes of Grasses. Grassland Research and Practice Series No. 13,* New Zealand Grassland Association, Dunedin, New Zealand, pp. 123–133. https://doi.org/10.33584/rps.13.2006.3078.

Clay, K. (1988) 'Clavicipitaceous fungal endophytes of grasses: coevolution and the change from parasitism to mutualism'. In Pirozynski, K. A. and Hawksworth, D. L. (Eds), *Coevolution of fungi with plants and animals*, Academic Press, San Diego, pp. 79–105.

Clay, K. and Cheplick, G. P. (1989) 'Effect of ergot alkaloids from fungal endophyte-infected grasses on fall armyworm (*Spodoptera frugiperda*)', *J. Chem. Ecol.*, 15(1), 169–182. https://doi.org/10.1007/BF02027781.

Clement, S. L., Elberson, L. R., Bosque Pérez, N. A. and Schotzko, D. J. (2005) 'Detrimental and neutral effects of wild barley-*Neotyphodium* fungal endophyte associations on insect survival', *Entomol. Exp. Appl.*, 114(2), 119–125. https://doi.org/10.1111/j.1570-7458.2005.00236.x.

Cooper, B. M. (1996) 'Evaluation of a meadow fescue (*Festuca pratensis*) population from Northland', *New Zealand Grassland Association Conference*, Whangarei, New Zealand, 69–72. https://doi.org/10.33584/jnzg.1995.57.2182.

Cooper, B. M., Hume, D. E., Panckhurst, K. A. and Popay, A. J. (2007) 'Agronomic responses of Italian ryegrasses infected with different *Neotyphodium* strains'. In Popay, A. J. and Thom, E. R. (Eds), *Proceedings of the 6th International Symposium on Fungal Endophytes of Grasses. Grassland Research and Practice Series No. 13,* New Zealand Grassland Association, Dunedin, New Zealand, pp. 297–300. https://doi.org/10.33584/rps.13.2006.3147.

Cosgrove, G. P., Anderson, C. B. and Berquist, T. R. N. (1996) 'Fungal endophyte effects on intake, health and liveweight gain of grazing cattle', *New Zealand Grassland Association Conference*, Waitangi, New Zealand, 43–48. https://doi.org/10.33584/jnzg.1995.57.2161.

Craig, A. M. (2009) 'Toxic effects of endophyte in seed straw'. In Fribourg, H. A., Hannaway, D. B. and West, C. P. (Eds), *Tall fescue for the twenty-first century. Agronomy Monograph 53,* American Society of Agronomy, Crop Science Society of America and Soil Science Society of America, Madison, Wisconsin, USA, pp. 327–335. https://doi.org/10.2134/agronmonogr53.c18.

Craig, A. M., Blythe, L. L. and Duringer, J. M. (2014) 'The role of the Oregon State University endophyte service laboratory in diagnosing clinical cases of endophyte toxicoses', *J. Agric. Food Chem.*, 62(30), 7376–7381. https://doi.org/10.1021/jf5027229.

Craven, K. D., Blankenship, J. D., Leuchtmann, A., et al. (2001) 'Hybrid fungal endophytes symbiotic with the grass *Lolium pratense*', *Sydowia*, 53(1), 44–73.

Crawford Jr., R. J., Forwood, J. R., Belyea, R. L. and Garner, G. B. (1989) 'Relationship between level of endophyte infection and cattle gains on tall fescue', *J. Prod. Agric.*, 2(2), 147–151. https://doi.org/10.2134/jpa1989.0147.

Cross, D. L. (2009) 'Toxic effects of the endophyte in horses'. In Fribourg, H. A., Hannaway, D. B. and West, C. P. (Eds), *Tall fescue for the twenty-first century. Agronomy Monograph 53,* American Society of Agronomy, Crop Science Society of America and Soil Science Society of America, Madison, Wisconsin, USA, pp. 311–325. 17. https://doi.org/10.2134/agronmonogr53.c17.

Cunningham, I. J. (1948) 'Tall fescue grass is poison for cattle', *N. Z. J. Ag.*, 77, 519.

Dang, X., Cao, G., Duan, D., et al. (1992) 'Studies on the toxic constituent of *Achnatherum inebrians*', *Acta Vet. Zootechn. Sin.*, 23(4), 366–371. (in Chinese with English abstract)

De Battista, J. (2005) '*Neotyphodium* research and application in South America'. In Roberts, C. A., West, C. P. and Spiers, D. E. (Eds), *Neotyphodium in cool-season grasses*, Blackwell Publishing Ltd Ames, Iowa, USA, pp. 65–71. https://doi.org/10.1002/9780470384916.ch1f.

De Battista, J., Altier, N., Galdames, D. R. and Dall'Agnot, M. (1997) 'Significance of endophyte toxicosis and current practices in dealing with the problem in South America'. In Roberts, C. A., West, C. P. and Spiers, D. E. (Eds), *Neotyphodium in cool-season grasses*, Blackwell Publishing Ltd Ames, Iowa, USA, pp. 383–387. https://doi.org/10.1007/978-1-4899-0271-9_60.

di Menna, M. E., Finch, S. C., Popay, A. J. and Smith, B. L. (2012) 'A review of the *Neotyphodium lolii / Lolium perenne* symbiosis and its associated effects on animal and plant health, with particular emphasis on ryegrass staggers', *N. Z. Vet. J.*, 60, 315–328. https://doi.org/10.1080/00480169.2012.697429.

di Menna, M. E., Mortimer, P. H., Prestidge, R. A., et al. (1992) 'Lolitrem B concentrations, counts of *Acremonium lolii* hyphae, and the incidence of ryegrass staggers in lambs on plots of *A. lolii*-infected perennial ryegrass', *N. Z. J. Agric. Res.*, 35, 211–217. https://doi.org/10.1080/00288233.1992.10417721.

Duringer, J. M., Blythe, L. L., Estill, C. T., et al. (2021) 'Determination of a sub-chronic threshold for lolitrem B and perennial ryegrass toxicosis in Angus cattle consuming endophyte-infected perennial ryegrass (Lolium perenne) straw over 64 days', *Livestock Science*, 250, 104570. https://doi.org/10.1016/j.livsci.2021.104570.

Dymock, J. J., Prestidge, R. A. and Rowan, D. D. (1989) 'The effects of lolitrem B on Argentine stem weevil larvae', *42nd New Zealand Weed and Pest Control Conference*, 73–75. https://doi.org/10.30843/nzpp.1989.42.10999.

Eady, C. (2021) 'The impact of alkaloid-producing *Epichloë* endophyte on forage ryegrass breeding: a New Zealand perspective', *Toxins*, 13(2), 158. https://doi.org/10.3390/toxins13020158.

Easton, H. S. (2007) 'Grasses and *Neotyphodium* endophytes: co-adaptation and adaptive breeding', *Euphytica*, 154(3), 295–306. https://doi.org/10.1007/s10681-006-9187-3.

Easton, H. S., Latch, G. C. M., Tapper, B. A. and Ball, O. J. P. (2002) 'Ryegrass host genetic control of concentrations of endophyte-derived alkaloids', *Crop Sci.*, 42(1), 51–57. https://doi.org/10.2135/cropsci2002.5100.

Easton, H. S., Lee, C. K. and Fitzgerald, R. D. (1994) 'Tall fescue in Australia and New Zealand', *N. Z. J. Agric. Res.*, 37, 405–417. https://doi.org/10.1080/00288233.1994.9513078.

Eerens, J. P. J., Lucas, R. J., Easton, S. and White, J. G. H. (1998) 'Influence of the endophyte (*Neotyphodium lolii*) on morphology, physiology, and alkaloid synthesis of perennial ryegrass during high temperature and water stress', *N. Z. J. Agric. Res.*, 41(2), 219–226. https://doi.org/10.1080/00288233.1998.9513305.

EFSA (2012) 'Scientific opinion on ergot alkaloids in food and feed', *EFSA Journal*, 10(7), 2798. https://doi.org/10.2903/j.efsa.2012.2798.

EFSA (2017) 'Human and animal dietary exposure to ergot alkaloids', *EFSA Journal*, 15(7), e04902. https://doi.org/10.2903/j.efsa.2017.4902.

Ekanayake, P. N., Hand, M. L., Spangenberg, G. C., et al. (2012) 'Genetic diversity and host specificity of fungal endophyte taxa in fescue pasture grasses', *Crop Sci.*, 52(5), 2243–2252. https://doi.org/10.2135/cropsci2011.12.0664.

Emile, J. C., Bony, S. and Ghesquiere, M. (2000) 'Influence of consumption of endophyte-infested tall fescue hay on performance of heifers and lambs', *J. Anim. Sci.*, 78(2), 358–364. https://doi.org/10.2527/2000.782358x.

Faeth, S. H., Gardner, D. R., Hayes, C. J., et al. (2006) 'Temporal and spatial variation in alkaloid levels in *Achnatherum robustum*, a native grass infected with the endophyte *Neotyphodium*', *J. Chem. Ecol.*, 32(2), 307–324. https://doi.org/10.1007/s10886-005-9003-x.

Faeth, S. H. and Saari, S. (2012) 'Fungal grass endophytes and arthropod communities: lessons from plant defence theory and multitrophic interactions', *Fungal Ecol.*, 5(3), 364–371. https://doi.org/10.1016/j.funeco.2011.09.003.

Faville, M. J., Briggs, L., Cao, M., et al. (2015) 'A QTL analysis of host plant effects on fungal endophyte biomass and alkaloid expression in perennial ryegrass', *Mol. Breed.*, 35, 161. https://doi.org/10.1007/s11032-015-0350-1.

Feuillet, C. and Muehlbauer, G. J. (Eds) (2009) *Genetics and genomics of the Triticeae*. Springer, Dordrecht, Netherlands.

Finch, S. C., Fletcher, L. R. and Babu, J. V. (2012) 'The evaluation of endophyte toxin residues in sheep fat', *N. Z. Vet. J.*, 60(1), 56–60. https://doi.org/10.1080/00480169.2011.634746.

Finch, S. C., Hawkes, A. D., Fitzgerald, J. M., et al. (2022a) 'Toxicological assessment of pure lolitrem B and ryegrass seed infected with the AR37 endophyte using mice', *J. Fungi*, 8(11), 1139. https://doi.org/10.3390/jof8111139.

Finch, S. C., Munday, J. S., Munday, R. and Kerby, J. W. F. (2016) 'Short-term toxicity studies of loline alkaloids in mice', *Food Chem. Toxicol.*, 94, 243–249. https://doi.org/10.1016/j.fct.2016.06.002.

Finch, S. C., Munday, J. S., Sprosen, J. M. and Bhattarai, S. (2019) 'Toxicity studies of chanoclavine in mice', *Toxins*, 11(5) 10.3390/toxins11050249.

Finch, S. C., Prinsep, M. R., Popay, A. J., et al. (2020) 'Identification and structure elucidation of epoxyjanthitrems from *Lolium perenne* infected with the endophytic fungus *Epichloë festucae* var. *lolii* and determination of the tremorgenic and anti-insect activity of epoxyjanthitrem I', *Toxins*, 12, 526. https://doi.org/10.3390/toxins12080526.

Finch, S. C., Thom, E. R., Babu, J. V., et al. (2013) 'The evaluation of fungal endophyte toxin residues in milk', *N. Z. Vet. J.*, 61(1), 11–17. https://doi.org/10.1080/00480169.2012.704626.

Finch, S. C., Vlaming, J. B., Sutherland, B. L., et al. (2018) 'Ergovaline does not alter the severity of ryegrass staggers induced by lolitrem B', *N. Z. Vet. J.*, 66(2), 93–97. https://doi.org/10.1080/00480169.2017.1416693.

Finch, S. C., Webb, N. G., Munday, J. S., et al. (2022b) 'Sub-acute feeding study of a tall fescue endophyte in a perennial ryegrass host using mice', *Toxicon*, 214, 30–36. https://doi.org/10.1016/j.toxicon.2022.04.019.

Fink-Gremmels, J. (2005) 'Mycotoxins in forages'. In Diaz, D. E. (Ed), *The mycotoxin blue book*, Nottingham University Press, Nottingham, UK, pp. 249–268.

Fisher, M. J., Bohnert, D. W., Ackerman, C. J., et al. (2004) 'Evaluation of perennial ryegrass straw as a forage source for ruminants', *J. Anim. Sci.*, 82(7), 2175–2184. https://doi.org/10.2527/2004.8272175X.

Fletcher, L. R. (1999) '"Non-toxic" endophytes in ryegrass and their effect on livestock health and production'. In Woodfield, D. R. and Matthew, C. (Eds), *Ryegrass Endophyte - An Essential New Zealand Symbiosis, Grassland Research and Practice Series No. 7,* New Zealand Grassland Association, Palmerston North, New Zealand, pp. 133–139. https://doi.org/10.33584/rps.7.1999.3393.

Fletcher, L. R. (2005) 'Managing ryegrass-endophyte toxicoses'. In Roberts, C. A., West, C. P. and Spiers, D. E. (Eds), *Neotyphodium in cool-season grasses*, Blackwell Publishing Ltd Ames, Iowa, USA, pp. 229–241. https://doi.org/10.1002/9780470384916.ch10.

Fletcher, L. R. (2012) 'Novel endophytes in New Zealand grazing systems: the perfect solution or a compromise?'. In Young, C. A., Aiken, G. E., McCulley, R. L., Strickland, J. R. and Schardl, C. L. (Eds), *Epichloae, endophytes of cool season grasses: implications, utilization and biology*, The Samuel Roberts Noble Foundation, Ardmore, Oklahoma, USA, pp. 5–13.

Fletcher, L. R. and Harvey, I. C. (1981) 'An association of a *Lolium* endophyte with ryegrass staggers', *N. Z. Vet. J.*, 29, 185–186. https://doi.org/10.1080/00480169.1981.34839.

Fletcher, L. R., Popay, A. J., Stewart, A. V. and Tapper, B. A. (2001) 'Herbage and sheep production from meadow fescue with and without the endophyte *Neotyphodium uncinatum*', *Fourth International Neotyphodium/ Grass Interactions Symposium*, Soest, Germany, 447–453.

Fletcher, L. R., Sutherland, B. L. and Fletcher, C. G. (1999) 'The impact of endophyte on the health and productivity of sheep grazing ryegrass based. In Woodfield, D. R. and Matthew, C. (Eds), *Ryegrass Endophyte - An Essential New Zealand Symbiosis, Grassland Research and Practice Series No. 7,* New Zealand Grassland Association, Palmerston North, New Zealand, pp. 11–17. https://doi.org/10.33584/rps.7.1999.3403.

Florea, S., Andreeva, K., Machado, C., et al. (2009) 'Elimination of marker genes from transformed filamentous fungi by unselected transient transfection with a Cre-expressing plasmid', *Fungal Genet. Biol.*, 46(10), 721–730. https://doi.org/10.1016/j.fgb.2009.06.010.

Florea, S., Jaromczyk, J. and Schardl, C. L. (2021) 'Non-transgenic CRISPR-mediated knockout of entire ergot alkaloid gene clusters in slow-growing asexual polyploid fungi', *Toxins*, 13(2), 153. https://doi.org/10.3390/toxins13020153.

Foot, J. Z., Heazlewood, P. G. and Cummins, L. J. (1988) 'The effect of high-endophyte perennial ryegrass pastures on reproduction in Merino ewes', *Aust. Adv. Vet. Sci.*, 146–147.

Fribourg, H. A., Hannaway, D. B. and West, C. P. (Eds) (2009) *Tall fescue for the twenty-first century. Agronomy monograph*, American Society of Agronomy, Crop Science Society of America, and Soil Science Society of America, Madison, Wisconsin, USA.

Fuchs, B., Krischke, M., Mueller, M. J. and Krauss, J. (2017) 'Plant age and seasonal timing determine endophyte growth and alkaloid biosynthesis', *Fungal Ecol.*, 29, 52–58. https://doi.org/10.1016/j.funeco.2017.06.003.

Fuchs, B., Kuhnert, E. and Krauss, J. (2020) 'Contrasting effects of grass – endophyte chemotypes on a tri-trophic cascade', *J. Chem. Ecol.*, 46(4), 422–429. https://doi.org/10.1007/s10886-020-01163-9.

Funk, C. R., Belanger, F. C. and Murphy, J. A. (1994) 'Role of endophytes in grasses used for turf and soil conservation'. In Bacon, C. W. and White, J. F. J. (Eds), *Biotechnology of endophytic fungi of grasses*, CRC Press, Inc., pp. 201–209. https://doi.org/10.1201/9781351070324-14.

Funk, C. R., Halisky, P. M., Johnson, M. C., et al. (1983) 'An endophytic fungus and resistance to sod webworms: association in *Lolium perenne* L.', *Nat. Biotechnol.*, 1(2), 189–191. https://doi.org/10.1038/nbt0483-189.

Funk, C. R., White, R. H. and Breen, J. P. (1993) 'Importance of *Acremonium* endophytes in turf-grass breeding and management', *Agric., Eco. & Envir.*, 44(1–4), 215–232. https://doi.org/10.1016/0167-8809(93)90048-T.

Galey, F. D., Barr, B. and Anderson, M. (1993) 'Grassland staggers of California', *N. Z. Vet. J.*, 41(4), 217.

Galey, F. D., Tracy, M. L., Craigmill, A. L., et al. (1991) 'Staggers induced by consumption of perennial ryegrass in cattle and sheep from northern California', *J. Am. Vet. Med. Assoc.*, 199(4), 466–470.

Gallagher, R. T., Hawkes, A. D., Steyn, P. S. and Vleggaar, R. (1984) 'Tremorgenic neurotoxins from perennial ryegrass causing ryegrass staggers disorder of livestock: structure elucidation of lolitrem B', *J. Chem. Soc., Chem. Comm.*, 614–616. https://doi.org/10.1039/c39840000614.

Garner, G. B. and Cornell, C. N. (1978) 'Fescue foot in cattle'. In Wylie, T. D. and Morehouse, L. G. (Eds), *Mycotoxic fungi, mycotoxins, mycotoxicoses. An encyclopedic handbook*, Marcel Dekker, New York, USA, pp. 45–82.

Gibbs Russell, G. E. and Ellis, R. P. (1982) 'The genus *Melica* L. (Poaceae) in southern Africa', *Bothalia*, 14, 37–44. https://doi.org/10.4102/abc.v14i1.1133.

Griffith, R. W., Grauwiler, J., Hodel, C., et al. (1978) 'Toxicologic considerations'. In Berde, B. and Schild, H. O. (Eds), *Ergot alkaloids and related compounds*, Springer Berlin Heidelberg, Berlin, Germany, pp. 805–851. https://doi.org/10.1007/978-3-642-66775-6_12.

Gundel, P. E., Bigaxzoli, F., Freitas, P. P., et al. (2020) 'Plant damage, seed production and persistence of the fungal endophyte *Epichloë occultans* in *Lolium multiflorum* plants under an herbivore lepidopteran attack and ozone pollution', *Ecol. Austral.*, 30(2), 321–330. https://doi.org/10.25260/ea.20.30.2.0.1034.

Hand, M. L., Cogan, N. O. I., Stewart, A. V. and Forster, J. W. (2010) 'Evolutionary history of tall fescue morphotypes inferred from molecular phylogenetics of the *Lolium-Festuca* species complex', *BMC Evol. Biol.*, 10, 303. https://doi.org/10.1186/1471-2148-10-303.

Heeswijck, R. V. and McDonald, G. (1992) '*Acremonium* endophytes in perennial ryegrass and other pasture grasses in Australia and New Zealand', *Aust. J. Agric. Res.*, 43(8), 1683–1709. https://doi.org/10.1071/AR9921683.

Hennessy, L. M., Popay, A. J., Finch, S. C., et al. (2016) 'Temperature and plant genotype alter alkaloid concentrations in ryegrass infected with an *Epichloë* endophyte and this affects an insect herbivore', *Frontiers Plant Sci.*, 7(JULY2016). https://doi.org/10.3389/fpls.2016.01097.

Hewitt, K. G., Popay, A. J., Hofmann, R. W. and Caradus, J. R. (2021) '*Epichloë* – a lifeline for temperate grasses under combined drought and insect pressure', *Grass Research*, 1, 7. https://doi.org/10.48130/GR-2021-0007.

Hoare, D. B. (2014) *Melica decumbens Thunb.* Available at: http://www.fao.org/ag/agp/AGPC/doc/gbase/Safricadata/meldec.htm (Accessed: 6 July 2014).

Hofmeister, F. (1892) 'The active constituents of *Lolium temulentum*', *Archiv Fur Experimentelle Pathologie Und Pharmacologie*, 30, 203–230.

Holmes, L. A., Frame, N. W., Frame, R. K., et al. (1999) 'Suspected tremorgenic mycotoxicosis (ryegrass staggers) in alpacas (*Llama pacos*) in the UK', *Vet. Rec.*, 145(16), 462–463. https://doi.org/10.1136/vr.145.16.462.

Huizing, H. J., Molen, W. v. d., Kloek, W. and Nijs, A. P. M. d. (1991) 'Detection of lolines in endophyte-containing meadow fescue in The Netherlands and the effect of elevated temperature on induction of lolines in endophyte-infected perennial ryegrass', *Grass Forage Sci.*, 46(4), 441–445. https://doi.org/10.1111/j.1365-2494.1991.tb02405.x.

Hume, D. E. and Barker, D. J. (2005) 'Growth and management of endophytic grasses in pastoral agriculture'. In Roberts, C. A., West, C. P. and Spiers, D. E. (Eds), *Neotyphodium in cool-season grasses*, Blackwell Publishing Ltd Ames, Iowa, USA, pp. 201–226. https://doi.org/10.1002/9780470384916.ch9.

Hume, D. E., Christensen, M. J. and Hickey, M. J. (1993) 'Changes over time in the incidence of *Acremonium* endophyte in annual and short-lived hybrid ryegrass swards', *Second International Symposium on Acremonium/Grass Interactions*, Palmerston North, New Zealand, 210–213.

Hume, D. E., Cooper, B. M. and Panckhurst, K. A. (2009) 'The role of endophyte in determining the persistence and productivity of ryegrass, tall fescue and meadow fescue in Northland', *New Zealand Grassland Association Conference*, Waitangi, New Zealand, 145–150. https://doi.org/10.33584/jnzg.2009.71.2754.

Hume, D. E., Drummond, J. B., Rolston, M. P., et al. (2018) '*Epichloë* endophyte improves agronomic performance and grain yield of rye (*Secale cereale*)', *10th International Symposium on Fungal Endophyte of Grasses*, Salamanca, Spain, 102.

Hume, D. E., Ryan, G. D., Gibert, A., et al. (2016) '*Epichloë* fungal endophytes for grassland ecosystems'. In Lichtfouse, E. (Ed), *Sustainable agriculture reviews*. Springer International Publishing, Switzerland, pp. 233–305. https://doi.org/10.1007/978-3-319-26777-7_6.

Hume, D. E. and Sewell, J. C. (2014) 'Agronomic advantages conferred by endophyte infection of perennial ryegrass (*Lolium perenne* L.) and tall fescue (*Festuca arundinacea* Schreb.) in Australia', *Crop & Pasture Sci.*, 65, 747–757. https://doi.org/10.1071/cp13383.

Hume, D. E., Stewart, A. V., Simpson, W. R. and Johnson, R. D. (2020) '*Epichloë* fungal endophytes play a fundamental role in New Zealand grasslands', *J. R. Soc. N. Z.*, 50(2), 279–298. https://doi.org/10.1080/03036758.2020.1726415.

Hunt, M. G., Rasmussen, S., Newton, P. C. D., et al. (2005) 'Near-term impacts of elevated CO_2, nitrogen and fungal endophyte-infection on *Lolium perenne* L. growth, chemical composition and alkaloid production', *Plant, Cell & Envir.*, 28(11), 1345–1354. https://doi.org/10.1111/j.1365-3040.2005.01367.x.

Iannone, L. J., Novas, M. V., McCargo, P. D., et al. (2021) 'Diversity, ecology, and applications of *Epichloë* fungal endophytes of grasses in South America'. In Rosa, L. H. (Ed), *Neotropical Endophytic Fungi*. Springer, Cham, Switzerland, pp. 11-36. https://doi.org/10.1007/978-3-030-53506-3_2.

Iannone, L. J., Novas, M. V., Young, C. A., et al. (2012) 'Endophytes of native grasses from South America: Biodiversity and ecology', *Fungal Ecol.*, 5, 357–363. https://doi.org/10.1016/j.funeco.2011.05.007.

Iannone, L. J., White Jr, J. F., Giussani, L. M., et al. (2011) 'Diversity and distribution of *Neotyphodium*-infected grasses in Argentina', *Mycological Progress*, 10(1), 9–19. https://doi.org/10.1007/s11557-010-0669-2.

Jaussaud, P., Durix, A., Videmann, B., et al. (1998) 'Rapid analysis of ergovaline in ovine plasma using high-performance liquid chromatography with fluorimetric detection', *J. Chromatogr.*, 815(1), 147–153. https://doi.org/10.1016/S0021-9673(98)00002-8.

Jensen, A. M. D. and Roulund, N. (2004) 'Occurrence of *Neotyphodium* endophytes in permanent grassland with perennial ryegrass (*Lolium perenne*) in Denmark', *Agric., Eco. & Envir.*, 104(3), 419–427. https://doi.org/10.1016/j.agee.2004.01.044.

Jensen, J. G. and Popay, A. J. (2004) 'Perennial ryegrass infected with AR37 endophyte reduces survival of porina larvae', *N. Z. Plant Prot.*, 57, 323–328.

Johnson, L. J., Bastías, D. A., Caradus, J. R., et al. (2021) 'The dynamic mechanisms underpinning symbiotic *Epichloë*–grass interactions: implications for sustainable and resilient agriculture'. In White, J., Kumar, A. and Droby, S. (Eds), *Microbiome Stimulants for Crops*. Woodhead Publishing, pp. 73–108. 6. https://doi.org/10.1016/B978-0-12-822122-8.00008-X.

Johnson, L. J. and Caradus, J. R. (2019) 'The science required to deliver *Epichloë* endophytes to commerce'. In Hodkinson, T. R., Doohan, F. M., Saunders, M. J. and Murphy, B. R. (Eds), *Endophytes for a growing world*. Cambridge University Press, Cambridge, UK, pp. 343–370. https://doi.org/10.1017/9781108607667.017.

Johnson, L. J., De Bonth, A. C. M., Briggs, L. R., et al. (2013) 'The exploitation of epichloae endophytes for agricultural benefit', *Fungal Diversity*, 60, 171–188. https://doi.org/10.1007/s13225-013-0239-4.

Johnson, L. J., Voisey, C. R., Faville, M. J., et al. (2019) 'Advances and perspectives in breeding for improved grass-endophyte associations', *Grass. Sci. Eur.*, 24, 351–363.

Johnson, M. C., Dahlman, D. L., Siegel, M. R., et al. (1985) 'Insect feeding deterrents in endophyte-infected tall fescue', *Appl. Environ. Microbiol.*, 49(3), 568–571. https://doi.org/10.1128/aem.49.3.568-571.1985.

Johnson, R., Voisey, C., Johnson, L., et al. (2007) 'Distribution of NRPS gene families within the *Neotyphodium/Epichloë* complex', *Fungal Genet. Biol.*, 44(11), 1180–1190. https://doi.org/10.1016/j.fgb.2007.04.009.

Johnson, R. D., Lane, G. A., Koulman, A., et al. (2015) 'A novel family of cyclic oligopeptides derived from ribosomal peptide synthesis of an in planta-induced gene, gigA, in *Epichloë* endophytes of grasses', *Fungal Genet. Biol.*, 85, 14–14. https://doi.org/10.1016/j.fgb.2015.10.005.

Johnson, R. D., Tsujimoto, H., Hume, D. E., et al. (2018) 'Alien chromatin from Hordeeae grasses enhances the compatibility of *Epichloë* endophyte symbiosis with the hexaploid wheat *Triticum aestivum*', *10th International Symposium on Fungal Endophyte of Grasses*, Salamanca, Spain, 104.

Jones, T. A., Ralphs, M. H., Gardner, D. R. and Chatterton, N. J. (2000) 'Cattle prefer endophyte-free robust needlegrass', *Journal of Range Management*, 53(4), 427–431. https://doi.org/10.2307/4003755.

Jung, G. A., Van Wijk, A. J. P., Hunt, W. F. and Watson, C. E. (1996) 'Ryegrasses'. In Moser, L. E., Buxton, D. R. and Casler, M. D. (Eds), *Cool-season forage grasses*, American Society of Agronomy, Crop Science Society of America and Soil Science Society of America, Madison, Wisconsin, USA, pp. 605–641. https://doi.org/10.2134/agronmonogr34.

Justus, M., Witte, L. and Hartmann, T. (1997) 'Levels and tissue distribution of loline alkaloids in endophyte-infected *Festuca pratensis*', *Phytochem.*, 44(1), 51–57. https://doi.org/10.1016/S0031-9422(96)00535-3.

Kauppinen, M., Saikkonen, K., Helander, et al. (2016) *Epichloë* grass endophytes in sustainable agriculture. *Nature Plants* 2, 15224. https://doi.org/10.1038/nplants.2015.224.

Kaur, J., Ekanayake, P. N., Tian, P., et al. (2015) 'Discovery and characterisation of novel asexual *Epichloë* endophytes from perennial ryegrass (*Lolium perenne* L.)', *Crop. Pasture Sci.*, 66, 1058–1070. https://doi.org/10.1071/cp14332.

Kellerman, T. S., Prozesky, L., Schultz, R. A. and Rabie, C. J. (1993) 'Neuromycotoxicoses of livestock on pastures in South Africa, with special reference to diploidosis', *N. Z. Vet. J.*, 41, 212–213.

Kennedy, C. W. and Bush, L. P. (1983) 'Effect of environmental and management factors on the accumulation of N-acetyl and N-formyl loline alkaloids in tall fescue', *Crop Sci.*, 23(3), 547–552. https://doi.org/10.2135/cropsci1983.0011183X002300030024x.

König, J., Fuchs, B., Krischke, M., et al. (2017) 'Hide and seek – Infection rates and alkaloid concentrations of *Epichloë festucae* var. *lolii* in *Lolium perenne* along a land-use gradient in Germany', *Grass Forage Sci.*, 73(2), 510–516. https://doi.org/10.1111/gfs.12330.

Lamichhane, J. R. (2017) 'Pesticide use and risk reduction in European farming systems with IPM: An introduction to the special issue', *Crop Protect.*, 97, 1–6. https://doi.org/j.cropro.2017.01.017.

Latch, G. C. M., Potter, L. R. and Tyler, B. F. (1987) 'Incidence of endophytes in seeds from collections of *Lolium* and *Festuca* species', *Ann. Appl. Biol.*, 111, 59–64. https://doi.org/10.1111/j.1744-7348.1987.tb01433.x.

Layton, D. L., Fletcher, L. R., Litherland, A. J., et al. (2004) 'Effect of ergot alkaloids on liveweight gain and urine lysergol level in ewe hoggets and heifers', *N. Z. Soc. Anim. Prod.*, 192–196.

Le Bars, J. and Le Bars, P. (1996) 'Recent acute and subacute mycotoxicoses recognized in France', *Vet. Res.*, 27(4–5), 383–394.

Leuchtmann, A., Bacon, C. W., Schardl, C. L., et al. (2014) 'Nomenclatural realignment of *Neotyphodium* species with genus *Epichloë*', *Mycologia*, 106, 202–215. https://doi.org/10.3852/106.2.202.

Lean, I. J. (2001) 'Association between feeding perennial ryegrass (*Lolium perenne* cultivar Grasslands Impact) containing high concentrations of ergovaline, and health and productivity in a herd of lactating dairy cows', *Aust. Vet. J.*, 79, 262–264. https://doi.org/10.1111/j.1751-0813.2001.tb11978.x.

Leuchtmann, A. and Clay, K. (1997) 'The population biology of grass endophytes'. In Tudzynski, C. (Ed.), *Mycota V. Plant Relationships*, Springer-Verlag, Berlin, Germany, pp. 185–202. https://doi.org/10.1007/978-3-642-60647-2_12.

Leuchtmann, A., Schmidt, D. and Bush, L. P. (2000) 'Different levels of protective alkaloids in grasses with stroma-forming and seed-transmitted *Epichloë/Neotyphodium* endophytes', *J. Chem. Ecol.*, 26(4), 1025–1036. https://doi.org/10.1023/a:1005489032025.

Leury, B., Henry, M., Kemp, S. and Webb Ware, J. (2014) *Developing increased understanding, awareness and potential mitigation strategies for perennial ryegrass toxicosis in sheep production systems*. Limited, M. L. A., North Sydney, Australia [Online]. Available at: http://www.mla.com.au/Research-and-development/Final-report-details?projectid=15461.

Lewis, G. C. (1997) 'Significance of endophyte toxicosis and current practices in dealing with the problem in Europe'. In Bacon, C. W. and Hill, N. S. (Eds), *Neotyphodium/Grass Interactions. Proceedings of the Third International Symposium on Neotyphodium/*

Grass Interactions, Athens, Georgia, USA, pp. 377–382. https://doi.org/10.1007/978-1-4899-0271-9_59.

Lewis, G. C. (2001) 'An overview of the symbiotic association between *Neotyphodium/Epichloë* fungal endophytes and forage grasses', *Asp. Appli. Biol.*, 63, 203–212.

Lewis, G. C. and Clements, R. O. (1986) 'A survey of ryegrass endophyte (*Acremonium loliae*) in the UK and its apparent ineffectuality on a seedling pest', *J. Agric. Sci., UK*, 107(3), 633–638.

Lewis, G. C., Ravel, C., Naffaa, W., et al. (1997) 'Occurrence of Acremonium endophytes in wild populations of *Lolium* spp. in European countries and a relationship between level of infection and climate in France', *Ann. Appl. Biol.*, 130(2), 227–238. https://doi.org/10.1111/j.1744-7348.1997.tb06828.x.

Leyronas, C. and Raynal, G. (2001) 'Presence of *Neotyphodium*-like endophytes in European grasses', *Ann. Appl. Biol.*, 139(1), 119–127. https://doi.org/10.1111/j.1744-7348.2001.tb00136.x.

Li, C., Wang, Z., Chen, T. and Nan, Z. (2021) 'Creation of novel barley germplasm using an *Epichloë* endophyte', *Chin. Sci. Bull.*, 66(20), 2608–2617. https://doi.org/10.1360/TB-2020-1587.

Liang, Y., Wang, H., Li, C., et al. (2017) 'Effects of feeding drunken horse grass infected with *Epichloë gansuensis* endophyte on animal performance, clinical symptoms and physiological parameters in sheep', *BMC Vet. Res.*, 13(1) 10.1186/s12917-017-1120-6.

Lin, W., Gao, C., Wang, J., et al. (2022) 'Effects of drought stress on peramine and lolitrem B in *Epichloë*-endophyte-infected perennial ryegrass', *Life*, 12(8). https://doi.org/10.3390/life12081207.

Little, P., Jennings, G., Skews, H. and Bobik, A. (1982) 'Bioavailability of dihydroergotamine in man', *Br. J. Clin. Pharmacol.*, 13(6), 785–790. https://doi.org/10.1111/j.1365-2125.1982.tb01866.x.

Liu, J., Nagabhyru, P. and Schardl, C. L. (2017) '*Epichloë festucae* endophytic growth in florets, seeds, and seedlings of perennial ryegrass (*Lolium perenne*)', *Mycologia*, 109(5), 691–700. https://doi.org/10.1080/00275514.2017.1400305.

Lyons, P. C., Plattner, R. D. and Bacon, C. W. (1986) 'Occurrence of peptide and clavine ergot alkaloids in tall fescue grass', *Science, USA*, 232(4749), 487–489. https://doi.org/10.1126/science.300832.

Malinowski, D., Leuchtmann, A., Schmidt, D. and Nösberger, J. (1997a) 'Growth and water status in meadow fescue is affected by *Neotyphodium* and *Phialophora* species endophytes', *Agron. J.*, 89(4), 673–678. https://doi.org/10.2134/agronj1997.00021962008900040021x.

Malinowski, D., Leuchtmann, A., Schmidt, D. and Nösberger, J. (1997b) 'Symbiosis with *Neotyphodium uncinatum* endophyte may increase the competitive ability of meadow fescue', *Agron. J.*, 89(5), 833–839. https://doi.org/10.2134/agronj1997.00021962008900050019x.

Malinowski, D. P. and Belesky, D. P. (2000) 'Adaptations of endophyte-infected cool-season grasses to environmental stresses: mechanisms of drought and mineral stress tolerance', *Crop Sci.*, 40, 923–940. https://doi.org/10.2135/cropsci2000.404923x.

Malinowski, D. P. and Belesky, D. P. (2019) '*Epichloë* (formerly *Neotyphodium*) fungal endophytes increase adaptation of coolseason perennial grasses to environmental stresses', *Acta Agrobotanica*, 72(2), 1767. https://doi.org/10.5586/aa.1767.

Matsukura, K., Shiba, T., Sasaki, T. and Matsumura, M. (2012) 'Enhanced resistance to four species of clypeorrhynchan pests in *Neotyphodium uncinatum* infected Italian ryegrass', *J. Econ. Entomol.*, 105(1), 129–134. https://doi.org/10.1603/ec11216.

McCulley, R. L., Bush, L. P., Carlisle, A. E., et al. (2014) 'Warming reduces tall fescue abundance but stimulates toxic alkaloid concentrations in transition zone pastures of the U.S.', *Frontiers in Chemistry*, 2(OCT), 88. https://doi.org/10.3389/fchem.2014.00088

Miles, C. O., Di Menna, M. E., Kellerman, T. S., et al. (1995a) 'Melica decumbens'. In Garthwaite, L. (Ed), *Toxinology and food safety research report 1992–1995*. AgResearch, Hamilton, New Zealand, p. 20.

Miles, C. O., Lane, G. A., di Menna, M. E., et al. (1996) 'High levels of ergonovine and lysergic acid amide in toxic *Achnatherum inebrians* accompany infection by an *Acremonium*-like endophytic fungus', *J. Agric. Food Chem.*, 44(5), 1285–1290. https://doi.org/10.1021/jf950410k.

Miles, C. O., Menna, M. E. d., Jacobs, S. W. L., et al. (1998) 'Endophytic fungi in indigenous Australasian grasses associated with toxicity to livestock', *Appl. Environ. Microbiol.*, 64, 601–606. https://doi.org/10.1128/aem.64.2.601-606.1998.

Miles, C. O., Munday, S. C., Wilkins, A. L., et al. (1994) 'Large-scale isolation of lolitrem B and structure determination of lolitrem E', *J. Agric. Food Chem.*, 42(7), 1488–1492. https://doi.org/10.1021/jf00043a018.

Miles, C. O., Uzal, F., Garthwaite, I., et al. (1995b) '*Poa huecu* and *Festuca argentina*'. In Garthwaite, L. (Ed), *Toxinology and food safety research report 1992–1995*. AgResearch, Hamilton, New Zealand, p. 20.

Miller, T. A., Hudson, D. A., Johnson, R. D., et al. (2022) 'Dissection of the epoxyjanthitrem pathway in *Epichloë* sp. LpTG-3 strain AR37 by CRISPR gene editing', *Front. Fungal Biol.*, 3. https://doi.org/10.3389/ffunb.2022.944234.

Mirlohi, A., Sabzalian, M. R. and Nekouei, M. K. (2004) 'Endophytic fungi, characteristics and their potential for genetic manipulation', *Iran. J. Biotechnol.*, 2(2), 75–83.

Miyazaki, S. (2003) 'Current subjects on endophyte toxicosis', *Mycotoxins*, 53(1), 57–62. https://doi.org/10.2520/myco.53.57.

Miyazaki, S., Fukumura, M., Yoshioka, M. and Yamanaka, N. (2001) 'Detection of endophyte toxins in the imported perennial ryegrass straw', *J. Vet. Med. Sci.*, 63(9), 1013–1015. https://doi.org/10.1292/jvms.63.1013.

Miyazaki, S., Ishizaki, I., Ishizaka, M., et al. (2004) 'Lolitrem B residue in fat tissues of cattle consuming endophyte-infected perennial ryegrass straw', *J. Vet. Diag. Invest.*, 16(4), 340–342. https://doi.org/10.1177/104063870401600416.

Moate, P. J., Williams, S. R. O., Grainger, C., et al. (2012) 'Effects of wild-type, AR1 and AR37 endophyte-infected perennial ryegrass on dairy production in Victoria, Australia', *Anim. Prod. Sci.*, 52(12), 1117–1130. https://doi.org/10.1071/AN12126.

Moon, C. D., Guillaumin, J. J., Ravel, C., et al. (2007) 'New *Neotyphodium* endophyte species from the grass tribes Stipeae and Meliceae', *Mycologia*, 99(6), 895–905. https://doi.org/10.3852/mycologia.99.6.895.

Moon, C. D., Miles, C. O., Järlfors, U. and Schardl, C. L. (2002) 'The evolutionary origins of three new *Neotyphodium* endophyte species from grasses indigenous to the Southern Hemisphere', *Mycologia*, 94, 694–711. https://doi.org/10.1080/15572536.2003.11833197.

Moon, C. D., Scott, D. B., Schardl, C. L. and Christensen, M. J. (2000) 'The evolutionary origins of *Epichloë* endophytes from annual ryegrasses', *Mycologia*, 92, 1103–1118. https://doi.org/10.1080/00275514.2000.12061258.

Moore, J. R., Pratley, J. E., Mace, W. J. and Weston, L. A. (2015) 'Variation in alkaloid production from genetically diverse *Lolium* accessions infected with *Epichloë* species', *J. Agric. Food Chem.*, 63(48), 10355–10365. https://doi.org/10.1021/acs.jafc.5b03089.

Mortimer, P. H., Fletcher, L. R., Menna, M. E. d., et al. (1982) 'Recent advances in ryegrass staggers', *Ruakura Farmers' Conference*, Hamilton, New Zealand, 71–74.

Moubarak, A. S., Piper, E. L., Johnson, Z. B. and Flieger, M. (1996) 'HPLC method for detection of ergotamine, ergosine, and ergine after intravenous injection of a single dose', *J. Agric. Food Chem.*, 44(1), 146–148. https://doi.org/10.1021/jf950425i.

Munday, J. S., Finch, S. C., Vlaming, J. B., et al. (2017) 'Pathological changes seen in horses in New Zealand grazing Mediterranean tall fescue (*Lolium arundinaceum*) infected with selected endophytes (*Epichloë coenophiala*) causing equine fescue oedema', *N. Z. Vet. J.*, 65(3), 147–151. https://doi.org/10.1080/00480169.2016.1248516.

Odriozola, E., Lopez, T., Campero, C. and Giminez Placeres, C. (1993) 'Ryegrass staggers in heifers: a new mycotoxicosis in Argentina', *Vet. Hum. Toxicol.*, 35(2), 144–146.

OECD (2022) *Guidelines for the testing of chemicals. Test No. 425: Acute oral toxicity: Up-and-down procedure.* https://doi.org/10.1016/j.yrtph.2023.105517.

Oliveira, J. A., Rottinghaus, G. E. and González, E. (2003) 'Ergovaline concentration in perennial ryegrass infected with a lolitrem B-free fungal endophyte in north-west Spain', *N. Z. J. Agric. Res.*, 46(2), 117–122. https://doi.org/10.1080/00288233.2003.9513537.

Panaccione, D. G. (2005) 'Origins and significance of ergot alkaloid diversity in fungi', *FEMS Microbiol. Lett.*, 251(1), 9–17. https://doi.org/10.1016/j.femsle.2005.07.039.

Panaccione, D. G., Cipoletti, J. R., Sedlock, A. B., et al. (2006a) 'Effects of ergot alkaloids on food preference and satiety in rabbits, as assessed with gene-knockout endophytes in perennial ryegrass (*Lolium perenne*)', *J. Agric. Food Chem.*, 54(13), 4582–4587. https://doi.org/10.1021/jf060626u.

Panaccione, D. G., Johnson, R. D., Wang, J., et al. (2001) 'Elimination of ergovaline from a grass-*Neotyphodium* endophyte symbiosis by genetic modification of the endophyte', *Proc. Nat. Acad. Sci. USA*, 98(22), 12820–12825. https://doi.org/10.1073/pnas.2211986.

Panaccione, D. G., Kotcon, J. B., Schardl, C. L., et al. (2006b) 'Ergot alkaloids are not essential for endophytic fungus-associated population suppression of the lesion nematode, *Pratylenchus scribneri*, on perennial ryegrass', *Nematology*, 8(4), 583–590. https://doi.org/10.1163/156854106778614074.

Parish, J. A., McCann, M. A., Watson, R. H., et al. (2003a) 'Use of nonergot alkaloid-producing endophytes for alleviating tall fescue toxicosis in sheep', *J. Anim. Sci.*, 81(5), 1316–1322. https://doi.org/10.2527/2003.8151316x.

Parish, J. A., McCann, M. A., Watson, R. H., et al. (2003b) 'Use of nonergot alkaloid-producing endophytes for alleviating tall fescue toxicosis in stocker cattle', *J. Anim. Sci.*, 81(11), 2856–2868. https://doi.org/10.2527/2003.81112856x.

Patchett, B., Gooneratne, R., Chapman, B. and Fletcher, L. (2011a) 'Effects of loline-producing endophyte-infected meadow fescue ecotypes on New Zealand grass grub (*Costelytra zealandica*)', *N. Z. J. Agric. Res.*, 54(4), 303–313. https://doi.org/10.1080/00288233.2011.608686.

Patchett, B., Gooneratne, R., Fletcher, L. and Chapman, B. (2011b) 'Seasonal changes in leaf and stem loline alkaloids in meadow fescue', *Crop & Pasture Sci.*, 62(3), 261–267. https://doi.org/10.1071/CP10266.

Patchett, B. J., Chapman, R. B., Fletcher, L. R. and Gooneratne, S. R. (2008a) 'Endophyte-infected *Festuca pratensis* containing loline alkaloids deters feeding by *Listronotus bonariensis*', *N. Z. Plant Prot.*, 61, 205–209. https://doi.org/10.30843/nzpp.2008.61.6843.

Patchett, B. J., Chapman, R. B., Fletcher, L. R. and Gooneratne, S. R. (2008b) 'Root loline concentration in endophyte-infected meadow fescue (*Festuca pratensis*) is increased by grass grub (*Costelytra zealandica*) attack', *N. Z. Plant Prot.*, 61, 210–214. https://doi.org/10.30843/nzpp.2008.61.6844.

Patterson, C. G., Potter, D. A. and Fannin, F. F. (1991) 'Feeding deterrency of alkaloids from endophyte-infected grasses to Japanese beetle grubs', *Entomol. Exp. Appl.*, 61(3), 285–289. https://doi.org/10.1111/j.1570-7458.1991.tb01561.x.

Pennell, C. G. L., Popay, A. J., Ball, O. J.-P., et al. (2005) 'Occurrence and impact of pasture mealybug (*Balanococcus poae*) and root aphid (*Aploneura lentisci*) on ryegrass (*Lolium* spp.) with and without infection by *Neotyphodium* fungal endophytes', *N. Z. J. Agric. Res.*, 48, 329–337. https://doi.org/10.1080/00288233.2005.9513663.

Pérez, L. I., Gundel, P. E., Ghersa, C. M. and Omacini, M. (2013) 'Family issues: fungal endophyte protects host grass from the closely related pathogen *Claviceps purpurea*', *Fungal Ecol.*, 6(5), 379–386. https://doi.org/10.1016/j.funeco.2013.06.006.

Pérez, L. I., Gundel, P. E., Zabalgogeazcoa, I. and Omacini, M. (2020) 'An ecological framework for understanding the roles of *Epichloë* endophytes on plant defenses against fungal diseases', *Fungal Biol. Reviews*, 34(3), 115–125. https://doi.org/10.1016/j.fbr.2020.06.001.

Petigrosso, L. R., Colabelli, M. N., Fernandez, O. N., et al. (2013) 'Incidence of the endophyte fungus *Neotyphodium coenophialum* in pastures of tall fescue differing in age and soil characteristics', *African J. Agric. Res.*, 8(22), 2655–2662. https://doi.org/10.5897/AJAR12.2106.

Petroski, R. J., Powell, R. G. and Clay, K. (1992) 'Alkaloids of *Stipa robusta* (sleepygrass) infected with an *Acremonium* endophyte', *Nat. Toxins*, 1(2), 84–88. https://doi.org/10.1002/nt.2620010205.

Pomilio, A. B., Rofi, R. D., Gambino, M. P., et al. (1989) 'The lethal principle of *Poa huecu* (coiron blanco): a plant indigenous to Argentina', *Toxicon*, 27(12), 1251–1262. https://doi.org/10.1016/0041-0101(89)90056-1.

Poore, M. H., Stratton, T., Roberts, C. A., et al. (2019) 'The Alliance for Grassland Renewal presents workshops to enhance the understanding and adoption of novel endophyte tall fescue', *J. Anim. Sci.*, 97(Supplement_1), 8–9. https://doi.org/10.1093/jas/skz053.018.

Popay, A. J. and Cox, N. R. (2016) '*Aploneura lentisci* (Homoptera: Aphididae) and its interactions with fungal endophytes in perennial ryegrass (*Lolium perenne*)', *Frontiers Plant Sci.*, 7, 1395. https://doi.org/10.3389/fpls.2016.01395.

Popay, A. J., Hickey, M. J. and Hume, D. E. (2008) 'Potential use of a selected strain of fungal endophyte for insect control in turf ryegrass', *1st European Turfgrass Society Conference*, Pisa, Italy, 151–152.

Popay, A. J., Hume, D. E., Davies, K. L. and Tapper, B. A. (2003) 'Interactions between endophyte (*Neotyphodium* spp.) and ploidy in hybrid and perennial ryegrass cultivars and their effects on Argentine stem weevil (*Listronotus bonariensis*)', *N. Z. J. Agric. Res.*, 46, 311–319. https://doi.org/10.1080/00288233.2003.9513559.

Popay, A. J., Jensen, J. G., Simpson, W. R., et al. (2023) 'Translocation of loline alkaloids in *Epichloë*-infected cereal and pasture grasses: what the insects tell us', *J. Fungi*, 9(1), 96. https://doi.org/10.3390/jof9010096.

Popay, A. J. and Lane, G. A. (2000) 'The effect of crude extracts containing loline alkaloids on two New Zealand insect pests', *Fourth International Neotyphodium/Grass Interactions Symposium*, Soest, Germany, 163–168.

Popay, A. J., Tapper, B. A. and Podmore, C. (2009) 'Endophyte-infected meadow fescue and loline alkaloids affect Argentine stem weevil larvae', *N. Z. Plant Prot.*, 62, 19–27. https://doi.org/10.30843/nzpp.2009.62.4801.

Popay, A. J. and Thom, E. R. (2009) 'Endophyte effects on major insect pests in Waikato dairy pasture'. In Mercer, C. F. (Ed.), *Pasture persistence symposium. Grassland Research and Practice Series No. 15*, New Zealand Grassland Association, Dunedin, New Zealand, pp. 121–126. https://doi.org/10.33584/jnzg.2009.71.2758.

Popay, A. J. and Wyatt, R. T. (1995) 'Resistance to Argentine stem weevil in perennial ryegrass infected with endophytes producing different alkaloids', *48th New Zealand Plant Protection Conference*, 229–236. https://doi.org/10.30843/nzpp.1995.48.11487.

Potter, D. A., Tyler Stokes, J., Redmond, C. T., et al. (2008) 'Contribution of ergot alkaloids to suppression of a grass-feeding caterpillar assessed with gene knockout endophytes in perennial ryegrass', *Entomol. Exp. Appl.*, 126(2), 138–147. https://doi.org/10.1111/j.1570-7458.2007.00650.x.

Poudel, S., Zeller, W. E., Fike, J. and Pent, G. (2023) 'Condensed tannins attributes: potential solution to fescue toxicosis?', *Agriculture (Switzerland)*, 13(3). https://doi.org/10.3390/agriculture13030672.

Prestidge, R. A. (1993) 'Causes and control of perennial ryegrass staggers in New Zealand', *Agric., Eco. & Envir.*, 44(1–4), 283–300. https://doi.org/10.1016/0167-8809(93)90051-P.

Pritchard, G. C. and Lewis, G. C. (1995) 'Ryegrass staggers', *The Veterinary Record*, 137(18), 471–472.

Pulsford, M. F. (1950) 'A note on lameness in cattle grazing on tall meadow fescue (*Festuca arundinacea*) in South Australia', *Aust. Vet. J.*, 26, 87–88. https://doi.org/10.1111/j.1751-0813.1950.tb04880.x.

Rasmussen, S., Parsons, A. J., Bassett, S., et al. (2007) 'High nitrogen supply and carbohydrate content reduce fungal endophyte and alkaloid concentration in *Lolium perenne*', *New Phytol.*, 173(March 2007), 787–797. https://doi.org/10.1111/j.1469-8137.2006.01960.x.

Raynal, G. (1991) 'Observations francaise sur les *Acremonium*, champignons endophytes des graminées fourragères', *Fourrages*, 126, 225–236.

Read, J. C. and Camp, B. J. (1986) 'The effect of the fungal endophyte *Acremonium coenophialum* in tall fescue on animal performance, toxicity, and stand maintenance', *Agron. J.*, 78(5), 848–850. https://doi.org/10.2134/agronj1986.00021962007800050021x.

Realini, C. E., Duckett, S. K., Hill, N. S., et al. (2005) 'Effect of endophyte type on carcass traits, meat quality, and fatty acid composition of beef cattle grazing tall fescue', *J. Anim. Sci.*, 83(2), 430–439. https://doi.org/10.2527/2005.832430x.

Realini, F. M., Escobedo, V. M., Ueno, A. C., et al. (2024) 'Anti-herbivory defenses delivered by *Epichloë* fungal endophytes: a quantitative review of alkaloid concentration variation among hosts and plant parts', *Ann. Bot.*, 10.1093/aob/mcae014.

Reed, K. F. M. (2014) 'Perennial pasture grasses: an historical review of their introduction, use and development for southern Australia', *Crop & Pasture Sci.*, 65(8), 691–712. https://doi.org/10.1071/cp13284.

Reed, K. F. M., Leonforte, A., Cunningham, P. J., et al. (2000) 'Incidence of ryegrass endophyte (*Neotyphodium lolii*) and diversity of associated alkaloid concentrations among naturalised populations of perennial ryegrass (*Lolium perenne* L.)', *Aust. J. Agric. Res.*, 51(5), 569–578. https://doi.org/10.1071/AR99182.

Reed, K. F. M., Nie, Z. N., Walker, L. V., et al. (2011) 'Weather and pasture characteristics associated with outbreaks of perennial ryegrass toxicosis in southern Australia', *Anim. Prod. Sci.*, 51(8), 738–752. https://doi.org/10.1071/AN11016.

Reed, K. F. M., Scrivener, C. J., Rainsford, K. A. and Walker, L. V. (2005) '*Neotyphodium* research and application in Australia'. In Roberts, C. A., West, C. P. and Spiers, D. E. (eds.), *Neotyphodium in cool-season grasses*, Blackwell Publishing Ltd Ames, Iowa, USA, pp. 43–54. https://doi.org/10.1002/9780470384916.ch1d.

Reed, K. F. M., Vaughan, J. L., Cummins, L. J. and Moore, D. D. (2010) 'Impact of mycotoxins and of a mycotoxin deactivator on alpacas grazing perennial ryegrass infected with wild endophyte (*Neotyphodium* spp.)', *Anim. Prod. Sci.*, 50(9), 902–908. https://doi.org/10.1071/an10068.

Repussard, C., Zbib, N., Tardieu, D. and Guerre, P. (2014) 'Ergovaline and lolitrem B concentrations in perennial ryegrass in field culture in southern France: distribution in the plant and impact of climatic factors', *J. Agric. Food Chem.*, 62(52), 12707–12712. https://doi.org/10.1021/jf504581y.

Ribeiro, M. A. M. d. V., Gurney, K. A. and Bush, L. P. (1996) 'Endophyte *Acremonium lolii* in ecotypes of perennial ryegrass (*Lolium perenne* L.) collected in old Irish pastures', *Irish J. Agric. & Food Res.*, 35(2), 151–157.

Riedell, W. E., Kieckhefer, R. E., Petroski, R. J. and Powell, R. G. (1991) 'Naturally-occurring and synthetic loline alkaloid derivatives: insect feeding behavior modification and toxicity', *J. Entomol. Sci.*, 26(1), 122–129. https://doi.org/10.18474/0749-8004-26.1.122.

Roberts, C. and Andrae, J. (2004) 'Tall fescue toxicosis and management', *Crop Manag.*, 3, 1–18. https://doi.org/10.1094/CM-2004-0427-01-MG.

Roberts, C. and Andrae, J. (2005) 'Public education in tall fescue toxicosis'. In Roberts, C. A., West, C. P. and Spiers, D. E. (Eds), *Neotyphodium in cool-season grasses*, Blackwell Publishing Ltd Ames, Iowa, USA, pp. 361–379. https://doi.org/10.1002/9780470384916.ch18.

Roberts, C. A. and Andrae, J. G. (2023) 'Operational model of the alliance for grassland renewal', *XXV International Grassland Congress*, Covington, Kentucky, USA, 1345–1348. https://doi.org/10.52202/071171-0328.

Rodriguez, R. J., White Jr, J. F., Arnold, A. E. and Redman, R. S. (2009) 'Fungal endophytes: diversity and functional roles', *New Phytol.*, 182(2), 314–330. https://doi.org/10.1111/j.1469-8137.2009.02773.x.

Rottinghaus, G. E., Garner, G. B., Cornell, C. N. and Ellis, J. L. (1991) 'HPLC method for quantitating ergovaline in endophyte-infected tall fescue: seasonal variation of ergovaline levels in stems with leaf sheaths, leaf blades, and seed heads', *J. Agric. Food Chem.*, 39(1), 112–115. https://doi.org/10.1021/jf00001a022.

Rowan, D. D. (1993) 'Lolitrems, peramine and paxilline: mycotoxins of the ryegrass/endophyte interaction', *Agric., Eco. & Envir.*, 44(1–4), 103–122. https://doi.org/10.1016/0167-8809(93)90041-M.

Rowan, D. D., Dymock, J. J. and Brimble, M. A. (1990) 'Effect of fungal metabolite peramine and analogs on feeding and development of argentine stem weevil (*Listronotus bonariensis*)', *J. Chem. Ecol.*, 16(5), 1683–1695. https://doi.org/10.1007/bf01014100.

Rowan, D. D. and Gaynor, D. L. (1986) 'Isolation of feeding deterrents against Argentine stem weevil from ryegrass infected with the endophyte *Acremonium loliae*', *J. Chem. Ecol.*, 12(3), 647–658. https://doi.org/10.1007/BF01012099.

Saari, S., Lehtonen, P., Helander, M. and Saikkonen, K. (2009) 'High variation in frequency of infection by endophytes in cultivars of meadow fescue in Finland', *Grass Forage Sci.*, 64, 169–176. https://doi.org/10.1111/j.1365-2494.2009.00680.x.

Saikkonen, K., Young, C. A., Helander, M. and Schardl, C. L. (2016) 'Endophytic *Epichloë* species and their grass hosts: from evolution to applications', *Plant Mol. Biol.*, 90(6), 665–675. https://doi.org/10.1007/s11103-015-0399-6.

Salminen, S. O. and Grewal, P. S. (2002) 'Does decreased mowing frequency enhance alkaloid production in endophytic tall fescue and perennial ryegrass?', *J. Chem. Ecol.*, 28(5), 939–950. https://doi.org/10.1023/A:1015201616013.

Saper, J. R. and Silberstein, S. (2006) 'Pharmacology of dihydroergotamine and evidence for efficacy and safety in migraine', *Headache: J. Head & Face Pain*, 46(s4), S171–S181. https://doi.org/10.1111/j.1526-4610.2006.00601.x.

Schardl, C. L. (2010) 'The epichloae, symbionts of the grass subfamily Poideae', *Ann. Missouri Bot. Garden*, 97, 646–665. https://doi.org/10.3417/2009144.

Schardl, C. L., Craven, K. D., Speakman, S., et al. (2008) 'A novel test for host-symbiont codivergence indicates ancient origin of fungal endophytes in grasses', *Syst. Biol.*, 57(3), 483–498. https://doi.org/10.1080/10635150802172184.

Schardl, C. L., Grossman, R. B., Nagabhyru, P., et al. (2007) 'Loline alkaloids: Currencies of mutualism', *Phytochem.*, 68, 980–996. https://doi.org/10.1016/j.phytochem.2007.01.010.

Schardl, C. L., Leuchtmann, A. and Spiering, M. J. (2004) 'Symbioses of grasses with seedborne fungal endophytes', *Annu. Rev. Plant Biol.*, 55, 315–340. https://doi.org/10.1146/annurev.arplant.55.031903.141735.

Schardl, C. L., Young, C. A., Faulkner, J. R., et al. (2012) 'Chemotypic diversity of *epichloae*, fungal symbionts of grasses', *Fungal Ecol.*, 5(3), 331–344. https://doi.org/10.1016/j.funeco.2011.04.005.

Schardl, C. L., Young, C. A., Hesse, U., et al. (2013a) 'Plant-symbiotic fungi as chemical engineers: multi-genome analysis of the Clavicipitaceae reveals dynamics of alkaloid loci', *PLoS Genet.*, 9(2), e1003323. https://doi.org/10.1371/journal.pgen.1003323.

Schardl, C. L., Young, C. A., Pan, J., et al. (2013b) 'Currencies of mutualisms: sources of alkaloid genes in vertically transmitted epichloae', *Toxins*, 5, 1064–1088. https://doi.org/10.3390/toxins5061064.

Schmidt, D. (1993) 'Effects of *Acremonium uncinatum* and a *Phialophora*-like endophyte on vigour insect and disease resistance of meadow fescue', *Second International Symposium on Acremonium/Grass Interactions*, Palmerston North, New Zealand, 185–188.

Schmidt, S. P. and Osborn, T. G. (1993) 'Effects of endophyte-infected tall fescue on animal performance', *Agric., Eco. & Envir.*, 44, 233–262. https://doi.org/10.1016/0167-8809(93)90049-U.

Schumann, B., Lebzien, P., Ueberschär, K.-H. and Dänicke, S. (2009) 'Effects of the level of feed intake and ergot contaminated concentrate on ergot alkaloid metabolism and carry over into milk', *Mol. Nutr. Food Res.*, 53(7), 931–938. https://doi.org/10.1002/mnfr.200800319.

Seddon, H. R. and Carne, H. R. (1926) 'Staggers in stock due to rough-bearded grass (*Echinopogon ovatus*)', *Agricultural Gazette New South Wales*, 37, 684–690.

Sepulveda, N., Rodriguez, F. and Jara, R. (1996) 'Poisoning in cattle from consumption of tall fescue (*Festuca arundinacea*) with the fungal endophyte (*Acremonium coenophialum*)', *Archivos de Medicina Veterinaria*, 28(1), 113–116.

Shi, T.-Q., Liu, G.-N., Ji, R.-Y., et al. (2017) 'CRISPR/Cas9-based genome editing of the filamentous fungi: the state of the art', *Appl. Microbiol. Biotechnol.*, 101(20), 7435–7443. https://doi.org/10.1007/s00253-017-8497-9.

Shiba, T. and Sugawara, K. (2009) 'Fungal loline alkaloids in grass-endophyte associations confer resistance to the rice leaf bug, *Trigonotylus caelestialium*', *Entomol. Exp. Appl.*, 130(1), 55–62. https://doi.org/10.1111/j.1570-7458.2008.00791.x.

Shiba, T., Sugawara, K. and Arakawa, A. (2011) 'Evaluating the fungal endophyte *Neotyphodium occultans* for resistance to the rice leaf bug, *Trigonotylus caelestialium*, in Italian ryegrass, *Lolium multiflorum*', *Entomol. Exp. Appl.*, 141(1), 45–51. https://doi.org/10.1111/j.1570-7458.2011.01162.x.

Shimada, N., Yoshioka, M., Mikami, O., et al. (2013) 'Toxicological evaluation and bioaccumulation potential of lolitrem B, endophyte mycotoxin in Japanese black steers', *Food Additives and Contaminants - Part A Chemistry, Analysis, Control, Exposure and Risk Assessment*, 30(8), 1402–1406. https://doi.org/10.1080/19440049.2013.790090.

Shymanovich, T., Saari, S., Lovin, M. E., et al. (2015) 'Alkaloid variation among Epichloid endophytes of sleepygrass (*Achnatherum robustum*) and consequences for resistance to insect herbivores', *J. Chem. Ecol.*, 41(1), 93–104. https://doi.org/10.1007/s10886-014-0534-x.

Siegel, M. R., Johnson, M. C., Varney, D. R., et al. (1984) 'A fungal endophyte in tall fescue: incidence and dissemination', *Phytopath.*, 74(8), 932–937. https://doi.org/10.1094/Phyto-74-932.

Siegel, M. R., Latch, G. C. M., Bush, L. P., et al. (1990) 'Fungal endophyte-infected grasses: alkaloid accumulation and aphid response', *J. Chem. Ecol.*, 16, 3301–3315. https://doi.org/10.1007/BF00982100.

Simpson, W. R., Faville, M. J., Moraga, R. A., et al. (2014) '*Epichloë* fungal endophytes and the formation of synthetic symbioses in Hordeeae (=Triticeae) grasses', *J. System. & Evol.*, 52, 794–806. https://doi.org/10.1111/jse.12107.

Simpson, W. R. and Mace, W. J. (2012) 'Novel associations between epichloid endophytes and grasses: possibilities and outcomes'. In Young, C. A., Aiken, G. E., McCulley, R. L., Strickland, J. R. and Schardl, C. L. (Eds), *Epichloae, endophytes of cool season grasses: Implications, utilization and biology*, The Samuel Roberts Noble Foundation, Ardmore, Oklahoma, USA, pp. 5–13.

Simpson, W. R., Schmid, J., Singh, J., et al. (2012) 'A morphological change in the fungal symbiont *Neotyphodium lolii* induces dwarfing in its host plant *Lolium perenne*', *Fungal Biol.*, 116, 234–240. https://doi.org/10.1016/j.funbio.2011.11.006.

Smith, G. W., Rotstein, D. S. and Brownie, C. F. (2004) 'Abdominal fat necrosis in a pygmy goat associated with fescue toxicosis', *J. Vet. Diag. Invest..*, 16(4), 356–359. https://doi.org/10.1177/104063870401600420.

Song, M., Li, X., Saikkonen, K., et al. (2015) 'An asexual *Epichloë* endophyte enhances waterlogging tolerance of *Hordeum brevisubulatum*', *Fungal Ecol.*, 13, 44–52. https://doi.org/10.1016/j.funeco.2014.07.004.

Song, H., and Nan, Z. (2015) 'Origin, divergence, and phylogeny of asexual *Epichloë* endophyte in *Elymus* species from western China', *PLoS ONE* 10, e0127096. https://doi.org/10.1371/journal.pone.0127096.

Stewart, A. V. (1987) *Plant-breeding aspects of ryegrass (Lolium spp.) infected with endophytic fungus*, Ph.D. thesis, Lincoln College, Canterbury, New Zealand.

Stewart, A. V. (2005) 'The commercial use of endophyte in turf ryegrass in New Zealand', *Inter. Turfgrass Soc. Res. J.*, 10, 641–644. https://doi.org/10.1002/its2.131.

Stewart, A. V., Barcellos, G. and Brilman, L. (2022) 'Use of endophytic fungi in turfgrasses: difficulties in delivery to the market', *Inter. Turfgrass Soc. Res. J.*, 1–4. https://doi.org/10.1002/its2.131.

Strahan, S. R., Hemken, R. W., Jackson Jr, J. A., et al. (1987) 'Performance of lactating dairy cows fed tall fescue forage', *J. Dairy Sci.*, 70(6), 1228–1234. https://doi.org/10.3168/jds.S0022-0302(87)80135-2.

Strickland, J. R., Spiers, D. E., Oliver, J. W. and Aiken, G. E. (2009) 'Physiological basis for fescue toxicosis'. In Fribourg, H. A., Hannaway, D. B. and West, C. P. (Eds), *Tall fescue for the twenty-first century. Agronomy monograph 53,* American Society of Agronomy, Crop Science Society of America and Soil Science Society of America, Madison, Wisconsin, USA, pp. 203–227. 12. https://doi.org/10.2134/agronmonogr53.c12.

Stuedemann, J. A. and Hoveland, C. S. (1998) 'Fescue endophyte: history and impact on animal agriculture', *J. Prod. Agric.*, 1(1), 39–44. https://doi.org/10.2134/jpa1988.0039.

Sugawara, K., Inoue, T., Yamashita, M. and Ohkubo, H. (2006) 'Distribution of the endophytic fungus, *Neotyphodium occultans* in naturalized Italian ryegrass in western Japan and its production of bioactive alkaloids known to repel insect pests', *Grass. Sci.*, 52, 147–154. https://doi.org/10.1111/j.1744-697X.2006.00060.x.

Sutherland, B. L., Hume, D. E. and Tapper, B. A. (1999) 'Allelopathic effects of endophyte-infected perennial ryegrass extracts on white clover seedlings', *N. Z. J. Agric. Res.*, 42(1), 19–26. https://doi.org/10.1080/00288233.1999.9513349.

Tadych, M., Bergen, M. S. and White Jr., J. F. (2014) '*Epichloë* spp. associated with grasses: new insights on life cycles, dissemination and evolution', *Mycologia*, 106, 181–201. https://doi.org/10.3852/106.2.181.

Takach, J. E. and Young, C. A. (2014) 'Alkaloid genotype diversity of tall fescue endophytes', *Crop Sci.*, 54(2), 667–678. https://doi.org/10.2135/cropsci2013.06.0423.

Thom, E. R., Popay, A. J., Hume, D. E. and Fletcher, L. R. (2012) 'Evaluating the performance of endophytes in farm systems to improve farmer outcomes: a review', *Crop & Pasture Sci.*, 63, 927–943. https://doi.org/10.1071/cp12152.

Thom, E. R., Waugh, C. D., Minnee, E. M. K. and Waghorn, G. C. (2013) 'Effects of novel and wild-type endophytes in perennial ryegrass on cow health and production', *N. Z. Vet. J.*, 61(2), 87–97. https://doi.org/10.1080/00480169.2012.715379.

Tor-Agbidye, J., Blythe, L. L. and Craig, A. M. (2001) 'Correlation of endophyte toxins (ergovaline and lolitrem B) with clinical disease: fescue foot and perennial ryegrass staggers', *Vet. Hum. Toxicol.*, 43(3), 140–146.

Valentine, S. C., Bartsch, B. D. and Carroll, P. D. (1993) 'Production and composition of milk by dairy cows grazing high and low endophyte cultivars of perennial ryegrass',

Second International Symposium on Acremonium/Grass Interactions, Palmerston North, New Zealand, 138–141.

van Zijll de Jong, E., Dobrowolski, M. P., Bannan, N. R., et al. (2008) 'Global genetic diversity of the perennial ryegrass fungal endophyte *Neotyphodium lolii*', *Crop Sci.*, 48, 1487–1501. https://doi.org/10.2135/cropsci2007.11.0641.

Vázquez-de-Aldana, B. R., Ciudad, A. G., Zabalgogeazcoa, I. and Criado, B. G. (2001) 'Ergovaline levels in cultivars of *Festuca arundinacea*', *Anim. Feed Sci. Technol.*, 93(3–4), 169–176. https://doi.org/10.1016/S0377-8401(01)00285-1.

Vázquez-de-Aldana, B. R., Zabalgogeazcoa, I., Garcia, C. A. and Garcia, C. B. (2003) 'Ergovaline occurrence in grasses infected by fungal endophytes of semi-arid pastures in Spain', *J. Sci. Food Agric.*, 83(4), 347–353. https://doi.org/10.1002/jsfa.1319.

Vikuk, V., Young, C. A., Lee, S. T., et al. (2019) 'Infection rates and alkaloid patterns of different grass species with systemic *Epichloë* endophytes', *Applied & Envir. Microbiol.*, 85(17), 1. https://doi.org/10.1128/AEM.00465-19.

Voisey, C. R. (2010) 'Intercalary growth in hyphae of filamentous fungi', *Fungal Biol. Reviews*, 24, 123–131. https://doi.org/10.1016/j.fbr.2010.12.001.

Waller, J. C. (2009) 'Endophyte effects on cattle'. In Fribourg, H. A., Hannaway, D. B. and West, C. P. (Eds), *Tall fescue for the twenty-first century. Agronomy Monograph 53,* American Society of Agronomy, Crop Science Society of America and Soil Science Society of America, Madison, Wisconsin, USA, pp. 289–310. https://doi.org/10.2134/agronmonogr53.c16.

Wang, T., Chen, T., White, J. F. and Li, C. (2022) 'Identification of three *Epichloë* endophytes from *Hordeum bogdanii* Wilensky in China', *J. Fungi*, 8(9). https://doi.org/10.3390/jof8090928.

Watson, R. H., Keogh, R. G. and McDonald, M. F. (1999) 'Ewe reproductive performance and growth rate of suckling-lambs on endophyte-infected perennial ryegrass pasture'. In Woodfield, D. R. and Matthew, C. (Eds), *Ryegrass endophyte: an essential New Zealand symbiosis, grassland research and practice series No. 7,* New Zealand Grassland Association, Palmerston North, New Zealand, pp. 19–26. https://doi.org/10.33584/rps.7.1999.3402.

White, J. F. (1988) 'Endophyte-host associations in forage grasses. XI. A proposal concerning origin and evolution', *Mycologia*, 80(4), 442–446. https://doi.org/10.2307/3807845.

Wiewióra, B. and Żurek, G. (2023) 'Amenity grasses-a short insight into species, their applications and functions', *Agron.*, 13(4), 1164. https://doi.org/10.3390/agronomy13041164.

Wolfe, B. A., Bush, M., Monfort, S. L., et al. (1998) 'Abdominal lipomatosis attributed to tall fescue toxicosis in deer', *J. Am. Vet. Med. Assoc.*, 213(12), 1783–1786.

Wolff, J., Richter, W. and Spann, B. (1995) 'Mutterkornalkaloide in der Milch?'. In Braun, C. (Ed), *VDLUFA-Schriftenreihe*. VDLUFAVDLUFA-Verlag, Darmstadt, Germany, pp. 521–524.

Xia, C., Li, N., Zhang, Y., et al. (2018) 'Role of *Epichloë* endophytes in defense responses of cool-season grasses to pathogens: a review', *Plant Dis.*, 102(11), 2061–2073. https://doi.org/10.1094/PDIS-05-18-0762-FE.

Yates, S. G., Fenster, J. C. and Bartelt, R. J. (1989) 'Assay of tall fescue seed extracts, fractions, and alkaloids using the large milkweed bug', *J. Agric. Food Chem.*, 37(2), 354–357. https://doi.org/10.1021/jf00086a018.

Young, C. A., Charlton, N. D., Takach, J. E., et al. (2014) 'Characterization of *Epichloë coenophiala* within the US: Are all tall fescue endophytes created equal?', *Frontiers in Chem.*, 2(NOV). https://doi.org/10.3389/fchem.2014.00095.

Young, C. A., Hume, D. E. and McCulley, R. L. (2013) 'Fungal endophytes of tall fescue and perennial ryegrass: Pasture friend or foe?', *J. Anim. Sci.*, 91, 2379–2394. https://doi.org/10.2527/jas.2012-5951.

Young III, W. C. and Silberstein, T. B. (2012) 'The Oregon grass seed industry'. In Young, C. A., Aiken, G. E., McCulley, R. L., Strickland, J. R. and Schardl, C. L. (eds.), *Epichloae, endophytes of cool season grasses: Implications, utilization and biology*, The Samuel Roberts Noble Foundation, Ardmore, Oklahoma, USA, pp. 49–52.

Zabalgogeazcoa, I. and Bony, S. (2005) '*Neotyphodium* research and application in Europe'. In Roberts, C. A., West, C. P. and Spiers, D. E. (eds.), *Neotyphodium in cool-season grasses*, Blackwell Publishing Ltd Ames, Iowa, USA, pp. 23–33. https://doi.org/10.1002/9780470384916.ch1b.

Zabalgogeazcoa, I., Vazquez de Aldana, B. R., Criado, B. G. and Ciudad, A. G. (1999) 'The infection of *Festuca rubra* by the fungal endophyte *Epichloë festucae* in Mediterranean permanent grasslands', *Grass Forage Sci.*, 54(1), 91–95. https://doi.org/10.1046/j.1365-2494.1999.00155.x.

Zbib, N., Repussard, C., Tardieu, D., et al. (2015) 'Toxicity of endophyte-infected ryegrass hay containing high ergovaline level in lactating ewes', *J. Anim. Sci.*, 93(8), 4098–4109. https://doi.org/10.2527/jas.2014-8848.

Zhang, W., Card, S. D., Mace, W. J., et al. (2017) 'Defining the pathways of symbiotic *Epichloë* colonization in grass embryos with confocal microscopy', *Mycologia*, 109(1), 153–161. https://doi.org/10.1080/00275514.2016.1277469.

Part 5

Case studies

Chapter 17

Grasslands in Europe: current status, emerging challenges, and pathways to sustainable futures

A. Van den Pol-van Dasselaar, Aeres University of Applied Sciences, The Netherlands; D. Hennessy, University College Cork, Ireland; and J. Isselstein, University of Göttingen, Germany

1 Introduction

Grasslands are a cornerstone of European agriculture, supporting both food security and a wide range of ecosystem services. Covering approximately one-third of the utilized agricultural area in Europe, grasslands contribute to a range of ecosystem services. Ruminants such as cattle, sheep, and goats can transform the feed source grass that is inedible to humans into high-quality protein-rich foods, in the form of meat and milk. Grasslands also support biodiversity conservation, carbon sequestration, nutrient cycling, water regulation, and cultural values, while sustaining rural economies and communities. Historically shaped by traditional farming practices such as grazing and haymaking, European grasslands have evolved into vital habitats and production landscapes.

Yet European grasslands are at a crossroads. Climate change, land-use competition, socio-economic pressures, changing dietary preferences, and evolving policy frameworks are reshaping how grasslands are managed and valued. At the same time, technological advances and collaborative innovations

http://dx.doi.org/10.19103/AS.2025.0154.26

offer new opportunities to enhance the multifunctionality of grassland-based systems.

This chapter provides a coherent, integrated overview of grasslands in Europe, addressing (1) their current status and use, (2) the ecosystem services they provide, (3) the major drivers and trends affecting them, and (4) promising innovations and research directions. These are illustrated by a case study from the project Grazing4AgroEcology that shows how innovative farmers, researchers, and other stakeholders are now jointly exploring new approaches. Throughout the chapter, we advance the central argument that grassland systems can, if appropriately managed and supported, simultaneously deliver nutritious food and multiple ecosystem services, thereby contributing to resilient agricultural futures in Europe.

2 Current status of grasslands in Europe

2.1 Grassland typology

European grassland-based livestock systems range from extensive, low-input systems relying on semi-natural pastures to highly intensive pastures managed for maximum productivity. Regarding the typology of grasslands, definitions have been proposed by Allen et al. (2011) and Peeters et al. (2014). In recent years, several EU joint research projects (e.g. MULTISWARD, Super-G) have attempted to more precisely and quantitatively capture the diversity of European grasslands in terms of environmental conditions, management, and comprehensive ecosystem services. The most complex typology to date has recently been presented as part of the Super-G project and takes into account a stronger differentiation of the intensity of grassland use (Tonn, 2024). This classification can serve as a reference, particularly for the precise spatial characterization of ecosystem services and for informing agricultural policy related to their classification and promotion.

In this chapter, we propose a simplified typology with four grassland types, characterized by vegetation, sward persistence, and grassland management (Isselstein and Komainda, 2021). They are built on widely used typologies and help frame ecosystem functions and management needs.

- Cultivated permanent grasslands: regularly and intensively managed grasslands aiming at a high yield and forage quality, i.e. renovated, fertilized, weeds controlled, sown species more or less dominating, not part of a crop rotation.
- Permanent grasslands: regularly and more or less intensively managed, moderately yielding, contain perennial or self-seeding forage species, persist indefinitely, sown species not or less dominant.

Published by Burleigh Dodds Science Publishing Limited, 2025.

- Temporary grasslands/leys: sown grasslands composed of forage species, integrated into arable crop rotations, sown species dominate.
- Semi-natural grasslands: extensively managed, low yield potential, permanent grasslands/habitats, vegetation dominated by indigenous and naturally occurring species.

This typology is used throughout the chapter to interpret differences in ecosystem service provision.

2.2 Grassland area and distribution

Grasslands are an essential part of agriculture in every European country. Permanent grasslands cover almost one-third of the utilized agricultural area in the European Union (Eurostat, 2025a) (Table 1). Permanent grassland is defined by Eurostat as "land used permanently (for several consecutive years, normally 5 years or more) to grow herbaceous fodder, forage or energy purpose crops, through cultivation (sown) or naturally (self-seeded), and which is not included in the crop rotation on the holding. The grassland can be used for grazing, mown for silage and hay, or used for renewable energy production. Grassland must have fodder interest, i.e. they include vegetal species of fodder interest." Temporary grasslands (leys) are typically counted under arable land because they alternate with crops in rotations. In total, permanent and temporary grasslands can be considered the main land use in Europe.

Fig. 1 shows the distribution of permanent grasslands in the different Member States of the European Union. In some countries, grasslands are the main land use (e.g. Ireland), while they are less prominent in other countries (e.g. Finland or Bulgaria). National averages often mask substantial within-country regional differences where grasslands are critical to local economies and ecological networks.

Historically, the area of grasslands in Europe has fluctuated (Küster, 1995; Poschlod, 2016). Forest clearing for agriculture over millennia expanded grasslands. Winkler et al. (2021) showed that grasslands are not only historically shaped but also currently undergoing regionally distinct transitions. In the last

Table 1 Grassland areas in the EU-27 (million ha)

	2012	2016	2020	2023
Permanent grassland	48	49	51	51
Arable land (including temporary grasslands)	99	100	98	97
Permanent crops	12	12	12	12
Utilized agricultural area	161	161	162	161

Source: Eurostat (2025a).

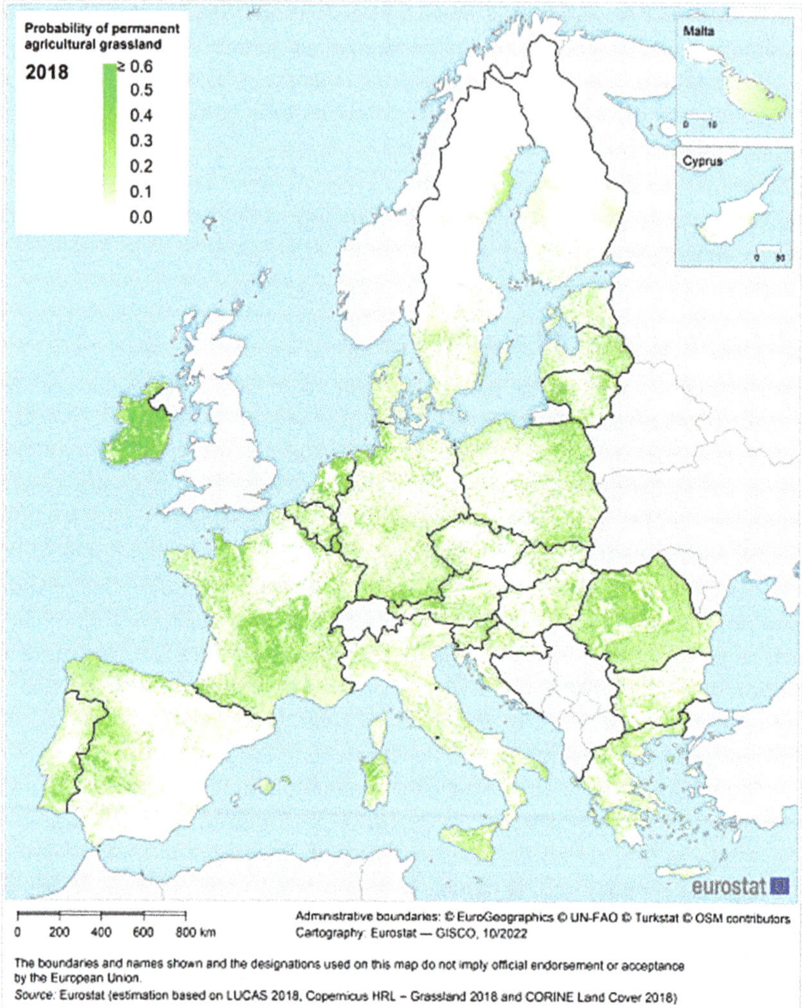

Figure 1 Probability of permanent agricultural grassland in Europe (https://ec.europa.eu/eurostat/statistics-explained/index.php?title=Permanent_agricultural_grassland_in_Europe; Eurostat, 2025b).

decades, the area of permanent grassland declined more sharply than the total utilized agricultural area (Huyghe et al., 2014; FAO, 2025). The main reasons for the loss of permanent grasslands were the conversion of grasslands into arable crops (32% of the lost area), urban expansion (30%), and land abandonment in rural areas (17%) (Erhard et al., 2016). Winkler et al. (2021) suggested that the regional mosaic underscores why policy and management must be tailored to local biophysical and socio-economic conditions.

2.3 Grazing in Europe

The primary and most significant function of European grasslands is the production of animal feed, i.e. grazed or conserved forage, particularly for ruminant livestock such as cattle, sheep, and goats. Ruminants possess a unique digestive system that enables them to convert fibrous plant material, which is indigestible to humans, into valuable food products for humans such as milk and meat. This process plays a crucial role in the agricultural economy and food supply chain, making grassland management a vital aspect of sustainable farming.

Traditionally, grazing was the main management practice on grasslands in Europe. Grazing by dairy cows, beef cattle, and small ruminants remains a defining feature of many European grassland systems. However, the proportion of dairy cows that graze has declined in several countries over the past two decades, with notable regional variation. Estimates derived from grazing experts throughout Europe showed a decrease in the period 2001–2019 in the percentage of dairy cows grazing (Van den Pol-van Dasselaar et al., 2020). More recent data from 2022 to 2024 show that this trend has continued, with a few exceptions (Table 2).

Some of the reasons for the decline in grazing across Europe include, in no particular order, the variability in herbage production and quality, lack of available labour, year-round calving patterns, use of high genetic merit cows, which require a steady supply of high-quality feed, increasing number of cows per farm and land fragmentation (Hennessy et al., 2020). There is a requirement for a robust type of cow, which can deal with the grazing conditions presented including topography, distance from the milking parlour, and variability in feed (grass) supply and quantity (Friggens et al., 2017; Delaby et al., 2021). Many breeding programmes across Europe are not designed to produce cows for pasture-based systems, rather cows fed indoors on concentrate-based diets with little variability in feed quality and with little emphasis on dairy cow fertility. Seasonal pasture-based systems require a level of seasonal calving to optimize the use of grazed grass in the diet and therefore demand a focus on fertility (Delaby et al., 2021). For example, in Ireland, the Economic Breeding Index focusses on fertility, maintenance/robustness, and milk solids production (Veerkamp et al., 2002; Berry et al., 2022).

The trends in grazing in Europe over the past decades are further illustrated in Fig. 2. This figure provides data on grazing from three countries, each with its own trends and challenges. While the percentage of grazing in Ireland is still very high, the percentage of grazing in Denmark has dramatically declined in the last 25 years. The percentage in the Netherlands lies in between.

Table 2 Grazing in Europe (% dairy cows) in six distinctive regions of Europe

Region and country	2010	2014	2016	2019	2022	2024
North						
Norway			90	80	75–95	95–100
Sweden	100	100	100	100	100	<100
Finland			70	80	60-80	60–70
West						
Ireland	99	98	95–100	95–100	90–100	95–100
UK		92	80–90	70–80	70–90	65–95
Central; grazing >50%						
The Netherlands[a]	74	69	65	73	75	69
Belgium[b]	85–95	75–80	60–85	30–95	40–90	50–90
Luxembourg	75–85	73	75		30–50	20–50
France	90–95	90	75–95	90	50–90	40–90
Switzerland	85–100	75–90	80–97	70–90	88–96	70–90
Central; grazing <50%						
Denmark	35–45	25–30	25	20–25	30	25–30
Germany	42		10–50	15–40	20–50	15–50
Austria	25		40	44	45	
East						
Poland			20	30	30	25
Lithuania		50–70				20
Czech Republic	20		3		4	2–5
Slovenia	25		20	20–40	20	20
Hungary			2–3	3–5	5–10	5
South						
Portugal	50			60		
Spain	20		10–30	20–30	10–50	10–60
Greece	15		10	10	5	5
Italy			10–20		2–70	35

[a]Statistical data in 2010-2024 (CBS, 2025).
[b]Large regional differences between Flanders (low) and Wallonia (high).
Source: Period 2010-2019: Van den Pol-van Dasselaar et al., 2020; period 2022-2024: expert input from the Working Group "Grazing" of the European Grassland Federation.

In Ireland, grazing remains consistently high. The country's mild climate and abundant rainfall create ideal conditions for grass growth, making pasture-based dairy and beef production highly cost-effective. Additionally, much of the land area in Ireland is not suited to the production of other feeds such as maize and barley due to topography, soil type, and rainfall. Compared to indoor feeding systems, grazing significantly reduces feed costs, as grass is the

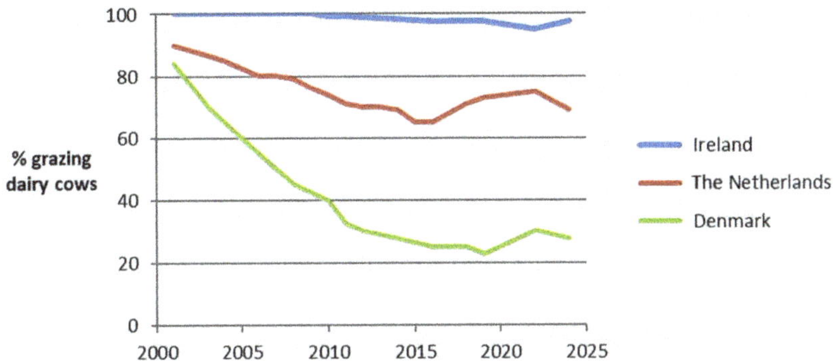

Figure 2 Grazing (% dairy cows) in Ireland (expert opinion), the Netherlands (statistical data), and Denmark (expert opinion). Expert opinions originate from the EGF Working Group "Grazing" and are partly reported by Van den Pol-van Dasselaar et al. (2019).

cheapest source of nutrition for livestock (Finneran et al., 2010). In addition, grass-fed milk and dairy products have been shown to be of high quality for human nutrition (e.g. O'Callaghan et al., 2016; Alothman et al., 2019; Timlin et al., 2023). Grassland-based systems are heavily promoted by science and advisory services, and grass contributes over 95% of a cow's annual diet as fed in Ireland (O'Brien et al., 2018). Additionally, Ireland's dairy sector relies heavily on exports, and grass-fed dairy products are marketed as high-quality and sustainable, further reinforcing the economic motivation to maintain high levels of grazing.

In Denmark, grazing has steadily declined over the past decades as dairy farming has become increasingly intensified. Dairy farms have shifted towards larger herds and higher milk production, prioritizing efficiency and productivity. As a result, many farmers have moved towards indoor feeding systems that allow for better control over nutrition and milk yields. This trend has led to a significant reduction in grazing, as pasture-based systems are often seen as less compatible with the high-output dairy model that dominates Danish agriculture today.

In the Netherlands, dairy cows usually graze only for part of the day. Grazed grass is supplemented indoors with conserved forages and concentrates. Intensification has led to a decrease in the percentage of grazing dairy cows since 2000. From 2005 onwards, societal concerns regarding the decline of grazing practices grew significantly (Runhaar et al., 2020). In response to these concerns, a "Grazing Covenant" was established in 2012, bringing together more than 80 organizations from across the entire dairy supply chain. To support the goal of halting the decline in grazing, several initiatives were introduced, such as financial incentives (e.g. grazing premiums) and grazing advice for farmers (grazing coaches). This led to a reversal in the trend between 2015 and 2022. Recent data, however, show that the percentage grazing is once again decreasing. This can be associated with a shift in the attention of dairy companies to broader sustainability goals (e.g. reducing carbon footprints, enhancing biodiversity), thereby putting less emphasis on grazing. These trajectories show how markets, advisory systems, infrastructure, and policy interact to shape the extent of grazing.

Country-specific insight in the barriers to grazing was obtained in the Horizon Europe project Grazing4AgroEcology (Table 3). Table 3 is based on results of an online survey among young farmers in eight European countries (n = 1410) (Van den Pol-van Dasselaar et al., 2024).

Next to the identification of barriers, research has also identified many options to improve grassland production, quality and utilization. Incorporating legumes can result in increased herbage quality, dry matter intake and milk production (e.g. Egan et al., 2018; McClearn et al., 2019). Improving grazing management (O'Donovan and Delaby, 2016) and developing grazing management decision support tools/systems that translate data on pasture growth and animal demand into actionable decisions can help provide

Table 3 Perceived barriers to grazing in France (FR), Germany (DE), Ireland (IE), Italy (IT, Sardinia = SA, South Tyrol = ST), the Netherlands (NL), Portugal (PT), Romania (RO), and Sweden (SE) as seen by young farmers (number of valid responses and % of respondents that chose a particular barrier in their top five barriers; only barriers with at least 40% are given for each country/region)

	FR	DE	IE	ITSA	ITST	NL	PT	RO	SE
Number of valid responses	458	116	58	33	59	456	43	111	76
Barriers									
Climate (no grass in dry periods, heat stress)	71	48	53	58		52	65	47	
Land fragmentation (not enough grazing land surrounding the farm)	62		70		41	42		42	50
Lack of knowledge/education	41		77	58			40	59	
Variability in grass quantity and quality	56		62			58			47
Predators like wolves		64			79				42
Consumer demands for low prices		61					40		67
Time/available labour		52				47	43		
Not enough grass available/low grass production when grazing	41							45	
Money, i.e. costs too high and/or benefit too low				42			43		
Unfavourable topography (e.g. the pastures are too steep)					50				

Source: Van den Pol-van Dasselaar et al. (2024).

confidence in decision-making (e.g. PastureBase Ireland, Hanrahan et al., 2017; MoSt Grass Growth model, Ruelle et al., 2018). Optimizing grazing systems means adjusting how livestock graze (timing, intensity, spatial patterns) to maximize benefits like food production and soil health, while minimizing harm to the surrounding ecosystem (e.g. avoiding overgrazing and soil erosion, and enhancing biodiversity).

3 Ecosystem services and multifunctionality

As mentioned in the previous section, the production of animal feed for ruminants leading to production of human food is the primary and most significant function of European grasslands. In many areas of Europe, this can occur without competing for land for crop production (Van Kernebeek et al., 2016) as grassland tends to occupy much of the land area unsuitable for crop production (access, topography, soil type, etc.). Grassland-based ruminant production systems can thus contribute to net human-edible protein when concentrates are minimized and diets are aligned with pasture supply (e.g. Mottet et al., 2018; Leroy et al., 2022). The net supply of edible protein by livestock can be considered using a number of metrics including feed conversion ratio (FCR; Wilkinson, 2011), total protein efficiency (Laisse et al., 2018), net protein efficiency (Laisse et al., 2018), edible protein conversion ratio (EPCR; Wilkinson, 2011; Laisse et al., 2019; Mosnier et al., 2021), and land-use ratio (LUR; van Zanten et al., 2016). Laisse et al. (2018) showed that the net protein efficiency of grassland-based dairy cattle production receiving low concentrate input is up to 2.57 in France, while Hennessy et al. (2021) reported a net protein efficiency of 4 for Irish grassland-based dairy cattle production systems (Fig. 3). This indicates that grassland- based ruminant production systems perform well in terms of converting feed protein to food protein and it highlights that grasslands (utilizing land unsuited to cropping) and arable lands are complementary in the production of human food.

Next to food production, grasslands provide many more ecosystem services, i.e. benefits to humans from nature (for recent reviews, see Bengtsson et al., 2019; Schils et al., 2022). Production services that play an essential role in supporting rural economies are closely related to grassland-based livestock farming, i.e. the products are milk, meat, and fibre (Isselstein and Kayser, 2014; Ryschawy et al., 2017). Other important services include the maintenance and enhancement of biodiversity (e.g. Habel et al., 2013; Isselstein, 2019), the purification and provision of clear ground and surface waters (Orr et al., 2016), the safeguarding of soil quality and reduction of soil erosion risks (Milazzo et al., 2023), climate regulation, including C sequestration (O'Mara, 2012; Bai and Cotrufo, 2022) and the provision of cultural services, i.e. recreation and landscape aesthetics (Pellaton et al., 2022; Jitea et al., 2024).

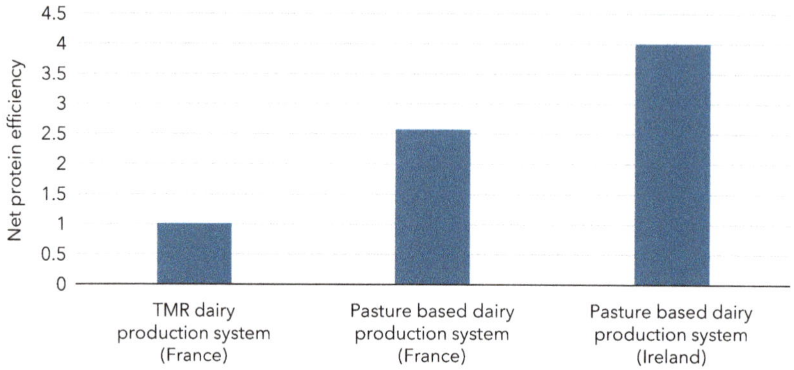

Figure 3 Net protein efficiency of dairy production systems (kg human edible protein produced/kg human edible protein consumed by livestock) in France (indoor feeding TMR and grassland-based; Laisse et al., 2018) and Ireland (grassland based; Hennessy et al., 2021) Source: Hennessy et al. (2024).

Bardgett et al. (2021) showed that the provision of these services and the ecological functioning of grasslands are at risk in situations of grassland degradation. Also, when grasslands are converted into arable or other crop systems, provisioning of ecosystem services is under threat (e.g. Schils et al., 2022). Schils et al. highlighted the importance of permanent grasslands in supporting different ecosystem services and concluded that land use change and intensification decreases the multifunctionality of permanent grasslands. Table 4 summarizes typical patterns of the ecosystem service provision of grasslands across the four grassland types introduced earlier.

In addition to the grassland type, the site and management conditions of grasslands affect the ecosystem services that these grasslands deliver. This is illustrated in Fig. 5. This figure provides a simplified, generalized representation of the conditions under which maximum performance of various ecosystem services can be assumed. On the vertical axis of the matrix, the characteristics

Table 4 Importance of ecosystem services provided from grasslands depending on the grassland type

Ecosystem service	Cultivated grasslands	Permanent grasslands	Temporary grasslands	Semi-natural grasslands
Herbage production	+++	++	+++	+
Soil-C/C-sequestration	++	+++	+	+
Nitrogen emission	++	++	++	+
Water provision	++	+++	++	++
Biodiversity/genetic resources	+	++	-	+++

For explanation of grassland types, see 2.1; importance/priority: "+++" highest, "++" moderate to high, "+" low, "−" no.
Source: Isselstein and Komainda (2021).

Published by Burleigh Dodds Science Publishing Limited, 2025.

or domains dependent on location and management are represented. On the horizontal axis, the various ecosystem services are presented. As an example, the highest production performance in terms of feed and food is achieved on fertile soils with adapted fertilization and frequent defoliation, making moderately high yields with very good feed quality possible.

The assessment in Fig. 4 can only be a rough representation of the complex relationships between location and management on the one hand, and ecosystem services on the other, and deviations from these generalized statements are possible. For example, with regard to climate impact of grassland farming, it is generally assumed that disturbance of the sward and/or the soil restricts carbon sequestration in the soil, which is therefore negatively assessed. However, a recent study shows that grassland improvement, which involved soil disturbance and reseeding, increased the productivity of the sward to such an extent that more carbon is captured and sequestered in the longer term (Feigenwinter et al., 2023). Recognizing such context dependencies is essential when designing policies and farm strategies that aim to increase multifunctionality rather than maximize a single service.

Ecosystem service	**Food**: produce feed and food	**Nature**: protect and promote biodiversity	**Climate**: reduce greenhouse gas emissions, sequester carbon	**Water**: provide sufficient and clean water	**Culture**: preserve typical cultural landscapes
Domain/ Characteristic					
site/soil	fertile, well supplied with basic nutrients	rather low fertility, low in basic nutrients	no tillage and sward disturbance	rather low fertility, high water permeability	small paddocks, diverse landscape structure.
soil water availability	moderate, no surplus, no drought	either dry or wet sites	wet on organic soils	no clear effect	diverse, from low to high
herbage quantity	Moderate	rather low	rather high	rather low for high water quantity, rather high for high water quality.	diverse, from low to high
herbage quality	High	often low, mature herbage	high to reduce methane emission by ruminants, low crude protein content to reduce N_2O emissions	low to moderate	diverse, from low to high
Fertilization	moderate to high	none to low	none to low	low to moderate	rather low
Defoliation	Frequent	infrequent	depending on methane or N_2O emission, either infrequent or frequent	rather frequent	diverse, grazing, mowing, frequent, infrequent

Figure 4 Simplified representation of the site and management conditions (domain/ characteristic) under which high or maximum performance is achieved for various ecosystem services that relate to food, nature, climate, water, and culture. Source: Isselstein (2021).

4 Drivers and trends of change

There are a number of major regional and global trends and challenges that may affect grasslands and grassland-based livestock production systems in Europe, shaping their productivity, sustainability, and role in society. These trends will not only lead to changes in the extent of grasslands but will also lead to changes in the way these grasslands are managed. This chapter discusses major drivers and trends of change and the implications they may have on grasslands in Europe.

4.1 Demographic trends

- *Changes in population size.* The United Nations (2024) predicts that the world's population will continue to grow, with projections indicating an increase from approximately 8 billion people today to a peak of more than 10 billion around 2084. The global population increase will heighten pressure on resources, food systems, and environmental sustainability. The growth is predicted to be unevenly distributed across regions. Global population growth is driven primarily by countries in Africa and parts of Asia; Europe already had its peak in 1992. Population increase is coupled with an increase in global demand for food and animal protein in diets (Alexandratos and Bruinsma, 2012). On a per capita basis, global meat consumption is expected to grow by 2% by 2032 compared to the base period of 2020–2022, driven by income and population growth (OECD-FAO, 2023). To meet the food demand of this growing population, increased food production is required from approximately the same land area and in some regions such as Europe, from a declining land area because of competition with other land uses such as bioenergy production and urbanization (O'Mara, 2012). This will require increases in efficiency of production from grassland, as well as other agriculture. Europe faces relatively stable or declining populations with ageing demographics but participates in global markets where demand for animal-source foods continues to evolve. These dynamics place a premium on efficient, multifunctional land use, an area where grasslands, including those in Europe, can excel.

4.2 Resource use and land-use change

- *Intensification.* In the last decades, roughly since the start of the CAP (Common Agricultural Policy of the European Union) in the early 1960s, grassland use in Europe has intensified: increased artificial fertilizer use, higher stocking rates, higher cutting frequency, more drainage and

Published by Burleigh Dodds Science Publishing Limited, 2025.

irrigation, more re-sowing and over-sowing, more weed control with herbicides (Peeters, 2009). As a result of intensification, the number of farms has decreased, while the average livestock numbers per farm and the land area per farm increased. The intensification of agriculture in Europe also led to the loss of grasslands on farms that were originally fully grass-based. To accommodate the increased energetic and nutritional needs of the animals, the feed ration has been supplemented with maize and concentrates. This led to conversion of grasslands into maize fields and arable lands to provide this supplementation. Ploughing grasslands for crops leads to severe biodiversity loss, soil carbon release, and hydrological changes that disrupt ecosystem functions. Grassland intensification boosts short-term productivity but often degrades soils, reduces biodiversity, and disrupts natural nutrient cycles. Over time, these systems can become less resilient to climate extremes and ecological stresses, threatening long-term sustainability.

- *Land abandonment*. This trend, that is in contrast with intensification, also occurred in Europe in the last decades (Peeters, 2009). In Europe, land abandonment was most common in marginal areas, which tend to be dominated by grassland, where farming was less economically viable, e.g. in mountainous or remote areas. Abandonment often triggers landscape changes, with grasslands gradually transitioning to shrubland or forest through natural succession. While this process can enhance carbon storage and woodland biodiversity, it reduces open-habitat species, reduces overall biodiversity, reduces agricultural productivity, and reduces the cultural value of traditional landscapes. These changes can disrupt rural communities, drive depopulation, and reshape local economies.

4.3 Societal expectations and market signals

- *Societal concerns*. Societal concerns regarding livestock systems have intensified in recent years, particularly with regard to greenhouse gas emissions, animal welfare, and the high land and water demands of these systems (Moran and Blair, 2021). Nutrient runoff and leaching, which degrade water and soil quality, and biodiversity loss linked to intensified production (Resare Sahlin et al., 2024) are also key issues. Growing public awareness of biodiversity loss, climate change, and animal welfare is reshaping expectations for grassland management. There is increasing pressure to adopt practices that enhance ecosystem services, improve animal welfare, and reduce environmental impacts. While this can support more sustainable and diversified management, it also challenges traditional practices and may require significant adaptation by farmers.

- *Dietary shifts towards plant-based alternatives.* The societal concerns and growing awareness of potential environmental problems related to livestock production in general are leading to dietary shifts. European consumers are increasingly replacing animal-based products with plant-based options (e.g. Mustapa et al., 2025). The rising demand for plant-based foods is reducing the reliance on livestock production in some regions, directly affecting grassland use. This trend can lead to lower grazing intensity and, in some areas, land abandonment or conversion to other uses. However, it may also create opportunities for grasslands to be managed for carbon storage, biodiversity, or other ecosystem services instead of meat or dairy production.
- *Socio-economic challenges.* Socio-economic challenges in agriculture are becoming increasingly prominent, with volatile prices creating financial uncertainty for farmers. A significant challenge is also the lack of collaboration between different actors in the food chain, which hinders the development of efficient and sustainable systems. Increasing price volatility and financial uncertainty make it difficult for farmers to invest in the long-term management of grasslands, often leading to under-management, intensification, or even abandonment. The lack of collaboration across the food chain further limits opportunities to develop value-added, sustainable grassland systems, such as premium pasture-fed products or ecosystem service markets. As a result, grasslands risk degradation or conversion, while their potential to support biodiversity, carbon storage, and resilient rural economies remains underutilized.
- *Policy instability.* In terms of policy, volatile politics can create an unstable regulatory environment, making it difficult for farmers to plan and invest in the long term. Shifting political priorities and policy changes can lead to uncertainty, impacting everything from subsidies to environmental regulations, and creating challenges for the agricultural sector in adapting to new demands and expectations. An example are the on-going, frequent changes to nitrogen and nitrates regulations in countries such as Ireland and the Netherlands. Shifting regulations make it difficult for farmers to adopt sustainable practices in grassland management or commit to infrastructure improvements. This instability can lead to reactive, short-term decision-making, increasing the risk of overuse, under-management, or even conversion of grasslands, and reducing their potential to deliver ecosystem services and long-term productivity.

4.4 Systemic trends

- *Path dependency* (a phenomenon whereby history matters; what has occurred in the past persists because of resistance to change) is a trend that is often seen in agricultural production systems. It is an economic and social concept that suggests that the decisions and outcomes in a system or process are heavily influenced by the past or the initial conditions, even when those past circumstances are no longer relevant. Once a particular course of action is chosen, it can be difficult or costly to change direction, leading to a "locked-in" trajectory (Goldstein et al., 2023). Farmers often feel that they are locked in to their current system. For farmers in indoor systems, examples of lock-ins include investment in housing and feeding equipment, loss of grazing management skills/knowledge, and poor grazing infrastructure. For grass-based farmers, examples of lock-ins may include inflexible regulatory frameworks, consumer preferences for low-cost food, constrained access to financial resources when investments in sustainability are requested, and limited labour availability. These "lock-ins" can prevent the adoption of more sustainable or diversified practices in grassland management. They slow innovation and adaptation, leaving grasslands vulnerable to degradation, reduced biodiversity, and missed opportunities for enhancing ecosystem services and resilience.

- *Climate change*. Climate change is a threat to grasslands (e.g. Gibson and Newman, 2019). As a largely uncontrollable phenomenon, it is altering precipitation patterns, temperature regimes, and the frequency of extreme weather events. These shifts affect productivity, species composition, and seasonal growth patterns, reducing the reliability and quality of forage. Increased droughts, flooding, and volatile weather also disrupt grazing management and can accelerate soil degradation or erosion, ultimately threatening the resilience, biodiversity, and long-term sustainability of grassland systems.

4.5 Trends in agricultural systems

- *Organic agriculture*. The European Commission has set ambitious targets for organic farming in Europe. The European Green Deal's Farm to Fork strategy sets the target that at least 25% of the European Union's agricultural area should be dedicated to organic farming by 2030. The share of the European Union's agricultural land under organic farming increased from 5.9% in 2012 to 10.5% in 2022 as a result of an increasing demand for organic products and policy support (EEA, 2024). Organic farming is generally lower input, and grasslands play an important role in supplying feed for livestock.
- *Regenerative agriculture*. Another trend, increasing in popularity, is regenerative agriculture. There is no agreed definition of regenerative agriculture (for example, see reviews of Newton et al., 2020 and Schreefel et al., 2020), but it is receiving much attention from producers, retailers,

researchers, and consumers, as well as politicians and the mainstream media (Newton et al., 2020). Regenerative agriculture has been proposed as an alternative means of producing food that may have lower, or even net positive, environmental and/or social impacts (Rhodes, 2017). In many cases grassland-based farming that includes grazing systems has many of the attributes associated with the various interpretations of regenerative agriculture like integration of livestock, reducing or eliminating tillage, increasing biodiversity, and sequestering carbon (Newton et al., 2020). An increase in regenerative agriculture could lead to an increase in grassland area.

4.6 Implications for grasslands

Grasslands are shaped by a complex interplay of environmental, socio-economic, policy, and societal drivers, each influencing their structure, function, and management. Intensification, conversion, and abandonment directly alter biodiversity, productivity, and carbon storage, while socio-economic pressures, policy instability, and path dependency constrain the ability of farmers to adopt sustainable practices in grassland management. Societal expectations, shifts toward plant-based diets, and climate change further reshape grazing patterns, management intensity, and landscape composition. Collectively, these trends create both risks, such as land degradation and soil erosion, species loss, and reduced resilience, and opportunities, including enhanced ecosystem services, and more diversified and sustainable grassland management. Understanding these interlinked drivers is essential for developing strategies that maintain the ecological, economic, and cultural value of grasslands in a rapidly changing world.

5 Case study: Grazing4AgroEcology

This case study from the project Grazing4AgroEcology (G4AE, EU funded under GA 101059626) illustrates how innovative farmers, researchers, and other stakeholders are now, at different levels, jointly exploring new approaches for sustainable grassland management to address the challenges they face.

5.1 The project Grazing4AgroEcology

To enhance the sustainability and productivity of grazing in grassland-based systems, the project G4AE (www.grazing4agroecology.eu) was established to foster agroecological innovations across Europe. G4AE aims to restore confidence among farmers and the agricultural industry in the performance and competitiveness of grazing and focuses on the five principles that Dumont et al. (2013) proposed to be optimized in animal production systems: (1) adopting management practices that aim to improve animal health, (2) reducing the inputs required for production, (3) reducing pollution by optimizing the biogeochemical functioning of farming systems, (4) enhancing diversity within animal production systems to strengthen their resilience, and (5) preserving biodiversity in agro-ecosystems by adapting management practices. A cornerstone of the G4AE approach is the implementation of a multi-actor strategy, recognizing that effective innovation arises from the collaboration of diverse stakeholders. The project brings together farmers' organizations, farmers, extension services, educational institutions, and research bodies from eight EU countries. In G4AE, they jointly explore new approaches for sustainable grassland management at different levels.

5.2 Farm level

Digital tools can benchmark and guide agroecological performance. G4AE has introduced a digital tool that helps farmers in all participating countries assess their agroecological performance. This so-called "Grazing4AgroEcology Self-Assessment Tool" that was developed in a multi-actor approach allows farmers to monitor the agroecological performance of their farms for the five principles: animal health, reducing inputs, reducing pollution, enhancing diversity and preserving biodiversity. Other examples of innovations at farm level are precision grazing techniques, where farmers have adopted precision grazing methods, utilizing GPS tracking devices and smart collars for cattle to guide grazing (virtual fencing). These technologies help optimize grazing patterns and improve pasture management, ensuring that livestock graze in an optimal way. Also, farmers have implemented rotational grazing systems where livestock are moved between different paddocks, giving each section of the pasture time to recover and regenerate. This method improves the overall productivity of the pasture while ensuring that soil fertility is maintained and that grasslands are not overgrazed.

5.3 Field level

Innovations have also taken place at field level. In many European countries, multispecies mixtures of grasses, legumes, and herbs are increasingly sown. As they come with complementary traits, they can deliver benefits such as reduced fertiliser requirements, increased resilience against drought, and a more stable production throughout the year. They also benefit biodiversity. Decision-support for nutrient applications, grounded in monitoring and modelling can lower costs and environmental losses while maintaining productivity.

5.4 Importance of multistakeholder approaches

Because contexts differ by climate, soils, infrastructure, markets, and policy, innovations must be co-designed with those who implement them. The G4AE project illustrates how a multi-actor approach, bringing together partners from different countries and with diverse expertise, can foster meaningful innovations in grassland-based production systems across Europe. By combining scientific knowledge, practical experience, and stakeholder perspectives, G4AE strengthens both the sustainability and productivity of grasslands, while highlighting their essential role in meeting broader environmental and societal objectives. The project demonstrates that economic viability and environmental management can go hand in hand, offering pathways towards more resilient and sustainable agricultural practices across Europe.

6 Conclusion: Future role of grassland systems in Europe

Earlier sections of this chapter have shown that grassland-based production systems contribute substantially to a wide range of ecosystem services. Yet, several major trends, including shifts in dietary preferences, evolving environmental regulations, and economic pressures, could lead to a reduction in livestock numbers across Europe. Such a decline may, in turn, decrease the area of managed grasslands. With lower demand for grazing, grasslands risk conversion to cropland, forests, or urban areas, or may be abandoned altogether. This could reduce the land available for agricultural production and diminish grassland-specific ecological benefits, such as carbon storage, wildlife habitat, soil fertility, and water infiltration. The extent of these impacts will vary depending on context, particularly the grassland type.

While reducing livestock numbers may address certain environmental concerns, it also introduces trade-offs that must be carefully considered in future land-use and agricultural policies. Taking multiple perspectives into account, the role of grasslands is likely to grow in importance as society increasingly relies on their capacity to deliver multiple ecosystem services. One key driver is the need for animal protein for human consumption, produced with minimal input of human-edible feed. Recent research has demonstrated that strictly grassland-based ruminant systems have substantial potential to meet this goal while relieving arable land from fodder production (e.g. Pfeifer and Winterberg, 2024; Wild et al., 2025). Achieving this requires improvements in both the efficiency of grassland management and the use of grasslands in livestock systems.

Technological and societal innovations will be essential to reduce trade-offs among competing objectives and enhance overall efficiency. These include diversifying grassland swards (Komainda et al., 2024), adopting smart farming and AI-based technologies (Horn and Isselstein, 2022), and engaging stakeholders in co-creation and co-development processes. Collaboration among farmers, researchers, policymakers, and industry experts is crucial to develop context-specific, sustainable solutions.

In conclusion, grassland dynamics are shaped by the complex interplay of natural processes and human interventions, which will continue to influence their structure, function, and productivity. We believe that grasslands are set to play a pivotal role in meeting the dual challenge of feeding a growing global population while safeguarding ecosystem health. Enhancing their efficiency and multifunctionality offers a promising pathway, enabling the production of high-quality food and net human-edible protein while maintaining biodiversity, soil fertility, and climate-regulating functions.

Ultimately, a holistic and interdisciplinary approach to grassland management is needed, recognizing the interconnectedness of agricultural productivity, environmental sustainability, and socio-economic factors. This approach should be supported by science, education and policies. By fostering collaboration among stakeholders, European grasslands can continue to provide essential ecosystem services while meeting the demands of a changing world.

7 Research priorities

Future grasslands research will require a holistic, multi-disciplinary approach that integrates knowledge from various spatial scales, actors along the value chain, and diverse scientific disciplines. Given the complexity of grassland ecosystems and the need to address multiple challenges in food production and sustainability, research will increasingly need a multistakeholder approach – an approach that fosters collaboration between farmers, advisors, researchers, policymakers, and industry stakeholders. This collective effort will be critical in developing solutions that are both locally relevant and scalable across different regions. It is crucial to bridge the gap between different scientific disciplines. Collaborative research between natural sciences (e.g. agronomy, ecology), production sciences (e.g. animal science, agribusiness), and socio-economic sciences (e.g. economics, policy studies) will provide a more comprehensive understanding of how to optimize grassland systems. Interdisciplinary research will ensure that innovations are not only technically sound but also economically viable, environmentally friendly and socially acceptable.

Some specific research needs are as follows:

- Technical research: a key focus area will be the development and optimization of grazing systems that enhance productivity while minimizing environmental impact. Research will continue to explore genetic improvements in livestock, such as breeding for improved disease resistance and enhanced nutrient use efficiency. Advanced systems management techniques will be central to improving the efficiency of grassland-based production systems. Additionally, technological solutions, such as sensor-based technologies, will be pivotal in improving labour conditions, increasing precision in grazing management, optimizing nutrient management planning, and monitoring livestock health in real-time. Research into multispecies swards, particularly in low-nutrient environments, will contribute to boosting grassland resilience and diversifying feed sources.

- Decision Support Tools (DSTs): the development of DSTs that combine data from various sources, such as satellite imagery, soil sensors, and livestock tracking devices, will empower farmers to make informed decisions that enhance productivity and sustainability. These tools will be essential in optimizing grazing patterns, managing pasture health, managing nutrient (fertilizer) use, and improving the overall efficiency of grassland-based systems. By using real-time data and predictive analytics, farmers can adjust their practices to maximize output while reducing environmental footprints. Digital tools, such as block chain for traceability and farm management software for planning, will help streamline operations, improve transparency, and enable more precise management of resources. Research will focus on how to integrate digital solutions in a way that is accessible and beneficial for farmers, especially those in smaller or less technologically advanced farms.
- Socio-economic and policy research: understanding the socio-economic context of grassland-based livestock production will be important for creating policies that support sustainability. Research will focus on developing effective strategies to increase societal support for grassland-based systems, including premiums, labels, and certification programs that incentivize sustainable practices. Research into short supply chains, and how they can be strengthened to improve profitability and food security, will be necessary. Additionally, communicating the value of grassland-based systems to the public will be increasingly important, with research focusing on improving how these systems are perceived and understood. Transparent communication and marketing strategies will play a key role in gaining public and political support.
- Finally, one of the most pressing research needs in grassland-based livestock production is research at the systems level as grassland-based systems are complex and influenced by many factors. Currently, much of the research focuses on isolated aspects of production (such as grazing management or incentives), but there is a need for more integrated, system-level approaches that also consider socio-economic effects. Research

at the system level requires examining how different components, such as grazing patterns, soil health, and biodiversity, interact and influence overall system performance. In addition, the role of societal interaction, good communication, and a comprehensive understanding of the broader context should be considered. This means exploring how stakeholders from science and practice, including farmers, researchers, policymakers, industry experts, and local communities, can work effectively together to share knowledge and coordinate efforts.

8 Where to look for further information

The European Grassland Federation (EGF) is a forum for students, research workers, advisors, teachers, industry representatives, farmers, policy-makers, and other people with active interest in all aspects of grassland in Europe. These aspects include utilization and management of all types of grasslands for production and other ecosystem services. The motto of the EGF is "*Connecting people to face the future challenges of grassland use in Europe*", and the objectives of EGF are:

- to facilitate and maintain close contact between Grassland Organizations in Europe;
- to promote the interchange of scientific and practical experience between grassland experts; and
- to initiate conferences and other meetings on all aspects of grassland production and utilization in Europe.

All papers of EGF conferences are peer-reviewed, edited and published in the proceedings that can be found at https://www.europeangrassland.org/.
Other sources of information are:

1. Encyclopedia pratensis (https://www.encyclopediapratensis.eu/). This is an online encyclopedia on grasslands built on the basis of international scientific literature and enriched by practices from farmers. It contains results of several international research projects.
2. Grassland use in Europe - a syllabus for young farmers (https://www.quae-open.com/produit/123/9782759231461/grassland-use-in-europe). The European project Inno4Grass (www.inno4grass.eu) provided an overview of grassland use in Europe, including grassland production, grazing management, hay and silage making, soil and nutrient management, environment and biodiversity, and quality of grassland-based animal products in the form of a syllabus for young farmers (Van den Pol-van Dasselaar et al., 2019). Understanding

and optimizing grassland use is key to ensuring the sustainability and resilience of European grassland farming systems in the face of environmental and economic challenges. The syllabus also provides country-specific information on grasslands in eight countries that are scattered throughout Europe (Sweden, Ireland, the Netherlands, Belgium, Germany, Poland, France, and Italy), thus providing a picture of grassland management in different pedoclimatic regions of Europe. The syllabus is currently being updated and will provide the latest insights with respect to grazing and information on grasslands in Sweden, Ireland, the Netherlands, Germany, France, Romania, Portugal and Italy.

3. Websites of European Union joint research projects on grasslands: Inno4Grass - https://www.inno4grass.eu/en/, SuperG - https://www.super-g.eu/, and Grazing4AgroEcology - https://grazing4agroecology.eu/.

9 References

Allen, V. G., et al. (2011). An international terminology for grazing lands and grazing animals. *Grass and Forage Science*, 66(1), 2.

Alothman, M., et al. (2019) The "grass-fed" milk story: understanding the impact of pasture feeding on the composition and quality of bovine milk. *Foods*, 8, 350. https://doi.org/10.3390/foods8080350.

Alexandratos, N. and Bruinsma, J. (2012). World agriculture towards 2030/2050: the 2012 revision. https://www.fao.org/4/ap106e/ap106e.pdf.

Bai, Y. and Cotrufo, M. F. (2022). Grassland soil carbon sequestration: Current understanding, challenges, and solutions. *Science*, 377(6606), 603–608.

Bardgett, R. D., et al. (2021). Combatting global grassland degradation. *Nature Reviews Earth & Environment*, 2(10), 720–735.

Bengtsson, J., et al. (2019) Grasslands–more important for ecosystem services than you might think. *Ecosphere*, 10(2), article e02582.

Berry, D. P., et al. (2022) The development of effective ruminant breeding programmes in Ireland from science to practice. *Irish Journal of Agricultural and Food Research*, 61, 38–54. https://doi.org/10.15212/ijafr-2020-0149.

CBS (2025). https://opendata.cbs.nl/statline (accessed on 17 March 2025)

Delaby, L., et al. (2021) Characteristics of robust animals for grass-based production systems. *Irish Journal of Agricultural and Food Research,* 59(2), 246–257. https://doi.org/10.15212/ijafr-2020-0147.

Dumont, B., et al. (2013). Prospects from agroecology and industrial ecology for animal production in the 21st century. *Animal*, 7(6), 1028–1043.

EEA (2024). *Agricultural area under organic farming in Europe*. European Environment Agency. https://www.eea.europa.eu/en/analysis/indicators/agricultural-area-used-for-organic (accessed on 1 September 2025)

Egan, M., et al. (2018). Incorporating white clover (*Trifolium repens* L.) into perennial ryegrass (*Lolium perenne* L.) swards receiving varying levels of nitrogen fertiliser:

effects on milk and herbage production. *Journal of Dairy Science*, 101, 3412–3427. https://doi.org/10.3168/jds.2017-13233.

Erhard, M., et al. (2016). *Mapping and assessment of ecosystems and their services.* 3rd Report – Final, March 2016. Technical Report - 2016 – 095. European Union, European Environment Agency, 183 pp.

Eurostat (2025a). Utilised Agricultural Area by Categories. https://doi.org/10.2908/TAG00025 (accessed on 17 March 2025)

Eurostat (2025b). Permanent agricultural grassland in Europe. https://ec.europa.eu/eurostat/statistics-explained/index.php?title=Permanent_agricultural_grassland_in_Europe (accessed on 17 March 2025)

FAO (2025). Europa: Entwicklung der Agrarfläche und der Waldfläche in den Jahren 2000 bis 2023 (Fläche in 1.000 Hektar). In Statista. https://de.statista.com/statistik/daten/studie/1258529/umfrage/agrarflaechen-und-waldflaechen-in-europa/ (accessed on 12 August 2025)

Feigenwinter, I., et al. (2023). Large inter-annual variation in carbon sink strength of a permanent grassland over 16 years: Impacts of management practices and climate. *Agricultural and Forest Meteorology*, 340, 109613.

Finneran, E., et al. (2010). Simulation modelling of the cost of producing and utilising feeds for ruminants on Irish farms. *Journal of Farm Management*, 14, 95–116.

Friggens, N. C., et al. (2017). Review: Deciphering animal robustness: a synthesis to facilitate its use in livestock breeding and management. *Animal*, 11, 2237–2251.

Gibson, D. J. and Newman, J. A. (2019). Grasslands and climate change: an overview. In: Gibson, D.J. and Newman, J. A. (eds.) *Grasslands and climate change. Ecological reviews*. Cambridge University Press, 3–18.

Goldstein, J. E., et al. (2023). Unlocking 'lock-in' and path dependency: a review across disciplines and socio-environmental contexts. *World Development*, 161, 106–116. https://doi.org/10.1016/j.worlddev.2022.106116.

Habel, J. C., et al. (2013). European grassland ecosystems: threatened hotspots of biodiversity. *Biodiversity and Conservation*, 22, 2131–2138.

Hanrahan, L., et al. (2017). PastureBase Ireland: a grassland decision support system and national database. *Computers and Electronics in Agriculture*, 136, 193–201. https://doi.org/10.1016/j.compag.2017.01.029.

Hennessy, D. P., et al. (2020). Increasing grazing in dairy cow milk production systems in Europe. *Sustainability*, 12(6), 2443.

Hennessy, D. P., et al. (2021). The net contribution of livestock to the supply of human edible protein: the case of Ireland. *The Journal of Agricultural Science*, 159, 463–471. https://doi.org/10.1017/S0021859621000642.

Hennessy, D., et al. (2024) What can we do to improve the contribution of grassland to net food security? *Grassland Science in Europe*, 29, 23–37.

Horn, J. and Isselstein, J. (2022). How do we feed grazing livestock in the future? A case for knowledge-driven grazing systems. *Grass and Forage Science*, 77(3), 153–166.

Huyghe, C., et al. (2014). *Grasslands and herbivore production in Europe and effects of common policies*. Quae Éditions, 287 pp.

Isselstein, J. (2019). *Protecting biodiversity in grasslands.* Burleigh Dodds Science Publishing, 16p, eBook.

Isselstein, J. (2021). Towards an informed grassland farming Shortfalls and perspectives. *Grassland Science in Europe*, 26, 2–10.

Isselstein, J. and Kayser, M. (2014). Functions of grassland and their potential in delivering ecosystem services. *Grassland Science in Europe*, 19, 199–214.
Isselstein, J. and Komainda, M. (2021). *Reconciling the goals–grass breeding for an improved grassland husbandry. Exploiting genetic diversity of forages to fulfil their economic and environmental roles*, 1–6. Palacký University Olomouc. https://doi.org/10.5507/vup.21.24459677.
Jitea, I. M., et al. (2024). Assessing the economic value of cultural ecosystem services from grasslands using choice experiments. *Grassland Science in Europe*, 29, 186–194.
Komainda, M., et al. (2024). Diverse grassland and diverse benefits: the different roles and purposes of grassland. *Grassland Science in Europe*, 29, 778–787.
Küster, H. J. (1995). *Geschichte der Landschaft in Mitteleuropa*. C.H. Beck, 424p.
Laisse, S., et al. (2018). The net contribution of ruminant production to the protein supply for humans. *Grassland Science in Europe*, 23, 718–720.
Laisse, S., et al. (2019). L'efficience nette de conversion des aliments par les animaux d'élevage: une nouvelle approche pour évaluer la contribution de l'élevage à l'alimentation humaine. *INRA Productions Animales*, 31, 269–288.
Leroy, F., et al. (2022) Animal board invited review: Animal source foods in healthy, sustainable, and ethical diets – An argument against drastic limitation of livestock in the food system. *Animal*, 16: 100457. https://doi.org/10.1016/j.animal.2022.100457.
McClearn, B., et al. (2019). Milk production per cow and per hectare of spring-calving dairy cows grazing swards differing in *Lolium perenne* L. ploidy and Trifolium repens L. composition. *Journal of Dairy Science*, 102, 8571–8585. https://doi.org/10.3168/jds.2018-16184.
Milazzo, F., et al. (2023). The role of grassland for erosion and flood mitigation in Europe: a meta-analysis. *Agriculture, Ecosystems & Environment*, 348, 108443.
Moran, D. and Blair, K. J. (2021). Review: sustainable livestock systems: anticipating demand-side challenges. *Animal*, 15, 100288.
Mosnier, C., et al. (2021). Evaluation of the contribution of 16 European beef production systems to food security. *Agricultural Systems*, 190, 103088. https://doi.org/10.1016/j.agsy.2021.103088.
Mottet, A., et al. (2018). Review: domestic herbivores and food security: current contribution, trends and challenges for a sustainable development. *Animal*, 12, s188–s198. https://doi.org/10.1017/S1751731118002215.
Mustapa, M. A. C., et al. (2025). Evolving appetites: current evidence and future perspectives in terms of meat substitutes in Europe. *Food Science & Nutrition*, 13(1), e4753. https://doi.org/10.1002/fsn3.4753.
Newton, P., et al. (2020). What is regenerative agriculture? A review of scholar and practitioner definitions based on processes and outcomes. *Frontiers in Sustainable Food Systems*, 4, 577723. https://doi.org/10.3389/fsufs.2020.577723.
O'Brien, D., et al. (2018). A national methodology to quantify the diet of grazing dairy cows. *Journal of Dairy Science*, 101, 8595–8604.
O'Callaghan, T. F., et al. (2016). Effect of pasture versus indoor feeding systems on raw milk composition and quality over an entire lactation. *Journal of Dairy Science*, 99, 9424–9440. https://doi.org/10.3168/jds.2016-10985
O'Donovan, M. and Delaby, L. (2016). Grazed grass in the dairy cow diet – how this can be achieved better! *Grassland Science in Europe*, 21, 350–365.

OECD and FAO (2023). *OECD-FAO agricultural outlook 2023-2032. Chapter 6. Meat.* OECD. https://www.fao.org/documents/card/en?details=CC6361EN

O'Mara, F. P. (2012). The role of grasslands in food security and climate change. *Annals of Botany*, 110(6), 1263–1270.

Orr, R. J., et al. (2016). The North Wyke Farm Platform: effect of temperate grassland farming systems on soil moisture contents, runoff and associated water quality dynamics. *European Journal of Soil Science*, 67(4), 374–385.

Peeters, A. (2009). Importance, evolution, environmental impact and future challenges of grasslands and grassland-based systems in Europe. *Grassland Science*, 55, 113–125.

Peeters, A., et al. (2014). Grassland term definitions and classifications adapted to the diversity of European grassland-based systems. *Grassland Science in Europe*, 19, 743–750.

Pellaton, R., et al. (2022). Cultural ecosystem services in European grasslands: a systematic review of threats. *Ambio*, 51(12), 2462–2477.

Pfeifer, C. and Winterberg, R. (2024). What is the role of grasslands under a feed-no-food scenario? *Grassland Science in Europe*, 29, 15–22.

Poschlod, P. (2016). *Geschichte der Kulturlandschaft*. 2. ed. 320p. Eugen Ulmer KG.

Resare Sahlin, K., et al. (2024). An exploration of biodiversity limits to grazing ruminant milk and meat production. *Nature Sustainability*, 7, 1160–1170.

Rhodes, C. J. (2017). The imperative for regenerative agriculture. *Science ProgressScience ProgressScience Progress*,Science Progress 100, 80–129. https://doi.org/10.3184/003685017X14876775256165.

Ruelle, E., et al. (2018). Development of the Moorepark St Gilles grass growth model (MoSt GG model): a predictive model for grass growth for pasture based systems. *European Journal of Agronomy*, 99, 81–90. https://doi.org/10.1016/j.eja.2018.06.010.

Runhaar, H., et al. (2020). Endogenous regime change: lessons from transition pathways in Dutch dairy farming. *Environmental Innovation and Societal Transitions*, 36, 137–150. https://doi.org/10.1016/j.eist.2020.06.001.

Ryschawy, J., C. et al. (2017). Assessing multiple goods and services derived from livestock farming on a nation-wide gradient. *Animal*, 11(10), 1861–1872.

Schils, R. L., et al. (2022). Permanent grasslands in Europe: land use change and intensification decrease their multifunctionality. *Agriculture, Ecosystems & Environment*, 330, 107891.

Schreefel, L., et al. (2020). Regenerative agriculture – the soil is the base. *Global Food Security*, 26, 100404. https://doi.org/10.1016/j.gfs.2020.100404.

Timlin, M., et al. (2023). Impact of varying levels of pasture allowance on the nutritional quality and functionality of milk throughout lactation. *Journal of Dairy Science*, 106, 6597–6622. https://doi.org/10.3168/jds.2022-22921.

Tonn, B. (2024). Permanent grassland typology – Atlas & field guide. Super-G Report, Deliverable 2.1, 198 p.

United Nations (2024). *World population prospects 2024: summary of results*. UN DESA/POP/2024/TR/NO. 9. United Nations.

Van den Pol-van Dasselaar, A., et al. (Eds) (2019). *Grassland use in Europe - a syllabus for young farmers*. Éditions Quae, 263 pp.

Van den Pol-van Dasselaar, A., et al. (2020). Grazing of dairy cows in Europe – An in-depth analysis based on the perception of grassland experts. *Sustainability*, 12, 1098.

Van den Pol-van Dasselaar A., et al. (2024). The perspectives of young European farmers and students on grazing. *Grassland Science in Europe*, 29, 896–898.

Van Kernebeek, H. R. J., et al. (2016). Saving land to feed a growing population: consequences for consumption of crop and livestock products. *International Journal of Life Cycle Assessment*, 21, 677–687. https://doi.org/10.1007/s11367-015-0923-6.

van Zanten, H. H. E., et al. (2016). Global food supply: land use efficiency of livestock systems. *The International Journal of Life Cycle Assessment*, 21, 747–758.

Veerkamp, R. F., et al. (2002). Dairy cattle breeding objectives combining yield, survival and calving interval for pasture-based systems in Ireland under different milk quota scenarios. *Livestock Production Science*, 76, 137–151.

Wild, M., et al. (2025). Feed the green for a sustainable and protein-efficient dairy production. *Agricultural Systems*, 223, 104216.

Wilkinson, J. M. (2011). Re-defining efficiency of feed use by livestock. *Animal*, 5, 1014–1022. https://doi.org/10.1017/S175173111100005X.

Winkler, K., et al. (2021). Global land use changes are four times greater than previously estimated. *Nature Communications*, 12, 2501. https://doi.org/10.1038/s41467-021-22702-2.

Published by Burleigh Dodds Science Publishing Limited, 2025.

Chapter 18

Challenges facing alpine/ mountain grasslands

Giovanni Peratoner, Laimburg Research Centre, Italy; Caren M. Pauler and Manuel K. Schneider, Agroscope, Switzerland; Vibeke Lind, NIBIO, Norway; Andreas Klingler and Andreas Schaumberger, AREC Raumberg-Gumpenstein, Austria; and Enrico Sturaro, University of Padova, Italy

1 Introduction

Mountain grasslands are the basis of a site-specific, locally based livestock production in a large part of the grassland area of the temperate zone. This chapter first outlines the site factors characterising mountain grasslands and the societal factors connected to them (Fig. 1). It then addresses the specificities of mountain grassland production in terms of yield and forage quality and links them to grassland-based animal production. Finally, it discusses relevant trade-offs between productive and environmental aspects with a focus on wildlife interaction and the expression of ecosystem services. Throughout this chapter, the mountain chain of the Alps, stretching from East to West from France to Slovenia and crossing eight countries, as well as the Scandinavian Mountains, running through the Scandinavian Peninsula, will serve as a primary case study. However, several issues addressed in this context are shared with other mountain areas characterised by comparable climatic and topographic

http://dx.doi.org/10.19103/AS.2025.0154.27

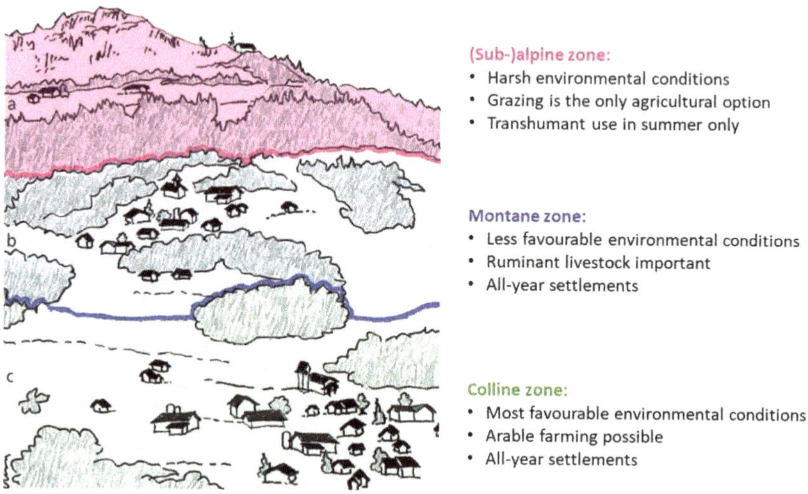

Figure 1 Type of agricultural activities depending on the altitudinal zone in the Alps (based on the artwork by U. Kaufmann, used with permission from Lauber et al. (2013), and modified by the authors).

conditions, e.g. other mountain chains in Europe like the Apennines or the Dinarides in Southern Europe, the Carpathians in Central Europe or the Pyrenees in Western Europe, as well as with other extra-European mountain chains.

2 Topography and climate

The dynamics and adaptability of grasslands to different site conditions, which vary significantly in mountain areas, are due to the diverse botanical composition and the interspecific interactions (Frei et al., 2014). As long as there is an equilibrium in terms of nutrient balance and coherence between site potential and management intensity, the adaptation of the plant community to the site conditions and to the management creates a stable and resilient ecosystem embedded in a sustainable circular economy (Pötsch et al., 2006). In mountain areas, abiotic site factors change on a comparatively small scale, resulting in a wide variety of utilisation forms and intensities in grasslands (Pötsch, 2012). Grassland farming in mountain areas is, therefore, generally small-scaled and strongly characterised by the following site factors:

- topography (elevation, slope, aspect);
- climate (temperature, precipitation, insolation); and
- soil (pH, depth, organic matter content, nutrient availability, texture, porosity).

Elevation causes a decrease in temperature according to a linear gradient, which largely determines the climatic conditions affecting grasslands management (Fig. 2a). Only through appropriate temperatures and radiation, assimilation processes are maintained and development phases such as germination or generative growth initiated (Keller et al., 1997). With increasing elevation, decreasing temperature reduces the duration of the growing season, thus limiting the number of growth cycles being potentially harvested, with decreasing performance in terms of forage yield. Slope (Fig. 2b) and aspect (i.e. the orientation faced by a terrain surface) usually change on a small scale, affecting the microclimate, which primarily depends on radiation and heat balance. This further increases the spatial variability of productivity and, in some cases, makes management considerably more difficult. Steeper areas also limit or complicate mechanisation and increase costs and workload, because of the need of special machineries or even manual work. Grazing in alpine grasslands requires livestock adapted to e.g. climatic conditions, terrain and wildlife, especially in steeper terrain. Light-weighted, robust breeds are particularly suited, as they are able to climb steep slopes and exert less pressure on the ground, thereby reducing the negative effects of trampling and the risk of erosion. The effect of aspect can be strong in mountain areas: south-facing areas experience an earlier start of the growing season due to longer and more intense radiation, causing earlier snowmelt and higher temperature but can be susceptible to summer drought spells due to higher evapotranspiration (Hua et al., 2022). Due to the mentioned environmental

a) Impact of increasing elevation = decreasing temperature on...		**b) Impact of increasing slope on...**		**c) Impact of increasing remoteness on...**	
Productivity	↓	Costs per product unit	↑	Conflicts with wildlife	↑
Duration of growing season	↓	Erosion	↑	Abandonment and encroachment	↑
Evapotranspiration and drought risk	↓	Physical workload	↑	Depopulation and missing generation turnover	↑
Duration of winter-feeding period	↑	Agricultural options	↓	Dependence on public payments	↑
Small-scale variability and abundance of specialised species	↑	Ease of mechanisation	↓	Communal ownership	↑
(Drinking) water supply	↑	Soil depth	↓	Infrastructure availability	↓

Figure 2 Impact of increasing mountain character of the environment (increasing elevation strongly correlated with decreasing temperature as well as increasing slope and remoteness) on different aspects of grassland farming, based on expert guesses of the authors. Upward pointing arrows indicate an increase, downward pointing arrows a decrease. The thickness of the arrows express the magnitude of the impact.

constraints, mountain grassland farming can only be maintained if there is a social interest and a political will to financially compensate these disadvantages or through innovative business models in which the price of the animal products economically rewards farmers for their added value contributions in delivering the ecosystem services provided by mountain grasslands (van den Pol-van Dasselaar et al., 2021). To this end, specially tailored subsidy models are already put into practice in various countries.

Physical and chemical soil properties, in conjunction with biotic factors, significantly influence the productivity of mountain grasslands. Intact soil life is essential for the mineralisation of organic material and the associated provision of nutrients in a form available to plants. The farmer can support productivity through fertilisation. In contrast, the consequences of agricultural activities, such as compaction due to mechanical stress, can impair soil fertility. In principle, soil type, nature, structure and depth are responsible for soil nutrient, water, air and heat balance (Keller et al., 1997). In former times, in a context of a subsistence economy, arable crops were widely diffused on a very small scale in mountain areas at the cost of huge amounts of working time and physical work, but this is no longer economically viable in the current context. Compared to arable land in favourable locations, permanent grasslands in mountain areas usually have lower soil fertility, in combination with topographical and climatic constraints. If basic requirements allowing for the mechanisation of ploughing are not met (e.g. where the slope exceeds about 20%), grassland and livestock farming are the only possible forms of agricultural utilisation in mountain regions. Moreover, grasslands provide several ecosystem services related to the soil. For example, permanent grasslands stabilise soil structural characteristics, prevent soil erosion, increase the organic matter content and improve water-retention capacity (Milazzo et al., 2023a,b; Saha and Kukal, 2015; Zhao et al., 2020b). Especially in connection with climatic changes and the trend towards drier conditions, the buffering function of the soil for a sufficient supply of nutrients and water is of great importance.

As for all crops, the climate and its interaction with topography and soil significantly influence grasslands. In high mountain regions, the growing season is short and plant growth is limited. The growing season depends less on the photoperiod itself than on the closely related annual temperature cycle and duration of snow cover (Filippa et al., 2019). Grassland vegetation is less demanding than other crops in terms of temperature but largely depends on precipitation as the primary factor determining grassland productivity, particularly in ecosystems that are either water-limited or under heath stress. Although solar radiation and temperature also affect productivity, their impact is secondary to the critical role of precipitation in driving grassland net primary productivity (Zhao et al., 2020a; Chen et al., 2023). Above all, a sufficient moisture at the beginning of the vegetation period is crucial for a high yield of the first growth cycle in spring. The Alps are particularly vulnerable to climate change and have already experienced twice as much warming since the beginning of the industrial age compared to the global average (Pepin et al., 2022; Kotlarski et al., 2023). This warming affects natural ecosystems and the cultivated landscape. Climate projections for the coming decades show an intensified trend towards further increases in mean temperature and unevenly distributed precipitation (IPCC, 2022). A decrease in moderate rainfall and an increase in heavy rainfall are already leading to recurrent intense drought spells in combination with high temperatures at various times of the year (Trnka et al., 2011; Pötsch et al., 2014). Although temperate grassland species have been shown to be largely robust and resilient against drought (Deléglise et al., 2015; Hofer et al., 2016), the uncertainty of forage production for farmers and the related need for purchasing off-farm forage jeopardises the profitability and the persistence of mountain grassland farming in the long term (Schaumberger et al., 2022). Because of the ongoing climate change, an increase in biomass provision is expected at low elevation in the Pre-Alps (i.e. the foothills at the base of the Alps), while a decrease in biomass provision is expected on the southern side of the Alps (Jäger et al., 2020). Moreover, a shift in the duration of the growing season through an earlier start and a later end is expected (Möhl et al., 2022; Wang et al., 2020).

Topography, soil and climate directly influence the dynamics and resilience of grassland ecosystems (Sebastià, 2004; Schirpke et al., 2017). Effective management and adaptation strategies must consider these complex relationships to ensure the sustainable utilisation of grasslands under a changing climate (Vogel et al., 2012; Schils et al., 2022). To a certain extent, the botanical composition can adapt to drier conditions, with drought-tolerant grass species and deep-rooted legumes or herbs increasing their proportion (Sanaullah et al., 2014; Tello-Garcia et al., 2020). However, active adaptation measures are required. This concerns the management (fertilisation and utilisation times, irrigation, etc.) and active adaptation of the plant communities through systematic weed control and reseeding with drought-tolerant forage species and cultivars within species.

3 Socioeconomic aspects

The topography of mountains increases the cost of travel from one point to another. This seemingly obvious characteristic has tremendous consequences for mountain livelihood systems and land use (Figs. 2b and 2c). As a first consequence, more energy or time is needed to reach a certain spot in mountain terrain than in a flat area (ESPON, 2019: Maps 5–7). This poorer accessibility increases transport costs of harvested biomass from field to farm and of manure and fertiliser from the farm to the field. As a consequence, the labour needs and the costs of forage production per area unit increase as a curvilinear, convex function of slope. If the costs of forage production are computed on the forage yield unit, the costs are a function of the interaction of slope and elevation and rise very rapidly as both factors increase (Peratoner et al., 2017; Peratoner et al., 2021). As slope steepness and remoteness of the grassland areas increase, it is therefore more advantageous to graze (i.e. lead animals to forage) than to feed in-house (bring forage to the animals). Grazing in remote places requires appropriate herding of these animals which is, due to the higher costs of travel, difficult for the owner of the animals himself besides work at the farmstead. Therefore, additional persons (e.g. family members, villagers or outsiders) are engaged during summer for the herding task for centuries in mountain areas (Calabrese et al., 2014) and the animal owner has time for work at the farmstead.

The low accessibility of mountain grasslands leads to a second particularity: the 'common property rights', which are frequent in mountain regions (Landolt and Haller, 2015; Galan et al., 2022). Communal ownership and communal organisation of work have several advantages in mountain grazing systems over private ownership, namely economies of scale in the workload, the hiring of external personnel and investments to access and housing. On the other hand, communal systems need a higher degree of agreement to ensure a fair and sustainable use of natural resources and complex decision processes for the adaptation to current and future changes (Baur and Binder, 2013).

Recent decades have seen important transitions in demographic and agricultural structures in mountain areas. In the Alps, there are strong regional differences, but several general trends can be observed: The overall population in the Alps (as defined by the perimeter of the Alpine Convention) increased by 7.8% from 1994 to 2004, but there are strong local differences among the 6000 alpine municipalities (Zanolla et al., 2007). In general, well-connected alpine towns are strongly gaining population at the expense of remote agricultural villages. Municipalities primarily devoted to agriculture are transformed into areas devoted to both agriculture and tourism or dominated by tourism only (Bätzing et al., 1996). The number of farms decreased by 35% between 1980 and 2000 (Streifeneder et al., 2007) due to a lack of successors (Flury et al., 2012) and this process is ongoing. In parallel with the reduced agricultural labour force, the importance of agriculture in municipal decision-making processes decreases and more importance is given to other stakeholders (e.g. tourism and power production).

Since the total agricultural area decreased less than the number of farms (by 7.6%; Flury et al., 2012), the surface area managed per farm increased. However, managing larger units and more land requires investments in infrastructure and machinery. These investments, in turn, necessitate the specialisation of animal production (unified herds) and forage conservation (silage or hay) as well as a rationalisation of work processes (including mechanisation). This has consequences for the livelihood system of mountain farmers: a recent study in a Swiss mountain region found that farmers increased commercial and administrative contacts at the expense of social contacts within their families and with friends (Junquera et al., 2022). This demonstrates the transition of family-run farms towards professional agri-food enterprises.

The professionalisation and rationalisation of farm operations have several consequences on mountain grassland use and maintenance. Forage production is concentrated on flat and productive sites, manageable with machinery (Marini et al., 2009). High productivity of livestock allows comparable output at lower numbers of heads leading to a substantial drop in animal numbers in certain regions (Battaglini et al., 2014). However, highly productive animals are less suited to graze low-productive and steep mountain pastures and are hence supplemented with concentrates or herded on the best sites only (van Dorland, 2007). Finally, the lower number of people employed in agriculture reduces the available labour force for manual work such as fencing or pasture maintenance (Baur et al., 2014).

4 Forage production

Grassland productivity in mountain areas is limited in the first instance by the decreasing duration of the growing season and heat balance with decreasing temperatures (Jäger et al., 2020). At high elevations, the period of indoor feeding far exceeds the grazing period. The limited duration of the growing season results in a need for harvesting and storing forage for the winter months. Especially under relevant topographic constraints, this leads to an additional burden in terms of workload and investments for special machineries and facilities (e.g. barn-drying systems) to perform this task.

Grassland management is closely linked to site factors; site-adapted management aims at taking them into account and establishes a robust and resilient utilisation system, which is able, in the long term, to cope with the climatic challenges of the mountain environment and to ensure a balance in terms of nutrients. Therefore, in mountain grasslands, the focus does not lie on the short-term maximisation of yields through the intensive use of external inputs, but on ensuring stable yields and high forage quality in line with the principles of circular agriculture. In mountain regions, shallow soils, steep slopes, and extreme weather events with significant annual fluctuations are often the main challenges. Climate change results in extended vegetation periods (Bellini et al., 2023) and opens up new utilisation possibilities but also requires adapted management to retain high forage quality. Increased management intensity through more cuts or intensive grazing, along with high fertiliser application, leads to higher evapotranspiration but better water-use efficiency. This applies above all to fertilisation, as this significantly enhances biomass production (Rose et al., 2012). Therefore, intensification should only occur in areas with sufficient water availability. Flexibility in management is essential to respond to annual fluctuations, e.g. by reducing the mowing frequency in unfavourable years.

Biodiversity is crucial for the ecological stability of mountain grassland habitats and their agricultural production, as biodiverse systems are expected to be more resilient, which is especially important in areas of high fluctuation in environmental conditions. Particularly, on less productive sites, it is essential to maintain or even promote plant and animal diversity to counteract the general loss of diversity observed in grasslands in the last decades (Isselstein, 2018).

Also, the economic viability of site-adapted farming methods requires careful evaluation. The direct production functions and the ecological and socio-economic functions of grassland production require equal consideration. Public payments and incentives for farmers who adopt sustainable practices are essential for implementing this approach. Nevertheless, many grassland areas in Europe, usually including the more favourable areas in the Alps, face a mismatch of site potential and management intensity due to overemphasising the production function. This focus often leads to disproportionate intensification – a practice that undermines the sustainability of these regions and diminishes the multifunctionality of permanent grasslands (Schils et al., 2022). Management intensity that exceeds the site optimum can lead to degeneration of the sward, whereby important species disappear. Undesirable, opportunist species often colonise the resulting gaps. Additionally, fertilisation is subject to stringent regulations within the European Union. Consequently, uniformly high nutrient levels across large-scale areas are not feasible, necessitating a differentiated approach to management intensity.

Conversely, many sites, particularly remote areas and those with challenging topographical conditions, including highly diverse habitats with a rich species variety, are subject to excessive extensification or abandonment. The resulting loss of valuable graduated agricultural mosaics leads to both agriculturally and ecologically detrimental consequences. The abandonment of utilisation leads to the establishment of tall grasses and perennial woody species, which increasingly reduce the biodiversity of moderately managed meadows and pastures (Dengler and Tischew, 2018). Another adverse effect of the reduction in utilisation is the increasing evapotranspiration in alpine grasslands, especially when the abandoned areas reach a higher leaf area index (LAI) than that of pastures (Inauen et al., 2013). Moreover, the additional water consumption following bush encroachment with alder species leads to a reduced rate of deep percolation and thus reduces the total amount of water available in a catchment area (van den Bergh et al., 2018).

Site-adapted management strives to create a resilient and comprehensive agricultural system that combines production goals with ecological and socio-economic functions. The aim is to recognise the site's potential and adjust management intensity to achieve stable yields that are resilient to biotic and abiotic stress factors, rather than focussing solely on short-term maximum yields. This requires a deep understanding of local conditions and flexible, forward-looking planning and implementation of management practices.

A further challenge is the evaluation of forage yield and quality. Its assessment is a challenging task in mountain areas, due to the small-scaled variability of topography and the lower predictability of the meteorological conditions. It requires an in-depth understanding of the many influencing factors and their complex interactions. The site conditions – including soil, climate and topography – form the basic framework for agricultural utilisation and are the starting point for all further management-related decisions. Management methods must be carefully selected and implemented in line with these natural conditions. The type and intensity of utilisation play a central role in ensuring stable yields and high forage quality. This includes practices such as mowing for hay or silage, grazing or combining both, coupled with fertilisation tailored to the site and utilisation aligned with maintaining closed nutrient cycles and avoiding both over-fertilisation and under-fertilisation.

The botanical composition of the sward, as a direct result of site factors and management, can be positively influenced by targeted interventions such as reseeding or weed control. To ensure long-term successful grassland management, continuously monitoring growth dynamics and regularly evaluating yields and forage quality are crucial.

Knowledge of grassland biomass yield is crucial for optimising management, but it is also a challenging task. Despite its importance, this data remains mostly unknown to farmers. In contrast, forage quality is assessed more frequently. However, the laboratory analysis of forage samples, a standard method for precisely determining the chemical composition, does not allow for timely adjustments in grassland management since it is only feasible post-harvest. Indirect indicators such as sward height and botanical composition of the vegetation provide valuable information even during the growth phase (Peratoner and Pötsch, 2019). In addition to simple methods, such as a scale, sward stick or a pasture plate meter (Murphy et al., 2021; Stewart et al., 2001), advanced techniques, such as creating 3D models using drone imagery (Bazzo et al., 2023) or ultrasonic sensors (Fricke et al., 2011), provide detailed insights into the yield situation. The phenology of dominating species also enables the estimation of forage quality; therefore, tabulated values based on empirical data deliver useful information in this sense (Resch et al., 2006; Daccord et al., 2007). The combination of automatically retrieved topographic and meteorological data with information about the management and a simple characterisation of the sward type provided by the users allows a further improvement of the estimate of forage quality by means of statistical predictive models driven by Growing Degree Days and other complementary information (Peratoner et al., 2016).

Modern technologies and innovations are increasingly important in grassland management, particularly in recording and analysing growth characteristics, including the start of the growing season (Dujakovic et al., 2024), cut detection (Dujakovic et al., 2025; Holtgrave et al., 2023; Schwieder et al., 2022; Siegmund et al., 2019; Taravat et al., 2019; Watzig et al., 2023) and yield (Castelli et al., 2023) and forage quality estimation. Combined with algorithms and artificial intelligence, satellite images and weather data provide valuable information on cut frequency and real-time yield forecasts. These technological advances enable comprehensive monitoring, even on small-structured inner-alpine meadows and pastures, and contribute significantly to the optimisation of grassland management.

5 Animal production

The advantage of ruminants is their ability to digest fibrous plant material and convert it into high-quality food for human consumption in terms of protein and energy. Thus, ruminants can be fed with forage from areas not suitable for the production of e.g. grain and other crops directly suited for human nutrition. In this way, mountain grasslands contribute to meeting the increasing food demand without competing with more efficient agricultural forms of food production. Depending on the length of the growing season, the animals can harvest more than 50% of their annual feed from mountain or alpine pastures (Bunger et al., 2018) which makes these areas highly relevant. In mountain areas, this often happens by altitudinal transhumance. Following the altitudinal gradient from the valley to alpine areas and the respective gradient of phenological advance, the animals graze plants at early phenological stages of high quality over weeks or even months. Due to this upward movement, the animals use the pastures at their optimum in terms of forage quality.

Food produced in mountain areas is perceived positively and considered sustainable (Hersleth et al., 2012). Sensitive consumers, however, may complain about off-odours and off-flavours and a variation in meat quality (Prache et al., 2022). Eating quality of meat is positively related to intramuscular fat content (Pannier et al., 2014). As an animal gets older, the muscle-to-fat proportion changes towards more fat and less muscle, which from a sensory point of view may be positive. Moreover, the highly diverse composition of grasses, shrubs, herbs and other species containing secondary metabolites may influence the taste of milk and meat, which are thus less uniform than food produced on lowland pastures (Ådnøy et al., 2005) or based on concentrate (Lind et al., 2009).

Some secondary metabolites, such as tannins, saponins, polyphenol oxidase, flavonoids and essential oils, may improve the fatty acid composition to a larger content of polyunsaturated fatty acids (Girard et al., 2016; Campidonico et al., 2016).

Environmental conditions of mountain grasslands can pose a challenge for grazing animals. Thus, site-adapted livestock is mandatory to cope with steep slopes, harsh climate and low forage quality at high elevation (Pauler et al., 2020a, b). Generally, local breeds seem to be well adapted to alpine vegetation systems (Willems et al., 2013). However, higher yielding dairy cows or goats depend on higher quality forage and additional concentrate (Sturaro et al., 2013) and may alter vegetation composition in an undesired way in the long term (Pauler et al., 2019). For highly productive breeds, pastures at high elevation are of less importance, as they may compromise both milk yield and animal health. Site-adapted management means matching livestock productivity level and pasture management intensity to the specific site conditions. However, due to the intensification of the farming structure in Europe, fewer animals ensure the same production level, as the production per animal increases. Therefore, more off-farm feed (including concentrate) is imported from lowland areas where grass yield and forage quality are higher and mechanisation is easier. Consequently, there is less livestock utilising alpine and mountainous pastures.

For instance, in the traditional production of summer milk in Norway, dairy cows are moved from the lowland farms to mountain farms for 2 months during summer. Today, the production system is endangered as the productivity of the cows is increasing. The pasture quality is insufficient, even when feeding additional concentrates to maintain the high yield of the cows. Those farms still producing summer milk are those that primarily have less intensive production in the lowland during the rest of the year. Native breeds (e.g. Sider-trønder Nordlandsfe) are more suited for the production of summer milk than high-yielding dairy cows (Norwegian Red Cattle). The native animals are smaller with lower milk yield and thus demand less feed. As the number of summer farms decreases, more areas are left abandoned with shrub encroachment as a result. The quality of the pasture decreases, and this makes it less likely that farmers will start summer farming again (Bele et al., 2018). A similar development is found in the Alps. Unique dairy and meat products produced in mountainous grasslands are often certified with a Protected Designation of Origin (PDO) label. Products must meet high quality to be certified, and the certification is followed by strict regulations for the production system and requires foresight for authentication and traceability criteria (Coppa et al., 2017). The distribution channels for these products often fall outside the established distribution channels and often must be developed alongside the products by the farmers themselves or smaller cooperatives. However, if successful, higher prices can be realised in the market due to the unique character of these products.

6 Interaction with predators

Pasture-based farming systems in mountain regions are threatened by an increasing number of large carnivore predators. Wolves are a commonly found predator in the Scandinavian mountains and the Alps. The recent natural recolonisation of many European areas by wolves has increased conflicts with humans (Boitani and Linnell, 2015). Conflicts arise particularly in regions where farmers are no longer used to protecting their livestock due to the absence of predators over centuries. There, livestock often is left grazing unattended and unprotected, even at night. Reducing conflicts due to predation on livestock therefore requires changes in farming practices and the adoption of protection methods (Dalmasso et al., 2011; Boitani and Linnell, 2015). In fact, damage compensation alone fails to reduce the conflicts between livestock farming and carnivores (Dalmasso et al., 2011). Sheep and goats are the livestock species most frequently killed by wolves in Europe (Reinhardt et al., 2012), but predation on cattle also occurs (Dalmasso et al., 2011). Additionally, lynx, wolverine, bears and eagles are seen in Scandinavia and the Alps. The predators predate in different ways and at different times of the grazing season. Preventive measures such as guarding dogs, early gathering of livestock and mortality collars are being used (Hansen et al., 2019). When farmers lose as much as 30–50% of their flock to predators, the production is endangered. While some farmers give up, others change their type of livestock to animals less exposed to predators (e.g. replacing sheep with beef cattle). Successful coexistence of large carnivores and humans depends on the effective mitigation of their impact on humans, especially through livestock predation. It is therefore essential for relevant stakeholders like livestock owners, conservation practitioners or carnivore managing authorities to know the effectiveness of interventions intended to reduce livestock predation by large carnivores: keeping livestock in enclosures, guarding or livestock guarding dogs, predator removal and using visual or auditory deterrents to frighten carnivores. Literature shows a general lack of scientific evidence of the effectiveness of any of these interventions: some interventions reduced the risk of depredation whereas other interventions did not result in reduced depredation. For example, livestock enclosures and livestock guarding (either by humans or dogs) generally appear to be effective interventions for protecting livestock from carnivores, but the effect of interventions is context-dependent and appears to vary with how well the actual problem is targeted (Eklund et al., 2017). Visual or auditory deterrents over time become familiar features in the wolves' environment, and as a consequence, their effectiveness decreases. For research, there is an urgent need to conduct studies of intervention effectiveness to support for the implementation of policies and strategies aiming at reducing conflicts between carnivores and livestock.

7 Ecosystem services

Especially in the valleys, the production of forage for ruminants is an important ecosystem service of mountain grasslands. Even at higher elevations, pastures contribute substantially to ruminant nutrition, although productivity decreases with elevation. In Switzerland, for instance, the transhumant summer pastures feed 10% of Swiss livestock while covering one-third of agricultural land (Stettler and Probst, 2023). As multifunctional ecosystems, mountain grasslands offer a broad bundle of services to society beyond food production, which become more important at higher elevations.

Mountain grasslands offer numerous cultural services, and the transhumant summer farming is even part of UNESCO's representative list of the intangible cultural heritage of humanity. For millennia, transhumant grazing of high-elevation areas has shaped mountain farming societies (Hafner and Schwörer, 2018). They developed an annual cycle in which the date of driving livestock up and down the mountain represents turning points setting the pace of life in the valley as well as on the mountain farms. Traditional costumes and rituals, like choosing the most beautiful cow in autumn, are closely related to mountain grazing traditions.

Grazing shaped not only culture in mountain areas but also the ecosystems themselves: over millennia, livestock opened forested or shrub-encroached areas by grazing, thereby creating semi-natural open grasslands and lowering the tree line. Grazing, hay cutting for winter forage and deforestation for firewood and timber accelerated this process and created large, open landscapes at the valley floor. The anthropogenic, extensive grasslands created by livestock farming host outstanding biodiversity (Kampmann et al., 2008). They are arks of diversity in times of declining species richness. Moreover, these wide and open landscapes are key for almost all types of mountain tourism (Parente and Bovolenta, 2012), including agritourism as well as hiking, mountain biking or skiing, and thus are the precondition of employment in the mountain tourism sector. Finally, mountain grasslands contribute to climate change mitigation by their mostly overlooked carbon sequestration potential (Ward et al., 2014) and enhance hazard protection. In this respect, grasslands provide better flood prevention and erosion mitigation and are almost comparable in these terms to forests (Milazzo et al., 2023b).

Multifunctional mountain grasslands and the ecosystem services they provide depend on continuous agricultural management. However, there is a two-sided trend in modern agriculture threatening the multifunctionality (Fig. 3): intensification and extensification, both leading to a loss of multifunctionality of ecosystem services (Tasser and Tappeiner, 2002).

On the one hand, the more favourable mountain grasslands are managed in increasing intensity. Farmers enhance productivity, for instance, by using

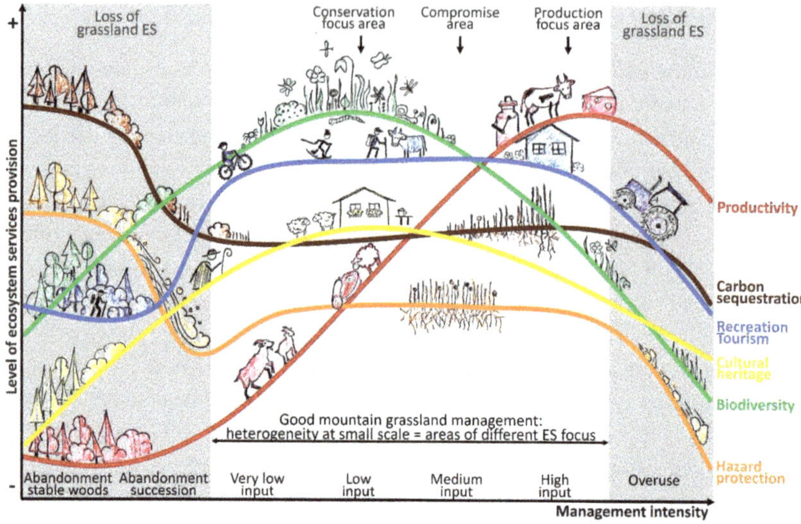

Figure 3 Change of the level of ecosystem services (ES) provision along a management intensity gradient, including abandonment and overuse at the lower and upper end of the gradient, based on expert guesses of the authors (artwork by Caren Pauler). As different ecosystem services peak at different management intensity levels, trade-offs are apparent.

machinery, applying fertilisation and by grazing more output-oriented livestock. Intensification causes uniformity of grasslands by promoting only a few highly competitive generalist species. Thus, intensified grasslands lose their specific biodiversity. Pastures grazed by a higher number of more productive, heavier animals suffer from increased trampling pressure (Pauler et al., 2019, 2020a) causing open and compacted soils. Consequently, intensification leads to less soil stability, more erosion, less water-holding capacity and less flood prevention.

On the other hand, if less favourable grasslands cannot be managed to cover costs, they are managed at very low grazing pressure or are even completely abandoned. This applies especially to steep and rocky pastures of shallow soil, low productivity and little infrastructure. As a result, succession begins, and in the long term, grasslands are overgrown by forest (below the natural tree line) or shrubs (above the natural tree line) and the majority of mountain grassland ecosystem services are lost (Prangel et al., 2023; Schirpke et al., 2020). For instance, cultural values as well as cheese and meat production are built on grasslands. Moreover, open grasslands and wood-grassland mosaics are generally more attractive than dense woodlands (Soliva et al., 2010) for tourism as well as for locals and provide a much better protection from forest fires (Conedera et al., 2024). However, avalanche protection (Mainieri et al., 2020) and carbon sequestration in above-ground biomass may benefit from reforestation. Concerning shrub encroachment, one shrub species has to be emphasised in particular due to its rapidly increasing distribution and multifarious impairment of ecosystem services: The green alder (*Alnus viridis*) is the most common invasive shrub in the European Alps (Pauler et al., 2022). It emits nitrous oxide, a highly potent greenhouse gas (Smith et al., 2021) probably negating the carbon sequestration in wood biomass. In general, dense shrub stands host much lower biodiversity than open mountain grasslands or patchy wood-grassland mosaics, but for grasslands encroached by *Alnus viridis*, the decline of diversity is much stronger than for other shrubs (Zehnder et al., 2020). Moreover, due to its high competitiveness, *A. viridis* even hinders forest reestablishment and the realisation of forest-related ecosystem services.

In summary, it is impossible to maximise all ecosystem services in the same places due to strong trade-offs among them, but the high small-scale variability of mountain grasslands allows for the realisation of a multitude of services in proximity. Overuse or underuse leads to a loss of most ecosystem services; only above-ground carbon sequestration and hazard protection benefit from abandonment and wood succession. However, during the transition phase – which can last for decades at high elevations – long, bent-over grass not foraged by animals can even exacerbate the risk of avalanches. During the last centuries and decades, additional functions of mountain grasslands have emerged, causing additional conflicts. For instance, the positive aspects of tourism, such as employment and improved infrastructure, come along with land-use conflicts, dangerous interactions between tourists and herd guard dogs, and a degeneration of mountain traditions into sheer touristic events. Moreover, in view of climate change, there is a legitimate interest in green energy produced in mountain areas. This, however, causes additional pressure on grasslands, which have been identified as ideal places for solar parks, wind farms or reservoirs for hydroelectric power. Today, mountain grasslands face the challenge of incorporating these new services while maintaining their productivity, species richness and their outstanding cultural value.

8 Future trends in research

Based on the current situation described in this chapter and the ongoing changes, the following research topics emerge as particularly relevant:

- Adaptation of mountain farming systems to changing climate, including the selection of plant species/varieties and livestock species/breeds adapted to mountain conditions and exploration of the genetic basis.
- Establishment of systems making use of the potential of digitalisation and technological advances to increase the efficiency of mountain farming

systems and to decrease the costs of production under disadvantaged conditions.
- Development of approaches to solve or mitigate the conflicts of interest between agricultural production, tourism, predators and energy production.
- Development and evaluation of sustainable mountain farming systems (with regard to ecological, economic and social dimensions).

9 Where to look for further information

In most countries (or regions within countries), including mountainous areas, associations dealing with grassland management in a mountain environment have been established. They provide at the local level a valuable source of information concerning relevant topics in the foreground and the results of research activities to address emerging issues, with a special focus on knowledge transfer. As their activities usually foresee the interaction of both scientists and practitioners, they contribute to effectively bridge the gap between research and practice. As an example, we mention here SoZooAlp (Società per lo Studio e la Valorizzazione dei Sistemi Zootecnici Alpini) in Northern Italy (https://www.sozooalp.it/), AGFF (Swiss Grassland Society) in Switzerland (https://www.agff.ch/), ÖAG (Österreichische Arbeitsgemeinschaft für Grünland und Viehwirtschaft) in Austria (https://gruenland-viehwirtschaft.at) and NLR (Norsk Landbruksraedgiving) in Norway (https://www.nlr.no/). The FAO-CIHEAM Mountain Pastures Sub-Network is an international platform for exchange between research institutions operating in mountain regions.

10 References

Ådnøy, T., et al. (2005). Grazing on mountain pastures – Does it affect meat quality in lambs? *Livestock and Production Science*, 94, 25–31.

Battaglini, L., et al. (2014). Environmental sustainability of Alpine livestock farms. *Italian Journal of Animal Science*, 13(2), 3155.

Bätzing, W., et al. (1996). Urbanization and depopulation in the Alps. *Mountain Research and Development*, 16(4), 335–350.

Baur, I. and Binder, C. R. (2013). Adapting to socioeconomic developments by changing rules in the governance of common property pastures in the Swiss Alps. *Ecology and Society*, 18(4), 60.

Baur, I., et al. (2014). Why do individuals behave differently in commons dilemmas? The case of alpine farmers using common property pastures in Grindelwald, Switzerland. *International Journal of the Commons*, 8(2), 657–685.

Bazzo, C. O. G., et al. (2023). A review of estimation methods for aboveground biomass in grasslands using UAV. *Remote Sensing*, 15(3), 639.

Bele, B., et al. (2018). Localized agri-food systems and biodiversity. *Agriculture*, 8, 22.

Bellini, E., et al. (2023). Impacts of climate change on european grassland phenology: A 20-year analysis of MODIS satellite data. *Remote Sensing*, 15(1), 218.

Boitani, L. and Linnell, J. D. (2015). Bringing large mammals back: large carnivores in Europe. In: Pereira, H. M. and Navarro, L. M. (eds.), *Rewilding European Landscapes*, Springer Open, New York, pp. 67–84.

Bunger, A., et al. (2018). *Småfenæringen - største sektoren i norsk jordbruk*. AgrAnalyse, Rapport 5 – 2018, Oslo.

Calabrese, C., et al. (2014). Alpine farming in Switzerland: discerning a lifestyle-driven labor supply. *Review of Social Economy*, 72(2), 137–156.

Campidonico, L., et al. (2016). Fatty acid composition of ruminal digesta and longissimus muscle from lambs fed silage mixtures including red clover, sainfoin, and timothy. *Journal of Animal Science*, 94(4), 1550–1560.

Castelli, M., et al. (2023). Insuring Alpine grasslands against drought-related yield losses using sentinel-2 satellite data. *Remote Sensing*, 15(14), 3542.

Chen, B., et al. (2023). The role of climatic factor timing on grassland net primary productivity in Altay, Xinjiang. *Ecological Indicators*, 157, 111243.

Conedera, M., et al. (2024). Linking the future likelihood of large fires to occur on mountain slopes with fuel connectivity and topography. *Natural Hazards*, 120(5), 4657–4673.

Coppa, M., et al. (2017). Grassland-based products: quality and authentication. *Grassland Science in Europe*, 22, 39–60.

Daccord, R., et al. (2007). *Bewertung von Wiesenfutter. Nährstoffgehalt für die Milch- und Fleischproduktion*. AGFF Merkblatt, 3. AGFF, Zürich.

Dalmasso, S., et al. (2011). An integrated program to prevent, mitigate and compensate Wolf (*Canis lupus*) damage in the Piedmont region (northern Italy). *Hystrix, the Italian Journal of Mammalogy*, 23, 54–61.

Deléglise, C., et al. (2015). Drought-induced shifts in plants traits, yields and nutritive value under realistic grazing and mowing managements in a mountain grassland. *Agriculture, Ecosystems & Environment*, 213, 94–104.

Dengler, J. and Tischew, S. (2018). Grasslands of Western and Northern Europe–between intensification and abandonment. In: Squires, V. R., et al. (eds.), *Grasslands of the world: diversity, management and conservation*, CRC Press, Boca Raton, pp. 27–63.

Dujakovic, A., et al. (2024). Growth unveiled: decoding the start of grassland seasons in Austria. *European Journal of Remote Sensing*, 57(1), 2323633.

Dujakovic et al., (2025). Enhancing grassland cut detection using Sentinel-2 time series through integration of Sentinel-1 SAR and weather data. *Remote Sensing Applications: Society and Environment*, 37, 101453.

Eklund, A., et al. (2017). Limited evidence on the effectiveness of interventions to reduce livestock predation by large carnivores. *Scientific Reports*, 7, 2097.

ESPON (2019). *Ministry of energy and spatial planning*. Department of Spatial Planning, Luxembourg.

Filippa, G., et al. (2019). Climatic drivers of greening trends in the Alps. *Remote Sensing*, 11(21), 2527.

Flury, C., et al. (2012). Future of mountain agriculture in the Alps. In: Mann, S. (ed.), *The future of mountain agriculture*. Springer Berlin Heidelberg, Berlin, Heidelberg, pp. 105–126.

Frei, E. R., et al. (2014). Plant population differentiation and climate change: responses of grassland species along an elevational gradient. *Global Change Biology*, 20(2), 441–455.

Fricke, T., et al. (2011). Assessment of forage mass from grassland swards by height measurement using an ultrasonic sensor. *Computers and Electronics in Agriculture*, 79(2), 142–152.

Galán, E., et al. (2022). The contribution of the commons to the persistence of mountain grazing systems under the Common Agricultural Policy. *Land Use Policy*, 117, 106089.

Girard, M., et al. (2016). Forage legumes rich in condensed tannins may increase n-3 fatty acid levels and sensory quality of lamb meat. *Journal of the Science of Food and Agriculture*, 96, 1923–1933.

Hafner, A. and Schwörer, C. (2018). Vertical mobility around the high-alpine Schnidejoch pass: indications of Neolithic and Bronze Age pastoralism in the Swiss Alps from paleoecological and archaeological sources. *Quaternary International*, 484, 3–18.

Hansen, I., et al. (2019). Impacts of Norwegian large carnivore management strategy on national grazing sector. *Journal of Mountain Science*, 16, 2470–2483.

Hersleth, M., et al. (2012). Lamb meat – Importance of origin and grazing system for Italian and Norwegian consumers. *Meat Science*, 90, 899–907.

Hofer, D., et al. (2016). Yield of temperate forage grassland species is either largely resistant or resilient to experimental summer drought. *Journal of Applied Ecology*, 53(4), 1023–1034.

Holtgrave, A.-K., et al. (2023). Grassland mowing event detection using combined optical, SAR, and weather time series. *Remote Sensing of Environment*, 295, 113680.

Hua, X., et al. (2022). Differential effects of topography on the timing of the growing season in mountainous grassland ecosystems. *Environmental Advances*, 8, 100234.

IPCC (2022). *Climate change 2022: impacts, adaptation and vulnerability. contribution of working group II to the sixth assessment report of the intergovernmental panel on climate change*. Press, Intergovernmental Panel on Climate Change, Cambridge and New York, 3068 pp.

Inauen, N., et al. (2013). Hydrological consequences of declining land use and elevated CO_2 in alpine grassland. *Journal of Ecology*, 101, 86–96.

Isselstein, J. (2018). Chapter 16. Protecting biodiversity in grasslands. In: Marshall, A., and Collins, R. (eds.), *Improving grassland and pasture management in temperate agriculture*, Burghley Dodds Science Publishing, Cambridge, pp. 381–396.

Jäger, H., et al. (2020). Grassland biomass balance in the European Alps: current and future ecosystem service perspectives. *Ecosystem Services*, 45, 101163.

Junquera, V., et al. (2022). Structural change in agriculture and farmers' social contacts: insights from a Swiss mountain region. *Agricultural Systems*, 200, 103435.

Kampmann, D., et al. (2008). Mountain grassland biodiversity: impact of site conditions versus management type. *Journal of Natural Conservation*, 16, 12–25.

Keller, E.R., et al. (1997). *Grundlagen der landwirtschaftlichen Pflanzenproduktion*. Handbuch des Pflanzenbaues, Verlag Eugen Ulmer, Stuttgart, 860 pp.

Kotlarski, S., et al. (2023). 21st Century alpine climate change. *Climate Dynamics*, 60(1), 65–86.

Landolt, G. and Haller, T. (2015). Alpine common property institutions under change: conditions for successful and unsuccessful collective action by alpine farmers in the canton of Grisons, Switzerland. *Human organization*, 74(1), 100–111.

Lauber, S., et al. (2013). *Zukunft der Schweizer Alpwirtschaft*. Eidgenössische Forschungsanstalt für Wald, Schnee und Landschaft WSL, Birmensdorf.

Lind, V., et al. (2009). Effects of concentrate or ryegrass-based diets (*Lolium multiflorum*) on the meat quality of lambs grazing on semi-natural pastures. *Acta Agriculturae Scandinavica, Section A – Animal Science*, 59(4), 230–238.

Mainieri, R., et al. (2020). Impacts of land-cover changes on snow avalanche activity in the French Alps. *Anthropocene*, 30, 100244.

Marini, L., et al. (2009). Impact of farm size and topography on plant and insect diversity of managed grasslands in the Alps. *Biological Conservation*, 142(2), 394–403.

Milazzo, F., et al. (2023a). An overview of permanent grassland grazing management practices and the impacts on principal soil quality indicators. *Agronomy*, 13, 1366.

Milazzo, F., et al. (2023b). The role of grassland for erosion and flood mitigation in Europe: a meta-analysis. *Agriculture, Ecosystems & Environment*, 348, 108443.

Möhl, P., et al. (2022). Growth of alpine grassland will start and stop earlier under climate warming. *Nature Communications*, 13(1), 7398.

Murphy, D.J., et al. (2021). Evaluation of the precision of the rising plate meter for measuring compressed sward height on heterogeneous grassland swards. *Precision Agriculture*, 22(3), 922–946.

Pannier, L., et al. (2014). Associations of sire estimated breeding values and objective meat quality measurements with sensory scores in Australian lamb. *Meat Science*, 96, 106–1087.

Parente, G. and Bovolenta, S. (2012). The role of grassland in rural tourism and recreation in Europe. *Grassland Science in Europe*, 17, 733–743.

Pauler, C. M., et al. (2019). Influence of Highland and production-oriented cattle breeds on pasture vegetation: a pairwise assessment across broad environmental gradients. *Agriculture, Ecosystems & Environment*, 284, 106585.

Pauler, C. M., et al. (2020a). Grazing allometry: anatomy, movement, and foraging behavior of three cattle breeds of different productivity. *Frontiers in Veterinary Science*, 7, 494.

Pauler, C. M., et al. (2020b). Choosy grazers: Influence of plant traits on forage selection by three cattle breeds. *Functional Ecology*, 34(5), 980–992.

Pauler, C. M., et al. (2022). Thinning the thickets: foraging of hardy cattle, sheep and goats in green alder shrubs. *Journal of Applied Ecology*, 59(5), 1394–1405.

Pepin, N. C., et al. (2022). Climate changes and their elevational patterns in the mountains of the world. *Reviews of Geophysics*, 60(1), e2020RG000730.

Peratoner, G., et al. (2016). webGRAS: a web application to estimate the potential forage quality of mountain permanent meadows. *Grassland Science in Europe*, 21, 203–205.

Peratoner, G. and Pötsch, E. M. (2019). Methods to describe the botanical composition of vegetation in grassland research. *Die Bodenkultur: Journal of Land Management, Food and Environment*, 70(1), 1–18.

Peratoner, G., et al. (2017). Effect of slope and altitude on the costs of forage production in mountain areas. *Grassland Science in Europe*, 22, 215–217.

Peratoner, G., et al. (2021). Arbeitszeitbedarf bei der Futterproduktion in Südtirol. *Laimburg Journal*, 3, 2021.008.

Pötsch, E. M. (2012). Optimale Grünlandbewirtschaftung in Bergregionen. In: Lehr- und Forschungszentrum für Landwirtschaft Raumberg-Gumpenstein (ed.), *39. Viehwirtschaftliche Fachtagung zum Thema "Milchproduktion - Status quo und Anpassung an zukünftige Herausforderungen"*, LFZ Raumberg-Gumpenstein, pp. 9–18.

Pötsch, E. M., et al. (2014). Impact of climate change on grassland productivity and forage quality in Austria. *Grassland Science in Europe*, 19, 139–141.

Pötsch, E. M., et al. (2006). Effect of different management systems on quality parameters of forage from mountainous grassland. *Grassland Science in Europe*, 11, 484–486.
Prache, S., et al. (2022). Review: Factors affecting sheep carcass and meat quality attributes. *Animal*, 16, 100330.
Prangel, E., et al. (2023). Afforestation and abandonment of semi-natural grasslands lead to biodiversity loss and a decline in ecosystem services and functions. *Journal of Applied Ecology*, 60, 825–836.
Reinhardt, I., et al. (2012). Livestock protection methods applicable for Germany – a country newly recolonized by wolves. *Hystrix, the Italian Journal of Mammalogy*, 23, 62–72.
Resch, R., et al. (2006). Futterwerttabellen für das Grundfutter im Alpenraum. *Der fortschrittliche Landwirt*, 84(24), 1–20.
Rose, L., et al. (2012). Effects of fertilization and cutting frequency on the water balance of a temperate grassland. *Ecohydrology*, 5(1), 64–72.
Sanaullah, M., et al. (2014). Effects of drought and elevated temperature on biochemical composition of forage plants and their impact on carbon storage in grassland soil. *Plant and Soil*, 374(1–2), 767–778.
Schaumberger, A., et al. (2022). Impact of drought stress and climate change on yield and forage quality of grassland. *Grassland Science in Europe*, 27, 472–474.
Schils, R. L. M., et al. (2022). Permanent grasslands in Europe: land use change and intensification decrease their multifunctionality. *Agriculture, Ecosystems & Environment*, 330, 107891.
Saha, D. and Kukal, S. S. (2015). Soil structural stability and water retention characteristics under different land uses of degraded lower Himalayas of North-West India. *Land Degradation & Development*, 26, 263–271.
Schirpke, U., et al. (2020). Spatio-temporal changes in ecosystem service values: effects of land-use changes from past to future (1860–2100). *Journal of Environmental Management*, 272, 111068.
Schirpke, U., et al. (2017). Future impacts of changing land-use and climate on ecosystem services of mountain grassland and their resilience. *Ecosystem Services*, 26, 79–94.
Schwieder, M., et al. (2022). Mapping grassland mowing events across Germany based on combined Sentinel-2 and Landsat 8 time series. *Remote Sensing of Environment*, 269, 112795.
Sebastiá, M.-T. (2004). Role of topography and soils in grassland structuring at the landscape and community scales. *Basic and Applied Ecology*, 5(4), 331–346.
Siegmund, R., et al. (2019). *Grassland monitoring based on Sentinel-1*. Proc. SPIE 11149, Remote Sensing for Agriculture, Ecosystems, and Hydrology XXI, 1114902.
Smith, C., et al. (2021). The Earth's energy budget, climate feedbacks, and climate sensitivity: supplementary material. In: Masson-Delmotte, V., et al. (Eds.), *Climate change 2021: The physical science basis. Contribution of working group I to the sixth assessment report of the intergovernmental panel on climate change*. Cambridge University Press, Cambridge.
Soliva, R., et al. (2010). Differences in preferences towards potential future landscapes in the Swiss Alps. *Landscape Research*, 35, 671–696.
Stettler, A. and Probst, S. (2023). How much livestock does Switzerland need for optimal land use? *Swiss Agricultural Research*, 14, 236–242.
Stewart, K. E. J., et al. (2001). An evaluation of three quick methods commonly used to assess sward height in ecology. *Journal of Applied Ecology*, 38(5), 1148–1154.

Streifeneder, T., et al. (2007). Selected aspects of agro-structural change within the Alps. A comparison of harmonised agro-structural indicators on a municipal level in the Alpine Convention area. *Journal of Alpine Research|Revue de géographie alpine*, 95(3), 41–52.
Sturaro, E., et al. (2013). Dairy systems in mountainous areas: farm animal biodiversity, milk production and destination, and land use. *Livestock Science*, 158, 157–168.
Taravat, A., et al. (2019). Automatic grassland cutting status detection in the context of Spatiotemporal Sentinel-1 imagery analysis and artificial neural networks. *Remote Sensing*, 11(6), 711.
Tasser, E. and Tappeiner, U. (2002). Impact of land use changes on mountain vegetation. *Applied Vegetation Science*, 5(2), 173–184.
Tello-García, E., et al. (2020). Drought- and heat-induced shifts in vegetation composition impact biomass production and water use of alpine grasslands. *Environmental and Experimental Botany*, 169, 103921.
Trnka, M., et al. (2011). Evaluating drought risk for permanent grasslands under present and future climate conditions. *Procedia Environmental Sciences*, 3, 50–57.
van den Bergh, T., et al. (2018). *Alnus* shrub expansion increases evapotranspiration in the Swiss Alps. *Regional Environmental Change*, 18(5), 1375–1385.
Van den Pol-Van Dasselaar, A., et al. (2021). Societal and economic options to support grassland-based dairy production in Europe. *Irish Journal of Agricultural and Food Research*, 59, 258–269.
Van Dorland, H. A., et al. (2007). Species-rich swards of the Alps: constraints and opportunities for dairy production. In: Elgersma, A., et al. (eds.), *Fresh herbage for dairy cattle*, Wageningen UR Frontis Series, 18, Springer, Dordrecht, pp. 27–43.
Vogel, A., et al. (2012). Grassland resistance and resilience after drought depends on management intensity and species richness. *PLoS One*, 7(5), e36992.
Wang, H., et al. (2020). Alpine grassland plants grow earlier and faster but biomass remains unchanged over 35 years of climate change. *Ecology Letters*, 23(4), 701–710.
Ward, A., et al. (2014). A global estimate of carbon stored in the world's mountain grasslands and shrublands, and the implications for climate policy. *Global Environmental Change*, 28, 14–24.
Watzig, C., et al. (2023). Grassland cut detection based on Sentinel-2 time series to respond to the environmental and technical challenges of the Austrian fodder production for livestock feeding. *Remote Sensing of Environment*, 292, 113577.
Willems, H., et al. (2013). Vegetation-type effects on performance and meat quality of growing Engadine and Valaisian Black Nose sheep grazing alpine pastures. *Livestock Science*, 151, 80–91.
Zanolla, G., et al. (2007). *Demographic dynamics in the Alpine arch: trends and future developments with special focus on Italy*. International Conference on Regional and Urban Modeling, Brussels.
Zehnder, T., et al. (2020). Dominant shrub species are a strong predictor of plant species diversity along subalpine pasture-shrub transects. *Alpine Botany*, 130, 141–156.
Zhao, W., et al. (2020a). Contributions of climatic factors to interannual variability of the vegetation index in Northern China Grasslands. *Journal of Climate*, 33, 175–183.
Zhao, Y., et al. (2020b). Grassland ecosystem services: a systematic review of research advances and future directions. *Landscape Ecology*, 35, 793–814.

Chapter 19

Grasslands in New Zealand: challenges and solutions

David R. Stevens[1], AgResearch Ltd, New Zealand

1 Introduction

Grassland farming has continued to support the lifestyle and growth of New Zealand since the first shipments of frozen meat and chilled cheese and butter to Britain in 1882. Its importance remains as relevant to the wealth of New Zealand now as it was then. This makes New Zealand unique in developed nations across the world, as it is a major trading nation in dairy, sheep meat and deer products, all from grassland systems, with little government support. Meat continues to be exported frozen or chilled, while the majority of dairy products are dried and exported as powders, with some cheese and butter exported.

1 E-mail: David.stevens@agresearch.co.nz

http://dx.doi.org/10.19103/AS.2025.0154.31

Grasslands in New Zealand comprise nearly 40% of the land area, or approximately 9.5 million hectares. These grasslands are predominantly introduced European grasslands with native tussock lands comprising approximately 2.1 million hectares (Statistics NZ 2023). Geo-climatic zones range from sub-tropical to cool temperate (Mackintosh 2001), and low (350 mm/annum) to high (>2500 mm/annum) rainfall as well as low to high fertility soils (Hewitt 1998). These resources produce a wide array of both seasonal and total pasture production profiles around New Zealand (Radcliffe 1974). Farm systems have been developed to fit the winter trough and late spring peak in pasture production (McCall & Sheath 1993), utilising the natural seasonal breeding and lactation cycles to match pasture supply with animal demand.

The pastures in New Zealand have a legacy of being a mixture of grasses and legumes (Sears 1951). This has often been referred to as a ryegrass (*Lolium perenne*)/white clover (*Trifolium repens*) mixture, but our hill country pastures have long been recognised as a complex mixture of temperate (C3) grasses and legumes (Cosgrove & Field 2016) that varies at a microsite level within paddocks (Gillingham & During 1973), seasonally through species differences in temperature optima (Mitchell 1956) and more generally between regions depending on the combination of soil fertility, soil moisture and temperature (Scott et al. 1985). Some tropical (C4) grasses such as kikuyu (*Cenchrus clandestinus*) and paspalum (*Paspalum dilatatum*) are present especially in the far north of the country.

A unique feature of those grassland systems is that they are a transplant from Europe (often established in the late 19th and early 20th centuries) and as such are an incomplete ecosystem, lacking many of the balances of predator and prey alongside the pasture resource (Goldson 2018). Indeed, even simple bacteria required for the fixation of nitrogen have had to be imported to inoculate legumes (Lowther and McDonald 1972). Fertiliser inputs are uniquely designed to match and enhance the current combination of imported plants and local soils. Grazing management systems have then been developed to optimise the use of the resultant seasonal nature of the pasture growth with animals that were also imported. This has seen the evolution of seasonally driven outdoor grazing systems. Animal production systems match pasture growth, with calving and lambing occurring in spring, while lactation ceases in summer and autumn.

Dairy farming, predominantly on flat land (Table 1), utilises a rotational grazing system with pastures being grazed for 12 to 24 hours before being rested for 18–25 days in spring, 25–45 days in summer and autumn (fall) and 60–90 days in winter. Sheep and beef breeding farms tend to use a continuous grazing system during spring (late pregnancy and early lactation) and revert to grazing periods of 4 to 10 days during summer and autumn, with resting periods of 30 to 60 days, and to a rotational grazing system in winter using

Table 1 Area for pastoral farming enterprises and number of animals in each livestock type in New Zealand, categorized by mean slope (Stevens et al. 2019).

		Area each Year (000 ha)			Numbers each year (000)[1]		
Farm enterprise	Slope Class	2008	2013	2018	2008	2013	2018
Beef	<7°	401	405	384	1195	1070	1069
Beef	7–15°	510	528	501	1891	1592	1563
Beef	15–25°	289	292	284	964	958	916
Beef	>25°	410	320	311	75	69	54
Dairy	<7°	1424	1558	1846	4746	5555	5471
Dairy	7–15°	377	393	499	770	851	972
Dairy	15–25°	47	46	49	59	76	86
Dairy	>25°	1	1	1	0	1	0
Deer	<7°	140	120	92	612	494	382
Deer	7–15°	84	77	92	417	364	303
Deer	15–25°	76	67	51	160	149	139
Deer	>25°	31	60	26	19	21	13
Sheep	<7°	697	619	519	9617	8205	6900
Sheep	7–15°	606	564	521	13814	12775	11575
Sheep	15–25°	591	512	499	9511	9003	8384
Sheep	>25°	319	241	176	762	737	578
Sheep and Beef	<7°	793	764	718	na[1]	na	Na
Sheep and Beef	7–15°	2491	2530	2490	Na	na	na
Sheep and Beef	15–25°	2848	2971	2953	Na	na	na
Sheep and Beef	>25°	847	871	728	Na	na	na

[1] Livestock numbers are reported under respective stock types and therefore no data is presented for livestock numbers for combined sheep and beef enterprises.

grazing periods of one to four days with rest periods of 90 to 120 days. Beef finishing farms often use an intensive rotational grazing system similar to dairy farming.

2 Livestock farming in New Zealand: an overview

Grazing enterprises contribute to red meat production (beef, sheep meat and venison) and dairy (cow, sheep and goat) as well as specialist products such as velvet antler (Beef + Lamb New Zealand 2022). These industries contribute approximately 43% to national export earnings and employ approximately 1 in 10 workers (Dalziel et al. 2018). The grassland area is managed by approximately 23 400 sheep and beef farmers, 11 100 dairy farmers and 780 red deer farmers (Table 2, Beef + Lamb New Zealand 2022), whose primary income is

Table 2 Industry demographics for ruminant farming on grassland in New Zealand

Enterprise	Stock numbers (000)	Farm numbers	Average farm size (ha)	Agricultural area (000 ha)
Sheep[1]	25 700	23 400[2]	375	8765
Beef[1]	4000			
Dairy cattle[1]	6200	11 100	220	2442
Red deer[1]	800	780	nd[5]	261
Dairy goats[3]	66	92	55	na[6]
Dairy sheep[4]	12	18	na[6]	na[6]

[1] Beef + Lamb New Zealand (2022).
[2] Sheep and beef are commonly farmed together, and so numbers are combined for reporting purposes.
[3] Scholtens et al. (2017).[4] Ministry of Primary Industries (2020).
[5] No data as deer farming often occurs as part of a sheep and beef farming operation.
[6] Not available.

farming. The number of ruminants farmed is approximately 25.7 million sheep, 6.2 million dairy cattle, 4.0 million beef cattle and 0.8 million red deer (Beef + Lamb New Zealand 2022). These populations produce enough high-quality protein-rich food for approximately 40 million people, while the population of New Zealand is approximately 5 million people. Thus between 80% and 99% of animal production is exported (Statistics NZ 2023), with New Zealand being the largest exporter of cow dairy (Whitfield 2023), venison and velvet antler (USDA 2023a), and the second largest exporter of sheep meat (USDA 2023b). Small populations of dairy goats (Scholtens et al. 2017) and dairy sheep (MPI 2020) are also farmed with the major emphasis on exporting high-quality milk powder for direct human consumption.

The distribution of pastoral farms in New Zealand by farm type (Figure 1), aligned with the distribution based on slope (Figure 2) demonstrates the significant clusters of dairy farming on land below 7° slopes, while sheep and beef farms tend to be on slopes greater than 7°. Sheep farming is more predominant in the South Island, while mixed sheep and beef farms occur throughout the country.

Grassland farming in many red meat enterprises in New Zealand comprises high stock numbers (2000 to 5000 sheep and 100 to 500 cattle) with low labour inputs (1 labour unit per 3000 sheep), with stocking rates of 7 to 10 sheep, or 2 cattle per hectare. Dairy farms manage approximately 440 cows, with stock rates of 2.5–3.5 cows/ha, with low labour inputs of 1 labour unit to 120 cows. Farms are predominantly family-owned and operated (92%) while providing the family income.

Internationally, production of dairy milk was approximately 810 million t/annum (2016) and sheep meat 69 million t/annum (2021). In contrast the production in New Zealand was approximately 21 million t/annum (Shadbolt and Apporao 2016) and 0.38 million t/annum (Beef + Lamb NZ 2021) or

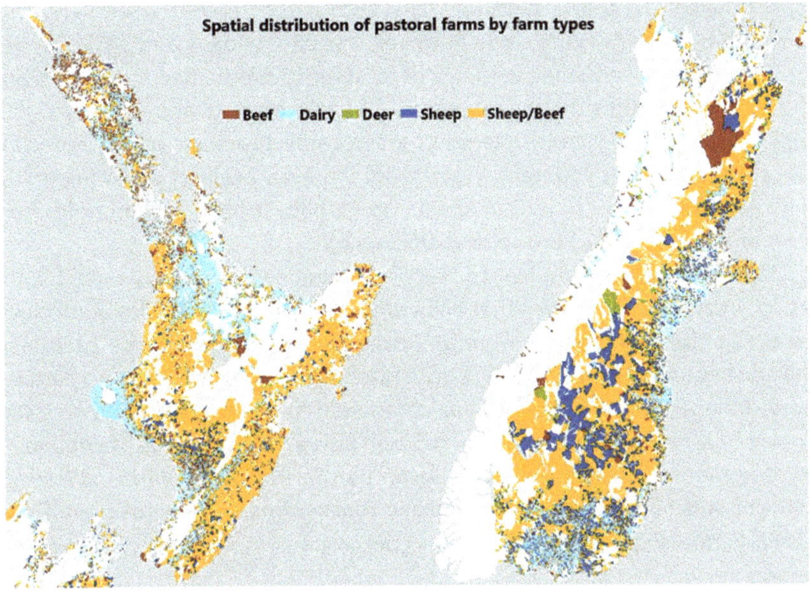

Figure 1 Spatial distribution of pastoral farms by farm types in New Zealand (Stevens et al. 2019). The North Island is on the left (latitude -34.4 to -41.5) and the South Island on the right (latitude -40.5 to -46.7).

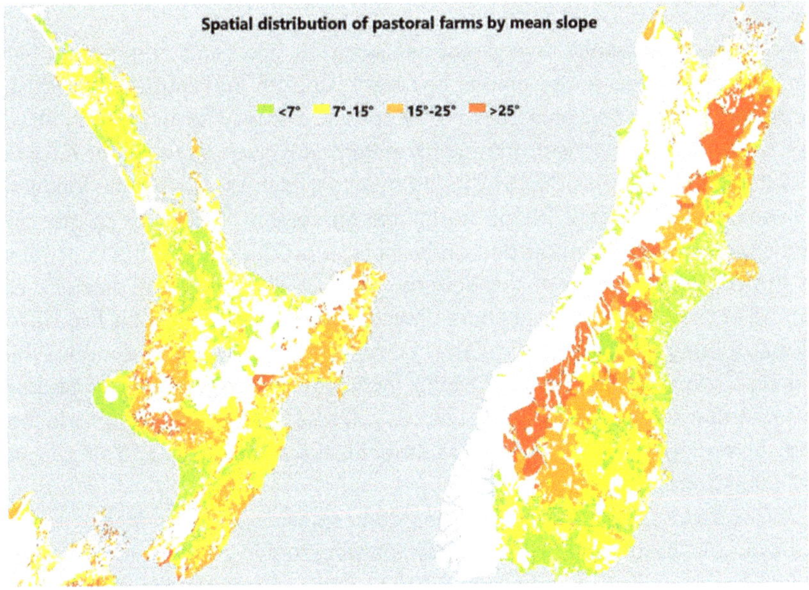

Figure 2 The spatial distribution of pastoral farms in New Zealand by mean slope (Stevens et al. 2019). The North Island is on the left (latitude -34.4 to -41.5) and the South Island on the right (latitude −40.5 to 46.7).

approximately 2.5% and 0.6% of dairy and sheep meat, respectively. New Zealand's position in world trade however, was 28% of dairy and 38% of sheep meat. These relatively small amounts of total world production are often being traded into countries that have a range of trade barriers and subsidies, thus prices received are directly related to the subsidy structures of the importing nations. Without any counter-support from the New Zealand government, the only recourse farmers have to maintain profitability is to capture benefits from economies of scale and production efficiency.

With exporting of animal protein as a major income generator for the nation, international treaties that allow the free trade of agricultural products are a major focus (Ministry of Foreign Affairs and Trade 2023). Since 2000, New Zealand has entered into free trade agreements with many of its trading partners around the world. New Zealand has committed to free trade agreements with China (2008), Hong Kong (2011) and South Korea (2015), as well as continuing trade agreements with Pacific and Asian partners through initiatives such as CER with Australia, ASEAN-Australia-New Zealand free trade agreement, Trans-Pacific Partnership and the Regional comprehensive economic partnership, among many.

Traditional trade agreements with the United Kingdom and the European Union have been recently updated to account for the results of the United Kingdom leaving the European Union. These trade agreements were often in the form of quotas and tariffs to control the entry of products and the price gained, to protect local producers.

These agreements have direct influence on grassland enterprises. New Zealand developed a free market approach to both agriculture and industry in the early 1980s (Wallace 2014). This opened the way for innovation across the agricultural sector with the rise of enterprises such as red deer farming, and the doubling of productivity in red meat and dairy sectors in the following 20 years (Fennessy et al. 2016). Traditional agreements which use quotas and tariffs place a significant burden on farmers to produce within the constraints of the pricing structure of the country which is importing the product. For example, the Common Agricultural Policy of the European Union has driven global trading prices of agricultural commodities since its inception. While this drives innovation and productivity increases in those farmers which have to meet that price barrier, it also has the adverse effect of driving production with increasing externalities such as nutrient leakage into water and air, causing pollution (Campbell et al. 2017).

The development of free trade has led to significant shifts in market supply. For example, before the China free trade agreement in 2008, exports to China were 5% of New Zealand's exports. By 2022, this had increased to 32% (Kiernan 2022), comprising dairy products and red meat. Similar developments in trade have been facilitated by the free trade agreement with South Korea, as both velvet antler production and export have increased by over 100% over the last decade (Deer Industry NZ 2020).

This shift from traditional markets has been perceived as a risk due to exposure to a dominant market (USDA 2020), yet product value also increases due to the lack of a trade price cap imposed by traditional markets. The opportunity in these changes is that product price can truly reflect production price, rather than the artificial pricing structures that may occur in markets with subsidised farming. For example, the European Union Common Agricultural Policy accounts for 33.1% of the 2021 EU budget or €55.71 billion (Milicevic and DuPont 2023). This supports agriculture on approximately 157 million hectares, or a support of €355/ha. To place this in context, the average net income for a red meat farmer in New Zealand is approximately €260/ha before wages of management, taxes, debt servicing and capital expenditure (Beef + Lamb NZ 2023). In New Zealand, government support to all agriculture, biosecurity, and food safety amounts to 0.6% of the total budget (NZ Treasury 2023) and equates to €42/ha. This inequity, due to agricultural protections, drives changes in efficiency, intensification and the mining of environmental capital, increasing emissions to air and water.

External forces will continue to shape the use of grasslands in New Zealand. The threats and opportunities of free trade have moulded the way we farm. Significant reductions in the sheep flock have been seen since the height of subsidised farming in New Zealand in the early 1980s (Wallace 2014) from nearly 70 million sheep down to approximately 25.7 million in 2021. This reduction has been accompanied by an increase in red meat productivity per ewe of 118% since 1990 (Beef + Lamb New Zealand 2022), resulting in a constant export of approximately 360 - 400 000 t sheep meat/annum (FigureNZ 2023). Alongside this, the dairy industry has expanded significantly, increasing total numbers from approximately 3.4 million head in 1990 to 6.5 million in 2015, settling at approximately 6.3 million in 2019 (Statistics NZ 2023). A concurrent outcome from the expansion of dairy cattle has been a gradual increase in beef exports as surplus calves are reared and slaughtered. Exports of beef and veal have increased steadily since 1990, increasing from under 300 000 t/annum to approximately 500 000 t/annum in 2023 (FigureNZ 2023). This increase has occurred while the dedicated beef herd has declined from approximately 5.0 million head to 3.8 million head (Beef + Lamb New Zealand 2023). These shifts in enterprise have been driven by the relative profitability and potential capital investment opportunities of each enterprise. Challenges to profitability will continue to face the grassland industries of New Zealand. Meeting consumer and societal expectations, adapting to climate change and reducing the externalities from ruminant farming are becoming central to maintaining the competitive advantage of our grassland farming systems.

3 Challenges facing pastoral livestock production in New Zealand

Challenges to grassland production come from many quarters. The first challenges are generated by the need to meet consumer and society expectations for ethical and environmentally sustainable food production. Then a range of challenges sit below this as farmers must also meet profitability metrics within changing climatic variability and the need to ensure production efficiencies.

3.1 Consumer requirements and impacts on livestock production

Consumers around the world hold standards of food safety, animal welfare and environmental stewardship in high regard, regardless of their country of origin (Saunders et al. 2015). While there are some differences in ranking, all agree that food production needs to meet these underlying standards. Consumer concerns will drive enterprise choice in grassland farming. The rising concern of global warming and climate change adds an extra challenge as consumers become aware of the impacts of methane emissions from ruminant livestock on global warming. Consumers have been exposed to a narrative that animals are bad for the environment, and that they are to blame for global warming (Hedberg 2023), deflecting attention away from the role of burning fossil fuels.

Translating consumer requirements into actionable processes that the farmer can manage produces the following themes, which are illustrated in Figure 3. These include the delivery of a high-quality product for consumers, reducing the impacts of grassland farming on losses to air and water, and maintaining and improving animal health and well-being.

3.2 Reduction of, or elimination of, the externalities of farming

Externalities are currently recognised as emissions to water and emissions to air (Figure 3). Emissions to water include nutrients, especially nitrogen and phosphorus, and contaminants, especially sediment and bacteria. Within New Zealand grazing systems, there is very little requirement for use of products such as antibiotics and other medicines, with New Zealand agriculture ranking as the third lowest user of antibiotics in the world (O'Neill 2015). Therefore, these products are not present in measurable quantities in water from grazing systems. Emissions to air from grassland systems are primarily the greenhouse gases, methane and nitrous oxide.

The nutrient requirements of an expanding population of the world have been met with remarkable increases in productivity, both from expanding land use, with conversion to agricultural use, and from increases in productivity, over

Figure 3 A depiction of the interconnectedness of New Zealand grassland farming to the environment.

the past 50 years. However, that golden era of expansion may be waning as upper boundaries of planetary stress are being reached (Campbell et al. 2017).

The need to constrain the impact of agricultural production exists across all enterprises. While grassland agriculture often uses resources that cannot be utilised for other forms of agriculture, such as grain or fruit production, it still must ensure that externalities such as emissions to water and air are brought within acceptable limits set through government policy. In New Zealand, for example, freshwater quality is under pressure with an estimated national nitrogen surplus in our pastoral industries of approximately 170 000 tonnes (Monaghan et al. 2021a). While this number appears large, it translates into an estimated surplus of approximately 12 kg N/ha on sheep and beef enterprises, and 49 kg N/ha on dairy enterprises (Monaghan et al. 2021b). National emissions to air from ruminants account for 38.8 million tonnes of CO_2 per annum (Ministry for the Environment 2022), though pastoral landscapes also store an estimated 1.48 billion tonnes of carbon in their soils nationally (Wall et al. 2021). These emissions are approximately 50% of the annual total greenhouse gas emissions of New Zealand (Ministry for the Environment 2022), and thus place a large responsibility on the farmers of New Zealand grasslands when meeting New Zealand's obligations agreed to in the Paris Agreement on climate change to reduce GHG emissions by 50% below 2005 levels by 2030.

3.3 Ensuring the highest ethical standards when raising and harvesting products from animals

New Zealand uses an animal ethics system that enshrines ethical standards in legislation using evidence-based methodologies (Ministry for Primary Industries 2022a), with animal welfare codes being constantly reviewed as new evidence emerges. This care extends to animal medicines through the agricultural medicines and veterinary medicines legislation (Ministry for Primary Industries 2022b). All exporters of animal products from grassland have quality assurance programmes which ensure that consumer requirements are met (e.g. NZFAP 2024). These programmes include origin and traceability, food safety and animal welfare standards, and each farm is audited every three years to ensure compliance.

It is recognised that ruminants grazing pasture generally have fewer health and welfare issues than housed or feedlot animals (Webster 2016). The complex environment of grazed pasture provides space, opportunities to graze and explore, as well as engagement in social activities that are associated with natural behaviours and provide a degree of control and choice (Webster 2016). Longevity of dairy cows, for example, is greater on NZ pasture-based systems at 6.8 years (Dairy NZ 2023) than in US housed systems, at 4.8 years (De Vries 2013).

Challenges still exist, however. The survival of young animals may be compromised, especially during the peri-natal period (McCoard 2017), when grazing animals have the ability to express natural birthing behaviours in open grassland. Nutritional challenges also still occur, depending on the ability of the soil and pasture to provide appropriate nutrients to the animal. For example, magnesium deficiency is common in dairy cattle during early spring (Roche et al. 2017) when low soil temperatures reduce the availability of magnesium to plants (Winter 1989). Exposure to disease challenge such as facial eczema, ryegrass staggers and internal parasites remains a threat (Smith and Towers 2002).

3.4 Safety of food harvest and processing

Consumers around the world rate food safety as a top priority when making buying decisions (Saunders et al. 2015).

New Zealand has a long track record of food safety excellence, based on the need to meet trading requirements with other nations. Food safety begins with the harvesting of products, with regulations in place from dairy parlour hygiene and velvet antler removal to slaughter processes and subsequent food storage and handling. Standards for food safety are enshrined in legislation (Ministry for Primary Industries 2022c). While food safety is an on-going potential threat, the very low inputs of chemicals and excellent food processing procedures ensure that this threat is constantly minimised.

3.5 Impacts of climate change and global warming

Severe weather events affect grassland agriculture by increasing the variation in feed production (mainly drought-related) and causing loss of soil and pastures (due to flooding, waterlogging and soil erosion). Increases in the base temperature alter potential pasture production and pasture species balance, in the most extreme, a shift from C3 to C4 pasture species, and promote disease and pest incursions.

Climatic variability is always a part of farming with farmers developing strategies to cope (Gray et al. 2008). However, climate variability is continuing to evolve with an increase in the frequency of drought at 13 of 30 regional sites, and drought intensity at 14 regional sites around New Zealand, since 1972 (Stats NZ 2020). Climate change has been having a direct effect on the productivity of perennial plant systems in New Zealand (Mackay et al. 2023). Downward trends in pasture production under a range of topographic conditions, and across several regions were reported, along with downward trends in forestry productivity (*Pinus radiata*). Mackay et al. (2023) hypothesised that the predicted benefits of enriched CO_2 were being negated by changes in rainfall amount and frequency, as well as potential changes in photosynthetic efficiency (Mackay et al. 2023). This observation means that future productivity gains that may be available through animal genetics and animal management systems may not be matched by increases in pasture and forage production. This then leaves the pastoral industry in the position of need to import supplements to match those animal productivity gains or to decrease livestock numbers.

As well as the impact of climate change on pasture productivity, further changes are being recorded in both disease threat (Philips et al. 2023) and potential insect pest spread (Hewitt et al. 2023). For example, New Zealand is unique in having a strain of the *Pseudopithomyces chartarum* fungus, which produces a mycotoxin, sporidesmin, in the spores. This mycotoxin forms in response to weather conditions and is most prevalent in late summer and autumn in warm environments (di Menna et al. 2009). Records in the recent past have noted an increase in the range of the disease, extending further south into traditionally lower temperatures and summer safe areas, with further spread predicted (Philips et al. 2023).

The impacts and spread of some internal parasites, such as *Haemonchus contortus* (barbers pole worm) are increasing. Barbers pole worm is highly pathogenic and can cause rapid death if undetected. Currently, they occur in the warmer, northern parts of the country but are predicted to have a greater range as temperatures increase (Johnson et al. 2021).

Pest outbreaks are also increasing. Outbreaks of the African black beetle (*Heteronchus arator*) are increasing (Ferguson et al. 2019). Spread of African black beetle beyond its range in the upper North Island is predicted (Hewitt et al. 2023), extending into the lower North Island and upper South Island over the next 75 years. Impacts of other pests, and the potential for new pests to emerge, have also been documented, with all scenarios having great impact through reduced pasture productivity and associated greater control costs being predicted (Mansfield et al. 2021).

3.6 Competition from alternate foods mimicking animal products

Biotechnology-based food developments, such as the impossible burger (impossiblefoods.com) are emerging. While this type of product has limited uptake at present, it demonstrates the potential to disrupt livestock enterprises. Many grassland red meat production systems in New Zealand are based in landscapes and slope classes that are unsuitable for other agricultural enterprises such as cropping (Figure 2, Moot and Davison 2021). Thus, their position in utilising grassland to generate high-quality protein and fats for human consumption will remain less threatened by these consumer trends.

The development of new ways of producing food via conversion of plant-sourced nutrients into pseudo-animal products will be one avenue to meet the future nutritional needs of further population expansions, and therefore may not be as great a threat as has been perceived (Manning et al. 2023).

A more immediate threat is the potential replacers for dairy products. The New Zealand dairy cattle industry relies predominantly on producing milk powder and associated products, which are used in the production of a range of manufactured food products. The potential for yeast-based systems to incorporate appropriate dairy cattle DNA to replicate these products has been proven, and the product is available for purchase (Powell et al. 2023). Mass production by upscaling is required to become a real threat to current animal-based milk production practices. International pressure for businesses such as

Nestle and Danone to reduce their GHG footprint will be significant in product choice. Therefore, the opportunity for large manufacturers to reduce the GHG footprint of their products by shifting to yeast-based or plant-based ingredients has the potential to be a much greater threat than consumer preference alone.

Transferring from grass-based systems to arable-based systems, needed to provide a plant-based source stock, will transfer GHG emissions from methane to carbon dioxide. Initial cultivation will create a potentially large loss in CO_2, though it will eventually reduce the total footprint, shifting the emissions from a recyclable product, methane, to a permanent product, CO_2.

This shift will also require farmland which is suitable for arable crops, and thus is a more direct competitor with dairy cattle than other grassland enterprises in New Zealand. Interestingly, the expansion of dairy farming in New Zealand has often been in regions where arable farming was predominant, indicating the greater profitability and return on capital that dairy farming currently has compared to arable farming. This suggests that to displace dairy farming, the profitability of arable farming would need to increase significantly before enterprise change would occur.

4 Addressing challenges: an overview

As trade delegations create space to ensure the access of products to markets who need it, we turn our attention back to the biophysical limitations of our livestock enterprises. Overall, the challenges that farmers face are common among many grazing nations. To remain as a farmer of livestock on grassland, farmers need to develop and implement solutions which reduce the externalities of their actions, while ensuring that the livestock under their management are well cared for. This needs to be framed within profitable farming enterprises. The greatest challenges that grassland farmers in New Zealand face are the losses to air and water produced by the farming of grassland, and the changes being seen from climate change.

Of the challenges presented, ethical standards and food safety are dealt with via government support and industry-led initiatives. These challenges are well recognised, and continual improvement is always sought. These initiatives ensure that market access is maintained and enhanced (via regulation, quality assurance programmes and trade agreements). Alternate foods remain a potential challenge, but issues of scaling, resource availability, processing and limited consumer acceptance are slowing progress. The need to increase global food production and adequate nutrition means that this challenge may never be realised.

The challenge of reducing externalities is about the balance of resource use and product output. Some of this balance depends on product price and is influenced by international trading conditions. However, other influences

Table 3 Animal management opportunities to mitigate or adapt to the externalities that impact on grassland farming in New Zealand both now and in the future.

	Current		Future	
Parameter	Mitigation	Adaptation	Mitigation	Adaptation
Species	Shift from cattle to sheep to lower N loss. Shift from sheep to cattle to reduce GHG outputs per unit product	Shift to more beef, less sheep as climate warming alters disease prevalence and feed quality	Shift from cattle to sheep to lower N loss. Shift from sheep to cattle to reduce GHG outputs per unit product	Replace Bos taurus with Bos indicus
Genetics	Choose low GHG emissions sires (sheep). Choose low N output sires (dairy cattle only)	Choose hair sheep varieties to reduce animal health issues and reduce costs. Choose lower micron wool sheep to increase product value. Choose sires bred for disease tolerance	Choose low GHG emissions sires (cattle and deer).	Choose deer sires bred for seasonal growth variation to match pasture production. Choose sires bred for disease tolerance
Stocking rate	Make culling decisions early. Lower replacement rate, increase per head performance to reduce emissions intensity	Use variable sales policies to adapt to climatic conditions. Increase per-head performance to reduce maintenance costs as a proportion of total energy requirements		Increase progeny survival
Nutrition	Reduce N intake through pasture fertiliser inputs and supplementary feed choice. Add methane inhibitors to the diet		Use plant varieties that reduce methane emissions	
Product output	Optimise lactation length. Reduce supplementary feeding	Vary milking intervals to match pasture supply. Vary animal growth rates to utilise feed as it grows	Increase product functionality to increase value and so reduce the need to produce as much	Integrate various perennial cropping systems with livestock systems
Product quality	Reduce supplementary feeding. Use alternative pasture species	Use alternative pasture species	Use alternative pasture species	Use alternative pasture species
Herd/flock management	Optimise disease control. Reduce culling. Use virtual herding technologies to exclude cattle from sensitive areas	Use virtual herding technologies (cattle) to improve feed utilisation	Use virtual herding technologies to exclude sheep and deer from sensitive areas	Use virtual herding technologies (sheep and deer) to improve feed utilisation. Change lambing and calving dates to match changes in pasture production

Table 4 Pasture/forage opportunities to mitigate or adapt to the externalities that impact on grassland farming in New Zealand both now and in the future .

	Current		Future	
Parameter	Mitigation	Adaptation	Mitigation	Adaptation
Species	Use high-growth-rate species to maximise available soil N use. Use plantain as a nitrification inhibitor. Use forage crops such as kale and fodder beet which have lower methane outputs	Use legume-dominant pastures. Add legumes to forage crops	Use new pasture species to reduce methane emissions	Use adapted pasture species, including C4 species
Genetics	Breed plants with greater abilities to uptake nutrients. Breed plants able to alter soil processes (e.g. plantain)	Breed plants with greater tolerance to disease and moisture and temperature extremes	Breed plants with greater abilities to uptake nutrients. Breed plants able to alter soil processes (e.g. plantain). Integrate novel endophytes to enhance overall animal performance while reducing emissions (e.g. high lipid ryegrass)	Breed plants with greater tolerance to disease and moisture and temperature extremes
Pasture mixtures	Choose pasture mixtures with low overall N content	Use functional mixtures which tolerate extremes in moisture and temperature	Choose pasture mixtures which influence soil microbiome function	Use functional mixtures which tolerate extremes in moisture and temperature
Nutrition	Optimise fertiliser inputs including fertiliser type and coatings. Add legumes to forage crops	Optimise nutrient additions to ensure maximum growth responses to sporadic water inputs.	Manipulate soil microbiome to access soil nutrient reserves	Manipulate soil microbiome to access soil nutrient reserves
Nutritive value	Use mixtures which have a low overall N content. Optimise trace element nutrition to reduce animal disease.	Use legume-dominant pastures. Add legumes to forage crops	See above, plus the ability of the soil microbiome to control potential diseases such as facial eczema	See above, plus the ability of the soil microbiome to control potential diseases such as facial eczema and internal parasites
Grazing management	Alter spelling durations to manipulate feed quality	Alter resting duration to stockpile feed		Alter the spelling duration to increase pasture longevity and soil quality
Water use	Irrigation management. Effluent disposal management	Irrigation. High water use efficiency plants		

Table 5 System opportunities to mitigate or adapt to the externalities that impact on grassland farming in New Zealand both now and in the future

	Current		Future	
Parameter	Mitigation	Adaptation	Mitigation	Adaptation
Decision support tools	Use of nutrient management software to minimise nutrient losses	Use of pasture growth forecasting software. Use of feed budgetting software, use of weather forecasting	Integration of data, such as landscape, animal feed requirements, soil types and climate, to manage animals in the landscape	Integration of data to manage animal movement between farms and regions
Enterprise choice	Shift from traditional beef operation to dairy beef operation. Shift from cow to sheep dairy production	Shift from traditional beef operation to dairy beef operation. Change to once-a-day milking systems	Integrate various annual cropping systems with livestock systems	Integrate various perennial cropping systems with livestock systems
Feed management	Use forage crops to support animal nutrition when pasture growth is low. Import low N supplementation to lower N emissions (from the arable sector or using international sources such as palm kernel expeller)	Match animal requirements to feed supply changes, such as altering buying and selling decisions, changing lambing and calving dates or drying off lactating animals	Use a range of forage/pasture species to reduce GHG emissions and nitrogen loss	Utilise tree fodder crops during climatic extremes
Land management	Restrict access to sensitive areas. Control grazing to minimise damage to sensitive areas	Use virtual herding technologies to manage variations in feed supply	Apply inhibitors to reduce nitrate and nitrous oxide emissions	Use virtual herding technologies (all ruminant species) to match variations in feed supply
Cultivation practices	Use of minimum tillage options to reduce potential soil and carbon loss	Minimum tillage to prevent damage in severe weather events		
Event management	Use off-paddock systems to restrict damage to land during severe conditions	Use virtual herding to control cattle if the infrastructure fails. Use virtual herding to optimise feed utilisation patterns		Use virtual herding to control animals if the infrastructure fails. Use virtual herding to optimise feed utilisation patterns. This would apply to all ruminant species.
Product flows	Increase early lactation performance to reduce time to finishing in sheep	Vary product delivery time to match pasture production		Implement on-farm processing to increase flexibility in product supply

include system choice, enterprise choice and input costs. These choices are influenced by both the availability and costs of capital, which are cyclically related to the relative profitability of different enterprises. For example, the shift from arable and sheep farming to dairy farming in New Zealand has been partly driven by the relative profitability of each enterprise and the willingness of lending institutes to encourage conversion to dairy (Copland and Stevens 2012).

At the individual farm-level solutions to reduce externalities, whether they be emissions to air or water, can be viewed as mitigations. These can be defined as practices which can alter impacts and are often implemented at operational or tactical scales. Many mitigations have been developed and are available for implementation, and potential future mitigations are being developed.

Responding to the influences of climate change can be viewed as adaptations. These are practices and systems choices which are implemented at tactical or strategic scales, with solutions evolving over time. Current options for adaptation are listed (Table 3), while some potential adaptations for the future are imagined (Table 4). There will be some cross-over between these two definitions and the solutions to these two major challenges of reducing externalities and responding to climate change.

In New Zealand, many potential mitigations and adaptations have been investigated in the recent past. These are grouped into animal (Table 3), pasture/forage (Table 4) and systems management approaches (Table 5). Each grouping is further divided to represent the range of targets that are available, along with future opportunities which may follow on from current initiatives. This list is not exhaustive, but a reflection of solutions available in New Zealand.

5 Improving animal management: from species and genetics to herd/flock management

5.1 Animal species

Shifting from sheep to cattle and vice versa can be used as a mitigation, most readily in red meat production, as the capital investment and farming structures are often already in place. However, the shifts are conflicting, depending on the emissions which are being targeted (Table 3). For example, if the farmer targets emissions to water, then a shift from cattle to sheep is warranted, as N loss from sheep is lower than cattle (Hoogendoorn et al. 2011). However, if a farmer targets emissions to the air, then a shift from sheep to cattle is warranted, as emissions from cattle (per kg product) are lower than sheep. This further interacts with the whole farm enterprise, as finishing cattle have a lower GHG footprint than a breeding herd.

When considering adaptation, then shifting from enterprise to enterprise may be forced upon the farmer. For example, as the climate warms, diseases that impact on the current sheep flock increase, causing farmers to shift towards beef production (McRae et al. 2018). As the climate continues to warm, a change from European cattle breeds (*Bos taurus*) to tropical cattle breeds (*Bos indicus*) may also be seen (Table 3).

5.2 Animal genetics

Genetic solutions are ongoing with significant investment in low GHG emissions sheep and cattle (Rowe et al. 2019). Low nitrogen-emitting sires are also available in the dairy industry (AmBreed 2023). Interactions between selecting for these traits and other production traits need to be carefully managed to prevent negative trade-offs (Hickey et al. 2022).

Expansion of animal breeding efforts to reduce methane outputs is being initiated in cattle (AmBreed 2023) and may become increasingly available in the near future (Table 3). Efforts in the deer industry are further away as the industry is small in scale, restricting potential investment in what are considered minor traits.

Adaptations using genetic solutions are being investigated, especially in the sheep industry (Pamu 2024). In these cases, a shift from traditional crossbred sheep based on European breeds is being considered with an increased focus on resilience with new breeding programmes in hair sheep and mid-micron wool sheep (Pamu 2024). This programme will potentially supply new genetics with greater disease resistance and greater heat tolerance, thus reducing costs while adding higher value wool. Some of these options are currently available, and genetic improvement programmes aim to increase the availability of appropriate, resilient genetics in the future. There is also some development of genetics which will utilise different environments and pasture feed profiles within the deer industry (Thompson et al. 2022). This may lead to animal genetics that are better adapted to future climatic conditions.

5.3 Stocking rate

Stocking rates on New Zealand grassland systems are primarily determined by the balance between low winter forage production and high spring forage production (Stevens et al. 2021a), to maximise the utilisation of pasture. Thus, changing relative stocking rate is most often done using tools such as increasing per head performance, culling decisions and sales policies to modify feed requirements to match pasture production (Table 3). Increasing progeny survival in the natural conditions of grassland provides a future opportunity (McCoard 2017) to improve feed use efficiency and optimise resource use.

5.4 Animal nutrition

Using low-N forages can potentially reduce nitrate leaching. This may be achieved by introducing some supplementation with low N feeds, though is confined to dairy systems due to logistic constraints. The same is true of using methane inhibitors as their use in the diet requires constant supplementation and this is uneconomic under current New Zealand farming conditions (Marmont et al. 2023). Plant varieties and species can be introduced to reduce methane emissions (see section on Plant/forage management). These choices, however, require significant changes in grazing systems to ensure integration into current grazing systems. Thus, they remain as potential options for the future (Table 3).

5.5 Product output

Altering product output can be used in the short term to mitigate some environmental impacts (Table 3). For example, many New Zealand dairy systems have implemented some level of supplementary feeding to augment pasture production, and as a result, have increased stocking rate (DairyNZ 2023). This trend has been a response to variations in pasture-based feed supply to generate greater resilience (Stevens et al. 2021a). Reverting to previous systems less reliant on imported supplements as a mitigation would reduce emissions to both water and air (Smith et al. 2023). However, this would conflict with the use of supplementary feeding as part of an adaptation strategy for future resilience (Stevens et al. 2021a).

Changing product output in the sheep and beef sector comes with trade-offs in potential market price. For example, lighter carcass weights could be achieved, reducing the amount of feed required and so reducing GHG emissions. However, market specifications determine the optimum carcass weight, and therefore, both product quality and returns to farmers would decline.

5.6 Product quality

Some elements of product quality could be influenced for the better with changes in the use of some feeds. For example, palm kernel expeller (PKE), a common supplement in the dairy industry, is also known to alter milk fat composition, altering its suitability for processing (DairyNZ 2023). Thus, removing PKE as a supplement would result in altered product quality while reducing the amount of feed used and so resulting in lower emissions.

Some pasture species may alter product quality. For example, the use of chicory and red clover in lamb finishing enhances intramuscular fat and omega 3 fatty acid deposition, improving eating experience (Lambe et al. 2017). These forages, with relatively low ADF concentrations, also have a lower GHG output per unit of intake (Waghorn et al. 2002), so their use would both alter product quality and reduce emissions to air. These options have had little uptake, even though the change in the functionality of the food may add value to the product. The resultant increase in value may then mean that the stocking rate could be lowered while maintaining profitability.

5.7 Herd/flock management

The management of the herd can influence both the mitigation of impacts as well as influence adaptation (Table 3). Optimising disease control of a herd or flock ensures that feed resources are utilised for productive purposes, rather than being used to fight the disease.

The use of sexed semen in the dairy industry provides the opportunity to maximise growth and shorten finishing times, while also providing only the right number of female replacements. Taken further, this can be used to ensure high-quality male dairy beef animals enter the value chain. This will reduce the GHG emissions per unit of product due to the greater efficiency of growing males for meat. However, a similar amount of feed is likely to be consumed and therefore no reduction in overall GHG emissions will occur.

New technologies such as virtual herding/fencing provide significant opportunities to control animals and thus reduce their impact on the environment (Stevens et al. 2023). Virtual herding has been commercialised by Gallagher Ltd (Campbell et al. 2018) and Halter USA Inc., creating the first commercially available GPS-based virtual fencing systems for cattle. The systems utilise GPS technology in conjunction with an audio cue to indicate the proximity to virtual fence boundaries, and use an electric pulse to deter the animal from crossing that boundary in a similar fashion to traditional electric fences, though with much lower energy delivery. This system has undergone extensive research on individual and small groups of cattle in Australia investigating the animal welfare implications, such as exposure to the pulse and audio stimuli, and learning behaviours such as the length of the learning process (Campbell et al. 2018 Campbell et al. 2019; Lee et al. 2008). Keeping cattle from entering waterways or eroding stream banks is a major mitigation option. However, fencing costs are very high and virtual herding technology is a much more cost-effective option, being between 40% and 80% less than conventional fencing, depending on the terrain (Box et al. 2023). Use of virtual herding to improve feed utilisation also provides the opportunity to increase future pasture growth and feed quality (Thompson et al. 2017). The continued deployment and development of the virtual herding technology (Stevens et al. 2023) will continue to increase its impact into the future (Table 4).

6 Improving pasture/forage management: from species to grazing management and water use

6.1 Plant species

Using pasture and forage species which have high growth rates can be used to help reduce losses of soluble soil nutrients such as nitrogen, as high growth will extract nutrients from the soil to support that growth (Table 4). Species such as Italian ryegrass (*Lolium multiflorum*) are known to take up soil nutrients in excess to their immediate growth requirements (Sandana et al. 2021). Using cool-tolerant grasses such as Italian ryegrass is promoted to reduce winter leaching losses by utilising nitrogen deposited as urine during winter (Maxwell et al. 2019).

Research has discovered that plantain (*Plantago lanceolata*) provides several environmental benefits. Most critically is the ability to reduce nitrogen losses (Herath et al. 2023). This highlights the potential for other pasture species to have attributes that may assist in further reductions in the environmental impacts of grassland use by ruminants (Table 4).

Greater use of legumes, like lucerne, red clover and *Lotus* spp, may assist in adapting to future climate change. This group of species often has a higher temperature growth threshold than grasses, and maintains feed quality, both as it matures and as ambient air temperature increases (Lambert and Litherland 2000). Legumes also fix nitrogen from the air, resulting in reduced needs for nitrogen fertiliser use and associated GHG production (Moot et al. 2019). However, legumes can have significant challenges which limit their use (Dodd and Sheath 2003). They can often only be used in situations where soil pH and phosphorus supply can be modified to optimum levels. This may be uneconomic in many farming circumstances. Many legumes also need rotational grazing with long rest intervals to maximise productivity and longevity, and are not suited to some grazing enterprises which rely on long periods of continuous grazing. When added to pasture mixtures they are often outcompeted by grasses. Thus, revised production systems are often required when implementing this strategy (Moot et al. 2019, Stevens et al. 2023a).

6.2 Plant genetics

Genetic programmes are in place to breed pasture and forage plants better adapted to future climatic conditions (Caradus et al. 2021). The need to accelerate selection for traits of resilience must be recognised to aid adaptation. New plant genetics may also have a direct impact on greenhouse gas emissions and nitrate losses. However, potential progress is currently restricted to standard breeding methods, as genetic modification is restricted by regulation in New Zealand (Caradus 2023). This means that technologies such as high lipid ryegrass, while developed in New Zealand, and with an ability to reduce methane emissions by 10–15% (Winichayakul et al. 2020), will be unable to be grown in New Zaland under current legislation.

Current commercial settings in the plant breeding industry see nearly all funding in pasture genetic improvement being funded by commercial companies (Chapman 2021). This means that investment in alternative pasture species may be lacking as minimum seed sales volumes drive product development (Rolston 2003). Investment at a public-good level is also relatively low (Chapman 2021) resulting in some potential solutions not making it to a commercial product (Monk et al. 2016).

Ryegrass (*Lolium perenne*) has been the major pasture grass in plant breeding programmes. Along with this programme has been the discovery and development of strains of an endophyte, *Epichloë*, which have conferred tolerance to a range of pasture pests (Popay and Easton 2006). These strains have provided opportunities to retain ryegrass production under changing climatic conditions (Ferguson et al. 2019). These endophyte products are integrated into a range of ryegrass cultivars and employed throughout New Zealand. Further strains are being developed.

6.3 Pasture mixtures

Pasture/forage mixtures use the integration of grasses, legumes and now forbs such as plantain (Herath et al. 2023) to provide high quality feed, natural N fixation and environmental improvements. Further research to increase the functionality of future pasture and forage mixtures will encompass interactions with the soil microbiome. However, this research needs funding and support to ensure delivery of these future benefits (Casey et al. 2022).

6.4 Plant nutrition

Optimising nutrient inputs may reduce potential losses from grazing systems (Monaghan et al. 2021b). This includes both N and P. However, this does interact with farming system and the influence of pasture production on whole farm productivity (Savage and Lewis 2005). The effectiveness of this approach will also vary depending on the current state. Sheep and beef farming, for example, has a much lower N surplus than dairy farming (Monaghan et al. 2021a) and therefore has a smaller opportunity for improvement.

Interestingly, this optimisation of nutrients for plant growth is an important tool for resilience and thus future adaptability (Stevens et al. 2021a). Thus, while altering plant nutrition may be an option to reduce current emissions, this may interact with future options for adaptability. There is an increasing use, for example, of legumes being sown with forage crops. Farmers are aiming to replace fertiliser N inputs with natural N fixation, maintaining forage yield while reducing the GHG emissions from fertiliser manufacture.

New fertiliser technologies such as coated fertilisers show promise, though their impact under pasture conditions continues to vary significant depending on conditions, thus reducing their reliability (Edmeades & McBride 2023).

6.5 Grazing management

Rotational grazing practices are widespread in New Zealand, especially in the dairy industry, and during autumn and winter in the sheep and beef industry. These practices have been developed around optimising feed supply and pasture growth to meet animal nutrition needs (McDonald and Roche 2023). Recent research has been revisiting rotational grazing practices (Stevens et al. 2011; Devantier et al. 2017, Donaghy et al. 2021). From this research, managements to defer grazing over periods of several weeks and months result in lower feed quality and nitrogen content (Devantier et al. 2017), thus reducing the potential for N leaching losses, when compared with common practises of continuous grazing or long rotations of 40 to 60 days which would be employed during summer and autumn.

Research into the grazing management of winter forage crops, such as kale and fodder beet, with dairy cattle was demonstrated to reduce sediment loss to waterways by an order of magnitude, nearly eliminating losses under normal winter conditions (Monaghan et al. 2017). This has been promoted and implemented across many areas using winter forage crop grazing (Beef + Lamb NZ 2023).

Changes in the use of rotational grazing techniques such as deferred grazing also provide the opportunity to manage feed supply when drought occurs, resulting in improved profitability resilience (Tozer et al. 2021). Understanding of further benefits of altering grazing intervals is on-going. As these benefits are documented, the practice of manipulating grazing interval will become more tailored to maximise the benefits, depending on circumstances and climatic influences (Table 4).

6.6 Water use

Using irrigation water efficiently provides the opportunity to mitigate environmental impacts through reducing potential for sediment loss and nitrate leaching (Smith et al. 2023). In the longer term, water use efficiency through the use of efficient plants such as legumes (Moot et al. 2019) will provide resilience as climate changes. The installation of irrigation is a known technique to manage climatic variability, though the availability of water may vary in future.

7 Improving production systems management

7.1 Enterprise choice

Managements at the systems level often combine the options present in animal and forage management (Table 5). These changes may also need to be accompanied by changes in capital, infrastructure and labour. One needs to keep in mind the scale of New Zealand farming enterprises and their capital investment when considering these changes. For example, changing from sheep farming to dairy farming may involve an investment of several million dollars to develop the appropriate infrastructure, as well as the employment of several staff.

Simple changes at the enterprise level, such as a shift from breeding cows to finishing dairy beef (Table 5) are associated with relatively little change in infrastructure or capital requirements, though does come with changing feed quality requirements. For example, breeding cows are often used to consume low-quality feed during autumn and winter on hill country farms, while finishing beef needs much higher pasture quality to be financially successful.

More significant changes, such as a shift from dairy cows to dairy sheep, would require capital inputs and retraining of staff to understand the requirements of dairy sheep. This change would also cause a revaluation of the farm, resulting in a change in the accessibility and costs of capital. These options may have impacts at both mitigation and adaptation scales. They are also influenced by a range of social factors which may drive change regardless of relative profitability merits (Kaye-Blake et al. 2014).

Often, animal production systems are integrated into annual cropping systems to allow for breaks between crops, nutrient management and crop cleanup. Integrating annual crops within animal production systems (Table 5) is much less common on a regular basis. This has been hypothesised to increase the potential resilience of farming systems in the future by increasing the diversity of income streams. However, present evidence suggests that this introduces extra complexity which is hard to manage, potentially leading to sub-optimal outcomes (Noe and Alroe 2012). This shift in thinking would be a significant departure from current practices for livestock farmers.

7.2 Feed management

Managing feed flows within a grassland system is paramount for overall efficiency and productivity (Stevens et al. 2021b). Rotational grazing management practices are one method used to vary feed supply throughout the year. As the variation between seasonal pasture production increases, forage crops can provide a buffer to maintain winter stocking rate while utilising the spring pasture surplus or fill summer feed deficits induced by drought (Lambert et al. 2000). Using imported supplements also provides this buffer within dairy farms (Stevens et al. 2021a) though may induce greater emissions to both air and water due to overall stocking rate increases.

7.3 Land management

Managing the land across the range of topography and aspect that is often present on sheep and beef farms presents a significant challenge due to the scale of farming operations. New technologies such as virtual herding can assist with this management (Box et al. 2023), reducing both labour and capital requirements.

Precision landscape management is a way of micro-managing the landscape to improve resource use and reduce environmental footprint. It is often associated with the application of digital tools and data collection. At present, it is limited to dairy farms as they often have limited variation, being predominantly on flat land (Figure 2). However, there are still variations in soil type and water courses, which can be managed by techniques such as temporary fencing or virtual herding.

Products such as nitrate inhibitors have been used in New Zealand in the past. These products had issues with residues being found in milk products and were therefore withdrawn from the market (MPI 2013). Future options would need to be free of such issues.

7.4 Event management

Off-paddock practices, such as feed pads or wintering barns, are currently used by dairy farmers (Smith et al. 2023) and some deer farmers (Stevens et al. 2023b) to mitigate the losses to water.

Virtual herding can also be used to manage the impacts from events such as floods and droughts by ensuring that the farmer can quickly place animals in the right part of the landscape (Stevens et al. 2023c) to prevent loss or damage.

7.5 Product flow

Managing product flow (Table 5) again utilises the incorporation of both animal and plant attributes at the systems level. For example, increasing flock fecundity to increase the number of lambs born in spring increases the feed demand of the ewe at a time when pasture supply is plentiful. This increased utilisation of spring feed also maintains feed quality and increases overall farm efficiency (MacKay et al. 2012), resulting in little change to nitrate leaching and total GHG emission, though an increase in production efficiency with more product being produced per unit of GHG (MacKay et al. 2012).

Using product flow as an adaptation tool is more problematic. However, individual farmers may be able to use this approach. For example, changing from twice daily milking to once daily milking as the season progresses is one way farmers manage variations in feed supply. Beef farmers will alter the slaughter date of animals both bringing it forward, slaughtering a lighter carcase weight, or postponing slaughter until carcass weight targets are met, depending on current feed supply and future expectations of feed supply and demand (Lambert et al. 2000).

Future options for on-farm product processing may include condensation of milk to reduce transport GHG emissions. Technology for the on-farm slaughter and processing of livestock is already available, though is not economic to implement at this time.

8 Implementing solutions: challenges and opportunities

The information provided in Tables 3, 4 and 5 highlights a range of potential solutions. The following section provides context regarding the status of the ability to implement these solutions. The solutions that exist affect a range of domains directly through to forage supply and systems changes (Tables 3, 4, 5). Each of these interventions may require application at different temporal and spatial scales. They also may require different labour and capital investment, as well as different levels of skill and technical support. They also apply differently to the livestock enterprises in New Zealand as the ability to apply them differs with labour, topography and management approaches. The outcomes in the red meat (sheep and beef) and dairy industries are outlined in the next sections.

Systems analysis of sheep and beef, and dairy enterprises (Stevens et al. 2021a) highlighted the importance of nutrient inputs, especially nitrogen, to ensuring both profitability and resilience of New Zealand farming systems. Nitrogen is one of the most limiting nutrients in New Zealand grassland systems, and inputs come from fertilizer or legume nitrogen fixation. In dairy farming, they were most commonly used as fertiliser inputs, at both operational and tactical levels. Nitrogen inputs in sheep and beef farms were mainly biological through the choice of legume-dominant pastures, and thus were tactical and strategic decisions. Potential environmental impacts of these approaches come from the difference in control of the supply of N. When fertiliser N is used, the control of use remains with the farmer. In the choice of legumes, N inputs are governed by biological feedback systems which turn N fixation off if soil N availability is high. This then naturally controls the amount of N in the system, while the farmer may continue to add N regardless of any potential surplus in the system.

It is important to acknowledge changes that have been made over time in the grassland enterprises of New Zealand. As previously stated, the sheep industry has had a significant decline in numbers while maintaining red meat production. This has seen GHG emissions decline from approximately 1 kg CH_4/kg meat to approximately 0.7 kg CH_4/kg meat between 1990 and 2016, respectively (Moot and Davison 2021). Overall, the total emissions from sheep enterprises have declined from approximately 14.6 million t CO_2 equivalents/annum in 1990 to approximately 8.3 million t CO_2 equivalents by 2020 (Ministry for the Environment 2022). While beef production has increased (FigureNZ 2023), the efficiency of production has also increased, with total emissions remaining constant at approximately 6.0 million t CO_2 equivalents/annum (Ministry for the Environment 2022). During the same period from 1990 to 2016, the expansion of dairy cattle numbers has increased total dairy enterprise emissions from 6.2 million t CO_2 equivalents/annum to 14.0 million t CO_2 equivalents/annum. At the same time, the efficiency of production has increased, resulting in a reduction in methane emissions per kg product from approximately 0.025 kg CH_4/kg milk to 0.021 kg CH_4/kg milk (Ministry for the Environment 2022).

Agricultural research in New Zealand has spent the last two decades shifting from a productivity basis towards an environmental focus, acknowledging that those externalities must be addressed. This firstly took the form of emissions to water, such as nitrogen, phosphorus and sediment loss from soil (Monaghan et al. 2021a). More latterly, it has focused more strongly on emissions to air in the form of methane and nitrous oxide (van der Weerden et al. 2020). Most of this research has focused on dairy cattle.

The intensive nature of dairy farming and its rapid expansion during the 1990s and 2000s highlighted potential environmental impacts (Julian et al. 2017). The realisation that dairy cow farming produces significantly more environmental challenge per stock unit has driven the industry to search for solutions. Nitrate leaching, for example, has been measured at 123 g mineral N leached per kg N intake for cattle compared with 75 g mineral N leached per kg N intake for sheep (Hoogendoorn et al. 2011).

The insidious effects of climate change are most often being addressed through genetic breeding programmes. For example, sthe pread of facial eczema and the impacts of other animal diseases, such as internal parasites, have ongoing animal breeding programmes to reduce both current and future impact (Johnson et al. 2021). However, these initiatives may need greater urgency placed on them as climate change expands the range of these threats (Johnson et al. 2021).

Further initiatives may also be required to counter future impacts such as increased heat stress (Johnson et al. 2021), and whole enterprise change may also be required if genetic solutions cannot be found. For example, current enterprise choice from north (warm) to south (cool) is driven by a combination of pasture quality (increasing as temperatures cool) and disease prevalence in sheep (decreasing as temperatures cool) (McRae et al. 2018). Increases in temperature then lead to greater number of beef cattle in northern New Zealand, and this trend spreading further south than currently (McRae et al. 2018), creating changes in both export products and potential emissions to air and water.

Solutions to reduce potential emissions to water and air for red meat farms are much harder to both develop and implement than for dairy farming. Monaghan et al. (2021b) estimated that, on red meat farms, land retirement was the most significant intervention in reducing environmental impact, and that this applied mostly to reducing sediment and P losses by 29% and 23%, respectively. The authors acknowledged the difficulties when scaling mitigations to represent the range of farm typologies and topographies, as well as underlying soil characteristics, when making such estimates at a national scale. However, relative reductions in losses are still relevant when trying to estimate progress.

Greenhouse gas emissions have also been declining in the red meat industry, with land retirement being a significant part of that change. Both production and carbon-capture forestry are viable and profitable alternative land uses for moderate to steep land. Furthermore, severe climate events in the form of drought and flood also impact grassland farming viability. While these events cannot be controlled, they may influence farming decisions and alter potential land use choices. For example, the erosion damage during cyclone events is greater on grazed pasture than under indigenous vegetation. Thus, to protect the landscape, the farmer may make decisions to further reduce the area of grassland.

Much of the impact of ruminants on the grazing landscape, and in GHG emissions, relates to the movement of animals in the landscape (Donovan and Monaghan 2020) and the forages which they access (Leahy et al. 2019). This interacts with the underlying soil properties and the topography of the landscape (Donovan and Monaghan 2020). Thus, interventions and mitigations must be able to control the animal and direct its access to appropriate feed sources under these scales and associated grazing management approaches. For example, the development of virtual herding technology for cattle may enable the implementation of quite sophisticated grazing management strategies (Box et al. 2023) that are currently unable to be implemented due to a lack of labour and too great a scale. Thus, technologies may provide opportunities to refine the management of herds and individuals both at the mitigation level and the adaptation level (Stevens et al. 2023c).

8.1 Interactions between mitigations affect the relative impact

Implementing solutions to restrict emissions to air and water have proved more problematic when developing solutions at a grazing systems level. Interactions, or trade-off's emerge when solutions are actioned.

For example, dairy cattle grazing systems over three sites across New Zealand experimented with a range of potential mitigations, including altering per head performance, stand-off facilities, varied cropping programmes and importation of supplements, mostly aimed at reducing nitrogen losses (van der Weerden et al. 2018). Total emissions as GHG are directly related to the amount of feed eaten and its nitrogen concentration. Thus, systems which decreased the total amount of feed eaten, while increasing per-head production, provided reductions in both nitrogen and GHG emissions. However, farm systems which increased productivity through increasing feed supply and stocking rate resulted in increasing emissions of all contaminants (van der Weerden et al. 2018). This demonstrates the interaction between maintaining profitability and reducing the externalities of ruminant livestock production if product prices are forced down by global market forces.

Trade-offs of potential contaminant mitigation strategies were assessed across a range of dairy farm typologies (Smith et al. 2023). New Zealand dairy farm typologies were developed using physical characteristics of soil drainage, soil wetness and slope (Monaghan et al. 2021). Mitigations evaluated included improved nitrogen management, off-paddock facilities to house cows in adverse soil conditions, effluent management practices, edge of field attenuation (artificial wetland development), and irrigation management practices (Smith et al. 2023). Of the nitrogen management practices chosen, reducing nitrogen fertiliser use and changing supplementary feed type to a low-N alternative had the greatest effect by reducing N, P and GHG losses. This is unsurprising as both feed supply and feed nitrogen concentration were reduced, resulting in direct reductions in losses. However, the magnitude of the reductions needs to be considered. The reductions in productivity were much lower than the concurrent reductions in losses to water and air. This indicates that current management practices chosen by the farmer may be exceeding optimal conditions, resulting in greater environmental impact than necessary.

Using off-paddock facilities to house cows when soil conditions are prone to damage transfers the problem of nitrogen deposition from soil to artificial bedding. The trade-off here means that effluent can be managed and potential sediment loss reduced, while losses to air in the form of both methane and nitrous oxide, from the interaction with the bedding (van der Weerden et al.

2014) are increased. So, losses to water may decrease, but losses to air increase (Smtih et al. 2023). However, several typologies did record either no change to N and P lost to water or even increases. These were generally related to increased effluent loadings to pasture, and to application along with irrigation, increasing the risk of losses (Smith et al. 2023). The capital and labour costs of building and using off-paddock facilities also need to be accounted for.

Effluent management was also relatively neutral in terms of impacts. Strategies such as not applying N fertiliser on effluent areas are already widespread, so reductions in potential nitrous oxide emissions are unlikely. Other management practices, such as storing effluent for longer to avoid saturated soil conditions result in the trade-off of increasing nitrous oxide emissions from the storage compared with reducing leaching losses to water (Smith et al. 2023).

Improved irrigation management applied to approximately 20% of dairy farms in the typologies tested (Smith et al. 2023). This is consistent with the proportion of dairy farm land under irrigation. The impacts of optimising irrigation use will depend on the current practices. Therefore, the maximum benefits of reducing both N and P loss will seldom be achieved. However, the implementation of full sensor-based irrigation optimisation had the potential to reduce N loss by between 67% and 85%, while P loss could be reduced by between 26% and 69%. Overall, more moderate improvements of 20% and 12% reductions in N and P losses, respectively, were estimated (Smith et al. 2023). The reduction in N loss in leaching drainage also had a small impact on reducing nitrous oxide losses to air (Smith et al. 2023).

Farming of red deer uses outdoor grazing, similar to other ruminants. Nitrate leaching losses can be slightly higher than sheep, though much lower than cattle (Hoogendoorn et al. 2011). Winter forage management can be a significant source of sediment loss as deer disrupt soil (McDowell and Stevens 2008). Initiatives to reduce this disruption have included farmer groups testing the use of indoor barn feeding as a mitigation option (Stevens et al. 2023b). Trade-offs are evident in substituting outdoor winter forage grazing with indoor barn feeding. Stevens et al. (2023b) calculated the potential GHG and nutrient loss footprints of red deer that either wintered on forage crops or confined to indoor barn feeding. While calculated feed intakes were 9.5% greater outdoors and losses to water were calculated to be up to 2.9 t sediment, and 60 kg N/ha, indoor wintering was calculated to increase GHG emissions by up to 14% due to a 21% reduction in methane emissions on high-quality forage crops and a 5% increase in nitrous oxide emissions in the barn. Losses to water in the form of sediment may be able to be mitigated using grazing management techniques documented by Monaghan et al. (2017), but losses of nitrogen to air or groundwater are directly related to N deposition on soil (rather than soil movement) and are therefore much harder to prevent.

These farm systems studies (van der Weerden et al. 2018, Smith et al. 2023, Stevens et al. 2023b) of potential interactions between the various externalities associated with ruminant livestock farming in grassland conditions demonstrate the trade-offs that must be accounted for when developing management plans and mitigation strategies. Also important in these calculations is the recognition that a wide range of variation is encountered. Farmers who have implemented the mitigations already will not get further benefits, while those who haven't will be able to record reductions in footprint. Early adopters of mitigations are unlikely to benefit from government incentives to reduce their footprint but are likely to meet consumer expectations of a low footprint. Therefore, the prospect of future government incentives may encourage a lack of progress in the near term, while consumer expectations, driven through pricing or outright refusal to purchase, hasten immediate change.

9 Conclusion

Grassland farming in New Zealand is centred around productive and profitable systems, with farms being 92% family-owned and providing most of the family income. This means that the scale of enterprises is significant, which influences the ability of the farmer to provide capital investment and the choice of solutions that may be available such as a shift between enterprises.

There is a significant body of research that has increased efficiency, reduced emissions to water and air and improved animal health and wellbeing. At the same time, developments in animal and plant genetics have supported improvements in efficiency, reduced the use of artificial chemicals such as fertiliser, anthelmintics, and pesticides, and increased productivity. The application of research has seen improvements across the board. Current research has highlighted a relative decline in grassland production, leading to the conclusion that future potential productivity gains in animal breeding may not be able to be supported by gains in pasture and forage production. This may lead to an increase in imported supplements to support future gains.

The unique year-round use of grasslands has created a grass-fed product of high quality, with a relatively low environmental footprint. There is a wide range of solutions available for both mitigation to reduce the externalities of grassland farming and adaptation to future climate change, with more being developed. However, work, both as research and practice adoption, needs to continue to ensure further reductions.

The application of solutions does come with potential trade-offs. These need to be understood when developing and applying solutions to reduce emissions, and adapt to future climate change while maintaining profitability.

10 Future trends in research

Casey et al. (2022) summarised the outcomes of a Symposium exploring the resilience of New Zealand grasslands. They recognised the challenges that were present and helped define future research needs. They identified the need to develop a greater understanding of both our current resources and our response to future challenges.

The soil microbiome, as a vast community of species, needs greater understanding. There has been significant investment in understanding the chemistry of soil and now there is a need to know more about the life within the soil. The development of analytical tools in recent years has enabled the development of soil microbiome research.

While there is a general understanding of energy and protein requirements of animals, the interactions between the micronutrient composition of the feeds and their impact on animal products is not well-understood, both for animal and human wellbeing.

More effort is required to develop plants and animals that reduce the impact of grassland farming on the environment, including applying genetic modifications and importing new plant germplasm. The role of plants in altering GHG emissions and reducing N leakage from grazing systems has only just begun with the research into the effects of plantain. More functional species, such as plantain, which reduces soil nitrate loss, will be out there.

The pasture microbiome must also be a target for further investment. Discoveries of the impacts of endophyte, a fungus living within pasture plants, have revolutionised our ability to manage pasture pests in ryegrass. This work will continue to expand into other species and has the potential to include pasture-based free living organisms.

An understand about integrating more annual and perennial crops into grassland systems to diversify production streams and create greater resilience in the face of increasing severity of weather events is required.

The animal genetics base needs to be broadened with a greater diversity of breeds and potential enterprises (such as dairy sheep and hair sheep) to create greater resilience as climates change.

Developments in other sectors may influence the inputs and outputs from grassland systems. These may include new technologies for producing and managing nitrogen, food production and human nutrition.

There is a range of digital technologies being developed which show promise for application to grassland farming. We need to continue to foster their development and find new ways of applying them.

Finally, we need to understand the interactions of each of these technologies when applied with the farm system, to ensure solutions are robust and have impact in the face of the complexity of a farming system.

11 Where to look for further information

There is a range of resources available to define the current status of grassland farming in New Zealand, challenges and current solutions. Many of these are industry good or government resources freely available to the public. These include:

Ministry for Primary Industries (MPI). The MPI website covers a wide range of support for grassland farming and agriculture in general. This support ranges from resources to aid recovery during and after severe climatic events such as droughts and cyclones through to animal welfare codes of practice, food safety and environmental protection and practice mitigation. The initial contact can be made through:

- https://www.mpi.govt.nz/.

with more refined information at:

- https://www.mpi.govt.nz/animals/.
- https://www.mpi.govt.nz/agriculture/.
- https://www.mpi.govt.nz/export/.
- https://www.mpi.govt.nz/funding-rural-support/.
- https://www.mpi.govt.nz/biosecurity/.
- https://www.groundrules.mpi.govt.nz/.

There are a number of bodies which producers support through a levy on production which is mandated by grassland farmers once every 5 years. These include Dairy NZ, Beef + Lamb New Zealand and Deer Industry NZ.

Dairy NZ: Dairy NZ provides support for dairy cattle farmers through a range of mechanisms. It houses a significant repository of information on its website. This includes general information regarding dairy cattle production as well as a range of tools for aiding practice change and decision making.

- https://www.dairynz.co.nz/.
- https://www.dairynz.co.nz/resources/.
- https://www.dairynz.co.nz/research/.
- https://www.dairynz.co.nz/regulation/.

Beef + Lamb New Zealand: B+LNZ collects a levy from sheep meat and beef to support programmes in maintaining industry statistics and productivity data, supporting science, coordinating and delivering on-farm improvement programmes and providing access to a wide range of knowledge and tools for industry.

- https://beeflambnz.com/.
- https://beeflambnz.com/knowledge-hub.
- https://beeflambnz.com/industry-data/.
- https://beeflambnz.com/science-genetics/.
- https://beeflambnz.com/programmes-partnerships/.
- https://beeflambnz.com/markets/.

Deer Industry NZ: The red deer industry collects levies on two products, venison and velvet antler. As such it supports research, farmer improvement, market advocacy and product safety across these products.

- https://www.deernz.org/.
- https://www.deernz.org/home/the-deer-difference/.
- https://www.deernz.org/home/deer-industry-new-zealand/.
- https://www.deernz.org/home/our-great-products/.
- https://www.deernz.org/deer-hub.
- https://www.deernz.org/nzdfa.

Other independent organisations also support the advancement of grassland agriculture in New Zealand. These include the New Zealand Grassland Association (NZGA), the New Zealand Society for Animal Production (NZSAP), the Agronomy Society of New Zealand (Agronomy NZ), New Zealand Society of Soil Science and the New Zealand Plant Protection Society. These organisations run annual conferences, produce an annual journal (or conference proceedings), most of which are open access, host occasional symposia and have active websites.

The New Zealand Grassland Association has a membership which includes farmers, science and agribusiness both national and internationally. It hosts an annual conference which moves from region to region. It is frequently co-host to Agronomy NZ and the New Zealand Society for Animal Production. Recent symposia have provided significant insight into challenges in Hill Country and to Future Resilience of grasslands in New Zealand. The resultant publications are a reflection of the current situation and potential future challenges and solutions to grassland farming.

- https://www.grassland.org.nz/.
- https://www.nzgajournal.org.nz/index.php/JoNZG/index.
- https://www.nzgajournal.org.nz/index.php/rps/issue/view/vol16.
- https://www.nzgajournal.org.nz/index.php/rps/issue/view/114.

The New Zealand Society of Animal Production is primarily a science organisation and hosts an annual conference which alternates between the main islands.

Recent special publications have updated the nutritional requirements for ruminants in New Zealand and the state of animal welfare in New Zealand.

- https://www.nzsap.org/.
- https://www.nzsap.org/content/publications.
- https://www.nzsap.org/pasture-and-supplements-grazing-animals.
- https://www.nzsap.org/animal-welfare-new-zealand.

The New Zealand Agronomy Society hosts an annual conference, often in conjunction with the NZGA. Recent special publications have investigated the need for future plant species and the seed production that may accompany that.

- https://www.agronomysociety.org.nz/.
- https://www.agronomysociety.org.nz/agronomy-nz-2020-.
- https://www.agronomysociety.org.nz/13.-seeds-for-futures.

The New Zealand Society of Soil Science co-host a conference with Soil Science Australia biennially and support a New Zealand conference every other year. Presentations are published as conference proceedings and much of the information is held on the Farmed Landscapes Research Centre website.

- https://www.soilscience.org.nz/.
- https://flrc.massey.ac.nz/.

The New Zealand Plant Protection Society hosts an annual conference and produces an annual journal. It is primarily a science organisation concentrating on the weeds and pests which impact production and threaten biosecurity.

- https://nzpps.org/.
- https://journal.nzpps.org/index.php/nzpp.

Independent industry support groups include the New Zealand Institute of Primary Industry Management, the Meat Industry Association and the Fertiliser Association. These are primarily professional networks supported by membership fees.

The New Zealand Institute of Primary Industry Management is primarily for working professionals, such as consultants, accountants and bankers, who support the agricultural industries. It hosts professional development events and produces a quarterly journal, available on subscription, which publishes management-related articles, as well as hosting an annual conference.

- https://www.nzipim.co.nz/.

The Meat Industry Association is a voluntary trade association representing New Zealand's red meat processors, marketers and exporters. Members are self-funding. It hosts a range of resources and publications regarding the red meat industry. As well as hosting an annual conference.

- https://www.mia.co.nz/.
- https://www.mia.co.nz/publications/.
- https://www.mia.co.nz/resources/.

The Fertiliser Association of New Zealand promotes and encourages responsible and scientifically-based nutrient management. Major fertiliser producers are members. Their website includes a range of resources regarding practices and nutrient requirements of the wide range of soils around New Zealand.

- https://www.fertiliser.org.nz/.
- https://www.fertiliser.org.nz/Site/code-of-practice/.
- https://www.fertiliser.org.nz/Site/resources/.

Individual companies also host websites that provide farmers and others with information regarding their business and support for implementing best practice in grassland farming. Of note are:

- Fonterra, New Zealand's largest milk processor and exporter - https://www.fonterra.com/nz/en.html.
- Silver Fern Farms, New Zealand's largest meat processor - https://silverfernfarms.com/.
- Barenbrug New Zealand and PGG Wrightson Seeds, producing pasture and forage varieties - https://www.barenbrug.co.nz/ and https://pggwrightsonseeds.com/.
- Ravensdown Ltd and Ballance Agrinutrients, the major fertiliser manufacturuers and importers - https://www.ravensdown.co.nz/expertise and https://ballance.co.nz/.
- New Zealand national farm assurance programmes - https://www.nzfap.com/.

12 Acknowledgements

Many thanks to Peter Pletnyakov for the generation of Figures 1 and 2 and Liz Wedderburn and Alice Bailey or the creation of Figure 3. Thanks also to Marie Casey for comments on the manuscript.

13 References

AmBreed. 2023. Genetic solutions https://crv4all.co.nz/nz/services/genetic-solutions accessed 1 February 2024.

Beef + Lamb New Zealand. 2022 Compendium of New Zealand farm facts. 2021, 45th edition. https://beeflambnz.com/sites/default/files/data/files/Compendium%202021_digital.pdf

Beef + Lamb New Zealand. 2022. New Zealand beef and lamb among the most carbon efficient in world. https://www.beeflambnz.co.nz/news/2022/11/8/lca-update accessed 16 January 2024

Beef + Lamb New Zealand. 2023. Winter grazing. https://beeflambnz.com/knowledge-hub/winter-grazing accessed 16 January 2024.

Beef + Lamb NZ. 2023 Sheep and beef farm survey. https://beeflambnz.com/industry-data/farm-data-and-industry-production/sheep-beef-farm-survey

Box L, et al. 2023. Developing a multiagent framework to explore the implications of fencing waterways using virtual fencing on New Zealand hill country farms. *Journal of New Zealand Grasslands*, 85, 101–108. https://doi.org/10.33584/jnzg.2023.85.3657

Campbell BM, et al. 2017. Agriculture production as a major driver of the Earth system exceeding planetary boundaries. *Ecology and Society*, 22(4), 8. https://doi.org/10.5751/ES-09595-220408

Campbell DLM, et al. 2018. Virtual fencing of cattle using an automated collar in a feed attractant trial. *Applied Animal Behaviour Science*, 200, 71–77. https://doi.org/10.1016/j.applanim.2017.12.002

Campbell DLM, et al. 2019. Virtual fencing is comparable to electric tape fencing for cattle behaviour and welfare. *Frontiers of Veterinary Science*, 6, 445. https://doi.org/10.3389/fvets.2019.00445

Caradus J. 2023. Genetic modification – benefits and risks for New Zealand grassland production systems. *Journal of New Zealand Grasslands*, 85, 39–49. https://doi.org/10.33584/jnzg.2023.85.3654

Caradus J, et al. 2021. Plant breeding for resilient pastures. *Resilient Pastures Symposium: New Zealand Grassland Association Research and Practice Series*, 17, 81–104. https://doi.org/10.33584/rps.17.2021.3441

Casey MJ, et al. 2022. Greener pastures: priorities for a low-footprint, high-value food producing future. A supplement to the Resilient Pasture Symposium. Grassland Research and Practice Series 17. 13pp. https://doi.org/10.33584/rps.17.2021.3604

Chapman DF. 2021. Resilient pastures. Resilient pastures symposium: New Zealand Grassland Association Research and Practice Series 17: 7–14. https://doi.org/10.33584/rps.17.2021.3522

Copland RJ and Stevens DR. 2012. The changing face of southern New Zealand farming: opportunities of land use change. *Proceedings of the New Zealand Grassland Association*, 74, 1–6. https://doi.org/10.33584/jnzg.2012.74.2876

Cosgrove GP and Field TRO. 2016. Hill country pastures, botanical composition and productive capacity. In: Thom ER. Ed. *Hill Country Symposium. Grassland Research and Practice Series 16*. Dunedin, New Zealand: New Zealand Grassland Association, pp. 281–288. https://doi.org/10.33584/rps.16.2016.3240

DairyNZ. 2023. New Zealand Dairy Statistics 2022–23. https://connect.dairynz.co.nz/2022-23_Dairy_Statistics/national-dairy-statistics.html

DairyNZ. 2023. Palm kernel expeller. https://www.dairynz.co.nz/feed/supplements/palm-kernel-extract/

Dalziel P, et al. 2018. The New Zealand food and fibre sector: a situational analysis. Agribusiness and Economics Research Unit, Lincoln University, 58pp. Accessed 26 April 2022. https://www.mpi.govt.nz/dmsdocument/33457-The-New-Zealand-Food-and-Fibre-Sector-A-Situational-Analysis-Report

Deer Industry NZ. 2020. Annual Report. https://www.deernz.org/assets/DINZ/Deer-Industry-of-New-Zealand/DINZ-Annual-Reports/2019-20-Annual-Report-DINZ-pages-web-version-reduced-size.pdf accessed 15 January 2024.

Devantier B, et al. 2017. The effect of deferred grazing in spring and early summer pastures on pasture growth rate and feed quality. *Journal of New Zealand Grasslands*, 79, 189–196. https://doi.org/10.33584/jnzg.2017.79.550

De Vries, A. 2013. Cow longevity economics: the cost benefit of keeping the cow in the herd. Cow Longevity Conference, DeLaval, Hamra Farm, Tumba, Sweden, pp. 22–52.

Di Menna ME, et al. 2009. A history of facial eczema (Pithomycotoxicosis) research. *New Zealand Journal of Agricultural Research*, 52(4), 345–376. https://doi/org/10.1080/00288230909510519

Dodd MB and Sheath GW 2003. Persistence: the key to legume performance in summer-dry hill country. Legumes for dryland pasture. *Grassland Research and Practice Series*, 11, 73–79. https://doi.org/10.33584/rps.11.2003.2993

Donaghy D, et al. 2021. Will current rotational grazing management recommendations suit future intensive pastoral systems? *Resilient Pastures Symposium. Grassland Research and Practice Series*, 17, 225–242. https://doi.org/10.33584/rps.17.2021.3464

Donovan M and Monaghan RM. 2020. Modelling grazing impacts on inherent soil loss risk for New Zealand landscapes; a novel geospatial approach to evaluating land use suitability. *Journal of Environmental Management*, V, 24.

Edmeades D and McBride R. 2023. An assessment of the agronomic effectiveness of N-(n-butyl) thiophosphoric triamide (nBTPT) - treated urea on the production of clover-based pastures, pastures, grasses and crops. *Journal of New Zealand Grasslands*, 85, 147–151. https://doi.org/10.33584/jnzg.2023.85.3637

Fennessey PF, et al. 2016. Innovations behind the farm gate that will influence performance of hill farming. In: Thom, ER. Ed. *Hill Country Symposium. Grassland Research and Practice Series 16*. Dunedin, New Zealand: New Zealand Grassland Association, pp. 15–20. https://doi.org/10.33584/rps.16.2016.3253

Ferguson CM, et al. 2019. Quantifying the economic cost of invertebrate pests to New Zealand's pastoral industry. New Zealand Journal of Agricultural Research, 62, 255–315. https://doi.org/10.1080/00288233.2018.1478860

FigureNZ. 2023. Exports of beef, veal, lamb and mutton from New Zealand. https://figure.nz/chart/L5XrXO6izARmQ7A2 accessed 16 January 2024.

Gillingham AG and During C. 1973. Pasture production and transfer on fertility within a long-established hill pasture. *New Zealand Journal of Experimental Agriculture*, 1, 227–232. https://doi.org/10.1080/03015521.1973.10427647

Goldson SL. 2018. Levy Oration 2017: reflections on a unique landscape. *Journal of New Zealand Grasslands*, 80, 11–14. https://doi.org/10.33584/jnzg.2018.80.353

Gray DI, et al. 2008. Strategies used to manage climatic risk: lessons from farmers with expertise in dryland farming. *Proceedings of the New Zealand Grassland Association*, 70, 59–68. https://doi.org/10.33584/jnzg.2008.70.2725

Hedberg RC. 2023. Bad animals, techno-fixes, and the environmental narratives of alternative protein. *Frontiers of Sustainable Food Systems*, 7, 1160458. https://doi.org/10.3389/fsufs.2023.1160458

Herath G, et al. 2023. Incorporating plantain into ryegrass-white clover mixed sward for an economically and environmentally sustainable dairy system: year one of a farm system study: plantain-based pasture for sustainable dairy systems. *Journal of New Zealand Grasslands*, 85, 321–329. https://doi.org/10.33584/jnzg.2023.85.3635

Hewitt A. 1998. *New Zealand Soil Classification*. 2nd edition. Landcare Research Series 1: 131pp.

Hewitt KG, et al. 2023. Drought impacts African black beetle feeding on perennial ryegrass. *Journal of New Zealand Grasslands*, 85, 239–251. https://doi.org/10.33584/jnzg.2023.85.3664

Hickey SM, et al. 2022. Impact of breeding for reduced methane emissions in New Zealand sheep on maternal and health traits. *Frontiers in Genetics*, 13, 910413. https://doi.org/10.3389/fgene.2022.910413

Hoogendoorn CJ, et al. 2011. Nitrogen leaching from sheep-, cattle- and deer-grazed pastures in the Lake Taupo catchment in New Zealand. *Animal Production Science*, 51, 416–425. https://10.1071/AN10179.1836-0939/11/050416

Julian JP, et al. 2017. River water quality changes in New Zealand over 26 years: response to land use intensity. *Hydrology and Earth System Sciences*, 21(2), 1149–1171. https://doi.org/10.5194/hess-21-1149-2017

Johnson PL, et al. 2021. A review of the current sheep industry in New Zaland and opportunities for change to meet future challenges. *New Zealand Journal of Animal Science and Production*, 81, 1–15.

Kaye-Blake W, et al. 2014. Validation of an agricultural MAS for Southland, New Zealand. *Journal of Artificial Societies and Social Simulation*, 17, 5. https://doi.org/10.18564/jasss.2564

Kiernan G. 2022, December 5. Why diversification from exporting to China must be a deliberate choice for NZ. Stuff. Why diversification from exporting to China must be a deliberate choice for NZ https://www.stuff.co.nz/business/the-monitor/300738276/why-diversification-from-exporting-to-china-must-be-a-deliberate-choice-for-nz accessed 24 January 2024.

Lambe NR, et al. 2017. Prediction of intramuscular fat content using CT scanning of packaged lamb cuts and relationships with meat eating quality. *Meat Science*, 123, 112–119.

Lambert MJ and Litherland AJ. 2000. A practitioner's guide to pasture quality. *Proceedings of the New Zealand Grassland Association*, 62, 111–115. https://doi.org/10.33584/jnzg.2000.62.2383

Lambert MJ, et al. 2000. How do sheep and beef farmers manage pasture quality? *Proceedings of the New Zealand Grassland Association*, 62, 117–121. https://doi.org/10.33584/jnzg.2000.62.2378

Leahy SC, et al. 2019. Mitigating greenhouse gas emissions from New Zealand pasture-based livestock farm systems. *Journal of New Zealand Grasslands*, 81, 101–110. https://doi.org/10.33584/jnzg.2019.81.417

Lee C, et al. 2008. The effect of low energy electric shock on cortisol, beta -endorphin, heart rate and behaviour of cattle. *Applied Animal Behaviour Science*, 113, 32–42. https://doi.org/10.1016/j.applanim.2007.10.002

Lowther WL and McDonald IR. 1972. Clover response to inoculation and pelleting on Te Anau soils. *Proceedings of the New Zealand Grassland Association*, 34, 122–128. https://doi.org/10.33584/jnzg.1972.34.1341

Macdonald KA and Roche JR. 2023. Farm systems research at Ruakura – a 60 year legacy underpinning profitable and sustainable pasture-based dairy systems. *New Zealand Journal of Agricultural Research*. https://doi.org/10.1080/00288233.2023.2168707

Mackay A, et al. 2023. Are observed rates of productivity compared to model predictions indicating negative climate impacts in perennial plants? *Journal of New Zealand Grasslands*, 85, 51–60. https://www.nzgajournal.org.nz/index.php/JoNZG/article/view/vol85

Mackay AD, et al. 2012. Has eco-efficiency of sheep and beef farms changed in the last 20 years? *Proceedings of the New Zealand Grassland Association*, 73, 119–124. https://doi.org/10.33584/jnzg.2012.74.2878

Mackintosh L. 2001. *Overview of New Zealand's climate*. Retrieved 15 June 2021 from https://niwa.co.nz/education-and-training/schools/resources/climate/overview

Manning L, et al. 2023. Threat or opportunity? An analysis of perceptions of cultured meat in the UK farming sector. *Frontiers of Sustainable Food Systems*, 7, 1277511. https://doi.org/10.3389/fsufs.2023.1277511

Mansfield S, et al.. 2021. Climate change impacts on pest ecology and risks to pasture resilience. *Resilient Pastures Symposium: New Zealand Grassland Association Research and Practice Series*, 17, 123–138. https://doi.org/10.33584/rps.17.2021.3477

Marmont B, et al. 2023. Farm-level cost-effectiveness analysis of in-paddock feeding of methane inhibitors in pasture-based dairy. *Journal of New Zealand Grasslands*, 85, 263–271. https://doi.org/10.33584/jnzg.2023.85.3655

Maxwell TMR, et al. 2019. Italian ryegrass swards reduce N leaching via greater N uptake and lower drainage over perennial ryegrass cultivars varying in cool season growth rates. *New Zealand Journal of Agricultural Research*, 62(1), 69–82. https://doi.org/10.1080/00288233.2018.1426615

McCall DG and Sheath GW. 1993. Development of intensive grassland systems: from science to practice. *Proceedings of the XVII International Grassland Congress*, 1993, 1257–1265.

McCoard SA. 2017. Issues and opportunities to capitalize on increased litter size in hill country sheep farming systems – a New Zealand perspective. *Animal Frontiers*, 17(3), 32–37. https://doi.org/10.2527/af2017-0126

McDowell RW and Stevens DR. 2008. Potential waterway contamination associated with wintering deer on pastures and forage crops. *New Zealand Journal of Agricultural Research*, 51, 287–290. https://doi.org/10.1080/00288230809510459

McRae KN, et al. 2018. Potential alterations in New Zealand sheep, cattle and deer numbers due to climate change: what can genetics offer? *New Zealand Journal of Animal Science and Production*, 78, 146–150.

Milicevic V, DuPont S. 2023. Financing of the CAP. https://www.europarl.europa.eu/factsheets/en/sheet/106/financing-of-the-cap#:~:text=The%20CAP%20accounts

%20for%2033.1,above%20%E2%80%93%20column%20(A)%5D. accessed 24 January 2024.
Ministry for the Environment. 2022. New Zealand's Greenhouse Gas Inventory 1990–2020. https://environment.govt.nz/assets/publications/GhG-Inventory/New-Zealand-Greenhouse-Gas-Inventory-1990-2020-Chapters-1-15.pdf accessed 16 January 2024.
Ministry of Foreign Affairs and Trade. 2023. Free trade agreements in force. https://www.mfat.govt.nz/en/trade/free-trade-agreements/free-trade-agreements-in-force/ Accessed 15 January 2024.
Ministry of Primary Industries. 2013. Withdrawal of DCD in New Zealand. https://www.mpi.govt.nz/dmsdocument/35-Withdrawal-of-DCD-in-New-Zealand-Letter-of-Assurance
Ministry of Primary Industries – Massey University. 2020. New Zealand Dairy Sheep survey. Wellington and Palmerston North New Zealand. 37 pp. https://www.mpi.govt.nz/dmsdocument/43183-Sheep-Dairy-Survey-Report-July-2020
Ministry of Primary Industries. 2022a. Animal Welfare. https://www.mpi.govt.nz/animals/animal-welfare/ accessed 25 January 2024.
Ministry of Primary Industries. 2022b. Animal compounds and veterinary medicines. https://www.mpi.govt.nz/agriculture/agricultural-compounds-vet-medicines/ accessed 24 January 2024.
Ministry of Primary Industries. 2022c. Food Safety. https://www.foodsafety.govt.nz/index.htm%20 accessed 24 January 2024.
Mitchell KJ. 1956. Growth of pasture species under controlled environment. 1. Growth at various levels of constant temperature. *New Zealand Journal of Science and Technology 37A:* 203–216.
Monaghan RM, et al. 2017. Grazing strategies for reducing contaminant losses to water from forage crop fields grazed by cattle during winter. *New Zealand Journal of Agricultural Research*, 60, 333–348. https://doi.org/10.1080/00288233.2017.1345763
Monaghan R, et al. 2021a. Quantifying contaminant losses to water from pastoral landuses in New Zealand I. Development of a spatial framework for assessing losses at a farm scale. *New Zealand Journal of Agricultural Research*, 64, 344–364. https://doi.org/10.1080/00288233.2021.1936572
Monaghan R, et al. 2021b. Quantifying contaminant losses to water from pastoral landuses in New Zealand II. The effects of some farm mitigation actions over the past two decades. *New Zealand Journal of Agricultural Research*, 64: 365–389. https://doi.org/10.1080/00288233.2021.1876741
Monk S, et al. 2016. Availability of seed for hill country adapted forage legumes. Hill Country Symposium. *New Zealand Grassland Association Grassland Research and Practice Series*, 16, 257–268. https://doi.org/10.33584/rps.16.2016.3237
Moot DJ, et al. 2019. Animal performance changes over 11 years after implementing a lucerne grazing system on Bog Roy Station. *Journal of New Zealand Grasslands*, 81, 75–80. https://doi.org/10.33584/jnzg.2019.81.390
Moot DJ and Davison R. 2021. Changes in New Zealand red meat production over the past 30 yr. Animal Frontiers 11: 26–31. https://doi.org/10.1093/af/vfab027
Noe E and Alroe HF. 2012. Observing farming systems: insights from social systems theory. Chapter 17. In *Farming Systems research into the 21st Century: The new*

dynamic. I Darnhofer, D Gibbon and B Dedieu (eds). pp. 387–403. https://doi/org/10.1007/978-94-007-4503-2_17

NZFAP. 2024. New Zealand's National Farm Assurance Programmes. https://www.nzfap.com/ accessed 24 January 2024.

NZ Treasury. 2023. The estimates of appropriations for the Government of New Zealand for the year ending 30 June 2024. https://www.treasury.govt.nz/publications/budgets/estimates-appropriations-government-new-zealand-year-ending-30-june-2024 accessed 15 January 2024.

O'Neill J. 2015. Antimicrobials in agriculture and the environment: reducing unnecessary use and waste. https://wellcomecollection.org/works/x88ast2u/items

Pamu. 2024. Sheep of the future. https://www.pamunewzealand.com/news/sheep-of-the-future-transformational-for-nz-sheep-farming. accessed 25 January 2024.

Phillips C, et al. 2023. Predicting facial eczema risks in a changing New Zealand climate. *Journal of New Zealand Grasslands*, 85, 61–71. https://doi.org/10.33584/jnzg.2023.85.3650

Popay AJ and Easton HS. 2006. Interactions between host plant genotype and Neotyphodium fungal endophytes affects insects. In *Breeding for Success: Diversity in Action. Proceedings of the 13th Australasian Plant Breeding Conference, Christchurch, New Zealand 18–21 April 2006*. pp. 561–567. Mercer CF. Ed. Advances in Pasture Plant Breeding. NZ Grassland Research and Practice Series 12. https://doi.org/10.33584/rps.12.2006.3020

Powell LJ, et al. 2023. Perceptions and acceptance of yeast-derived dairy in British Columbia, Canada. *Frontiers in Sustainable Food Systems*, 7, 1127652. https://doi/10.3389/fsufs.2023.1127652

Radcliffe JE. 1974. Seasonal distribution of pasture production in New Zealand. I Methods of measurement. *New Zealand Journal of Experimental Agriculture*, 2, 337–340. https://doi.org/10.1080/03015521.1974.10427692

Roche JR, et al. 2017. A 100-year review: a century of change in temperate grazing dairy systems. *Journal of Dairy Science*, 100, 10189–10233.

Rolston M. 2003. Seed production issues that limit supplies or result in high market prices of dryland legume species. *New Zealand Grassland Association. Research and Practice Series*, 11, 161–167. https://doi.org/10.33584/rps.11.2003.3012

Rowe SJ, et al. 2019. Selection for divergent methane yield in New Zealand sheep – A ten-year perspective. *Association for the Advancement of Animal Breeding and Genetics*, 23, 306–309.

Sandana P, et al. 2021. Nitrogen nutrition index and forage yield explain nitrogen utilization efficiency in hybrid ryegrasses under different nitrogen availabilities. *Field Crops Research*, 265, 108101. https://doi.org/10.1016/j.fcr.2021.108101

Saunders C, et al. 2015. Consumer preferences in developing and developed country markets and the impact on EU and New Zealand. *Asia Pacific Journal of EU Studies*, 13(2), 1–24.

Savage J and Lewis C. 2005. Applying science as a tool for dairy farmers. *Proceedings of the New Zealand Grassland Association*, 67, 61–66. https://doi.org/10.33584/jnzg.2005.67.2594

Scholtens MR, et al. 2017. The current state of the New Zealand goat industry. *Proceedings of the New Zealand Society of Animal Production*, 77, 164–168.

Scott D, et al. 1985. Limitations to pasture production and choice of species. In: Burgess RE, Brock JL. Eds. *Using herbage cultivars. Grassland Research and Practice Series*

3. Palmerston North, New Zealand: New Zealand Grassland Association, pp. 9–16. https://www.grassland.org.nz/viewpublication.php?pubID=318

Sears PD. 1951. Potentialities of intensive grassland farming. *Proceedings of the New Zealand Grassland Association*, 13, 66–76. https://doi.org/10.33584/jnzg.1951.13.969

Shadbolt NM and Apparao D. 2016. Factors influencing the dairy trade from New Zealand. *International Food and Agribusiness Management Review*, Special Issue 19(B), 241–255.

Smith BL and Towers NR. 2002. Mycotoxicosis of grazing animals in New Zealand. *New Zealand Veterinary Journal*, 50, 28–345.

Smith C, et al. 2023. Co-benefits and trade-offs of water quality mitigation measures on greenhouse gas emissions from New Zealand dairy systems. *Journal of New Zealand Grasslands*, 85, 283–300. https://doi.org/10.33584/jnzg.2023.85.3667

Stats NZ. 2020. Drought: key findings - Severity trends for meteorological droughts 1972–2019. Retrieved 20 October 2020 from https://www.stats.govt.nz/indicators/drought

Statistics NZ. 2023. Livestock Numbers. https://www.stats.govt.nz/indicators/livestock-numbers accessed 16 January 2024.

Stevens D, et al. 2023a. Responses of Lotus pedunculatus to sulphur fertiliser and defioliation regime in acid soils at low temperature. *Journal of New Zealand Grasslands*, 84, 49–56. https://doi.org/10.33584/jnzg.2022.84.3570

Stevens D, et al. 2023b. Effects of water-quality management mitigations on greehouse-gas emissions from deer farms. *Animal Production Science,* https://doi.org/10.1071/AN22436

Stevens D, et al. 2021a. A systems approach to understanding the connection between farm systems resilience and pasture resilience. *Resilient Pastures Symposium. Grassland Research and Practice Series*, 17, 421–432. https://doi.org/10.33584/rps.17.2021.3467

Stevens DR, et al. 2011. Winter feeding- changing labour requirements and productivity. *Proceedings of the New Zealand Grassland Association*, 73, 51–56. https://doi.org/10.33584/jnzg.2011.73.2850

Stevens DR, et al. 2019. Review of livestock walking distance and implications for the New Zealand Greenhouse Gas Inventory. MPI Technical Paper No. 2021/07. https://www.mpi.govt.nz/dmsdocument/46246-Review-of-Livestock-Walking-Distances-and-Implications-for-the-New-Zealand-Greenhouse-Gas-Inventory- accessed 16 January 2024.

Stevens DR, et al. 2021b. Using new insights in grazing management to buffer the impacts of climatic variability on pasture resilience. *Resilient Pastures Symposium. Grassland Research and Practice Series*, 17, 379–386. https:/doi.org/10.33584/rps.17.2021.3439

Stevens D, et al. 2023c. Understanding pathways of digital technology development to improve farm sustainability and resilience. *Journal of New Zealand Grasslands*, 85, 91–100. https://doi.org/10.33584/jnzg.2023.85.3661

Thompson BR, et al. 2017. Winter grazing and dead material effects on early spring pasture production, New Zealand Journal of Agricultural Research. https://doi.org/10.1080/00288233.2017.1283633

Thompson BR, et al. 2022. Deer seasonal growth pathways. *New Zealand Journal of Animal Science and Production*, 82, 50–56.

Tozer K, et al. 2021. Impact of deferred grazing on profitability of a hill country farm in north-western Waikato. *Resilient Pastures Symposium. Grassland Research and Practice Series*, 17, 393–400. https://doi.org/10.33584/rps.17.2021.3442

USDA. 2020. New Zealand Agricultural exports to China continue to surge. https://apps.fas.usda.gov/newgainapi/api/Report/DownloadReportByFileName?fileName=New+Zealand+Agricultural+Exports+to+China+Continue+to+Surge_Wellington_New+Zealand_07-20-2020.pdf. accessed 23 January 2024.

USDA. 2023a. New Zealand deer production and trends. https://fas.usda.gov/data/new-zealand-new-zealand-deer-production-and-trends. accessed 25 January 2024.

USDA. 2023b. New Zealand sheep production and trends. https://fas.usda.gov/data/new-zealand-new-zealand-sheep-production-and-trends. accessed 24 January 2024.

van der Weerden TJ, et al. 2014. Addition of straw or sawdust to mitigate greenhouse gas emissions from slurry produced by housed cattle: a field incubation study. *Journal of Environmental Quality*, 43, 1345–1355. https://doi.org/10.2134/jeq2013.11.0452

van der Weerden T, et al. 2018. The *effects of system changes in grazed dairy farmlet trials on greenhouse gas emissions. Animals (Basel)*, 8, 234.

van der Weerden TJ, et al. 2020. Meta-analysis of New Zealand's nitrous oxide emission factors for ruminant excreta supports disaggregation based on excreta form, livestock type and slope class. *Science of the Total Environment*, 732, 139235.

Waghorn GC, et al. 2002. Methanogenesis from forages fed to sheep. *Proceedings of the New Zealand Grassland Association*, 64, 167–171.

Wall A, et al. 2021. Importance of resilient pastures for New Zealand's agricultural soil carbon stocks. *Resilient Pasture Symposium. New Zealand Grassland Association Grassland and Practice Series*, 17, 191–200. https://doi.org/10.33584/rps.17.2021.3455

Wallace N. 2014. *When the Farm Gates Opened: The Impacts of Rogernomics on Rural New Zealand*. 160pp. Otago University Press, Dunedin, New Zealand.

Webster J. 2016. Animal welfare: freedoms, dominions and "a life worth living". Animals https://doi.org/10.3390/ani6060035

Whitfield. 2023. Top milk exporting countries in 2023. https://www.linkedin.com/pulse/top-milk-exporting-countries-2023-spencer-whitfield/ accessed 15 January 2024.

Winichayakul S, et al. 2020. In vitro gas production and rumen fermentation profile of fresh and ensiled genetically modified high-metabolizable energy ryegrass. *Journal of Dairy Science*, 103. https://doi.org/10.3168/jds.2019-16781

Winter DS. 1989. The mineral composition of pasture in Southland based on a 5-year monitoring programme. *Proceedings of the New Zealand Grassland Association*, 50, 89–94. https://doi.org/10.33584/jnzg.1989.50.1872

Part 6

Conclusion

Chapter 20

Conclusion: pathways to resilient temperate grasslands

Agnes van den Pol-van Dasselaar, Aeres University of Applied Sciences, The Netherlands

1 Introduction

Temperate grasslands face a range of challenges (e.g. Delaby et al., 2020) that often occur simultaneously and therefore need to be addressed in a coherent and integrated way. Examples of these challenges include intensification, land abandonment, land-use change, and climate change. At the same time, research (e.g. Bengtsson et al., 2019) suggests that temperate grasslands have enormous opportunities to support the transition to more resilient, sustainable, and circular agricultural systems. This book reviews the wealth of recent research addressing these challenges and opportunities.

The individual chapters endorse the potential of grasslands and highlight the unique opportunities of grasslands in delivering multiple benefits to farmers and to wider society. The book summarises key research on the role of grasslands in agricultural systems (Chapters 1–3) and on grassland dynamics and management (Chapters 4–9). The book also draws attention to grasslands as a feed source, linking this to animal health and product quality (Chapters 10–12), and describes recent developments in the botanical composition of grasslands (Chapters 13–16). Furthermore, it provides several key case studies reviewing major challenges in temperate grasslands in specific regions (Chapters 17–19). This concluding chapter does not attempt to summarise the entire book but aims to highlight key insights in relation to temperate grasslands

http://dx.doi.org/10.19103/AS.2025.0154.32

and translate the findings from the individual chapters into recommendations for practice, science, and policy.

2 Cross-cutting insights

The following key cross-cutting insights emphasise the interplay between ecological processes, management practices, human decision-making, and policy in temperate grasslands:

- Multifunctionality is the rule, not the exception. Grasslands provide multiple services simultaneously: provisioning (forage, meat, milk), regulating (carbon sequestration, water regulation), supporting (soil formation, nutrient cycling), and cultural (recreation, landscape beauty). Management decisions inevitably involve trade-offs between these services. Yet, with well-designed grazing and cutting regimes, carefully chosen species mixtures, and thoughtful landscape planning, it is possible to enhance several benefits at once.
- The biodiversity of temperate grasslands underpins resilience and productivity. Diverse plant communities, soil organisms, and associated fauna improve nutrient cycling, disease resistance, forage stability, and ecosystem resilience, supporting both ecological and production outcomes.
- Soil health is foundational. Soil structure, organic matter, and microbial activity are critical for long-term productivity, water retention, carbon storage, and the overall sustainability of temperate grassland systems.
- Grazing management is central. The timing, intensity, and type of grazing animals strongly influence outcomes. Well-timed grazing can boost nutrient cycling, maintain forage quality, and support botanical diversity. In contrast, mismatched grazing pressure can accelerate degradation and nutrient losses.
- Species composition matters. Multi-species swards, including grasses, legumes, and forbs such as plantain, often show benefits for resilience and nitrogen-use efficiency. Plant secondary metabolites and beneficial endophytes offer opportunities to reduce environmental losses and improve product quality, though knowledge gaps remain. Semi-natural and natural grasslands continue to play a key role in sustaining biodiversity across temperate regions.
- Scale and time horizons matter. Many ecological processes and climate-related impacts in grasslands unfold over decades and across landscapes. Short-term experiments or single-field metrics can overlook key dynamics. Long-term, spatially explicit monitoring, combined with adaptive management is therefore fundamental.

- The farmer drives day-to-day decisions in grassland management. These decisions determine the sustainability and productivity of grassland-based systems. The socio-economic viability of these systems is therefore crucial, as it shapes what farmers can realistically implement. The mindset of the farmer, shaped by knowledge, experience, values, and preferences, ultimately drives management outcomes. Farmers are the key agents of change, translating insights and technologies into practical action on the ground.
- Technology enables, people decide. Precision livestock farming, including pasture monitoring, remote sensing, and decision-support systems, can improve efficiency and reduce environmental impacts. Adoption, however, depends on human factors: cost, usability, trust, and fit with routines of farmers. Effective extension, co-development, and participatory approaches are therefore essential for successful implementation.
- Knowledge alone does not determine practice; policy frameworks are decisive. Financial incentives, regulations, and market signals largely determine whether innovations are adopted, scaled, and sustained. Aligning policy frameworks to reward multifunctional outcomes is a prerequisite for desired outcomes.

3 Practical recommendations for farmers and advisers

Grass-based farming systems in temperate regions can benefit from pragmatic, evidence-based suggestions drawn from the chapters in this book, including the following:

- Design swards for function, not just yield. Where possible, shift from single-species ryegrass leys towards mixtures that include legumes and strategic forbs to stabilise production, improve nitrogen fixation, and support animal health.
- Prioritise co-benefits in trait selection. Select for multiple traits in plant breeding and favour cultivars and mixtures that contribute to multiple objectives: resilience (e.g. to drought or frost), nutritional quality of grasslands as livestock feed, and positive effects on ecosystem services such as soil carbon sequestration and biodiversity.
- Match grazing to plant growth curves. Adjust stocking rate and rotation timing to avoid the extremes of overgrazing and undergrazing. This preserves sward quality and soil structure.
- Optimise nutrient management. Calibrate fertilisation to soil tests, close manure loops, and cover manure storage to reduce losses.

- Optimise forage conservation. Plan cutting windows, optimise settings of machinery, ensure proper compaction and sealing, and monitor silage stability at feed-out.
- Monitor what matters. Track, e.g., forage quality, animal performance, soil organic matter, and simple biodiversity indicators (e.g. sward species richness). Use these data to guide adaptive management decisions.
- Build collaborative learning networks. Participate in or establish farmer-led trials, knowledge exchange groups, and regional demonstration sites that accelerate uptake of new knowledge and practical innovations. Share both failures and successes.

4 Policy and institutional priorities

Policymakers influence temperate grassland management by shaping incentives, information, and markets. Key priorities to support temperate grasslands are as follows:

- Reward multifunctionality and public goods. Implement payment schemes that recognise and compensate for carbon sequestration, biodiversity, and landscape services alongside productivity. Ensure these schemes use robust and cost-effective verification methods.
- Foster cross-sectoral coherence. Align agricultural, environmental, and climate policies to avoid contradictory incentives, such as encouraging high-input feed production while aiming to restore biodiversity-rich grasslands.
- Enable market recognition. Promote labelling and procurement standards that reward pasture-based quality or local production conditions. Where grassland management enhances product quality (e.g. milk fatty acid profiles), these schemes can add value for farmers adopting sustainable practices.
- Invest in advisory capacity and peer learning. Adequately strengthen extension programmes to translate scientific knowledge into context-specific advice. Digital tools are useful but cannot replace human facilitation. Support farmer-to-farmer networks to accelerate learning.
- Provide fair pathways for change. Offer targeted support to vulnerable regions (e.g. mountain areas) and smallholders to adopt improved practices without compromising livelihoods.

5 Research agenda

The chapters in this book identify many promising research directions for temperate grasslands, closely linked to the content of each individual chapter.

The following overarching priorities are proposed for a research agenda in the coming decades. These priorities are listed in an order going from fundamental research design through to practical adoption, monitoring, and synthesis:

- Long-term and landscape-scale research. Fund and maintain not only short-term grassland trials but explicitly also long-term trials (spanning up to decades) that capture interactions among grazing regimes, species mixtures, nutrient management, and climate variability, to study slow and interacting ecological processes.
- From process understanding to system-level insights. Bridge mechanistic plant and soil studies with whole-farm life-cycle assessments and socio-economic models to quantify trade-offs and synergies at scales relevant to practice and policy.
- Mindset and motivation of farmers. Evidence from participatory research across Europe shows that farmers' values, experiences, preferences, and habits strongly shape management choices. Any transition effort must respect these drivers, engaging farmers and their families in ways that align with their lived realities and goals.
- Adoption of innovations. Investigate social, economic, and behavioural barriers to adopting innovations in temperate grasslands to ensure that investments translate into practice.
- Stakeholder engagement in grassland science. Involve farmers, advisers, and local communities from trial design through to interpretation. Participatory approaches ensure relevance, practicality, and higher adoption rates.
- Socio-ecological resilience indicators. Identify metrics that capture both ecological stability (e.g. biodiversity, soil health) and community vitality (e.g. farmer income stability, generational renewal). Combine simple field metrics with higher-tier measures where justified.
- Verification of ecosystem services. Develop robust, cost-effective monitoring, and remote sensing approaches to enable credible claims and payments for ecosystem services such as carbon sequestration and biodiversity.
- Standardised and accessible data management. Ensure datasets are standardised, openly available where possible, and managed to protect farmer privacy while enabling cross-site synthesis.

6 Conclusion: shaping tomorrow's temperate grasslands

Temperate grasslands are socio-ecological systems, where people and nature interact and evolve together. Ecological science offers valuable tools for better outcomes, but these only succeed when paired with trust, local knowledge, supportive institutions, and fair markets. Durable change depends on helping farmers make sound decisions under uncertainty, aligning incentives with public goods, and building the skills and networks that sustain good management.

Transition management provides a practical way forward: set a clear long-term direction, enable diverse experiments, evaluate results openly, and adapt policies and practices in response. Inclusion of research is essential. Long-term studies reveal slow processes, species dynamics, and climate responses, while shorter-term trials drive innovation in grazing, sward diversity, and nutrient management. Research quantifies trade-offs and synergies, informs policy, and supports monitoring and verification of ecosystem services.

Education and research are closely linked. The students and young farmers of today will be the practitioners and decision-makers of tomorrow. Engaging them in research, participatory projects, and innovation networks ensures that scientific insights become practical, locally adapted solutions. Continuous, co-designed learning builds technical skills, strengthens stewardship values, and increases confidence in adopting new practices. In this way, education amplifies the impact of research and secures the next generation of grassland stewards.

The future of temperate grasslands will be shaped where science meets practice, when researchers, teachers, farmers, advisers, and policymakers work together to co-develop solutions, test them in real conditions, and share lessons widely. By pursuing multifunctionality, resilience, and equitable livelihoods as connected goals and by recognising the societal value of grassland-based production, we can ensure that landscapes with temperate grasslands remain productive, climate-resilient, and rich in biodiversity for decades to come.

7 References

Bengtsson, J., et al. (2019). Grasslands–more important for ecosystem services than you might think. *Ecosphere* 10, e02582. https://doi.org/10.1002/ecs2.2582.

Delaby, L., et al. (2020). Pasture-based dairy systems in temperate lowlands: challenges and opportunities for the future. *Frontiers in Sustainable Food Systems* 4, 543587. https://doi.org/10.3389/fsufs.2020.543587.

Index

www.ingramcontent.com/pod-product-compliance
Ingram Content Group UK Ltd.
Pitfield, Milton Keynes, MK11 3LW, UK
UKHW021824100926
8556IPUK00003B/137

9 781801 469029